T0272022

Automation in Construction toward Resilience

Robotics, Smart Materials, and Intelligent Systems

Edited by
Ehsan Noroozinejad Farsangi
Mohammad Noori
Tony T.Y. Yang
Paulo B. Lourenço
Paolo Gardoni
Izuru Takewaki
Eleni Chatzi
Shaofan Li

CRC Press
Taylor & Francis Group
Boca Raton London New York

CRC Press is an imprint of the
Taylor & Francis Group, an **informa** business

Designed cover image: Shutterstock

First edition published 2024
by CRC Press
2385 Executive Center Drive, Suite 320, Boca Raton, FL 33431

and by CRC Press
4 Park Square, Milton Park, Abingdon, Oxon, OX14 4RN

CRC Press is an imprint of Taylor & Francis Group, LLC

© 2024 selection and editorial matter, Ehsan Noroozinejad Farsangi, Mohammad Noori, Tony T.Y. Yang, Paulo B. Lourenço, Paolo Gardoni, Izuru Takewaki, Eleni Chatzi, and Shaofan Li; individual chapters, the contributors

---

Library of Congress Cataloging-in-Publication Data

Names: Farsangi, Ehsan Noroozinejad, editor.
Title: Automation in construction toward resilience: robotics, smart materials & intelligent systems/
edited by Ehsan Noroozinejad Farsangi, Mohammad Noori, Tony T.Y. Yang, Paulo B. Lourenço,
Paolo Gardoni, Izuru Takewaki, Eleni Chatzi, and Shaofan Li.
Description: First edition. | Boca Raton, FL: CRC Press, [2024] | Series: Resilience and sustainability in
civil, mechanical, aerospace and manufacturing engineering systems; book 9 | Includes bibliographical
references and index.
Identifiers: LCCN 2023004188 (print) | LCCN 2023004189 (ebook) | ISBN 9781032350868 (hardback) |
ISBN 9781032350899 (paperback) | ISBN 9781003325246 (ebook)
Subjects: LCSH: Building–Automation.
Classification: LCC TH1000 .A866 2024 (print) | LCC TH1000 (ebook) | DDC 690–dc23/eng/20230216
LC record available at https://lccn.loc.gov/2023004188
LC ebook record available at https://lccn.loc.gov/2023004189

ISBN: 978-1-032-35086-8 (hbk)
ISBN: 978-1-032-35089-9 (pbk)
ISBN: 978-1-003-32524-6 (ebk)

DOI: 10.1201/9781003325246

Typeset in Times
by KnowledgeWorks Global Ltd.

Contents

Preface

Advancements in smart and innovative building materials and systems, artificial intelligence, augmented reality, robotics, nanotechnology, building information modeling, use of drones, additive manufacturing (3D printing), and blockchain construction management technology have resulted in transformative changes in the construction industry and the adoption of novel technologies that are defining a new era in this domain. Massive and radical changes have surged owing to the possibilities created by big data, the advent of the Internet of Things, and the emergence of data-centric engineering, along with the technological advances driving down the cost of sensors, data storage, and computational services.

Furthermore, life cycle building information modeling (BIM), extended virtual reality, drone usage, smart and innovative building materials, and blockchain-based construction management techniques and technologies are increasingly utilized in standardized, modularized, or prefabricated products and components. In addition, on-site fully autonomous robotic 3D printing, utilization of AI and machine learning, and advanced sensing technologies have demonstrated the onset of significant advancements in construction engineering and the potential of a transformation of the construction industry to increase its productivity and, at the same time, to address emerging global challenges such as housing crisis, resource shortages, climate change, and the global population growth.

This book project covers refereed material on various aspects of technological advancements in the construction industry pertaining to the use of information technologies in design, engineering, construction technologies, and maintenance and management of constructed facilities. The scope of the book is broad, encompassing a variety of emerging topics toward smart and resilient structures and infrastructure.

This book addresses several issues in the development and implementation of novel technologies in construction automation in a collection of 30 chapters. We are confident that this book delivers practical and state-of-the-art materials to a broad group of industry professionals and researchers interested in gaining a better understanding of automation in construction and the potential opportunities and challenges.

Editors
Ehsan Noroozinejad Farsangi
Mohammad Noori
Tony T.Y. Yang
Paulo B. Lourenço
Paolo Gardoni
Izuru Takewaki
Eleni Chatzi
Shaofan Li
January 2023

About the Editors

Dr. Ehsan Noroozinejad Farsangi is currently a senior researcher at UBC Smart Structures, Canada. Concurrently, he is a tenured faculty member at the Graduate University of Advanced Technology, Iran. He is also an adjunct associate professor at Southeast University, China. Dr. Noroozinejad is the director of the Resilient Structures Research Group consisting of tens of top-ranked international researchers. Besides, he is the founder and chief editor of the *International Journal of Earthquake and Impact Engineering*, the associate editor of the *ASCE Practice Periodical on Structural Design and Construction*, the associate editor of the *IET Journal of Engineering*, the associate editor of *Frontiers in Built Environment: Earthquake Engineering Section*, the editor of the *Journal of Reliability Engineering and Resilience*, and the engineering editor of *ASCE Natural Hazards Review*. He has published over 120 high-impact journal papers in indexed journals and five books with reputed publishers in his field of expertise. His main research interests include smart structures, resilience-based design, reliability analysis, artificial intelligence, construction robotics, intelligent infrastructure, and digital twins in construction. Dr. Noroozinejad is also the recipient of many national and international awards, including the prestigious Associate Editor Award in 2022 by the American Society of Civil Engineers (ASCE), because of his consistent and exemplary service to enhance ASCE's publication activities.

Dr. Mohammad Noori is a professor of mechanical engineering at Cal Poly, San Luis Obispo, a fellow and life member of the American Society of Mechanical Engineers, and a recipient of the Japan Society for Promotion of Science Fellowship. Noori's work in modeling the complex hysteretic behavior of structural systems, including pinching phenomenon, cited in the literature as the Bouc-Wen-Baber-Noori model, is an original contribution that has been widely utilized in nonlinear random vibrations for seismic response analysis of concrete structures and has been incorporated in Open Sees seismic analysis program. His work in non-zero mean, non-Gaussian response analysis, and the first passage of hysteretic systems are also original contributions. Over the past 25 years, he has also carried out extensive work in seismic isolation of secondary systems, and the application of artificial intelligence methods for structural health monitoring, which have been widely cited. He has authored over 300 refereed papers, including over 150 journal articles, 12 scientific books, and has edited 25 technical and special journals and volumes. Noori has supervised over 100 graduate students and post-doc scholars and has presented over 100 keynotes, plenaries, and invited talks. He is the founding executive editor of an international journal and has served on the editorial boards of over ten other journals and as a member of numerous scientific and advisory boards. He has been a Distinguished Visiting Professor at several highly ranked global universities and directed the Sensors Program at the National Science Foundation in 2014. He has been a founding director or co-founder of three industry-university research centers and held chair professorships at two major universities. He served as the dean of engineering at Cal Poly for five years, served as the chair of the National Committee of Mechanical Engineering Department heads, and was one of seven co-founders of the National Institute of Aerospace, in partnership with NASA Langley Research Center. Noori also serves as the chief technical advisor for several scientific organizations and industries.

Dr. Tony T.Y. Yang is a professor at the University of British Columbia (UBC), Vancouver, Canada. He is a world-renowned researcher in structural and earthquake engineering, with over 200 high-impact refereed papers, books, and reports with over 2,300 citations. He has developed numerous novel technologies that have enhanced the practice of smart and sustainable infrastructure internationally. He is an active member of numerous national and international code committees, including one of the 19 voting members of the Standing Committee for Earthquake Design, which is responsible for writing the seismic design provision of 2020/2025 National Building Code of Canada

(NBCC). He has supervised over 170 highly qualified personnel and received over ten coveted international awards. He is the recipient of the 2014 CISC H.A. Krentz award, 2019 Technology award from the New Zealand Concrete Society, and 2020 Meritorious achievement award from Engineers & Geoscientists British Columbia, Canada, the most prestigious award given to a professional engineer (P.Eng.) in the province of British Columbia, Canada. He has worked with multiple engineering firms in Canada and around the world. His exemplary contributions to structural and earthquake engineering have significantly impacted earthquake engineering communities internationally.

Dr. Paulo B. Lourenço is a professor at the Department of Civil Engineering, University of Minho, Portugal, and head of the Institute in Sustainability and Innovation in Structural Engineering, with 250 researchers. Experienced in the fields of non-destructive testing, advanced experimental and numerical techniques, innovative repair and strengthening techniques, and earthquake engineering. He is a specialist in structural conservation and forensic engineering, with work on 100+ monuments, including 17 UNESCO World Heritage; leader of the revision of the European masonry code (EN 1996-1-1); coordinator of the European Master on Structural Analysis of Monuments and Historical Constructions, with alumni from 70+ countries and the European Heritage/Europa Nostra Award (the most prestigious in Europe); editor of the *International Journal of Architectural Heritage* and advisor of the Conference Series on Structural Analysis of Historical Constructions. He has supervised more than 60 PhD theses and coordinated multiple national and international research projects; awarded an Advanced European Research Council Grant to develop an integrated seismic assessment approach for heritage buildings; and coordinator of an Innovative Training Network on sustainable building lime applications via circular economy and biomimetic approaches with 15 PhD students across Europe.

Dr. Paolo Gardoni is the Alfredo H. Ang Family Professor and an Excellence Faculty Scholar in the Department of Civil and Environmental Engineering at the University of Illinois at Urbana-Champaign. He is also a professor in the Department of Biomedical and Translational Sciences at the Carle Illinois College of Medicine, and a fellow of the Office of Risk Management & Insurance Research in the Gies College of Business at the University of Illinois at Urbana-Champaign. Prof. Gardoni is the director of the MAE Center, which focuses on creating a Multi-hazard Approach to Engineering, the editor-in-chief of the journal *Reliability Engineering and System Safety*, and the founder and former editor-in-chief of the journal *Sustainable and Resilient Infrastructure*. His research interests include probabilistic mechanics; reliability, risk, and life cycle analysis; decision-making under uncertainty; performance assessment of deteriorating systems; modeling of natural hazards and societal impact; ethical, social, and legal dimensions of risk; optimal strategies for natural hazard mitigation and disaster recovery; and engineering ethics. Prof. Gardoni is the 2021 recipient of the Alfredo Ang Award on Risk Analysis and Management of Civil Infrastructure from the American Society of Civil Engineers for his contributions to risk, reliability, and resilience analysis, and his leadership in these fields.

Dr. Izuru Takewaki is a professor of building structures at Kyoto University and was the 56th president of the Architectural Institute of Japan (AIJ) during 2019–2021. He is the field chief editor of *Frontiers in Built Environment (Frontiers SA in Switzerland),* which includes 15 specialty sections related to various fields of built environment. His main interests are passive structural control, robust and resilient structural design of buildings, structural optimization, inverse problem in vibration, soil-structure interaction, critical excitation method for worst-case analysis, etc. He has published over 230 international journal papers and six monographs in English. The most recent one is 'An Impulse and Earthquake Energy Balance Approach in Nonlinear Structural Dynamics' from CRC Press in 2021. He was awarded numerous prizes, e.g., the Research Prize of AIJ (2004), the 2008 Paper of the Year in J. of The Structural Design of Tall and Special Buildings, the Prize of AIJ for Book (2014).

Dr. Eleni Chatzi is an associate professor and chair of Structural Mechanics and Monitoring at the Department of Civil, Environmental and Geomatic Engineering of ETH Zurich, Switzerland. Her research interests include the fields of structural health monitoring (SHM) and structural dynamics, nonlinear system identification, and intelligent life cycle assessment for engineered systems. She has authored more than 300 papers in peer-reviewed journals and conference proceedings, and further serves as an editor for international journals in the domains of dynamics and SHM. She led the recently completed ERC Starting Grant WINDMIL on the topic of "Smart Monitoring, Inspection and Life-Cycle Assessment of Wind Turbines". Her work in the domain of self-aware infrastructure was recognized with the 2020 Walter L. Huber Research prize, awarded by the American Society of Civil Engineers (ASCE).

Dr. Shaofan Li is currently a professor of applied and computational mechanics at the University of California, Berkeley. Dr. Li graduated from the Department of Mechanical Engineering at the East China University of Science and Technology (Shanghai, China) with a bachelor of science (B.S.) degree in 1982; he also holds master of science (M.S.) degrees in applied mechanics (Huazhong University of Science and Technology, Wuhan, China) and in aerospace engineering (the University of Florida, Gainesville, FL, USA) in 1989 and 1993, respectively. In 1997, Dr. Li received a PhD degree in mechanical engineering from Northwestern University (Evanston, IL, USA), and he had been a post-doctoral researcher at Northwestern University during 1997–2000. In 2000, Dr. Li joined the faculty of the Department of Civil and Environmental Engineering at the University of California, Berkeley. Dr. Shaofan Li is the recipient of the Fellow Award of the International Association of Computational Mechanics (IACM), the U.S. Association of Computational Mechanics (USACM) Fellow Award (2013), and the U.S. National Science Foundation (NSF) Career Award (2003). Dr. Li has published more than 200 technical papers in peer-reviewed scientific journals with an h-index of 53 (Google Scholar), and he also co-authored three research monographs/graduate textbooks.

Contributors

Naida Ademović
University of Sarajevo
Sarajevo, Bosnia and Herzegovina

Ali Akbari
Kharazmi University
Karaj, Alborz, Iran

Abdalla Alhashmi
Dalhousie University
Halifax, NS, Canada

Hamed Alizadeh
Kharazmi University
Karaj, Alborz, Iran

Harish Chandra Arora
AcSIR-Academy of Scientific and Innovative
    Research
Ghaziabad, India

Ersin Aydin
Nigde Omer Halisdemir University
Nigde, Turkey

Mohsen Azimi
The University of British Columbia (UBC)
Vancouver, BC, Canada

Yingnan Bao
The University of British Columbia (UBC)
Vancouver, BC, Canada

Behrouz Behnam
Amirkabir University of Technology
Tehran, Tehran, Iran

A. Bogdanovic
Institute of Earthquake Engineering and
    Engineering Seismology
Skopje, N. Macedonia

J. Bojadjieva
Institute of Earthquake Engineering and
    Engineering Seismology
Skopje, N. Macedonia

Weijia Cai
The University of British Columbia (UBC)
Vancouver, BC, Canada

Claudia Casapulla
University of Naples Federico II
Naples, Italy

Huseyin Cetin
Nigde Omer Halisdemir University
Nigde, Turkey

Paula Couto
Laboratório Nacional de Engenharia Civil
Lisboa, Portugal

Daniel Davies
University of Nottingham
Nottingham, England

Sevilay Demirkesen
Gebze Technical University
KOCAELİ, Turkey

K. Edip
Institute of Earthquake Engineering and
    Engineering Seismology
Skopje, N. Macedonia

Saeideh Farahani
Amirkabir University of Technology
Tehran, Tehran, Iran

Ehsan Noroozinejad Farsangi
The University of British Columbia (UBC)
Vancouver, BC, Canada

D. Filipovski
Institute of Earthquake Engineering and
    Engineering Seismology
Skopje, N. Macedonia

Simona Fontul
Laboratório Nacional de Engenharia Civil
Lisboa, Portugal

Kohei Fujita
Kyoto University
Kyoto, Japan

Vahidreza Gharehbaghi
The University of Kansas
Lawrence, KS, USA

I. Gjorgjeska
Institute of Earthquake Engineering and
    Engineering Seismology
Skopje, N. Macedonia

Erhan Güneyisi
Harran University
Şanlıurfa, Turkey

Esra Mete Güneyisi
Gaziantep University
Gaziantep, Turkey

Saurabh Gupta
Indian Institute of Technology Kanpur
Kalyanpur, Uttar Pradesh, India

Ming Hu
University of Maryland
College Park, Maryland, USA

Li Huan
Curtin University
Bentley, Perth, Australia

Lei Huang
The University of British Columbia (UBC)
Vancouver, BC, Canada

Koshin Iguchi
Kyoto University
Kyoto, Japan

Süleyman İpek
Bingol University
Bingöl, Turkey

D. Ivanovski
Institute of Earthquake Engineering and
    Engineering Seismology
Skopje, N. Macedonia

Hashem Jahangir
University of Birjand
Birjand, Iran

Muhammed Zain Kangda
REVA University
Bangalore, India

Nishant Raj Kapoor
AcSIR-Academy of Scientific and Innovative
    Research
Ghaziabad, India

Ali Khansefid
KN Toosi University of Technology
Tehran, Tehran, Iran

Koosha Khorramian
Norlander Oudah Engineering Limited
    (NOEL)
Calgary, Alberta, Canada

Soliman Khudeira
Illinois Institute of Technology
Chicago, Illinois, USA

T. Kitanovski
Institute of Earthquake Engineering and
    Engineering Seismology
Skopje, N. Macedonia

Denise-Penelope N. Kontoni
University of the Peloponnese
Patras, Greece

Aman Kumar
AcSIR-Academy of Scientific and Innovative
    Research
Ghaziabad, India

Tian Lan
RMIT University
Melbourne, Melbourne, Australia

H.H. Lavasani
Kharazmi University
Karaj, Alborz, Iran

Jianchun Li
University of Technology Sydney
Ultimo, NSW, Australia

Shuai Li
RMIT University
Melbourne, Melbourne, Australia

Yancheng Li
University of Technology Sydney
Ultimo, NSW, Australia

Ali Maghsoudi-Barmi
KN Toosi University of Technology
Tehran, Tehran, Iran

Arman Mamazizi
University of Kurdistan
Sanandaj, Iran

F. Manojlovski
Institute of Earthquake Engineering and
    Engineering Seismology
Skopje, N. Macedonia

I. Markovski
Institute of Earthquake Engineering and
    Engineering Seismology
Skopje, N. Macedonia

Priyan Mendis
RMIT University
Melbourne, Melbourne, Australia

Pravin R. Minde
MIT World Peace University
Kothrud, Pune, India

Arsalan Mousavi
University of Kurdistan
Sanandaj, Iran

Elham Mousavian
University of Naples Federico II
Naples, Italy

N. Naumovski
Institute of Earthquake Engineering and
    Engineering Seismology
Skopje, N. Macedonia

Mohammad Noori
California Polytechnic State University
San Luis Obispo, CA, USA

Fadi Oudah
Dalhousie University
Halifax, NS, Canada

Baki Ozturk
Hacettepe University
Ankara, Turkey

Xiao Pan
The University of British Columbia (UBC)
Vancouver, BC, Canada

Dipak Patil
MIT World Peace University
Kothrud, Pune, India

Mehrdad Piri
Kharazmi University
Karaj, Alborz, Iran

A. Poposka
Institute of Earthquake Engineering and
    Engineering Seismology
Skopje, N. Macedonia

Arash Karimi Pour
University of Texas at El Paso (UTEP)
El Paso, Texas, USA

Rohan Raikar
REVA University
Bangalore, India

Z. Rakicevic
Institute of Earthquake Engineering and
    Engineering Seismology
Skopje, N. Macedonia

Milad Roohi
University of Nebraska–Lincoln
Omaha, NE, USA

Filipa Salvado
Laboratório Nacional de Engenharia Civil
Lisboa, Portugal

Bijan Samali
Western Sydney University
Penrith, NSW, Australia

Abhaysinha G. Shelake
MIT World Peace University
Kothrud, Pune, India

V. Sheshov
Institute of Earthquake Engineering and
    Engineering Seismology
Skopje, N. Macedonia

Ali Shojaeian
University of Oklahoma
Norman, Oklahoma, USA

A. Shoklarovski
Institute of Earthquake Engineering and
    Engineering Seismology
Skopje, N. Macedonia

Maria João Falcão Silva
Laboratório Nacional de
    Engenharia Civil
Lisboa, Portugal

Miroslaw J. Skibniewski
University of Maryland
College Park, Maryland, USA

Izuru Takewaki
Kyoto University
Kyoto, Japan

Manzoor Tantray
National Institute of Technology
    Srinagar
Jammu and Kashmir Srinagar, India

Sina Tavasoli
The University of British
    Columbia (UBC)
Vancouver, BC, Canada

Serik Tokbolat
University of Nottingham
Nottingham, England

Selcuk Toprak
Gebze Technical University
KOCAELİ, Turkey

Phuong Tran
RMIT University
Melbourne, Melbourne, Australia

Zubair Rashid Wani
Indian Institute of Science Bangalore
Bangalore, India

Yifei Xiao
The University of British Columbia (UBC)
Vancouver, BC, Canada

Tony T.Y. Yang
The University of British Columbia (UBC)
Vancouver, BC, Canada

Yang Yu
UNSW
Sydney, NSW, Australia

Hao Xuan Zhang
The University of British Columbia (UBC)
Vancouver, BC, Canada

Zhengbo Zou
The University of British Columbia (UBC)
Vancouver, BC, Canada

1 Autonomous Inspection and Construction of Civil Infrastructure Using Robots

Yifei Xiao, Xiao Pan, Sina Tavasoli, Mohsen Azimi,
Yingnan Bao, Ehsan Noroozinejad Farsangi, and Tony T.Y. Yang

CONTENTS

1.1 INTRODUCTION

In recent years, with the advancement of control and sensing hardware as well as computational capabilities, robotic technologies have gained more traction in research developments and field applications in various scientific fields (Bock and Linner, 2016; Contreras, Wilkinson, and James, 2021). The use of robotic technologies is meant to enhance the degree of autonomy to reduce and eventually eliminate dependency on human labor, thus enhancing the work efficiency, lowering the monetary and time cost, and creating a more economical and sustainable built environment.

Robotic technologies have been adopted into civil and structural engineering fields for construction and inspection of civil infrastructures. Robots can be divided into different categories based on their mobility, such as unmanned aerial vehicle (UAV) (flying), unmanned ground vehicle (UGV) (ground/surface-locked), climbing robots, robotic arms, and a collaborative scheme of these. Research on automated construction methods has been attempted using various robotic technologies. Earlier, UAVs have been used for construction of structures (e.g., Lindsey, Mellinger, and Kumar, 2011; Augugliaro et al., 2014; Braithwaite et al., 2018). These studies have shown that the UAVs generally have very limited payload and short flight times, which are not suitable to construct structural components with a moderate to heavy weight. Robotic arms can be adopted for automated construction procedures (Koerner-Al-Rawi et al., 2020; Liang, Kamat, and Menassa, 2020; Qiao et al., 2021). However, these studies mainly focus on small-scale structure construction in laboratory environments. More research about large-scale structure construction should be conducted.

DOI: 10.1201/9781003325246-1

1

The use of climbing robots and ground-locked robots for construction tasks was also attempted (e.g., Allwright et al., 2014; Cucu, Rubenstein, and Nagpal, 2015; Soleymani et al., 2015). These studies were mainly carried out on planar structures in a well-controlled laboratory environment. Their real-world applicability requires further investigation. On the other hand, automated inspection methods using robots have been investigated on civil structures (Azimi, Eslamlou, and Pekcan, 2020; Ayala-Alfaro et al., 2021). In general, compared to ground vehicles, the advantages of UAVs are their flexibility in flying at different altitudes and their ease in flying over ground obstacles, which makes them an ideal candidate to inspect the upper portion of large-scale civil structures such as high-rise buildings and bridges. UGVs, on the other hand, have much higher payload than UAVs, which allows them to carry more types of sensory and control units, and have a prolonged battery life. Therefore, they are one of the best candidates for interior investigation for buildings and pipelines.

1.2 LITERATURE REVIEW

1.2.1 Automatic Construction and Robot Control

Compared with traditional construction methods, automatic construction can reduce construction cost, improve working efficiency, ensure construction safety, and control construction progress and quality. Robotic cranes can be used to hoist and place heavy-weight and large-scale structural components and robotic arms can be implemented to grasp and install light-weight and small-scale structural components.

For heavy-weight and large-scale structural components construction, robotic cranes are more suitable than robotic arms because the capability and working range of robotic arms are very limited. So far, a number of research investigations have been conducted for robotic cranes. Dutta et al. (2020) implemented computer-aided lift planning (CALP) systems for robotic cranes. This CALP system provides optimal solutions for path planning and lifting work through computer graphics and simulations supported by intelligent decision-making and planning algorithms. Zi, Lin, and Qian (2015) introduced a cooperative localization scheme and an improved localization algorithm for mobile cranes. They also used a global path planning method for obstacle avoidance and adopted a four-point collaborative leveling method for automatic leveling control of mobile cranes. Kang et al. (2011) designed a robotic-crane-based automatic construction system for high-rise steel structures. This system is expected to improve the construction efficiency and alleviate problems related to the lack of skilled work. Zhao et al. (2022) proposed a trajectory prediction approach to address the interference handling and crane scheduling problems for robotic twin-crane systems. This method is proved to be able to increase accuracy by 20% through simulation. Lei et al. (2013) introduced a generic lifting binary path-checking method for mobile cranes. This method considers the minimum and maximum crane lift radius and the modules' erection orders. Through the literature review on robotic crane, it is found that most studies focus on trajectory planning and obstacle avoidance. More research toward automatic construction using robotic cranes should be carried out in the future.

The robotic arm is more suitable for light-weight and small-scale structural components construction due to its flexibility and high accuracy. Qiao et al. (2021) used a robotic arm to construct a new mortise and tenon timber structure. Koerner-Al-Rawi et al. (2020) proposed an efficient workflow for robotic assembly of a discrete timber structure. These two studies proved that construction efficiency and precision were improved by automatic construction using robotic arm. Apolinarska et al. (2021) applied Ape-X DDPG reinforcement learning algorithms for the assembly of timber joints using robotic arms. The robotic arms were trained in simulation environments, and then the control policy was successfully deployed in reality. The experiments showed that the control policy also considered tolerances and shape variations of the joints. Liang, Kamat, and Menassa (2020) applied a human demonstration method to teach a robotic arm how to install ceiling tiles. A number of virtual and real demonstration videos were used to teach the robotic arm in a simulation environment, and the simulation results presented a 78% success rate in installing ceiling tiles. Ying et al. (2021) developed a deep learning-based optimization algorithm for path planning and collision avoidance of robotic arms. The developed algorithm was proved to be able to reduce the path length with shorter computational time when the robotic arm is operating.

The joints of a robot are usually powered by electric motors, hydraulic systems, or pneumatic systems. All of these power methods rely on control algorithms to achieve high-accuracy performance. Control algorithms have been widely used in various engineering fields, such as shake table control (Yang et al., 2015; Xiao, Pan, and Yang, 2022), robotic arm control (Wai and Muthusamy, 2013), and flight control (Ignatyev, Shin, and Tsourdos, 2020), etc. Among all the control algorithms, the proportional-integral-derivative (PID) controller is the most commonly used control method in industry. The PID control algorithm is simple and the selection of parameters is relatively easier compared to other control algorithms. It also has good adaptability and strong robustness, and can be applied to various industrial applications. However, PID control is not suitable for a system with high nonlinear properties and other complicated cases. Therefore, more advanced control methods have been proposed to improve the control performance of robotic arms. For instance, Wang (2016) introduced two adaptive control schemes to consider the uncertain kinematics and dynamics of robotic arms. The control performance of these two schemes was examined through numerical simulations. Poignet and Gautier (2000) proposed a nonlinear model predictive control method to consider the nonlinear properties of a robotic arm. Yang, Fukushima, and Qin (2011) implemented a decentralized adaptive robust controller with disturbance observer for trajectory tracking of robotic arms. Hence, when PID control performance is not desired and cannot meet the requirements for robot applications, more advanced control methods can be implemented to control the joints of robotic arms.

1.2.2 AUTONOMOUS INSPECTION WITH DIFFERENT ROBOTIC TECHNOLOGIES

Recent advances of image processing techniques (IPTs) have demonstrated their success for vision-based structural inspection. In the past, traditional IPTs were usually developed using handcrafted features, which are prone to fail in real-world situations where there exist many external disturbances and noise. More recently, convolutional neural network (CNN)-based vision methods have been shown to achieve a substantial increase in accuracy, and higher robustness against image noise and environmental disturbances. In the past few years, CNN-based vision methods have been widely attempted to inspect various structural systems and components (Spencer, Hoskere, and Narazaki, 2019), such as reinforced concrete (RC) structures (Gao and Mosalam, 2018; Liang, 2019; Ni, Zhang, and Chen, 2019; Pan and Yang, 2020), steel structures (Yeum and Dyke, 2015; Yun et al., 2017; Cha et al., 2018; Kong and Li, 2018; Pan and Yang, 2022a, b), masonry structures (Wang et al., 2018; Wang et al., 2019), and structural bolted assemblies (Ramana, Choi, and Cha, 2019; Pan and Yang, 2022). Although these studies have well demonstrated the effectiveness of the vision methods in damage inspection, there are several limitations: (1) most of the studies assume image data were manually collected, without further attempts in automating the data collection process; (2) most of the studies were built upon 2D computer vision, which has shown promising results in the qualitative judgment of the structural damage state, but cannot provide precise damage localization and quantification in 3D space.

More recently, research has been shifted toward automating the data collection and evaluating structural conditions from 2D space to 3D space. Automation of the data collection process is typically done by the use of advanced robotic technologies. For example, recent research has attempted to design autonomous single-agent UAV or UGV systems for data collection and damage detection purposes. The UAVs and UGVs equipped with various types of sensing and controlling units can greatly enhance the degree of automation in data collection process, and the level of damage evaluation in 3D space. However, the majority of the existing studies have been conducted in a controlled environment and they are inefficient if occluded objects and obstacles make navigation challenging for certain robots. Each type of robot has limitations, for example, UAVs are constrained by payload and flight time, which are not an issue for UGVs. Since UGVs operate on the ground, their payloads are much larger than UAVs and can operate for many hours and days. In addition, UGVs can serve UAVs as portable charging stations and can carry high-performance processing units for real-time decision-making. On the other hand, UGVs have limitations in speed, field of view, and obstacle avoidance.

These days UAVs are controlled by iPhone or tablet and can be equipped with cameras, sensors, or other intelligent devices providing useful information for different applications such as agriculture, inspection of infrastructures, mapping, and constructional monitoring. In construction, they are used for various goals such as monitoring the performance of the construction site (Howard, Murashov, and Branche, 2018; Fernandez Galarreta, Kerle, and Gerke, 2015; Freimuth and König, 2015; Lin, Han, and Golparvar-Fard, 2015), safety (Feifei et al., 2012; Yamamoto, Kusumoto and Banjo, 2014; De Melo et al., 2017; Moeini et al., 2017; Álvares, Costa, and de Melo, 2018), 3D modeling and photogrammetric applications for reconstruction of buildings (Eschmann et al., 2012; Irizarry, Gheisari, and Walker, 2012; Álvarez et al., 2014; Gheisari, Irizarry, and Walker, 2014; Zollmann et al., 2014; Shang and Shen, 2017), and monitoring of damages and cracks in buildings and bridges (Rodriguez-Gonzalvez et al., 2014; Zhu, 2015). Most inspections using UAVs are basically done using a pilot. This requires an expert force to do the monitoring and piloting which might be tedious and sometime subjective to the pilot expertise and knowledge. Elimination of this role provides more efficiency and higher level of autonomy in inspections and assessments. More recently, research has shown that UAVs can operate autonomously in both indoor and outdoor environments. Since GPS signals are usually blocked by the structure, the autonomous navigation is usually achieved by simultaneous localization and mapping (SLAM) methods, using advanced sensors such as Light Detection and Ranging (LiDAR) and Radio Detection and Ranging (RADAR). This method depends on advanced modules and sensors to localize the UAV and generate a map of the exploring area. Using LiDAR and other sensors for construction monitoring and SLAM provides high accuracy. However, it creates limitations such as high cost and high payload for UAVs, which requires the use of medium- or large-size UAVs. Application of the medium- or large-size UAVs for construction assessments or structural monitoring might be difficult, especially when dealing with interior areas where there are narrow passages such as openings and doors. This limits the application of UAV for indoor assessments. Therefore, small-size UAVs, named micro aerial vehicles (MAVs), are more suitable candidates for interior assessment. Autonomous assessment and monitoring of the construction using MAVs is the emerging area that requires more research and field validations.

On the other hand, UGVs are similar to MAVs in many ways, with limited spatial maneuverability, but much longer battery life. Modern UGVs play important roles in many industries and vary in chassis typologies (differential drive, mecanum wheel, tracked, etc.) and equipment installed on them (camera, sonar, robotic arm, etc.), depending on the usage scenes. Compared to MAVs, UGVs don't need to lift themselves against gravity, a much more powerful battery package and computer can be easily installed. Benefiting from these features, UGVs can be deployed for autonomous construction assessment and inspection. With 5G and other telecom technologies (MAKINO et al., Voigtländer et al., 2017), using multiple UGVs on-site with distributed computing (Voigtländer et al., 2017) is possible. With the powerful onboard computer, signal processing (Manzano et al., 2019) and machine vision (Moemen et al., 2020) can be embedded and implemented into the robot. In recent years, research on UGVs for structural engineering applications has been attempted. Phillips and Narasimhan (2019) employed a Jackal ground robot to automatically obtain the 3D point cloud data of a bridge using a predetermined inspection plan which was manually created. Automating the inspection route remains challenging for full-scale bridges, which is an active area of research.

McLaughlin, Charron, and Narasimhan (2020) employed a Husky ground robot to detect spalls and delamination for a concrete bridge, using vision and LiDAR-based methods. The methods were limited to two damage types, and further research is needed to detect other damage types such as concrete cracks and steel reinforcement exposure. Yuan et al. (2022) designed a robot using various types of sensors such as stereo camera, LiDAR, and inertial measurement unit (IMU), to inspect concrete damages. While their robot was deployed to inspect a local portion of a laboratory concrete specimen, the capability of the robot in the field applications still remains unexplored. In addition, due to limited sensing abilities and 3D reconstruction accuracy, concrete cracks less than 10 mm cannot be quantified.

To address these limitations, a multi-robot system that consists of various types of robots such as MAVs, UGVs, or climbing robots, equipped with different sensors with high enough precision, should be first developed. Second, advanced autonomous navigation, control, sensing, and their respective processing algorithms should be developed in the virtual environment to examine their potential for navigation and inspection. Third, the multi-robot system should be validated on different real-world infrastructures such as buildings and bridges.

1.3 METHODOLOGY

1.3.1 Automatic Construction

In this section, the robotic crane is used to illustrate methodologies of achieving automatic construction. Similar methodologies are also applicable for robotic arms. Forward kinematics and inverse kinematics are first introduced. Then, the PID control algorithm used to control the joints of robots is presented. Finally, visual grasping using OpenCV, YOLOv3-tiny, and reinforcement learning is illustrated.

1.3.1.1 Forward Kinematics (FK) and Inverse Kinematics (IK)

In the process of robot modelling, the robotic crane is expressed as a series of links and joints. These links can be straight or bent and the joints can be prismatic or revolute. In robot kinematics, forward kinematics (FK) is used to calculate the position and the rotation of the end joint of a robotic crane given the angles of each joint. On the contrary, inverse kinematics (IK) is used to calculate the required angles of each joint given the position and rotation of the end of a robotic crane.

In FK, the pose (including position and rotation) of each joint relative to the previous joint should have six degrees of freedom (three positions and three rotations). This pose relationship can be expressed by a transformation matrix, which will be multiplied in turn to solve the pose of the end joint of the robotic crane in the robot base coordinate system. In order to describe the translational and rotational relationship between two interconnected links, the Denavit–Hartenberg (DH) convention (Denavit and Hartenberg, 1955) is used when analyzing robotic kinematics. Figure 1.1 shows the description of links and joints using DH parameters, where α, a, d, and θ are the link twist, link length, link offset, and joint angle respectively. $\hat{X}$, $\hat{Y}$, and $\hat{Z}$ are the X-axis, Y-axis, and Z-axis, respectively. i-1 represents a notation for the last link or joint while i represents a notation for the current link or joint. Then, the transformation matrix (T_i^{i-1}) of a joint i relative to the previous joint i-1 can be expressed in Equation (1.1).

$$T_i^{i-1} = \begin{bmatrix} \cos\theta_i & -\sin\theta_i & 0 & a_{i-1} \\ \sin\theta_i\cos\alpha_{i-1} & \cos\theta_i\cos\alpha_{i-1} & -\sin\alpha_{i-1} & -\sin\alpha_{i-1}d_i \\ \sin\theta_i\sin\alpha_{i-1} & \cos\theta_i\sin\alpha_{i-1} & \cos\alpha_{i-1} & \cos\alpha_{i-1}d_i \\ 0 & 0 & 0 & 1 \end{bmatrix} \qquad (1.1)$$

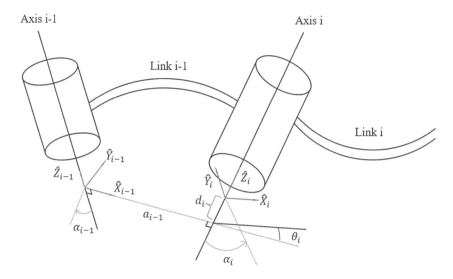

FIGURE 1.1 Description of links and joints using DH parameters.

where the rotation matrix (R) is shown as Equation (1.2) and translation matrix (T) is shown in Equation (1.3)

$$R = \begin{bmatrix} \cos\theta_i & -\sin\theta_i & 0 \\ \sin\theta_i \cos\alpha_{i-1} & \cos\theta_i \cos\alpha_{i-1} & -\sin\alpha_{i-1} \\ \sin\theta_i \sin\alpha_{i-1} & \cos\theta_i \sin\alpha_{i-1} & \cos\alpha_{i-1} \end{bmatrix} \quad (1.2)$$

$$T = \begin{bmatrix} a_{i-1} \\ -\sin\alpha_{i-1} d_i \\ \cos\alpha_{i-1} d_i \end{bmatrix} \quad (1.3)$$

Therefore, the FK of a robotic crane from its base coordinate system to its nth joint coordinate system can be expressed using Equation (1.4).

$$T_n^{base} = T_1^{base} T_2^1 T_3^2 \cdots T_{n-1}^{n-2} T_n^{n-1} \quad (1.4)$$

This FK process is straightforward and has a unique solution. Compared to FK, IK is more complicated and sometimes it may have multiple solutions. There are three types of methods to solve an FK problem, which are the geometric method, algebraic method, and numerical method. For the algebraic method and numerical method, a number of methods have been widely used in robot kinematics problem, such as KDL-IK (ROS Wiki, 2014), TRAC-IK (Beeson and Ames, 2015), and IK-FAST (OpenRAVE, 2011), etc. Compared to geometric methods, algebraic and numerical methods are better when a complicated IK problem needs to be solved (e.g., IK for multi-degree of freedom robots). Here, the geometric method is taken as an example for concept illustration. Figure 1.2 shows a two-degree-of-freedom simplified robot. The target coordinate of the end joint of this robot is already known and expressed as (x, y). According to the law of cosines, Equations (1.5) and (1.6) can be obtained:

$$x^2 + y^2 = l_1^2 + l_2^2 - 2l_1l_2 \cos\left(180° - \theta_2\right) \quad (1.5)$$

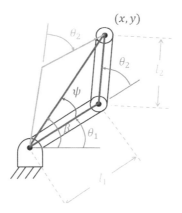

FIGURE 1.2 Two-degrees-of-freedom robot.

$$l_2^2 = l_1^2 + \left(x^2 + y^2\right) - 2l_1\sqrt{x^2 + y^2}\cos\psi \tag{1.6}$$

Then, the angle θ_2 and angle ψ can be calculated as:

$$\theta_2 = \arccos\left(\frac{x^2 + y^2 - \left(l_1^2 + l_2^2\right)}{2l_1l_2}\right) \tag{1.7}$$

$$\psi = \arccos\left(\frac{l_2^2 - \left(x^2 + y^2\right) - l_1^2}{-2l_1\sqrt{x^2 + y^2}}\right) \tag{1.8}$$

Therefore, the angle θ_1 can be calculated using Equation (1.9):

$$\theta_1 = \begin{cases} arctan2\left(y,x\right) + \psi, & \theta_2 < 0 \\ arctan2\left(y,x\right) - \psi, & \theta_2 < 0 \end{cases} \tag{1.9}$$

1.3.1.2 Low-Level Control Algorithms of Robotic Crane

The joints of a robotic crane can be simply controlled by the PID control algorithm. PID is a close-loop control framework, which calculates the error between a desired input signal and a feedback signal measured from sensors. The error will be minimized through proportional term (P gain), integral (I gain) term, and derivative (D gain) term. Figure 1.3 shows the block diagram of the PID control framework and it can be expressed in Equation (1.10). The rotation angles of electric motors are measured by sensors and this measured rotation angle is the feedback signal in the PID control algorithm. When required angle signals are sent to the electric motors of the robotic crane, the input angle signals will be subtracted from the measured feedback signals to obtain the error. Finally, the error will be minimized through P gain, I gain, and D gain.

$$u(t) = K_p e(t) + K_i \int_0^t e(\tau)d\tau + K_d \frac{de(t)}{dt} \tag{1.10}$$

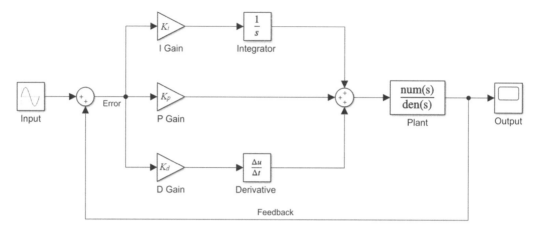

FIGURE 1.3 PID control framework.

1.3.1.3 Visual Grasping and Camera Calibration

In Section 1.3.1.1, FK and IK are introduced to let the end joint of a robotic crane move to a desired position. If an end effector, such as a gripper, is installed to the end joint, the robotic crane will be able to grasp the object based on FK or IK results. In a real situation, if a robotic crane is required to grasp an object, IK is applied more frequently than FK because it is easier to obtain the coordinates of the object rather than the required joint angles of the robotic crane. However, if only IK is implemented to control the robotic crane, the position and the orientation of the object needs to be defined manually. This also indicates that the robotic crane cannot localize the object by itself if the position of the object is changed and the corresponding pose is not given. Hence, in this case, vision technologies need to be implemented to solve this problem. By utilizing vision technologies, robotic cranes will be able to identify and localize the object by themselves, and then achieve grasping and placement automatically.

1.3.1.3.1 Camera Calibration: Intrinsic and Extrinsic Matrices

In visual grasping, cameras are often used as "eyes" of robotic cranes and these "eyes" can help robotic cranes find the position and orientation of the object automatically. The robotic cranes will first obtain the coordinate of an object in the image coordinate system. By utilizing the camera matrix (P), this image coordinate will then be transformed to a coordinate in a world coordinate system, which is normally selected as the base of a robotic crane. This transformation is presented as Equation (1.11)

$$x = \mathbf{PX} \tag{1.11}$$

where x is the coordinate in the image coordinate system and $\mathbf{X}$ is the coordinate in the world coordinate system. The camera matrix (P) is defined as Equation (1.12).

$$P = \begin{bmatrix} f_x & 0 & x_0 \\ 0 & f_y & y_0 \\ 0 & 0 & 1 \end{bmatrix} \begin{bmatrix} \mathbf{R} & \mathbf{T} \\ 0 & 1 \end{bmatrix} \tag{1.12}$$

where f_x and f_y are the focal length of camera, x_0 and y_0 are the principal point offset. $\mathbf{R}$ is the rotation matrix and $\mathbf{T}$ is the translation matrix. The camera matrix P includes intrinsic matrix (P_{in}) and extrinsic matrix (P_{ex}), which are shown in Equations (1.13) and (1.14), respectively. The intrinsic matrix determines the coordinate transformation from the image coordinate system to the camera coordinate system. The extrinsic parameters describe the geometrical relation between the camera coordinate system and the world coordinate system.

$$P_{in} = \begin{bmatrix} f_x & 0 & x_0 \\ 0 & f_y & y_0 \\ 0 & 0 & 1 \end{bmatrix} \tag{1.13}$$

$$P_{ex} = \begin{bmatrix} R & T \\ 0 & 1 \end{bmatrix} \tag{1.14}$$

where $R = \begin{bmatrix} r_{11} & r_{12} & r_{13} \\ r_{21} & r_{22} & r_{23} \\ r_{31} & r_{32} & r_{33} \end{bmatrix}$, $T = \begin{bmatrix} t_x & t_y & t_z \end{bmatrix}^T$. r_{11} to r_{33} describe the orientation relation between two coordinate systems. t_x, t_y, and t_z illustrate the position relation between two coordinate systems.

Camera calibration is usually conducted before visual grasping to ensure the accuracy of a camera, and the camera matrix P is obtained through camera calibration. Zhang's chessboard camera calibration method (Zhang, 1999) is widely used in the world. This calibration method is already built into the robot operating system (ROS) so that people can easily calibrate their camera using ROS. The general steps are summarized below:

1. Prepare a chessboard where the size of the chessboard is known. Use the camera, which needs to be calibrated, to take a number of pictures of it from different angles.
2. The feature points in the image, such as the corners of the chessboard, are detected to obtain the pixel coordinates. The world coordinates of the corners of the chessboard are calculated according to the known chessboard size and the origin of the world coordinate system.
3. Solve the intrinsic matrix (P_{in}) and extrinsic matrix (P_{ex}) based on Zhang's chessboard camera calibration method.

1.3.1.3.2 Visual Grasp Using Computer Vision, YOLOv3 and Reinforcement Learning

In order to position the structural components at the right location, it is required to first localize the structural components. For this purpose, real-time computer vision-based object detectors such as You-Only-Look-Once (YOLO) can be used. In this section, a real-time specific variant of YOLO named YOLOv3-tiny, developed by Pan and Yang (2022) can be used and re-trained to localize different types of structural components. The YOLOv3-tiny has only 44 layers, which is significant less than the original YOLOv3, leading to high efficiency while still maintaining satisfactory accuracy. Figure 1.4 depicts the architecture of the YOLOv3-tiny where it takes an input image of 416×416 in pixel dimensions and processes the image with a series of convolution (Conv), batch normalization (BN), rectified linear unit (ReLU), and depth concatenation. The output of YOLOv3-tiny includes the bounding boxes for the localization of targeted structural components in an image. After the localization of the structural components, the pixel coordinates of the bounding boxes can be obtained in 2D space. The pixel coordinates are then mapped to the 3D world coordinates in the real world with the camera calibration outcomes as shown in Section 3.1.3, where the robotic cranes can be automatically controlled to localize the structural components using the 3D coordinates.

By implementing robot IK, OpenCV and the YOLOv3-tiny, the robots are able to conduct some simple construction tasks, such as object identification, localization, grasp and placement. For more complex tasks, the reinforcement learning algorithms can be used to train the robot and let it know how to successfully complete a complicated task. The Actor-Critic reinforcement learning algorithm (Konda and Tsitsiklis, 1999), which includes a policy network and a critic network, is the foundation of many advanced reinforcement learning algorithms. The policy network is used to

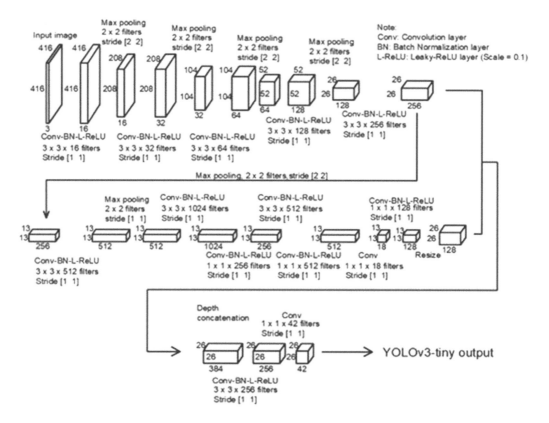

FIGURE 1.4   Architecture of the YOLOv3-tiny.

generate actions of the robotic crane while the critic network is used to evaluate how good the generated actions are. Figure 1.5 shows the framework of the Actor-Critic reinforcement learning method. Deep Deterministic Policy Gradient (DDPG) (Silver et al., 2014), which is an improved Actor-Critic reinforcement learning algorithm, can enable the robot to learn more effectively in continuous actions. Compared to the Actor-Critic, a target network is added to the policy network and critic network respectively to improve the training efficiency and the training success rate. During training, the parameters of policy network, critic network, and two target networks are updated so that the policy network can generate better action for robots for a specific state and the critic network can evaluate the actions more accurately. After training, only the policy network will be implemented and it will generate an optimal action to control the robot for a specific state.

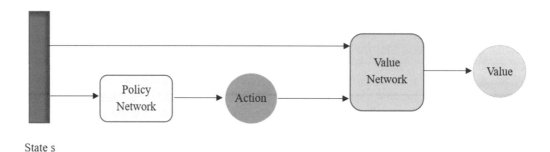

FIGURE 1.5   Framework of Actor-Critic reinforcement learning method.

1.3.2 Autonomous Inspection and Assessment

As described in Section 2, multi-robot systems should be considered to achieve a comprehensive inspection of civil structures. For this purpose, this section provides a detailed description of key algorithms, software platforms, and hardware units required for MAVs and UGVs to achieve autonomous navigation and inspection.

1.3.2.1 Collision-Free Simultaneous Localization and Mapping (SLAM) Using UGV

The goal of a SLAM algorithm is to create a map of the environment using the mounted vision-based sensors, such as RGB-D or LiDAR cameras for creating 3D SLAM and 2D Occupancy Grid Mapping, respectively. The LiDAR is a time-of-flight (ToF) sensor that is used for measuring the distance of objects by emitting laser pulses and measuring the time for the reflected pulses from an object. Because LiDAR-based SLAM has a relatively high precision compared to RADAR and vision-based SLAM, they are popular for autonomous and robotic applications, particularly in low-light conditions.

One of the challenging problems in robotics is the kidnapped robot problem (KRP), when a robot is placed in an unknown environment and forced to move instantly to create a map of its surroundings (Yu et al., 2020). Therefore, the localization and mapping algorithm is a critical component of an autonomous robot system. There are several path planning algorithms to solve collision-free navigation problems. The rapidly exploring random tree (RRT) algorithm (LaValle, 1998) is a sampling-based path planning algorithm that efficiently searches non-convex high-dimensional spaces to create a graph of routes until it finds a path between 'start' and 'goal' points. The pseudo-code as well as an example of the RRT algorithm are given in Figure 1.6, and more details about the algorithm are available from Karaman and Frazzoli (Karaman and Frazzoli, 2011). Details of the mapping algorithm can be accessed from the ROS website (wiki.ros.org/gmapping).

1.3.2.2 Autonomous Navigation Using MAV

Implementation of the MAVs for indoor navigation and inspection in a single pipeline is assessed using a new navigation algorithm and vision techniques (Tavasoli, Pan and Yang, 2023). The

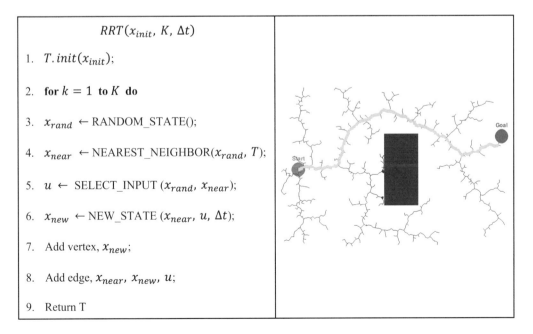

FIGURE 1.6 Collision-free navigation using the RRT algorithm.

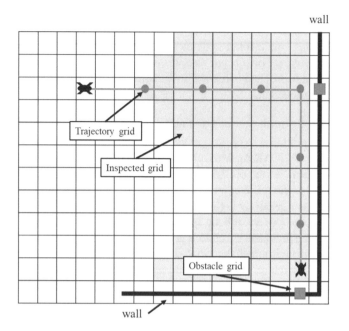

FIGURE 1.7 Grid representation of the floor plan.

navigation algorithm converts the scanning area into grids (Figure 1.7) and tries to move away from the formerly explored obstacle and trajectory grids in order to maximize the exploring area. For this purpose, when the MAV faces an obstacle, the exploring area will be divided into four zones with respect to the current location of the MAV. Then the least explored area will be determined using Equation (1.15). In this equation, T_i, O_i, and I_i are the distance between the center of the ith trajectory, obstacle, and inspected grids in a specific zone with respect to the MAV's current location, respectively. a is the power coefficient to weigh each grid and assess its effect on the navigation. The MAV avoids obstacles during the navigation using a time-of-flight (ToF) distance sensor installed on the front side of the MAV. The scanning is being done using a RGB camera installed on the front side of the MAV. In order to keep the MAV sufficiently small and economical, only one ToF sensor and one RGB camera are used in the presented methodology.

$$Explored\ factor = \sum_{1}^{n} \frac{1}{T_i^a} + \sum_{1}^{m} \frac{1}{O_i^a} + \sum_{1}^{p} \frac{1}{I_i^a} \tag{1.15}$$

1.3.2.3 Vision-Based Structural Damage Inspection

Images acquired by UAVs or UGVs can be automatically processed by computer vision methods to detect structural components and their damages. In this section, reinforced concrete (RC) columns and two of the severe and common damage types (i.e., spalling and steel reinforcement) are considered. Identification of these two selected damage types is crucial as they are closely related to the most severe damage state of the RC structures (Federal Emergency Management Agency [FEMA], 2012). Within this context, two CNN vision-based detection methods are developed and applied. On one hand, a YOLOv3-tiny architecture established by Pan and Yang (2022) is adopted and trained to localize RC columns and exposed steel reinforcement. Second, the DeepLabv3+ algorithm (Chen et al., 2018) is used to detect concrete spalling. As the original DeepLabv3+ algorithm (Chen et al., 2018) contains a relatively deep and complex backbone network to achieve high accuracy in segmenting multiple classes. In this study, only one class (i.e., concrete spalling) is considered, and the

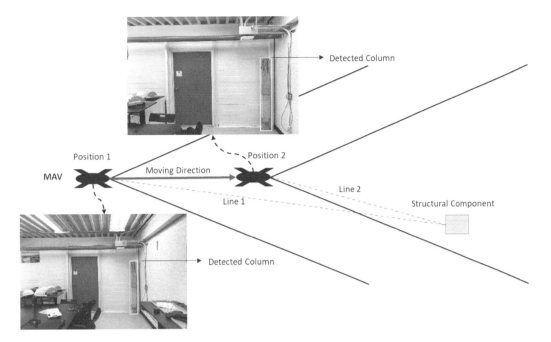

FIGURE 1.8   Illustration of structural component localization.

use of a complex backbone network is unnecessary. Therefore, to maintain a good balance between speed and accuracy, the DeepLabv3+ incorporates a lightweight backbone, ResNet-18 (He et al., 2016), to achieve real-time efficiency.

Once the structural component is identified by the vision-based algorithms, the detected bounding boxes within two consecutive image frames can be used to estimate the location of the structural components (Figure 1.8). The algorithm uses the relative position of the detected component in each of the frames to estimate the position of the component with respect to the current UAV's location. The global location of the MAV is known because of the embedded inertia device mounted on MAV. Hence, the global location of the damaged component can be calculated.

1.4   IMPLEMENTATION

1.4.1   Automatic Construction Using Robotic Cranes

The robot operating system (ROS) is an open-source platform, which provides lots of software libraries and tools that can help people to develop their own robotic applications. In this section, a mobile robotic crane is used and simulated in the ROS virtual environment for a small floor system construction. In ROS, this robotic crane model is generated by writing codes in URDF format files. Gazebo software is a 3D physical simulation platform, which is used to create a virtual simulation environment in ROS. Rviz is a 3D visualization tool that contains many plugins that can be used to display camera images, robotic models and robot movement trajectories, etc. In this floor construction demo, Gazebo is used to create ground, wooden columns and floors, and Rviz is implemented to provide plugins for visualization. In addition, the MoveIt plugin is also used combined with Gazebo and Rviz for path planning, as well as FK and IK analysis. For IK analysis, the TRAC-IK method (Beeson and Ames, 2015) is selected because it is easier to install than IK-FAST (OpenRAVE, 2011) and has higher stability than KDL-IK (ROS Wiki, 2014). PID controllers, introduced in Section 3.1.2, are implemented for all the joints of the robotic crane.

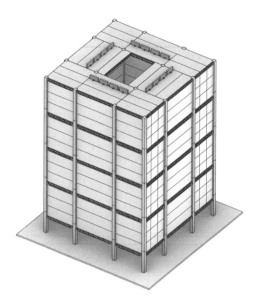

FIGURE 1.9   Wooden modular structure.

Figure 1.9 shows a wooden modular structure, and the target of this construction demo is to construct a scaled-down modular floor system for the first story using the robotic crane. The entire construction procedure can be divided into two stages. The first stage is to grasp, lift, and install the column on the ground (Figure 1.10(a)) and the second stage is to install the floors on the columns (Figure 1.10(b)). Visual grasping is implemented so that the robotic crane is able to localize and grasp columns automatically. The simulation demo presented in this paper only demonstrates the concept of how the algorithm could be used to control the robot to lift and place objects. Specific details during the construction process, such as the hoist rope installation on the columns or floors, and swing of the hoist rope during object transportation, are all ignored. General procedures of these two stages are summarized below:

- Stage 1: column installation
  Figure 1.11(a) shows the initial construction site before column installation. Due to the range limitations of robotic crane, only eight columns can be installed when the mobile robotic crane is located at the one side of the ground. After installing eight columns, the robotic crane will move to the other side of the ground and complete the installation of

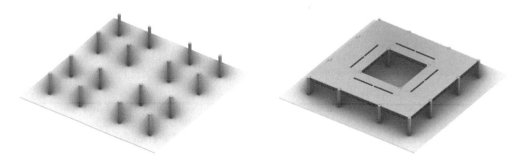

FIGURE 1.10   Two stages for modular construction of the first story.

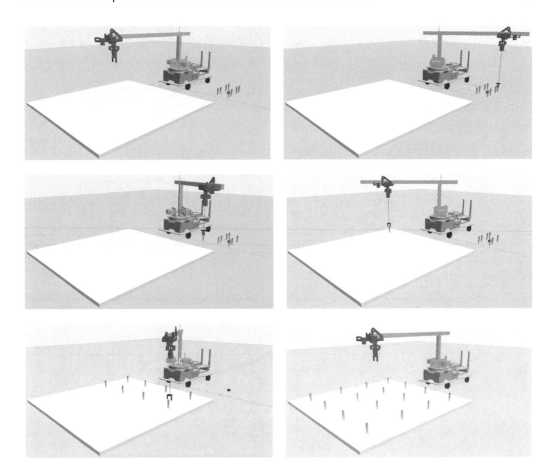

FIGURE 1.11 Construction procedures in stage 1.

the rest of eight columns. A camera is installed above the construction site for visual grasp-ing purposes. There are two options to achieve column localization. The first method is to use YOLOv3-tiny introduced in Section 3.1.3 to train the robotic crane so that the robotic crane can recognize wooden columns and find the grasping point automatically. The sec-ond approach is to use color recognition by using OpenCV. The color of the wooden col-umn is earthy yellow. Therefore, the HSV color threshold can take [26, 43, 46] as the lower limit and [34, 255, 255] as the upper limit. Figure 1.11(b) shows that the robotic crane has already found the position of the first column and moved the gripper to grasp the first column. The position where the column should be placed has been pre-defined for the robotic crane. By implementing the TRAC-IK method introduced in Section 3.1.1 for inverse kinematics and using the MoveIt plugin for path planning, the robotic crane will find an execution path and move the column to the target position. Figure 1.11(c) and (d) shows that the robotic crane has moved the first column to the target position and placed it on the ground. By repeating the procedures mentioned above, eight columns can be installed at the one side of the ground (Figure 1.11(e)). Finally, the robotic arm will be moved to the other side of the ground and will finish the installation of the remaining eight columns (Figure 1.11(f)).

- Stage 2: floor installation
  In stage 2, the wooden floors, which have been prefabricated in the factory, will be installed on the existing 12 columns. Figure 1.12(a) shows the initial construction site before floor

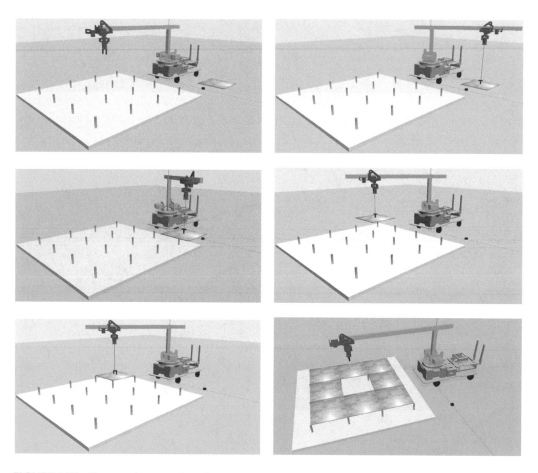

FIGURE 1.12   Construction procedures in stage 2.

installation. In order to lift the floor, hooks need to be temporarily installed on the floor surface, or a vacuum gripper may be used if the floor weight is within the capacity of gripper. In this demo, hooks are pre-installed at the middle of the floor so that the gripper can be used to lift the floor. Figure 1.12(b) shows that the robotic crane has already found the position of the floor and moved the gripper to grasp the floor. Then, by implementing the TRAC-IK method introduced in Section 3.1.1 for inverse kinematics and using the MoveIt plugin for path planning, the robotic crane will find an execution path and move the floor to the target position. Figure 1.12(c)–(e) demonstrates that the robotic crane has moved the floor to the target position and place it on four columns located at the upper-left corner of the ground. Finally, the robotic crane will be moved around the ground and will finish the installation of the remaining seven floors. Figure 1.12(f) shows the construction site after all columns and floors are installed.

1.4.2   Autonomous Inspection and Assessment Using UGVs and MAVs

In this demo, a collaborative UGV and MAV scheme is proposed, where the UGV is responsible for SLAM, while the MAV is used for structural inspection. The proposed scheme is validated on a reinforced concrete (RC) building in the structural laboratory at The University of British Columbia. First, a custom-built autonomous ground vehicle platform is designed and assembled. An NVIDIA Jetson Nano (4GB) with Ubuntu 18.04 LTS and ROS Melodic is used to deploy the launch

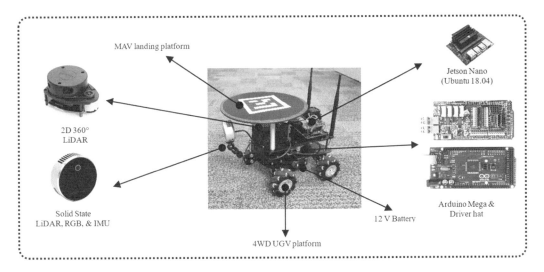

FIGURE 1.13 Hardware components of the ground robot.

the ROS nodes for sensors data collection and transmission. An Arduino Mega 2560 board is used to deploy the low-level C++ codes for parsing the navigational commands sent from Jetson Nano via a USB port. Such signals are finally decoded to run the 4 servo DC motors in the desired direction and speed. The Jetson Nano on the UGV is connected to the internet using a Wi-Fi module. The hardware components of the ground robot are shown in Figure 1.13.

With the help of modern SLAM technology, UGV can easily locate itself using mapped data or the provided map as well as navigating to a given destination. However, as with all sensors, noise always exists and the environment is not as ideal as theoretical. In order to enhance the localization accuracy, fiducial markers may be used to provide an extra layer of accuracy to the utilization of UGV. In this case, data fusion can be introduced to dynamically combine data from multiple sources and a more accurate data set can then be concluded. Fiducial markers are ideal in such cases, especially when the target is known and small in form. Figure 1.14 shows a UGV tracking a fiducial marker of 20mm × 20mm, while maintaining the spatial relationship between the UGV and the target. This is crucial in scenes where the UGV and robotic arm needs to localize and grasp a structural component, and install it to a destined location. Moreover, fiducial markers can have more information embedded into itself and can be used to provide specific information about the component the robot is processing, such as the ID number, geometry data, weight, etc. There are multiple algorithms that could achieve these goals, such as the open-source AR tag tracking library

FIGURE 1.14 Fiducial marker is being used for locating.

FIGURE 1.15 Fiducial marker is being used for enhanced locating accuracy.

in ROS. Figure 1.15 shows that using the fiducial markers, the UGV can navigate from marker A to marker B more accurately.

The UGV can be simulated using Gazebo, which is a software available in ROS and used for creating realistic simulation environment for development purposes. The simulation environment used in this study is the publicly available TurtleBot3 House, which includes connected small and large rooms and obstacles that are ideal for testing RRT-based exploration and GMapping algorithms that are described in Section 3.2.1. This simulation environment is shown in Figure 1.16.

Rviz, another ROS software, is used to visualize the mapping results. Figure 1.17 shows the different stages of RRT-based exploration and mapping. In this figure, the blue area shows the local cost map around the robot. The current laser scans are shown in red and the detected frontiers are shown with green markers. After detection of the first frontiers, the algorithm sets the frontiers as the new goal and uses the RRT algorithm to send the navigation command to the robot. This process continues until the map is generated.

On the other hand, an autonomous inspection pipeline for interior assessment of a building using a customized MAV has been developed and tested. The algorithm turns the map into grids and

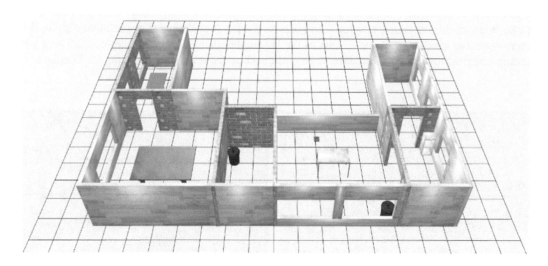

FIGURE 1.16 The simulation environment in gazebo.

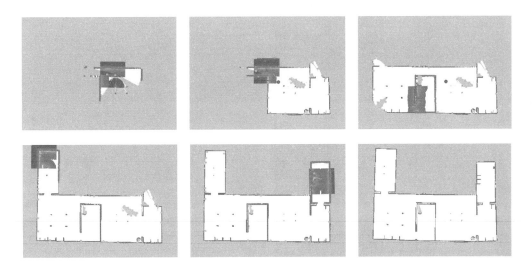

FIGURE 1.17   Different stages of RRT-based autonomous exploration and mapping.

uses the information of the previously inspected areas and obstacles to navigate into new areas. It uses vision to automatically detect and localize structural components and damages (i.e., concrete spalling, exposure of steel reinforcement) during its navigation. The developed pipeline is suitable for small-size to medium-size MAVs equipped with a minimum of a single camera for data collection and a single-distance sensor for obstacle avoidance. The end cost and the capabilities of the proposed pipeline make it a competitive solution for autonomous inspection of the interior areas with narrow passages.

Figure 1.18 demonstrates the navigation results using the developed pipeline in the RC building. Due to the limited battery life of the MAVs, two rooms in the RC building are selected as the inspection environment, where each room includes several desks and objects as obstacles. The developed inspection pipeline has successfully inspected the whole interior of the rooms to detect and localize the damaged structural components, as shown in Figure 1.19. In this case, the RC columns, concrete spalling, and steel reinforcement exposure have been successfully detected.

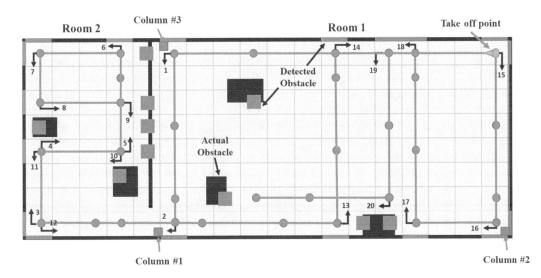

FIGURE 1.18   Result of navigation in two office rooms.

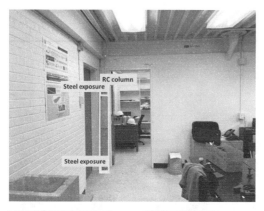 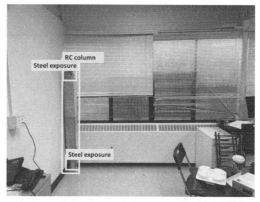

 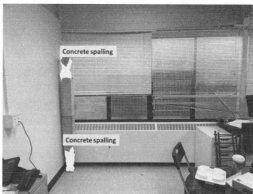

FIGURE 1.19   Column detection and steel exposure.

1.5   CONCLUSION

Recent years have witnessed the success of robotic technology applications in many different fields. In civil engineering, earlier attempts using robotic technologies have been made for automated construction and inspection. In this book chapter, first, some of the related works are reviewed to highlight their contributions and limitations. Second, a detailed methodology for robotic construction and inspection is described to provide a theoretical basis to develop these smart technologies. Third, three case studies are presented to demonstrate the automated construction and inspection procedures. The first case study is about the use of a mobile crane to achieve efficient construction of modular timber buildings. The second and the third studies focus on the use of MAVs and UGVs for autonomous navigation and inspection of real-world buildings. The second case focuses on the use of a MAV to navigate autonomously inside an RC building. The MAV successfully navigated inside the building and localized and geotagged damage components based on the explained algorithm. By implementing PID controllers for robot joint control, YOLOv3-tiny or color recognition for visual grasping, the TRAC-IK method for inverse kinematics, and the MoveIt plugin for path planning, the robotic crane can automatically localize and then install the columns and floors. The results indicate that the use of robots with other smart technologies (e.g., computer vision and LiDAR, etc.) can successfully achieve autonomous navigation, mapping, construction, and inspection of civil structures. By integrating autonomous exploration and mapping algorithms, a ground robot can autonomously create an occupancy map of surroundings. Although robot based autonomous inspection and construction has shown its potential in civil engineering, several limitations of current research can be observed, and future studies will be conducted.

1.5.1   FUTURE STUDY OF AUTOMATED CONSTRUCTION

Although existing studies have shown robotic technologies have capabilities for automated construction of civil structures, several limitations can be observed, and future studies are recommended accordingly. To wit:

- Many existing studies were conducted in the virtual ROS simulation. These studies assume perfect control and sensing capabilities, without considering local control or sensor failure which may happen occasionally in the real world. Therefore, their robustness against local failure remains as a further study.
- Many of the virtual simulation studies assume a simple construction environment without external disturbances. However, a real construction environment tends to be highly complicated, with discrete structural components, machineries, temporary houses, and people moving around at different locations. Therefore, the algorithms developed in any virtual simulation should be further developed and validated to account for these complex scenarios.
- There are some real-world demonstrations according to the literature review. However, most of these demonstrations were conducted in the laboratory environment. Most of them were focused on a small-scale simple structure (e.g., one-story and one-bay frame), with simple connections that are not rigorously validated on their force capacity in resisting lateral or vertical loads. Therefore, future study should be conducted for quality assurance and performance assessment of these robotically constructed structures.
- A cooperative robot working framework needs to be proposed, which includes a robotic crane used for hoisting large-scale structural components and a robotic arm used for installing small-scale structural components and connections.
- In order to enable robots to achieve more complicated construction tasks, reinforcement learning algorithms, such as DDPG, asynchronous advantage actor-critic (A3C), and proximal policy optimization (PPO), need to be implemented.

1.5.2   FUTURE STUDY OF AUTOMATED INSPECTION

Despite the promising results made by existing studies, there are several limitations. Therefore, recommendations for future studies are summarized herein:

- Many of the existing studies about robot-based structural damage inspection are only validated on local structural components or a small portion of full-scale structural systems. Although these studies have demonstrated the potential of robots in small-scale applications, their full potential on a complete inspection of large-scale structures still remains unexplored. Future research will be conducted on full-scale structural systems in the field.
- Most of the robot-based structural damage inspection research was mainly attempted using a single drone or ground robot. Although a single robot may be effective for local applications, it is inefficient or even impossible to inspect a full-scale structural system. In the future, multi-robot systems should be developed for full-scale civil structures to facilitate more efficient and comprehensive multi-tasking.
- Many of the existing studies were conducted in a well-controlled laboratory environment. While some studies conducted small-scale field applications, there still lack more rigorous validations in dealing with environmental disturbances, such as different weather conditions, winds, rains, complex terrains (e.g., a post-disaster site full of collapsed buildings or debris), which may significantly hamper the effective navigation of UGVs or MAVs. Therefore, future research should be conducted to validate these robotic navigation and inspection methods on full-scale civil structures in the field under different environmental conditions.

- Many of the existing studies relied on the use of LiDAR and heavy modules for assessment of the interior. These modules cannot be installed on small MAVs. Hence, more research is required to address the challenges of navigation in GPS-denied areas without using heavy or expensive modules.
- Stronger MAVs with novel design, more battery lifetime, and more payload are required to increase the versatility of navigation within a big post-disaster scenario with numerous pieces of debris.
- Robots are usually controlled by an operator remotely. Using connected robots within a local network, it is possible to fully automate collaboration of multiple robots as a system.
- Ground robots are capable of carrying heavy sensors and power supply. Therefore, they can be used for initial exploration and mapping, carrying aerial robots, and charging platforms.

REFERENCES

Allwright, M., Bhalla, N., El-faham, H., Antoun, A., Pinciroli, C., & Dorigo, M. (2014, September). SRoCS: Leveraging stigmergy on a multi-robot construction platform for unknown environments. In *International Conference on Swarm Intelligence* (pp. 158–169). Springer, Cham.

Álvares, J. S., Costa, D. B., & de Melo, R. R. S. (2018). Exploratory study of using unmanned aerial system imagery for construction site 3D mapping. *Construction Innovation*.

Álvarez, C., Roze, A., Halter, A., & Garcia, L. (2014). Generating highly accurate 3D data using a senseFly eXom drone. *White Paper*.

Apolinarska, A. A., Pacher, M., Li, H., Cote, N., Pastrana, R., Gramazio, F., & Kohler, M. (2021). Robotic assembly of timber joints using reinforcement learning. *Automation in Construction*, 125, 103569.

Augugliaro, F., Lupashin, S., Hamer, M., Male, C., Hehn, M., Mueller, M. W., & D'Andrea, R. (2014). The flight assembled architecture installation: Cooperative construction with flying machines. *IEEE Control Systems Magazine*, *34*(4), 46–64.

Ayala-Alfaro, V., Vilchis-Mar, J. A., Correa-Tome, F. E., & Ramirez-Paredes, J. P. (2021). Automatic Mapping with Obstacle Identification for Indoor Human Mobility Assessment. arXiv preprint arXiv:2111.12690.

Azimi, M., Eslamlou, A. D., & Pekcan, G. (2020). Data-driven structural health monitoring and damage detection through deep learning: State-of-the-art review. *Sensors*, *20*(10), 2778.

Beeson, P., & Ames, B. (2015, November). TRAC-IK: An open-source library for improved solving of generic inverse kinematics. In *2015 IEEE-RAS 15th International Conference on Humanoid Robots (Humanoids)* (pp. 928–935). IEEE.

Bock, T., & Linner, T. (2016). *Construction Robots: Volume 3: Elementary Technologies and Single-Task Construction Robots*. Cambridge University Press.

Braithwaite, A., Alhinai, T., Haas-Heger, M., McFarlane, E., & Kovač, M. (2018). Tensile web construction and perching with nano aerial vehicles. In *Robotics Research* (pp. 71–88). Springer, Cham.

Cha, Y. J., Choi, W., Suh, G., Mahmoudkhani, S., & Büyüköztürk, O. (2018). Autonomous structural visual inspection using region-based deep learning for detecting multiple damage types. *Computer-Aided Civil and Infrastructure Engineering*, *33*(9), 731–747.

Chen, L. C., Zhu, Y., Papandreou, G., Schroff, F., & Adam, H. (2018). Encoder-decoder with atrous separable convolution for semantic image segmentation. In *Proceedings of the European conference on computer vision (ECCV)* (pp. 801–818).

Contreras, D., Wilkinson, S., & James, P. (2021). Earthquake reconnaissance data sources, a literature review. *Earth*, 2(4), 1006–1037.

Cucu, L., Rubenstein, M., & Nagpal, R. (2015, May). Towards self-assembled structures with mobile climbing robots. In *2015 IEEE International Conference on Robotics and Automation (ICRA)* (pp. 1955–1961). IEEE.

De Melo, R. R. P. S. P. S., Costa, D. B., Álvares, J. S., & Irizarry, J. (2017). Applicability of unmanned aerial system (UAS) for safety inspection on construction sites. *Safety Science*, 98, 174–185.

Denavit, J., & Hartenberg, R. S. (1955). A kinematic notation for lower-pair mechanisms based on matrices.

Dutta, S., Cai, Y., Huang, L., & Zheng, J. (2020). Automatic re-planning of lifting paths for robotized tower cranes in dynamic BIM environments. *Automation in Construction*, *110*, 102998. https://doi.org/10.1016/j.autcon.2019.102998

Eschmann, C., Kuo, C. M., Kuo, C. H., & Boller, C. (2012). Unmanned aircraft systems for remote building inspection and monitoring.

Federal Emergency Management Agency (FEMA). (2012). Seismic Performance Assessment of Buildings Volume 1-Methodology. *Rep. No. FEMA P-58-1.*

Feifei, X., Zongjian, L., Dezhu, G., & Hua, L. (2012). Study on construction of 3D building based on UAV images. *The International Archives of the Photogrammetry, Remote Sensing and Spatial Information Sciences, 39,* B1.

Fernandez Galarreta, J., Kerle, N., & Gerke, M. (2015). UAV-based urban structural damage assessment using object-based image analysis and semantic reasoning. *Natural Hazards and Earth System Sciences, 15*(6), 1087–1101.

Freimuth, H., & König, M. (2015, October). Generation of waypoints for UAV-assisted progress monitoring and acceptance of construction work. In *15th International Conference on Construction Applications of Virtual Reality.*

Gao, Y., & Mosalam, K. M. (2018). Deep transfer learning for image-based structural damage recognition. *Computer-Aided Civil and Infrastructure Engineering, 33*(9), 748–768.

Gheisari, M., Irizarry, J., & Walker, B. N. (2014). UAS4SAFETY: The potential of unmanned aerial systems for construction safety applications. In *Construction Research Congress 2014: Construction in a Global Network* (pp. 1801–1810).

He, K., Zhang, X., Ren, S., & Sun, J. (2016). Deep residual learning for image recognition. *In Proceedings of the IEEE conference on computer vision and pattern recognition* (pp. 770–778).

Howard, J., Murashov, V., & Branche, C. M. (2018). Unmanned aerial vehicles in construction and worker safety. *American Journal of Industrial Medicine, 61*(1), 3–10.

Ignatyev, D. I., Shin, H. S., & Tsourdos, A. (2020). Two-layer adaptive augmentation for incremental backstepping flight control of transport aircraft in uncertain conditions. *Aerospace Science and Technology, 105,* 106051.

Irizarry, J., Gheisari, M., & Walker, B. N. (2012). Usability assessment of drone technology as safety inspection tools. *Journal of Information Technology in Construction (ITcon), 17*(12), 194–212.

Kang, T. K., Nam, C., Lee, U. K., Doh, N. L., & Park, G. T. (2011). Development of robotic-crane based automatic construction system for steel structures of high-rise buildings. In 28th International Symposium on Automation and Robotics in Construction, ISARC 2011.

Karaman, S., & Frazzoli, E. (2011). Sampling-based algorithms for optimal motion planning. *The International Journal of Robotics Research, 30*(7), 846–894.

Koerner-Al-Rawi, J., Park, K. E., Phillips, T. K., Pickoff, M., & Tortorici, N. (2020). Robotic timber assembly. *Construction Robotics, 4*(3), 175–185.

Konda, V., & Tsitsiklis, J. (1999). Actor-critic algorithms. Advances in *Neural Information Processing Systems,* 12.

Kong, X., & Li, J. (2018). Vision-based fatigue crack detection of steel structures using video feature tracking. *Computer-Aided Civil and Infrastructure Engineering, 33*(9), 783–799.

LaValle, S. M. (1998). Rapidly-exploring random trees: A new tool for path planning.

Lei, Z., Taghaddos, H., Hermann, U., & Al-Hussein, M. (2013). A methodology for mobile crane lift path checking in heavy industrial projects. *Automation in Construction, 31,* 41–53.

Liang, X. (2019). Image-based post-disaster inspection of reinforced concrete bridge systems using deep learning with Bayesian optimization. *Computer-Aided Civil and Infrastructure Engineering, 34*(5), 415–430.

Liang, C. J., Kamat, V. R., & Menassa, C. C. (2020). Teaching robots to perform quasi-repetitive construction tasks through human demonstration. *Automation in Construction, 120,* 103370.

Lin, J. J., Han, K. K., & Golparvar-Fard, M. (2015). A framework for model-driven acquisition and analytics of visual data using UAVs for automated construction progress monitoring. In *Computing in civil engineering 2015* (pp. 156–164).

Lindsey, Q., Mellinger, D., & Kumar, V. (2011). Construction of cubic structures with quadrotor teams. *Proc. Robotics: Science & Systems, VII,* 7.

Manzano, S. A., Hughes, D. T., Simpson, C. R., Patel, R., Heckman, C., & Correll, N. (2019, October). Embedded neural networks for robot autonomy. In *The International Symposium of Robotics Research* (pp. 242–257). Springer, Cham.

McLaughlin, E., Charron, N., & Narasimhan, S. (2020). Automated defect quantification in concrete bridges using robotics and deep learning. *Journal of Computing in Civil Engineering, 34*(5), 04020029.

Moeini, S., Oudjehane, A., Baker, T., & Hawkins, W. (2017). Application of an interrelated UAS-BIM system for construction progress monitoring, inspection and project management. *PM World Journal, VI*(VIII), 1–13.

Moemen, M. Y., Elghamrawy, H., Givigi, S. N., & Noureldin, A. (2020). 3-d reconstruction and measurement system based on multimobile robot machine vision. *IEEE Transactions on Instrumentation and Measurement, 70,* 1–9.

Ni, F., Zhang, J., & Chen, Z. (2019). Pixel-level crack delineation in images with convolutional feature fusion. *Structural Control and Health Monitoring, 26*(1), e2286.

OpenRAVE. "IKFast: The Robot Kinematics Compiler. 2011. Available from: http://openrave.org/docs/latest_stable/openravepy/ikfast/#ikfast-the-robot-kinematics-compiler.

Pan, X., & Yang, T. Y. (2020). Post-disaster image-based damage detection and repair cost estimation of reinforced concrete buildings using dual convolutional neural networks. *Computer-Aided Civil and Infrastructure Engineering, 35*(5), 495–510.

Pan, X., & Yang, T. Y. (2022a). 3D vision-based out-of-plane displacement quantification for steel plate structures using structure-from-motion, deep learning, and point-cloud processing. Computer-Aided Civil and Infrastructure Engineering.

Pan, X., & Yang, T. Y. (2022b). Image-based monitoring of bolt loosening through deep-learning-based integrated detection and tracking. *Computer-Aided Civil and Infrastructure Engineering, 37*(10), 1207–1222.

Phillips, S., & Narasimhan, S. (2019). Automating data collection for robotic bridge inspections. *Journal of Bridge Engineering, 24*(8), 04019075.

Poignet, P., & Gautier, M. (2000, March). Nonlinear model predictive control of a robot manipulator. In *Proceedings of the 6th International Workshop on Advanced Motion Control.* (Cat. No. 00TH8494) (pp. 401–406). IEEE.

Qiao, W., Wang, Z., Wang, D., & Zhang, L. (2021). A new mortise and tenon timber structure and its automatic construction system. *Journal of Building Engineering, 44*, 103369.

Ramana, L., Choi, W., & Cha, Y. J. (2019). Fully automated vision-based loosened bolt detection using the Viola–Jones algorithm. *Structural Health Monitoring, 18*(2), 422–434.

Rodriguez-Gonzalvez, P., Gonzalez-Aguilera, D., Lopez-Jimenez, G., & Picon-Cabrera, I. (2014). Image-based modeling of built environment from an unmanned aerial system. *Automation in Construction, 48*, 44–52.

ROS Wiki. (2014). Kinematics and Dynamics Library (KDL). Available from: http://wiki.ros.org/kdl.

Shang, Z., & Shen, Z. (2017). Real-time 3D reconstruction on construction site using visual SLAM and UAV. *arXiv preprint arXiv:1712.07122.*

Silver, D., Lever, G., Heess, N., Degris, T., Wierstra, D., & Riedmiller, M. (2014, January). Deterministic policy gradient algorithms. In *International Conference on Machine Learning* (pp. 387–395). PMLR.

Soleymani, T., Trianni, V., Bonani, M., Mondada, F., & Dorigo, M. (2015). Bio-inspired construction with mobile robots and compliant pockets. *Robotics and Autonomous Systems, 74*, 340–350.

Spencer, B. F. Jr, Hoskere, V., & Narazaki, Y. (2019). Advances in computer vision-based civil infrastructure inspection and monitoring. *Engineering, 5*(2), 199–222.

Voigtländer, F., Ramadan, A., Eichinger, J., Lenz, C., Pensky, D., & Knoll, A. (2017, October). 5G for robotics: Ultra-low latency control of distributed robotic systems. In *2017 International Symposium on Computer Science and Intelligent Controls (ISCSIC)* (pp. 69–72). IEEE.

Wai, R. J., & Muthusamy, R. (2013). Design of fuzzy-neural-network-inherited backstepping control for robot manipulator including actuator dynamics. *IEEE Transactions on Fuzzy Systems, 22*(4), 709–722.

Wang, H. (2016). Adaptive control of robot manipulators with uncertain kinematics and dynamics. *IEEE Transactions on Automatic Control, 62*(2), 948–954.

Wang, N., Zhao, Q., Li, S., Zhao, X., & Zhao, P. (2018). Damage classification for masonry historic structures using convolutional neural networks based on still images. *Computer-Aided Civil and Infrastructure Engineering, 33*(12), 1073–1089.

Wang, N., Zhao, X., Zhao, P., Zhang, Y., Zou, Z., & Ou, J. (2019). Automatic damage detection of historic masonry buildings based on mobile deep learning. *Automation in Construction, 103*, 53–66.

Xiao, Y., Pan, X., & Yang, T. T. (2022). Nonlinear backstepping hierarchical control of shake table using high-gain observer. *Earthquake Engineering & Structural Dynamics, 51*(14), 3347–3366.

Yamamoto, T., Kusumoto, H., & Banjo, K. (2014). Data collection system for a rapid recovery work: using digital photogrammetry and a small unmanned aerial vehicle (UAV). In *Computing in Civil and Building Engineering (2014)* (pp. 875–882).

Yang, Z. J., Fukushima, Y., & Qin, P. (2011). Decentralized adaptive robust control of robot manipulators using disturbance observers. *IEEE Transactions on Control Systems Technology, 20*(5), 1357–1365.

Yang, T. Y., Li, K., Lin, J. Y., Li, Y., & Tung, D. P. (2015). Development of high-performance shake tables using the hierarchical control strategy and nonlinear control techniques. *Earthquake Engineering & Structural Dynamics, 44*(11), 1717–1728.

Yeum, C. M., & Dyke, S. J. (2015). Vision-based automated crack detection for bridge inspection. *Computer-Aided Civil and Infrastructure Engineering, 30*(10), 759–770.

Ying, K. C., Pourhejazy, P., Cheng, C. Y., & Cai, Z. Y. (2021). Deep learning-based optimization for motion planning of dual-arm assembly robots. *Computers & Industrial Engineering, 160,* 107603.

Yu, S., Yan, F., Zhuang, Y., & Gu, D. (2020). A deep-learning-based strategy for kidnapped robot problem in similar indoor environment. *Journal of Intelligent & Robotic Systems, 100*(3), 765–775.

Yuan, C., Xiong, B., Li, X., Sang, X., & Kong, Q. (2022). A novel intelligent inspection robot with deep stereo vision for three-dimensional concrete damage detection and quantification. *Structural Health Monitoring, 21*(3), 788–802.

Yun, J. P., Kim, D., Kim, K., Lee, S. J., Park, C. H., & Kim, S. W. (2017). Vision-based surface defect inspection for thick steel plates. *Optical Engineering, 56*(5), 053108.

Zhang, Z. (1999, September). Flexible camera calibration by viewing a plane from unknown orientations. In *Proceedings of the Seventh IEEE International Conference on Computer Vision* (Vol. 1, pp. 666–673). IEEE.

Zhao, N., Lodewijks, G., Fu, Z., Sun, Y., & Sun, Y. (2022). Trajectory predictions with details in a robotic twin-crane system. *Complex System Modeling and Simulation, 2*(1), 1–17.

Zhu, B. (2015, August). The application of the unmanned aerial vehicle remote sensing technology in the FAST project construction. In *Remote Sensing of the Environment: 19th National Symposium on Remote Sensing of China* (Vol. 9669, pp. 187–194). SPIE.

Zi, B., Lin, J., & Qian, S. (2015). Localization, obstacle avoidance planning and control of a cooperative cable parallel robot for multiple mobile cranes. *Robotics and Computer-Integrated Manufacturing, 34,* 105–123.

Zollmann, S., Hoppe, C., Kluckner, S., Poglitsch, C., Bischof, H., & Reitmayr, G. (2014). Augmented reality for construction site monitoring and documentation. *Proceedings of the IEEE, 102*(2), 137–154.

Tavasoli, S., Pan, X., & Yang, T. Y. (2023). Real-time autonomous indoor navigation and vision-based damage assessment of reinforced concrete structures using low-cost nano aerial vehicles. *Journal of Building Engineering, 68,* 106193.

2 Robotics in 3D Concrete Printing
Current Progress & Challenges

Shuai Li, Tian Lan, Priyan Mendis, and Phuong Tran

CONTENTS

2.1 INTRODUCTION

Large-scale additive manufacturing (AM) with cementitious material is commonly referred to as 3D concrete orinting (3DCP). Most 3DCP fabrication processes are based on material extrusion, which typically involves phase transitions of material from liquid to solid [1]. In contrast to polymer AM that only involves physical changes, 3DCP involves changes in both physical properties and chemical reactions in the phase transition [2]. The additional requirement for suitable rheological properties poses challenges for 3DCP projects. To overcome this major barrier, more than 80% of existing literatures focused on exploring material properties and developing printable mix designs [3]. While the development in printable materials laid the foundation for 3DCP research, digital design frameworks and robotic controls play important roles in the advancement of automatic construction.

As shown in Figure 2.1, a typical 3DCP workflow starts with creating a digital model in computer aided design (CAD) software. Slicer software with dedicated toolpath generation algorithms is required to convert CAD models to a toolpath. After that, the slicer software will translate the toolpath to a machine control code that can be executed by a 3D concrete printer. The code contains instructions for setting the parameters of the 3D printer and positions of the printhead [4]. However, as pointed out by Bikas, Stavridis [5], the digital design phase of AM is usually disconnected from the machine programming and production phase.

Various types of 3DCP systems have been developed. They can be classified as gantry, frame, robot-arm, polar, and delta based on their motion systems [2]. The motion and extrusion systems have direct effects on the deposition of material and the quality of 3DCP. Consequently, the machine code that controls the printing process plays an equally important role as the material designs. There is a need for reviews considering the creation of digital models and control of printers to complement the existing studies on printable materials. A snapshot of current CAD modelling techniques, robotic design, and manipulation of 3DCP process will be presented in this chapter.

DOI: 10.1201/9781003325246-2

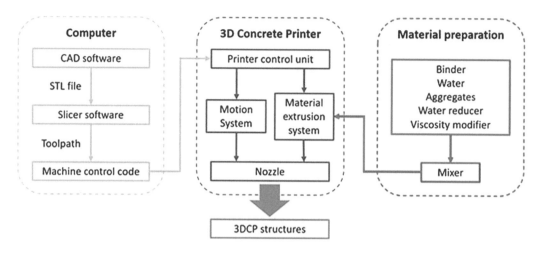

FIGURE 2.1   A typical 3DCP workflow.

2.2   DIGITAL DESIGN FRAMEWORK

2.2.1   FORM FINDING WITH CAD SYSTEMS

2.2.1.1   Background

CAD software plays a significant role in the workflow of 3DCP because the construction process of 3DCP is automated by computer numerical control (CNC) systems. With CNC, first introduced in the early 1970s, computers are integrated into the machines to replace their digital control boards. The integration of computers enabled accurate process control with servo loops and interpolator loops [6]. Needs for sophisticated programs can be observed in the development of CNC towards the modern systems deployed by 3DCP. Therefore, CAD software was utilized to generate the codes for controlling CNC systems. Those codes contain instructions on how to make, and what path to move, 3D printers. The machine control codes were initially known as an automatically programmed tool, and then recognized as ISO 6983, RS-274D, or G&M codes [7]. Moreover, the dramatic development and declining costs of CAD software have made it more appealing and accessible for designers [8]. CAD software became an essential step of the 3DCP workflow because of the needs to control CNC machines and create complex designs.

Khoshnevis and Hwang [9] have foreseen the necessity of a complete comprehensive feasibility evaluation of 3DCP based on geometry, physics, chemistry, and construction process. One of the main benefits of CAD systems is the ability to create complex designs within material constraints [10]. For example, as the result of low mechanical strength of fresh concrete, 3DCP can fail due to tension induced by overhang or cantilever, and compression from upper layers [11]. Those unstable structures can be identified in CAD models and redesigned. 3DCP also enables the fabrication of optimised structures with complex geometries. While the integration of computational modelling is important for 3DCP, commercial CAD packages do not consider the 3DCP's constraints and processing factors [10]. Therefore, developing form design strategies suitable for 3DCP is one of the challenges faced by researchers of 3DCP.

2.2.1.2   Parametric Designs and Shape Optimization

Parametric design tools have made a significant contribution to the research of 3DCP. They can provide a visual scripting environment and the ability to program customised components with programming languages (e.g., Python, C Sharp, Visual Basic) according to specific research objective. For examples, Ooms, Vantyghem [12] and Vantyghem, Ooms [13] developed a plug-in with

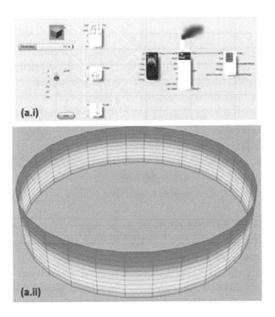

FIGURE 2.2 CAD models generated by different design methods, (a.i) & (a.ii) FEA created in Grasshopper for evaluating the deformation on a cylinder, (b) parametric design of a column with trigonometric functions.

Grasshopper in Rhinoceros, a parametric design software, to automatically pre-process models for finite element analysis (FEA) in SIMULIA Abaqus. This plug-in enabled designers to preview their 3DCP model as virtual printing processes, where the deformation and structural stability can be predicted quantitatively. Moreover, FEA plug-ins [14, 15] have been developed to facilitate designs of complex geometries within CAD environment. As shown in Figure 2.2.a, the deformation of a 3DCP cylinder can be simulated at the completion of each layer, where the time-dependent material properties are defined parametrically [16].

The design freedom and geometric complexities realised by 3DCP make it compatible with the topology optimization (TO) method. TO is the field of study that seeks for optimal structural designs under a prescribed condition by progressive removal of inefficient materials [17]. However, the resulting geometries can be too complex to be fabricated by conventional methods. Vantyghem, De Corte [18] created a post-tensioned girder bridge designed by TO using 3DCP. Due to the manufacturing constraints of 3DCP, the resulting bridge design with large overhangs needed to be post-processed and subdivided into several printable segments. The individual segments and rein-forcement were assembled manually, and grout mortar was poured into the cavities. Furthermore, the constraints of overhangs, toolpath continuities, and anisotropic material properties were added to the design conditions of the TO algorithm developed by Bi, Tran [17]. This TO algorithm also included a segmentation function to bypass the overhang limitation and assign different printing directions to each segment.

Computational design tools provide architects the ability to explore design spaces and robotic fabrications. Anton, Yoo [19] designed 3DCP columns using mathematical functions. As shown in Figure 2.2.b, points on a layer are displaced periodically according to the values of trigonomet-ric functions. The complexity of geometries can be controlled by varying the frequencies, ampli-tudes, and phases of the trigonometric functions. Another example of 3DCP structure designed with similar trigonometric functions is the sinusoidal wall by XtreeE [20], a company that designs and manufactures large-scale 3D concrete printers. In addition to Boundary Representations (BRep) as mentioned ealier, Function Representation (FRrep) can also be ultilized to create intricate geom-etries without compromising their printablity [21].

2.2.2 Slicing and Toolpath Generation

2.2.2.1 Slicing Strategies

The 3D models created in CAD software must be translated to machine control codes for 3D printers. This process typically starts with slicing 3D models to 2D layers. 3D shapes can be sliced into flat layers of constant thickness by creating contour lines or extracting cross sections with a series of horizontal planes. This approach is well-established for small-scale 3D printing with polymers and widely used by standard slicer software, such as Slic3r [22], Cura [23], and Simplify3D [24]. The resulting toolpath can be printed by a 3-axis printer because the normals of the layers are constantly vertical. However, it is inappropriate for large-scale 3DCP because the manufacturing constraints are not considered by this slicing strategy [25]. As shown in Figure 2.3.a, slicing with constant thickness can result in locally cantilevered layers, which are unstable and can result in large deformation in 3DCP.

The second strategy generates layers with constant distance between them. This method was developed by Carneau, Mesnil [26] based on the algorithm to create a bricklaying pattern on shells using geodesic coordinates [27]. Similarly, Gosselin, Duballet [25] proposed a constant tangent method that creates a non-planar toolpath with locally varying layer thickness, with the nozzle needing to be oriented perpendicularly to the surface of the previous layer. As shown in Figure 2.3.b, the advantage of this strategy is that a constant contact surface can be maintained between layers, which avoids the local cantilever issues created by the horizontal slicing method. As orientations of the nozzle need to be varied at different positions of the toolpath, six-axis printers are needed. Moreover, a 3D shape can be sliced by a series of planes inclined to a certain angle with the horizontal. Carneau, Mesnil [26] used this method to generate the toolpath by slicing the model of a barrel vault with planes inclined at 40°. As shown in Figure 2.3.c, this method can reduce local overhangs of layers and therefore improve the stability. A six-axis printer is also required for 3DCP the layers created with this method.

2.2.2.2 Infill Pattern Generation

The aforementioned slicer algorithms generate a toolpath from the boundary of a 3D model, which can only fabricate thin-shell structures. The cavities can be filled either with mortar manually as in [18] and [19], or by filaments printed with specific infill patterns. Two strategies commonly used for 3DCP are direction-parallel and contour-parallel strategies. The direction-parallel strategy

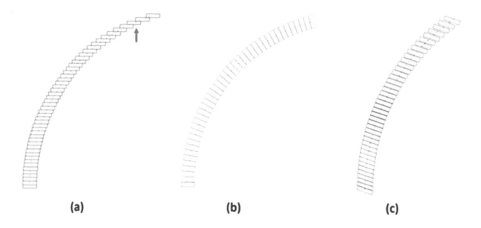

(a)   **(b)**   **(c)**

FIGURE 2.3 Cross-sectional view of layers created by different slicing strategies, (a) horizontal slicing strategy will create local overhangs as indicated by the arrow; printable by both 3-axis and 6 axis printer, (b) constant layer thickness strategy; only printable by 6-axis printer, (c) constant angle slicing strategy; only printable by 6-axis printer.

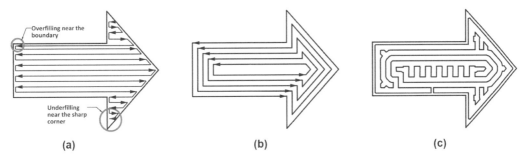

FIGURE 2.4 Infill patterns created by different toolpath strategies, (a) direction-parallel strategy, (b) contour-parallel strategy, (c) hybrid strategy.

creates a toolpath by arranging linear segments aligned with a specific reference line and altering directions [28]. As shown in Figure 2.4.a, the toolpath created with this strategy has a zig-zag pattern. Contour-parallel toolpaths are created by offsetting the boundary using various algorithms, such as the Voronoi diagram technique [28] or edge offset method [29]. While the contour-parallel strategy (Figure 2.4.b) creates a toolpath with less underfillings near the boundary, the direction-parallel strategy requires less computational power. The two strategies were combined by Xia, Ma [30] into a hybrid method (Figure 2.4.c) with the purpose of reducing underfilling, overfilling, and toolpath discontinuities.

The ability to design the printing procedure is attractive for 3DCP designers, because the printing processes can have direct effects on the printing outcomes. Gleadall [31] developed a generic toolpath generation algorithm. It can create a customised toolpath by replications and parametric variations of mathematically defined toolpath segments such as lines and arcs. Moreover, toolpath-based design for 3DCP was also investigated by Breseghello and Naboni [32]. In addition to toolpath optimization based on FEA, the proposed framework was able to control the printing speed and extrusion speed to create a woven pattern. Reinforcement can be placed accurately in the tensional regions of this toolpath pattern.

2.2.3 3D Printer's Execution of Machine Control Codes

Information needs to be extracted from the digital models and translated into codes that can be executed by 3D printers. Conversion from CAD model to machine control codes can be performed by the aforementioned slicer software [22–24]. Researchers have also developed algorithms using programming tools such as MATLAB [33] or the parametric design software Grasshopper [34] to meet specific research objectives. After receiving codes, 3D printers' controllers will interpret them and create the signals for individual manipulators. For example, the coordinates of a target point in a move command can be converted to the rotation angles of motors based on inverse kinematics. Interested readers should refer to [35] and [36] for a deeper understanding of motion planning for robots.

Various types of industrial robots are used in 3DCP systems, and the differences between them are discussed in Section 1. Most commercial manufacturers of industrial robots require proprietary programs to operate their 3D concrete printers. For example, KUKA uses a domain-specific programming language called KUKA robot language (KRL) for their 3D printers. As shown in Figure 2.5.a, KRL provides typical programming statements such as "FOR", "IF", and "WHEN", as well as robotic-specific statements for controlling motions and tools [37]. Those programming statements are also available in RAPID, the programming language developed for ABB industrial robots [38]. As shown in Figure 2.5.b, the tools and coordinates of target points are defined prior to instructing the robot to perform a specific action, such as robot move.

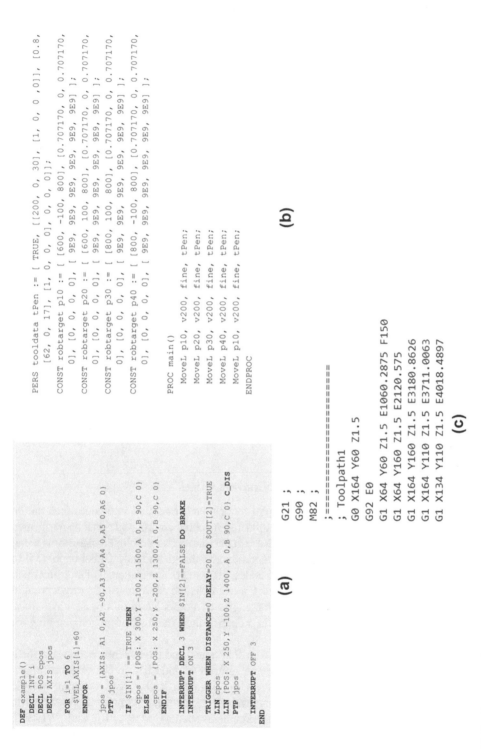

FIGURE 2.5 Examples of machine control codes, (a) KUKA robot language (KRL), (b) RAPID code for ABB industrial robots, (c) GCode contains relatively low-level information compared with its commercial counterparts.

GCode is a universal and classic machine programming tool with more than half a century of history [39]. As shown in Figure 2.5.c, a typical GCode command contains information about the motion type (G1), coordinates of the point to move to, speed, and extrusion volume. The auxiliary tools on 3D printers can be controlled by dedicated commands called M codes. Despite the popularity of GCode in 3DCP, collecting data and establishing feedback loops for machines programmed with it is challenging [40].

2.2.4 Real-time Control Systems for 3DCP

While robot motions are accurately controlled by the well-established machine control codes, the actual printing process can be affected by external factors: environmental conditions and geometric deviations due to cumulative deformations of fresh concrete. Therefore, a physical printing process can deviate significantly from the expected outcomes in its digital model [41]. The effects of changes in processing parameters can be characterized by inspecting the printing quality using the techniques discussed in Section 2.4. As shown in Figure 2.6, real-time measurements can be integrated into 3DCP systems to establish feedback loops to actively control printing processes.

Yuan, Zhan [42] proposed a real-time toolpath and extrusion control method to actively adjust the layer width during printing. The pump's motor speed during printing was monitored by a Hall effect sensor and then converted to materials' flowrate using a computer. FUROBOT, a Grasshopper plug-in, was used to communicate between their KUKA Printer's controller and the computer. The toolpath is real-time generated in Grasshopper and sent to the printer in 50 ms intervals.

Furthermore, a gantry 3D concrete printer with CNC-based Sinumerik numerical control unit was used by Wolfs, Bos [43]. Both the robot movement and the mixer-pump are controlled by GCode. A 1D time of flight distance sensor was mounted on the nozzle to measure the distance between the nozzle and previous layers, and the measurements were collected by an Arduino microcontroller. The robot movements prescribed in GCode are compared with the measurements to adjust the nozzle's positions in each interpolation cycle of the printer.

Panda, Lim [44] used a shear vane apparatus to measure the shear stress of fresh concrete in the pump's hopper during printing. The pumping system was integrated into the printer with a robotic operating system that accepts GCode as inputs. Real-time control of the flowrate was achieved by setting a potentiometer through an Arduino microcontroller and digital-to-analog converter circuit.

A vision-based control system was installed on a 3D concrete printer by Barjuei, Courteille [45] to monitor the variations in a filament's geometries during printing. The images acquired by the industrial camera mounted with the printhead are sent to a master controller. The velocity of the

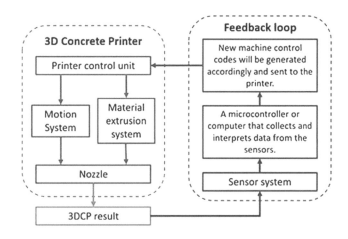

FIGURE 2.6 Integration of sensor system and feedback loop into the 3DCP system.

TABLE 2.1

IDS Developed by Companies to Facilitate 3DCP Designs

IDS Name	Company Names	Key Functionalities	Ref
RAP CAM	RAP Technologies	• Management cloud access • Planar and non-planar slicing • Robot simulation • Environment management	[46]
XtreeE Platform	XtreeE	• Multi-material printing • Custom robotic configurations • Design-to-production workflow • Real-time process monitoring • Non-planar slicing	[47]
Chisel and Artysan	Cybe	• Slicing and toolpath generation • Robot motion and statistics tracking • Robot control during printing	[48]
Kuka.Sim	Kuka	• CAD model reader • Python support • Slicing, toolpath, and KRL generation • Swept volume function • Robot simulation	[49]
RobotStudio	ABB	• Slicing, toolpath, and RAPID code generation • Translating GCode into RAPID code • Filtering GCode points for smooth robot movement • Interpolation of external axis • Extrusion control as an integrated robot axis • Robot simulation	[50]

printhead is set by the embedded control algorithm which works in parallel with the vision system. The time delays caused by communication, image acquisition, and vision processing are important for the stability of this close-loop control system, and they need to be synchronized with the 3DCP system's operating frequency.

2.2.5 INTERACTIVE DESIGN SYSTEMS FOR 3DCP

As discussed above, the design and control of the 3DCP process require high levels of expertise in interdisciplinary knowledge between CAD software, programming, robotics, and printable materials. It is necessary to develop a holistic design framework that allows users to directly control individual parameters prior and during printing processes. For non-expert users, an interactive design system (IDS) with simplicity and flexibility is needed. Like the programming languages developed by the manufacturers of industrial robots, a number of IDS have been developed by the companies specialize in 3DCP. Information about those systems is summarized in Table 2.1.

2.3 MOTION SYSTEMS

In recent years, progress in robot-assisted manufacturing techniques has expended the potential of automatic construction. There are two mainstreams in the robot-assisted manufacturing techniques: AM and the robotic assembly. In AM, the construction is facilitated by the deposition of materials layer by layer along the toolpath in the 3D space. During the printing process, the printer is fixed whereas the nozzle unit is carried by the mechanical parts to move along the toolpath. For the robotic assembly technique, the prefabricated components are assembled sequentially by the end effector

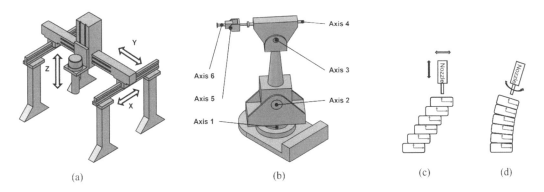

FIGURE 2.7 Typical motion systems in automatic construction: (a) gantry system, (b) robotic arm system, (c) printing layers in gantry system, (d) printing layers in robotic arm system.

to finish the construction. In this section, different motion systems in robot-assisted construction are reviewed. From the mechanical mechanism perspective, motion systems that are widely implemented in AM can be categorized into gantry robot [51, 52] and robotic arm [18, 25]. Typically, the motion in the gantry system is along the x-, y-, and z-axes whereas the robotic arm mainly supports more than three axes. In addition, motion systems including multi-robot and mobile robot systems are implemented for increasing efficiency and building volume.

Gantry motion systems, shown in Figure 2.7.a, are commonly used in the AM. As Figure 2.7.a shows, the motion in the x- and y- directions are facilitated by the movable frame whereas the motion in the z- direction is facilitated by the length-adjustable printhead [51, 52]. The gantry system can support both lab- and large-scale concrete printing. As Figure 2.7.a shows, the large-scale concrete printer can a print bridge segment with a dimension of 3440 mm × 920 mm × 1080 mm. Figure 2.7.b shows the robotic arm system, where the printhead is mounted on the end effector of the 6-axis industrial ABB IRB6650 robot [18]. The concrete girder segments with length of 800 mm are fabricated.

The simple structure of gantry systems provides advantages including easy setup, simple manipulation, low cost, and large build volume. The robotic arm system is more complex to utilize and expensive. However, the extra degree of freedom provided by the robotic arm system can improve the bonding between adjacent printing layers when the inclined wall needs to be fabricated. As Figures 2.7.c and d show, the rotation of the printhead in a robotic arm system can adjust the printing direction based on the curvature where the gantry system only supports single direction [25].

A wall with large curvature changes between layers can be fabricated with high quality by applying the strategy shown in Figure 2.7.d. The build volume of the gantry system cannot extend to the region outside of the frame. For the robotic arm system, the build volume is limited to the reachability of the end effector. In addition, large-scale construction is time-consuming with a single robot. Co-working robotic systems and mobile robotic systems are also applied in robot-assisted construction techniques. In addition, novel robots are designed in robot-assisted construction including flying robots and climbing robots. Research on mobile robotic construction and the co-working between multiple robots for different types of material are reviewed.

The printing-while-moving design proposed by Tiryaki et al. [53] illustrates a typical frame-of-motion system in mobile robot-assisted construction. As shown in Figure 2.8.a, most mobile robotic systems consist of an industrial robotic arm and a mobile platform. The robotic arm is mounted on the mobile base in most cases. Therefore, motion control can be separated into two parts including the control of robot joints and the control of the platform. As Figure 2.8.b shows, the trajectory generated by the motion planner is the input of both mobile base and robotic arm.

Multiple types of sensors, e.g., camera and laser sensors, have been embedded into the motion system so that the mobile base can be localized and controlled. The control of the robotic arm can

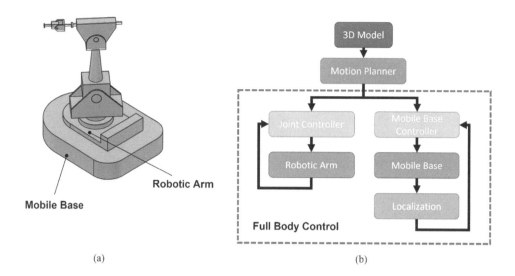

(a)   (b)

FIGURE 2.8 Single mobile robotic system in 3DP: (a) single mobile robotic system in concrete printing, (b) typical motion control system in mobile robotic 3DP system.

be facilitated by the built-in industrial program. The joints of robots can precisely follow the commands. However, the motion of joints is influenced by the mobile base. Uneven terrain conditions and the soft connection between mechanical parts can induce disturbance into the system. In most cases, model predictive control (MPC) is applied to mitigate the disturbance. The digital construction platform (DCP) designed by Keating et al. [54] demonstrated the ability of robot-assisted construction at large scale. The hydraulic robot arm at macro level and the electric robot arm at micro level contributes to the automatic construction of formwork. The formwork illustrated in DCP can be further utilized in the on-site construction of other architecture.

Mobile robotic arm system is also utilized in mesh mould construction. The stay-in-place scheme proposed by Hack et al. [55] required steel formwork fabrication for further non-standard concrete construction. The mesh mould, which is orthogonally crossed mesh, is composed of steel wire aligned along both horizontal and vertical directions. The on-site construction mechanism is to fabricate the mesh mould at first, where the fabricated mesh mould can be utilized as reinforcement with further concrete infill. The tool is mounted on the end effector of a robotic arm, where the discrete steel bar feed is implemented separately in horizontal and vertical direction. The cutting, welding, and bending tools are integrated for the continuous operations on steel bars. The bending, welding, and cutting of steel wire contribute to more geometric freedom of the mesh mould formwork, where the load-bearing concrete wall with non-standard shape can be constructed.

Figure 2.9 illustrates the track-based mobile robot system for 3DCP proposed by Xu et al. [56]. As shown in Figure 2.9, the robotic arm is mounted on the movable platform. Below the movable table, a platform can facilitate the movement along the x-axis track. The movements along the y-axis and z-axis are facilitated by the movable table. As illustrated in Figure 2.9, the track below the platform is fixed during on-site construction. Due to the existence of fixed track, the track-based printing system can adapt for complex terrains.

The cooperations between multiple robots can improve the ability of large-scale construction. Multiple robots can cooperate simultaneously to complete a construction task. Communication between different agents, therefore, needs to be implemented to avoid potential collision. The localization and motion planning for each robot is an essential component of a multi-robot construction system. The commonly applied localization algorithm is the simultaneous localization and mapping (SLAM). To provide the pose information of multiple agents, global reference systems such as laser reflectors and optical markers are installed. The external reference provides the precise positional

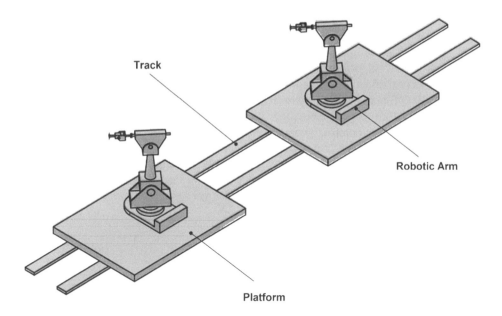

FIGURE 2.9 Track-based mobile robot-assisted 3DP system.

information for an absolute localization. As Figure 2.10.a shows, each robot can localize with an optical marker and broadcast its own position [57]. In addition, localization can also be achieved by the purely on-board sensing [58], where the 3D LiDAR is utilized to reconstruct the 3D building model and further provide pose information. As presented in Figure 2.10.b, the dot markers made by a robot are utilized as a building model to test the localization accuracy [58].

Cooperation between robots can automate the assembly of discrete blocks, where one of the robots can assemble blocks, while the other robot provides support for the structure [59]. The extrusion of material and assembly of concrete blocks can be sequentially applied to fabricate the specific infilling patterns, e.g., the polystyrene foam with concrete infillings and truss pattern can be

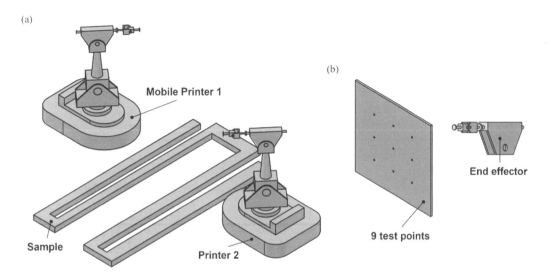

FIGURE 2.10 Multi-robot system in 3DCP: (a) Mobile multi-robot system with global reference, (b) mobile multi-robot system with point mark.

constructed with a multi-robot system [60]. During the assembly, one of the robots can transport the polystyrene foam blocks while another robot extrudes the mortar. The assembly of polystyrene foam can be regarded as temporary formwork where the extruded mortar can provide fully connected infillings with truss patterns.

This section has examined robotic systems in construction including 3D printing and assembly of discrete blocks. Different motion systems including mobile base and track systems are applied to facilitate the mobility of robotic construction. A robotic system also contributes to the on-site construction of structures with complex geometric shape. The utilization of a multi-robot system can increase the efficiency and scale of construction. Robots can facilitate different tasks sequentially for automatic construction. The utilization of a multi-robot system contributes to the automation of complicated construction. In the future, the autonomy of the robot-assisted construction system, especially the collaborations between robots and human-robot collaborations, could be further developed.

2.4   INSPECTION TECHNIQUES

Inspection techniques for checking the quality of robot-assisted construction at different stages including prior to construction, during construction, and post construction are vital due to challenges such as complex geometric features. The most common techniques in the inspection of construction quality include 3D laser scanning and photogrammetry-based methods [61].

3D laser scanning, which can obtain precise 3D point clouds, is a common technique in many engineering fields. 3D laser scanning has high accuracy, long measurement ranges, and low dependency on environmental conditions. With comparison to 3D laser scanning, the photogrammetry-based methods are cheaper due to the inexpensive cameras. However, these cameras require good lighting conditions. For the quality check of 3DCP features, multiple cameras need to be deployed so that images from multiple angles can be acquired. The applications of inspection techniques in construction are further summarized in Table 2.2.

With the comparison of the as-designed model and the 3D reconstructed model from laser scanner, light scanner, and cameras, the construction quality can be evaluated by the model deviation [42, 62, 64, 65]. The reconstructed model can also be utilized as the guidance for further construction process [63]. As discussed in Section 2, the inspection techniques in real-time tend to adjust the extrusion of material so that the surface quality can be improved during construction [42, 43, 67].

TABLE 2.2
Inspection Techniques Applied in Automatic Construction

Sensing System	Measurement	Application	Ref
Light scanner	Point cloud of flat surfaces	Evaluation of manufacturability	[62]
1D time of flight (ToF) distance sensor	Nozzle height	Real-time nozzle position correction	[43]
Laser scanner Multiple digital cameras	3D point cloud of printed components	Reconstruction of 3D model for further fabrication process	[63]
Laser scanner 4K sensor	3D point cloud of assembled structure	Evaluation of assembly quality	[64]
Laser scanner	3D point cloud of extruded structure	Evaluation of 3DCP quality	[65]
Commercial camera Laser distance sensor	Surface defects on extruded filament Vertical deformation of extruded concrete	Evaluation of extrudability and buildability for 3DCP	[66]
Hall effect sensor	Motor speed of pump	Real-time nozzle speed adjustment	[42]
Laser scanner	3D point cloud of extruded structure	Evaluation of 3DCP quality	
Camera	Width of extruded structure	Real-time extrusion speed control	[67]

The real-time monitoring of extruded material is also utilized to analyze the extrudability and buildability for 3D concrete printing [66].

Laser sensors and cameras can be utilized in the real-time inspection of construction quality. The laser distance sensor is utilized to measure the deformation of fresh concrete after printing where the commercial camera is utilized to capture surface images [66]. Cameras can capture images undergoing post-processing to detect printing failures. The surface imperfections and deformations are analyzed to improve the extrudability and buildability of concrete printing. The camera can be easily set up to continuously capture images during printing [67]. The processed images can be categorized as over-extrusion, correct, and under-extrusion. With the information from processed images, the speed of extrusion can be controlled through a feedback loop, in real-time, for the layer width correction.

This section has examined inspection techniques in construction including laser scanner, cameras, and laser distance sensor. Inspections at different stages including prior to construction, during construction, and post-construction are reviewed. Material properties such as extrudability and buildability can be examined via camera and laser scanner. For the real-time manipulation of a printer, the extrusion speed, nozzle speed, and nozzle height can be corrected according to the information captured from sensors.

2.5 CONCLUSION AND OUTLOOK

In conclusion, CAD software provides researchers the ability to create complex geometries and program robots to explore and expand 3DCP's abilities to automate the construction industry. While numerous design frameworks and control systems have been studied, they tend to be developed from established algorithms and commercially available sensors. Moreover, innovative robotic systems based on multi-robot collaboration have shown great potential for automating the construction process. The remaining challenges for digital designs and robotics of 3DCP can be summarized as below:

- A holistic digital design framework with comprehensive feasibility analysis is needed to simplify the complex decision-making process. Such a digital framework can be developed in the form of an interactive design system, which can make 3DCP more acceptable for non-expert users.
- The mechanical performance of structures fabricated as real-time control systems, parametric design tools, and multi-robot systems is underexplored. More research effort can be devoted to studies on the correlations between them.
- The degree of freedom provided by the gantry system has limitations for the construction of complex structures. However, the extra degree of freedom provided by the robotic arm system can support the construction of complex structures. In the future, more efforts can be focused on integrating multiple types of tools to support multi-process construction.
- A mobile robotic system provides more possibilities of on-site construction at a large scale. Localization and full-body control makes collaborations between robots feasible. In the future, more efforts can be focused on the automation of complex collaborations between robots and on human-robot collaborations.
- Multi-robot systems can be integrated into the existing 3DCP to fully automate the construction process. Tasks such as filling cavities with mortar and placement of reinforcement can be assigned to them to further reduce the reliance on manual work.

REFERENCES

1. Buswell, R.A., et al., *A process classification framework for defining and describing digital fabrication with concrete.* Cement and Concrete Research, 2020. **134**: p. 106068, doi: https://doi.org/10.1016/j.cemconres.2020.106068.

2. Cao, X., et al., *3D printing devices and reinforcing techniques for extruded cement-based materials: A review.* Buildings, 2022. **12**(4): p. 453.

3. Ma, G., et al., *Technology readiness: A global snapshot of 3D concrete printing and the frontiers for development.* Cement and Concrete Research, 2022. **156**: p. 106774, doi: https://doi.org/10.1016/j.cemconres.2022.106774.

4. Bryła, J. and A. Martowicz, *Study on the importance of a slicer selection for the 3D printing process parameters via the investigation of g-code readings.* Machines, 2021. **9**(8): p. 163, doi: https://doi.org/10.3390/machines9080163.

5. Bikas, H., et al., *A design framework to replace conventional manufacturing processes with additive manufacturing for structural components: A formula student case study.* Procedia Cirp, 2016. **57**: p. 710–715, doi: https://doi.org/10.1016/j.procir.2016.11.123.

6. Liang, S.Y., R.L. Hecker, and R.G. Landers, *Machining process monitoring and control: The state-of-The-art.* J. Manuf. Sci. Eng., 2004. **126**(2): p. 297–310, doi: https://doi.org/10.1115/1.1707035.

7. Latif, K., et al., *A review of g code, STEP, STEP-NC, and open architecture control technologies based embedded CNC systems.* The International Journal of Advanced Manufacturing Technology, 2021. **114**(9): p. 2549–2566, doi: https://doi.org/10.1007/s00170-021-06741-z.

8. Shivegowda, M.D., et al., *A review on computer-aided design and manufacturing processes in design and architecture.* Archives of Computational Methods in Engineering, 2022: p. 1–8, doi: https://doi.org/10.1007/s11831-022-09723-w.

9. Khoshnevis, B., et al., *Mega-scale fabrication by contour crafting.* International Journal of Industrial and Systems Engineering, 2006. **1**(3): p. 301–320, doi: http://dx.doi.org/10.1504/IJISE.2006.009791.

10. Liu, J., et al., *Additive manufacturing of sustainable construction materials and form-finding structures: A review on recent progresses.* 3D Printing and Additive Manufacturing, 2021, doi: https://doi.org/10.1089/3dp.2020.0331.

11. Ko, C.-H., *Constraints and limitations of concrete 3D printing in architecture.* Journal of Engineering, Design and Technology, 2021, doi: https://doi.org/10.1108/JEDT-11-2020-0456.

12. Ooms, T., et al., *A parametric modelling strategy for the numerical simulation of 3D concrete printing with complex geometries.* Additive Manufacturing, 2021. **38**: p. 101743, doi: https://doi.org/10.1016/j.addma.2020.101743.

13. Vantyghem, G., T. Ooms, and W. De Corte, *VoxelPrint: A grasshopper plug-in for voxel-based numerical simulation of concrete printing.* Automation in Construction, 2021. **122**: p. 103469, doi: https://doi.org/10.1016/j.autcon.2020.103469.

14. Preisinger, C. and M. Heimrath, *Karamba—A toolkit for parametric structural design.* Structural Engineering International, 2014. **24**(2): p. 217–221, doi: https://doi.org/10.2749/101686614X13830790993483.

15. Michalatos, P. *Millipede.* 2014 [cited 2022 09/10]; Available from: https://grasshopperdocs.com/addons/millipede.html#:~:text=Millipede%20is%20a%20structural%20analysis,forces%2C%20and%203d%20volumetric%20elements.

16. Jordy Vos, S.W. *Buckling Simulation for 3D Printing in Fresh Concrete.* [cited 2022 09/10]; Available from: https://www.karamba3d.com/examples/moderate/buckling-simulation-for-3d-printing-in-fresh-concrete.

17. Bi, M., et al., *Topology optimization for 3D concrete printing with various manufacturing constraints.* Additive Manufacturing, 2022. **57**: p. 102982, doi: https://doi.org/10.1016/j.addma.2022.102982.

18. Vantyghem, G., et al., *3D printing of a post-tensioned concrete girder designed by topology optimization.* Automation in Construction, 2020. **112**: p. 103084, doi: https://doi.org/10.1016/j.autcon.2020.103084.

19. Anton, A., et al., *Vertical Modulations.* 2019.

20. XtreeE. *Double sine wall.* 2016 [cited 2022 09/10]; Available from: https://xtreee.com/en/project/mur-sinusoidal/.

21. Bhooshan, S., T. Van Mele, and P. Block. *Morph & Slerp: Shape description for 3D printing of concrete.* in *Symposium on Computational Fabrication.* 2020. p. 1–10, doi: https://doi.org/10.1145/3424630.3425413.

22. Slic3r. *Slic3r: Open source 3D printing toolbox.* [cited 2022 09/10]; Available from: https://slic3r.org/.

23. 3DGBIRE. *Ultimaker Cura.* [cited 2022 09/10]; Available from: https://3dgbire.com/pages/ultimaker-cura?utm_term=&utm_campaign=gs-2019-08-20&utm_source=google&utm_medium=smart_campaign&hsa_acc=4134625929&hsa_cam=18311660389&hsa_grp=&hsa_ad=&hsa_src=x&hsa_tgt=&hsa_kw=&hsa_mt=&hsa_net=adwords&hsa_ver=3&gclid=Cj0KCQjw4omaBhDqARIsADXULuXfzjPu4plPNcQzod8Q0w1SS58XaMhjlBafFigFZSvc0117Ol8YgwaAkVWEALw_wcB.

24. Simplify3D, I., *Simplify3d.* 2019; Available from: https://www.simplify3d.com/.

25. Gosselin, C., et al., *Large-scale 3D printing of ultra-high performance concrete a new processing route for architects and builders.* Materials & Design, 2016. **100**: p. 102–109, doi: https://doi.org/10.1016/j.matdes.2016.03.097.

26. Carneau, P., et al., *Additive manufacturing of cantilever-from masonry to concrete 3D printing.* Automation in Construction, 2020. **116**: p. 103184, doi: https://doi.org/10.1016/j.autcon.2020.103184.

27. Adiels, E., M. Ander, and C.J. Williams, *Brick patterns on shells using geodesic coordinates.* in *Proceedings of IASS Annual Symposia.* 2017. International Association for Shell and Spatial Structures (IASS), p. 1–10.

28. Tang, K., S.-Y. Chou, and L.-L. Chen, *An algorithm for reducing tool retractions in zigzag pocket machining.* Computer-Aided Design, 1998. **30**(2): p. 123–129, doi: https://doi.org/10.1016/S0010-4485(97)00064-X.

29. Lin, Z., et al., *Smooth contour-parallel tool path generation for high-speed machining through a dual offset procedure.* The International Journal of Advanced Manufacturing Technology, 2015. **81**(5): p. 1233–1245, doi: https://doi.org/10.1007/s00170-015-7275-z.

30. Xia, L., et al., *Globally continuous hybrid path for extrusion-based additive manufacturing.* Automation in Construction, 2022. **137**: p. 104175, doi: https://doi.org/10.1016/j.autcon.2022.104175.

31. Gleadall, A., *FullControl GCode designer: Open-source software for unconstrained design in additive manufacturing.* Additive Manufacturing, 2021. **46**: p. 102109, doi: https://doi.org/10.1016/j.addma.2021.102109.

32. Breseghello, L. and R. Naboni, *Toolpath-based design for 3D concrete printing of carbon-efficient architectural structures.* Additive Manufacturing, 2022. **56**: p. 102872, doi: https://doi.org/10.1016/j.addma.2022.102872.

33. Lim, J.H., Y. Weng, and Q.-C. Pham, *3D printing of curved concrete surfaces using adaptable membrane formwork.* Construction and Building Materials, 2020. **232**: p. 117075, doi: https://doi.org/10.1016/j.conbuildmat.2019.117075.

34. Costanzi, C.B., et al., *3D printing concrete on temporary surfaces: The design and fabrication of a concrete shell structure.* Automation in Construction, 2018. **94**: p. 395–404, doi: https://doi.org/10.1016/j.autcon.2018.06.013.

35. Kucuk, S. and Z. Bingul, Robot kinematics: Forward and inverse kinematics. 2006: INTECH Open Access Publisher London, UK.

36. Aspragathos, N.A. and J.K. Dimitros, *A comparative study of three methods for robot kinematics.* IEEE Transactions on Systems, Man, and Cybernetics, Part B (Cybernetics), 1998. **28**(2): p. 135–145, doi: 10.1109/3477.662755.

37. Mühe, H., et al., *On reverse-engineering the KUKA Robot Language.* arXiv preprint arXiv:1009.5004, 2010, doi: https://doi.org/10.48550/arXiv.1009.5004.

38. Robotics, A., *Operating manual–introduction to Rapid.* Document ID: 3HAC0966-50 Revision, 2004. **1**.

39. Živanović, S.T. and G.V. Vasilić, *A new CNC programming method using STEP-NC protocol.* FME Transactions, 2017. **45**(1): p. 149–158, doi: https://doi.org/10.5937/fmet1701149Z.

40. Rauch, M., et al., *An advanced STEP-NC controller for intelligent machining processes.* Robotics and Computer-Integrated Manufacturing, 2012. **28**(3): p. 375–384, doi: https://doi.org/10.1016/j.rcim.2011.11.001.

41. Buswell, R., et al. *Inspection methods for 3D concrete printing.* in *RILEM International Conference on Concrete and Digital Fabrication.* 2020. Springer, p. 790–803, doi: https://doi.org/10.1007/978-3-030-49916-7_78.

42. Yuan, P.F., et al., *Real-time toolpath planning and extrusion control (RTPEC) method for variable-width 3D concrete printing.* Journal of Building Engineering, 2022. **46**: p. 103716, doi: https://doi.org/10.1016/j.jobe.2021.103716.

43. Wolfs, R.J., et al., *A real-time height measurement and feedback system for 3D concrete printing*, in High tech concrete: Where technology and engineering meet. 2018, Springer. p. 2474–2483.

44. Panda, B., et al. *Automation of robotic concrete printing using feedback control system.* in ISARC. Proceedings of the international symposium on automation and robotics in construction. 2017. IAARC Publications.

45. Barjuei, E.S., et al., *Real-time vision-based control of industrial manipulators for layer-width setting in concrete 3D printing applications.* Advances in Industrial and Manufacturing Engineering, 2022: p. 100094, doi: https://doi.org/10.1016/j.aime.2022.100094.

46. Technologies, R. *RAPCAM: Software for large format 3D printers.* [cited 2022 09/10]; Available from: https://www.raptech.io/rapcamam.

47. XtreeE. *XtreeE: The large-scale 3D printers.* [cited 2022 09/10]; Available from: https://xtreee.com/en/ecosystem/.

48. CyBe. *CyBe Software.* [cited 2022 09/10]; Available from: https://cybe.eu/3d-concrete-printing/software/.

49. *KUKA.Sim: features overview.* [cited 2022 09/10]; Available from: https://www.kuka.com/en-au/products/robotics-systems/software/simulation-planning-optimization/kuka_sim/kuka_sim_updates.

50. ABB. *ABB: RobotStudio 3D Printing PowerPac*. [cited 2022 09/10]; Available from: https://new.abb.com/products/robotics/application-software/3d-printing-powerpac.

51. Rahul, A. and M. Santhanam, *Evaluating the printability of concretes containing lightweight coarse aggregates*. Cement and Concrete Composites, 2020. **109**: p. 103570, doi: https://doi.org/10.1016/j.cemconcomp.2020.103570.

52. Salet, T.A., et al., *Design of a 3D printed concrete bridge by testing*. Virtual and Physical Prototyping, 2018. **13**(3): p. 222–236, doi: https://doi.org/10.1080/17452759.2018.1476064.

53. Tiryaki, M.E., X. Zhang, and Q.-C. Pham. *Printing-while-moving: a new paradigm for large-scale robotic 3D Printing*. in *2019 IEEE/RSJ International Conference on Intelligent Robots and Systems (IROS)*. 2019. IEEE, p. 2286–2291, doi: 10.1109/IROS40897.2019.8967524.

54. Keating, S.J., et al., *Toward site-specific and self-sufficient robotic fabrication on architectural scales*. Science Robotics, 2017. **2**(5): p. eaam8986, doi: 10.1126/scirobotics.aam8986.

55. Hack, N., et al., *Structural stay-in-place formwork for robotic in situ fabrication of non-standard concrete structures: A real scale architectural demonstrator*. Automation in Construction, 2020. **115**: p. 103197, doi: https://doi.org/10.1016/j.autcon.2020.103197.

56. Xu, W., et al., *Toward automated construction: The design-to-printing workflow for a robotic in-situ 3D printed house*. Case Studies in Construction Materials, 2022. **17**: p. e01442, doi: https://doi.org/10.1016/j.cscm.2022.e01442.

57. Zhang, X., et al., *Large-scale 3D printing by a team of mobile robots*. Automation in Construction, 2018. **95**: p. 98–106, doi: https://doi.org/10.1016/j.autcon.2018.08.004.

58. Gawel, A., et al. *A fully-integrated sensing and control system for high-accuracy mobile robotic building construction*. in *2019 IEEE/RSJ International Conference on Intelligent Robots and Systems (IROS)*. 2019. IEEE, p. 2300–2307, doi: 10.1109/IROS40897.2019.8967733.

59. Parascho, S., et al., *Robotic vault: A cooperative robotic assembly method for brick vault construction*. Construction Robotics, 2020. **4**(3): p. 117–126, doi: https://doi.org/10.1007/s41693-020-00041-w.

60. Duballet, R., O. Baverel, and J. Dirrenberger. *Space truss masonry walls with robotic mortar extrusion*. in Structures. 2019. Elsevier, p. 41–47, doi: https://doi.org/10.1016/j.istruc.2018.11.003.

61. Kim, M.-K., Q. Wang, and H. Li, *Non-contact sensing based geometric quality assessment of buildings and civil structures: A review*. Automation in Construction, 2019. **100**: p. 163–179, doi: https://doi.org/10.1016/j.autcon.2019.01.002.

62. Kinnell, P., et al., *Precision manufacture of concrete parts using Integrated Robotic 3D Printing and Milling*. 2021. Proceedings of the 21st International Conference of the European Society for Precision Engineering and Nanotechnology, EUSPEN 2021. 2021. https://www.euspen.eu/resource/precision-manufacture-of-concrete-parts-using-integrated-robotic-3d-printing-and-milling/

63. Lindemann, H., et al. *Development of a shotcrete 3D-printing (SC3DP) technology for additive manufacturing of reinforced freeform concrete structures*. In *RILEM International Conference on Concrete and Digital Fabrication*. 2018. Springer, p. 287–298, doi: https://doi.org/10.1007/978-3-319-99519-9_27.

64. Grasser, G., et al. *Complex architecture in printed concrete: the case of the Innsbruck University 350 th Anniversary Pavilion COHESION*. In *RILEM International Conference on Concrete and Digital Fabrication*. 2020. Springer, p. 1116–1127, doi: https://doi.org/10.1007/978-3-030-49916-7_106.

65. Anton, A., et al., *A 3D concrete printing prefabrication platform for bespoke columns*. Automation in Construction, 2021. **122**: p. 103467, doi: https://doi.org/10.1016/j.autcon.2020.103467.

66. Ting, G.H.A., et al., *Extrudable region parametrical study of 3D printable concrete using recycled glass concrete*. Journal of Building Engineering, 2022. **50**: p. 104091, doi: https://doi.org/10.1016/j.jobe.2022.104091.

67. Kazemian, A., et al., *Computer vision for real-time extrusion quality monitoring and control in robotic construction*. Automation in Construction, 2019. **101**: p. 92–98, doi: https://doi.org/10.1016/j.autcon.2019.01.022.

3 Reinforcement Learning-Based Robotic Motion Planning for Conducting Multiple Tasks in Virtual Construction Environments

Weijia Cai, Lei Huang, and Zhengbo Zou

CONTENTS

3.1 INTRODUCTION

The construction industry employs 7% of the working-age labor force globally, and approximately $10 trillion is spent annually on construction-related goods and services (Barbosa et al. 2017). Despite its size, the construction industry has been slow to implement innovative technologies, resulting in flattened productivity and limited profit margins. Most importantly, the construction industry claims the highest percentage (21%) of total fatal injuries among all industries (U.S. BLS 2020). Surveys showed the major cause (45%) of disability in construction is musculoskeletal injuries (Arndt et al. 2005), since many common construction tasks such as bricklaying, ceiling installation, and earthmoving require repetitive and physically demanding actions by human workers. Robotizing construction tasks has long been considered a promising solution in the construction industry to reduce the risks of occupational injuries and diseases by releasing human workers from the physically demanding tasks (Bock 2015).

Construction tasks involving operations of reaching, moving, and placing objects are considered the fundamental tasks for automation in construction (Everett and Slocum 1994). The major bottleneck of robotizing these tasks is optimal motion planning (Lundeen et al. 2019), which refers to construction robots attempting to obtain an optimal trajectory to safely move onsite objects from an initial placement to a final placement. For example, a ceiling installation task requires robots to first identify the location of a prefabricated ceiling panel to plan a series of actions to reach that panel, then pick it up and plan a safe trajectory to move it to an opening with a predefined position and orientation. To allow construction robots to plan their motions in a dynamic construction environment, reinforcement learning (RL) has been used to generate a series of optimal actions given observations from the environments (Haarnoja et al. 2018). Since an RL agent chooses its actions according to what it observes from the environment, it has the potential to enable robots to adapt to unstructured construction sites in which the observations might be diverse (Eysenbach et al. 2018).

However, exploring such diverse observations in the real world is time-consuming, costly, and possibly harmful to the robots (Gu et al. 2016). Therefore, it is beneficial to simulate the construction site and train the RL agent virtually in a physics-based game engine (James and Johns 2016). The game engine can be used to simulate motions of the robots and onsite objects (e.g., ceiling panels) to provide a realistic feedback loop for the construction robot (You et al. 2018). By leveraging virtual environments, an RL control agent can be trained in a variety of scenarios with different settings to test its generality without being deployed on a real construction robot.

There are a number of emerging applications of RL-based robot control in manufacturing and construction (Thomas et al. 2018; Dharmawan et al. 2020; Liang et al. 2020; Matsumoto et al. 2020; Apolinarska et al. 2021). However, these applications mostly focus on designing and training RL agents for a single task, which results in highly specialized RL agents that are not flexible to onsite uncertainties or slight changes to the trained operation. In fact, RL has the potential to transfer knowledge from a trained agent to accelerate the training of multiple similar downstream tasks (Zhu et al. 2020), using transfer learning (TL) (Yosinski et al. 2014). Considering that many construction tasks (e.g., ceiling installation, window installation, and flooring) can be seen as a specific case of a source task (e.g., pick and place), the transferability of RL can reduce computational cost and training effort for these similar construction tasks.

In this study, we proposed an RL-based robotic motion planning approach to conduct three construction tasks in a virtual construction environment (VCE). We first utilized Unity3D to build a realistic VCE including a 6 degree-of-freedom (DoF) robot arm and target objects (i.e., windows, ceiling panels, and floor panels) around a two-story unfinished building. Then we trained RL agents to conduct three similar construction tasks (i.e., window installation, ceiling installation, and flooring). Training of the RL agents can be divided into two stages: 1) pre-train an RL agent for a source task (i.e., pick-and-place task) using a four-step training plan, where the agent in this stage is referred to as the upstream agent; and 2) fine-tune the upstream agent to output three downstream agents for three specific construction tasks. Finally, we evaluated the robustness of the downstream agents using the final cumulative rewards and the success rate of each task.

The contribution of this study is twofold:

- First, we proposed a simulation-based approach for training construction robots controlled using RL in a game engine, where realistic construction environments and robotic control operations can be represented.
- Second, we introduced a staged training method for RL-based robotic control, so that the RL agent can transfer the learned knowledge from source tasks to downstream tasks without the time-consuming retraining process.

3.2 RELATED WORKS AND BACKGROUND

3.2.1 TRADITIONAL ROBOTIC MOTION PLANNING IN CONSTRUCTION

The main challenge of robotic motion planning for construction robots is that the unstructured nature of construction sites requires robots to be flexible and work under uncertainties (Feng et al. 2014).

In achieving this, previous works in traditional robotic motion planning focus on two directions. The first direction is to carefully constrain the workspace of the robots by designing new robotic systems and thereafter manually preprogram the motions of robots. For example, Lee et al. (2007) designed a ceiling installation robot composed of an industrial robot arm mounted on a vehicle with an aerial lift. The motion planning of their designed robot was then preprogrammed to enable transportation of the ceiling panels between certain locations. Thereafter, human workers still need to adjust the final position and orientation of the panel to complete the installation. Similarly, Iturralde et al. (2020) designed a cable driven parallel robot (CDPR) integrated with an industrial robot arm to automate curtain wall installations. The CDPR provided mobility for the industrial robot along the wall to conduct drilling and panel installation; however, CDPR can only work under a constrained workspace and is not generalizable for other construction tasks. Researchers have also created new tooling for existing robots. For example, Kumar et al. (2017) designed a new tool for an industrial robot to automate the construction of non-standard steel-reinforced steel meshes. The designed tool integrated a series of components for sub-tasks such as bending and cutting, which simplified the motion planning of complex manipulations. However, their robot tooling can only be used for a specific task.

Another direction is to innovate on classic robotic motion planning methods (Villumsen and Kristiansen 2017; Yang and Kang 2021). Sampling-based motion planning (SMP) methods such as the probabilistic roadmaps (PRM) method (Kavraki et al. 1996) and rapid-exploring random trees (RRT) (LaValle et al. 1998) are commonly used in optimizing robot motions. SMP attempts to use only restricted information about the robot and obstacles to optimize trajectories between an initial placement and a final placement considering collision avoidance (Elbanhawi and Simic 2014). The main drawback of SMP is that it needs to constantly check if collisions might occur in a sampled trajectory, which is an NP-complete problem requiring expensive computation for the robot (Yu and Gao 2021). For example, Yang and Kang (2021) implemented a PRM-based motion planning method for modular home prefabrication. The authors first sampled a series of robot configurations (i.e., the distance between an initial position and a final position, the shape of the target component, and the loading capacity of the robot). Then they implemented a path searching algorithm to obtain the optimal trajectory considering collision checking. This collision checking dictated whether a robot configuration can be accepted. However, the shapes of construction robots and onsite objects tend to be diverse and complex (Lundeen et al. 2019), which causes the sampling processing of SMP to be computationally expensive and infeasible.

3.2.2 ROBOTIC MOTION PLANNING USING REINFORCEMENT LEARNING

Unlike the preprogrammed motion planning and classic motion planning methods such as SMP, RL-based methods have the potential to improve the robustness and flexibility of construction robots (Konidaris and Barto 2006; Gu et al. 2016). This flexibility has been achieved in fundamental robotic control tasks such as pick-and-place (James and Johns 2016) and push-and-grasp (Zeng et al. 2018). In general, RL-based motion planning aims to train an RL agent that produces an optimal policy that maps the states (e.g., the robot arm joints' positions) to the actions (e.g., forces applied to the joints) with the guidance of a designed reward function. Stimulated by the reward, the RL agent can learn to work in dynamic environments such as a construction site (Jaakkola et al. 1994).

To elaborate, an RL agent's goal is to maximize the return G_t (i.e., the summation of the discounted future reward) over time under the learned policy model $\pi(A \mid S)$ from its interactions with the environment (see Equation 3.1).

$$max_{\pi(A|S)}G_t = max_{\pi(A|S)}\sum_{k=t+1}^{T}\delta^{k-t-1} \times R_k = max_{\pi(A|S)}\left(\delta G_{t+1} + R_{t+1}\right) \tag{3.1}$$

In Equation 3.1, R_k is the reward applied to the agent starting from time step $k = t+1$ and δ is a discount factor that determines the present value of the future rewards. *A* refers to the actions produced by the agent. In this paper, the action A is composed of a discrete vector, which selects the specific joints to rotate, and a continuous vector, which decides the values of rotation for the selected joints. S refers to the state observed from the environment. In this paper, S is composed of the target object posture q_t^{target}, the gripper posture $q_t^{gripper}$, the gripper velocity $v_t^{gripper}$, the final placement posture $q_t^{placement}$, and the joint position p^{joint}. Here, we define position as the 3D coordinates (x, y, z), and the posture as the combination of position and orientation (x, y, z, r_x, r_y, r_z).

The optimization problem from Equation 3.1 can be seen as a Markov decision process (MDP) with reward (Sutton and Barto 2018). A Q value function $q(s,a)$ is commonly used to represent the expected return for G_t at state s, taking an action a, and following a policy π, as described in Equation 3.2:

$$q(s,a) = E_\pi \left(\delta G_{t+1} + R_{t+1} \mid S_t = s, \; A_t = a \right) \tag{3.2}$$

By interacting with the environment over time, the agent keeps updating Equation 3.2 for each instance of state and action pair until it converges to the form of Equation 3.3:

$$q^*(s,a) = \sum_{s, r} \pi^*(a \mid s) f(\overline{s} \mid s,a) \left(r + max_{\overline{a}} q^* (s,\overline{a}) \right) \tag{3.3}$$

In Equation 3.3, f is the transition function mapping state s and a to the next state $\overline{s}$; r is the reward from the environment with the agent being at the state s; π^* is the optimal policy model. Due to the complexity of exploration for updating the Q value function, modern RL methods such as policy gradient (PG) (Sutton et al. 1999; Schulman et al. 2015; Schulman et al. 2017) use model-free surrogate functions to approximate the Q value function and the policy function. In this paper, we implemented the proximal policy optimization (PPO) algorithm (Schulman et al. 2017) for our robot motion planning task. We chose PPO because it guarantees a stable improvement of the policy model.

Despite the flexibility of RL, implementations of RL-based robotic motion planning in the construction domain are insufficient (Liang et al. 2020; Matsumoto et al. 2020). The main reason lies in the unstructured nature of construction sites (Feng et al. 2014), which hinders the real-world implementation of onsite construction robots. To avoid this issue, we trained our RL agents in a virtual construction environment in which onsite uncertainties can be simulated to enable sufficient explorations of the RL agents.

3.2.3 Transfer Learning in Deep Reinforcement Learning for Robotic Motion Planning

Deep reinforcement learning (DRL) combines RL and deep neural network (DNN), which enable the RL agents to reuse the learned model parameters of the trained agents, including the policy model and Q value function (Gu et al. 2016). This form of knowledge transferability has been widely discussed in the domain of transfer learning (TL) (Yosinski et al. 2014; Long et al. 2015). TL techniques are used to accelerate DNN model training by incorporating learned model parameters trained on similar tasks. The learned parameters are hypothetically close as their task goals are similar. For example, window installation and ceiling installation are similar tasks that both aim to pick up an object starting from a random placement and place it at another placement with a specific posture (e.g., windows are commonly vertically installed while the ceiling panels are horizontally installed), for which the agents of both tasks will have similar model parameters producing similar motion planning actions. Therefore, we can train an RL agent for a pick-and-place task as a source task and then reuse the learned agent in the training for downstream tasks, such as window installation and ceiling installation.

However, construction tasks require the robots to make accurate manipulations for safety reasons (Lundeen et al. 2019), i.e., placing the object with specific orientation within a small tolerance. This is challenging to achieve because the agents and the related reward functions must be carefully designed, which requires additional experiments in searching for the best design. To make stable improvement of RL training, curriculum learning (CL), as another form of TL, schedules the model training by a sequence of training criteria (e.g., distance tolerance of reaching an object) that ascend by difficulty (e.g., decreasing distance tolerance) (Bengio et al. 2009). An agent in a curriculum plan can utilize the transferable parameters of the models from the former criteria, hence accelerating the training for subsequent criteria. To this end, the upstream agent can be trained following a designed curriculum plan with increasing difficulty. Details of the implementation of CL in this paper will be discussed in the next section.

3.3 METHODOLOGY AND EXPERIMENT DESIGN

In this section, we introduce the proposed approach that enables a virtual robot to execute multiple tasks using RL with a CL-based training plan. We first introduce the setup of a VCE in Unity3D including the physical characteristics of the environment, the virtual sensors deployed for goal detection, and the link between the RL algorithm and VCE. Next, we explain the process of training an upstream RL agent and then fine-tune the agent for three specific construction tasks. The training process is composed of two stages (see Figure 3.1). In stage 1, we pre-trained an upstream agent that learns to complete a source task composed of picking up the target object at an initial placement and placing the object at a final placement with pre-defined posture (i.e., position and orientation of the object). In stage 2, additional trainings will be conducted to fine-tune the upstream agent to output three downstream agents for three specific tasks, namely, window installation, ceiling installation, and flooring.

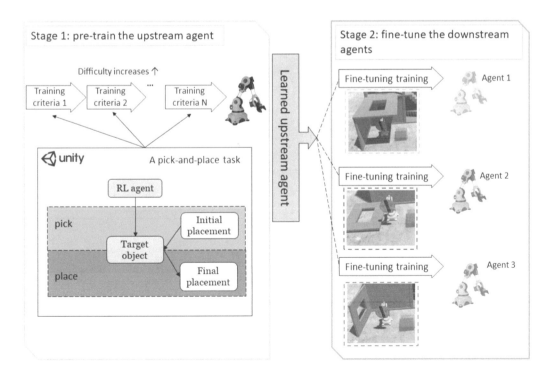

FIGURE 3.1 An overview of the proposed approach.

3.3.1 VIRTUAL CONSTRUCTION ENVIRONMENT DEVELOPMENT

The development of our VCE contains three main steps: 1) set up physical characteristics (e.g., weight and dimension) of the objects in the VCE; 2) set up virtual sensors for detecting whether the RL agents have completed the tasks; and 3) link the RL agent to the robot arm in VCE. In the first step, we designed our VCE as a construction site with an unfinished two-story building (see Figure 3.2), a 6 DoF industrial robot, a target object (i.e., either a window, a ceiling panel, or a floor panel), three types of openings as the final placement (i.e., final position and orientation of the target object), and several barriers that constrain the workspace of the robot. We set the force limit of each joint of the robot to be 200 N, which enables lifting a target object up to 15 kilograms. There are three opening examples for construction tasks, and each is within the reachable distance of the robot (see Figure 3.2b).

Next, we set up virtual sensors to determine whether the robot has completed the following two goals: 1) picking up the target object, and 2) placing the object with any predefined posture. For the first goal, we used a ray-cast sensor, which is similar to a sonic sensor that sends pulses and receives a signal if the pulses hit an object. The ray-cast sensor is designed to have a limited range as a distance tolerance, which is used as one of the training criteria in CL. For the second goal, we used the ray-cast sensor combined with an angle detector that limits the angle between the gripper plane and the final placement plane within a predefined angle tolerance. We also used the angle tolerance as one of the training criteria in CL. Finally, we linked the RL agents to the robot in the VCE to control motions of the robot's joints. During training of the general RL agent, we randomized the postures of the target and the final placements to force the robot to be flexible and robust in different scenarios.

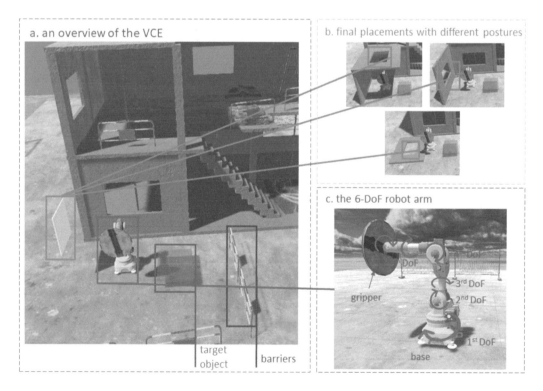

FIGURE 3.2   VCE used for RL-based robot motion planning: (a). an overview of the VCE including a 6-DoF robot arm, a cube-shaped target object, and several barriers; (b). three examples of the possible final placement: the top left is for ceiling installation, the top right is for window installation, the bottom is for flooring; (c). a 6-DoF robot arm with a vacuum gripper is shown.

3.3.1.1 Stage 1: Pre-train the Upstream Agent

In stage 1, the upstream agent aims to learn a pick-and-place task as the source task, which includes two goals: 1) picking up the target, and 2) placing the target with predefined postures. For the first goal, the agent learns to approach and pick up the target object within a distance tolerance. As described in Section 3.1, we used a ray-cast sensor to detect the distance between the robot's gripper and the target object. We assume the initial orientation of the target object is horizontal and its initial position is randomized. Thereafter, the agent is required to pick up the target from any position within the distance tolerance. For the second goal, the agent learns to place the target object to a final placement with a predefined posture (i.e., position and orientation). More specifically in Figure 3.1, for ceiling installation, Agent 1 aims to place a ceiling panel with a horizontal orientation above the robot; for window installation, Agent 2 aims to place a window vertically next to the robot; for flooring, Agent 3 aims to place a floor panel with a horizontal orientation on the ground. Therefore, for the upstream agent, the environment randomly samples one of the pre-defined postures from the above three construction tasks for every training episode.

During training, RL agents attempt to achieve the goals by maximizing the return of designed reward functions. Therefore, designing a proper reward function is one of the most fundamental problems in RL (Abbeel and Ng 2004). In this study, we designed our reward function R (final line in Figure 3.3) as the summation of an intrinsic reward R^I obtained from the exploration of the agent (Pathak et al. 2017) and an extrinsic reward R^E obtained from successful maneuvers that help achieve the goals. For the intrinsic reward function, we implemented a curiosity-based reward to encourage the agents to explore unfamiliar states using the intrinsic curiosity module (ICM) (Pathak et al. 2017). ICM includes a feature extractor, an inverse model, and a forward model. The feature extractor encoded states S_t and S_{t+1} into features $\omega(S_t)$ and $\omega(S_{t+1})$ that are trained to predict action A_t as $\widehat{A}_t$ in the inverse model. Then the forward model takes $\omega(S_t)$ and A_t as input and predicts the feature of S_{t+1} as $\hat{\omega}(S_{t+1})$. Finally, the intrinsic reward is calculated as the Euler distance between $\omega(S_{t+1})$ and $\hat{\omega}(S_{t+1})$ shown as a part of Algorithm 1.

Algorithm 1: Reward function

1 **for** *episode e = 1 to N* **do**
2   **for** *training step t = 1 to MaxStep* **do**
         # get intrinsic reward#
         $R^I_{e,t} = \frac{\delta}{2} \times \|\omega(S_{t+1}) - \hat{\omega}(S_{t+1})\|^2$
         # get extrinsic reward#
         $R^E_{e,t} = -\frac{1}{MaxStep}$
         if *Grasped* **then**
             $R^E_{e,t} = 1$ (only assigned for the first time)
             if *Placed* **then**
                 $R^E_{e,t} = 2$
                 # get total reward#
                 $R_{e,t} = R^I_{e,t} + R^E_{e,t}$
                 End current episode
         else
             continue
         # get total reward#
         $R_{e,t} = R^I_{e,t} + R^E_{e,t}$

FIGURE 3.3   Reward function design.

TABLE 3.1

Curriculum Plan (Cp) for the Upstream Agent

Training Criteria	τ_{pick} (m)	τ_{place} (m)	τ_{angle} (degree)
cp1	0.40	0.50	90.00
cp2	0.30	0.25	50.00
cp3	0.30	0.13	30.00
cp4	0.30	0.10	20.00

For the extrinsic reward design, we simply assigned a positive reward for the completion of the goals and a negative reward as a time penalty for not achieving any of the goals or taking too long to achieve them (Matulis and Harvey 2021). The extrinsic reward is also shown as a part of Algorithm 1 in Figure 3.3.

To pursue a stable improvement during the training of the agent, we adopted curriculum learning (CL) (Bengio et al. 2009; Matulis and Harvey 2021) to design the training plan for the upstream agent (see Table 3.1). As seen in Table 3.1, the training criteria consists of three parts: 1) distance tolerance for reaching target τ_{pick}, 2) distance tolerance for placing the target object τ_{place}, and 3) angle tolerance for limiting the orientation for placing the target object τ_{angle}. Each row in Table 3.1 refers to the training of an agent named as cp1 to cp4. The model parameters (i.e., policy model, Q value function, and the learned ICM) of an agent starting from cp2 are initiated from its former agent.

3.3.1.2 Stage 2: Fine-tune the Downstream Agents

In stage 2, we conducted additional fine-tuning of the upstream agent to output three downstream agents for three specific construction tasks: ceiling installation, window installation, and flooring. The downstream agents utilized the learned model parameters from the trained upstream agent (i.e., cp4) in stage 1 to expedite their training. We used the same criteria from the training of cp4 to compare the results between the upstream agent and the downstream agents, shown in Table 3.2. During fine-tuning, we used the same reward function as stage 1. Table 3.2 contains three conditions to distinguish each task: 1) whether to utilize the pre-trained upstream agent, 2) orientation for placing the target object r_{place}, and 3) position for placing the target object p_{place}.

To better evaluate the effectiveness of CL, we also trained three RL agents as a control group (i.e., cp5.1c, cp5.2c, cp5.3c, "c" for "control") to achieve these three construction tasks without inheriting the model parameters from the cp4. These agents are referred to as the control group agents. We compared the final cumulative rewards and the success rates of picking and placing the target objects between the downstream agents and the control group agents in the next section.

TABLE 3.2

Curriculum Plan for the Downstream Agents and the Control Group Agents

Downstream Task	Training Criteria	Inherited the Upstream Agent	r_{place}	p_{place}
Ceiling installation	cp5.1	Yes	Horizontal	Above the robot
	cp5.1c	No		
Window installation	cp5.2	Yes	Vertical	Around the robot
	cp5.2c	No		
Flooring	cp5.3	Yes	Horizontal	On the floor
	cp5.3c	No		

3.4 RESULTS AND DISCUSSION

In this section, we show the training results of the upstream agent and evaluate the robustness of the downstream agents using the success rates of picking and placing. Following the curriculum plan in Tables 3.1 and 3.2, the evaluation is divided into three sections: 1) performance of the pre-trained upstream agent, 2) a comparison between the upstream agent and the downstream agents, 3) and a comparison between the downstream agents and the control group agents.

3.4.1 PRE-TRAINING RESULTS OF THE UPSTREAM AGENT

The results of pre-training for the upstream agent are shown in Figure 3.4. Each line with the same color represents the cumulative extrinsic reward over 2 million action steps taken for each experiment. Transparent lines refer to the raw data while solid lines represent the moving average. It can be observed that all RL agents converged to a stable cumulative reward. To elaborate, cp1 converged to a reward around 1.6, while cp2 converged to around 2.1. For cp3 and cp4, they converged to around 1.2, which is lower than the results of cp1 and cp2. This is because the acceptable motions of the robot overlap among different predefined final placements with large angle tolerance in cp1 and cp2 (see Figure 3.1) so the agents can complete placing using similar motion plans. However, the angle tolerance is so small that the acceptable motions in cp3 and cp4 become distinct, so the agents have to plan different motions for distinct placing postures.

We evaluated the training results by the success rate of picking $N_{pick}/N_{episode}$ and the success rate of placing $N_{place}/N_{episode}$. The success rates of placing for cp4 are relatively low because the similarity among the three types of final placement is decreasing, which requires the policy model to distinguish the three types of final placements.

3.4.2 COMPARISON BETWEEN THE UPSTREAM AGENT AND THE DOWNSTREAM AGENTS

In this section, we evaluate the robustness of the downstream agents by comparing their results with the upstream agent. The training results are shown in Figure 3.5. The downstream agents (i.e., cp5.1

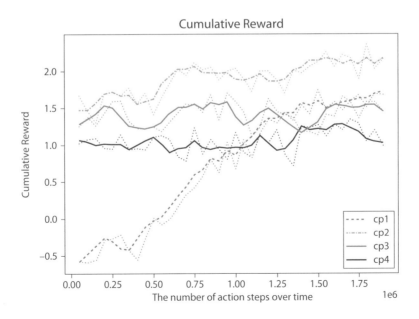

FIGURE 3.4 Cumulative reward over the number of action steps taken for the upstream agent.

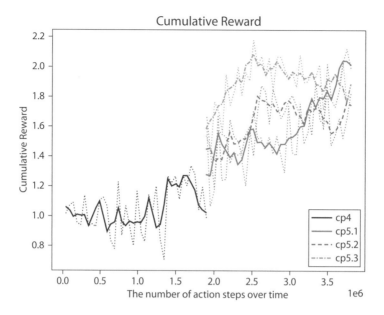

FIGURE 3.5 Cumulative reward over the number of action steps taken for the comparison between the upstream agent and the downstream agents.

to cp5.3) all have solid improvements from the upstream agent cp4. The three downstream agents all achieved the final cumulative reward of around 1.8.

The success rates of the downstream agents are shown in Table 3.3. For picking, it can be observed that the downstream agents achieved an average success rate of around 91%. For placing, the success rates of the downstream agents are in 60.92%, 62.45%, and 72.47%, which all achieved solid improvements from the result of the cp4 (i.e., 41%). The results also show that Agent 3 (i.e., cp5.3) has a much higher success rate of placing than the other downstream agents. This is because the orientation of the final placement for Agent 3 is similar to the orientation of the initial placement, which allowed Agent 3 to complete its task by simply rotating the base joint after picking up the target object. However, the final placements for Agent 1 and Agent 2 are more distinct from that of Agent 3, requiring these two agents to explore more complex motion paths. For example, for ceiling installation, Agent 1 is supposed to flip the robot's gripper to place the ceiling panel.

3.4.3 COMPARISON BETWEEN THE THREE DOWNSTREAM AGENTS AND THE CONTROL GROUP AGENTS

The training results of the control group agents are shown in Figure 3.6. Unlike the downstream agents, the control group agents (i.e., cp5.1c, cp5.2c, and cp5.3c) did not inherit the model

TABLE 3.3

Comparison of the Success Rates between the Downstream Agents and the Upstream Agent

Training Criteria	$N_{pick}/N_{episode}$ (%)	$N_{place}/N_{episode}$ (%)
cp5.1/cp4	94.84/90.57	60.92/42.76
cp5.2/cp4	92.19/90.57	62.45/42.76
cp5.3/cp4	86.06/90.57	72.47/42.76

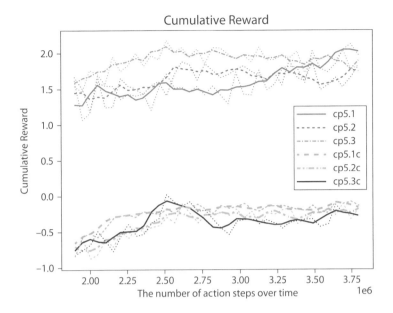

FIGURE 3.6 Cumulative reward over the number of action steps taken for the comparison between the downstream agents and the control group agents.

parameters from the upstream agent, instead, they are trained from scratch using the same RL algorithm and model structure, without the transferred knowledge from cp4. It can be seen that all control group agents resulted in negative final cumulative rewards, indicating that it is too difficult to learn to achieve their goals under the same training criteria as the downstream agents, without any pre-training.

The success rates of the control group agents and the downstream agents are shown in Table 3.4. For picking, the control group agents have an average success rate of 63%, as compared to 91% from the downstream agents. For placing, the control group agents have a low average success rate of around 1%, as compared to 63% from the downstream agents. These results show the effectiveness of acceleration introduced by transfer learning, which uses the learned model parameters from the upstream agent.

3.5 CONCLUSION AND FUTURE WORKS

In this study, we explored the performance of RL-based robotic motion planning for conducting multiple construction tasks in a virtual construction environment. We concluded that RL has the potential to control robots in completing construction tasks including complex manipulations such

TABLE 3.4

Comparison of the Success Rates between the Control Group Agents and the Downstream Agents

Training Criteria	$N_{pick}/N_{episode}$ (%)	$N_{place}/N_{episode}$ (%)
cp5.1c/**cp5.1**	74.79/**94.84**	0.45/**60.92**
cp5.2c/**cp5.2**	61.56/**92.19**	0.77/**62.45**
cp5.3c/**cp5.3**	56.67/**86.06**	1.96/**72.47**

as placing the target object with a random posture. Then we further concluded that our approach could reduce the efforts of designing and training RL agents for multiple construction tasks using transfer learning. We first built a realistic VCE in Unity3D and pre-trained the upstream RL agent for a pick-and-place task using a designed curriculum plan. Next, we fine-tuned the upstream agent to output three downstream agents for the three specific construction tasks, i.e., ceiling installation, window installation, and flooring. In addition, we conducted experiments using a control group, which did not use model parameters from the upstream agent, proving the effectiveness of the proposed training method. Furthermore, our approach is flexible as the pre-trained agent can be reused to accelerate the training of agents for other similar tasks such as paneling and framing. Researchers and practitioners aiming to use construction robots in their own work can use the proposed approach and the trained upstream agents to achieve flexible robotic control for various construction tasks, which is the first step in solving the problem of generality in construction robotics and eventually deploying these robotic solutions onsite.

We noticed that the downstream agents' improvements gained from the upstream agent vary among the three construction tasks due to the difference of the difficulties. One possible cause for this phenomenon is that the downstream agents are required to distinguish different tasks using same the RL agent design. To address this issue, we propose two streams of future work. First, it might be necessary to adopt a more flexible RL agent design, such as multi-task reinforcement learning (Wilson et al. 2007), which uses one single agent to perform multiple tasks without additional fine-tuning. Next, it may be beneficial to use visual observations which are more semantically rich as the training input for the RL agent. For example, the visual observations may contain the additional geometric information of the target and placements to enable RL agents to make smarter decisions.

ACKNOWLEDGEMENT

We acknowledge the support of the Natural Sciences and Engineering Research Council of Canada (NSERC), [funding reference number: ALLRP 570442-2021].

REFERENCES

Abbeel, P., and A. Y. Ng. 2004. "Apprenticeship learning via inverse reinforcement learning". In *Proceedings of the Twenty-First International Conference on Machine Learning,* 1 Presented at the Banff, Alberta, Canada.doi: https://doi.org/10.1145/1015330.1015430.

Apolinarska, A. A., M. Pacher, H. Li, N. Cote, R. Pastrana, F. Gramazio, and M. Kohler. 2021. "Robotic assembly of timber joints using reinforcement learning". *Automation in Construction* 125:103569.

Arndt, V., D. Rothenbacher, U. Daniel, B. Zschenderlein, S. Schuberth, and H. Brenner. 2005. "Construction work and risk of occupational disability: A ten year follow up of 14,474 male workers". *Occupational and Environmental Medicine* 62(8):559–566.

Barbosa, F., J. Woetzel, and J. Mischke. 2017. "Reinventing construction: A route of higher productivity". Technical report. Bengio, Y., J. Louradour, R. Collobert, and J. Weston. 2009. "Curriculum learning". In *Proceedings of the 26th Annual International Conference on Machine Learning,* 41–48.

Bengio, Y., Louradour, J., Collobert, R., and Weston, J. 2009. "Curriculum learning". In *Proceedings of the 26th Annual International Conference on Machine Learning,* 41–48. Presented at the Montreal, Quebec, Canada. doi: https://doi.org/10.1145/1553374.1553380.

Bock, T. 2015. "The future of construction automation: Technological disruption and the upcoming ubiquity of robotics". *Automation in Construction* 59:113–121.

Dharmawan, A. G., Y. Xiong, S. Foong, and G. S. Soh. 2020. "A model-based reinforcement learning and correction framework for process control of robotic wire arc additive manufacturing". In *2020 IEEE International Conference on Robotics and Automation (ICRA),* 4030–4036. IEEE.

Elbanhawi, M., and M. Simic. 2014. "Sampling-based robot motion planning: A review". *IEEE Access* 2:56–77.

Everett, J. G., and A. H. Slocum. 1994. "Automation and robotics opportunities: Construction versus manufacturing". *Journal of Construction Engineering and Management* 120(2):443–452.

Eysenbach, B., A. Gupta, J. Ibarz, and S. Levine. 2018. "Diversity is all you need: Learning skills without a reward function". arXiv preprint arXiv:1802.06070.

Feng, C., Y. Xiao, A. Willette, W. McGee, and V. Kamat. 2014. "Towards autonomous robotic in-situ assembly on unstructured construction sites using monocular vision". In Proceedings of the 31st International Symposium on Automation and Robotics in Construction, 163–170.

Gu, S., T. Lillicrap, I. Sutskever, and S. Levine. 2016. "Continuous deep q-learning with model-based acceleration". In *International Conference on Machine Learning*, 2829–2838. PMLR.

Haarnoja, T., V. Pong, A. Zhou, M. Dalal, P. Abbeel, and S. Levine. 2018. "Composable deep reinforcement learning for robotic manipulation". In *2018 IEEE International Conference on Robotics and Automation (ICRA)*, 6244–6251. IEEE.

Iturralde, K., M. Feucht, R. Hu, W. Pan, M. Schlandt, T. Linner, T. Bock, J. Izard, I. Eskudero, and M. Rodriguez et al. 2020. "A cable driven parallel robot with a modular end effector for the installation of curtain wall modules". In *Proceedings of the International Symposium on Automation and Robotics in Construction*, Volume 37, 1472–1479. IAARC Publications.

Jaakkola, T., S. Singh, and M. Jordan. 1994. "Reinforcement learning algorithm for partially observable Markov decision problems". In *Proceedings of the 7th International Conference on Neural Information Processing Systems*, 345–352. Presented at the Denver, Colorado. doi: https://doi.org/10.5555/2998687.2998730.

James, S., and E. Johns. 2016. "3d simulation for robot arm control with deep q-learning". *arXiv preprint* arXiv:1609.03759.

Kavraki, L. E., P. Svestka, J.-C. Latombe, and M. H. Overmars. 1996. "Probabilistic roadmaps for path planning in high-dimensional configuration spaces". *IEEE Transactions on Robotics and Automation* 12(4):566–580.

Konidaris, G., and A. Barto. 2006. "Autonomous shaping: Knowledge transfer in reinforcement learning". In *Proceedings of the 23rd International Conference on Machine Learning*, 489–496.

Kumar, N., N. Hack, K. Doerfler, A. N. Walzer, G. J. Rey, F. Gramazio, M. D. Kohler, and J. Buchli. 2017. "Design, development and experimental assessment of a robotic end-effector for non-standard concrete applications". In *2017 IEEE International Conference on Robotics and Automation (ICRA)*, 1707–1713. IEEE.

LaValle, S. M. et al. 1998. "Rapidly-exploring random trees: A new tool for path planning".

Lee, S., M. Gil, K. Lee, S. Lee, and C. Han. 2007. "Design of a ceiling glass installation robot". In Proceedings of the 24th International Symposium on Automation and Robotics in Construction, 247–252.

Liang, C.-J., V. R. Kamat, and C. C. Menassa. 2020. "Teaching robots to perform quasi-repetitive construction tasks through human demonstration". *Automation in Construction* 120:103370.

Long, M., Y. Cao, J. Wang, and M. Jordan. 2015. "Learning transferable features with deep adaptation networks". In *International Conference on Machine Learning*, 97–105. PMLR.

Lundeen, K. M., V. R. Kamat, C. C. Menassa, and W. McGee. 2019. "Autonomous motion planning and task execution in geometrically adaptive robotized construction work". *Automation in Construction* 100:24–45.

Matsumoto, K., A. Yamaguchi, T. Oka, M. Yasumoto, S. Hara, M. Iida, and M. Teichmann. 2020. "Simulation-based Reinforcement Learning Approach towards Construction Machine Automation". In Proceedings of the International Symposium on Automation and Robotics in Construction, Volume 37, 457–464. IAARC Publications.

Matulis, M., and C. Harvey. 2021. "A robot arm digital twin utilising reinforcement learning". *Computers & Graphics* 95:106–114.

Pathak, D., P. Agrawal, A. A. Efros, and T. Darrell. 2017. "Curiosity-driven exploration by self-supervised prediction". In *International Conference on Machine Learning*, 2778–2787. PMLR.

Schulman, J., F. Wolski, P. Dhariwal, A. Radford, and O. Klimov. 2017. "Proximal policy optimization algorithms". arXiv preprint arXiv:1707.06347.

Schulman, J., S. Levine, P. Abbeel, M. Jordan, and P. Moritz. 2015. "Trust region policy optimization". In *International Conference on Machine Learning*, 1889–1897. PMLR.

Sutton, R. S., and A. G. Barto. 2018. *Reinforcement Learning: An Introduction*. MIT press.

Sutton, R. S., D. McAllester, S. Singh, and Y. Mansour. 1999. "Policy gradient methods for reinforcement learning with function approximation". *Proceedings of the 12th International Conference on Neural Information Processing Systems*, 1057–1063. Presented at the Denver, CO. doi: https://doi.org/10.5555/3009657.3009806.

Thomas, G., M. Chien, A. Tamar, J. A. Ojea, and P. Abbeel. 2018. "Learning robotic assembly from cad". In *2018 IEEE International Conference on Robotics and Automation (ICRA)*, 3524–3531. IEEE.

U.S. BLS 2020. "Fatal Occupational Injuries by Industry and Event or Exposure, All United States, 2020". https://www.bls.gov/iag/tgs/iag23.htm. (Accessed April 20, 2022)

Villumsen, S. L., and M. Kristiansen. 2017. "PRM based motion planning for sequencing of remote laser processing tasks". *Procedia Manufacturing* 11:300–310.

Wilson, A., A. Fern, S. Ray, and P. Tadepalli. 2007. "Multi-task reinforcement learning: a hierarchical bayesian approach". In *Proceedings of the 24th International Conference on Machine Learning*, 1015–1022.

Yang, C. H., and S. C. Kang. 2021. "Collision avoidance method for robotic modular home prefabrication". *Automation in Construction* 130:103853.

Yosinski, J., J. Clune, Y. Bengio, and H. Lipson. 2014. "How transferable are features in deep neural networks?". *Advances in Neural Information Processing Systems* 27.

You, S., J. H. Kim, S. Lee, V. Kamat, and L. P. Robert Jr. 2018. "Enhancing perceived safety in human–robot collaborative construction using immersive virtual environments". *Automation in Construction* 96:161–170.

Yu, C., and S. Gao. 2021. "Reducing collision checking for sampling-based motion planning using graph neural networks". *Advances in Neural Information Processing Systems* 34:4274–4289. https://proceedings.neurips.cc/paper_files/paper/2021/file/224e5e49814ca908e58c02e28a0462c1-Paper.pdf.

Zeng, A., S. Song, S. Welker, J. Lee, A. Rodriguez, and T. Funkhouser. 2018. "Learning synergies between pushing and grasping with self-supervised deep reinforcement learning". In *2018 IEEE/RSJ International Conference on Intelligent Robots and Systems (IROS)*, 4238–4245. IEEE.

Zhu, Z., K. Lin, and J. Zhou. 2020. "Transfer learning in deep reinforcement learning: A survey". arXiv preprint arXiv:2009.07888.

4 The Impact of Smart Materials on Structural Vibration Control

Bijan Samali, Yang Yu, Yancheng Li, Jianchun Li, and Huan Li

CONTENTS

4.1 INTRODUCTION

Because the idea of structural vibration control in civil engineering was introduced in 1972 [1], structural vibration control has been significantly put forward from theory to practice. Each form of control has its own advantages and disadvantages. Passive control devices are simple, easy to implement and have a good damping effect, and have been used to some extent in civil engineering, especially for foundation isolation, but their control effect and practical scope are still somewhat limited.

In recent years, the research and application of smart materials have developed considerably, and according to their functional characteristics, they can be divided into two categories: smart sensing materials and smart driving materials. In civil engineering, research is mainly focused on the real-time monitoring and diagnosis of structural vibrations, self-adaptation, self-healing and self-control of structures [2]. The development and application of smart materials have opened up new horizons in vibration control and mitigation of civil structures.

In this chapter, first of all, four types of smart materials, namely, shape memory alloys (SMA) [3–5], piezoelectric [6–9], magnetorheological (MR) [10–14] and electrorheological (ER) materials [15, 16] are comprehensively reviewed, including the fundamentals of materials and their application in structural vibration control and mitigation. Then, a case study is presented to demonstrate how smart materials–based seismic response suppression of civil structures is implemented. In this case study, a prototype rotary damper made of MR fluids developed by the authors' group is employed. The dynamic tests of this innovative smart device under different loading conditions of amplitude, frequency and applied currents are conducted to investigate its dynamic behaviour. The test results show that the rotary MR damper exhibits highly nonlinear hysteresis responses, so a reliable and precise model needs to be proposed first to describe unique behaviour of rotary MR damper before the corresponding controller is designed. To deal with this issue, a novel sigmoid function-based model is developed to characterise hysteretic shear force response of a rotary smart damper. Then, particle swarm optimisation (PSO) is employed to identify model parameters, which is regarded as resolving a global minimum optimisation problem. The optimisation objective is set as mean square

DOI: 10.1201/9781003325246-4

error (MSE) between experimental responses and model predictions. Subsequently, on the basis of a parameter identification result, a generalised rate-dependent model is built up for control application of a rotary smart damper, in which parameters of a hysteresis model are correlated with control currents. A numerical study is conducted on a smart building structure composed of a three-storey building model and several rotary MR dampers subject to scaled benchmark earthquakes, including El-Centro and Northridge, for performance evaluation of current control algorithm of rotary MR dampers. The results describe that the proposed smart rotary MR damper-based control system could effectively diminish inter-storey drift and floor acceleration of structures, demonstrating a promising application perspective. Finally, the challenges and future work in this area are discussed in the conclusion section.

4.2   SMART MATERIALS FOR STRUCTURAL VIBRATION CONTROL

4.2.1   PIEZOELECTRIC MATERIALS

Since the discovery of piezoelectric materials, their variety has been greatly enriched. At present, the main types include piezoelectric ceramics, single piezoelectric crystals, piezoelectric polymers and piezoelectric fibre composites. Among them, piezoelectric ceramics and piezoelectric fibre composites are increasingly used as two important intelligent control materials in vibration control.

When using piezoelectric materials for vibration control of structures, they act as sensors and actuators based on their positive and negative piezoelectric effect. The positive piezoelectric effect of converts mechanical energy into electrical energy, thus transforming vibration information of the structure into identifiable electrical signals. The controller calculates the required control voltage based on the electrical feedback signal in combination with the control algorithm and then applies electrical fields to piezoelectric material, which deforms the material based on inverse piezoelectric effect, thereby suppressing the structural vibration and achieving structural vibration control. Depending on the presence or absence of external energy input in controlled structures, structural vibration control can be broadly categorised into three types: passive, active and semi-active control.

Piezoelectric passive control involves the implantation of piezoelectric materials inside the structure being controlled. When the structure is deformed by external loads, the piezoelectric material uses its positive piezoelectric effect to sense the strain generated by the structure and to transform the mechanical energy produced by the vibration into electrical energy, which is transmitted via an external circuit, thereby dissipating or absorbing mechanical energy produced by structural vibration.

Piezoelectric active vibration control uses the positive and negative piezoelectric effect of piezoelectric materials, which are arranged inside or on the surface of a structure as sensors and actuators for the controlled structure. When the structure vibrates and deforms under external loads, the sensing converts the vibration signal into a voltage signal. The control signal is then applied to the actuator by a power amplifier, which deforms the actuator to produce control force and converts electrical energy into mechanical energy, thus realising the vibration control.

Piezoelectric semi-active control is an evolution of piezoelectric active vibration control and piezoelectric passive vibration control. Instead of feeding energy directly to the actuator to generate the control force, the energy is used in the semi-active control loop to change the structural parameters such as stiffness, damping and inertia of the system by varying the parameters of the piezoelectric element placed inside the controlled member, thus achieving the control purpose.

Numerous research has been carried out on piezoelectric devices for structural vibration control. Kamada et al. [17] established a total of 32 piezoelectric ceramic actuators at the bottom of four columns on the ground floor of a steel frame, and used an active control algorithm to achieve the desired objective by controlling bending moments and axial forces, resulting in a 10% increase in the damping ratio of the first and second vibration patterns of the frame. In addition, by combining piezoelectric materials with passive friction dampers, they have designed piezoelectric material smart friction dampers for the semi-active control of wind vibration responses in towering structures [18].

4.2.2 SHAPE MEMORY ALLOYS (SMA)

SMA is defined as alloy material with a large deformation that can be returned to pre-deformed shape after hearting. Shape memory effect (SME) and superelasticity effects (SE) are the main excellent effects of SMA, which have self-resetting properties and high energy consumption. SME plays a significant role in active control and repair of structures by providing recovery of deformed plastic SMA through heating. In addition, due to the unique thermodynamic and phase change properties of SMAs, their stiffness and damping properties can be adjusted by varying the working temperature, which is critical in structural semi-active control.

The SE effect of SMA can be used to make energy dissipators. By heating the SMA to a suitable temperature and loading the SMA to a suitable stress level, the energy dissipator can consume a large amount of the structural vibration energy when the structure vibrates, thus effectively suppressing the structural vibration. In addition, using the property of SME of SMA, the SMA is pre-buried into the structure, and by heating and cooling the SMA, the configuration of the SMA is changed, and the variation of modulus of elasticity of the SMA is controlled, so that the local or overall stiffness of structure could be changed and structural natural frequencies can be changed to avoid resonance.

Based on unique characteristics of SMA, numerous research has been carried out on its application in structural vibration suppression. Motogi et al. [19] studied a composite plate embedded with SMA filaments. By heating the SMA filaments with electric current, the modulus of elasticity of the SMA filaments was drastically changed, thus changing the modulus of elasticity of the composite plate to avoid structural resonance. Liang and Rogers [20] studied the design of SMA springs for vibration control. The temperature-dependent properties of the SMA modulus of elasticity can be used to vary the spring constant of SMA springs by a factor of 3 to 4. Epps and Chandra [21] investigated adjusting of resonant frequencies of composite beams by means of pre-strained SMA filaments. Numerical studies show that resonant frequency of beam can be controlled by the restoring force produced by the SME of SMA. Bidaux et al. [22] utilised the SME of SMA to adjust stiffness of composite beams to actively change the natural frequency of composite beams. Shahin et al. [23] studied an approximate model that simulates the vibration control of SMA cables and multi-storey buildings. SME of SMA was used to actively control the vibration of a multi-storey building by heating the SMA cables with electric current. Choi et al. [24] investigated the active control of vibration in two types of hybridised intelligent structure, one containing piezoelectric film actuators and current variant actuators, and the other containing piezoelectric ceramic and SMA actuators. The test result shows that a hybrid actuator system is much more capable of controlling the vibration of smart structures than a single-type actuator system.

4.2.3 ELECTRORHEOLOGICAL (ER) AND MAGNETORHEOLOGICAL (MR) FLUIDS

ER and MR fluids, collectively known as "controlled fluids", can be transformed from Newtonian fluids into Bingham plastic or viscoelastic bodies in milliseconds under the impact of applied electrical or magnetic fields.

ER fluid is the suspension of non-conductive mother liquor (often silicone oil or mineral oil) and solid electrolyte particles (inorganic non-metallic materials, organic semiconductor materials, polymeric semiconductor materials) uniformly dispersed in it. The current-variable fluid is transformed from a Newtonian fluid with good fluidity and a certain degree of viscosity to a viscoplastic fluid with a certain yield shear stress subjected to action of electrical fields, which is called the ER effect.

The MR fluids are high-quality soft magnetic materials with high saturation magnetic induction and low magnetic coercivity, uniformly distributed in a non-permeable carrier medium (silicone oil or mineral oil). When a magnetic field is applied, the suspended particles are induced to change from magnetically neutral to strongly magnetic and interact with each other to form a "chain" of bridges between the magnetic poles, causing them to change instantaneously from Newtonian fluid to viscoplastic fluid with a certain yield shear stress, which is called the magnetorheological effect.

The electro/magnetic rheological effect is similar and has the following three characteristics:

1. Continuity. The yield shear stress varies continuously as the electromagnetic field strength changes.
2. Reversibility. The electromagnetic rheology fluid can "harden" as the electromagnetic field strength increases and "soften" as it decreases.
3. Short frequency response time. The time required to change the yield shear stress of an electromagnetic rheology fluid in the forward and reverse directions as the electromagnetic field strength varies in the order of 10^{-4} s and 10^{-3} s, respectively.

ER/MR fluids have been used to make many forms of vibration damping drives, representative of which have been successfully used in civil engineering structures, including ER seismic isolator with a damping coefficient of 2215 kNs/m, designed by Mcmahon and Makris [25], and MR fluid damper with the maximum damping force of 200 kN, developed by Calson and Spencer [26]. In addition, a large amount of research in this area has been conducted and reported. Shahali et al. [27] designed a smart sandwich cylinder shell composed of ER fluids and thick FG face sheet. The numerical investigation was conducted to prove that changing electric field applied is able to adjust the natural frequency and modal parameters of the structure. Hoseinzadeh and Rezaeepazhand [28] studied the effect of ER fluids damper on composite plate in terms of dynamic response. In this study, nonlinear behaviour of a smart damper was characterised by the Bingham model. Pei et al. [29] devised a novel hybridised control method for earthquake-induced vibration suppression of MR dampers incorporated smart structure, where fuzzy logic and NSGA-II-based multi-objective optimisation were employed. Mousavi Gavgani et al. [30] employed a MR damper-based smart system in offshore infrastructure under wave excitations, with excellent performance in vibration mitigation. Şahin et al. [31] compared three semi-active control algorithms, i.e., sliding mode, PID and energy-based control for the structures with a MR damper installed between the 1st level and the ground. The comparison result validated that the sliding mode control achieves the best performance in base shear reduction.

4.3   CASE STUDY OF MR DEVICE APPLICATION

4.3.1   Design and Dynamic Test of MR Materials-Based Smart Device

A prototype of a rotary damper based on MR fluids was developed and fabricated by the authors [32], consisting of a thin rotational plate with a shaft connected to transmit shear force, MR smart materials, two housings to generate cavity and circular coil to generate magnetic fields to energise materials, as described in Figure 4.1. The distance between the housing and plate is set to 1 mm to keep the smart materials inside the damper.

Then, dynamic tests are designed and carried out to evaluate the performance of this novel intelligent device. Figure 4.2 presents the experimental setup, where a shaking table generates horizontal excitation to a smart damper. In the test, a smart damper is attached to a supporter on the floor. The local cell with a capacity of 300 N is adopted to gauge shear force generated to the smart damper. The LVDT is employed to gauge horizontal deflection due to excitations from the shaking table. In this experiment, three levels of excitation frequencies are considered, namely, 0.5 Hz, 1 Hz and 2 Hz. Four levels of excitation amplitude are set, namely 7 mm, 18 mm, 28 mm and 35 mm. The

FIGURE 4.1   Prototype of rotary MR damper.

current applied to a smart damper ranges from 0 to 2 A, with increments of 0.5 A. During the test, sampling duration and frequency are 1 s and 2048 Hz, respectively. The data acquisition (DAQ) can record displacement response of a smart device under harmonic loading directly, while the velocity response of the device could be obtained by differentiating the displacement response.

Figure 4.3 (a) describes shear force-displacement response of a smart damper and Figure 4.3 (b) depicts shear force-velocity response of a smart damper, when the device is excited with a harmonic load of 7 mm amplitude and 2 Hz frequency, under different levels of current. The results in both figures show that with the increment of the current level, the shear force ascends gradually. If there

FIGURE 4.2   Experiments for dynamic test of rotary smart damper.

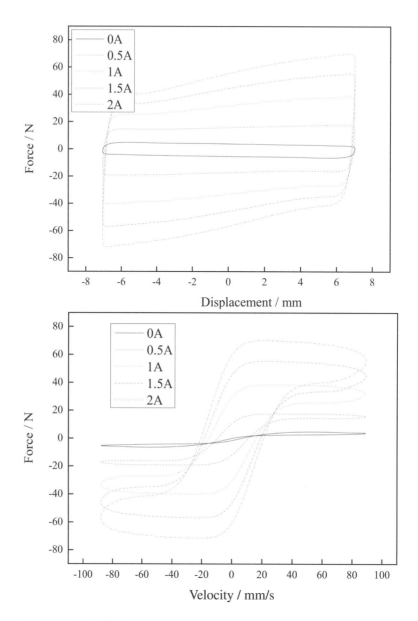

FIGURE 4.3 One example of device responses.

is no current supplied, a smart damper exhibits apparent viscous characteristics due to an elliptical shear force-displacement loop. When the current level is increased, the loop slope gradually ascends. Hence, if we would like to adopt this smart damper for control application, a precise model that can characterise these behaviours needs to be proposed first.

4.3.2 Proposed Model and Parameter Estimation

To characterise this innovative device, this book chapter devises a novel mathematical model for interpreting nonlinear and hysteretic behaviour of smart damper. In this model, a linear damping component, a linear stiffness component and a hysteresis component indicated by sigmoid function are connected in parallel, with five parameters, as shown in Equation (4.1) and Equation (4.2).

Compared to the classical Bouc-Wen hysteresis model, the proposed one contains fewer parameters and excludes a differential equation.

$$F = k_0\theta + c_0\dot{\theta} + az + F_0 \tag{4.1}$$

$$z = \frac{b\dot{\theta} + csign(\theta)}{\sqrt{1 + \left[b\dot{\theta} + csign(\theta)\right]^2}} \tag{4.2}$$

where k_0 is the stiffness parameter (N/m), and c_0 is the damping parameter (Ns/m); a, b and c are three parameters to adjust the scale, shape and width of the loop; F_0 is the initial shear force. Given the excitation condition, a group of model parameters can be identified based on experimental data.

In essence, model parameter identification is a global minimisation optimisation problem, where mean square error (MSE) between experimental force and predicted force sequence in one cycle is defined as an optimisation target. Subsequently, particle swarm optimisation (PSO) [33, 34] is employed to achieve this target by minimising the MSE value. If the MSE value is equal to 0, and the corresponding solution is optimal to this model. The expression of optimisation target is given in Equation (4.3):

$$Obj\_F = \frac{1}{N}\sum_{i=1}^{N}\left[F_{exp}(i) - F(i)\right]^2 \tag{4.3}$$

where F_{exp} and F represent experimental shear force and predicted shear force; N denotes the entire number of samples used for parameter identification.

Figure 4.4 compares experimental force and predicted forces in terms of time-historical plots, when the load of 28 mm amplitude, 2 Hz frequency and 1 A current is applied to a smart damper. In the figure, the straight line indicates the error between experimental force and predicted force. Apparently, the newly developed hysteresis model is capable of conducting precise estimation of the force response of a smart damper for most data points. The estimation error of shear force of the device maintains a low level, satisfying the modelling task in this research.

The comparisons of displacement/speed-force loops of a smart damper between experimental results and predicted results are presented in Figure 4.5, where the loading case is 2 Hz frequency and 2 A current with various levels of amplitude. As can be observed in Figure 4.5 (a), abnormal quartering shapes with blunt angles exist in displacement-shear force hysteresis loops, indicating the impact of device speed on shear force. In addition, speed-force loops in Figure 4.5 (b) indicate

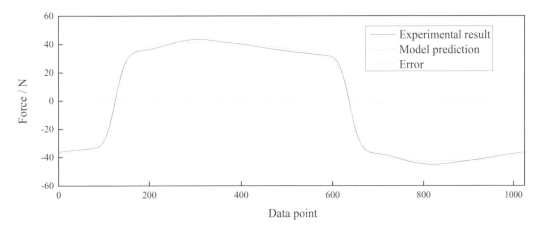

FIGURE 4.4 One example of device responses.

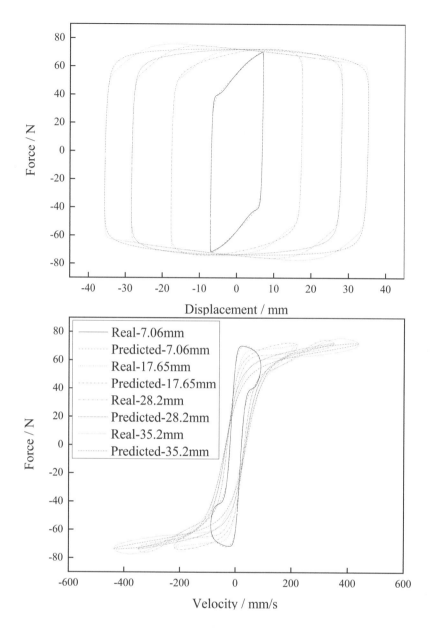

FIGURE 4.5 Comparison of device response between real and predicted results with different loading amplitudes (2 Hz-2 A).

the important impact of speed on the shape of loops. Outstanding agreement between experimental force and model predictions sufficiently validates the capability of this sigmoid function-based model in the demonstration of these features.

Figure 4.6 illustrates the capability of proposed model to demonstrate the impact of applied current level on the output force of a smart damper, where the excitation condition is 17 mm amplitude and 2 Hz frequency. If the magnetic fields are not applied to a smart damper (current is 0), the hysteretic loops of the smart device exhibit viscous characteristics of a smart damper, represented by elliptical loops. When the applied magnetic fields continuously ascend, the loop slope gradually increases with an expandable enclosed area. In particular, when applied current level is 2 A,

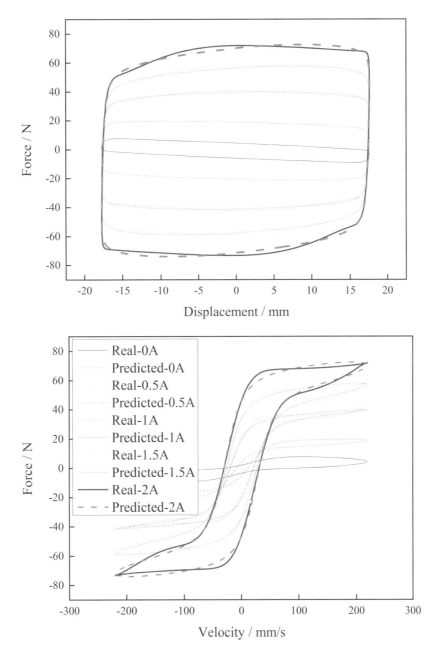

FIGURE 4.6 Comparison of device response between real and predicted results with different applied current levels (2 Hz-2 A).

displacement shear force loop is anomalous. In accordance with comparisons between experimental forces and model predictions, it can be concluded that the sigmoid function-based hysteretic model is a promising solution to the optimal design of controllers for adaptive control of MR fluids damper incorporated smart structures against external vibration.

A large amount of research has revealed that smart material responses rely on loading amplitude, frequency and supplying current level. However, in the practical condition, loading amplitude and frequency are always unknown. Hence, in the chapter, the current dependence of parameters of

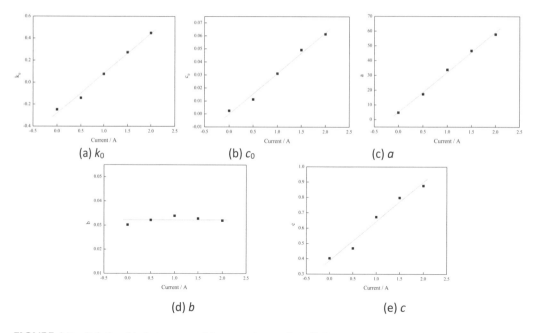

FIGURE 4.7   Relationship between model parameters and applied current.

sigmoid function is investigated to better apply the proposed model in the development and realisation of the adaptive control algorithm for a rotary MR damper. Figure 4.7 gives the relationships between parameters of the model and variable current. It is noticeable that except b, all the model parameters are linearly increased with the addition of the current, indicating that the linear relationship could be elaborated by polynomial function with first order. On the other hand, the value of parameter b fluctuates with the varied current, but the change amplitude is relatively small. As a result, in this study, the value of parameter b is fixed to 0.030781, which is the parameter mean value under different current conditions. In summary, the mathematical expressions of the current-dependent model parameters are given in Equation (4.4) through Equation (4.8).

$$k_0 = 0.3613 \cdot I - 0.2791 \tag{4.4}$$

$$c_0 = 0.03125 \cdot I - 0.000111 \tag{4.5}$$

$$a = 27.13 \cdot I + 4.953 \tag{4.6}$$

$$b = 0.030781 \tag{4.7}$$

$$c = 0.2552 \cdot I + 0.3881 \tag{4.8}$$

4.3.3   CONTROL APPLICATION OF MR DEVICE FOR STRUCTURAL SEISMIC MITIGATION

In this section, a real-time control strategy on the basis of a sliding mode approach is developed and the corresponding current control strategy is illustrated in Equation (4.9):

$$u(t) = \begin{cases} u_1 & \text{if } s \neq 0 \text{ and } \kappa(t) \leq 2 \\ u_2 & \text{if } s = 0 \text{ and } \kappa(t) \leq 2 \\ 0 & \text{if } \kappa(t) > 2 \end{cases} \tag{4.9}$$

TABLE 4.1

Mass, Stiffness and Damping of Three-Storey Shear Structure

Floor No.	Mass (kg)	Stiffness (kN/m)	Damping (N•s/m)
1	98.3	516	125
2	98.3	684	50
3	98.3	684	50

where $u_1 = \min\left[\rho, \max\left\{-\rho, -\xi\left(\dot{s}|s|^{-0.5} + \vartheta sign(s)\right)\right\}\right]$, and $u_1 = -\rho sign(\dot{s})$. Three constant parameters, ξ, ρ and ϑ, satisfy the following condition: $\xi\,\vartheta > \rho$, $\rho v_m - \tau > 0.5\vartheta^2$.

In order to assess the capability of a designed semi-active control algorithm on the basis of a sliding mode approach and rate-dependent sigmoid function–based model, a simulation investigation is carried out on a three-storey shear structure model. In the simulation, the smart structure is established by adding four rotary MR dampers at each level of the structure model. Table 4.1 shows the parameter values of the building model. The properties of mass, damping and stiffness of a smart damper are available from Equation (4.4) to Equation (4.8). To validate the effectiveness of the proposed sliding model control on structural vibration suppression, this smart damper–incorporated shear structure is numerically loaded by benchmark earthquakes, namely El-Centro and Northridge. In addition, both seismic inputs are narrowed by 0.5 to ensure that natural frequencies of a smart structure are within the main earthquake frequency spectrum. In this regard, the vibration response of a smart structure without control current will be greatly reinforced, which can further verify the capacity of a smart structure with the proposed control strategy.

Figure 4.8 records the evaluation of current signals applied to smart MR dampers with time for different earthquake loads. As can be seen in the figure, the current keeps continuous change during the event of an earthquake, and the value is in the range of [0, 2] throughout, satisfying the requirement of the upper limit of current (2 A). This is mainly due to the fact that when the current in smart damper varies, the coils in the device produce a large amount of heat, resulting in an unstable performance.

Acceleration is a major metric to evaluate the intensity of vibration response of the structure due to an external load, which is used to assess the capacity of the proposed control algorithm in this study. In general, the floor acceleration rises with the structure height because of the first mode of

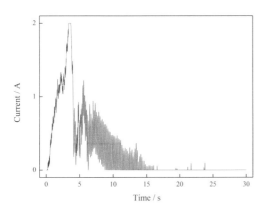
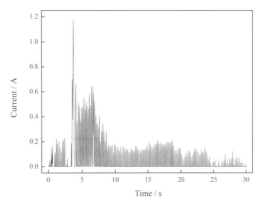

FIGURE 4.8 Control current signals.

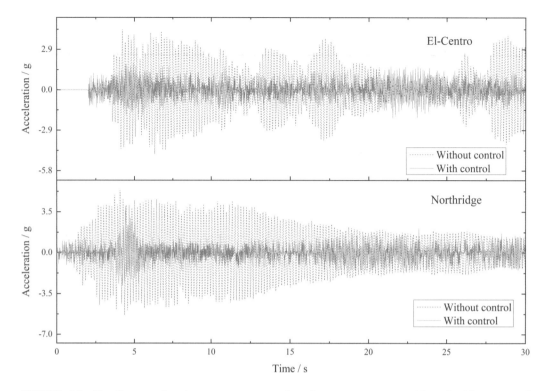

FIGURE 4.9 Top floor acceleration response comparison between structures with and without control method under seismic events.

a seismically excited structure. Consequently, with the increment of level number of the structure, floor acceleration gradually rises, indicating the fact that maximal acceleration often occurs on the top floor. Therefore, this chapter utilises change in acceleration of the top floor when acceleration reaches its peak as an evaluation indicator to evaluate the degree of deformation and vibration of the shear structure. A comparison of time historical top-level acceleration of a shear structure without and with control method under El-Centro and Northridge seismic excitations is conducted, as shown in Figure 4.9. The comparison results clearly demonstrate that the structure is capable of significantly reducing the acceleration response in both seismic situations based on an adaptive control method of the proposed model.

Besides time series top-level acceleration, maximum acceleration of all levels and inter-storey drift are important indicators to describe the performance of a developed semi-active controller. Inter-storey drift is able to provide deformation intensity between neighbour levels, while the maximum acceleration of each level roughly generates the architecture of a structure in terms of most severe vibrations. Figure 4.10 (a) shows the comparison of inter-storey drift of all the levels of the structure with and without a proposed control algorithm subjected to scaled El-Centro and Northridge earthquakes, and Figure 4.10 (b) portrays the maximum acceleration of all the levels of the structure with and without a proposed control algorithm under two scaled seismic inputs. Apparently, in both seismic scenarios, peak floor acceleration often increases with structural height, while inter-storey drift shows a decreasing relationship with structural level. It is worth mentioning that the structure with a proposed control algorithm shows relatively little variation in both indicators for all levels compared to a passive smart structure. The promising comparative results amply demonstrate an excellent application of a smart damper and proposed control methods on the basis of a sliding mode for seismic protection and response suppression of civil structures.

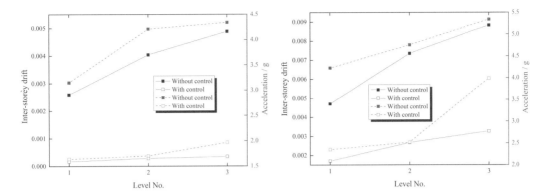

FIGURE 4.10 Maximum acceleration and inter-storey drift comparisons between structures with and without proposed control method under seismic events.

4.4 CONCLUSION, CHALLENGE AND FUTURE WORK

This chapter mainly illustrated the impact of smart materials on structural vibration control. In particular, SMA, piezoelectric, ER and MR materials are deeply reviewed in terms of the fundamental and potential applications in structural vibration mitigation. A case study on a MR fluid-based damper was presented to demonstrate how the smart materials can be used for structural vibration suppression. The detailed conclusions are summarised as follows.

- Smart materials and relevant devices assume highly nonlinear and hysteretic behaviours under external excitations.
- A novel sigmoid function–based hysteresis model is proposed for characterising nonlinear responses of a rotary MR damper, with high accuracy.
- PSO is employed for model parameter identification, which contributes to optimal values of model parameters of the model under each loading condition.
- A generalised rate-dependent model is established by correlating the model parameters with applied current for the control application.
- A simulation investigation is carried out to assess the capability of smart device–incorporated structures against earthquakes. The result indicates outstanding capacity of the proposed system in suppressing inter-storey drift and floor acceleration.

Even though current research has shown great potential of smart materials in structural vibration control, several challenges should be well addressed for their practical applications. The first problem is material design. For MR materials, how to achieve maximum MR effect by material optimisation design is still challenging. In addition, the current smart devices are mainly developed for increasing stiffness with the increment of applied current. However, the low stiffness is preferable to the structure under seismic loading. In future work, designing a smart device with adjustable negative stiffness may be a potential solution to this problem. Moreover, the problems due to practical implementation of smart devices should be considered. For example, this case study employs a shear building model to demonstrate the proposed method. However, in some cases, the structures perhaps are asymmetric, where the torsion will be generated during the earthquake event. It is critical to resolve this challenge in future work.

REFERENCES

1. Yao JTP. Concept of structural control. ASCE Journal of the Structural Division. 1972;98(ST7):1567–1574.
2. Li J, Li Y, Askari M, et al., editors. Future intelligent civil structures: Challenges and opportunities. 31st International Symposium on Automation and Robotics in Construction and Mining, ISARC 2014-Proceedings; 2014.

3. Das S, Chakraborty A, Barua I. Optimal tuning of SMA inerter for simultaneous wind induced vibration control of high-rise building and energy harvesting. Smart Materials and Structures. 2021;30(2).
4. Qiu C, Gong Z, Peng C, et al. Seismic vibration control of an innovative self-centering damper using confined SMA core. Smart Structures and Systems. 2020;25(2):241–254.
5. Tian L, Luo J, Zhou M, et al. Research on vibration control of a transmission tower-line system using SMA-BTMD subjected to wind load. Structural Engineering and Mechanics. 2022;82(5):571–585.
6. Yin D, Yi K, Liu Z, et al. Design of cylindrical metashells with piezoelectric materials and digital circuits for multi-modal vibration control. Frontiers in Physics. 2022;10.
7. Hashemi A, Jang J, Hosseini-Hashemi S. Smart Active Vibration Control System of a Rotary Structure Using Piezoelectric Materials. Sensors. 2022;22(15).
8. Sharma A. Effect of porosity on active vibration control of smart structure using porous functionally graded piezoelectric material. Composite Structures. 2022;280.
9. Sharma S, Kumar A, Kumar R, et al. Active vibration control of smart structure using poling tuned piezoelectric material. Journal of Intelligent Material Systems and Structures. 2020;31(10):1298–1313.
10. Yu Y, Li Y, Li J, et al. Nonlinear characterization of the MRE isolator using binary-coded discrete CSO and ELM. International Journal of Structural Stability and Dynamics. 2018;18(08):1840007.
11. Yu Y, Hoshyar AN, Li H, et al. Nonlinear characterization of magnetorheological elastomer-based smart device for structural seismic mitigation. International Journal of Smart and Nano Materials. 2021;12(4):390–428.
12. Yu Y, Yousefi AM, Yi K, et al. A new hybrid model for MR elastomer device and parameter identification based on improved FOA. Smart Structures and Systems. 2021;28(5):617–629.
13. Yu Y, Royel S, Li Y, et al. Dynamic modelling and control of shear-mode rotational MR damper for mitigating hazard vibration of building structures. Smart Materials and Structures. 2020;29(11):114006.
14. Yu Y, Li J, Li Y, et al. Comparative investigation of phenomenological modeling for hysteresis responses of magnetorheological elastomer devices. International Journal of Molecular Sciences. 2019;20(13): 3216.
15. Choi SB, Han YM, Sung KG. Vibration control of vehicle suspension system featuring ER shock absorber [Conference Paper]. International Journal of Applied Electromagnetics and Mechanics. 2008;27(3):189–204.
16. Sung KG, Han YM, Cho JW, et al. Vibration control of vehicle ER suspension system using fuzzy moving sliding mode controller. Journal of Sound and Vibration. 2008;311(3-5):1004–1019.
17. Kamada T, Fujita T, Hatayama T, et al. Active vibration control of frame structures with smart structures using piezoelectric actuators (vibration control by control of bending moments of columns). Smart Materials and Structures. 1997;6(4):448.
18. Wu J, Chen B, Song X. Wind-induced response control of a television transmission tower by piezoelectric semi-active friction dampers. International Journal of Structural Stability and Dynamics. 2022;22(06):2250064.
19. Shinya M, Motohiro T, Takehito F. Stiffness change with temperature in glass fiber reinforced composite laminates embedded with SMA wires. Nippon Kikai Gakkai Ronbunshu, C Hen/Transactions of the Japan Society of Mechanical Engineers, Part C. 1997;63(615):3772–3777.
20. Liang C, Rogers C. Design of shape memory alloy springs with applications in vibration control. 1993.
21. Epps J, Chandra R. Shape memory alloy actuation for active tuning of composite beams. Smart Materials and Structures. 1997;6(3):251.
22. Bidaux J-E, Månson J-A, Gotthardt R. Active stiffening of composite materials by embedded shape-memory-alloy fibres. MRS Online Proceedings Library (OPL). 1996;459.
23. Shahin AR, Meckl PH, Jones JD. Modeling of SMA tendons for active control of structures. Journal of Intelligent Material Systems and Structures. 1997;8(1):51–70.
24. Choia S, Park Y, Fukuda T. A proof-of-concept investigation on active vibration control of hybrid smart structures. Mechatronics. 1998;8(6):673–689.
25. McMahon S, Makris N, editors. Large-scale ER-damper for seismic protection. Smart Structures and Materials 1997: Passive Damping and Isolation; 1997: SPIE.
26. Dyke SJ, Spencer Jr B, Sain M, et al. Modeling and control of magnetorheological dampers for seismic response reduction. Smart materials and structures. 1996;5(5):565.
27. Shahali P, Haddadpour H, Shakhesi S. Dynamic analysis of electrorheological fluid sandwich cylindrical shells with functionally graded face sheets using a semi-analytical approach. Composite Structures. 2022:115715.
28. Hoseinzadeh M, Rezaeepazhand J. Vibration suppression of composite plates using smart electrorheological dampers. International Journal of Mechanical Sciences. 2014;84:31–40.

29. Pei P, Peng Y, Qiu C. An improved semi-active structural control combining optimized fuzzy controller with inverse modeling technique of MR damper. Structural and Multidisciplinary Optimization. 2022;65(9):1–25.
30. Gavgani SAM, Jalali HH, Farzam MF. Semi-active control of jacket platforms under wave loads considering fluid-structure interaction. Applied Ocean Research. 2021;117:102939.
31. Şahin Ö, Adar NG, Kemerli M, et al. A comparative evaluation of semi-active control algorithms for real-time seismic protection of buildings via magnetorheological fluid dampers. Journal of Building Engineering. 2021;42:102795.
32. Li Y, Li J. Dynamic characteristics of a magnetorheological pin joint for civil structures. Frontiers of Mechanical Engineering. 2014;9(1):15–33.
33. Yu Y, Zhang C, Gu X, et al. Expansion prediction of alkali aggregate reactivity-affected concrete structures using a hybrid soft computing method. Neural Computing and Applications. 2019;31(12):8641–8660.
34. Yu Y, Li Y, Li J. Parameter identification of a novel strain stiffening model for magnetorheological elastomer base isolator utilizing enhanced particle swarm optimization. Journal of Intelligent Material Systems and Structures. 2015;26(18):2446–2462.

5 Green Construction Workforce Training Using Virtual Reality

Ming Hu and Miroslaw J. Skibniewski

CONTENTS

5.1 OVERVIEW

5.1.1 BACKGROUND

Rapid adoption of energy-efficient technologies is a key tenet of the architecture, engineering, and construction industry (AEC) sustainably practices and underpins the critical path to create and sustain American leadership in the transition to a global clean energy economy. To best accelerate this adoption, future professional training options must combine the rich, high interactivity of real-world training with low-cost, on-demand offerings that boost participation and access. Available AEC continuing-education options based on traditional training methods characterize two opposing poles: *low fidelity, low cost, remote training* versus *high fidelity, expensive, in-situ training*. Increasingly, as evidenced by the robust American Institute of Architects (AIA) and United States Green Building Council (USGBC), catalogues of online, continuing-education resources [1], low-fidelity trainings predominate.

These conventional text-and-recorded-lecture-based trainings and 2D diagrams are insufficient in representing the full complexity and interoperability of 3D objects and systems [2] that characterize energy-efficient building systems. In-parallel, in-situ training is becoming more expensive as the cost of next-generation equipment/techniques exacerbates costs and safety hazards on active construction sites. What is needed are options that allow trainees to visualize and understand building components and assemblies in both static and dynamic models [3] while conveying real-world scale and perspective. Virtual reality (VR) simulation is the gold standard training tool that can fill this

DOI: 10.1201/9781003325246-5

need. Immersive environments immediately make scale-dependent relationships clear. However, creating effective VR trainings remains challenging, especially for collaborative, highly interactive immersive environments. There are three challenges that need to be addressed: (1) improving rendering protocols for multiple participants; (2) making transitions between multiple 360-degree images as seamless as possible; and (3) reconciling depth and location of 3D objects in relation to 360-degree images.

Expediency in developing prototype training material that can supplant traditional pedagogies is a vital step to energy independence through technology adoption. The University of Maryland, College Park (UMD) is uniquely well-suited to tackle this challenge. As a research-oriented university, UMD has discipline expertise in architecture, construction engineering, and VR. Long-term industry collaborations give the research team an exceptional background in bridging training/ applied needs. New curriculum expansions—such as immersive media design (IMD, a computer science and arts-based undergraduate major)—provide experienced VR-curriculum experts. The team will draw on their previous experience in VR training [4] and research [5] that highlight the effectiveness of VR in enhancing *spatial intelligence* and *spatial memory*-dependent learning outcomes to prototype *immersive, high fidelity, mid-range cost trainings*.

This chapter describes a proposed VR-enhanced energy efficiency training curriculum leveraging content and infrastructure from the UMD School of Architecture, Planning, and Preservation (ARCH); the School of Engineering, Project Management Center for Excellence (PMCE); the Maryland Blended Reality Center (MBRC); and the IMD Program in the Department of Computer Science and Department of Art. We envision the proposed training curriculum as a mid-tier training option, one where training infrastructure costs could easily be borne by individual companies or individual practitioners who invest in VR-based technologies (e.g., high-performance computer, VR headsets). The proposed two-module curriculum will focus on the high-priority subject areas of (1) energy-efficient building systems and (2) sustainable construction materials selection. Each training module will consist of written materials, videos, tests, and immersive VR experiences. The curriculum's target population will be architects, engineers, and contractors with 2–5 years' experience who self-select for training on advanced/sustainable building systems.

5.1.2 THEORETICAL FOUNDATION: SPATIAL INTELLIGENCE

An important aspect afforded by virtual environments is the subjective experience of being virtually present at a location, even when one is physically elsewhere. This notion of presence has long been considered central to virtual environments, for evaluation of their effectiveness as well as their quality [6]. More precisely, Slater developed the idea of place illusion (PI), referring to the aspects of presence "constrained by the sensorimotor contingencies afforded by the virtual reality system." [7]. Sensorimotor contingencies are actions used in the process of perceiving the virtual world, such as moving the head and eyes to change gaze direction or looking around occluding objects to gain an understanding of the space [8]. Slater therefore concluded that establishing the presence or "being there" for lower-order immersive systems such as desktops is not feasible [9]. In contrast, the sensorimotor contingencies of walking and looking around facilitated by head-mounted displays (HMDs) contribute to their higher-order immersion and establishing a presence. Recent research in cognitive psychology [10] suggests that the mind is inherently embodied. The way we create and recall mental constructs is influenced by the way we perceive and move [11, 12], known as embodied cognition. The most common straightforward definition is "states of the body modify states of the mind" [13]. The early cognitive theory, the disembodied theory of mind proposed in the 17th century believed our behaviour was mediated by something internal to the organism, our brain. According to Cartesian dualism, the mind is entirely distinct from the body, which is exactly opposite to the current embodied cognition theory that treats the body as the primary condition for experience since it comprises a collection of active meaning about the surroundings and its objects [14]. Since the 1950s, researchers have collected empirical data through experiments to understand how cognitive functions such as learning and memorizing are influenced by the physical body and the physical environment where the body is immersed [15].

VR's potential to augment AEC trainings derives from its ability to enhance spatial intelligence-dependent learning outcomes, a fundamental cognitive task/ability for architects and engineers [16]. Defined as "the ability to generate, retain, retrieve, and transform well-structured visual images" [17], it is associated with a heightened sense of relational and situational awareness [18]. In general, immersive VR environments have been shown to improve students' spatial intelligence thus enhancing their understanding of 3D objects [19] and increasing memory recall ability [4]. Sorby et al. (2018) tested more than 3000 first-year engineering students over a five-year period and found that VR training had a particularly positive impact on problem-solving and analysis skills [20]. The improved learning outcomes derived from enhanced spatial intelligence are related to how the human memory system works. The memory system that encodes, stores, recognizes, embodies, and recalls spatial information about the environment is called *spatial memory* [21]. Several studies have found that embodied navigation and memory are connected [22, 23]. Workforce training for green construction, experienced in an immersive virtual environment, could enhance learning and recall by leveraging the integration of vestibular and proprioceptive inputs (overall sense of body position, movement, and acceleration) [24]. Studies have shown that several aspects of training in virtual environments can be transferred to the real world [9, 25–27]. These include natural physical and psychological reactions, reflex response, heightened emotional response, as well as the transfer of knowledge and skills. Of particular interest are the design of VR environments for workforce training [28] including those to provide training for first responders [29], aircraft cabin safety procedures [30], and automotive manufacturing [31]. VR training environments that create close-to-real conditions are likely to invoke/support use of the trainees' spatial intelligence. They may also increase the retention of spatial memory-dependent features including object size, orientation, height, and placement in the environment. Combined, these features may increase better comprehension and recall of energy-efficient building systems' complex features.

5.2 TECHNICAL DESCRIPTION AND INNOVATION

5.2.1 RELEVANCE AND OUTCOMES

VR Options: Suitable simulation options for AEC industry training can be characterized by the visualization platform (see Table 5.1): desktop-based, immersive, 3D game-based, and building information modelling (BIM)-enabled.

1. Desktop-based VR displays a 3D virtual environment with objects that can be rotated [32]. It is displayed on a screen and viewed without a head-mounted device (HMD).
2. Immersive VR requires HMD and sensor gloves for sensations of "spatial immersion" that provide authenticity or realness to the VR environment [33]. Research on immersive VR in the context of building/built environment design is scarce [19].
3. 3D game-based VR creates a game-like training scene through integrating visual, interactive, network, and multi-user operating technologies [34]. It has been used to develop fire safety training [35] and construction process management training [36].
4. BIM-enabled VR augments BIM, a software and technology-dependent process that involves making and managing digital representations of physical and functional place characteristics. BIM-generated objects (vs. conventional CAD-based 3D objects) include supplementary data such as a building material's physical and structural properties, construction schedule, associated cost, and operation and maintenance requirements. This content allows users to analyze/contrast multiple factors (such as material types or cost) simultaneously to develop effective design solutions in real time [37]. This method can be immersive if HMD's are used.

TABLE 5.1

VR Training Options

	VR Type	Immersive	Cost	Production Time	AEC Adoption	Multi-user	Mobility
1	Desktop	No	Low	Short	High	No	Low
2	Immersive	Yes	High	Long	Low	Yes	Low
3	3D game-based	No	High	Long	Low	Yes	Low
4	BIM-enabled	Possible	Medium	Medium	Medium	Yes	Medium

5.2.1.1 VR Training Approach

In the AEC industry, training videos that utilize 3D objects without immersive VR have limitations. These demonstration videos, with cutaway views of 3D objects and animations of building systems, cannot demonstrate the 3D object *in relation to* humans, scale, location/distance, and spatial organization. Immersive VR environments immediately make these relationships clear.

Two advantages recommend immersive BIM-enabled VR for the proposed curriculum. (1) It enables the use of building information data [38] as trainees move from conventional 2D to 3D VR environments. Other conventional CAD-based 3D objects lose embedded building data during the translation to the VR environment. (2) Users may initiate building model changes while in the program that are then immediately reflected [39]. Both advantages increase efficiency as design changes/modifications can be made pre-build. This approach confers high-fidelity information to boost both the analytical understanding (spatial intelligence) and recall (spatial memory) of complex systems. Leveraging existing course content and infrastructure developed at UMD, AIA, and USGBC, the proposed simulation training will focus on: (1) energy-efficient building systems, and (2) sustainable building construction materials. Table 5.2 provides the rationale, system requirements, and expected outcomes.

5.2.1.2 Expected Module Outcomes

After the modules, trainees should more effectively be able to:

1. Name, list and label building system components; equipment operation steps and procedures; material composition and characteristics.
2. Describe and demonstrate understanding of: building system functions and characteristics; size and orientation of equipment; sustainable materials characteristics and applications.
3. Demonstrate facility at spatial skills: accuracy in identifying geometry, rotations, and patterns.
4. Demonstrate enjoyment and appreciation of the learning process.

TABLE 5.2

VR Application in Proposed Teaching Module

	System	VR Rationale	Equipment	Expected Outcomes	Assessment
1	*Energy-efficient building systems*	Complexity and size, integration with BIM data	Desktop/ Oculus DK2 HMD	1,2,3,4	Quiz and spatial test (refer to 3.2)
2	*Sustainable building construction materials*	Quantity, integration with BIM data	180-degree Projection screen/ Oculus DK2 HMD	1, 2, 4	Quiz and spatial test (refer to 3.2)

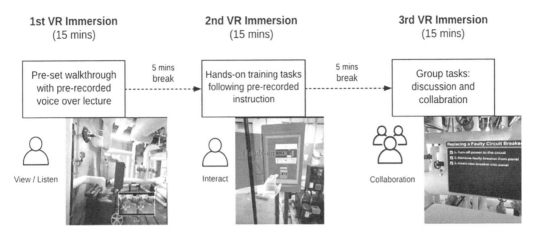

FIGURE 5.1 VR immersion.

5.2.2 Training Modules/Technology Focus

The two training modules will be developed and tested at UMD with ~80 trainees (AEC professionals with 2–5 years' experience). Each module will include three, 15-minute VR immersions in a 4.5-hour/day training session (refer to Figure 5.1). The first 15-minute VR immersion will situate the trainee in a real site environment (for example, for Module 1, the trainee will be immersed in a detached single-family house), and a pre-recorded, voice-over lecture will introduce information regarding the specific building system. The voice-over lecture will be accompanied by a preset walkthrough in the house. In the first 15-minute VR immersion, trainees will only be exposed to the building system/components visually (they will only view); they will not interact with objects and surroundings. The second 15-minute VR immersion will contain a series of hands-on training tasks. The trainee will follow pre-recorded, voice-over instructions to touch, open or move parts of the system, to enhance their understanding of the topics explained in the first immersion. The second immersion will allow the trainee to view the 3D objects and systems, and interact with their surroundings. In the third 15-minute VR immersion, two trainees will be immersed in a group environment and interact via virtual avatars (refer to Section 5.3 for more information). They will follow the pre-recorded, voice-over instruction to perform certain tasks to test their understanding and memory of the contents taught in the previous two VR immersions. For example, a task for Module 2 may be to select a building's exterior facade materials including weather barriers, insulation, and structure support layers that provide the highest thermal resistance value in a cold climate. They will then be asked to assemble these materials in the correct spatial organization. The two trainees will have the opportunity to discuss options and make a collective decision; this collaboration and interaction may enhance learning outcomes.

VR-equipped laboratory facilities at UMD will be used so that small groups of trainees may be simultaneously fully immersed in the VR environment, while the project team will monitor individual and collaborative performances and collect data for assessment/evaluations (refer to Section 3.2). The control group will train concurrently using traditional 2D lecture/video materials that parallel the proposed training modules. In total, the training session will take ~4.5 hours including preparation, pre- and post-assessments, training, breaks, and an end-of-session survey.

5.2.2.1 Module 1: <u>Energy-Efficient Building Systems</u> Are Essential for Achieving America's Energy Efficiency Goals

Module 1 focuses on an advanced mechanical system: the ground source, or geothermal heat pump system. This highly efficient renewable energy technology can be applied in both residential and

 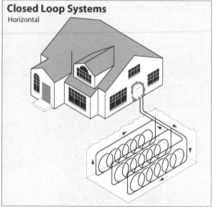

FIGURE 5.2   Geothermal heat pump in VR.

commercial buildings for space heating/cooling and water heating uses [40]. Efficiency is based on the exploitation of naturally existing heat, rather than by producing heat through a secondary source. Although the technology has been in use since the 1940s [41], widespread adoption continues to be limited in the United States. While a higher installation cost is a contributing factor, professional practitioner deficits in applied knowledge of this system-based technology also contribute to its more limited use. We will demonstrate the detailed assembly, installation, and operation of the geothermal heat pump (see Figure 5.2 and Table 5.2).

5.2.2.2   Module 2: Sustainable Construction Materials

The use of basic and advanced building materials including concrete, wood, metal, and plastic will be taught from the life cycle perspective. The simulation environment will enable trainees to visualize a building materials' impact across time, to establish a whole system thinking mindset. This training module will help trainees integrate the environmental impact of material selection across systems. The trainee will be able to walk through and physically explore the entire simulated environment that will be based on a built LEED building (see Figure 5.3).

5.2.3   Educational and Technical Feasibility

5.2.3.1   Curriculum Infrastructure and Online Teaching Capabilities

We will leverage the curriculum from the UMD Project Management Center in the School of Engineering (PMCE) and the School of Architecture (ARCH). This scaffold will allow the rapid integration of new VR environments designed specifically for pre-existing course content. PMCE offers degrees/certifications to engineers, architects, and any professional with a technical background. It is the first program in an engineering school to be accredited by the Project Management Institute's Global Accreditation Center. PMCE is equipped with technology-enhanced classrooms where lectures will be filmed with dedicated technical staff. Sustainability is the heart of ARCH's curriculum/programs including 2 undergraduate degrees, 4 standalone master's, 1 PhD, 3 minors, 3 certifications, and 16 interdisciplinary dual-masters. The proposed curriculum will leverage content from an existing graduate-level course: Carbon Neutral Development through Net Zero Building Design (taught by author).

5.2.3.2   VR Learning Outcome Research Capabilities

UMD's Maryland Blended Reality Center (MBRC) is dedicated to innovative VR, AR, and XR research and training for professionals in high-impact areas. Virtual reality displays, such as

FIGURE 5.3 Simulated VR environment for module 2.

HMDs, afford a superior spatial awareness by leveraging vestibular and proprioceptive senses (see Figure 5.4). On this theme, people have long used memory palaces as spatial mnemonics to help remember information by organizing it spatially in an environment and associating it with salient features in that environment. MBRC has explored whether using virtual memory palaces in an

FIGURE 5.4 UMD virtual memory palaces study.

HMD will allow a user to be able to recall information better than when using a traditional desktop display [11]. In a 2019 study, the research team had 40 participants recall the identity and location of faces on both displays. The research team found that virtual memory palaces in an HMD provide a superior memory recall ability compared to traditional desktop displays. Specifically, researchers found a statistically significant difference of 8.8% between the recall accuracies in an HMD compared to a desktop.

5.2.3.3 Cyberinfrastructure and Facilities

MBRC's well-equipped VR/AR laboratory and education space will facilitate the proposed module's different simulation configurations per the following equipment (see Figure 5.5):

- VR/AR headsets, including those from Oculus, MetaAR, and Vuzix
- Interaction sensors (Kinects, LeapMotion), infra-red cameras, EEG sensors, and eye-trackers
- Multi-camera lightfield arrays for multiscopic video capture
- Rapid prototyping facilities with 3D laser scanner, 3D printer
- High-end 3D graphics workstations

5.2.3.4 VR Headsets

Standalone HMD: Fully wireless immersion is enabled via VR standalone devices for both, single-trainee and multi-trainee simulations. These self-contained headsets include built-in processors, GPUs, sensors, batteries, memory, and displays thus freeing the participants from tethered computers and enabling movement in the physical environment that simulates behaviour taking place in the VR environment.

Collaboration enhancing technology: Oculus headsets are configured for group VR demonstrations. A single powerful workstation with an NVIDIA K6000 GPU drives the headset and displays

FIGURE 5.5 MBRC lab.

to a Samsung DM75D LED panel. In this configuration, a single user can wear and experience the headset, while others can watch the experience on a large LED display panel. Eight additional Oculus headsets can then be deployed to support additional trainees as needed.

5.2.4 INNOVATION AND POTENTIAL IMPACT

5.2.4.1 Innovation: Addressing Challenges in Collaborative Immersive Training

Collaborative workforce training requires innovative systems for compression, streaming, and rendering of multiple participants and their shared state from individual headsets to a central server and back. Achieving a low-latency and highly interactive shared virtual environment enables new modes of immersive experiential training that engages natural modes of interaction in the real world such as gestures, pose, and gaze. Building upon Microsoft's Holoportation project, UMD has developed a prototype software system for acquiring and fusing multiple dynamic video light fields from a system of 8 cameras with Kinects. The results, shown in Figure 5.6, are superior to the Holoportation project's quality and still run at 90 frames per second [42]. While such a method can fuse video light fields for the 8 cameras, the challenge is in the compact representation and streaming of such content over wireless networks.

A UMD research team has created virtual environments by capturing building and manufacturing floors with multiple 360-degree cameras at high-resolution (6K stereo or 8K mono) images. The research team has integrated them into a persistent 3D environment that can be used in Unity to develop a navigable 3D environment. For the application of workforce training for green construction, the challenge will be to ensure that as a person navigates multiple 360-degree images, the transitions are as seamless as possible. UMD research team has also developed methods of acquiring and reproducing standard operating procedures (SOPs) for training in virtual or augmented reality. Specifically, we have been experimenting with the addition of virtual 3D CAD objects mixed with real 360-degree images to augment virtual scenes with objects, that will be needed for various SOPs. One of the research challenges here is developing novel methods to reconcile the depth and location of 3D CAD objects, to merge seamlessly with 360-degree image panoramas and relight them so that they appear to be an integral part of the 3D immersive experience. This will require accurate depth computation of 360 images for an appropriate placement, as well as the incorporation of techniques such as spherical harmonic lighting for relighting 3D CAD objects in virtual scenarios. In parallel with resolving the above technical challenges, the research team plans to film

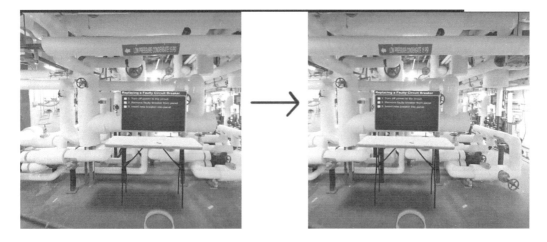

FIGURE 5.6 Multiview-fusion: Microsoft's Holoportation results (left) to our Montage4D (right).

various SOPs using multiple 360 cameras to create a VR training library. The research team also plans to explore the creation of an intuitive and easy-to-use user interface for selecting SOPs from the video library, playback control of the SOP, as well as fully flexible navigation and teleportation in the SOP environment.

5.2.4.2   Industry Impact

Accelerating ways to meet AEC industry's changing preferences for energy-efficient technologies and systems drives our project, innovation, and choice of the target audience. While new graduates may have some green building exposure during pre-licensing education, early career professionals have much less decision power in the marketplace. The burden to drive forward energy-efficient practices will continue to fall within the continuing education training space.

The innovation of this proposed training module is to utilize immersive VR training technologies to meet known training needs. Augmenting traditional pedagogical curriculums with simulation-based approaches and accessory information (BIM-data objects) will enable high fidelity, middle-tier cost, and remote training opportunities. We will accomplish this by addressing key technical challenges currently hindering the adoption of collaborative VR environments. Advantages and the intended impacts of the proposed VR training include (a) 3D spatial representation of complex systems; (b) training at the edge of realism; (c) increase in trainee focus, retention, and engagement; (d) advancing research to improve collaborative immersive training environments; (e) allowing trainers to directly observe and evaluate learning outcomes; and (f) data feedback to validate training experiences. Outcomes/utilization of the POC are envisioned as:

Short term (0–3 years post-project period):

- Use of POC at UMD with potential replication at other higher education institutions via promotion of the research and best practices at discipline-specific conferences and via promotions through our professional organization partners (AIA and USGBC).
- Immediate distribution to professional organization partners (AIA and USGBC) for inclusion in continuing education courses. AEC professionals with access to VR standard equipment will be able to experience fully immersive VR environments. An embedded survey will track use.

Medium term (3–7 years post-project period):

- Future scaling to high school level certificate, transitioning worker, and two-year community college programs.

5.3   PROPOSED WORK PLAN AND OUTCOMES

5.3.1   Technical Scope Summary

The technical scope of the plan tasks is given below.

- Develop, test, and validate the training module: two training modules described in Section 2.1 (Yr 1,2).
- Extend VR content via self-guided VR walkthrough lectures (Module 1), and 360-degree videos of case studies (Module 2). Students may opt to view accompanying lectures through VR headsets, via computer (~3D game), or online video (Yr 1, 2).
- Execute pilot in 2nd year summer (Yr 2).
- Collect feedback and assess the achievement of the learning outcomes (Yr 2).

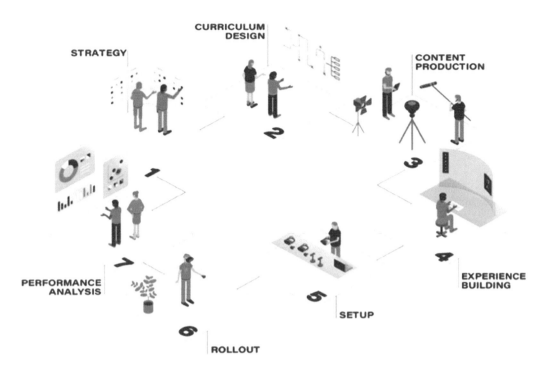

FIGURE 5.7   Workflow and tasks.

5.3.2   Work Breakdown Structure (WBS) and Task Description

Tasks and deliverables are described below (refer to Figure 5.7).

Task 1: Plan strategy with partners to determine the course content and schedule (Yr 1).

The initial curriculum design will be based on UMD's ARCH and PMCE's ongoing curriculum (see Section 2.3) with the input from USGBC and AIA experts. A one-day virtual workshop will be organized with the team, professional organization experts, and industry stakeholders (approximately 20 participants) to determine the specific course content, module level, and targeted learning objectives and outcomes. The course evaluation methods will be defined. Participants will be invited to participate in the pilot (in Yr 2).

Subtask 1.1: Review existing ARCH and PMCE course content.

Subtask 1.2: Review and select case studies to be included (Module 2).

Subtask 1.3: Conduct workshop to solicitate recommendations from AEC industry stakeholders on specific training pain-points/limitations; Evaluate specific procedures used in real-world trainings that will translate into VR simulations.

Subtask 1.4: Collaborate with AIA and USGBC to conduct a survey on the accessibility and cost of VR course. Results will provide information on access/demand/cost constraints.

Task 2: Create the training modules (Yr 1). Diagrams and course narratives will be created and prepped for future VR content creation (see Task 4). Initial content such as 3D models of components will be constructed using BIM software (Autodesk Revit).

Subtask 2.1: Develop Module 1 – *Energy efficient building systems.*

This module will focus on the geothermal heat pump—an energy-efficient building technology that contributes to carbon neutral goals (refer to Section 2.2.1).

<u>Subtask 2.2</u>: Develop Module 2 - *Sustainable building construction materials.*
This module will focus on the visualization of sustainable materials' impact through the whole building life cycle (refer to Section 2.2.2).

Task 3: Produce content; course lectures for both the control group and VR group will be recorded using Media Site digital technology in UMD technology-enhanced classrooms (Yr 1, 2).

<u>Subtask 3.1</u>: Record instructor lectures/narratives.
<u>Subtask 3.2</u>: Transfer BIM models to VR environment for merge per task 4.

Task 4: Design and build the VR experiences. These will include instructor voice-over narratives. Commercial (Oculus) HMD's will be used for the pilot (Yr 1, 2).

<u>Subtask 4.1</u>: Integrate and merge VR BIM objects to create a smooth and realistic experience.
<u>Subtask 4.2</u>: Create self-guided VR walkthrough lectures for Module 1.
<u>Subtask 4.3</u>: Create 360-degree VR videos of case studies for Module 2.

Task 5: Set up the testing space and website; *Canvas™* will be the Learning Management System used (Yr 2).

<u>Subtask 5.1</u>: Create the online course website on Canvas for pilot course rollout.
<u>Subtask 5.2</u>: Set up physical space for pilot course roll out in MBRC VR lab.

Task 6: Roll out the experimental pilot to test the POC (in person, no fee) in 2nd-year summer for 80 trainees at the MBRC VR lab. The pilot will be run four times; each run will occur on a separate day for 20 trainees over approximately 4.5 hours. The course target population will be licensed architects, engineers, and contractors with two to five years' experience who self-select for training on advanced/sustainable building systems. Prior knowledge and experience with VR will not be required. Figure 5.8 illustrates the pilot schedule and training process.

During each pilot run, both Module 1 and 2 will be tested. The pilot is designed as an experimental study. During each run, the 20 trainees will be randomly assigned to one of two groups (10 per group): the control group will take the modules in the conventional lecture/video method; the experimental group will take the course in full VR immersion. Trainees will consent to participate in the pilot course/study without knowing the placement. Prior to each module, groups will complete a pre-course quiz, consisting of ten multiple choice questions and five building or system anatomy questions related to the module topic.

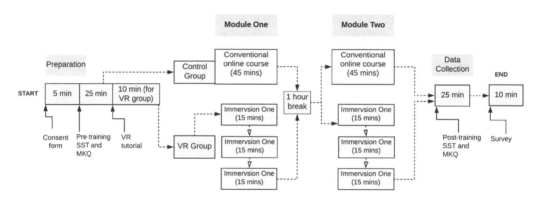

FIGURE 5.8   Pilot course experimental design/schedule.

The control group will take the conventional online pre-recorded lecture (45 mins) on individual desktops. The collaboration will take place online (through *Canvas*™) to simulate future online conditions. For the VR group, 10 trainees will be outfitted with VR HMD's and controllers. They will be given 5-minute tutorial on how to view and interact with the course platform (virtual manual); how to interact with the building system and components; and how to use the controller to manipulate the building components in three dimensions.

Module 1: After the tutorial, ten VR trainees will be given three, 15-minute immersions (see Section 2.2 for description). The project team will intervene only if a participant requests clarification or assistance. Post-module, both groups will be given a second quiz with ten multiple-choice and five anatomy questions. The post-module quiz will be similar in design but assess different content than the pre-module quiz.

Module 2: After a one-hour break, repeat the same process.

Control group option: After training, control group participants will be invited to experience a VR immersion (either Module 1 or 2).

Questionnaire: Post Module 1 and 2, a subjective questionnaire will be administered to both control and experimental groups to assess the trainee's perceived engagement, knowledge acquisition, and effectiveness of teaching pedagogy used (either video lessons or VR immersion).

Subtask 6.1: Trainee recruitment will be open to all PMCE and ARCH enrolled students. The typical PMCE graduate student is a full-time working AEC professional. PMCE has ~120 graduate students per year. ARCH enrols +400 students per year. In addition, all professionals who participated in the Yr 1, 1-day virtual workshop will be invited to participate (see Task 1). If there is difficulty in filling the quota of 80 students for the pilot, an invitation will be sent via listserv to the UMD Architecture and Engineering alumni networks.

Subtask 6.2: Data Collection: Pre- and post-module learning assessment data will be collected through the Spatial Skill Test (SST) and Module Knowledge Quiz (MKQ). SST employs items from two psychometric tests of spatial skills: the Differential Aptitude Test: Space Relation (DAT:SR) [43] and Purdue Spatial Visualization Test Rotations (PSVT:R) [44]. MKQ will contain ten multiple-choice and five building or system anatomy questions related to the module training content. Data collection will be administrated by the project team.

Subtask 6.3: Run the pilot course four times during 2nd-year summer.

Subtask 6.4: Collect training results.

Task 7: Analyze/validate the learning outcomes and modify the content as needed.

Subtask 7.1: Analyses and evaluation of training data per learning outcomes (refer to Table 5.2).

Subtask 7.2: Revise module training content based on the above assessment.

5.3.3 PROJECT RISK AND MITIGATION MANAGEMENT

Table 5.3 outlines risks from three key technological and pedagogical categories. (1) Course content (CC): Information needed to build the course content may be difficult to find and/or render. (2) Technical (T): VR requires intensive graphics capabilities that are not always possible with standard computer equipment. Investments and technical advances may be required to achieve smooth implementation, immersion, and interaction. (3) Trainee experience (TE): VR trainees must master the VR interface and accompanying interaction tools; this can be more difficult for non-digital natives. In addition, VR training times may be difficult to estimate and vary widely across students based on discipline differences, workforce/cultural characteristics.

TABLE 5.3

Risk Assessment of Key Challenges and Mitigation Strategies

Risk	Description	Prob	Impact	Risk Factor	Approach/Mitigation
Case-building information (CC)	Cannot access the information	1	3	3	Case-building information is typically publicly accessible via internet. If this risk occurs, we will contact the project manager directly through PI's network/contacts.
BIM model building (CC/T)	Large model/size	3	3	9	Realistic VR experiences are best created with high levels of detail conferred into the simulated environment. However, higher-resolution objects create bigger digital files. If files get too big, we may need to reduce our scope or utilize UMD high-performance computing resources, such as Deep thought 2 with 2.8 GHz processor speeds and 128 GB memory.
High fidelity of VR environment (T)	Develop methods to reconcile the depth and location of 3D objects, to merge seamlessly with 360-degree image panoramas.	5	3	15	Preliminary work (1) fusing 360-degree imagery from Google Street View and Microsoft Street Side and (2) adding virtual 3D CAD objects mixed with real 360-degree images has been successful. We will employ currently developed SOP's and make modifications as needed.
Cybersickness (TE)	Duration that trainee can endure VR determines the length of modules	3	5	15	During the design of the training modules, we will execute several preliminary tests to determine the appropriate VR duration. We will also design all training modules as 15-minute segments to mitigate potential cybersickness. This design modification will be tested in the pilot to determine if it is adequate for the full course or needs to be further refined.
Realism of training experience (TE) Scalable and measurable hands-on experience (TE)	Insufficiently detailed VR content and scale issues may hinder the immersive experience	3	3	9	The research team has extensive experience developing education and training content for VR environments. However, due to the complexity of the building systems/components, there is a risk that the pilot VR experience may not be comparable to in-situ training experiences. We plan to conduct detailed assessments and evaluations from the data collected from the pilot course to better understand any inherent limitations. This feedback will be used to adjust and modify the VR environment for a future full launch.
The effectiveness of VR training	VR training may not prove more effective than traditional pedagogy	3	3	9	A major objective of the study is to elucidate the effectiveness of the VR simulations in furthering the learning outcomes. Our detailed assessments and evaluation will look at factors (trainee experience, outcome, and more) influencing effectiveness thus providing information for adaptation.

Score: Low (1), Medium (3), High (5).

Risk Factor: >15 Red (highest risk), 15–10 Orange, 3–9 Yellow, <3 Green.

Risk factor = Probability (Prob) × Impact.

5.4 CONCLUSION

If the proposed curriculum pilot and evaluation of learning outcomes prove successful, the two validated VR immersive modules will be disseminated through AIA and the USGBC as online courses to satisfy required licensure training hours. For USGBC members, the two modules will be integrated into their Green Builder [45] Program that trains professionals in all areas of green building, from material selection to building data analytics. Further funding for the development of additional VR-immersive training modules on other sustainable building technologies will be solicited from interested industrial partners and professional organizations.

Future applications of the validated training module beyond professional workforce training are likely. The team expects to fully integrate the VR modules into existing undergraduate and graduate courses within ARCH and PMCE. Another potential scaler at UMD will be to offer the curriculum infrastructure development process and lessons learned as a "best practices" guide to other undergraduate majors. The research team will also collaborate with local community colleges to reach out to underrepresented students and provide free training to these students as well as instructors.

REFERENCES

1. AIA https://www.aia.org/continuing-education and https://www.usgbc.org/education accessed on January 2, 2021.
2. Brown, Jonathan R., Irina Kuznetcova, Ethan Kirk Andersen, Nick H. Abbott, Deborah M. Grzybowski, and Christopher Douglas Porter. "Full Paper: Implementing Classroom-Scale Virtual Reality into a Freshman Engineering Visuospatial Skills Course." In *2019 FYEE Conference*. 2019.
3. Whyte, Jennifer, Nouran Bouchlaghem, Antony Thorpe, and R. McCaffer. "From CAD to virtual reality: Modelling approaches, data exchange and interactive 3D building design tools." *Automation in construction* 10, no. 1 (2000): 43–55.
4. Krokos, Eric, Catherine Plaisant, and Amitabh Varshney. "Virtual memory palaces: Immersion aids recall." *Virtual reality* 23, no. 1 (2019): 1–15.
5. Simon, Madlen, and Ming Hu. "Mind The Perception and Emotional Response To Design." In *ARCC Conference Repository*. 2019.
6. Skarbez, Richard, Frederick P. Brooks Jr, and Mary C. Whitton. "A survey of presence and related concepts." *ACM computing surveys (CSUR)* 50, no. 6 (2017): 1–39.
7. Slater, Mel. "Place illusion and plausibility can lead to realistic behaviour in immersive virtual environments." *Philosophical transactions of the royal society b: Biological sciences* 364, no. 1535 (2009): 3549–3557.
8. O'regan, J. Kevin, and Alva Noë. "A sensorimotor account of vision and visual consciousness." *Behavioral and brain sciences* 24, no. 5 (2001): 939–973.
9. Kozak, J. J., Peter A. Hancock, E. J. Arthur, and Susan T. Chrysler. "Transfer of training from virtual reality." *Ergonomics* 36, no. 7 (1993): 777–784.
10. Repetto, Claudia, Silvia Serino, Manuela Macedonia, and Giuseppe Riva. "Virtual reality as an embodied tool to enhance episodic memory in elderly." *Frontiers in psychology* 7 (2016): 1839.
11. Barsalou, Lawrence W. "Grounded cognition." *Annual review of psychology* 59, no. 1 (2008): 617–645.
12. Shapiro, Lawrence. *Embodied cognition*. Routledge, 2010.
13. Dove, Guy. "On the need for embodied and dis-embodied cognition." *Frontiers in psychology* 1 (2011): 242.
14. Merleau-Ponty, Maurice. *Phenomenology of perception*. Routledge, 2013.
15. Wilson, Andrew D., and Sabrina Golonka. "Embodied cognition is not what you think it is." *Frontiers in psychology* 4 (2013): 58.
16. Wolfartsberger, Josef. "Analyzing the potential of virtual reality for engineering design review." *Automation in construction* 104 (2019): 27–37.
17. Lohman, D. F. Spatial ability and g. *Human abilities: Their nature and measurement*. (pp. 97–116). Lawrence Erlbaum Associates, Inc., 1996.
18. Gardner, Howard. "Multiple intelligences (Vol. 5, p. 56)." *Minnesota Center for arts education* (1992).
19. Paes, Daniel, Eduardo Arantes, and Javier Irizarry. "Immersive environment for improving the understanding of architectural 3D models: Comparing user spatial perception between immersive and traditional virtual reality systems." *Automation in construction* 84 (2017): 292–303.

20. Sorby, Sheryl, Norma Veurink, and Scott Streiner. "Does spatial skills instruction improve STEM out-comes? The answer is 'yes'." *Learning and individual differences* 67 (2018): 209–222.

21. Madl, Tamas, Ke Chen, Daniela Montaldi, and Robert Trappl. "Computational cognitive models of spatial memory in navigation space: A review." *Neural networks* 65 (2015): 18–43.

22. Leutgeb, Stefan, Jill K. Leutgeb, Carol A. Barnes, Edvard I. Moser, Bruce L. McNaughton, and May-Britt Moser. "Independent codes for spatial and episodic memory in hippocampal neuronal ensembles." *Science* 309, no. 5734 (2005): 619–623.

23. Buzsáki, György, and Edvard I. Moser. "Memory, navigation and theta rhythm in the hippocampal-entorhinal system." *Nature neuroscience* 16, no. 2 (2013): 130–138.

24. Hartley, Tom, Colin Lever, Neil Burgess, and John O'Keefe. "Space in the brain: How the hippocampal formation supports spatial cognition." *Philosophical transactions of the royal society b: Biological sciences* 369, no. 1635 (2014): 20120510.

25. Witmer, Bob G., John H. Bailey, Bruce W. Knerr, and Kimberly C. Parsons. "Virtual spaces and real world places: Transfer of route knowledge." *International journal of human-computer studies* 45, no. 4 (1996): 413–428.

26. Rose, F. David, Elizabeth A. Attree, Bennett M. Brooks, David M. Parslow, and Paul R. Penn. "Training in virtual environments: Transfer to real world tasks and equivalence to real task training." *Ergonomics* 43, no. 4 (2000): 494–511.

27. Hamblin, Christopher James. *Transfer of* training from virtual reality environments*. Wichita State University, 2005.

28. Carruth, Daniel W. "Virtual reality for education and workforce training." In *2017 15th International Conference on Emerging eLearning Technologies and Applications (ICETA)*, pp. 1–6. IEEE, 2017.

29. Mossel, Annette, Christian Schoenauer, Mario Froeschl, Andreas Peer, Johannes Goellner, and Hannes Kaufmann. "Immersive training of first responder squad leaders in untethered virtual reality." *Virtual reality* 25, no. 3 (2021): 745–759.

30. Buttussi, Fabio, and Luca Chittaro. "Effects of different types of virtual reality display on presence and learning in a safety training scenario." *IEEE transactions on visualization and computer graphics* 24, no. 2 (2017): 1063–1076.

31. Ordaz, Néstor, David Romero, Dominic Gorecky, and Héctor R. Siller. "Serious games and virtual simulator for automotive manufacturing education & training." *Procedia computer science* 75 (2015): 267–274.

32. Mawlana, Mohammed, Faridaddin Vahdatikhaki, Ahmad Doriani, and Amin Hammad. "Integrating 4D modeling and discrete event simulation for phasing evaluation of elevated urban highway reconstruction projects." *Automation in construction* 60 (2015): 25–38.

33. Waly, Ahmed F., and Walid Y. Thabet. "A virtual construction environment for preconstruction planning." *Automation in construction* 12, no. 2 (2003): 139–154.

34. Wang, Peng, Peng Wu, Jun Wang, Hung-Lin Chi, and Xiangyu Wang. "A critical review of the use of virtual reality in construction engineering education and training." *International journal of environmental research and public health* 15, no. 6 (2018): 1204.

35. Guo, Hongling, Heng Li, Greg Chan, and Martin Skitmore. "Using game technologies to improve the safety of construction plant operations." *Accident analysis & prevention* 48 (2012): 204–213.

36. Le, Quang Tuan, Akeem Pedro, Hai Chien Pham, and Chan Sik Park. "A virtual world based construction defect game for interactive and experiential learning." *Int. J. Eng. Educ* 32 (2016): 457–467.

37. Goulding, Jack Steven, Farzad Pour Rahimian, and Xiangyu Wang. "Virtual reality-based cloud BIM platform for integrated AEC projects." *Journal of information technology in construction* 19 (2014): 308–325.

38. Boton, Conrad. "Supporting constructability analysis meetings with immersive virtual reality-based collaborative BIM 4D simulation." *Automation in construction* 96 (2018): 1–15.

39. Petrova, Ekaterina Aleksandrova, Mai Rasmussen, Rasmus Lund Jensen, and Kjeld Svidt. "Integrating Virtual Reality and BIM for End-user Involvement in Building Design: a case study." In *The Joint Conference on Computing in Construction (JC3) 2017*, pp. 699–709. The Joint Conference on Computing in Construction (JC3) 2017, 2017.

40. Self, Stuart J., Bale V. Reddy, and Marc A. Rosen. "Geothermal heat pump systems: Status review and comparison with other heating options." *Applied energy* 101 (2013): 341–348.

41. Department of Energy. "Geothermal Heat Pumps". https://www.energy.gov/energysaver/heat-and-cool/heat-pump-systems/geothermal-heat-pumps

42. Du, Ruofei, David Li, and Amitabh Varshney. "Interactive Fusion of 360 Images for a Mirrored World." In *2019 IEEE Conference on Virtual Reality and 3D User Interfaces (VR)*, pp. 900–901. IEEE, 2019.

43. Connor, Jane M., and Lisa A Serbin. "Visual-spatial skill: Is it important for mathematics? Can it be taught." *Women and mathematics: Balancing the equation* (1985): 151–174. https://books.google.com/books?id=vjyZAgAAQBAJ&pg=PT214&lpg=PT214&dq=Connor,+Jane+M.,+and+Lisa+A+Serbin.+%E2%80%9CVisual-spatial+skill:+Is+it+important+for+mathematics?+Can+it+be+taught.%E2%80%9D+Women+and+mathematics:+Balancing+the+equation&source=bl&ots=2NRfOiWjYu&sig=ACfU3U3L8i3dFEoIndiGykMdGRa9MvnpiQ&hl=en&sa=X&ved=2ahUKEwizp4abir7-AhWrmWoFHYXvCcQQ6AF6BAgHEAM#v=onepage&q=Connor%2C%20Jane%20M.%2C%20and%20Lisa%20A%20Serbin.%20%E2%80%9CVisual-spatial%20skill%3A%20Is%20it%20important%20for%20mathematics%3F%20Can%20it%20be%20taught.%E2%80%9D%20Women%20and%20mathematics%3A%20Balancing%20the%20equation&f=false

44. Bodner, George M., and Roland B. Guay. "The Purdue visualization of rotations test." *The chemical educator* 2, no. 4 (1997): 1–17.

45. https://www.usgbc.org/education/sessions/be-green-builder-12846203 accessed on January 2, 2021.

6 Building Information Modeling (BIM) in Geotechnics and Infrastructures

Selcuk Toprak and Sevilay Demirkesen

CONTENTS

6.1 INTRODUCTION

The construction industry is complex and dynamic in nature (Chan and Chan, 2004; Fahmy et al., 2020). Especially, the involvement of different parties makes projects even more complex and hard to manage (Ward and Chapman, 2008). Therefore, it is essential that construction companies use effective methods and tools to avoid the problems brought by complexity and dynamicity. Being one of the most effective practices, technology adoption and integration is of utmost importance to enhance performance in construction projects as well as reduce waste. In recent years, building information modeling (BIM) has become an effective means of managing construction projects in a cooperative and collaborative environment (Lu et al., 2013; Zhao et al., 2015). Certain benefits have already been achieved in different projects such as time and cost savings, better inter-organizational skills, and better communication with subcontractors and other project partners (Bryde et al., 2013; Lu et al., 2013). Despite various efforts, it is still a challenge that BIM has not fully been understood by industry practitioners in terms of its potential benefits (Neff et al., 2010; Merschbrock, 2012). Considering the lack of emphasis on potential benefits and usage, this chapter aims to promote BIM applications in terms of their usage in geotechnical engineering and infrastructure applications. In this respect, the chapter first points out the purpose of BIM and its benefits for construction projects. Then, widely used BIM software has been discussed in terms of its functionalities and capabilities. In this context, various examples are presented and discussed accordingly.

DOI: 10.1201/9781003325246-6

6.2 BIM PURPOSE AND BENEFITS

The most widely used definition of BIM is made by the U.S. National Institute of Building Science (NIBS). "BIM is a digital representation of physical and functional characteristics of a facility and a shared knowledge resource for information about a facility forming a reliable basis for decisions during its life-cycle; defined as existing from earliest conception to demolition" (NIBS, 2007). Autodesk (2023) defines BIM as "the holistic process of creating and managing information for a built asset. Based on an intelligent model and enabled by a cloud platform, BIM integrates structured, multi-disciplinary data to produce a digital representation of an asset across its lifecycle, from planning and design to construction and operations."

BIM is one of the most commonly used tools for designing, building, and maintaining buildings. Succar (2009) defines it as an interactive set of policies, processes, and technology that help manage building design and project data. BIM is further preferred by a broader community for the fact that it provides a common digital resource for all members of a project during the lifecycle of a building from design to facilities management (Sabol, 2008). Besides promoting the use of new software, BIM also introduces a new way of thinking and a new approach to project delivery (Smith, 2014). It mainly targets the transition to a business model, where participants can work together on the BIM model eliminating the burden of working on separate information channels and incompatible software.

According to Masood et al. (2014), BIM is a 3D digital representation of an intelligent building with key components. Eadie et al. (2015) further implied that BIM is a tool for virtual design appearing as a platform for information and data sharing among stakeholders leading to better communication through the project lifecycle. Khemlani (2009) implies that BIM is an innovative technology that extends far beyond the 3D design phase to manage processes throughout the project's life cycle. BIM can be used to estimate cost and quantity in the pre-construction phase as well as to sustain integration and coordination between structural components and schedules. Li et al. (2020) implied that BIM can improve the effectiveness of quality control in construction projects. Aibinu and Venkatesh (2014) indicated that BIM is sometimes defined as a product itself facilitating collaboration, integration, and simultaneous work by multiple design disciplines during the project delivery process. In this respect, as a parametric digital model, BIM describes all the physical and functional characteristics of a project.

Apart from the above-listed functionalities, BIM can simulate a building's construction process in a virtual environment. Hence, BIM is an important guide for planners when making important decisions by enabling the details of forward-looking work to be visualized at any time (Chau et al., 2004). BIM further allows the creation a workable building model, where information provided by different disciplines is available to the project team including all members. The ability of BIM to combine graphical and non-graphical data models enables more accurate cost information for cost consultants (McCuen, 2008). To better understand the functionalities of BIM, it is of utmost importance to mention some terms such as dimensions, level of development, and maturity level, which are presented in below sections.

6.2.1 DIMENSIONS OF BIM

The different subsets in BIM are described by the dimensions. The three-dimensional (3D) model refers to the geometric shapes and information providing a basis for integrating information in the model. BIM is not limited to the 3D model but also allows information to be used in the graphic model for different purposes (Aibinu and Venkatesh, 2014). Different researchers define BIM's multi-dimensional capability as "nD" modeling since it has almost infinity of dimensions (Eastman et al., 2011; Smith, 2014). On the other hand, there has been no consensus achieved regarding the BIM dimensions in recent years. For example, Saxon (2018) mentions that 3D is the geometric model, 4D time, and 5D cost, whereas 6D and 7D definitions have still been questioned. Arnal (2018) defined 6D as sustainability and 7D as operation and maintenance, whereas 8D health and safety, 9D Lean

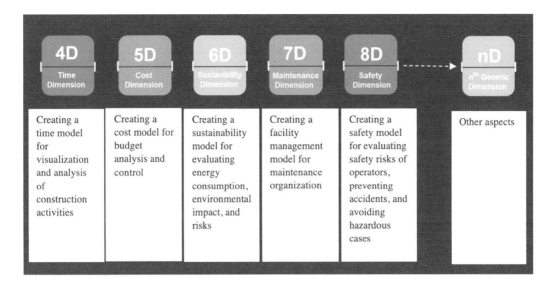

FIGURE 6.1 BIM dimensions and descriptions.

management, and 10D digitization. The dimensions defined by Smith (2014) are 3D (object model), 4D (time), 5D (cost), 6D (as-built operation), 7D (sustainability), and even 8D (safety). Given this background, Figure 6.1 presents the generic BIM dimensions along with definitions.

According to Figure 6.1, 4D planning relates to the time dimension, which means that 3D models and construction processes on timelines are combined to produce a real-time graphical description of the project's progress. All project participants can observe, analyze, and assess progress in terms of sequential, spatial, and temporal features with the help of 4D, which makes workflow planning and constructability evaluation possible throughout time. Improved site design and logistics are also made possible by 4D scheduling. In order to show a financial representation of the model in real-time, costs are incorporated into the BIM model in 5D. The BIM sustainability dimension, or 6D, enables models to be assessed for their energy use, environmental effects, and pollution hazards. The BIM facility management, or 7D, enables a project's management and maintenance during its entire life cycle to be assessed. The 8D model incorporates safety elements into design and construction to help enhance safety management. An addition to BIM, nD models include all the design data required throughout the life cycle of a construction facility (Lee et al., 2005).

6.2.2 LEVELS OF DEVELOPMENT (LOD)

The United BIM (2022) has defined level of development (LOD) as "a set of specifications that gives professionals in the AEC industry the power to document, articulate and specify the content of BIM effectively and clearly. Serving as an industry standard, LOD defines the development stages of different systems in BIM." Thanks to LOD, the data such as geometric information of building elements or non-geometric material components, insulation properties, color, acoustic properties, and application workmanship details can be added to the models. LOD has been developed to allow better communication among BIM users in terms of the data quality of building elements (United BIM, 2022). The American Institute of Architects (AIA) defines six different LOD, where LOD specifies design requirements at each stage.

- LOD 100 is a representation detail level of the model that specifies the general size, shape, and placement of the building pieces during the preliminary design stage. Items with a LOD of 100 are not geometric representations.

- LOD 200 represents the level of detail of the model and contains a general system, item, or assembly, comprising approximations of quantities, dimensions, form, position, and direction, for which non-geometric data is also associated with structural parts.
- LOD 300 is a representation of the model's detail level and includes a generic system, item, or assembly with precise measurements, forms, positions, and directions that are ready for manufacture and application where non-geometric data is also linked to structural components.
- LOD 350, in addition to the LOD 300 detail level, includes interfaces and parts necessary for the coordination of the element with nearby or attached elements in the model.
- As a representation of the detail level of the model, LOD 400 contains dimensions, form, position, quantity, and orientation for a particular system, object, or assembly as well as details, fabrication, assembly, and installation information, along with non-geometric information associated with building elements. To have a representation of production, a LOD 400 element is modeled in the required detail.
- LOD 500 represents a field verification of dimensions, form, position, quantity, and direction information whereas non-geometric data are also associated with building elements. Figure 6.2 presents the LOD summary with respect to the construction stages.

6.2.3 THE BIM MATURITY LEVELS

Succar (2010) defines BIM maturity as "the quality, repeatability, and degree of excellence within a BIM capability." BIM maturity levels represent the technological progress in the AEC industry with respect to the degree of collaboration and information sharing among different stakeholders in a project. There are basically four fundamental levels defining this extent. Level 0 refers to low collaboration, where the user works in 2D with computer aided design (CAD). Level 1 represents partial collaboration, where the user works in 2D or 3D with digital files. Level 2 presents full collaboration, where the user works in 3D. The final level, Level 3, represents full integration (BibLus, 2022). Figure 6.3 presents the BIM maturity map proposed by the Department of Business, Innovation, and Skills (2011). The figure shows each level with respect to the degree and extent of

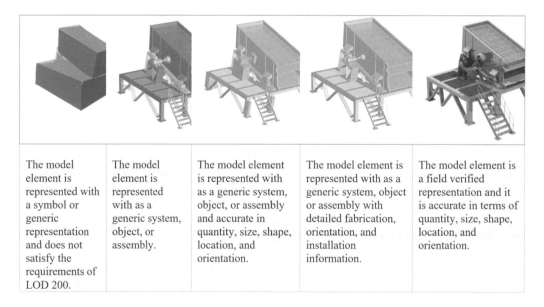

| The model element is represented with a symbol or generic representation and does not satisfy the requirements of LOD 200. | The model element is represented with as a generic system, object, or assembly. | The model element is represented with as a generic system, object, or assembly and accurate in quantity, size, shape, location, and orientation. | The model element is represented with as a generic system, object or assembly with detailed fabrication, orientation, and installation information. | The model element is a field verified representation and it is accurate in terms of quantity, size, shape, location, and orientation. |

FIGURE 6.2 LOD summary.

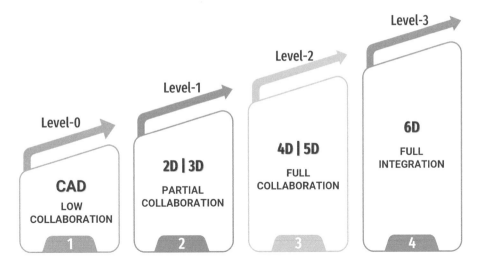

FIGURE 6.3 BIM maturity map.

collaboration. Barlish and Sullivan (2012) indicated that organizations' different levels of maturity should be considered carefully to compare one organization's BIM case to another.

6.3 BIM TOOLS AND SOFTWARE

In construction projects, BIM can be easily implemented using a variety of software. Therefore, it is important to list the most widely used software features and capabilities. The purpose of this chapter is to summarize the software used mainly during the design and construction stages, although there is a variety of software that can be mentioned. Accordingly, selected software is categorized into four main groups based on their prominent features:

- BIM 3D modeling/design software core features can be defined as architectural, structural, and MEP modeling tools and libraries. BIM software can create geometric models as well as incorporating non-geometric information such as cost, time, and other essentials.
- Collaboration BIM software enables the simultaneous review and visualization of a group of models, as well as the management of information requests and cloud sharing.
- BIM verification and control software offers multi-disciplinary coordination, clash detection, and other control features. Additionally, these programs offer features like cost estimation and quantification.
- For the construction phase, BIM 4D/5D software primarily offers 4D and 5D simulation, budgeting, and cost control capabilities. Additionally, it offers tools for managing safety, quality, documents, and post-construction processes. With their features, some BIM software can meet the needs of more than one group.

Besides, it may also be possible to classify BIM software according to the features they offer. For example, software such as Revit and ArchiCAD can be called BIM 3D modeling/design Software with the design tools they offer. Similarly, some software focuses on Collaboration, some are BIM Validation and Checking and some are 4D and 5D construction process.

Table 6.1 provides a comparison of features for BIM tools and software. The features offered are primarily taken from their promotional materials. It should be noted that BIM software has sometimes used different nomenclature for the solutions it offers on its promotional materials. The details of the BIM tools and software are presented below along with their comparisons.

TABLE 6.1

Comparing Features of BIM Tools and Software (Yilmaz, 2021)

BIM Software	BIM 3D Modeling/Design Software					Collaborative BIM Software				BIM Verification and Control Software				BIM 4D/5D Software		
	Revit	Archicad	AllPlan	Tekla Structures	SkechUp	BIM 360	Trimble Connect	BIMx	Allplan BIMplus	Navisworks	Bexel Manager	Solibri	iTWO costX	Vico Office	Synchro	MTWO iTWO 4.0
Provider	Autodesk	Nemetschek	Nemetschek	Trimble	Trimble	Autodesk	Trimble	Nemetschek	Nemetschek	Autodesk	Bexel	Nemetschek	RIB Software	Trimble	Bentley	RIB Software
Architectural Design	+	+	+	-	+	-	-	-	-	-	-	-	-	-	-	-
Structural Design	+	+	+	+	-	-	-	-	-	-	-	-	-	-	-	-
MEP Design	+	+	-	-	-	-	-	-	-	-	-	-	-	-	-	-
Parametric Modeling	+	+	+	+	+	-	-	-	-	-	-	-	-	-	-	-
Scheduling-Time input	-	-	+	+	-	+	-	-	-	+	+	-	+	+	+	+
Cost input	+	-	-	+	+	+	-	-	-	+	+	-	+	+	-	+
Multi-discipline Coordination	+	+	+	-	+	+	+	+	+	+	+	+	+	+	+	+
Design Reviews and Collaboration	+	+	+	-	+	+	+	+	+	+	+	+	+	+	+	+
Change Visualization Management	-	+	+	-	-	+	+	+	+	-	+	-	+	+	+	+
RFI Management	-	+	-	-	-	+	+	+	+	-	-	-	+	+	-	+
Clash Detection	+	+	+	+	+	+	-	-	+	+	+	+	-	+	+	+
Quantity Takeoff	+	+	+	+	+	-	-	-	-	+	+	+	+	+	+	+
Cost Estimation	+	-	-	+	+	-	-	-	-	+	+	-	+	+	+	+
4D Construction Simulation	-	-	+	+	+	+	-	+	+	+	+	-	+	+	+	+
Cost Budgeting	-	-	-	-	-	+	-	-	-	-	+	-	+	+	-	+
5D Cost Control Simulation	-	-	-	-	-	+	-	-	-	+	+	-	+	+	-	+
Cost - Budget Control	-	-	-	-	-	+	-	-	-	-	+	-	+	+	-	+
Purchase and Pay Application	-	-	-	-	-	+	-	-	-	-	+	-	-	+	-	-
Safety Management	-	-	-	-	-	+	-	-	-	-	-	-	-	-	+	+
Quality Management	-	-	-	-	-	+	+	-	-	-	-	+	-	-	+	+
Doc. Management	+	+	-	-	+	+	-	-	+	-	+	+	-	+	+	+
6D Facility Management	-	-	-	-	-	-	-	-	-	-	+	-	-	-	+	+

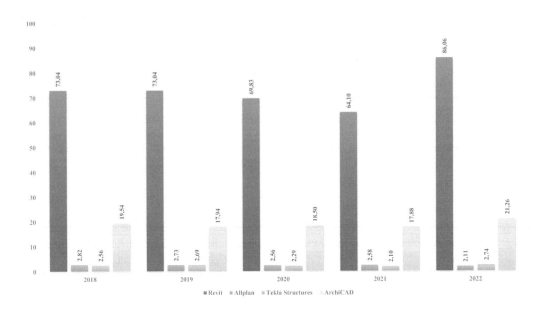

FIGURE 6.4 Comparing Revit, ArchiCAD, Allplan, and Tekla Structures based on the interest they attracted over time.

6.3.1 BIM 3D MODELING/DESIGN SOFTWARE

In BIM, 3D software is commonly used to support multi-disciplinary collaboration. A majority of its use is for architectural design. Among these, Revit software, which combines structural and MEP design elements, has recently gained popularity. Figure 6.4 indicates that Autodesk Revit is the most popular BIM design software in recent years among a set of others such as Allplan, Tekla Structures, and ArchiCAD based on Google Trends (2022) data.

6.3.1.1 Revit

Revit is a software of the Autodesk company. It is a multi-discipline software, which includes advanced capabilities for architectural design, MEP, structural engineering, and construction management. Revit is now being used by a broad community including but not limited to MEP engineers, architects, designers, contractors, and landscape architects. Revit allows users to design with parametric modeling elements with Dynamo. Revit creates a single database of files that can be shared among multiple users. The plans, sections, and elevations of the 3D model, as well as quantity, schedules, and costs are all linked, when a change was made on one view, the other views and outputs are updated automatically. As a result, the Revit model, which includes drawings, quantities, schedules, and cost data, is fully coordinated in terms of the architectural items shown in the drawings. There may be engineers and architects from multiple disciplines working together on the same project. Each discipline creates its own project databases and connects to other consultants' databases for verification. Clash detection, which identifies whether various components of the building occupy the same physical space, is possible in Revit. Additionally, Revit can import, export, and link data in the industry's most popular file formats, including DWG, IFC, and DGN (Autodesk, 2022a).

6.3.1.2 ArchiCAD

ArchiCAD is software developed by Nemetschek Group. In the design and modeling process, ArchiCAD is a 3D BIM tool primarily used by urban planners, architects, engineers, and designers. The user-friendly interface of ArchiCAD makes it one of the easiest BIM softwares to learn. Known as one of the first BIM implementations, ArchiCAD is a leading CAD tool for creating both 2D and 3D geometry (Zigurat Global Institute of Technology, 2019).

ArchiCAD has become a holistic BIM design program by integrating architectural drawing capabilities as well as structural and MEP design tools in its latest versions. It's a sophisticated platform designed to meet the needs of today's 3D design and BIM functions from planning to project implementation and facilities management. It also offers a BIM cloud, which is a real-time BIM collaboration environment, as well as BIMx, a strong mobile BIM visualization tool. ArchiCAD is an open BIM, which means it can write and read BIM data in a vendor-independent (IFC) format. This ensures that data is exchanged seamlessly with other project participants, regardless of the design tools they use (Financesonline, 2022a; Graphisoft, 2022a).

6.3.1.3 Allplan

Allplan is also software of Nemetschek Group. Allplan describes itself as a BIM-oriented CAD system. Allplan is the multi discipline platform for architects, engineers and entrepreneurs that integrate design and construction in all phases of the project. Allplan offers quick, easy work on bigger and more complicated projects with tough geometry, high details and numerous partnerships. Allplan's cloud technology enables to take full advantage of BIM using efficient workflows. Allplan Bimplus provides additional possibilities for combining and analyzing submodels from other areas. The task pane provides simple access to the Allplan Bimplus system, which provides efficient task management and localization-related tasks with mobile devices (Allplan, 2022a).

6.3.1.4 Tekla Structures

Tekla Structures is a 3D structural design software owned by Trimble Solutions Corp. It offers all the physical information needed to build and maintain structures in addition to allowing the creation, fusion, management, and sharing of multi-material 3D models. Tekla models can deliver LOD 500, which enables constructible structures. It only offers 3D structural design tools, unlike other design programs. Data from Tekla Structures can be imported, exported, and linked with other software solutions, digital construction tools, and fabrication machinery to streamline workflows. Collaboration with project members and third parties is easy with Trimble Connect. It is possible to integrate Tekla Structures with SketchUp, an architectural design program. Additionally, MEP and facility design tools can be incorporated into 3rd party applications (Tekla, 2022).

6.3.1.5 SketchUp

SketchUp is 3D architectural software that was purchased and developed by Trimble Solutions Corp. SketchUp is a popular 3D design solution known for its ease of use. Although it has an extremely fast interface for creating simple elements with a pull and push motion, it is well designed to rival the most powerful CAD software. Impressive 3D models of architecture, interior design, construction, landscape architecture, graphic design, and more can be created in a snap. For designers, drafters, architects, or engineers, SketchUp offers powerful, professional-grade features such as lighting effects, textures, layer managers, animations, and access to Trimble's 3D warehouse (SketchUp, 2022).

6.3.2 COLLABORATIVE BIM SOFTWARE

6.3.2.1 Bim 360

Autodesk BIM 360 is a potent cloud-based project management software for the construction sector. Project, site, and BIM management can expedite project delivery through the use of this program while remaining under budget and in accordance with industry standards, safety regulations, and specifications (Autodesk, 2022b).

BIM 360 is a versatile project management solution that includes almost all construction industry participants. The software provides a common platform for owner, project managers, subcontractors, design teams to improve coordination and collaboration (Financesonline, 2022b). General features of the software include but not limited to BIM and Multi-discipline coordination, design reviews and collaboration, work assignment and tracking real-time project status, project control, quality and safety operations, document control, and reporting.

6.3.2.2 Trimble Connect

Trimble Connect is a cloud-based collaboration tool available to engineers, subcontractors, architects, enabling informed decision-making and increasing project efficiency. Communication on-site and off-site, design coordination, and project management are some of the main features (Trimble Connect, 2022). General features of the software include but not limited to aligning models, combining and viewing selected models, multi-dimensional collaboration, improved workflows, task management, quality control, and custom reports and exports.

6.3.2.3 BIMx

BIMx is a Graphisoft software developed by Nemetschek Group. BIMx, the most widely used tool for project coordination and presentation, connects the design studio and the building site. The BIMx platform includes the "BIM Hyper Model," a tool similar to a game that allows everyone to explore and understand the building model. In the field, BIMx provides real-time model transitions, in-context measurements, and project markups within the context of the model. The latest version of BIMx offers seamless use and quality performance for projects with both complex 3D models and extensive 2D documents. The application allows team members to access all construction documents of the buildings, comprehensive functions that are easily accessible via mobile devices anytime and anywhere (Graphisoft, 2022b). Some of the general features that the software has are connection between 2D and 3D models, real-time 3D cross-section, position feedback during navigation, and measurements in 2D drawings or directly in a BIM model (Financesonline, 2022c).

6.3.2.4 Allplan Bimplus

Allplan Bimplus is a company operating under Nemetschek Group. It is an open BIM platform for different disciplines to collaborate efficiently on projects in the cloud. Throughout the entire building lifecycle, BIM model information, documentation, and tasks are centrally managed. Visualization and control of BIM projects, effective project management, incorporating project data into BIM software, merging, coordinating, and reviewing in a central model, detection of discrepancies or clashes, coordinating task board quickly and coordinate tasks, and accessing the project from different devices are some general features that the software offers (Allplan, 2022b).

6.3.3 BIM VERIFICATION AND CONTROL SOFTWARE

6.3.3.1 Navisworks

AEC teams and experts can review and manage projects using the Autodesk program Navisworks. During the pre-construction phase, it enables stakeholders to take part in a thorough examination of integrated models and data, improving control over the project's result (Autodesk, 2022c). Through its manage and simulate capabilities, Navisworks manages and simulates projects, offering greater coordination, simulation, and full project analysis for integrated project review. While Navisworks Manage offers tools for clash detection, advanced coordination, 5D analysis, and simulations, Navisworks Simulate emphasizes reviewing and communicating project details. Each package has unique review and management features that allow projects to be completed earlier than planned with increased profitability (Financesonline, 2022d). Some primary properties of the software are model coordination integration with BIM 360 and AutoCAD, BIM 360 integration, IFC file reader, real-time navigation, photorealistic model rendering, reality capture capabilities, 4D/5D project scheduling, and collected data in a single model.

6.3.3.2 BEXEL Manager

BEXEL Manager is an integrated suite of BIM tools that makes it easy for construction companies to manage large projects. It also recognizes many structural elements that have different tasks in the programming system of the platform. It can create custom quantity purchases with 4D and 5D BIM tools and easily export those. In addition, BEXEL Manager may work with any BIM model classification system. This allows you to quickly switch between tools to satisfy the needs of each project's regulations and customers (Bexel Manager, 2022; Financesonline, 2022e). Some features of the software are innovative 3D visualization, importing/exporting multi-source model, IFC formats support, 4D visualization, 5D cost comparison and reporting, and 6D project documentation integration.

6.3.3.3 Solibri

Solibri is software provided by Nemetschek. Solibri enables BIM managers, engineers, designers, and other interested parties to collaborate and solve problems by combining models from different disciplines for advanced quality control and quality assurance tools. The program includes tools for validating BIM, checking compatibility, coordinating the design process, conducting design reviews, analyzing the designs, and checking codes. With Solibri, designers can improve BIM quality and optimize the entire design process. It is free for site managers, subcontractors, and other interested parties to use, and the information they need can be viewed anywhere, anytime (Solibri, 2022). The system can combine and view multiple models by IFC, use markups, dimensions, classifications, create and comment on issues, and extract quantity take-offs.

6.3.3.4 iTWO costX

iTWO costX is an RIB Software International tool that includes fully integrated 2D and 3D estimation software. It enables the integration of 2D and 3D takeoffs, estimating, and customized reporting all in one application. This software helps to unify disparate processes, increase transparency, and improve communication for smooth performance. With iTWO costX, users can analyze and takeoff data from 3D or BIM models, including IFC, as well as support for 5D BIM with live links to rate libraries and workbooks. A range of drawing files and external rate information can also be imported and exported by iTWO costX, bringing interoperability to a whole new level (iTWO costX, 2022). It supports all major 3D design software, provides advanced viewing tools for 2D and 3D, enables users to define properties, and generates cost plan reports on 2D and 3D models.

6.3.4   BIM 4D/5D Software

6.3.4.1 Vico Office

Vico Office is a software that was acquired by Trimble Solutions Corp. The Vico Office module adds constructability analysis and coordination, quantity takeoffs, 4D production control, and 5D estimating to the basic 3D model. Vico Office provides a BIM workflow for construction projects that are fully integrated. Vico Office can be used for cost studies and planning purposes in 2D and 3D models. The software checks and compares 2D drawings and 3D models and automatically processes changes if desired. Vico Office provides the ability to manage the partial or entire project and edit user privileges. BIM models created in different BIM software and in different formats can be combined with Vico office (Vico Office, 2022). Some features are location-based planning, 5D financial management, location-based calculations, risk management with BIM, risk analysis, quantities, and schedules, and change management.

6.3.4.2 Synchro Pro

Synchro Pro is a simple 4D BIM virtual design and construction software platform powered by Bentley. It enables design teams to increase the quality, safety, productivity, and efficiency of projects.

One of Synchro's popular solutions, using real-time visualization, Synchro Pro provides greater insight and control over projects throughout all phases of development while enhancing quality, security, and productivity from BIM (Bentley, 2022). Some basic features are progress monitoring, taking advantage of existing technology, alternative sequences testing for lower cost, CPM-based planning, custom reporting, resource management and reporting, viewing different files in open viewer, and file storage with workgroup project.

6.3.4.3 MTWO for iTWO 4.0

MTWO is a cloud-based 5D BIM construction management solution provided by RIB Software International for connecting all project contributors, processes, and data. By utilizing artificial intelligence in the cloud, MTWO assists with virtual to physical construction planning. Contractors, developers, and owners looking to accelerate their digital transformation can take advantage of MTWO's 5D BIM construction management solution by integrating RIB's iTWO 4.0 solution with Microsoft Azure. With MTWO, project participants collaborate and streamline workflow, improve communication, and increase productivity. As the building and construction industry moves into Industry 4.0, iTWO 4.0 provides a cloud-based 5D BIM enterprise platform that redefines management. iTWO 4.0 provides enterprise-wide digital project management throughout the entire construction lifecycle by integrating cloud computing, 5D BIM, big data, and modular construction (MTWO Cloud, 2022). BIM model management, quantity takeoff and cost estimating, visualized schedule and 5D BIM simulation, bid & tender management, procurement management, facility management, building lifecycle, and document and data management are some of the generic features.

6.4 BIM APPLICATIONS IN GEOTECHNICS AND INFRASTRUCTURES

The usage of BIM in Architecture-Engineering-Construction (AEC) has recently received support from both public and private entities. Governments provide specific directives for the application of BIM in this regard, primarily in public projects. To take advantage of BIM's benefits, such as improved project quality and efficiency, the various governments and public institutions have already begun to push for mandating its usage. Such BIM adoption requirements have proven successful in numerous large-scale building and transportation infrastructure projects (Yigiter, 2020). Successful BIM implementation trends have lately been put into practice in wealthy nations. Australia, Canada, the EU, Hong Kong, Singapore, South Korea, and the USA are among the nations that have developed the most effective implementation techniques (Bhatti et al., 2018). The data provided in Panuwatwanich et al. (2013) data indicates that 51% of construction companies are using BIM in their projects. In late 2008, the BIM council was established in Canada (Jin et al., 2015). In 2014, the EU parliament supported BIM, and nations like the UK, Denmark, and Finland encouraged its usage in major projects (Kolaric et al., 2016). The German government announced that BIM would be a requirement for all transportation projects until 2020. (Cobuilder, 2017). For the plan submissions of all projects in Singapore with a floor area of more than 5000 square meters, BIM was required. The federal government of the United States began a five-year campaign to promote the adoption of BIM (Jin et al., 2015). By requiring BIM use in all public construction projects, the South Korean government promoted the use of BIM in the construction industry (HM Government, 2012). Government regulations requiring the use of BIM in public sector projects, according to Smith (2014), could hasten the introduction of BIM and prevent waiting for client demands. Despite the fact that BIM is required in many nations, its use still faces difficulties in some nations where practitioners are hesitant to use it because the benefits of BIM are either poorly understood or because there is a lack of laws, guidelines, or legislation pertaining to its use. For instance, Turkey still lacks a national standard for the application of BIM. Even though BIM has been used in numerous projects thus far, there are still not enough standards or people who are aware of it. On the other hand, in 2014 the Turkish Ministry of Transport required the use of BIM in railway projects. As a result, there was a significant increase in BIM awareness and a decrease in change resistance (Toklu and Mayuk, 2020). The Mecidiyekoy-Mahmutbey Metro Line, which was constructed between 2014 and 2018, has a total construction area of 27,400 m², and is one of these projects that was first developed with BIM (Acar, 2019). Construction of homes, infrastructure, businesses, and airports are just a few examples of the project types that continue this trend that began with railroad projects. According to Jiang et al. (2022), mandates can be effective motivators for promoting BIM implementation. Governments have a crucial role in the adoption of BIM. Given this context, it is clear that recommending rules, guidelines, and mandates can encourage and accelerate the use of BIM.

BIM finds applications in a variety of projects allowing users a different set of features and dimensions. In addition to its standard applications in buildings structures, BIM has been gaining momentum recently in the implementation for geotechnical engineering and infrastructures applications. Because almost every structure is in contact with soil and gets support from the ground against the static and dynamic loadings, geotechnical applications like foundations, excavations, retaining walls are an integral part of the complete project process. Indeed, geotechnical considerations for any structure start from the planning stage and continue throughout the design, construction, monitoring, and life cycle of the structure. A complete cycle of geotechnical assessment involves: i) field investigations with laboratory and field experiments to characterize the soil properties and design parameters; ii) design for short term and long term according to codes and regulations supported by analysis including empirical, theoretical, and numerical modeling approaches; iii) the construction of geotechnical elements (e.g., piles) and system (e.g. whole foundation) according to design and specifications complying with quality control and assurance steps; iv) monitoring during the geotechnical structure and upper structure construction as well as during the lifetime of the structures against future hazards; and v) modifications as required by future demands caused by different factors such as nearby construction, natural hazards, climate change, etc. Integration of the whole process with BIM provides flexibility and control not only during the construction period but for the lifetime of the structures.

There are several recent studies about the use of BIM for protection against geotechnical and construction hazards. For example, Wang et al. (2015) discussed the use of BIM in terms of identifying fall and cave-in hazards related to excavation pits and promoting the use of required fall protection equipment. In this respect, they used 3D range point clouds from the excavated pits to measure the geometric properties of the pits. Then, an algorithm was developed to extract height information automatically to locate fall hazards. This integrated approach resulted in a BIM including the installation of safety equipment. They concluded that this informative model has the potential to assist project managers, safety experts, and workers to recognize hazards in the workplace and develop mitigation strategies accordingly. They further mentioned that this information model has the potential to increase communication among project participants.

Geographical information systems (GIS) have been used extensively in infrastructure construction, management and assessments. Figure 6.5 shows some examples from different cities in Los Angeles, U.S.A., and Christchurch, New Zealand. Toprak et al. (2018) illustrated and presented a comparison of horizontal ground displacements in the Avonside area, Christchurch from air photos, LiDAR, and satellite measurements regarding pipeline damage assessment. Data about the new or existing components on an existing system in GIS can be added by locating them using the global positioning system (GPS). Costin et al. (2018) show how integrating GPS data into geographic information systems (GIS) increases the quality of communication and decision-making. Furthermore, Costin et al. (2018) state that combining GIS and BIM capabilities is very efficient to control and monitor roadside utilities in ground transportation infrastructure including geotechnical, structural, and drainage data. Indeed, BIM offers a cooperative environment, where the accuracy of data is enhanced through better visualization and communication. Wu et al. (2021) explained that BIM is an effective way to accelerate the informatization of geotechnical engineering. They further implied that Industry Foundation Classes (IFC) in BIM are a good source for the exchange of information.

FIGURE 6.5   GIS data of water pipeline systems for a) Los Angeles, USA (Toprak, 1998), b) Christchurch, New Zealand (Toprak et al. 2019).

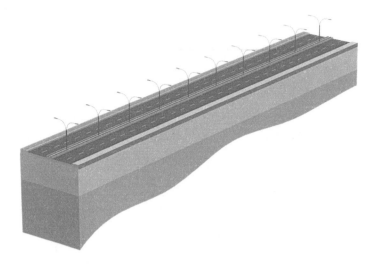

FIGURE 6.6   Conceptual ground model.

BIM was used in Geological Survey Organizations (GSO) in order to have 3D geological models accessible to the civil engineering community (Kessler et al., 2015). This way efficiency gains are possible and risk reduction is achieved. In their study, they presented an example project, which is the 3D modeling project of a 28km of railway line between Leeds and York (Burke et al., 2015). The final ground model (Figure 6.6) aimed to reveal areas needing ground investigations in terms of early assessment of the design of deep or shallow foundations for route electrification. According to Figure 6.6, it was detected that the client was able to integrate the conceptual ground model within their in-house workflow, which helped to overcome certain challenges. Considering the unpredictable ground conditions, the BIM integration in geotechnical workflows is essential to avoid delays (Kessler et al., 2015).

BIM has already been applied to numerous projects, including tunnel construction. Construction of tunnels, in particular, differs greatly from those of buildings and above-ground civil infrastructure projects. BIM has been employed as a useful method to efficiently handle data information since tunnel projects are complicated. However, drill and blast tunneling has not yet fully embraced BIM. By merging interconnected data models and performance data for drill and blast tunnel construction, Sharafat et al. (2021) created a BIM-based multimodal tunnel information modeling (TIM) framework to address this issue and improve project management, construction, and delivery performance. BIM's visualization capabilities can also be used in infrastructure projects to quickly uncover complicated relationships. IFC is flexible and simple to extend, and it is based on an object-oriented data model. IFC allows for the hierarchical modeling of things in terms of spatial areas (such as building floors) before other elements are connected to the spatial objects. Although IFC was initially created for modeling buildings, its use has recently been expanded to a number of civil engineering fields, including bridges and highways (Koch et al., 2017). Shield tunnel modeling has already been used in a number of studies using the IFC-based multi-scale product model (Amann et al., 2013; Yabuki et al., 2013; Borrmann et al., 2015). These models allowed the modeling of multiple geometric spaces to characterize the inside of a tunnel in a hierarchical fashion and supplied a specific number of new IFC classes for describing tunnels and their alignment. Koch et al. (2017) created a framework for tunnel information modeling to enable management, simulation, and visualization in mechanized tunneling projects in light of the significant role that BIM plays in tunnel projects. They also provided two case studies in which the framework was effectively used.

In another study, Lee et al. (2021) explored the application of geotechnical building information modeling in Malaysia. In this respect, they presented a set of case studies using BIM with

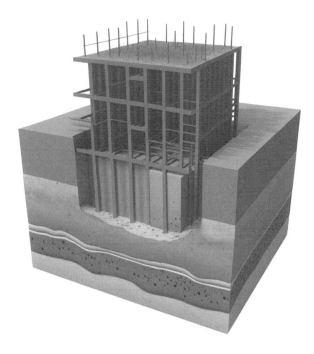

FIGURE 6.7 A representative 3D subsurface model based on geotechnical investigation data.

geotechnical data. Figure 6.7 presents a 3D model that was created incorporating geotechnical investigation data. The modeling helped to identify site areas in terms of containing different materials and foresee potential challenges.

The use of BIM in geotechnical engineering is growing but there still exist some challenges. On the other hand, digitization of the data contributes to easy workflow and better visualization. Therefore, it is foreseen that the use of BIM in geotechnical engineering and infrastructure applications will be more common soon and the benefits will be clearer. Vanicek et al. (2021) emphasized the utilization of the BIM model during lifetime structure expectancy which brings a huge potential for future applications in geotechnics and infrastructures.

6.5 FUTURE TRENDS

This chapter aimed at showing the use of BIM in geotechnical engineering and infrastructure applications. In this context, the purpose and benefits of BIM were presented. Then, commonly used BIM software is presented providing a comparative list of features and capabilities. The presented software was further classified based on the purpose category like collaboration, coordination, and so on. Then, the uses of BIM in geotechnical engineering and infrastructure applications were given based on the outputs of recent research. Some case studies were presented being part of these implementations. Given this background, one can conclude that BIM use in geotechnical engineering and infrastructure applications is critical and certain benefits can be achieved such as time and cost savings along with better coordination and collaboration. Moreover, it is considered that better visualization with BIM has the potential to avoid safety risks.

Recently, BIM has been fueled by several auxiliary components like GIS. It was suggested that combining GIS and BIM capabilities can inspire fresh ideas for infrastructure management while also boosting effectiveness and performance (Costin et al., 2018). Numerous academics underlined the importance of using geotechnical, structural, and drainage data to monitor and control roadside utilities in ground transportation in order to improve the planning of pipeline, port, and airport infrastructure (Bradley et al., 2016; Chong et al., 2016; Aziz et al., 2017). In a different study, Kim et al. (2015) noted that the integration of BIM and GIS permitted a quicker and more accurate calculation of cut and fill procedures. Kurwi et al. (2017) investigated the advantages of combining GIS and BIM for railways and concluded that doing so could improve decision-making and integrity during the building process. Given this context, integrating BIM with additional tools results in significant benefits.

Utilizing Lean construction methods and techniques in conjunction with BIM implementation is another trend to increase the potential of BIM. In several studies, the usage of BIM was cited as a means of decreasing waste and fostering both Lean and green practices (Arayici et al., 2011; Ahuja et al., 2017). Production is a significant aspect in the connection between Lean and BIM, Bayhan et al. (2022) noted, highlighting the strong synergy between the two disciplines. As part of lean construction, BIM enables improved information flow among a group of project participants, improving real-time management (Al Hattab and Hamzeh, 2015). Lean construction and BIM integration also make design and construction cooperation simple, encourage end-user participation, and optimize the entire system (Alarcon et al., 2013; Tauriainen et al., 2016). Additionally, some Lean techniques like significant prefabrication, just-in-time (JIT) delivery, last-planner-system (LPS), and push/pull planning are supported by BIM (Hamdi and Leite, 2012). Dong et al. (2013) suggested that incorporating BIM and Lean into supply chain management (SCM) has significant advantages, including proactive periodic reporting, real-time quality checks, time savings, and accurate production data. Given these advantages, it is clear that Lean and BIM work well together to facilitate construction management procedures and streamline the construction process.

REFERENCES

Acar, B. (2019). Implementation problems of building information modeling: Case study of Istanbul Airport project, Master's Thesis, MSGÜ, Institute of Science, Istanbul.

Ahuja, R., Sawhney, A., and Arif, M. (2017). Driving lean and green project outcomes using BIM: A qualitative comparative analysis." *International Journal of Sustainable Built Environment, 6*(1), 69–80. https://doi.org/10.1016/j.ijsbe.2016.10.006.

Aibinu, A., and Venkatesh, S. (2014). Status of BIM adoption and the BIM experience of cost consultants in Australia. *Journal of Professional Issues in Engineering Education and Practice, 140*(3), 04013021.

Al Hattab, M., and Hamzeh, F. (2015). Using social network theory and simulation to compare traditional versus BIM–Lean practice for design error management. *Automation in Construction*, 52, 59–69. https://doi.org/10.1016/j.autcon.2015.02.014.

Alarcon, L. F., Mesa, H., and Howell, G. (2013). Characterization of Lean Project Delivery. *Proc. 21st Ann. Conf. of the Int'l Group for Lean Construction.*

Allplan (2022). Allplan AEC Buildability at Its Best. Retrieved from https://www.allplan.com/us_en/products/allplan-aec

Allplan (2022). Allplan Bimplus. https://www.allplan.com/us_en/products/allplan-bimplus/

Amann, J., Borrmann, A., Hegemann, F., Jubierre, J. R., Flurl, M., Koch, C., and König, M. (2013). A refined product model for shield tunnels based on a generalized approach for alignment representation. Proc. of the ICCBEI.

Arayici, Y., Coates, P., Koskela, L., Kagioglou, M., Usher, C., and O'Reilly, K. (2011). Technology adoption in the BIM implementation for lean architectural practice. *Automation in Construction, 20*(2): 189–195. https://doi.org/10.1016/j.autcon.2010.09.016.

Arnal, P. I. (2018). Why don't we start at the beginning? The Basics of a Project: Lean Planning and Pre-Construction. Retrieved from https://www.bimcommunity.com/news/load/490/why-don-t-we-start-at-the-beginning on September 5, 2022.

Autodesk (2022). Design and build with BIM.

Autodesk (2022a). Revit: BIM software for designers, builders, and doers. https://www.autodesk.com/products/revit/overview?term=1-YEAR&tab=subscription

Autodesk (2022b). BIM 360. https://www.autodesk.com/bim-360/

Autodesk (2022c). Features of Navisworks 2023. https://www.autodesk.com/products/navisworks/features.

Autodesk (2023). What is BIM. https://www.autodesk.com/industry/aec/bim.

Aziz, Z., Riaz, Z., and Arslan, M. (2017). Leveraging BIM and big data to deliver well maintained highways. *Facilities*, *35*(13/14), 818–832.

Barlish, K., and Sullivan, K. (2012). How to measure the benefits of BIM—A case study approach. *Automation in Construction*, *24*, 149–159.

Bayhan, H. G., Demirkesen, S., Zhang, C., and Tezel, A. (2022). A lean construction and BIM interaction model for the construction industry. *Production Planning & Control*. https://doi.org/10.1080/09537287.2021.2019342.

Bentley (2022). https://www.bentley.com/en/products/brands/synchro

Bexel Manager (2022). https://bexelmanager.com/software/

Bhatti, I. A., Abdullah, A. H., Nagapan, S., Bhatti, N. B., Sohu, S., and Jhatial, A. A. (2018). Implementation of building information modeling (BIM) in Pakistan construction industry. Engineering *Technology & Applied Science Research*, *8*(4), 3199–3202.

BibLus (2022). BIM Maturity Levels: BIM Level 3. Retrieved from https://biblus.accasoftware.com/en/bim-maturity-levels-bim-level-3/ on September 25, 2022.

Borrmann, A., Kolbe, T. H., Donaubauer, A., Steuer, H., Jubierre, J. R., and Flurl, M. (2015). Multi-scale geometric-semantic modeling of shield tunnels for GIS and BIM applications. *Computer-Aided Civil and Infrastructure Engineering*, *30*(4), 263–281.

Bradley, A., Li, H., Lark, R., and Dunn, S. (2016). BIM for infrastructure: An overall review and constructor perspective. *Automation in Construction*, *71*, 139–152.

Bryde, D., Broquetas, M., and Volm, J. M. (2013). The project benefits of building information modeling (BIM). *International Journal of Project Management*, *31*(7), 971–980.

Burke, H. F., Hughes, L., Wakefield, O. J. W., Entwisle, D. C., Waters, C. N., Myers, A., ... & Horabin, C. (2015). A 3D geological model for B90745 North Trans Pennine Electrification East between Leeds and York.

Chan, A. P., and Chan, A. P. (2004). Key performance indicators for measuring construction success. *Benchmarking: An International Journal*, *11*(2), 203–221. https://doi.org/10.1108/14635770410532624.

Chau, K. W., Anson, M., and Zhang, J. P. (2004). Four-dimensional visualization of construction scheduling and site utilization. *Journal of Construction Engineering and Management*, *130*(4), 598–606.

Chong, H. Y., Lopez, R., Wang, J., Wang, X., and Zhao, Z. (2016). Comparative analysis on the adoption and use of BIM in road infrastructure projects. *Journal of Management in Engineering*, *32*(6), 05016021.

Cobuilder (2017). BIM in Germany. Retrieved from: https://cobuilder.com/en/bim-in-germany/.

Costin, A., Adibfar, A., Hu, H., and Chen, S. S. (2018). Building information modeling (BIM) for transportation infrastructure–Literature review, applications, challenges, and recommendations. *Automation in Construction*, *94*, 257–281.

Department of Business, Innovation, and Skills (2011). A report for the Government Construction Client Group Building Information Modeling (BIM) Working Party Strategy Paper. Retrieved from https://www.cdbb.cam.ac.uk/system/files/documents/BISBIMstrategyReport.pdf

Dong, N. T., Khanzode, A., & Lindberg, H. (2013). Applying lean principles, BIM, and quality control to a construction supply chain management system. In *Proceedings of the 30th CIB W78 International Conference*, 535–544.

Eadie, R., Browne, M., Odeyinka, H., McKeown, C., and McNiff, S. (2015), "A survey of current status of and perceived changes required for BIM adoption in the UK", *Built Environment Project and Asset Management*, *5*(1), 4–21. https://doi.org/10.1108/BEPAM-07-2013-0023

Eastman, C. M., Eastman, C., Teicholz, P., Sacks, R., and Liston, K. (2011). *BIM handbook: A guide to building information modeling for owners, managers, designers, engineers and contractors.* John Wiley & Sons.

Fahmy, A., Hassan, T., Bassioni, H., and McCaffer, R. (2020). Dynamic scheduling model for the construction industry. *Built Environment Project and Asset Management*.

Financesonline (2022a). ARCHICAD Review. Retrieved from https://reviews.financesonline.com/p/archicad/ on September 1, 2022.

Financesonline (2022b). Autodesk BIM 360 Review. https://reviews.financesonline.com/p/autodesk-bim-360/

Financesonline (2022c). BIMx Review. https://reviews.financesonline.com/p/bimx/

Financesonline (2022d). Navisworks Review. Retrieved from https://reviews.financesonline.com/p/navisworks/.

Financesonline (2022e). https://reviews.financesonline.com/p/bexel-manager/

Google Trends (2022). Interest on BIM software over the years. https://trends.google.com/trends/explore?date=today%205-y&q=%2Fm%2F0260j49,%2Fm%2F0c0tdp,%2Fg%2F1211j65g,%2Fm%2F0127x3j4#TIMESERIES, (Erişim Tarihi: 10/02/2021).

Graphisoft (2022a). Archicad – BIM by architects for architects.

Graphisoft (2022b). BIMx-Explore, engage, mobilize. https://graphisoft.com/solutions/bimx

Hamdi, O., and Leite, F. (2012). BIM and Lean Interactions from the Bim Capability Maturity Model Perspective: A Case Study. *IGLC 2012-20th Conference of the International Group for Lean Construction*. The International Group for Lean Construction.

HM Government (2012). *Building information modeling - industrial strategy: Government and industry in partnership*. HM Government, London, UK.

iTWO costX (2022). https://www.itwocostx.com/costx/products/

Jiang, R., Wu, C., Lei, X., Shemery, A., Hampson, K. D., and Wu, P. (2022). Government efforts and roadmaps for building information modeling implementation: lessons from Singapore, the UK and the US. *Engineering, Construction and Architectural Management*.

Jin, R., Tang, L., and Fang, K. (2015). Investigation into the current stage of BIM application in China's AEC industries. *WIT Transactions on The Built Environment, 149*, 493–503.

Kessler, H., Wood, B., Morin, G., Gakis, A., McArdle, G., Dabson, O., …, and Dearden, R. (2015). Building Information Modeling (BIM): A route for geological models to have real world impact.

Khemlani, L. (2009). Visual estimating-extending BIM to construction: AECbytes. *Building the Future "Article." AECbytes: Analysis, Research, and Reviews of AEC Technology. Web, 21.*

Kim, H., Chen, Z., Cho, C. S., Moon, H., Ju, K., and Choi, W. (2015). Integration of BIM and GIS: Highway cut and fill earthwork balancing. In *Computing in Civil Engineering 2015* (pp. 468–474).

Koch, C., Vonthron, A., and König, M. (2017). A tunnel information modeling framework to support management, simulations and visualisations in mechanised tunnelling projects. *Automation in Construction, 83*, 78–90.

Kolaric, S., Vukomanovic, M., Radujkovic, M., and Pavlovic, D. (2016). Perception of building information modeling within the Croatian construction industry. In *International Scientific Conference People, Buildings and Environment 2016 (PBE2016)* (Vol. 29).

Kurwi, S., Demian, P., and Hassan, T. M. (2017). Integrating BIM and GIS in Railway Projects: A Critical Review In: Chan, P W and Neilson, C J (Eds.). *Proceeding of the 33rd Annual ARCOM Conference, 4-6 September 2017*, Cambridge, UK, Association of Researchers in Construction Management, 45–53.

Lee, A., Wu, S., Marshall-Ponting, A. J., Aouad, G., Cooper, R., Tah, J. H. M., and Barrett, P. S. (2005). nD modeling road map: A vision for nD-Enabled construction.

Lee, M. L., Lee, Y. L., Goh, S. L., Koo, C. H., Lau, S. H., and Chong, S. Y. (2021). Case Studies and Challenges of Implementing Geotechnical Building Information Modelling in Malaysia. *Infrastructures, 6*(10), 145.

Li, H., Zhang, C., Song, S., Demirkesen, S., and Chang, R. (2020). Improving tolerance control on modular construction project with 3D laser scanning and BIM: A case study of removable floodwall project. *Applied Science, 10*, 8680. https://doi.org/10.3390/app10238680

Lu, W., Zhang, D., and Rowlinson, S. M. (2013). BIM collaboration: A conceptual model and its characteristics. In *Proceedings of the 29th Annual Association of Researchers in Construction Management (ARCOM) Conference*. Association of Researchers in Construction Management.

Masood, R., Kharal, M. K. N., and Nasir, A. R. (2014). Is BIM adoption advantageous for construction industry of Pakistan? *Procedia Engineering, 77*, 229–238.

McCuen, T. L. (2008). Scheduling, estimating, and BIM: A profitable combination. *AACE International Transactions*, BIM11.

Merschbrock, C. (2012). Unorchestrated symphony: The case of inter-organizational collaboration in digital construction design. *Journal of Information Technology in Construction (ITcon), 17*(22), 333–350.

MTWO Cloud (2022). https://www.mtwocloud.com/features-overview.html

Neff, G., Fiore-Silvast, B., and Dossick, C. S. (2010). A case study of the failure of digital communication to cross knowledge boundaries in virtual construction. *Information, Communication & Society, 13*(4), 556–573.

NIBS, N., (2007), "United States national building information modeling standard version 1—Part 1: Overview, principles, and methodologies," NIBS und buildingSMART alliance Final Report. https://buildinginformationmanagement.files.wordpress.com/2011/06/nbimsv1_p1.pdf

Panuwatwanich, K., Wong, M. L., Doh, J. H., Stewart, R. A., and McCarthy, T. J. (2013). Integrating building information modeling (BIM) into Engineering education: an exploratory study of industry perceptions using social network data.

Retrieved from https://graphisoft.com/solutions/products/archicad on September 3rd, 2022.

Sabol, L. (2008). Challenges in cost estimating with building information modeling. *IFMA World Workplace, 1*, 1–16.

Saxon, R. (2018). Getting the dimensions of BIM into focus. Retrieved from https://www.bimplus.co.uk/getting-dimensions-bim-focus/ on September 6, 2022.

Sharafat, A., Khan, M. S., Latif, K., and Seo, J. (2021). BIM-based tunnel information modeling framework for visualization, management, and simulation of drill-and-blast tunneling projects. *Journal of Computing in Civil Engineering*, *35*(2), 4020068.

SketchUp (2022). 3D Modeling for Professionals? We've got you. https://www.sketchup.com/plans-and-pricing/sketchup-pro

Smith, P. (2014). BIM & the 5D project cost manager. *Procedia-Social and Behavioral Sciences*, *119*, 475–484.

Solibri (2022). https://www.solibri.com/our-offerings

Succar, B. (2009). Building information modeling framework: A research and delivery foundation for industry stakeholders. *Automation in Construction*, *18*(3), 357–375.

Succar, B. (2010, May). The five components of BIM performance measurement. In *CIB World Congress* (Vol. 14). Salford: United Kingdom.

Tauriainen, M., Marttinen, P., Dave, B., and Koskela, L. (2016). The effects of BIM and lean construction on design management practices. *Procedia Engineering, 164*, 567–574. https://doi.org/10.1016/j.proeng.2016.11.659.

Tekla (2022). Why Tekla Structures? Retrieved from https://www.tekla.com/products/tekla-structures on September 1, 2022.

Toklu, S., and Mayuk, S. G. (2020, October). The Implementation of Building Information Modeling (BIM) in Turkey. In *ICONARCH International Congress of Architecture and Planning* (pp. 87–100).

Toprak, S., Nacaroglu, E., and Koc, A. C. et al. (2018). Comparison of horizontal ground displacements in Avonside area, Christchurch from air photo, LiDAR and satellite measurements regarding pipeline damage assessment. *Bulletin of Earthquake Engineering, 16*, 4497–4514. https://doi.org/10.1007/s10518-018-0317-9.

Toprak, S., Nacaroglu, E., and Van Ballegooy, S. et al. (2019). Segmented pipeline damage predictions using liquefaction vulnerability parameters, *Soil Dynamics and Earthquake Engineering, 125*, 105758, ISSN 0267-7261, https://doi.org/10.1016/j.soildyn.2019.105758.

Toprak, S. 1998. *Earthquake effects on buried lifeline systems*, Ph.D. Thesis, Cornell University, Ithaca, NY, USA.

Trimble Connect (2022). Capabilities. https://connect.trimble.com/capabilities

United BIM (2022). BIM Level of Development | LOD 100, 200, 300, 350, 400, 500. Retrieved from https://www.united-bim.com/bim-level-of-development-lod-100-200-300-350-400-500/

Vanicek, I., Pruska, J., and Jirásko, D. (2021). BIM – An application in geotechnical engineering. *Acta Polytechnica CTU Proceedings*, *29*, 24–28, 47th Conference Foundation Engineering 2019. https://doi.org/10.14311/APP.2020.29.0025.

Vico Office (2022). https://www.itwocostx.com/costx/products/

Wang, J., Zhang, S., and Teizer, J. (2015). Geotechnical and safety protective equipment planning using range point cloud data and rule checking in building information modeling. *Automation in Construction*, *49*, 250–261.

Ward, S., & Chapman, C. (2008). Stakeholders and uncertainty management in projects. *Construction management and economics*, *26*(6), 563–577.

Wu, J., Chen, J., Chen, G., Wu, Z., Zhong, Y., Chen, B., …, and Huang, J. (2021). Development of data integration and sharing for geotechnical engineering information modeling based on IFC. *Advances in Civil Engineering*, *2021*, 1–15.

Yabuki, N., Aruga, T., and Furuya, H. (2013). Development and application of a product model for shield tunnels. In *ISARC. Proceedings of the International Symposium on Automation and Robotics in Construction* (Vol. 30, p. 1). IAARC Publications.

Yilmaz, S. (2021). Facilitating cost control practices through BIM applications, MSc Thesis, Gebze Technical University, Kocaeli.

Yigiter, F. (2020). An assessment of building information modeling (BIM) implementation for the Turkish transportation infrastructure industry, MSc Thesis, Istanbul Technical University, Istanbul).

Zhao, D., McCoy, A. P., Bulbul, T., Fiori, C., and Nikkhoo, P. (2015). Building collaborative construction skills through BIM-integrated learning environment. *International Journal of Construction Education and Research*, *11*(2), 97–120.

Zigurat Global Institute of Technology (2019). BIM software tools for all occasions. Retrieved from https://www.e-zigurat.com/blog/en/bim-software-tools-for-all-occasions/ on September 9, 2022.

7 Building Information Modelling Implementation in the UK Construction Industry
A Commercial Perspective

Daniel Davies and Serik Tokbolat

CONTENTS

7.1 INTRODUCTION

This chapter delves into the commercial side of Building Information Modelling (BIM) in the United Kingdom, the technology is helping to drive the construction towards a leaner construction. Literature has been reviewed to establish the industries' adaptation to BIM, the improvements in efficiency of usage that the industry has and will continue to have to make to continue to implement the information model. The segments of the literature review looked into the increase of collaboration deriving from BIM, its effects on change management, contracts and cost reduction before finally delving into the current challenges being faced when trying to integrate BIM within the industry. The research strategy was set out via the methodology chapter, showcasing the research objectives, giving the reader an outlook on the overall purpose of the piece of research. Once this has been set out, analysis of the primary data has been undertaken, examining respondents' thoughts on the set-out questions, giving an up-to-date snapshot of the current state of BIM commercially within the industry. Finally, conclusions were made on the research objectives, reviewing possible

DOI: 10.1201/9781003325246-7

limitations and them ultimately giving recommendations on what should happen next regarding the push of BIM usage and possible further research.

7.2   LITERATURE REVIEW

This literature review looked into the commercially related aspects which BIM can possibly assist with, along with the current challenges, bringing the research to a place where further research is needed to progress the research further.

7.3   OVERVIEW OF BUILDING INFORMATION MODELLING

BIM has been construction's big step of the 21st century, looking to substantially improve the delivery of projects, as at the start of the century they were becoming substantially harder to manage (Alshawi and Ingirige, 2003). BIM is best described as a methodology that integrates building objects and their relationships to others in a precise manner, therefore enabling stakeholders within the contract the ability to put to query, simulate and estimate activities and their certain implications on the project delivery process (Amudu-Yusuf, 2018). BIM is put in place with interest to interlink all aspects of a project as a platform for integrated collaboration and communication through a single model, to try and deliver as lean a project as possible (Sacks et al., 2010).

Before delving into the different aspects of BIM, it would be wise to outline further what it is, and how it is used by the industry currently. BIM can be described as the process of both creating and managing the digital information throughout a construction project's life cycle, or otherwise the digital description of every aspect of the built asset within a contract. The model has potential use at all stages of a project life cycle: At the preconstruction phase, it can be utilised by the client to gain an understanding of project needs. Currently, it is used heavily by the design team to analyse, design and develop the project during the preconstruction phase. The main contractor and their supply chain should have usage during the construction phase and then lastly BIM can be utilised by the facility manager during the operation and decommissioning phases of a project (Grilo and Jardim-Goncalves, 2010).

To go into more detail on the construction phase, the aim should be to involve all parties from the subcontractor to the client in the BIM model; this will stimulate and improve the efficiency of the overall performance of a project through increased collaboration and communication, all in interest to achieve optimal results in a cost effective and timely manner, bringing stakeholders within a project together. This combines resources, abilities and knowledge to complete tasks that a lone party may find more challenging to undertake by themselves (Hughes, 2012). BIM also allows for the automation of many of the slower, more tedious tasks that are required within project management, reducing the number of repetitive tasks that have to be undertaken usually (Peterson et al., 2011).

Regarding the actual technology behind BIM, there are many companies who are trying to push the use of their programme. The current leader in terms of usage is AutoDesk's Collaborate Pro, a progression of the AutoDesk 360 software; this empowers teams from design to civil engineering with the ability to build and manage projects as they are happening. AutoDesk claims that "BIM Collaborate Pro is the only Civil 3D, Revit, and AutoCAD Plant 3D collaboration solution on the market that connects your teams, data, and workflows, and delivers powerful insights" (AutoDesk, 2021).

The UK government is a big advocator for the increased implementation of BIM, with the first major strategy being set out in 2011. This saw several "mobilisation and implementation" plans set out in the overall BIM strategy published by the Department for Business Innovation & Skills (CDBB, 2011). The strategy composed of a push element as well as a pull element. Push was defined as the element that looks at the supply chain and different processes by which the implementation and adaptation of BIM could be made easier and more viable for firms. Pull, the second element to the strategy, regards the client "pull" and delves into how a client within a project should be very specific and consistent about what it specifies in terms of the aspects of BIM. The strategy also introduced formal terms such as BIM Maturity Levels and BIM Deliverables. BIM Maturity Levels range from 0–3 currently, with plans for further levels in the future, with Level 0 being an unmanaged CAD that is basic, utilising simple processes and 2D drawings to Level 3, which is a much more extensive data exchange within a project, holding data from the operational, delivery and performance phases of a project (Government, 2015). With regard to BIM Deliverables, this refers to a list of BIM-related questions that have a specific answer (Deliverable) that is required to pass a phase of a project, along with what BIM Maturity Level a project is at in order to reach that deliverable, providing companies with a tool to measure how well they are adapting to BIM (CDBB, 2011). Since the BIS Strategy was released, the push of BIM usage across centrally procured public construction projects has been the most prevalent, along with even more long-term aims to open up the industry (small to medium-sized firms).

7.4 BUILDING INFORMATION MODELLING CURRENTLY IN THE UK

BIM in the UK has progressed massively in terms of usage since 2011, since the government initiative in 2011 mentioned above, the usage has increased from 13% to 73% as of 2020 (NBS, 2020). This proves that the increased encouragement of the use of BIM towards a leaner construction has been effective to date and going by year-to-year stats will furthermore become more prevalent within the industry. Along with the strategy set out by the UK government, there are now standards being put set out to provide people and organisations with a rulebook on the do's and don'ts for the tool (BSI, 2019). To give a brief outlook on the standards put in place, currently there are two International Standards in place, first drafted in 2017 and then published in January 2019 under the names BS EN ISO 19650-1 and BS EN ISO 19650-2. Part 1 outlines the principles, concepts and provides recommendations on how to manage building information (Seyedhabibollah, 2019). Part 2 supplies information management requirements in the delivery phase of assets (Winfield, 2020). Parts 3, 4 and 5 of the standards are also going through the same process as Parts 1 and 2 ISO 19650 went through in order to be developed into international standards. Part 3 provides specification management for the operational phase of assets using BIM, whilst Part 5 provides a security minded approach to information management, both are due in Q4 and Q2 of 2021, retrospectively. Part 4 is about fulfilling employer's information exchange requirements using the Construction Operations Building Information Exchange (COBie); this part is currently in the early stages of development. From all of the above, it is wise to assume that the UK industry is now beginning to adapt well to BIM but also there is still much work to be done.

7.5 COMMERCIALLY RELATED ASPECTS OF BUILDING INFORMATION MODELLING

7.5.1 COLLABORATION AND COMMUNICATION

A lack of collaboration in construction projects often results in misunderstandings, misinterpretations of data and poor communication, consequently resulting in increased rework (Oraee et al., 2019). An improvement of collaboration deriving from BIM is a necessary change to the old traditional way in which a project is delivered, comprised of a set of related linear work activities that are divided between functional specialties within design, management and construction. This dated process of running a project is what can possibly be described as an "over the wall" approach to information (Matthews et al., 2018). This can also be described as an "unstructured stream of text or graphic entities" (BSI, 2010). Not only is this attitude towards information inefficient but it also derives issues with communication between disciplines and project teams; a commercial example of a breakdown of communication is valuations of items not being agreed on, being sent back and forth between quantity surveyors of both the client and main contractor, rather than working together on the valuation until an agreement is made. If there were collaboration between these parties enabled by BIM, then the completion of work would be much more efficient due to the push to work with each other, rather than against. One of the staples of BIM is the improved ability to share information with parties integrated within the BIM of a specific project, this naturally will encourage collaboration as the communication of information becomes significantly more efficient. The increased integration of teams is seen as something that that stimulates innovations inside of a collaborative environment (Wamelink, 2014). Along with this, the increase in communication is seen to increase the trust between parties as they will be a lot more familiar with the way each other work, again increasing efficiency of teams as it will be known which tasks can be done by oneself and which will need a collaborative integrated approach.

This force comes from having the centralised source of project specific information in BIM, derives from responsibilities which are set out in the Stage 3 BIM Responsibility Matrix – RACI (Responsible, Authorising, Contributing, Informed) set out in the ISO 19650-1 mentioned previously. This sets out who is responsible for management and updating of information within a model, giving more of an impetus for an individual to be collaborative and complete work to the highest standard possible (Kassem et al., 2014).

7.5.1.1   Change Management and Disputes

Due to the complexity of construction and civil engineering projects, change is inevitable. From this often disputes may arise which are one of the main factors which prevent the successful and efficient completion of a project (Cakmak and Cakmak, 2014). To obtain an understanding of the impact of BIM on change, it is important to be aware of the causes and how BIM helps to reduce or eliminate them. As change is inevitable, it is how it is managed that is most important. The management focuses on adapting and controlling the changing factors within a project. BIM supports teams and individuals with change when fully integrated with a project, enabling foreseeing change before it happens, for example with clash detection, which finds design issues that would have previously been unforeseen and resultantly would have had to be solved on-site, resulting in increased cost from delays, rework and potential redesign (Akponeware and Adamu, 2017). Also, as mentioned above, BIM's collaborative platform will encourage commercial teams to work together. This will naturally encourage teams to work together when realising change, also massively reducing the probability of disputes as stakeholders within a contract will more often be working together than against each other; this is especially true with Integrated BIM Contracts (Manderson et al., 2015).

Disputes arising often means that conflict can occur, BIM can result in reduced conflict and overall project team benefits (Ghaffarianhoseini et al., 2017). From this, it is feasible to see that even if a significant change does arise, disputes that may arise can be avoided; this goes back to technology being a collaborative tool. The reduce in change orders as well as disputes comes down to too many BIM-related processes, one being the decision support benefits that it provides, enabling simplified knowledge management. This means that the project-related data is continuously and progressively collated, stored and maintained throughout the building life cycle, streamlining, tracking and evaluating project details, giving less room for inter-project confrontation. This multidisciplinary integration between teams allows for the identification and resolution of issues during the pre-construction phase; the effects of this mean that conflicts and disputes are kept to a minimum, including other factors such as reducing cost and programme impacts by reducing errors along with requests for information which in turn reduces reworks (Demian and Walters, 2014).

7.5.1.2   BIM Integration with Contracts

For BIM to continue to become more prevalent within the industry, standardised contracts will have to continue to be adapted to allow for the use of a fully collaborative BIM. There has been much progress in recent years to integrate BIM into contracts since the Construction Industry Council issued their first BIM Protocol in 2013 (CIC, 2013), stating that BIM were to be incorporated into contracts of all those imputing to BIM. New Engineering Contract's (NEC3) approach to BIM was to provide general guidance on BIM using technical content to form part of Works Information and Parties Rights and Responsibilities in the form of Z clauses, which are clauses used to amend NEC contracts, adding terms and conditions (Elliott, 2013). This brings to light that BIM is not incorporated in the standard version of the NEC3 contract, meaning further adaptation will be needed down the line to fully adapt contracts to BIM. Although with this being said, NEC have been credited with proficient use of BIM terminology, with it being claimed that this is something that other contracts have yet to implement (Kirby-Turner, 2021). Differently to NEC's BIM Guidance, the Joint Contracts Tribunal (JCT) public sector supplement includes schedules of modifications for all of the JCT main contracts (Pantry, 2020), which also includes forms of contracts used or maintenance works. JCT themselves as an entity believe that the majority of benefits from BIM will be accrued in the design phase of a project, but also see that this may not be the case if "as built" data within the Information Model is meaningful and accurate, also enabling for further maintenance (Joyce and Houghton, 2014).

Further to the NEC and JCT contracts, the Partnering Contract for a single Project (PPC2000) contract was chosen as the contract by the Ministry of Justice to Support to be used on the Cookham Wood Young Offenders Institution Trial Project, the first of the UK government's projects to test the application and benefits of BIM Level 2, which is the level that is distinguished by collaborative working and requires "an information exchange process which is specific to that project and coordinated between various systems and project participants" (Ganah and John, 2014). BIM Level 2 is the highest current level of the model which has been fully defined by the government, so it is only possible for current contracts to adapt to this level of the technology. For this project, the PPC2000 Contract was used with no amendments in respect to BIM, the contract is a multi-party contract which is used with interest to improve collaboration. Through the contract dates were agreed between all parties for integrating deadlines together, it was stated that having this type of contract provided a bolder approach to the implementation of BIM as it didn't limit each party's obligations (Mosey, 2014).

7.5.1.3 Cost Reduction and Analysis

As BIM is a tool that looks to improve efficiency, working towards a leaner construction; one of its main project-based benefits is being able to reduce costs. A report that further backs up this point stated that out of 35 projects, 21 of them claimed BIM to be a successful cost reduction tool with 2 stating negative effects financially (Bryde et al., 2013). Another more recent study also reiterates the point of cost reduction stating that the tool reduces the capital costs of construction and whole life cost of built assets by 33%, a figure that seems a bit too impressive to believe this early in the technology's maturity, but still shows that cost reduction is definitely a major advantage of BIM (Georgiadou, 2019). Another way in which BIM allows for cost reduction is due to the enabling of cost modelling, which is defined as "techniques used for forecasting the estimated cost of a proposed construction project". This more accurate cost forecast from BIM means there will be much less unforeseen costs arising, allowing for project teams to analyse and manage expected costs and evaluate the impact of those outlays over time. Also adding to this, if a decision is made on a forecast which is wrong then it could also mean a cost increase; this in a long-term perspective will save costs as well (Rooke, 2021).

The forecasting aspect can be assisted via BIM by using technologies for the take-off of quantities, such as AutoDesk Revit, Vico and Assemble. These programmes have the intelligence to accelerate the speed of take-off, along with increasing the accuracy of the estimate; this saves costs both from a man-hour perspective and also that it offers more reliable figures for the commercial team to utilise (Olsen and Taylor, 2017).

When BIM is at a level where the model can assist with budgetary and cost considerations associated to a project it is considered to be a 5D BIM, this is the second highest BIM dimension (7D currently the highest) which means at that point the technology is very integrated with a project. 5D BIM allows for The New Rules of Measurement to be integrally linked with BIM, meaning there is a consistent approach to estimating and cost planning, again improving efficiency (Smith, 2016). Further to the take-off benefits mentioned above, there are other possibilities that can come with a 5D BIM including analysing cash flows through the use of automated processes and producing operational and lifecycle costs which can be analysed before a building is even built (Mayouf et al., 2019).

7.5.1.4   Current Limitations to BIM Commercially

Although BIM is seen as the tool which will ultimately continue to be used more and more through-out the industry, literature suggests that the model is still mainly designer-driven meaning that the commercial use of BIM may still be limited, meaning that it has not fully integrated with a good majority of stakeholders within a project (Li, 2020). Another issue with implementation that has become prevalent is conflicting expectations of what performance the model is supposed to be working at, this is due to a lack of awareness in the actual capabilities of BIM, with some simply unaware of the massive improvement in efficiency it can offer (Vass and Gustavsson, 2017). Another similar issue with implementation is that workers do not feel comfortable with having to change the way that they work (cultural barrier); this means that the specific attributes of some roles needed are not there (Sebastian, 2011).

Due to the dynamic of a construction project, subcontractors will always be required for works that the main contractor does not specialise in. This can be a problem as usually subcontractors are smaller companies with a lot less capital to be able to invest in new technologies (Chan et al., 2019). This can result in barriers such as high set-up costs and also the subcontractors becoming being the weakest link, which means that poor performance by a subcontractor becomes a limiting force in a set of supply chain relationships (Robson et al., 2014).

There are also legal issues that come with BIM due to topics such as ownership issues regarding the BIM, this is seen as increased risk exposure which discourages companies. The legal issues in the majority come down to who will own the model after the completion of the works, the loss of control over the design also becomes a legal barrier to the implementation of BIM (Eadie et al., 2015)

7.5.2   METHODOLOGY

7.5.2.1   Introduction

A research methodology is the detailed description of what techniques/methods are going to be used to collect and analyse data in a research project. There are two types of research, firstly primary data, which is first-hand data that is collected by a researcher, adding to existing overall social knowledge. This data is then in circulation for use as research data that people can reuse within the general research community, therefore making it secondary data (Hox and Boeije, 2005).

7.5.3   DEFINITIONS

Research Methodology - When first looking into research methodology, it is important to know that the term refers to a lot more than the research methods which are adopted within a particular study. The research methodology can be described as what encompasses the rationale behind a piece of research, along with the philosophical assumptions which attribute to emphasising the study (Chynoweth, 2008). Another way in which a methodology can be defined is "a contextual framework" for research, a coherent and logical scheme based on views, beliefs and values, that guides the choices in which researchers make (Helen, 2015). From this it is clear that the researcher must have good understanding of the science in which research is obtained, allow-ing for the most suitable research methods within the methodology to be utilised. All of these choices around the suitable methods are in aim to answer the overall problem or question of the research wholly. There are a few methodologies that could be applicable to this research, which will mostly be analysis on a quantitative basis with some qualitative aspects. From research already undertaken by the researchers, a conclusion has been made that although not as com-mon with quantitative research as compared to qualitative, grounded research will be the most suitable. Systematically, it is expected to present the data collected in an efficient way; it can be described as the methodology that involves the construction of hypothesis and theories through the collecting and analysis of data (Chun Tie et al., 2019). This research methodology will be evident in the research design and strategy as this will show how the up-to-date theories on commercial BIM will be reached.

Research Method and Research Paradigm - Research methods rather than the broader methodology, are the actual methods in which data will be collected by the researcher (Kothari, 2004.) Upon the initiation of collecting data, it is vital for the researcher to understand which research paradigm the data lies in, this will in turn allow for easier data analysis, as there are methods attributed to each type of data paradigm. When collecting data under the grounded theory, it is a very efficient approach to involve existing literature in the post data collection discussion as it produces a richer data set, meaning a more reliable theory can be constructed by the researcher (Chynoweth, 2008). The methods of research chosen will involve both qualitative and quantitative data, these methods together can be described as a mixed method approach. The two different data types are different research approaches that sit in different paradigms, although they provide a good basis for discussion.

7.5.3.1   Quantitative Research

Quantitative research can be defined as social research which employs empirical methods and empirical statements. With empirical statements being defined as what is the case in the "real world" rather than what is ought to be the case (Sukamolson, 2007). These empirical statements come in the form of data that are numerical values; the numerical values create a data set that can then be quantified and analysed by the researcher (Punch, 2013). As the researcher is collecting quantitative data, it is said that the researcher is acting as a positivist, this in short means the researcher has to act as a realist, utilising the quantitative logic to come to conclusions on topics that are being researched. For this piece of research, the quantitative form of research that has been selected is the Likert scale which will be the major aspect of the survey that will be the main form of primary research, the scale will be discussed further in Section 7.5.4.6 Questionnaire: Adopted Method.

7.5.3.2   Qualitative Research

Qualitative research is more of a holistic approach in comparison to quantitative, involving discovery rather than the selection of premeditated answers. The process allows for the researcher to develop more a higher-level detail on answers from a high involvement from the respondent if they've had personal experience in a chosen topic (Williams, 2007). When collecting qualitative data, rather than the researcher operating in a positivism paradigm, they are acting in an interpretivist paradigm. This leaves room for more subjective answers from the respondents, along with subjective interpretation from the researcher (Fellows and Liu, 2015).

7.5.4   RESEARCH DESIGN

Research design as a term describes the process in which data will be collected, will aim to answer the research question proposed at the start of the piece of research (Bell et al., 2018). In order to be able to evaluate the found data and conclude on it, the design framework in question needs to be sophisticated enough as to where the majority of the data collected can be analysed in a way where it gives the best results for the piece of research (Chynoweth, 2008). If the research collected integrates well with the research done in the literature review, then they will complement each other well to form a high-quality discussion and therefore, good quality results.

7.5.4.1   Secondary Data

As Figure 7.1 shows, the research design is very linear, opposing for the use of triangulation and favouring to be more direct with the research. This linear approach allows for the questionnaire which will be directly based off the literature review, this allows for the data obtained to be updates with up-to-date theories on the state of BIM commercially. The secondary data will take up the

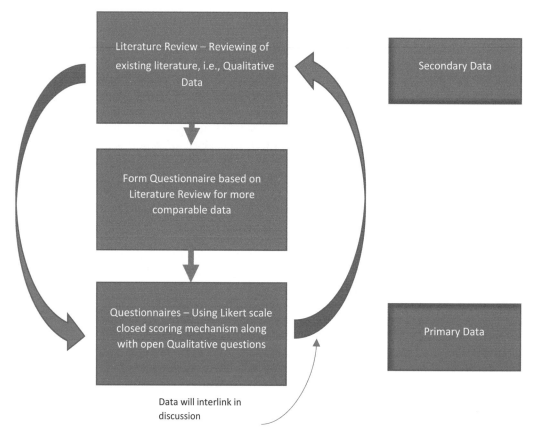

FIGURE 7.1   The adopted research design.

majority of the qualitative data with the primary data coming from the Likert Scale questionnaire; secondary data isn't collected directly by the researcher but rather gathered indirectly via the findings of others (literature review).

7.5.4.2   Primary Data

Figure 7.1 again shows how the primary data will be gathered through the questionnaire, with the questions being created from the literature review for more comparable data. With limited direct research on BIM commercially, overall the comparisons will have to be to the smaller subtopics of the literature review such as change orders, contracts and cost reduction.

7.5.4.3   Research Aim

Give an up-to-date commercially based perspective on the implementation of BIM within the construction industry and highlight the challenges that are currently being faced.

7.5.4.4   Research Objectives

The research objectives for this piece of research are as follows:

1. To develop a good understanding of the current state of BIM and then give an overview.
2. To review the selected commercial aspects of BIM, preparing sufficiently for the primary research.
3. To investigate the current status of BIM as a commercial tool.

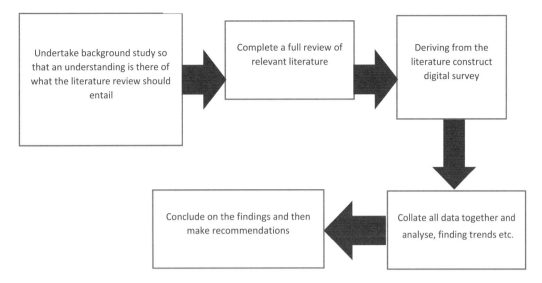

FIGURE 7.2 Adopted research strategy.

7.5.4.5 Adopted Research Strategy

Figure 7.2 gives a visual representation of the workflow required from start to finish so that the research objectives can be met, ranging from the literature review to collecting data all the way to evolving the raw data into finally coming to conclusions and recommendations:

7.5.4.6 Questionnaire – Adopted Method

Questionnaires have many beneficiaries for the researcher, the first being that the research method enables the mass gathering of research, allowing for a larger and more reliable possible pool of data that can possibly be collected (Bell et al., 2018). This will be useful when trying to analyse the tool in BIM that has such a large potential, hopefully producing up-to-date theories on its state and in turn giving advice to the rest of the industry. There are certain terms that give researchers or readers an idea of what type of document it is that is being used for a piece of research, open or closed. Open questionnaires are either interviews where answers are given verbally or open text box questions on digital surveys. Closed questions on the other hand are normally designed to be completed without any interaction with the researcher, therefore being more time efficient to complete as there is not a back-and-forth type of conversation for data to be produced. A closed internet survey also allows for a more personal response, as the respondent is anonymous. The adopted method of research by the researchers is a mostly closed questionnaire with open questions for more personalised opinions nearer the end of the survey. The closed questions were issued through basic yes or no questions as well as Likert Scale–type questions, in aim of gaining opinions identifying levels of agreement or disagreement on statements made on subjects such as collaboration, change orders and contracts, the topics discussed heavily in the literature. The Likert Scale can be defined as similar to the rating scale, except the questions consist of attitudinal statements with respondents answering on numerical scale from one extreme of favourableness to another (Naoum, 2007), this allows for direct opinions on the selected topics, which gives more opportunity for a high-quality analysis.

7.5.4.7 Sampling

Sampling within quantitative research can be a challenge due to the anonymity, meaning if the aim is for a relatively high number of respondents, otherwise the demographic of respondents may become irregular and not the type that was first intended (Delice, 2010). The objective of sampling has been described as a practical means of enabling the data collection and processing components of research to be carried out, while ensuring that the sample provides a high-quality representation of the desired population (Fellows, 2008). For this piece of research, the target sample was commercially based workers who work for large (250+ employees) or very large (1500+) companies as this group is best suited to rating the performance of BIM as their companies are expected to have had the capital to invest in the technology. The broader approach to smaller firms was not needed as from the literature review it is already known that these companies are still yet to be widely integrated into using the model. The type of probability sampling in terms of theory that would be most applicable is expert sampling (Etikan and Bala, 2017); this is in aim of constructing the views of individuals in the area, i.e., quantity surveying professionals so that the views that are given on the chosen topic are as reputable as possible.

7.5.5 Findings and Discussion

The basis of this section is the presentation of the data collected, done through the primary research that has been undertaken by the study through the questionnaires, along with the secondary data collected through the literature review. The structure of this section is based off of the structure of the questionnaire, which was based off the literature review. This structure is an aim for easier discussion and possible findings of correlation between the primary and secondary research. Another aim regarding the analysis is to identify trends and discrepancies between a previously discovered theory through the literature review and the fresh theory deriving from the primary research. Once each question has been analysed individually, there is a conclusion to see if the research objectives set out have been met.

7.5.5.1 Introduction

From the questionnaire, there were 52 respondents from a commercially based background, following the set-out demographic for this piece of research. As the questionnaire was sent out with intentions to only be answered by commercially based workers, there was no question asking for job roles, in hopes of more security of anonymity, meaning more open and confident answers from the respondents.

7.5.6 Analysis – Quantitative Results

7.5.6.1 Question 1

This question was instilled into the questionnaire to give certification that the results obtained, derived from the demographic that was first intended. From the results shown in Figure 7.3, it can be seen that 93% of respondents were from the required demographic set out in the methodology, workers from firms with 250+ employees. These companies realistically will have the sufficient amount of capital to implement BIM to a level as to where it can be analysed properly. This gives more assurance that the research question along with the research objectives is met as the questionnaire was answered, for the majority by the right people. From the 52 respondents, there were 4 respondents (or 7%) from demographics that were not sought after in the methodology. The impact of this 7% in terms of systematic response bias (Moors et al., 2014) is not seen as significant by the study; therefore, the results will be analysed as they are on face-value.

7.5.6.2 Question 2

The overall aim of question 2 was to gain data on whether the workers themselves regard themselves as knowing the 3D model-based process well or not. The most common response to the scale of 1–6 was "3 – Below Average" (38%) as shown in Figure 7.4; this means using the mode average, most of the respondents have the opinion that their knowledge is below the medium point on the scale, closer to a low understanding than a high understanding. The mean average for this question is 3.85; this gives an answer stating that the average respondent had an above average understanding of BIM. This correlates with the increased usage from 13% to 73% within the industry, proving

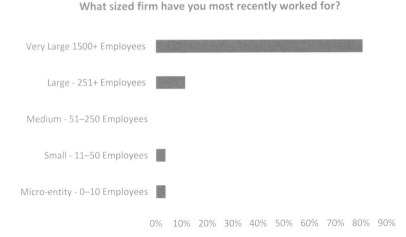

FIGURE 7.3 Respondent firm sizes.

that with the increased usage comes increased knowledge (NBS, 2020). From the data it can also be seen that there were 4 answers (8%) of "6 – Very High"; this is promising for the industry as more commercial workers become highly knowledgeable with the tool, making the tool more integrated within projects, as the tool moves from being purely design based. This increase in knowledge also allows for designers to be integrated with the commercial side of projects that can mean there is a more efficiently implemented model, which caters to both party's needs (Farnsworth et al., 2015).

7.5.6.3 Question 3

The aim of this question was to find out the actual applied effectiveness for the respondents person- ally, giving first-hand views on the efficiency of the tool. The most common answer (mode average) for this question was "2 – Ineffective" (46%) as shown in Figure 7.5; this portrays that the tool even within civil engineering and construction firms with 250+ employees and sufficient capital to invest in the technology sees a lack of effectiveness commercially for workers. The mean average

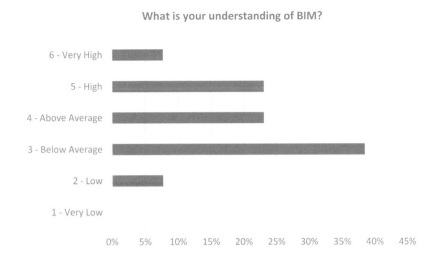

FIGURE 7.4 Understanding of BIM.

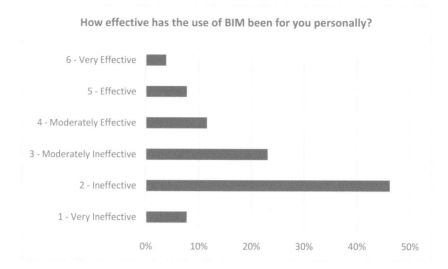

FIGURE 7.5   Personal effectiveness of BIM usage.

answer for this question sat at 2.77, which rounds to "3 – Moderately Effective" (23%); this gives
the impression that the tool is slightly more effective than the answer that was given in majority,
but still does not show BIM as a fully effective and integrated tool. This may again point towards
BIM still being designer-driven (Li, 2020). The answers on the ineffective side of the Likert Scale
give a narrative that a 5D BIM can assist with cost and budgetary considerations associated to a
project (Smith, 2016) is yet to be consistently met. Looking into the moderately effective and above
answers, massive potential is shown that BIM can be an effective and very effective tool for the
commercial side of projects, giving enthusiasm for further investment and usage of the tool.

7.5.6.4   Question 2 vs 3 – Understanding of BIM vs Effectiveness

If Figures 7.4 and 7.5 are compared as shown in Figure 7.6, correlation between the respondents is
understood and the effectiveness of usage of BIM is found. The patterns of both sets of data are very

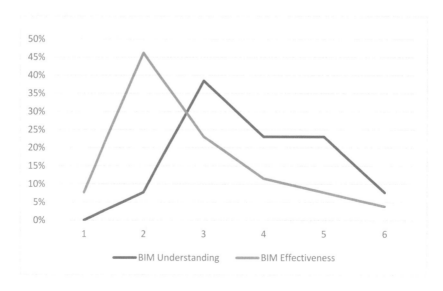

FIGURE 7.6   BIM understanding vs effectiveness.

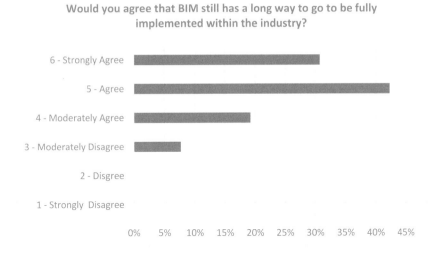

FIGURE 7.7 Does BIM have a long way to go until it's fully implemented?

similar with around 1 number difference; this helps to illustrate within the commercial side of the industry. The lack of understanding directly correlates with the amount of effectiveness that BIM can have, suggesting that if more money were to be invested into awareness and different ways of usage, then the effectiveness would rise with the understanding.

7.5.6.5 Question 4

Question 4 follows the theme of 2 and 3 in terms of the attitudes and opinions given on BIM, asking whether the commercial workers within the industry feel that the tool has a long way to go in terms of being fully implemented within the industry. The most common answer to the question was "5 – Agree" (42%) as shown in Figure 7.7, showing strong agreement with the statement and another 16 strongly agreeing (31%); this illustrates that on a commercial front it is still felt that there is a long way to go to get the technology in a place where it can be used efficiently, throughout the industry. Another interesting statistic that derives from this question is that there were no "1 – Strongly Disagree" or "2 – Disagree" answers, meaning that from Figures 7.4 and 7.5, even those who have a very high understanding as well as experiencing a very effective usage, still have the opinion that there is a long way to go with the implementation of BIM. This correlates with the 2020 National BIM Report statistic that only 27% of workers from the industry in its entirety (not exclusively commercially based workers) feel that BIM will be used for a majority of projects and 6% for all projects (NBS, 2020).

7.5.6.6 Question 5

Question 5 asks the respondents if they agree with the statement "Would you agree that BIM is now at a stage where it efficiently helps with the commercial side of projects?". From the resultant data set derived from the respondents it illustrates that there may be an attitudinal divide of opinions on whether BIM assists the commercially related aspects of projects. The most common answer to the question was actually "4 – Moderately Agree" (33%) as shown in Figure 7.8, meaning actually the majority of votes agree that BIM efficiently helps with projects, going against the correlation of data obtained in questions 2–4. The divide becomes visible with only 3 respondents answering that they "3 – Moderately Disagree" (6%). This data may start to highlight the effectiveness set out in the literature review of the technology's advantages, such as an increase in collaboration with all stakeholders within a project (Wamelink, 2014). When analysing the data from a mean average

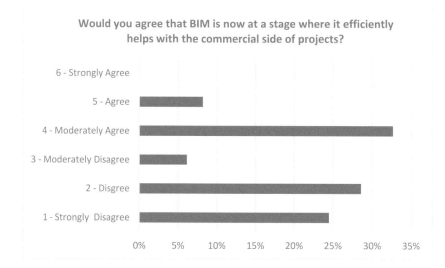

FIGURE 7.8   BIM's effectiveness commercially.

perspective, normal correlation starts to reappear stating that the respondents disagree that BIM is at a stage where it efficiently helps with projects.

The study does question the results from this question as they are not too dissimilar to previous questions, meaning there may be an aspect of respondent fatigue (Naoum, 2007), although 16 respondents all with the same fatigue to the same question this early on in the survey may say otherwise.

7.5.6.7   Question 6

Question 6 allowed for more than one answer, as BIM has many benefits that may be felt as equally as important by the respondents; the overall aim of this question was to find out what aspect of BIM commercial workers see as the most beneficial. The most common answer (32 times picked) to this question was that the "Shared Building Model" aspect of the technology was the most important as shown in Figure 7.9; from this it can be said that the centralised

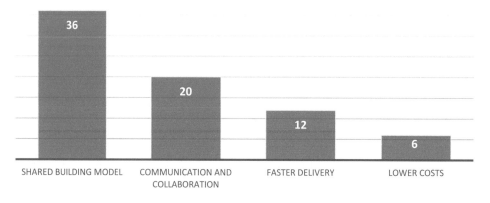

FIGURE 7.9   Commercial workers' most important aspects of BIM.

source of information for an entire project is proving to be the most efficient aspect of BIM for commercial workers, allowing projects to achieve optimal results by combining resources, knowledge and abilities to complete tasks correlating with the theory set out in the literature (Hughes, 2012). The second most common answer with 20 responses was increased communication and collaboration; this points towards the technology encouraging parties to work together efficiently to solve problems rather than adopting a "over the wall" mentality when working with information with stakeholders within a contract (Matthews et al., 2018). The "Faster Delivery" answer was inserted into the question by the study with expectancy of it being the least picked by the respondents, but it actually had double the number of answers as "Lower Costs". From this derives the theory that commercial workers are less interested in the tool cutting costs, but otherwise, more efficiently delivering a project with fewer mistakes (this could in turn also decrease costs as a resultant).

7.5.6.8 Question 7

Question 7 was implemented into the survey to gain data on the commercial effectiveness of BIM when it comes to using the tool as a centralised source of information. The most common answer for the statement made in the question "BIM helps as a centralised source of information to improve collaboration with parties involved in a project" was "5 – Agree" (42%) as shown in Figure 7.10. This answer gives the overall opinion that commercial workers understand the effectiveness of this aspect of BIM as a centralised tool, helping to involve every party in situations, such as reviewing design that has cost implications or keeping all relevant cost information in one digital storage space (Hughes, 2012). The mean average answer for Q7 was 4.42, rounding down to "4 – Moderately Agree" (19%); this gives the interpretation that commercially based workers on average see the collaborative benefits of BIM, showing that the majority understand one of the main benefits of the technology. This correlated with the data in Figure 7.9, with respondents regarding the most important use of BIM is the shared model. Other interesting statistics deriving from this question include that 4 respondents chose "2 – Disagree" (8%); this could be down to bad personal experience with BIM, where the tool has resulted in negative collaboration with parties involved in a project. Inversely, 8 respondents (15%) chose the answer "6 – Strongly Agree", giving connotations that they either fully understand the potential advantages of BIM or have had a high quality and efficient use of BIM.

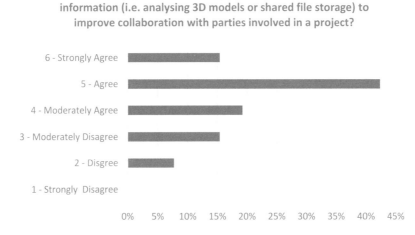

FIGURE 7.10 BIM's effectiveness as a collaboration tool.

When a Change Order has arisen, has BIM helped to solve an
issue with a party who is integrated within the Model?

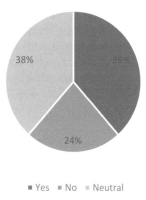

FIGURE 7.11   BIM's effectiveness at resolving change orders.

7.5.6.9   Question 8

The aim of question 8 was to find out if on the occurrence of a change order arising on a project, has the use of BIM assisted with resolving an issue with a change order. Question 7 proved that the respondents are aware of its effectiveness, but this question will find if it's actually helped with situations personal to the respondent. Firstly, it is worth mentioning that the "Neutral" answer was inserted into this question as typically with projects that are using BIM, change orders are much less common, leaving the respondents the option of no answer (Ghaffarianhoseini et al., 2017). Another reasoning for the "Neutral" answer is that some of the respondents may have not have had the opportunity to use BIM to help with this issue at all, meaning they will not be able to give an opinion on its effectiveness. Looking at the two answers that either give positive or negative impressions towards BIM regarding this topic, the most common answer with 20 votes was "Yes" (38%) as shown in Figure 7.11, showing that BIM can and has been an efficient tool for assisting with change orders, giving further backup to the secondary research gained through the literature review. There were also 12 "No" (24%) answers given to this question, showing that even when the tool has been there to use, it hasn't yet been able to help, showing that there is still more investment and awareness needed for the tool to be at an efficient enough level commercially.

7.5.6.10   Question 9

As findings show in the literature review, the ISO 19650 Series setting out International Standards for the use of BIM, with Parts 2 and 3 being released in 2019 and Parts 4 and 5 due to be published in 2021 (Winfield, 2020). As International Standards are there to be met, this question was issued to grasp how many commercially based workers within the industry are actually aware of the standards, giving way for an opinion on whether more needs to be done to increase this awareness or not. From the graph above it can be seen that is still a lot that needs to be done to increase commercial workers' awareness of the ISO 19650 Series of standards so that they can be met, only 38% of respondents considered themselves aware of the standards as shown in Figure 7.12, showing that over 62% were not actually aware that there were certain international standards that have to be met. This result may come from potentially unexperienced respondents who haven't had much experience with BIM itself, naturally meaning that they wouldn't have to be knowledgeable of the standards or they have had experience with BIM and simply didn't know of the standards entirely.

Are you aware of the ISO 19650 Series, setting out standards for
managing information over a whole project life cycle?

■ Yes ■ No

FIGURE 7.12   Awareness of international standards.

7.5.6.11   Question 10

This question's aim was to find out the current perspective from commercial workers on the current state of standardised contracts within the industry, establishing where they have the opinion that they have been adapted enough to withstand the use of BIM yet or not. As in the previous question, there were only 5 options included in the Likert Scale to leave a "Neutral" answer available for those with inefficient experience and knowledge in this subject. Looking into the data deriving from this question, it is a very spread-out set, giving no real leniency towards agreeing or disagreeing, showing a spread out set of both opinions and knowledge in standardised contracts. The most common non-neutral answer was "12 – Disagree" (23%) as shown in Figure 7.13; this may be a resultant from some of the points mentioned in the literature review such as the NEC3 contract only offering amendments to contracts rather than actual selectable clauses inserted into the overall contract for use (Elliott, 2013). Although the amendments are still as effective, on face value this makes the contract seem like it isn't as committed to contracts as JCT's schedule of modifications to adapt all of their contracts to BIM (Joyce and Houghton, 2014). Opposing to this, some still feel that the

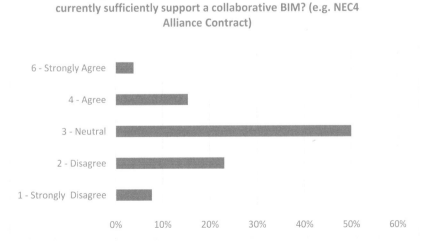

FIGURE 7.13   Standardised contracts.

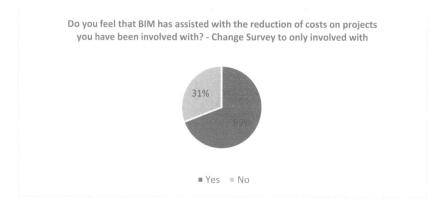

Do you feel that BIM has assisted with the reduction of costs on projects
you have been involved with? - Change Survey to only involved with

31%

■ Yes   ■ No

FIGURE 7.14   BIM's assistance with cost reduction.

standardised contracts in place do sufficiently support the use of BIM with 8 "4 – Agree" (15%) and
2 "5 – Strongly Agree" (4%) answers, showing that the mechanisms in place can be effective and
rather it may be attitudes that aren't enabling the use of some contracts effectively.

7.5.6.12   Question 11

Questions 11, 12 and 13 were set out to gain an understanding on BIM's use as a cost reduction tool
for commercial workers. Question 11's aim was to gain the overall experiences from the respondents using BIM for these purposes; from the results, it can be seen that the tool has helped to cut
costs with its efficiency, as well as other cost-saving aspects. Sixty-nine percent of respondents
stated that BIM assisted as a cost-saving tool as shown in Figure 7.14; this shows correlation and
follows the same trend to the secondary data, collected through the literature review stating that
21/35 projects claimed the technology was a successful cost reduction tool, with only 2 stating
negative effects financially (Bryde et al., 2013). These are quite positive and encouraging results
considering the earlier data collected regarding the adaptation of BIM commercially, showing that
even with the integration to the commercial side of projects in its early stages, the effects have
still been there for workers to see, again proving the potential is definitely there for an effective
commercial use of BIM. Opposed to this, 31% respondents to this question did not feel that BIM
had assisted with projects that they have been involved with, highlighting the youth of BIM commercially once more.

7.5.6.13   Question 12

From the secondary research via the literature review, it was found that there are efficient programmes in place to assist with the take-off process, aiding for more accurate forecasting of projects such as AutoDesk Revit, Vico and Assemble (Olsen and Taylor, 2017). This question was put in
place to discover whether these modern methods are being utilised by commercial workers or not.
From the data received, it is clear to see that although the technology is there, the usage has not yet
picked up. Showing that the take-off process may be lagging behind other commercial topics, with
only 15% of the 52 respondents (as shown in Figure 7.15) claiming experience of the use of BIM for
this process as compared to Q3's results, showing that 12 (23%) respondents had had "Moderately
Effective" to "Very Effective" use of BIM, an 8% differential. If this data were to be analysed from
a more positive perspective, 65% of the respondents who have had effective use of BIM (Q3), have
had experience using it for the take-off process, showing that once usage and effectiveness increases
through the industry, the take-off process will also become more BIM prominent, making way for
the increases in efficiency by undertaking the take-off manner using this process, as mentioned in
the literature review (Olsen and Taylor, 2017).

Have you had experience using BIM technologies for the take-off process?

■ Yes ■ No

FIGURE 7.15 Personal experience with BIM during the take-off process.

7.5.6.14 Question 13

As from the previous questions, the impression has been gathered that BIM is still in its early stages commercially; this question was to put in place to establish how many of the respondents have had experience with a full 5D BIM, which can assist with budgetary and cost considerations to a project, with interest to discovering if there has been any experience of this as of yet. From the results, it can be seen that a fully integrated 5D is very hard to achieve and has only as of yet been fulfilled personally by 10% of the 52 respondents as shown in Figure 7.16; this result was to be expected to be this way as a full 5D BIM would naturally have less respondents agreeing with this statement as compared to those 23% mentioned in Q3 who have had basic effective use of the technology. From this, it can be grasped that some of the major advantages analysed in the literature, such as analysing cash flows through the use of automated processes, along with producing operational and lifecycle costs can be analysed before a building is even built (Mayouf et al., 2019).

7.5.6.15 Question 14

Incorporating subcontractors into a BIM model in past research has proven to be an issue due to a reduced amount of capital that is there to utilise due to their smaller company size (Chan et al., 2019). This is a mostly commercially based problem due to the management becoming increasingly difficult due to differing processes due to the required smaller subcontractors not being able to be integrated into the model, meaning they are unable to input their own relevant data efficiently. The aim of the question was to give an up-to-date commercial opinion on whether it is still a problem to

A 5D BIM is where the model can assist with budgetary and cost considerations to a project - do you feel you have had experience of full 5D BIM as of yet?

■ Yes ■ No

FIGURE 7.16 Experience of 5D BIM.

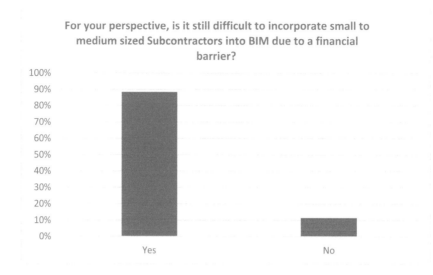

FIGURE 7.17   Difficulty of incorporating small firms into BIM.

incorporate the subcontractors, and the data given back still gave a massive feeling that this is still an issue within the industry, with 88% respondents giving this perspective as shown in Figure 7.17. This means that BIM needs to become more cost efficient to use, along with opening up more sources of training for subcontractors to learn the ins and outs of the technology, all in aim of producing more integrated use of the technology within the industry.

7.5.6.16   Question 15

The secondary research undertaken through the literature review showcased that the use of BIM is being pushed by the UK government and has been since 2011 (CDBB, 2011). The aim of this question was to establish if the sample of respondents had the feeling that there could still be more done to progress the technology further. From the results shown in Figure 7.18, it can be seen that 27% of respondents had the opinion of "5 – Strongly Agree" showing that commercially based workers

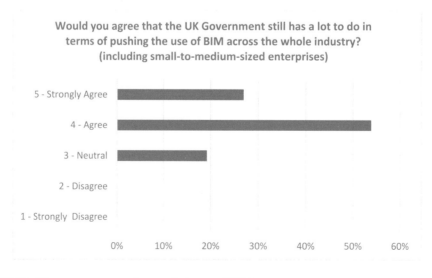

FIGURE 7.18   The government needs to push the use of BIM.

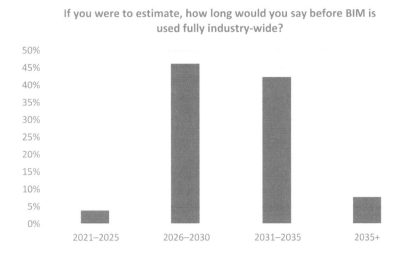

FIGURE 7.19   Full BIM implementation estimation.

feel strongly that more can be done to push the use of BIM. The most common answer with 54% of respondents choosing this answer was "4 – Agree" with the mean answer being close to this at 4.07, showing that the heavy majority have the opinion that there needs to be further push from the government, be that through initiatives or further strategies that are more expansive, detailed and applicable by more companies. Lastly for this question, there were no answers for "1 – Strongly Disagree" or "2 – Disagree", again showing BIM is still early in its implementation.

7.5.6.17   Question 16

As has already been established, the general thought from respondents is hat BIM is yet to be integrated industry wide, this question was constructed to grasp opinion from the respondents on when they think BIM will be used industry wide. The data shown in Figure 7.19 gives the most common answer of "2026–2030" with 46% votes, showing that with the government push in Q15 along with the increased affordability which is felt is needed deriving from Q14, then the UK construction and civil industries are as close as 5 years away from BIM being used fully industry-wide. The next most common answer was "2031–2035' with 43% votes, a close second. From the previous results within the survey, the study feels that this answer is the most realistic, considering the amount of data that isn't in favour of BIM commercially, with realistically there being 10 years minimum before an integrated fully industry-wide BIM is in place, progressing to a leaner construction industry (Dave et al., 2013).

7.5.7   Analysis – Qualitative Results

7.5.7.1   Qualitative Results

Q: Do you feel there are any challenges with BIM that currently hinder its effect commercially? (Implementation could be better etc.)

A: To give the chance of more personal opinion on BIM from commercial workers, this question, along with following (Q18), will complete the qualitative data sub-chapter; the question asks the respondents their own thoughts on what hinders the progress of commercial implementation for BIM.

One of the first responses obtained gave the opinion that it was very much design led, with the "information" required to help with commercial decisions has been inefficient or missing. From this it can be seen that a fully integrated BIM, which includes all parties within a project, including the commercial staff, is still a while away, along with the actual information/data that the commercial team is obtaining from the model being either wrong or not put into a format in which it is easily interpretable, meaning less efficiency when it comes to performance. Another similar response that correlates with this stated that a challenge that hinders its effect commercially is the "Integration with commercial systems and the time it would take to get cost and schedule included in BIM". This shows projects with BIM in place are struggling to integrate with the commercial systems that are used to implement cost and programmes into BIM, two massive aspects of a commercial position such as a quantity surveyor.

Another interesting response was "At the moment having a shared source of information is a struggle commercially. Not having everyone uploading information to a shared data storage area really hinders the likes of using up to date cost sheets. If I am 3 periods behind, then my costs will be way out which isn't a good standing point commercially". From this derives the theory that some workers within projects are adapting to the technology and some aren't, which is becoming a problem as there are different levels of BIM intelligence. This shows that for the tool to work properly there needs to be a full commitment from all members of a project team; this relates back to the secondary research that states that "workers do not feel comfortable with having to change the way that they work (cultural barrier), this means that the specific attributes of some roles needed are not there" (Sebastian, 2011).

A response which correlates with data obtained from the literature review was "Hard to get small contractors on board as the capital investment to get individuals trained up". This follows the theory such as BIM barriers such as high set-up costs and also being the weakest link, which means that poor performance by a subcontractor becomes a limiting force in a set of supply chain relationships (Robson et al., 2014). Showing that the scalability of BIM financially is still to be acquired, backed up by another answer stating, "Masking the software, hardware and training available at a cost that does render any of these things as a barrier to use" shows that the tool is still overly expensive. Similarly, another respondent further substantiates this, with their opinion being "the cost-effectiveness of smaller firms implementing it. If subcontractors don't use it, it doesn't benefit the scheme", showing that if subcontractors can't integrate it then it reduces the effectiveness of BIM.

Another challenge which was brought up by a respondent was that "client demand needs to be there; the positive effects of BIM need to be communicated better". This produces the theory that the technology in some situations is not being driven by the client properly; this reiterates one of the challenges indicated in the literature review stating that the lack of client demand reduces the effectiveness of the tool, meaning that the use of BIM ends up becoming lacklustre (Vass and Gustavsson, 2017); second to this, the respondent mentioned how some workers simply aren't aware of the effectiveness and therefore weren't committing to BIM. This will come with staff training, as another respondent stated with the answer of, "Staff training and ability to use the system are playing a big part in the BIM's use, or lack thereof." In similar relation to the above, other responses highlighting their opinions on the challenges said, "Cost of implementation and client demand" and "There has to be 100% buy in by all parties to ensure that BIM is a useful tool." Showing that for the tool to be efficient commercially, the commitment has to be there by all involved.

Another standpoint on the challenges of BIM stated that a challenge is "Security of the system and the information contained within it", which the study relates back to the secondary research found in the literature review regarding legal issues that come with the data collected within a model, with increased risk exposure, which discourages companies (Eadie et al., 2015); this fresh data shows that this is still seen as an issue.

Lastly, the most extensive answer given was "BIM use needs a better business case to support its adoption and/or a better selling job regards the improvements it has already, or in the future will, genuinely deliver. To date, the main feedback we tend to get (my personal experience being on the project) is it adds cost and is an 'add-on'. I have a feeling we may not have fully committed to its implementation (you get out what you put in?) but in my limited experience the jury remains 'out'." This gives light to the not only the need for training to have suitable professionals who can use BIM, but also the need for each project to be adjusted properly to the tool, so it can function more efficiently. In this case, it seems as if the respondent's project has had experience of the usage of BIM, but as it has not been implemented properly it has had a negative cost effect, rather than the positives mentioned in the literature review (Bryde et al., 2013).

7.5.7.2   Question 18

Q: "Any extra thoughts/comments on BIM would be much appreciated!"

A: The final question of the survey was left as open as possible in an aim of obtaining some results that were in envisaged at the time when the study constructed the questionnaire. This question was not required to be answered by the respondents, giving may for an answer to be given only if they felt they were needed.

Firstly, an interesting comment from one of the first respondents said, "BIM-FM is a new and valuable aspect of the interface, & needs to be implemented more, the UK government have to increase the compulsory BIM level on centralised projects imminently." BIM-FM wasn't reviewed in the secondary research and from brief secondary research done after seeing this answer by the study, BIM FM stands for BIM Facility Management and the use of it by the respondent in their opinion is clearly effective; it allows for the transfer of data regarding the model of a project along with assisting on-site workers with technical sheets. It also provides the user an up-to-date real-time vision of the facilities that is all controlled by an operator. From this, BIM FM appears to be another usable programme/tool that can be as effective as the processes mentioned in the literature review such as AutoDesk Collaborate (AutoDesk, 2021).

Another comment showed that a lack of up-to-date or powerful enough technology may be an issue, saying, "Integrated 3d/4d models show promise but the ability to run from a standard corporate laptop is lacking." This gives the theory that although the programmes are there to use, the company laptops that are given to workers for usage are not yet fast enough to run the BIM technology, showing that companies may need to invest more money into more powerful laptops for a more effective BIM implementation.

Two comments made by the respondents relate back to the cultural barrier mentioned in the literature review (Sebastian, 2011); two respondents said, "Industry needs to be more aware of it, especially the older generation" and "we are a slow industry to adopt new ideas!", showing that the adaptation of BIM commercially may be under a hindrance due to dated attitudes on new technologies.

Lastly, a comment which reiterates the point of the need for increased government initiative (CDBB, 2011) was put by a respondent with them saying, "Structured procedures needs to be in place to deal with design changes and fabrication drawings"; these structured procedures could possibly be put in place either through government initiatives which would put in place more rules and regulations for BIM or even it could be made compulsory to use certain structured procedures through updated standardised contracts formulated specifically for BIM (Elliott, 2013).

7.6   DISCUSSION AND CONCLUSION

A review of the selected aspects of BIM that have commercial relations was undertaken via the literature review, making way for the primary research, with many interesting findings. Firstly, in terms of collaboration and communication, there are increases in efficiency through an increased integration of project teams, removing the "over the wall" approach to information (Matthews et al., 2018), providing space for more efficient ideas due to the more collaborative environment produced by BIM (Wamelink, 2014). Change orders and disputes were also found to decrease heavily due to this, with rather than project teams such as clients and main contractors working against each other for profit (Ghaffarianhoseini et al., 2017), they're now put into environment where they all have the same common goal, therefore reducing these costly altercations from arising (Demian and Walters, 2014). The next of the commercial topics to be reviewed was BIM-based contracts within the industry, with findings that there have been adaptations from contract producing organisations such as JCT, with supplying schedules of modifications for all of their contracts (Pantry, 2020). NEC also has been adapting relatively well to the integration of BIM as highlighted in The BIM Alliance report notes, noting the NEC's proficient use of BIM terminology, something that other contracts have lacked on (Kirby-Turner, 2021), although the only current way to adapt NEC's contracts is through Z clauses, used to amend their contracts. The last commercial aspect of BIM that was reviewed was its use as a cost reduction tool, highlighting secondary research, which shows that throughout 35 projects using the technology, 21 stated that BIM was a successful cost reduction tool, with only 2 stating negative effects (Bryde et al., 2013). Other studies stated that the tool reduces the capital costs of construction by 33%, a very promising figure (Georgiadou, 2019).

The primary research which was used to investigate the current status of BIM commercially was deemed to be successful by the study, with many interesting findings. Overall, from the results it can be seen that BIM is still in its early stages commercially, with on average the workers finding it a moderately ineffective tool, which is still predominantly design based, with commercial workers mostly disagreeing with the statement that BIM is now at a stage where it can efficiently help with the commercial side of projects. The use of BIM as a cost reduction tool is known, but the experience of a 5D BIM is still sought after and by the looks of it still 5–10 years away. Other data deriving from the primary research suggested that the biggest and most important aspect of BIM commercially is its use as a collaboration tool, using the shared building model that is required for the usage of BIM, but only when it is integrated properly throughout a project from the smaller subcontractors all the way up to the client; otherwise the implementation is not able to be effective. The heavy majority felt that for a fully integrated and implemented BIM to be available more commonly in projects, the UK government still has a lot to do push and make BIM more cost efficient, with the need for increased awareness of the effectiveness of the tool (along with increased awareness of the newly produced international standards) and also increased training, so there is an increased amount of BIM experts within the commercial sectors of all companies throughout the industry. Overall, the potential for an effective commercial BIM is there to be seen by most, but it seems the tool is still early in its maturity with still lots of work to be done regarding its implementation.

The questionnaire derived a lot of interesting data and theories, giving updates to the literature review, even sometimes in contrast. The first finding was that the understanding of BIM is still at an average state with the actual personal effectiveness low, although these two seem to be connected going by the data, meaning the effectiveness will increase with the understanding. The commercial workers who responded in the majority have the opinion that BIM still has a long way to go to be fully implemented within the industry with the data showing that it can assist with collaboration as well as reducing costs when it is integrated efficiently within a project team. In terms of using BIM to manage costs and forecasts the respondents stated that there have been instances where it has helped as a cost reduction tool, but they have had little experience with a 5D BIM, which involved cost and budgetary considerations, along with little use for the take-off process, which is used for forecasting. Furthermore, from the results it is now understood that for BIM to genuinely deliver, it needs to be integrated between all parties within a project, with it being difficult to incorporate small firms into the BIM model, in turn leading to decreased efficiency overall of project use of BIM. Client demand is also essential for the integration between all parties as in some situations it is not being driven by the client properly. Also, ultimately it is felt that the UK government still has a lot to do to push the usage of BIM, with the current need for an increase in training and funding to increase the awareness and personal usage effectiveness of BIM; this along with a introduction of specific structured and standardised procedures could be a viable option. Overall, at the point of data collection, BIM is still in an early stage of its maturity commercially within the industry with a lot of progress needed in various different areas.

7.6.1 LIMITATIONS

Limitations which were felt by the study are as follows:

- COVID-19 resulted in a limiting variety in ways in which research could be undertaken – e.g., only realistic research options were online questionnaires and video call interviews.
- Limited number of responses; a larger sample size naturally means a more reliable data set.
- Sample could have been biased towards certain projects; the study contacted large projects for respondents, meaning overall there could have been some aspects of response bias.
- Lack of triangulation in the research; only using one research method may have hindered the diversity of results.

7.7 RECOMMENDATIONS

7.7.1 RECOMMENDATIONS FOR GOVERNMENT INITIATIVES AND SCHEMES

Deriving from this research, there are a few recommendations; firstly, a recommendation is for there to be a government initiative to fund the sufficient training of all workers within the industry, so that they are aware of the effectiveness of BIM, therefore increasing the usage of the tool, meaning it can integrate properly within every team involved in a project. Along with this, there needs to be an increase in funding from the government to make BIM more affordable for smaller firms, so that commercial teams from all companies are able to collaborate around shared building models, resulting in more cost-efficient projects and long-term savings.

7.7.2 RECOMMENDATIONS FOR FURTHER RESEARCH

In terms of further research off of the back of this piece, the study would recommend further research into the effectiveness of BIM as a cost reduction tool; the literature reviewed (Georgiadou, 2019) and (Bryde et al., 2013) showed that it can be effective but the results seemed to differ between both pieces of research. The primary research showcased that commercial teams have seen first-hand its effectiveness, but haven't used it for forecasting as much. Further research into the effectiveness, if positive, will result in an increased push towards a 5D BIM.

REFERENCES

Akponeware, A.O. and Adamu, Z.A., 2017. Clash detection or clash avoidance? An investigation into coordination problems in 3D BIM. Buildings, 7(3), p. 75.

Alshawi, M. and Ingirige, B., 2003. Web-enabled project management: an emerging paradigim in construction. Automation in Construction, 12(4), pp. 349–364.

Amudu-Yusuf, G., 2018. Critical success factors for building information modelling implementation. Construction Economics and Building, 18(3), pp. 55–73.

AutoDesk, 2021. BIM Collaborate Pro – Deliver better projects, faster. [Online] Available at: https://www.autodesk.co.uk/campaigns/aec-collaboration [Accessed 12 April 2021].

Bell, E., Bryman, A. and Harley, B., 2018. Business research methods. Oxford University Press.

Bryde, D., Broquetas, M. and Volm, J.M., 2013. The project benefits of building information modelling (BIM). International Journal of Project Management, 31(7), p. 977.

BSI, 2010. Constructing the Business Case: Building Information Modelling. British Standards Institution and Building. SMART UK, London and Surrey, UK.

BSI, 2019. BIM – Building Information Modelling – ISO 19650. [Online] Available at: https://www.bsigroup.com/en-GB/iso-19650-BIM/ [Accessed March 2021].

Cakmak, E. and Cakmak, P.I., 2014. An analysis of causes of disputes in the construction industry using analytical network process. Procedia-Social and Behavioral Sciences, 109, pp. 183–187.

CDBB, 2011. Building Information Modelling (BIM) Working Party Strategy Paper.

Chan, D.W., Olawumi, T.O. and Ho, A.M., 2019. Perceived benefits of and barriers to building information modelling (BIM) implementation in construction: The case of Hong Kong. Journal of Building Engineering, 25, p. 100764.

Chun Tie, Y., Birks, M. and Francis, K., 2019. Grounded theory research: A design framework for novice researchers. SAGE Open Medicine, 7, 2050312118822927.

CIC, 2013. Building Information Modelling (BIM) Protocol. Standard Protocol for use in projects using Building Information Modelling, Volume 1.

Dave, B., Koskela, L., Kiviniemi, A., Tzortzopoulos, P. and Owen, R., 2013. Implementing lean in construction: Lean construction and BIM [CIRIA Guide C725].

Delice, A., 2010. The sampling issues in quantitative research. Educational Sciences: Theory and Practice, 10(4), pp. 2001–2018.

Demian, P. and Walters, D., 2014. The advantages of information management through building information modelling. Construction Management and Economics, 32(12), pp. 1153–1165.

Eadie, R., McLernon, T. and Patton, A., 2015. An investigation into the legal issues relating to building information modelling (BIM). In Rics Cobra Aubea 2015. Royal Institution of Chartered Surveyors.

Elliott, F. (2013). May 2013. [Online] Available at: https://www.fenwickelliott.com/research-insight/newsletters/insight/23 [Accessed 24 Apr. 2023].

Etikan, I. and Bala, K., 2017. Sampling and sampling methods. Biometrics & Biostatistics International Journal, 5(6), p. 00149.

Farnsworth, C.B., Beveridge, S., Miller, K.R. and Christofferson, J.P., 2015. Application, advantages, and methods associated with using BIM in commercial construction. International Journal of Construction Education and Research, 11(3), pp. 218–236.

Fellows, R.F., 2008. Research methods for construction. Blackwell, pp. 1–300.

Fellows, R.F. and Liu, A.M., 2015. Research methods for construction. John Wiley & Sons.

Ganah, A.A. and John, G.A., 2014. Achieving Level 2 BIM by 2016 in the UK. Computing in Civil and Building Engineering (2014), pp. 143–150.

Georgiadou, M., 2019. An overview of benefits and challenges of building information modelling (BIM) adoption in UK residential projects. Construction Innovation, 19(3), pp. 298–320.

Ghaffarianhoseini, A., Tookey, J., Ghaffarianhoseini, A., Naismith, N., Azhar, S., Efimova, O. and Raahemifar, K., 2017. Building information modelling (BIM) uptake: Clear benefits, understanding its implementation, risks and challenges. Renewable and Sustainable Energy Reviews, 75, pp. 1046–1053.

Government, H., 2015. Level 3 Building Information Modelling – Strategic Plan. Digital Built Britain, p. 15.

Grilo, A. and Jardim-Goncalves, R., 2010. Value proposition on interoperability of BIM and collaborative working environments. Automation in Construction, 19(5), pp. 522–530.

Helen, K., 2015. Creative research methods in the social sciences: A practical guide. Policy Press, p. 5.

Hox, J.J. and Boeije, H.R., 2005. Data collection, primary versus secondary, pp. 593–599.

Hughes, W. R., 2012. Differing perspectives on collaboration in construction. Construction Innovation, 12(3), pp. 355–368.

Joyce, R. and Houghton, D.,2014. Briefing: Building information modelling and the law. Proceedings of the Institution of Civil Engineers-Management, Procurement and Law, 167(3), pp. 114–116.

Kassem, M., Iqbal, N., Kelly, G., Lockley, S. and Dawood, N., 2014. Building information modelling: Protocols for collaborative design processes. Journal of Information Technology in Construction (ITcon), 19, pp. 126–149.

Kirby-Turner, C., 2021. NEC4, Collaboration and BIM. [Online] Available at: https://constructionmanagement.co.uk/bim-collaboration-and-nec4/ [Accessed 23 March 2021].

Kothari, C., 2004. Research methodology: Methods and techniques. New Age International.

Li, Y., 2020. Brief analysis of the application and limitation of BIM in project life cycle management. IOP Conference Series: Materials Science and Engineering, 780(5), p. 052001.

Manderson, A., Jefferies, M. and Brewer, G., 2015. Building information modelling and standardised construction contracts: A content analysis of the GC21 contract. Construction Economics and Building, 15(3), pp. 72–84.

Matthews, J., Love, P.E., Mewburn, J., Stobaus, C. and Ramanayaka, C., 2018. Building information modelling in construction: Insights from collaboration and change management perspectives. Production Planning & Control, 29(3), pp. 202–216.

Mayouf, M., Gerges, M. and Cox, S., 2019. 5D BIM: an investigation into the integration of quantity surveyors within the BIM process. Journal of Engineering, Design and Technology, 17(3), pp. 537–553.

Moors, G., Kieruj, N.D. and Vermunt, J.K., 2014. The effect of labeling and numbering of response scales on the likelihood of response bias. Sociological Methodology, 44(1), pp. 369–399.

Mosey, D., 2014. BIM and related revolutions: A review of the Cookham Wood Trial Project. Society of Construction Law, Hinckley, UK.

Naoum, S., 2007. Dissertation research & writing for construction students. Elsevier, pp. 1–80.

NBS, 2020. 10th Annual BIM Report. p. 23.

NBS, 2020. National Building Specification. NBS' 10th National BIM Report.

Olsen, D. and Taylor, J.M., 2017. Quantity take-off using building information modeling (BIM), and its limiting factors. Procedia Engineering, 196, pp. 1098–1105.

Oraee, M., Hosseini, M.R., Edwards, D.J., Li, H., Papadonikolaki, E. and Cao, D., 2019. Collaboration barriers in BIM-based construction networks: A conceptual model. International Journal of Project Management, 37(6), pp. 839–854.

Pantry, M. E. F., 2020. Working with BIM and JCT Contracts. [Online] Available at: https://corporate.jctltd.co.uk/working-with-bim-and-jct-contracts/ [Accessed 21 March 2021].

Peterson, F., Hartmann, T., Fruchter, R. and Fischer, M., 2011. Teaching construction project management with BIM support: Experience and lessons learned. Automation in Construction, 20(2), pp. 115–125.

Punch, K., 2013. Introduction to social research: Quantitative and qualitative approaches. SAGE.

Robson, A., Boyd, D. and Thurairajah, N., 2014. UK construction supply chain attitudes to BIM. In 50th ASC Annual International Conference.

Rooke, R., 2021. How Construction Forecasting Software Can Benefit Your Company and Power Your Growth. https://www.viewpoint.com/blog/how-construction-forecasting-software-can-benefit-your-company-and-power-your-growth.

Sacks, R., Koskela, L., Dave, B.A. and Owen, R., 2010. Interaction of lean and building information modeling in construction. Journal of Construction Engineering and Management, 136(9), pp. 968–980.

Sebastian, R., 2011. Changing roles of the clients, architects and contractors through BIM. Engineering, Construction and Architectural Management, 18(2), pp.176–187.

Seyedhabibollah, S., 2019. Requirement Management in a Life-Cycle Perspective Using Asset Information Based on ISO 19650-1 and CoClass.

Smith, P., 2016. Project cost management with 5D BIM. Procedia-Social and Behavioral Sciences, 226, pp. 193–200.

Sukamolson, S., 2007. Fundamentals of quantitative research. Language Institute Chulalongkorn University, 1, pp. 2–3.

Vass, S. and Gustavsson, T.K., 2017. Challenges when implementing BIM for industry change. Construction Management and Economics, 35(10), pp. 597–610.

Wamelink, J.W.F. and Heintz, J.L., 2014. Innovating for integration: clients as drivers of industry improvement. Construction Innovation, pp. 149–164.

Williams, C., 2007. Research methods. Journal of Business & Economics Research (JBER), 5(3).

Winfield, M., 2020. Construction 4.0 and ISO 19650: A panacea for the digital revolution? Management, Procurement and Law, 173(4), pp. 175–181.

8 Towards Automated Quality Assurance
Generating Synthetic Images of Building Components for Vision-Based Semantic Segmentation

Hao Xuan Zhang, Lei Huang, Weijia Cai, and Zhengbo Zou

CONTENTS

8.1 INTRODUCTION

The construction industry is facing challenging demands from rapid population growth and an accelerating urbanization rate (Idrees et al., 2017). It is expected that the urban population will reach seven billion by the year 2050, consisting of more than 65% of the global population (Ritchie and Roser, 2018), which lays immense pressure on the current infrastructure and drive up the need for new infrastructure. During construction, discrepancies between as-built structures and as-designed models can lead to schedule delays and cost overruns (Josephson and Hammarlund, 1999). This issue is exacerbated by the recent trend of increased offsite manufacturing of building components, since the assembly of the prefabricated components demands high precision and a tight margin of error (Yin et al., 2019). Therefore, quality assurance (QA) is essential throughout the life cycle of a construction project.

DOI: 10.1201/9781003325246-8

139

Currently, onsite QA is primarily conducted through visual inspection to catch problems including geometric defects such as dimensional mismatch, mispositioning of structural components; and surface defects, such as spalling, corrosion, and cracks (Kim et al., 2019). Another form of onsite QA relies on inspectors to manually measure discrepancies between the design model and the finished structure, which can be time-consuming, costly and error-prone (Phares et al., 2004). Furthermore, research showed that manually comparing as-built structures with 2D drawings while touring the building creates a heavy mental workload for building inspectors, which also leads to errors and omissions (Shi et al., 2019). Thus, there is a need to develop automated QA approaches that can replace the manual measurements for increased accuracy and reduced inspection time (Kim et al., 2019).

The main challenge for automating onsite QA is to provide fast and accurate representation of the condition of the built structures, which sparked innovation and adaptation of vision-based sensors (e.g., laser scanners, RGB-D sensors) and computer vision algorithms (e.g., convolutional neural networks) in the QA process. Indeed, research studies in the architecture, engineering, and construction (AEC) domain explored possibilities of using point clouds and RGB images as input data to help QA through as-built model generation (Czerniawski and Leite, 2020), construction progress monitoring (Kopsida and Brilakis, 2020), discrepancy detection (Rahimian et al., 2020), and surface defect detection (Liu et al., 2019). Though the specific inspection task varies in these studies, one foundational step remains common, as the semantic segmentation of point clouds or images into building components (e.g., a pixel-by-pixel labeling of a construction site image into construction equipment, worker, etc.). Therefore, in this study, we focus on improving the performance of automated semantic segmentation of images into building components as the first step of achieving automated vision-based QA.

Currently, two challenges must be addressed before automated semantic segmentation of images and point clouds can be achieved at scale for construction projects. First, it is challenging to collect large-scale real-world data of images or point clouds in buildings and/or on construction sites. Despite efforts in the AEC domain (Agapaki and Brilakis, 2020; Czerniawski and Leite, 2020) aiming to provide public datasets of indoor environments and construction sites, the scale of these datasets is small when compared to those in the computer science domain, which plays an essential role for creating high performing algorithms for semantic segmentation (Chen et al., 2017; Zhou et al., 2017). The more pressing challenge lies in the labeling of these datasets. Currently, human labelers draw boundaries of objects within an image and then label each object manually as a building component class, which is time-consuming and error-prone (Hua et al., 2021; Ros et al., 2016; Zhu et al., 2019).

Given these challenges, leveraging synthetic datasets has been of particular interest among researchers in the AEC domain (Chen et al., 2020; Ma et al., 2020; Ros et al., 2016; Soltani et al., 2016), since data generated from virtual environments can be easily scaled up and has the potential for creating a diverse dataset of building components. However, merely generating synthetic images or point clouds of buildings or construction sites does not solve the issue of lacking an automated method for labeling these synthetic datasets. In this study, we propose a novel approach to overcome this issue by combining building information modeling (BIM) and game engines to simultaneously generate and label synthetic datasets of building components.

The proposed approach starts with a BIM authoring tool (e.g., Revit), which served as the model geometry and building information provider. Next, in a game engine (e.g., Unity 3D), we generated synthetic building images by randomizing the position and rotation of the camera and taking a "photo" of the building consisting of various building components. Thereafter, a linking module custom-built in the game engine was used to trace back to the BIM software and retrieve semantic information of the building components encompassed in that image in real time. Finally, we applied a convolutional neural network (CNN) model to conduct semantic segmentation of the images generated by the game engine.

To validate the approach, we used five building information models from a variety of building types (i.e., residential, commercial, and educational) as testbeds. In total, 50,000 images were generated from these five models as a training set for the CNN model, and another 10,000 images were generated as a testing set. The performance of the CNN-based segmentation model was measured by average pixel-wise accuracy and mean intersection over union (IoU) across all building component classes.

The contribution of this study is twofold:

- First, we proposed an automated method for generating diverse images of building components in a game engine to provide a wealth of training and testing images for semantic segmentation of building components.
- Second, we introduced an automated method for labeling the synthetically generated images from the game engine by tracing back to the original building information model in the BIM authoring tool. Through this study, we aim to address the problem of lacking training data in the AEC domain for semantic segmentation of building components, which is a concrete step towards automated QA using computer vision.

8.2 BACKGROUND

This study is built on the intersection of (1) vision-based quality assurance in AEC, and (2) existing efforts for creating synthetic datasets in AEC.

8.2.1 VISION-BASED QUALITY ASSURANCE IN AEC

Quality assurance is essential for construction projects to avoid deviations, nonconformance, and defects, which are common factors leading to rework. Previous studies found the direct cost of rework can constitute more than 5% of the total construction cost (Hwang et al., 2009), leading to cost and schedule overruns. The use of remote-sensing instruments such as total stations, and the application of BIM, eliminated the need for the complete manual approach for QA. However, the current practice is still labor intensive since inspectors need to manually compare measurements from the collected data with BIM. With the development of cost-effective vision-based sensors (e.g., RGB cameras, RGB-D camera, LiDAR) and computer vision algorithms, researchers and practitioners started to explore the opportunity of using images, videos, and point clouds for automated QA to replace the existing labor-intensive and error-prone practice of manual comparisons.

Several types of vision-based tasks can be implemented as a fundamental first step for QA, including image/point cloud classification, object detection, and semantic segmentation. Image classification refers to the task of identifying the content in an image and then assigning a class from a group of known labels to the image, e.g., classifying construction objects such as studs, electrical outlets, and insulation onsite to ensure construction quality (Hamledari et al., 2017). Object detection refers to the task of classifying and localizing multiple objects of different classes in an image, e.g., detecting cracks in concrete columns (Dinh et al., 2016). Semantic segmentation refers to the task of classifying each pixel of an image with a label, which is essentially a pixel-level classification of an image, e.g., segmenting point clouds and images into cracks and non-cracks for pavement management and structural health monitoring for concrete columns (Gopalakrishnan et al., 2017; Liu et al., 2019). These vision-based tasks have been extensively applied in AEC for QA to achieve a higher level of automation.

The most studied application area of computer vision-based QA in construction is surface defect detection (Martinez et al., 2019), which is also an important aspect of consideration for construction safety. Defect detection has been implemented in assessing the condition of horizontal civil infrastructures, including reinforced concrete bridges, underground concrete pipes, and asphalt pavements, and vertical structures such as steel frames, reinforced concrete columns and exterior masonry walls (Koch et al., 2015). For example, an integrated vision-based system was developed to automatically inspect the quality of slate slabs for defects and ensure the adherence to construction guidelines (Iglesias et al., 2018). An image processing pipeline used to inspect external wall insulation quality was developed to provide real-time feedbacks for automated QA for remodeling (Cho et al., 2019). Vision-based QA was also applied for robotic construction using an extrusion system to print wall frames, where tight tolerance is made possible by measuring the extruded layer width in real time to detect conditions of over-extrusion or under-extrusion (Kazemian et al., 2019).

Aside from defect detection, studies also explored the use of computer vision for detecting dimensional, geometrical, and positional variations for QA. For example, a vision-based approach was proposed to inspect the alignment of tile installation by analyzing the geometric characteristics of the tile finishes (Lin and Fang, 2013). Furthermore, semantic segmentation of building point clouds was achieved by conducting multi-scale feature detection for the curvature of objects to improve the quality of recreated 3D models of as-built environments (Dimitrov and Golparvar-Fard, 2015). In addition, geometric and relationship-based reasoning was implemented to find discrepancies between as-built and as-planned models (Maalek et al., 2019). Furthermore, a real-time vision-based framework was developed to inspect steel frame assemblies for validating and correcting errors in the manufacturing stage (Martinez et al., 2019). Concluding from previous studies, although vision-based QA has been implemented in a variety of application areas (e.g., defect detection, geometric discrepancy detection) for many tasks (e.g., object detection, semantic segmentation), it was concluded from a recent survey (Martinez et al., 2019) that research and applications of vision-based quality assurance and control are still limited, and there is a need to explore different vision-based techniques to automatically evaluate built structures for defects and non-compliance.

8.2.2 Existing Efforts for Creating Synthetic Datasets in AEC

A significant bottleneck of developing vision-based methods for AEC, cited by many recent reviews (Guo et al., 2021; Martinez et al., 2019; Zhong et al., 2019), is the lack of open training and testing data. There have been noticeable efforts in creating open data that can be accessed by all researchers to benchmark their algorithms. For example, the first workshop of computer vision in the built environment set out to provide an open dataset for the task of scan-to-BIM, where point clouds are given as inputs and the output is an automatically created building information model (Han et al., 2021; Wu and Xue, 2021). Other similar efforts (Czerniawski and Leite, 2020) include releasing scanned point clouds of educational and commercial buildings for as-built BIM generation and discrepancy detection. Agapaki and Brilakis (2020) also released a dataset including scans from an industrial factory for segmentation, object detection, and classification of point clouds for the purpose of creating a digital twin of the factory. Despite these efforts, a challenge persists as the collected real-world data need to be labeled manually to provide ground truths. These manual inputs, such as drawing boundaries on an image for semantic segmentation or classification, are often costly and time-consuming, preventing the wide adoption of existing open datasets for developing and validating vision-based algorithms in AEC.

One approach to solving the lack of open data is to create synthetic datasets. Synthetic data generation has seen success in several areas (e.g., autonomous driving, object segmentation, and text recognition), by increasing the performance for vision-based methods without the need for large-scale manually labeled datasets (Chen et al., 2020; Danielczuk et al., 2019; Hou et al., 2021; Jaderberg et al., 2014; Yue et al., 2018). One of the most widely applied fields of synthetic data is autonomous driving. For instance, a simulation engine-based environment was created by (Johnson-Roberson et al., 2016) to render realistic urban environments for the training of autonomous vehicles, including simulated weather and traffic patterns. The simulation allowed pixel-level segmentation of objects in an urban environment. Similarly, (Ros et al., 2016) created an open dataset of synthetic urban environments, aiming to train deep neural network (DNN) algorithms for segmenting objects from a collection of diverse urban images for autonomous driving. Eleven classes of objects were included in the virtual environments. More recently, (Yue et al., 2018) proposed a LiDAR point cloud generator capable of providing labeled synthetic data of cars in various realistic virtual environments. The generator enabled changes of car model, color, orientation, weather, time of day, and background. Similarly, Chen et al. (2020a, 2020b) explored the possibility of extracting semantic information from aerial point clouds to generate interactive, photorealistic virtual environments.

Researchers also explored synthetic data generation for indoor environments. Notable efforts from (Handa et al., 2015) showed the possibility of creating synthetic images of indoor scenes by randomly placing a virtual camera in 3D models, which included offices, bedrooms, kitchens, living rooms and bathrooms. The proposed dataset was used to train the navigation of indoor robot and unmanned aerial vehicles (UAV). More recently, synthetic point clouds of indoor scenes were generated by (Ma et al., 2020) for the segmentation building components. 3D models of indoor objects were created in SketchUp and then imported into FME Workbench for point cloud data generation, with an eventual goal of generating building information models from point clouds for existing buildings without BIM.

Despite significant strides in creating synthetic datasets in the AEC domain to combat the issues of manual data acquisition and labeling, a main research gap exists, as the connection between BIM and synthetic data is rarely examined. Even when BIM is used as the base model to generate synthetic data (Ma et al., 2020), the process is usually partially manual, with BIM only being used as the 3D geometric input. Motivated by the previous research and the research gap, we attempt to propose an automated approach to generate diverse synthetic images from BIM for QA. Specifically, we focused on using a convolutional neural network (CNN) to solve the problem of semantic segmentation of synthetic images into indoor building components.

8.3 METHODOLOGY

In this study, we proposed an automated approach to generate synthetic images of indoor environments and applied a CNN-based algorithm to conduct semantic segmentation of the generated images. On a high level, the approach, as seen in Figure 8.1, is composed of three main components: a BIM, a game engine, and a CNN model. First, the BIM served as the model geometry and building information contributor, where the 3D model of a building is created and later transferred to the

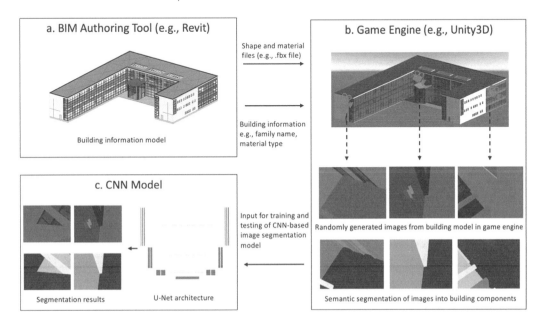

FIGURE 8.1 Three main components of the approach for automatic semantic segmentation of synthetic images into building components.

game engine. Next, in the game engine, we generated synthetic building images by randomizing the position and rotation of the camera to take "photos" of the building components. Thereafter, a linking module built in the game engine was used to trace back to the BIM software and retrieve semantic information of the building components encompassed in the images. Finally, we applied a CNN model to conduct semantic segmentation of the generated synthetic images.

8.3.1 Connecting BIM and a Game Engine for Automated Retrieval of Building Information for Semantic Segmentation

In this first step, we implemented a cloud-based approach to connect a BIM authoring tool (i.e., Revit) with a game engine (i.e., Unity 3D). The reason for choosing a game engine over a commercial BIM software to generate images is twofold. First, existing software suites (e.g., IrisVR, InsiteVR) lack the capability to render high-resolution and realistic images of indoor environments. Second, existing software lacks programming APIs which enable flexible randomization of camera positions and generating pixel-level class labels for building components. Therefore, we first imported 3D building models from Revit to Unity 3D (Du et al., 2018), including model geometry and materials. Next, a cloud-based linking module (as seen in Figure 8.2) was created to establish server-client connections to allow the flow of semantic building information from Revit to Unity 3D.

In Revit, we created a custom plugin to find all building components in the model, and retrieve their parameters (e.g., family type, location) to store in a Hashtable (i.e., a C# data structure based on key-value pairs). Next, we initialized a TCP/IP connection from Revit to the cloud server to upload the Hashtable of BIM objects and parameters. If the model has been changed, we can run the plugin again to reflect any changes to the cloud server. On the server, we maintained a hashtable of all objects and their parameters sent from Revit, and updated the existing data when a new request was sent from the Revit plugin. Finally, we created a plugin in Unity to retrieve the detailed building information (i.e., hashtable of objects and their parameters) from the cloud server and matched the objects with the game objects in the scene. The matching process utilizes the Element ID, which is a unique identifier for each BIM object, and is shared between the game objects in Unity 3D and BIM objects in Revit.

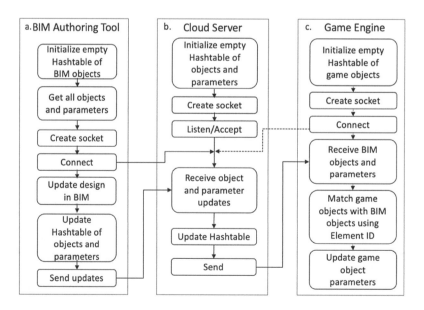

FIGURE 8.2 Cloud-based linking module for transmitting objects in BIM and their parameters to the game engine.

TABLE 8.1

Classes of Building Components Included for Semantic Segmentation Task

Building Component Group	Type of Building Components
Architectural	Ceiling, Floors, Walls, Doors, Windows, Furniture, Lighting, Stairs
Structural	Columns, Framing, Truss
Mechanical Electrical and Plumbing	Duct, Pipes, Conduits, Plumbing

The essential information to transfer from Revit to Unity 3D is the type of an object, which was used to provide true labels of the game objects in Unity 3D. To this end, we adopted a combination of the widely used Revit family classification system and the Uniformat building component classification system, following similar setups of previous studies (e.g., Czerniawski and Leite, 2020). Two types of families were used from the Revit classification system, namely, system families and loadable families. System families are basic elements that are assembled or constructed onsite, such as walls, ducts, and floors. Other the other hand, loadable families are building and system components that would usually be purchased, delivered, and installed in and around a building, such as windows, doors, boilers, water heaters, and plumbing fixtures. Uniformat is a widely used standard for classifying construction information in North America, developed in collaboration between the Construction Specification Institute in the U.S., and Construction Specifications Canada (Charette and Marshall, 1999; Kasi and Chapman, 2011). Uniformat separates construction objects using different levels of groups, including major group elements (level 1), group elements (level 2), and individual elements (level 3). After combining the two classification standards, a total of 15 classes were included in our semantic segmentation task (as seen in Table 8.1).

8.3.2 SYNTHETIC IMAGE GENERATION IN A GAME ENGINE

In this step, we focused on bridging the 'reality gap' between simulated built environments and physically built environments to improve data availability for vision-based algorithms in AEC tasks. To this end, we explored domain randomization, which is a technique used for improving the performance of computer vision algorithms by training on simulated images with randomized rendering. The goal of domain randomization is to generate enough variations through simulation so that the real world will appear as another simulated variation (Tobin et al., 2017). In this study, we focused on randomizing the location and rotation of the camera in the game engine to generate synthetic images. As seen in Figure 8.3, for each 3D model rendered in the game engine (i.e., Unity 3D), we randomly generated the location and rotation parameters of the camera in the scene, and then assigned the camera to that location and rotation to create an image, until we reached the maximum of the number of images needed. This automated approach is fast and cost-effective as compared to physically taking photos in buildings. Moreover, by using BIMs of a variety of building types, this approach is well-suited for providing a diverse dataset of images containing a variety of building components.

Since the game objects in Unity3D are connected to the cloud server, which holds building information for each object, we can easily retrieve the semantic information and assign it to the corresponding game object. Since we focused on semantic segmentation, a plugin was created in Unity 3D to link the true labels (shown in colors in Figure 8.4) of game objects in a scene to the 15 classes of building components identified in step one. As seen in Figure 8.4, we assigned a unique color to each class of objects and then used the colored image as the ground truth of class labels. Eventually, for each randomly generated image (i.e., images shown in Figure 8.3), we obtained a companion image with pixel-level classification labels (i.e., images shown in Figure 8.4). These images were then separated into training and testing sets for developing the semantic segmentation algorithms.

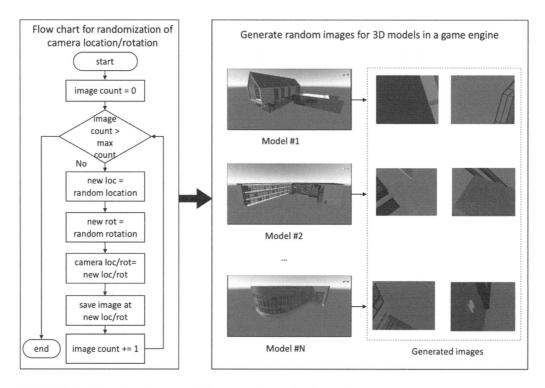

FIGURE 8.3 Flow chart of automatically generating random images in a game engine.

8.3.3 SEMANTIC SEGMENTATION OF SYNTHETIC IMAGES INTO BUILDING COMPONENTS USING U-NET

In this step, we focused on addressing the task of semantic segmentation of images into building components. To this end, we used a CNN-based approach, as CNN has been proven effective for semantic segmentation (Ros et al., 2016; Zhou et al., 2017; Zhu et al., 2019). A typical architecture of CNN includes an input layer, a sequence of alternating convolutional layers and pooling layers, several fully connected layers, and finally the output layer. At the heart of a CNN architecture are

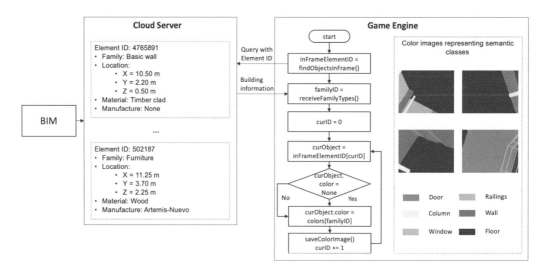

FIGURE 8.4 Retrieving class labels of building components from the cloud server and assigning colors.

the convolutional layers. A convolutional layer applies a set of filters (i.e., a set of tensors, each with equal width and height while the depth is the same as the number of channels of the input) to convolve the input (as shown in Equation 8.1), and the result of which would be passed through an activation function such as rectified linear unit (ReLu) to generate a set of feature maps (i.e., the output of the convolutional layer that contains the information collected from the input). The stacked feature maps are subsequently passed to the next layer.

Unlike that of a regular fully connected neural network, each neuron in the convolutional layer is connected to only a local region (i.e., a small patch of the input image), which helps extract local features (i.e., the information of local regions such as edges and corners) from the input. As the network grows deeper, higher-level features of the input (e.g., contours of objects) will be detected. The convolution operation on a two-dimensional image with one channel is defined as follows:

$$O[m, n] = (I * K)[m, n] = \sum_{i}\sum_{j} I[i, j] \cdot K[m - i, n - j] \tag{8.1}$$

where O represents the output of a convolutional layer; I represents the input; K represents a filter; m and n are the indices. A pooling layer typically follows a convolutional layer to compress similar features in each local region. Pooling layers decrease the size of feature maps, which boosts the computation speed and prevents overfitting. After pairs of convolutional and pooling layers, the result (i.e., the detected features) is passed to fully connected layers, which take all features passed from previous layers into consideration and make a final prediction based on a global view. In this study, we adopted U-Net (Ronneberger et al., 2015), as the underlying CNN architecture, since U-Net was designed for semantic segmentation tasks so it requires less training data while being able to achieve high performance. The basic architecture of U-Net is shown in Figure 8.5.

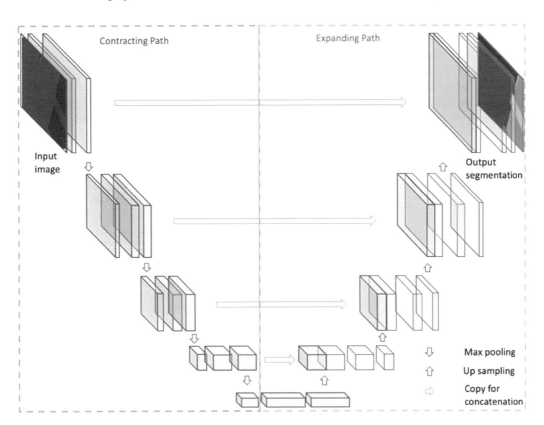

FIGURE 8.5 The U-Net structure used for semantic segmentation of images.

The original U-Net structure is divided into two halves (Ronneberger et al., 2015). As seen in Figure 8.5, the left half is the contracting path, with four successive blocks, each consisting of two convolution operations and a max pooling operation for down-sampling. The right half is the expansive path, with four successive blocks, each consisting of an up-convolution (i.e., transposed convolution) for up-sampling, a concatenation with the corresponding feature map from the contracting path, which contains the feature details lost during down-sampling, and two convolution operations. Every convolution operation in U-Net is activated by ReLu. For our application, we adopted the original U-Net architecture with a simple modification of changing the output classes to 15 in accordance with the total number of classes identified in step one.

8.4   RESULTS AND DISCUSSIONS

The main objective of this study is to prove the effectiveness of using synthetic images generated from BIM and a game engine for the task of semantic segmentation of images into building components. Therefore, we held the semantic segmentation algorithm as a control variable, using only variations of the U-Net architecture. In the following sections, we will first introduce the BIMs used for generating synthetic images, then the distributions of components in the 15 classes, and finally the performance metrics and results of our U-Net model on the segmentation task.

8.4.1   BIMs Used to Generate Synthetic Images

In this study, we used five building information models from a variety of building types and sizes as the base from which to generate synthetic images. Figure 8.6 shows the 3D geometry of the BIMs. In summary, we had two educational buildings, two commercial buildings, and one residential building. For each building, the final model used was merged from architectural, structural, and MEP (mechanical, electrical, and plumbing) models.

To generate the synthetic images, we first connected each building information model to the cloud server, where five repositories of building information were created, one for each building. Next, we transferred the five models into Unity 3D, and ran the plugin to retrieve the building information from the cloud server, and then matched the game objects with building components using Element IDs. Finally, we ran the script to randomly place the camera in Unity 3D to take "photos" of building components, with semantic information labeled simultaneously using unique colors.

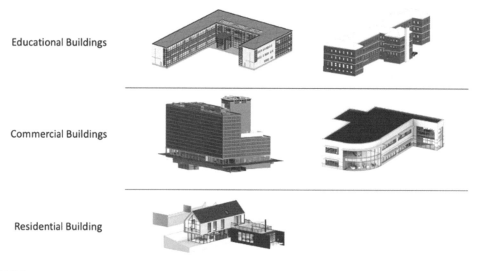

FIGURE 8.6   BIMs used for generating synthetic images.

TABLE 8.2

Distribution of Building Components in Generated Images

Ceilings	35.71%	Furniture	11.67%	Truss	2.00%
Floors	40.10%	Lighting	6.67%	Duct	5.30%
Walls	42.85%	Stairs	3.59%	Pipes	4.75%
Doors	8.33%	Columns	4.10%	Conduits	1.99%
Windows	5.17%	Framing	3.09%	Plumbing	4.17%

Finally, we took 10,000 images for each building for the training dataset, 50,000 images in total. Then, we generated 2,000 images for each building for the test set, 10,000 images in total. The distribution for the 15 classes of building components in the training and testing datasets is shown in Table 8.2. Because the camera was randomly put in the scene, the distribution of the building components reflects the proportion of each building component class in the five BIMs. From Table 8.2, it can be seen that ceiling, floor, and wall occurred the most in the synthetic datasets, while conduits, truss, and framing appeared the least. If an object did not fall into the 15 classes of building component categories, we labeled it as background.

8.4.2 Model Architecture and Parameters for U-Net

The U-Net architecture is detailed in Figure 8.5. The essence of the U-Net architecture lies in the contracting and expansive paths on the left and right hand side (Ronneberger et al., 2015), which serve as a pair of encoding and decoding operations. The contracting path encodes the "knowledge" of the input image to a set of condensed feature maps in the middle, and then the expansive path decodes the condensed feature maps to produce pixel-level image segmentation results (Ibtehaz and Rahman, 2020). Since the original U-Net, multiple advances have been made to apply more complex CNN architectures as the encoder and decoder (Howard and Gugger, 2020; Ibtehaz and Rahman, 2020), such as VGG (Simonyan and Zisserman, 2014) and ResNet (He et al., 2016). In this study, we used the original U-Net architecture, and the modified architectures with ResNet18, ResNet34, and ResNet50 as the encoder and decoder. The model parameters for these architectures are shown in Table 8.3.

8.4.3 Performance of Semantic Segmentation Using U-Net

In this study, we used categorical cross entropy as the training and validation loss to evaluate our models. For 15 classes, the cross entropy was calculated as, $L = -\frac{1}{N}\Sigma_{i=1}^{N}\Sigma_{k=1}^{K}y_{i,k}logp_{i,k}$, where K is the number of building component classes (i.e., 15); N is the number of training examples (i.e., 50000); and $y_{i,k}$ is the either one, when the training sample i has the class label k or zero otherwise; and $p_{i,k}$ is the probability of $y_{i,k}$ being one.

TABLE 8.3

U-Net Architectures and Number of Parameters

U-Net Architecture	Number of Parameters
Original U-Net	8 million
U-Net with ResNet18	31 million
U-Net with ResNet34	41 million
U-Net with ResNet50	339 million

TABLE 8.4

Performance of Semantic Segmentation Models on Testing Images

U-Net Architecture	Average Pixel-wise Accuracy	Percentage Increase for Average Pixel-wise Accuracy over Original U-Net	Mean IoU	Percentage Increase for Mean IoU over Original U-Net
Original U-Net	0.86	-	0.42	-
U-Net with ResNet18	0.89	3.49%	0.53	26.19%
U-Net with ResNet34	0.92	6.98%	0.61	45.24%
U-Net with ResNet50	0.94	9.30%	0.62	47.62%

Once the model converged on the training set, we deployed the trained model on the testing set. To evaluate the model performance, we used average pixel-wise accuracy and mean intersection-over-union (IoU), as testing metrics. Pixel-wise accuracy is the percent of pixels in an image that was correctly classified, which gives an overall evaluation of the segmentation performance. However, pixel-wise accuracy is not representative when the image has imbalanced classes. For example, when an image is only 5% window and 95% wall, classifying the whole image as wall will result in 95% pixel-wise accuracy, which does not represent the segmentation performance for all classes. During image generation, because we randomly placed the camera with different locations and rotations, it is likely for class imbalance to occur in some of the training images. Therefore, we adopted mean IoU as another performance metric, which is essentially the overlap between the predicted segmentation and the ground truth, divided by the union between the predicted segmentation and the ground truth. Mean IoU considers all classes in an image, and has been proven effective in dealing with class imbalance (Thoma, 2016).

Table 8.4 shows performance of the models built using U-Net architectures detailed in Table 8.3. It can be seen that the original U-Net had the lowest average pixel-wise accuracy and mean IoU, and more complex models performed better. Specifically, the performance increased as the complexity of the model increased, with U-Net with ResNet50 as the encoder/decoder achieving the highest average pixel-wise accuracy (0.94) and mean IoU (0.62).

To visualize the performance, we ranked the segmentation results from the test set by mean IoU, and randomly sampled ten images each from the best-segmented 100 and worst-segmented 100 images. As shown in Figures 8.7 and 8.8, the trained model generally performs well when the image is composed of simple contours from building component classes such as walls, ceilings, and columns. On the other hand, the model struggles when the contours of the building components are complex, including classes such as furniture and railings. This is expected since furniture, although only represented as one class in the training and testing data, has various shapes, increasing the difficulty of correctly labeling pixels that belong to furniture. This can be potentially improved by having detailed categories of various types of furniture such as tables, chairs, and sofas.

8.4.4 LIMITATIONS

High performance of the segmentation model showed the viability of using synthetic data set for automated building component segmentations from images. However, several limitations of the study exist and shed light on future research. First, the imbalance of building component classes can be an issue if the training data is small, since the segmentation model will not have enough examples from a certain class of building component, which causes misclassification. This problem can potentially be solved by generating more images for the training set, given it is efficient to obtain synthetic data, and it is possible to obtain images that contain specific classes of building components in game engines. Second, the generalization capability of models trained using synthetic data

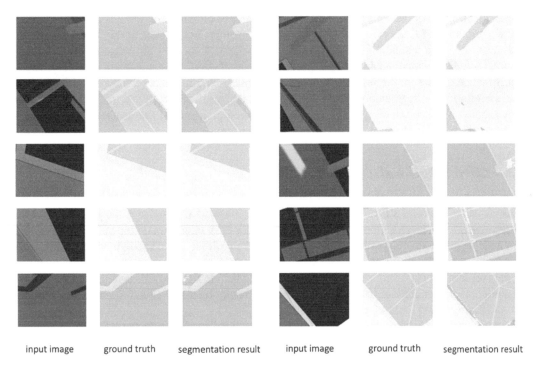

input image   ground truth   segmentation result   input image   ground truth   segmentation result

FIGURE 8.7 Sampled Best performing segmentation results based on mean IoU.

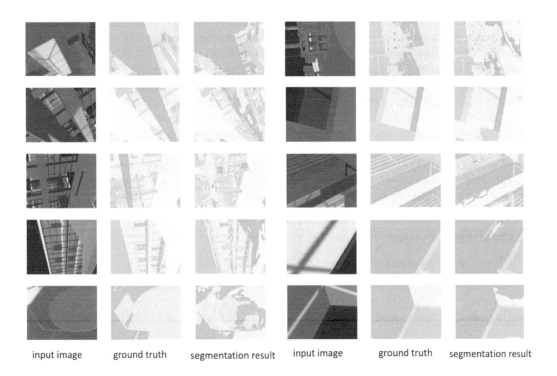

input image   ground truth   segmentation result   input image   ground truth   segmentation result

FIGURE 8.8 Sampled worst-performing segmentation results based on mean IoU.

needs further validation. Previous studies showed the use of synthetic point clouds can improve the performance of CNN-based models for building component segmentation (Ma et al., 2020). However, further research is needed to validate if similar effects exist for image-based segmentations of building components. Finally, we used average pixel-wise accuracy and mean IoU as the performance metrics for the segmentation models. While these are standard metrics widely used for the task of semantic segmentation, they are not designed specifically for problems in AEC. Further investigation is needed to design metrics that are tailored for semantic segmentation in AEC, such as geometrical and topological connections between building classes.

8.5   CONCLUSION AND FUTURE WORK

In this study, we proposed an automated approach for generating synthetic images of building components using BIM and a game engine. The approach is aimed at addressing the pressing need for large, diverse, and labeled image data sets of an indoor built environment for training and testing semantic segmentation algorithms. The approach links building information models with a game engine to randomly generate "photos" of building components at various locations and rotations with pixel-level labels. The proposed approach lays a concrete step towards automating the process of quality assurance for construction projects using imagery inputs by allowing automated segmentation of images into building components.

Several conclusions can be drawn from the study. First, results from the building component class distribution showed the viability of the approach for generating diverse sets of images containing multiple classes of building components. Second, it was shown that a connection between BIM and a game engine can enable the automated retrieval of building component classes in a "photo" taken in the game engine, which solves the issue of automated labeling of the image. Next, the synthetic images were used to train four U-Net-based neural networks, and the best model using ResNet 50 as the underlying encoder/decoder architecture achieved 0.62 mean IoU, which rivals the current state of the art in image segmentation for indoor environments (Liu et al., 2021). From the segmentation performance, it is safe to conclude that synthetic images can be effective in providing the training and testing images for indoor building component semantic segmentation.

For future work, we plan to deploy the developed model to real-world building images and test the performance of a model trained purely on synthetic data. Previous studies investigating the performance of segmentation models trained on synthetic point clouds of building components (Ma et al., 2020) showed improvement of the model when synthetic data was used together with real-world data. However, it is unclear how the performance of image segmentation models will change when mixing synthetic and real data. Another thread of future work is to apply the segmentation model for geometric QA. As the performance of the segmentation model continues to improve through training from larger data sets, we will deploy the model for detecting dimensional, geometrical, and positional variations for as-built structures.

ACKNOWLEDGEMENT

We acknowledge the support of the Natural Sciences and Engineering Research Council of Canada (NSERC), [funding reference number: ALLRP 570442-2021].

REFERENCES

Agapaki, E., & Brilakis, I. (2020). CLOI-NET: Class segmentation of industrial facilities' point cloud datasets. *Advanced Engineering Informatics*, 45, 101121.
Charette, R. P., & Marshall, H. E. (1999). *UNIFORMAT II Elemental Classification for Building Specifications, Cost Estimating, and Cost Analysis*. US Department of Commerce, Technology Administration, National Institute of Standards and Technology.

Chen, M., Feng, A., Hou, Y., McCullough, K., Prasad, P. B., & Soibelman, L. (2021). Ground material classification and for UAV-based photogrammetric 3D data A 2D-3D Hybrid Approach. *arXiv preprint. arXiv:2109.12221.*

Chen, M., Feng, A., McCullough, K., Prasad, P. B., McAlinden, R., & Soibelman, L. (2020a). Generating synthetic photogrammetric data for training deep learning based 3D point cloud segmentation models. *arXiv preprint. arXiv:2008.09647.*

Chen, M., Feng, A., McCullough, K., Prasad, P. B., McAlinden, R., Soibelman, L., & Enloe, M. (2020b). Fully automated photogrammetric data segmentation and object information extraction approach for creating simulation terrain. *arXiv preprint arXiv:2008.03697.*

Chen, L. C., Papandreou, G., Schroff, F., & Adam, H. (2017). Rethinking atrous convolution for semantic image segmentation. *arXiv preprint. arXiv:1706.05587.*

Cho, S. H., Lee, K. T., Kim, S. H., & Kim, J. H. (2019). Image processing for sustainable remodeling: Introduction to real-time quality inspection system of external wall insulation works. *Sustainability, 11*(4), 1081.

Czerniawski, T., & Leite, F. (2020). Automated segmentation of RGB-D images into a comprehensive set of building components using deep learning. *Advanced Engineering Informatics, 45*, 101131.

Danielczuk, M., Matl, M., Gupta, S., Li, A., Lee, A., Mahler, J., & Goldberg, K. (2019, May). Segmenting unknown 3D objects from real depth images using mask R-CNN trained on synthetic data. In *2019 International Conference on Robotics and Automation (ICRA)* (pp. 7283–7290). IEEE.

Dimitrov, A., & Golparvar-Fard, M. (2015). Segmentation of building point cloud models including detailed architectural/structural features and MEP systems. *Automation in Construction, 51*, 32–45.

Dinh, T. H., Ha, Q. P., & La, H. M. (2016, November). Computer vision-based method for concrete crack detection. In *2016 14th International Conference on Control, Automation, Robotics and Vision (ICARCV)* (pp. 1–6). IEEE.

Du, J., Shi, Y., Zou, Z., & Zhao, D. (2018). CoVR: Cloud-based multiuser virtual reality headset system for project communication of remote users. *Journal of Construction Engineering and Management, 144*(2), 04017109.

Gopalakrishnan, K., Khaitan, S. K., Choudhary, A., & Agrawal, A. (2017). Deep convolutional neural networks with transfer learning for computer vision-based data-driven pavement distress detection. *Construction and Building Materials, 157*, 322–330.

Guo, B. H., Zou, Y., Fang, Y., Goh, Y. M., & Zou, P. X. (2021). Computer vision technologies for safety science and management in construction: A critical review and future research directions. *Safety Science, 135*, 105130.

Hamledari, H., McCabe, B., & Davari, S. (2017). Automated computer vision-based detection of components of under-construction indoor partitions. *Automation in Construction, 74*, 78–94.

Han, J., Rong, M., Jiang, H., Liu, H., & Shen, S. (2021). Vectorized indoor surface reconstruction from 3D point cloud with multistep 2D optimization. *ISPRS Journal of Photogrammetry and Remote Sensing, 177*, 57–74.

Handa, A., Patraucean, V., Badrinarayanan, V., Stent, S., & Cipolla, R. (2015). SceneNet: understanding real world indoor scenes with synthetic data. *arXiv preprint. arXiv preprint arXiv:1511.07041.*

He, K., Zhang, X., Ren, S., & Sun, J. (2016). Deep residual learning for image recognition. In *Proceedings of the IEEE Conference on Computer Vision and Pattern Recognition* (pp. 770–778).

Hou, Y., Chen, M., Volk, R., & Soibelman, L. (2021). An approach to semantically segmenting building components and outdoor scenes based on multichannel aerial imagery datasets. *Remote Sensing, 13*(21), 4357.

Howard, J., & Gugger, S. (2020). Fastai: A layered API for deep learning. *Information, 11*(2), 108.

Hua, Y., Marcos, D., Mou, L., Zhu, X. X., & Tuia, D. (2021). Semantic segmentation of remote sensing images with sparse annotations. *IEEE Geoscience and Remote Sensing Letters, 19*, 1–5.

Hwang, B. G., Thomas, S. R., Haas, C. T., & Caldas, C. H. (2009). Measuring the impact of rework on construction cost performance. *Journal of Construction Engineering and Management, 135*(3), 187–198.

Ibtehaz, N., & Rahman, M. S. (2020). MultiResUNet: Rethinking the u-net architecture for multimodal biomedical image segmentation. *Neural Networks, 121*, 74–87.

Idrees, M. D., Hafeez, M., & Kim, J. Y. (2017). Workers' age and the impact of psychological factors on the perception of safety at construction sites. *Sustainability, 9*(5), 745.

Iglesias, C., Martínez, J., & Taboada, J. (2018). Automated vision system for quality inspection of slate slabs. *Computers in Industry, 99*, 119–129.

Jaderberg, M., Simonyan, K., Vedaldi, A., & Zisserman, A. (2014). Synthetic data and artificial neural networks for natural scene text recognition. *arXiv preprint. arXiv:1406.2227.*

Johnson-Roberson, M., Barto, C., Mehta, R., Sridhar, S. N., Rosaen, K., & Vasudevan, R. (2016). Driving in the matrix: Can virtual worlds replace human-generated annotations for real world tasks? *arXiv. preprint arXiv:1610.01983.*

Josephson, P. E., & Hammarlund, Y. (1999). The causes and costs of defects in construction: A study of seven building projects. *Automation in Construction, 8*(6), 681–687.

Kasi, M., & Chapman, R. E. (2011). Proposed UNIFORMAT II Classification of Bridge Elements.

Kazemian, A., Yuan, X., Davtalab, O., & Khoshnevis, B. (2019). Computer vision for real-time extrusion quality monitoring and control in robotic construction. *Automation in Construction, 101,* 92–98.

Kim, M. K., Wang, Q., & Li, H. (2019). Non-contact sensing based geometric quality assessment of buildings and civil structures: A review. *Automation in Construction, 100,* 163–179.

Koch, C., Georgieva, K., Kasireddy, V., Akinci, B., & Fieguth, P. (2015). A review on computer vision based defect detection and condition assessment of concrete and asphalt civil infrastructure. *Advanced Engineering Informatics, 29*(2), 196–210.

Kopsida, M., & Brilakis, I. (2020). Real-time volume-to-plane comparison for mixed Reality–Based progress monitoring. *Journal of Computing in Civil Engineering, 34*(4), 04020016.

Lin, K. L., & Fang, J. L. (2013). Applications of computer vision on tile alignment inspection. *Automation in Construction, 35,* 562–567.

Liu, Z., Cao, Y., Wang, Y., & Wang, W. (2019). Computer vision-based concrete crack detection using u-net fully convolutional networks. *Automation in Construction, 104,* 129–139.

Liu, Z., Hu, H., Lin, Y., Yao, Z., Xie, Z., Wei, Y., & Guo, B. (2021). Swin Transformer V2: Scaling Up Capacity and Resolution. *arXiv preprint. arXiv:2111.09883.*

Ma, J. W., Czerniawski, T., & Leite, F. (2020). Semantic segmentation of point clouds of building interiors with deep learning: Augmenting training datasets with synthetic BIM-based point clouds. *Automation in Construction, 113,* 103144.

Maalek, R., Lichti, D. D., & Ruwanpura, J. Y. (2019). Automatic recognition of common structural elements from point clouds for automated progress monitoring and dimensional quality control in reinforced concrete construction. *Remote Sensing, 11*(9), 1102.

Martinez, P., Al-Hussein, M., & Ahmad, R. (2019). A scientometric analysis and critical review of computer vision applications for construction. *Automation in Construction, 107,* 102947.

Phares, B. M., Washer, G. A., Rolander, D. D., Graybeal, B. A., & Moore, M. (2004). Routine highway bridge inspection condition documentation accuracy and reliability. *Journal of Bridge Engineering, 9*(4), 403–413.

Rahimian, F. P., Seyedzadeh, S., Oliver, S., Rodriguez, S., & Dawood, N. (2020). On-demand monitoring of construction projects through a game-like hybrid application of BIM and machine learning. *Automation in Construction, 110,* 103012.

Ritchie, H., & Roser, M. (2018). Urbanization. Our world in data.

Ronneberger, O., Fischer, P., & Brox, T. (2015, October). U-net: Convolutional networks for biomedical image segmentation. In *International Conference on Medical Image Computing and Computer-Assisted Intervention* (pp. 234–241). Springer, Cham.

Ros, G., Sellart, L., Materzynska, J., Vazquez, D., & Lopez, A. M. (2016). The synthia dataset: A large collection of synthetic images for semantic segmentation of urban scenes. In *Proceedings of the IEEE Conference on Computer Vision and Pattern Recognition* (pp. 3234–3243).

Shi, Y., Du, J., Ahn, C. R., & Ragan, E. (2019). Impact assessment of reinforced learning methods on construction workers' fall risk behavior using virtual reality. *Automation in Construction, 104,* 197–214.

Simonyan, K., & Zisserman, A. (2014). Very deep convolutional networks for large-scale image recognition. *arXiv preprint. arXiv:1409.1556.*

Soltani, M. M., Zhu, Z., & Hammad, A. (2016). Automated annotation for visual recognition of construction resources using synthetic images. *Automation in Construction, 62,* 14–23.

Thoma, M. (2016). A survey of semantic segmentation. *arXiv preprint. arXiv:1602.06541.*

Tobin, J., Fong, R., Ray, A., Schneider, J., Zaremba, W., & Abbeel, P. (2017, September). Domain randomization for transferring deep neural networks from simulation to the real world. In *2017 IEEE/RSJ International Conference on Intelligent Robots and Systems (IROS)* (pp. 23–30). IEEE.

Wu, Y., & Xue, F. (2021). FloorPP-Net: Reconstructing Floor Plans using Point Pillars for Scan-to-BIM. *arXiv preprint. arXiv:2106.10635.*

Yin, X., Liu, H., Chen, Y., & Al-Hussein, M. (2019). Building information modelling for off-site construction: Review and future directions. *Automation in Construction, 101,* 72–91.

Yue, X., Wu, B., Seshia, S. A., Keutzer, K., & Sangiovanni-Vincentelli, A. L. (2018, June). A lidar point cloud generator: from a virtual world to autonomous driving. In *Proceedings of the 2018 ACM on International Conference on Multimedia Retrieval* (pp. 458–464).

Zhong, B., Wu, H., Ding, L., Love, P. E., Li, H., Luo, H., & Jiao, L. (2019). Mapping computer vision research in construction: Developments, knowledge gaps and implications for research. *Automation in Construction, 107*, 102919.

Zhou, B., Zhao, H., Puig, X., Fidler, S., Barriuso, A., & Torralba, A. (2017). Scene parsing through ade20k dataset. In *Proceedings of the IEEE Conference on Computer Vision and Pattern Recognition* (pp. 633–641).

Zhu, Y., Sapra, K., Reda, F. A., Shih, K. J., Newsam, S., Tao, A., & Catanzaro, B. (2019). Improving semantic segmentation via video propagation and label relaxation. In *Proceedings of the IEEE/CVF Conference on Computer Vision and Pattern Recognition* (pp. 8856–8865).

9 Truss Bridge Construction Using Self-Propelled Modular Transporters Compared to Traditional Bridge Building Methods

Soliman Khudeira

CONTENTS

DOI: 10.1201/9781003325246-9

9.1   INTRODUCTION

A self-propelled modular transporter (SPMT) is a computerized flat vehicle used to move bridges or other heavy structure with precision (Figure 9.1). Assembly of a bridge is done off-site in an adjacent staging area under safe conditions and then followed by moving the structure to its final location [1]. Traffic impacts on roads and railroads are measured in hours compared to the many months when conventional bridge construction techniques are used.

Utilizing SPMTs will reduce the project's construction time. This is due to the reduction of the multiple sequential and time-consuming processes of conventional bridge construction methods. The SPMT reduces the various steps in conventional methods to only one step: moving the prefabricated structure from the nearby assembly area to the final position. This technology needs be considered and utilized for all projects where on-site construction time needs to be reduced, stakeholders' convenience recognized, and enhanced safety during construction is a priority. Value Engineering (Value Methodology) should be implemented early in the design

FIGURE 9.1   Self-propelled modular transporters (SPMTs).

phase to assess a project-specific feasibility, cost, and benefits analysis of using SPMTs compared to traditional methods.

Utilizing SPMTs to move prefabricated bridges has a direct positive correlation with reducing construction time, minimizing disturbance to traffic, minimize traffic detour time, improving safety to all stakeholders, reducing negative environmental impact, expanding the options for constructability, improving final quality of the constructed facility, and lowering project cost [1].

The use of SPMTs in bridge projects was the recommended by the Prefabricated Bridge Elements and Systems International Scan sponsored by the Federal Highway Administration (FHWA), the American Association of State Highway and Transportation Officials (AASHTO), and the Transportation Research Board's National (TRB) [1, 2].

Off-site prefabrication of bridges close to the ground and under controlled conditions, then moving the bridge to the final position on-site, can substantially improve the quality in significantly less time compared to conventional bridge construction. The feasibility of using SPMTs should be investigated and implemented in all bridge projects, if found to be feasible, especially in projects where the road has high traffic volumes of more than 40,000 vehicles per day. Conventional methods of bridge construction include various sequential steps all done over live road and rail traffic. Using SPMT technology, the conventional sequence is reduced to only one step, which is to move the prefabricated bridge to its final position.

Using SPMTs can result in time savings due to the change from the sequential step-by-step processes of conventional bridge construction to concurrent processes, followed by using SPMTs to move the bridge to the final on-site location. Assembly of the bridge can begin before site preparation has been completed. The new bridge can be constructed at the adjacent staging area at a lower elevation, which substantially improves construction safety to the worker by avoiding severe environmental factors, such as high wind at higher elevations.

9.2 DETAILS OF THE SPMT's EQUIPMENT

SPMTs are computerized platform used to move structures and components in various industries including petrochemical, offshore, power, and bridges in civil engineering (Figure 9.2). A single SPMT unit has multiple axels. SPMTs can travel on uneven ground with surface variations up to 18 inches. They also can be operated on longitudinal grades with slopes up to 8%. All axles of SPMT are individually controllable. SPMT axles can telescope independent of each other for the

FIGURE 9.2   SPMTs carrying a fully assembled truss bridge.

load to remain horizontal on the platform. SPMTs move slowly at a maximum of 1 mile per hour when the load is heavy and require more care and stability.

9.2.1 BENEFITS OF USING SPMTS

Removing and replacing a bridge on its original horizontal alignment using conventional construction methods typically requires closing the bridge to traffic for a long period of time and detouring the traffic that previously used the bridge. The ability of SPMT technology to move bridges in hours can significantly reduce traffic disruption, restore the use of the highways in significantly less time, improve work zone safety for workers, minimize environmental impact, improve quality of work, improve constructability, lower costs of the project, reduced on-site construction time, and increased contractor options are possible when SMPTs are used to move the structure into final position.

Quality is enhanced by using SPMTs for several reasons. First, the bridges are built in the controlled conditions of the staging area, allowing continuous operations without the need to accommodate traffic. Second, the bridge can be built as a unit in the staging area and moved as a unit to the bridge site, reducing the number of deck joints typically required in prefabricated bridge systems. Bridge owners in Europe have demonstrated this with their multiple-span continuous bridges, both with and without substructures, constructed off-site and moved into final position with SPMTs. Third, work can begin early in the staging area, independent of on-site construction status, allowing adequate time for proper construction of the system and proper curing of the concrete to ensure proper long-term performance. These advantages can lead to improved reliability in achieving the desired 100-year service life compared to conventional construction.

The SPMT driving method increases the contractor's options. Its use allows the new bridge to be assembled at an off-site location and quickly driven in any direction to its final location. Precision movement to position the bridge within a fraction of an inch is possible with SPMTs. Driving also eliminates many of the issues related to overhead height restrictions that impact crane lifting operations, and the supported SPMT loads provide added safety assurance relative to suspended crane loads.

Benefits to contractors also include being able to use conventional construction methods to build spans at staging areas near the installation site. The contractor further benefits from the flexibility of being able to work on the bridge as many hours as needed during the day, as well as from the improved safety and reduced risks of working at a lower height away from traffic. In addition, since foundation and possibly substructure work is being done on-site concurrently with prefabricated construction activities at the near-site staging area, an on-site delay (for example, because of unexpected utilities) will not affect construction at the staging area. Also, since more of the work is done off-site, fewer workers are required over or near traffic, further enhancing safety. Fewer and shorter lane closures also improve safety since any deviation from normal traffic patterns can result in accidents.

9.2.2 PLANNING AND COST

Companies supplying the SPMTs should be included in discussions from the initial project's planning stages. Their inclusion will facilitate the development of contract plans and specifications. If federal funds are used as part of the project funds, the SPMT company that works with the owner in the planning stage will not be allowed to bid on the project, unless all SPMT companies are invited to participate in the planning stage of the project.

The cost of using SPMTs could be high. However, this cost should be weighed against the reduced costs that result from significantly reducing the on-site construction time, the higher final quality, reduced inconvenience to the stakeholders, increased safety to the works and the public, and reduced negative environmental factors because of a shorter project duration.

FIGURE 9.3   Bridge staging area adjacent to the project site used to assemble the truss.

9.2.3   TRAFFIC IMPACT: CONVENTIONAL AND SPMT CONSTRUCTION

The following bridge construction activities are not allowed above active traffic: bridge demolition, beam erection, deck form erection and removal, and concrete deck pouring operations.

Bridge demolition and beam erection typically require full closure of the roadway under the bridge. For conventional construction, concrete deck form erection and removal and concrete pouring operations typically require traffic to be moved from underneath these operations using lane shifts, lane closures, or both. Lane closure restrictions for a specific site may require construction operations to be performed only at night, increasing labor and traffic control costs, and decreasing production rates.

9.2.4   STAGING SITE REQUIREMENTS AND PREPARATION

Sufficient land area (staging area) for bridge assembly should be available for SPMT construction methods (Figure 9.3). The staging area should be within a suitable distance from the bridge site. A feasible path must be available from the staging area to the bridge site, or from the staging area to barges that can go to the bridge site. The ground surface at the staging area and along the path should be relatively flat and have adequate ground-bearing capacity or steel plates provided to resist the loads.

9.2.5   DESIGN REQUIREMENTS PRIOR TO UTILIZING THE SPMTs

Various design factors need to be investigated and properly design for a safe and efficient move. These factors include clearly stating the design assumptions, ground-bearing capacity, temporary shoring bents at staging area, stresses and deflection during the moving operation, and wind loading. Some of these factors are discussed below.

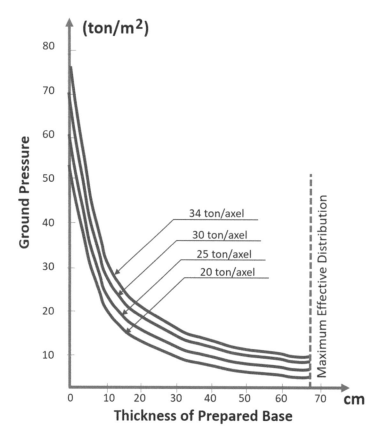

FIGURE 9.4   Relationship between ground pressure, base thickness, and load per axel.

9.2.6 Ground-Bearing Capacity

The contractor must prepare the ground for adequate bearing at the staging area for construction of the bridge on temporary supports, and along the path from the staging area to the final bridge location to resist the weight of the loaded SPMTs. Graphs providing the relationship between ground pressure (ton/m²), required base thickness (cm), and axel load (ton per axel) are provided by the manufactures of the SPMTs. Figure 9.4 is a representative relationship between these variables.

9.2.7 Allowable Temporary Stresses and Deflections

Lifting and moving a bridge with a SPMT usually results in a condition in which the bridge will be temporarily supported at locations that are not the final support points. The designer should check the bridge under these temporary support conditions. Structural members will likely see stress reversals that may be detrimental. An impact factor should be added to the dead load to account for movement on uneven surfaces; a value of at least 15% is recommended.

The analysis should also account for the SPMT system, which will essentially provide equal bearing at each support. This is especially true for larger bridges where multiple sets of interconnected SPMTs are proposed. The SPMTs will provide soft support for the bridge. Appropriate spring coefficients need to be applied at each support so that the bridge will experience essentially equal reactions.

This will result in a redistribution of dead load stresses in the bridge. The designer should calculate and specify an accurate bridge weight, especially for large bridges. For example, a steel fabricator's estimates are not for the finished product because cutting waste has not been included; the final weight of each piece is not determined until the shop drawings are complete. Simple rule-of-thumb weight estimates for items such as cross frames and diaphragms may not be good enough, and the weight of a coating system alone can be substantial. Accuracy is needed to provide an adequate lifting operation.

The contractor should provide the jacking forces and procedures for lifting a superstructure from its temporary supports to its setting height before installation, and the jacking procedures for lowering a removed superstructure from its setting height to a lower height for demolition. The contractor should provide analysis of jacking, moving, and erection loads to ensure that temporary stresses and deflections are within allowable values for the bridge and that the bridge is adequately reinforced to handle the calculated temporary stresses without cracking (for example, deck tensile stress).

The analysis should include the number of jacking towers at the ends of the span, the jacking force per jack, and the locations of the jacks along the beam length. Jacks at each end are to be connected to each other to push and support equal loads. The SPMTs will typically lift the bridge span as near to the final support positions as possible to reduce installation stresses. Lifting points for the move will be controlled by SPMT width and available clearance between obstacles.

Temporary overstresses on the bridge should not be allowed during lifting and moving. Geometric tolerances should avoid excessive stresses while permitting an optimum speed of movement. The contractor should verify that the bridge is adequately braced during the move (for example, end and intermediate diaphragms of superstructure spans are constructed before transporting).

9.2.8   LOAD AND LOAD COMBINATIONS FOR STRUCTURES DURING CONSTRUCTION

An American Society of Civil Engineers (ASCE) document for loading and load combination during construction is available [4], which specifies that structures and their components shall be designed so that their strength exceeds the effects of factored loads.

Structures shall resist the effects of the following loads and load combinations:

Final loads:

D – dead load
L – live load

Construction loads:

Weight of temporary structures
C_D – construction dead load

Material loads:

C_{FML} – fixed material load
C_{VML} – variable material load

Construction procedure loads:

C_P – personnel and equipment load
C_H – horizontal construction loads
C_F – erection and fitting forces
C_R – equipment reactions
C_C – lateral pressure of concrete
Lateral earth pressure
C_{EH} – lateral earth pressures

Environmental loads:

W – wind load
T – thermal load
S – snow load
E – earthquake load
R – rain load
I – ice load

9.2.9   LOAD COMBINATIONS AND LOAD FACTOR FOR STRENGTH DESIGN

The contractor will provide analysis of the loads and stresses on the bridge at its temporary location and during the move. The owner should specify a design wind load and earthquake loads for movement of the bridges. The wind speed should be based on statistical analysis that includes the probability of occurrence. Thunderstorms can occur during the move; therefore, reviewing the weather report before the move needs to be done. An American Society of Civil Engineers (ASCE) document provides guideline on loads and load combinations during construction [4]. The load combinations form this document for strength design and allowable stress design are summarized below.

9.2.9.1   Strength Design

Structures and their components shall be designed so that their strength exceeds the effects of factored loads in the following basic load combinations:

1. $1.4D + 1.4\,C_D + 1.2\,C_{FML} + 1.4C_{VML}$
2. $1.2D + 1.2\,C_D + 1.2\,C_{FML} + 1.4C_{VML} + 1.6C_P + 1.6C_H + 0.5L$
3. $1.2D + 1.2\,C_D + 1.2\,C_{FML} + 1.3W + 1.4C_{VML} + 0.5C_P + 0.5L$
4. $1.2D + 1.2\,C_D + 1.2\,C_{FML} + 1.0E + 1.4C_{VML} + 0.5C_P + 0.5L$
5. $0.9D + 0.9\,C_D + (1.3W\ or\ 1.0E)$

The most unfavorable effects from wind and earthquake loads shall be considered, but they need not be assumed to act simultaneously. Similarly, C_H need not be assumed to act simultaneously with wind or seismic loads.

9.2.9.2   Allowable Stress Design

Specified loads shall be combined to obtain the maximum design load effects for members and systems. The following basic load combinations shall be used:

1. $D + C_D + C_{FML} + C_{VML}$
2. $D + C_D + C_{FML} + C_{VML} + C_P + C_H + L$
3. $D + C_D + C_{FML} + C_{VML} + W + C_P + L$
4. $D + C_D + C_{FML} + C_{VML} + 0.7E + C_P + L$
5. $D + C_D + (W\ or\ 0.7E)$

Reference [4] requires that the load combinations for design loads on bridges during construction to be also in accordance with the AASHTO, AREMA, or other applicable specifications.

9.3   REVIEW OF AREMA-RELATED DESIGN CRITERIA

The following are selected sections from AREMA's design criteria [3]. These sections are selected as they are pertinent to the design of this railroad truss bridge.

TABLE 9.1
Unit Weight of Various Materials

Type	Pounds per Cubic Foot
Steel	490
Concrete	150
Sand, gravel, and ballast	120
Asphalt-mastic and bituminous macadam	150
Granite	170
Paving bricks	150
Timber	60

9.3.1 LOADS AND FORCES

Bridges shall be proportioned for the following loads: (1) dead load, (2) live load, (3) impact load, (4) wind forces, (5) centrifugal force, (6) forces from continuous welded rail, (7) other lateral forces, (8) longitudinal forces, and (9) earthquake forces.

9.3.2 DEAD LOAD

In estimating the weight for the purpose of computing dead load stresses, the unit weights found in Table 9.1 shall be used.

The track rails, inside guard rails, and their rail fastenings shall be assumed to weigh 200 lb per linear foot for each track.

9.3.3 LIVE LOAD

The recommended live load in lb per axle and uniform trailing load for each track is the Cooper E 80 load shown in Figure 9.5, or the Alternate Live Load on four axles spaced as shown in Figure 9.6, whichever produces the greater stresses.

For bridges on curves, provisions shall be made for the increased proportion carried by any truss, girder, or stringer due to the eccentricity of the load.

For members receiving load from more than one track, the design live load on the tracks shall be as follows:

1. For two tracks, full live load on two tracks
2. For three tracks, full live load on two tracks and one-half on the other track
3. For four tracks, full live load on two tracks, one-half on one track, and one-quarter on the remaining one.
4. For more than four tracks, as specified by the engineer.

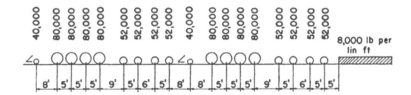

FIGURE 9.5 Cooper E 80 load.

FIGURE 9.6   Alternate live load on four axels.

The selection of the tracks for these loads shall be such as will produce the greatest live load stress in the member.

9.3.4   Distribution of Live Loads

The maximum wheel load on each rail is distributed equally to all ties within a length of 4 feet, but not to exceed three ties, and is applied without impact. For the design of beams or girders, the live load shall be considered as a series of loads, as shown in Figure 9.5 or 9.6. No longitudinal distribution of such loads shall be assumed.

9.3.5   Longitudinal Forces

The longitudinal force for E-80 loading shall be taken as the larger of:

   a. Force due to braking, as prescribed by the following equation, acting 8 feet above top of rail
      Longitudinal braking force (kips) = 45 + 1.2 L
   b. Force due to traction, as prescribed by the following equation, acting 3 feet above top of rail
      Longitudinal traction force (kips) = 25√L

where
   L is length in feet of the portion of the bridge under consideration.

For design loads other than E-80, these forces shall be scaled proportionally. The points of force application shall not be changed.

The longitudinal force shall be distributed to the various components of the supporting structure, considering their relative stiffness. The soil resistance of the backfill behind the abutments shall be utilized where applicable. The mechanisms (rail, bearings, load transfer devices, etc.) available to transfer the force to the various components shall also be considered.

For members receiving load from more than one track, the design load on the tracks shall be as follows: For two tracks, apply the full load on two tracks, and for three tracks, apply the full load on two tracks and one-half on the other track.

Longitudinal force due to braking acts at the center of gravity. The center of gravity height is taken as 8 feet above the top of rail. This force is transferred from vehicle to rail as a horizontal force at the top of rail and a vertical force couple transmitted through the wheels.

Longitudinal forces transmitted by tractive effort of locomotives or the braking action of trains will be distributed to bridge members in accordance with their relative stiffness and orientation with respect to the force path between the applied longitudinal force and the supporting substructure.

In bridges with stringer and floor-beam floor systems, longitudinal forces are first applied to the stringers. The force must then be transferred to the members to which stringers are connected, usually the floor beams. Traction bracing can be used to directly transfer the longitudinal force from the stringers to truss or girder panel points.

In bridges with transverse floor-beam floor systems (such as through girder spans), traction bracing can be used to transfer the longitudinal force to the bridge members supporting the floor beams (typically girders).

It is generally considered good practice to design traction bracing to be the same depth as the member being braced. However, when traction bracing isn't used, the floor beams should be designed for transverse bending and torsion where applicable. It is generally considered good practice to provide traction bracing rather than design floor beams or transverse members for lateral bending and torsion.

Fixed bearings and their anchorages should be designed to transfer the longitudinal force from superstructure to substructure. In addition to designing the fixed bearings to take all the longitudinal forces, it is the practice of some engineers, given that bearings tend to become frozen or stuck with time, to design the area around and below expansion bearings for a εpercentage of the longitudinal forces going through those bearings as though the expansion bearings were partially fixed.

9.3.6 Serviceability Criteria: Limiting LL Deflection

The deflection of the structure shall be computed for the live loading plus impact loading condition producing the maximum bending moment at midspan for simple spans.

The structure shall be so designed that the computed deflection shall not exceed L/640 of the span length center to center of bearings for simple spans.

9.3.7 Camber

The camber of trusses shall be equal to the deflection produced by the dead load plus a live load of 3,000 lb per foot of track.

9.3.8 Thermal Movements (Expansion Joints)

The deformation caused by a change in temperature (ΔL) depends upon the extreme bridge design temperatures (ΔT) in the area, and it is determined as:

$$\Delta L = (\alpha)(L)(\Delta T)$$

where
ΔL= deformation caused by a change in temperature (inch)
α = coefficient of thermal expansion (in./in./°F) = 0.0000065 (for steel and concrete)
L = distance between the expansion joints (inch)
ΔT = change in temperature = (final T − initial T). Temperature range is from −30°F to 130°F, with normal installation of 50°F

9.3.9 Impact Loads

Impact load, due to the sum of vertical effects and rocking effect created by passage of locomotives and train loads, shall be determined by taking a percentage of the live load and shall be applied vertically at top of each rail.

Impact load due to vertical effects: This is expressed as a percentage of live load applied at each rail. For a truss span, this force = [15 + (4,000/(L + 25))], where L = length, feet, center to center of supports for stringers, transverse floor beams without stringers, longitudinal girders, and trusses (main members).

Impact load due to rocking effect, RE, is created by the transfer of load from the wheels on one side of a car or locomotive to the other side from periodic lateral rocking of the equipment. RE shall

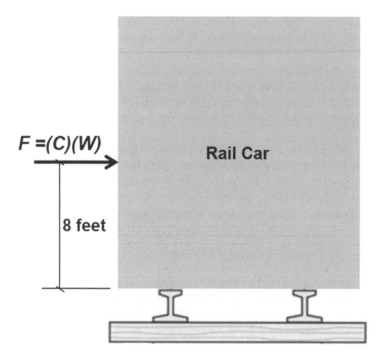

FIGURE 9.7   Location of the centrifugal force above the rail.

be calculated from loads applied as a vertical force couple, each being 20% of the wheel load without impact, acting downward on one rail and upward on the other. The couple shall be applied on each track in the direction that will produce the greatest force in the member under consideration.

9.3.10   CENTRIFUGAL FORCE

On curves, a centrifugal force shall be applied horizontally through a point 8 feet above the top of the rail (Figure 9.7) measured along a line perpendicular to the plane at the top of the rails and equidistant from them.

When a maximum design speed is not specified by the engineer, the centrifugal force shall correspond to 15% of each axle load without impact. The superelevation of the outer rail used in determining the point of application of the force shall be assumed as 6 inches.

When the maximum design speed and superelevation are specified by the engineer, the resulting centrifugal force shall correspond to the percentage of each axle load, without impact, determined by the following formula:

$$C=(0.00117)(S^2)(D)$$

where
    C = centrifugal factor, percent
    S = speed, miles per hour
    D = degree of curve (central angle of curve subtended by a chord of 100 ft.)

9.3.11   LATERAL FORCES FROM EQUIPMENT

A single moving concentrated lateral force equal to one-quarter of the weight of the heaviest axle of the specified live load, without impact, shall be applied at the base of rail in either direction and at any point along the span.

9.3.12   Wind Forces on Loaded Bridge

In general, the wind force shall be considered as a moving load acting in any horizontal direction. As a minimum, the bridge shall be designed for laterally and longitudinally applied wind forces acting independently as follows.

9.3.13   Wind Forces on the Train

The lateral wind force shall be taken at 300 lb. per linear foot applied normal to the train on one track at a distance of 8 feet above the top of the rail.

9.3.14   Wind Forces on the Bridge

Lateral wind pressure shall be taken at 30 lb. per square foot normal to the following surfaces: For truss spans, the vertical projection of the span plus any portion of leeward trusses not shielded by the floor system.

The lateral wind force on girder and truss spans, however, shall not be taken as less than 200 lb. per foot for the loaded chord or flange and 150 lb. per foot for the unloaded chord or flange, neglecting the wind force on the floor system.

The longitudinal wind force on spans shall be taken for truss spans as 50% of the lateral wind force.

9.3.15   Wind Forces on Unloaded Bridge

In general, the wind force shall be considered as a moving load acting in any horizontal direction. As a minimum, the bridge shall be designed for laterally and longitudinally applied wind forces acting independently as follows:

a. The lateral wind force on the unloaded bridge shall be taken as 50 lb. per square foot of surface.
b. The longitudinal wind force on the unloaded spans shall be taken for truss spans as 50% of the lateral wind force.

The specified basic wind force of 50 lb. per square foot on an unloaded structure has a long historic background in railroad specifications. It was assumed that a hurricane wind, during which train operations would not be attempted, could produce a load of 50 lb. per square foot on such surfaces.

9.3.16   Stability Check

For wind, nosing, and centrifugal forces, the vertical weight of a train on a tower or pier usually improves the lateral stability of the structure, so it is prudent to model the least weight train that would be present with the applicable lateral overturning load. A uniform vertical loading of 1,200 lbs/ft applied to the leeward track represents a consist of empty cars. For multiple track structures supported by the same pier(s), only the leeward track is loaded. This stability check is designed to ensure that a load equal to half the full design load on the verge of incipient roll will not cause the span to roll over. It is not intended to prevent damage to the structure, nor is it intended for deck design.

9.4   CASE STUDY – TRUSS BRIDGE DESIGN AND CONSTRUCTION USING SPMTs

A 400-foot-long, 5-million-pound (2,500 ton) steel railroad truss bridge was rolled into place using self-propelled modular transporters (SPMTs). This bridge serves as the top tier of a multi-level grade separation project on the southeast side of Chicago, IL, USA. The 130th Street and Torrence

FIGURE 9.8   Main span of the existing CSS&SB bridge over NS tracks and torrence avenue.

Avenue intersection improvement project is an extremely complex, $200-million program by the Chicago Department of Transportation.

Using the Accelerated Bridge Construction (ABC) techniques was an innovative solution to facilitating construction in a busy, urban environment. ABC was incorporated into the design contract documents. This resulted in transporting the steel truss 800 feet from its staging area to its final position in less than four hours. The commuter-freight rail truss now makes a striking silhouette. In the following sections, the author will discuss the evolution of the truss bridge configuration during design and the use of ABC techniques as a means to facilitate construction of the truss with minimal impacts to the operating railroads and motorists.

The new railroad truss bridge is a key component of the 130th Street and Torrence Avenue project. This area, located on the far southeast side of Chicago, serves 39,000 vehicles per day. Over 52 freight trains cross at-grade Norfolk Southern Railway (NS) tracks nearby at 130th Street and at Torrence Avenue. This site is further complicated with the recent expansion of the adjacent Ford Motor Company Manufacturing Campus.

Eliminating conflicts caused by two at-grade railroad crossings resulted in realigning and grade-separating 130th Street and Torrence Avenue below the existing Norfolk Southern Railroad (NSRR) tracks. The existing CSS&SB bridge (Figure 9.8) is located over the existing NSRR and Torrence Avenue and carries over 40 passenger trains per day. Making sure all the project components fit in this complex puzzle while maintaining all rail traffic required the CSS&SB railroad truss span to be constructed first.

9.4.1   KEY PROJECT COMPONENTS

The project consists of a three-tiered grade separation, including a total of six new bridges (three railroad, one roadway, and two pedestrian/bicyclist bridges); realignment of three major roadways; mixed-use path; over 10,000 linear feet of retaining walls; new drainage system (including underground detention chamber and pump station); street lighting; traffic signals; roadway pavement; extensive landscaping; and other elements (Figure 9.9).

9.4.2   VALUE ENGINEERING (VALUE PLANNING)

During the early stages of the design, alternatives were identified for the assembly and erection of the CSS&SB truss bridge. Results of the evaluation analysis demonstrated that the assembly of the

FIGURE 9.9   Rendering of the completed project showing the truss and other project elements.

truss in a staging area (Figure 9.10) and then transporting it using self-propelled modular transporters (SPMTs) was a cost-effective solution.

The value engineering (VE) sessions explored all possible approaches and methods for the assembly and erection of the CSS&SB truss span. Twenty-five (25) ideas were generated during the VE sessions. Of the 25 ideas, two (2) alternatives were developed and further evaluated: Alternative 1 – Off-Site Assembly with Roll-in, and Alternative 2 – Build in Place. Results of the performance and

FIGURE 9.10   Off-site assembly of the truss bridge at ground elevation.

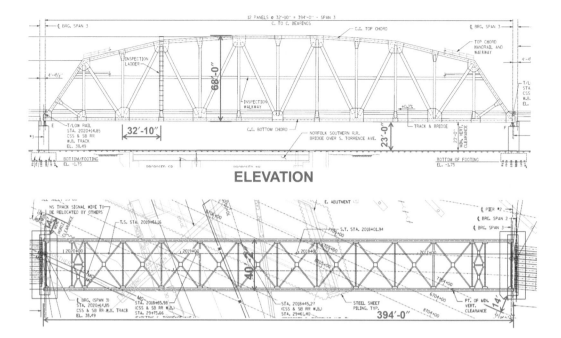

FIGURE 9.11 Final plan and elevation of CSS&SB truss bridge.

acceptance evaluation analysis demonstrated that the Off-Site Assembly with Roll-in Alternative #1 would be safer than Alternative # 2, best address all stakeholders' needs by minimizing impacts to the motoring public, the NS freight operations, and NICTD/CSS&SB commuter operations and by simplifying the assembly and erection of the truss while adding value to the project.

9.4.3 EVOLUTION OF THE TRUSS DESIGN

A clear-span structure was determined to be the most feasible alternative; and at the end of preliminary design, the proposed CSS&SB structure consisted of a 368-foot span truss bridge with abutments skewed at 45 degrees. The 45-degree skew was suggested to have the shortest bridge span. The detailing and fabrication of a skewed portal frames of the truss is more complex than square frames and could add 15% premium to the fabrication cost. This premium is a result of the complex geometry of the portal frames and the loss of economy in the fabrication of the end segments of the floor system.

The elimination of the skew also had two other advantages beyond the impacts to the steel and installation. The volume of concrete required at the abutments was reduced by approximately 30% due to the reduced width of the abutments. The end floor-beam span was reduced from approximately 57'8″ to 40'2″, eliminating the need for an intermediate bearing for the floor beam. The revised and final layout of the truss resulted in a 394′ span center to center of bearings with supports perpendicular to the structure.

The end result is a truss with a span length of 394 feet and a maximum height of 67'4″ at its tallest point. The substructure consists of full height concrete piers supported on driven steel piles. Figure 9.11 shows the final configuration of the truss span.

9.5 TRUSS DESIGN CRITERIA

The truss is a single-span double-track ballast-deck structure. It was designed to support Cooper E-80 live load and other design provisions in accordance with the American Railway Engineering and Maintenance of Way Association Manual (AREMA) [3]. A summary of these design provisions follows:

9.5.1 SERVICEABILITY CRITERIA

Serviceability criteria of limiting live load deflection of L/640; including provisions for 12 inches of ballast.

9.5.2 IMPACT LOAD – ECCENTRIC LOAD

Eccentric load on the truss due to the spiral and super-elevation resulted in additional 10.4% more load on the truss (see Figure 9.12)

9.5.3 IMPACT LOAD – ROCKING EFFECT

Rocking effect, which is an impact load created by the transfer of load from the wheels on one side of a car or locomotive to the other side from periodic lateral rocking of the equipment. As 20% of the wheel load is applied downward on one rail and upward on other rail this results in a 2.5% increase to the impact load (see Figure 9.13).

9.5.4 TOTAL IMPACT LOAD

The total impact load is the sum of the vertical and rocking effects reduced by 10% as AREMA allows for reduction for ballast conditions. Total impact vertical load is: [17.65% (vertical effect) + 2.5% (rocking effect)] (0.90) = 18.14% of the live load applied at each rail.

9.5.5 CENTRIFUGAL FORCE – LATERAL FORCE

Centrifugal force must be accounted for since the tracks are on a spiral at each end of the truss (Figure 9.14): 0.00117(S)(D); where S = 35 mph and D = 4 degrees. This results in a 3.38% increase to the live load.

9.5.6 CENTRIFUGAL FORCE CALCULATION

Per AREMA, on curves, a centrifugal force shall be applied horizontally through a point 8 feet above the top of the rail (Figure 9.15) measured along a line perpendicular to the plane at the top of the rails and equidistant from them.

The centrifugal force is:

$$F = C\,W$$

where
    C = factor (in percentage) = $0.00117\ V^2\ D$
    D = degree of curve = 4 degrees
    W = train weight = 568 kips
    V = speed (mph) = 40 mph

Therefore,
    Factor C = $(0.00117)(40)^2(4°) = 7.49\% = 0.0749$
    The Centrifugal Force:

$$F = (C)\,(W)$$
$$F = (0.0749)(568\ \text{kips})(2\ \text{trains})$$
$$F = 85.0\ \text{kips}$$

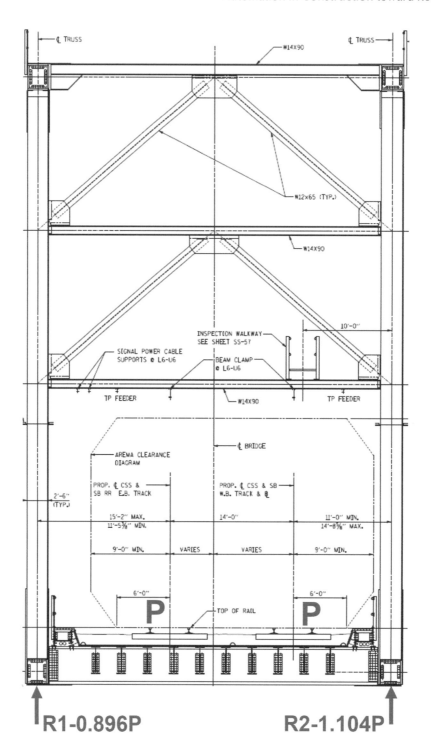

FIGURE 9.12 Eccentric load on truss.

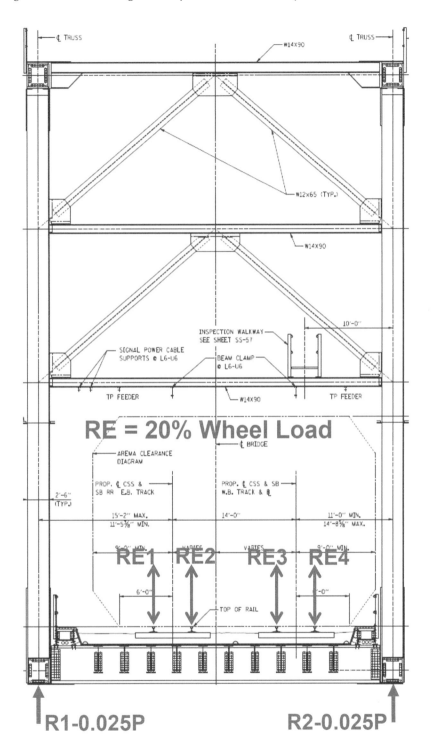

FIGURE 9.13   Rocking effect.

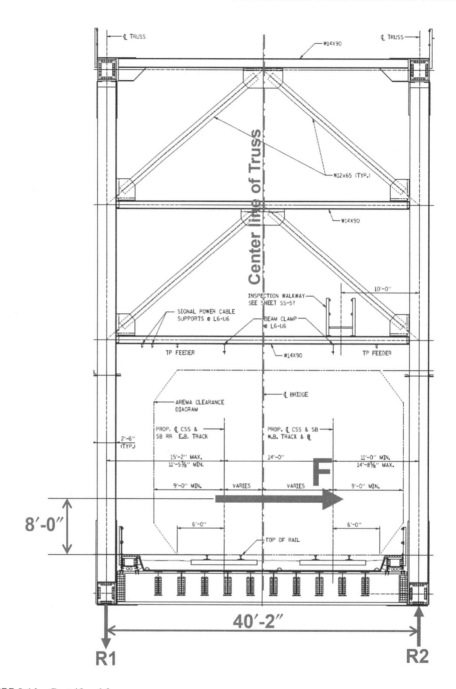

FIGURE 9.14 Centrifugal force.

This force is then distributed to the truss as follows:

$$F = 85.0 \text{ kips}(11.88'/40.16') = 25.2 \text{ kips}$$

9.5.7 Nosing Forces – Lateral Force

Nosing forces are lateral force due to equipment. Nosing forces are calculated by applying 25% of wheel load at the rail base; this results in an additional load of 1.6 kips to each internal panel point (see Figure 9.16).

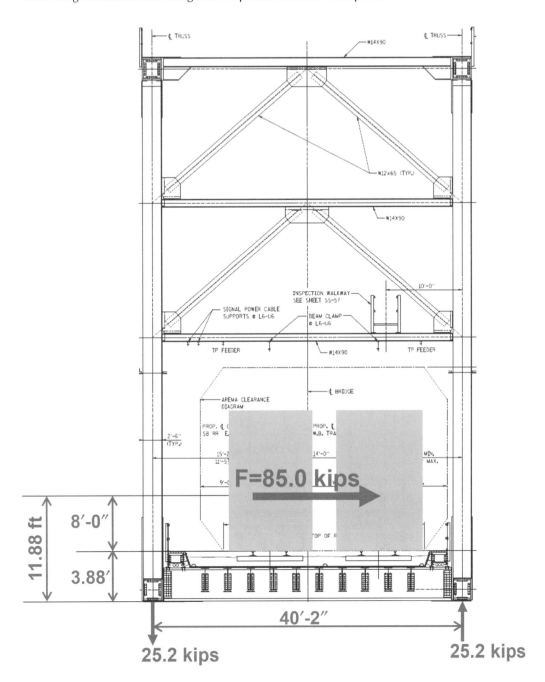

FIGURE 9.15 Centrifugal force on the truss.

9.5.8 WIND LOADS – LATERAL FORCE

Wind load in the lateral direction is calculated for the unloaded and loaded truss.

For the unloaded truss: 50 psf was used in the windward and leeward directions.

For the loaded truss: wind load of 300 lb./ft on the train was used, applied 8 feet above the top of the rail. A wind load of 30 psf on the truss was applied. The surface area includes the vertical projections of the span and any portion of the leeward trusses not shielded by the floor system.

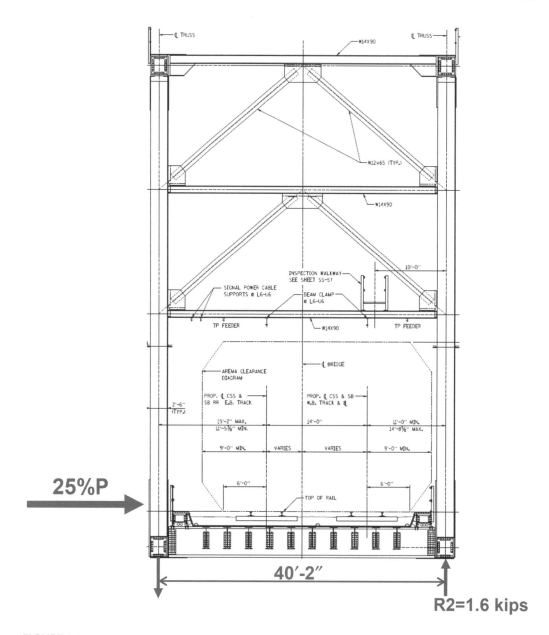

25%P →

40'-2"

R2=1.6 kips

FIGURE 9.16   Nosing force.

9.5.9   LONGITUDINAL FORCES FROM BRAKING AND TRACTION

Other unusual forces include the large longitudinal forces from braking and traction due to stopping and starting of the trains. The larger of the two values presented below is resisted by the tracking bracing located in the plane of the stringers.

- Breaking $45 + 1.2 L$ at 8 ft. above top of rail
- Tracking: $25\sqrt{L}$ at 3 ft. above top of rail

Loads calculated are per track and include an eccentricity factor (10.4%).

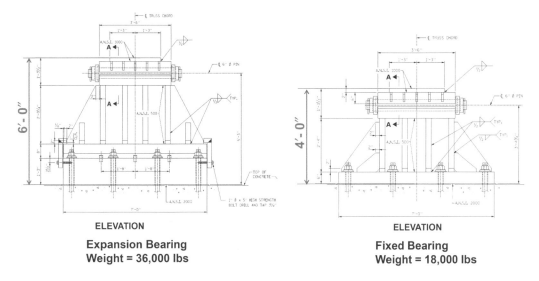

ELEVATION

**Expansion Bearing
Weight = 36,000 lbs**

ELEVATION

**Fixed Bearing
Weight = 18,000 lbs**

FIGURE 9.17   Truss expansion bearing and fixed bearing.

9.5.10 Bridge Bearings

The bridge bearings were designed for the following conditions and forces:

- Expansion and contraction (temperature): 1 ¼ inches per 100 ft., which for 394 ft. length results in 5 inches at 50 degrees
- Jacking provisions at the end floor beams
- Lateral forces
- Longitudinal braking force of 519 kips per track governs, which is resisted by the fixed bearing

These forces resulted in large bearings: the expansion bearings are nearly 6 feet tall and fixed bearings are just over 4 feet (see Figure 9.17).

Grade 50 steel was used for the beams and girders. To keep plate sizes of the box chords to reasonable thicknesses, Grade 70W HPS steel was used for the most highly stressed members. The maximum plate thickness of any box member is 2.25 inches. To facilitate inspections, handholes are spaced every 6 feet and detailed with bird screens.

The truss was cambered for dead load (actual) plus 3,000 lb./ft. (for each track), which resulted in a 5 ½ inches at midspan. The load combination of DL + LL + I + CF controlled member sizes.

Fatigue controlled the tension member size. The truss has an inspection walkway system to allow inspectors to safely access most of the members of the truss. The benefits gained with eliminating the skew and lengthening the truss span included simplified geometry, reduced abutment size, and optimized steel shapes using Grade 70 steel that resulted in significant savings.

9.5.11 Thermal Movements (Expansion Joints)

The deformation caused by a change in temperature (ΔL) for the truss is calculated to determine the width of the expansion joint at the expansion bearing.

$$\Delta L = \alpha \, L \, \Delta T = (0.0000065)(394')(12)(80) = 2.46 \text{ inch}$$

where
  ΔL = deformation caused by a change in temperature (inch)
  α = coefficient of thermal expansion (in./in./°F) = 0.0000065 (for steel and concrete)
  L = distance between the expansion joints (inch) = (394′) (12)
  ΔT = change in temperature = (final T – initial T). Temperature range is from –30°F to 130°F, with normal installation of 50°F. Therefore, ΔT = (130°F – 50°F) = 80°F

9.6 TRUSS STAGING AREA

The results of the analysis demonstrated that the assembly of the truss in a staging (Figures 9.18 and 9.19) area and then transporting it using SPMTs was an efficient and cost-effective solution. The

FIGURE 9.18 Members of the truss being assembled in the truss staging area.

FIGURE 9.19 Truss staging area.

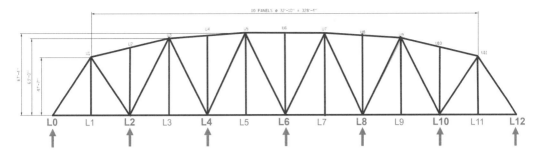

FIGURE 9.20   Temporary foundation supports at L0, L2, L4, L6, L8, L10, and L12.

use of SPMTs requires the availability of an adjacent area to be used for the off-site assembly of the structure. Location away from the roadway and railroad gave the contractor much more flexibility in terms of construction operations and schedule.

The bridge was assembled and painted in four months off-site at the staging area. Once complete, the 43-foot-wide, 67-foot-high bridge was ready for the 800-ft. move to its final location. Four SPMTs contorted by one operator made the move in two hours to the edge of roadway. Another two hours were required to align and set the bearings in their final locations.

9.6.1   PREPARING FOR THE TRUSS ROLL-IN USING SPMTs

During assembly, the truss was supported at L0, L2, L4, L6, L8, L10, and L12 as detailed in Figure 9.20. The truss was assembled and painted in the staging area.

Right before transferring the truss to the SPMTs, the truss was jacked onto temporary supports (at L2 and L10). Figures 9.21 and 9.22 show the jacking operations at L2. Two operators controlled the hydraulic jacks and were in communications to lift all four points the same amount at same time. The SPMTs carried the truss at L3, L4, L8, and L9 and moved it to the final location.

In preparation of the roll-in, the contractor prepared a detailed schedule for the eight-hour track shutdown window. The eight hours were divided into 15-minute increments and detailed the move operation and showed where each crew was to be and what task had to be accomplished.

9.6.2   JOINT STRENGTHENING

Local joint strengthening reinforcement was incorporated at L3 and L9 for support on the SPMTs (Figures 9.23 and 9.24). With the truss supported on the SPMTs, it was ready to make its 800-foot journey from the staging area to its final location.

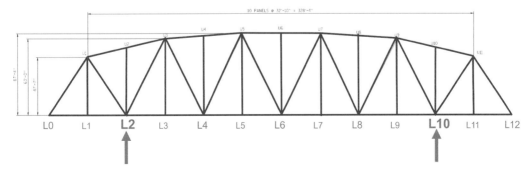

FIGURE 9.21   Truss jacked up and supported on towers (at L2 and L10) to prepare it to be carried by the SPMTs.

FIGURE 9.22   Jacking operation at L2.

9.6.3   COORDINATION WITH THE RAIL COMPANY

The specifications allowed for the bridge to be brought within 25 feet of the centerline of the NS track before the move. The bridge was supported by four SPMT units, with each unit capable of rotating 360 degrees. All were controlled by one single operator connected to the units by a single cable. The SPMTs were also capable of lifting and lowering the truss, which eliminated the need for

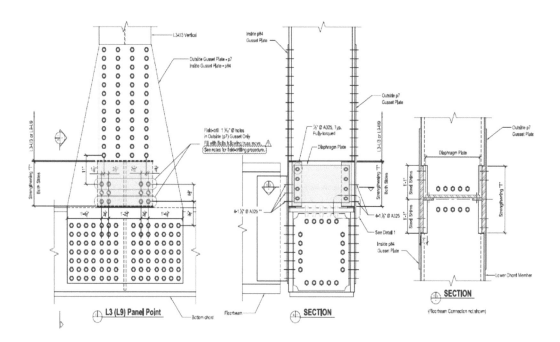

FIGURE 9.23   Joint strengthening at L3 and L9 for support on SPMTs.

FIGURE 9.24   Jacking towers at L2 and L10, and SPMTs supporting the truss at L3 and L4 and at L8 and L9.

using cranes. Moving the truss the night before (to the edge or the road) allowed for a practice run to make sure the SPMTs were operating as planned.

The NS tracks were closed at 8:00 a.m. on a Saturday morning, and then started to move the truss to its final location. It took two hours to get the truss lowered onto the piers and another two hours to get the bridge bearings adjusted properly. Figures 9.25 shows the SPMTs moving the truss over the NS existing tracks and toward the new truss piers. The SPMTs were removed and the track signals and crossing were restored well within the eight-hour shutdown window.

Figure 9.26 shows the new truss alongside the existing bridge. Shifting the CSS&SB on a new alignment and utilizing ABC techniques allowed commuter rail service to remain operational during assembly and installation of the truss.

Once the truss was in place, the contractor and railroad forces could continue work (Figure 9.27) that included placing the ballast and ties on the bridge and installing the catenary cables that provide the electricity to the trains.

FIGURE 9.25   SPMTs moving truss over the existing NS tracks.

FIGURE 9.26 New truss alongside the existing bridge allowing commuter rail service to continue during assembly and installation of truss.

FIGURE 9.27 Project site under construction. The three rail bridges completed and functional, while roadway construction continues safely under the rail bridges.

REFERENCES

1. Prefabricated Bridge Elements and Systems in Japan and Europe International Technology Exchange Program FHWA/US DOT (HPIP) 400 Seventh Street, SW Washington, DC, international@fhwa.dot.gov, www.international.fhwa.dot.gov, Publication No. FHWA-PL-05-003, HPIP/01-05(3M)EW.

2. Ralls, M.L., B.M. Tang, and H.G. Russell, "Self-Propelled Modular Transporters for Bridge Movements in Europe and the U.S.". Proceedings, Transportation Research Board 6th International Bridge Engineering Conference, July 2005. https://www.semanticscholar.org/paper/Self-Propelled-Modular-Transporters-for-Bridge-in-Ralls-Tang/bc19b17e78c7a325d74cfc4c23e66e1912db23e4
3. American Railway Engineering and Maintenance of Way Association (AREMA). "Manual for Railway Engineering". American Railway Engineering and Maintenance of Way Association (AREMA), 2012.
4. American Society of Civil Engineers (ASCE). "Design Loads on Structures during Construction". American Society of Civil Engineers (ASCE), 2019.

10 Bacteria-Based Self-Healing Concretes for Sustainable Structures

Süleyman İpek, Esra Mete Güneyisi, and Erhan Güneyisi

CONTENTS

10.1 INTRODUCTION

In structural applications, concrete is the most widely used construction material in the world due to its durability, strength, easy fabrication and application, and reasonable price (Wiktor and Jonkers, 2011). Concrete is principally made of cement, water, and aggregate is a composite material; therefore, it may show widely changeable final features and has characteristics of high compressive strength and comparatively low tensile strength (Jonkers, 2007). Because of this relatively low tensile strength characteristic, cracking in concrete is a regular occurrence (Van Tittelboom et al., 2010). However, concrete's comparatively low tensile characteristics can be compensated by implementing steel or other reinforcing materials (Jonkers, 2007). In this way, it is expected of the concrete to resist substantial tensile stresses originating from deformations, plastic shrinkage and settlement, external loads, and expansive reactions (Van Tittelboom et al., 2010). Even taking all precautions, avoiding micro-crack formations in traditional concrete may not be possible (Wiktor and Jonkers, 2011). Micro-cracks have a tendency to expand and grow when timely and effective treatment is not applied, which leads to a formation of a continuous crack network causing durability problems (Schlangen and Joseph, 2009; Van Tittelboom et al., 2010). The expanded and grown cracks offer a simple pathway for the transportation of gases and liquids that may contain aggressive substances (Schlangen and Joseph, 2009). In addition to the concrete, the reinforcement will suffer from being corroded by the presence of water, oxygen, and possibly carbon dioxide and chlorides in the medium as cracks propagate to the reinforcement. Consequently, micro-cracks are a sign of impending structural collapse (De Belie and De Muynck, 2008).

On the other hand, it must be emphasized that not all initially formed micro-cracks progress to become unstable and harmful cracks (Wiktor and Jonkers, 2011). Numerous investigations have shown that tiny cracks in concrete may be repaired when particular circumstances are satisfied (Edvardsen, 1999; Aldea et al., 2000; Dry, 2000; Reinhardt and Jooss, 2003; Zhong and Yao, 2008; Parks et al., 2010), which is referred to concrete's "self-healing" (Wiktor and Jonkers, 2011). At that point, it is worthy to state that there are two main strategies of self-healing technologies taken into account in cement-based concrete: autogenous healing strategy consists of the addition of internal curing agents, minerals, fibers, and nanofillers and autonomous healing strategy involving microbial,

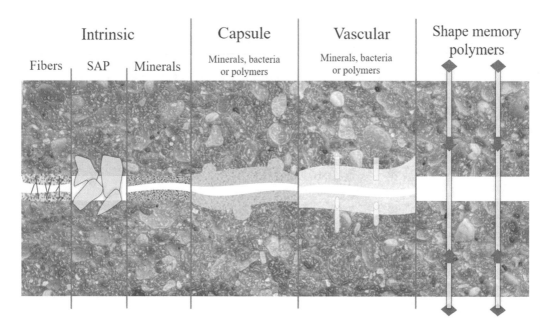

FIGURE 10.1 Schematical indication of self-healing mechanism of typical agents. (Figure adapted from Souza, 2017.)

capsule, electrodeposition, vascular, and SMA technologies. Self-healing mechanisms carried out by the application of these strategies are schematically indicated in Figure 10.1 (Souza, 2017).

Although it is believed that physical, chemical, and mechanical processes are the main reasons for autogenic healing, calcium carbonate precipitation has been identified as having the greatest impact on concrete's autogenic healing (Edvardsen, 1999). Neville (2002) stated that the cement, which constitutes the integrity of concrete, continues to hydrate in the presence of moisture and maintains the development of concrete on the basis of strength and durability, and in this context, defined autogenous healing as an extension of this living medium. Mainly, autogenous healing is a process taking place by carbonation of the portlandite (calcium hydroxide; $Ca(OH)_2$) released by the hydration of the alite (tricalcium silicate; $3CaO.SiO_2$), and belite (dicalcium silicate; $2CaO.SiO_2$) of the cement and the free calcium oxide (CaO) in the cement. The carbon dioxide (CO_2) initiating the carbonation may be in the surrounding air, water, and soil. The reaction formula for this process is given below:

$$Ca(OH)_2 + CO_2 = CaCO_3 + H_2O \qquad (10.1)$$

Specific healing agents can be employed to repair the cracks on their own in addition to healing them autogenously. For that purpose, in order to improve concrete's potential for self-healing, a variety of healing agents have been suggested in addition to chemical-based (Dry, 2000; Li and Yang, 2007; Schlangen and Joseph, 2009) and mineral-based (Jiang et al., 2015; Hung and Su, 2016; Pang et al., 2016) self-healing agents. The potential implementation of bacteria-based self-healing agents has widely been investigated in recent times (De Muynck et al., 2010; Jonkers et al., 2010; Sierra-Beltran et al., 2014; Wong, 2015; Alazhari et al., 2018; Jang et al., 2020).

Igarashi et al. (2009) have published a technical committee report on behalf of the Japan Concrete Institute (JCI TC-075B), in which they have classified and defined the self-healing/repairing phenomena using the Venn diagram shown in Figure 10.2. De Rooij et al. (2013) have mathematically described this Venn diagram as: (i) X – autogenous healing, (ii) Y – engineered healing/repairing, (iii) X∪Y – self-healing/repairing, (iv) [A]=X∩Y′ – natural healing, (v) [B]=X∩Y – autonomic healing, and (vi) [C]=X′∩Y – activated repairing. In order for a detailed explanation of each

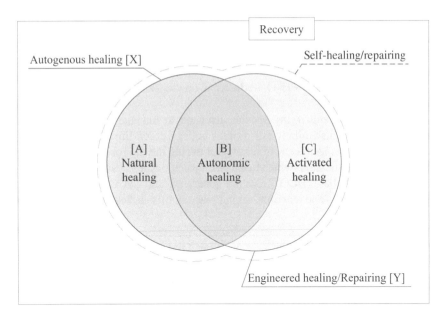

FIGURE 10.2 Classification and Venn diagram presentation of self-healing/repairing phenomena according to JCI TC-075B. (Figure adapted from De Rooij et al., 2013.)

classification, see the report published by Igarashi et al. (2009) and the book chapter published by De Rooij et al. (2013).

Nowadays, the bacteria-based self-healing agents also named microbially generated $CaCO_3$ precipitations are applied to repair the crack of in situ concrete (Phillips et al., 2013; Xu and Yao, 2014; Wang et al., 2014c), stabilize the soil (DeJong et al., 2010), reduce leakage from geologically trapped CO_2, bioremediate the metals (Mitchell and Ferris, 2005; Achal et al., 2009), and achieve the seal in gas and oil wells (Phillips et al., 2016). This book chapter will discuss the healing/repairing mechanism, bacteria species implemented and their effects, and structural applications.

10.2 HEALING MECHANISMS

Mainly and briefly, the healing mechanism operating by the application of bacteria to seal the cracks in concrete is based on the reaction of calcium ions (Ca^{2+}) with the carbonate ions (CO_3^{2-}), and leading to the $CaCO_3$ precipitation (Justo-Reinoso et al., 2021). At first sight, the concrete matrix having an alkaline and very dry environment may appear to be inhospitable for bacteria; however, there are systems where bacteria thrive and exist in nature resembling the concrete matrix, e.g., deserts, inside rocks, and ultra-basic environments as well as at even more than 1 km depth of the earth's crust (Jonkers, 2007). These bacteria can withstand high levels of mechanical and chemical stresses by generating spores, which are specialized cells (Sagripanti and Bonifacino, 1996). Spores also have very extensive lifespans and little metabolic activity; certain species have even been reported to create spores that can live up to 200 years (Schlegel, 1993). One of the significant factors affecting the precipitation of $CaCO_3$ is the bacteria species. In this context, generally, three different bacteria species (e.g., ureolytic, denitrifying, and aerobic non-ureolytic) that are operating the healing process with different strategies have been employed as bacteria-based self-healing materials in the literature (Justo-Reinoso et al., 2021). In addition to the bacteria species, the amount of dissolved inorganic carbon, pH, the concentration of calcium ions, and the existence of nucleation sites are some of the parameters determining the microbial precipitation of $CaCO_3$ (Wu et al., 2012).

Basically, in the case of utilization of ureolytic species, the urea ($CO(NH_2)_2$) reacted with water is broken down into ammonia (NH_3)/ammonium (NH_4^+) and CO_3^{2-} by bacteria using urease activity.

In the case of the presence of sufficient Ca^{2+}, $CaCO_3$ is precipitated (Phillips et al., 2012). This phenomenon is also known as the degradation of urea and its reaction can be formulated as follows (Stocks-Fischer et al., 1999; Justo-Reinoso et al., 2021):

$$CO(NH_2)_2 + H_2O \rightarrow \text{Bacterial urease} \rightarrow NH_4^+ + CO_3^{2-} \tag{10.2}$$

On the other hand, when denitrifying species are used to manage the precipitation of $CaCO_3$, the process is operated by dissimilation of nitrates under oxygen-limited conditions (Erşan et al., 2015b). Nitrate (NO_3^-) serves as the electron acceptor during the microbial oxidation of organic carbon in the course of denitrification, which takes place when oxygen is absent (De Belie et al., 2018). Thereby, CO_3^{2-} and bicarbonate (HCO_3^-) ions are generated, and $CaCO_3$ is precipitated when there is dissolved calcium present. The reactions taking place in this strategy can be formulated through Equations (10.3)–(10.5):

$$5HCOO^- + 2NO_3^- \rightarrow N_2 + 3HCO_3^- + 2CO_3^{2-} + H_2O \tag{10.3}$$

$$Ca^{2+} + 3HCO_3^- + OH^- \longleftrightarrow CaCO_3(precip) + H_2O \tag{10.4}$$

$$Ca^{2+} + CO_3^{2-} \longleftrightarrow CaCO_3(precip) \tag{10.5}$$

The last most frequently utilized bacteria species for the purpose of self-healing of concrete is the (aerobic) non-ureolytic bacteria that provide the $CaCO_3$ precipitation by oxidation of organic carbon (Justo-Reinoso et al., 2021). The act of the bacteria as a nucleation site for the precipitation of $CaCO_3$ and elevation of the pH of the environment because of the HCO_3^- production are probably two primary factors managing the $CaCO_3$ precipitation process when non-ureolytic bacteria are present (Rivadeneyra et al., 1998; Hoffmann et al., 2021). The Ca^{2+} ions are attracted by the bacterial cells that are charged negatively, and then promote the precipitation of calcium in the form of $CaCO_3$ (Hoffmann et al., 2021; Justo-Reinoso et al., 2021). Afterward, the bodies of bacteria are covered by the gradually accumulated $CaCO_3$ precipitations, and as a result, the cell engulfs in the $CaCO_3$ crystals (Castanier et al., 1999). On the other side, non-ureolytic bacteria utilize organic acids as the energy and carbon source, which causes an increase in pH. As a consequence of the utilization of organic acid, alkaline HCO_3^- ions are generated, and because of that, the pH of the environment elevates (Castanier et al., 1999; Hoffmann et al., 2021; Justo-Reinoso et al., 2021). Due to the alkaline environment and abundance of calcium in concrete, the $CaCO_3$ are precipitated by further reaction of the generated HCO_3^- ions. Here, Equation 10.6 presents the formula of the reactions taking place in this strategy:

$$Ca^{2+} + 3HC_3^- + OH^- \rightarrow CaCO_3 + H_2O \tag{10.6}$$

10.3   BACTERIA SPECIES AND EFFECTS

The portlandite (calcium hydroxide; $Ca(OH)_2$), which is quantitatively the second-most significant hydration product in traditional concrete, is the primer cause of high alkalinity in the matrix of fresh concrete. Therefore, bacteria introduced to the concrete mixture must be capable of withstanding both mechanical loads and high alkalinity for extended periods of time. Alkaliphilic (namely alkali-resistant) spore-forming bacteria consequently seem to be the most suitable bacterial agents for inclusion in the concrete matrix. In addition, bacteria must also have tolerance to oxygen due to the toxic medium of the concrete matrix, which is because of the intrusion of oxygen by diffusion through the matrix capillaries (Jonkers et al., 2010).

As stated in the previous section, there are three bacteria species incorporated into the concrete for the purpose of self-healing. They are ureolytic, denitrifying, and aerobic non-ureolytic species.

The ureolytic and denitrifying species involve the nitrogen cycle, whereas the (aerobic) non-ureolytic species involve the carbon cycle. In the self-healing applications of cementitious materials, ureolytic species are the most often employed bacteria among these species (Justo-Reinoso et al., 2021) since ureolytic bacteria exhibit the quickest $CaCO_3$ precipitation rates compared to other bacteria species (De Belie et al., 2018), whereas non-ureolytic bacteria proceed with the process of $CaCO_3$ precipitation more slowly (Reeksting et al., 2020). However, there are several substantial concerns even though the effectiveness of using ureolytic species such that ions causing nitrogen oxide emissions into the atmosphere are produced, raising implications for both the environment and health. However, even with the high effectiveness of using ureolytic species, there are several substantial disadvantages such that producing NH_4^+ ions causing nitrogen oxide (NO_x) emissions into the atmosphere raises both environment and health problems (De Belie et al., 2018; Zhang et al., 2020). Additionally, too much ammonium presence in the concrete matrix raises the possibility of salt damage due to its conversion to nitric acid (Seifan et al., 2016), which might accelerate the corrosion of the reinforcing bars (Zamani et al., 2020). Moreover, it has been stated that the process of $CaCO_3$ precipitation is hindered by ammonia deposition, occurring in less-than-ideal precipitation (Jung et al., 2020). Besides, it has been reported that a sudden decrease in the survival of ureolytic bacteria resulted from possibly the production of a high amount of ammonia was observed and it would suggest that the mineralization process can only occur for a finite amount of time (Reeksting et al., 2020; Justo-Reinoso et al., 2021). The most popular types of each bacterial species for use in the self-healing of concrete are listed in Table 10.1 of the literature. It should be noted that in addition to the bacteria species and types, the growth medium, embedment method, concentration, and

TABLE 10.1
The Most Popular Strains of Each Bacterial Species Used for Self-Healing of Concrete

Bacteria Species	Bacteria Types	Studies
Ureolytic	*Bacillus licheniformis*	Krishnapriya et al. (2015), Seifan et al. (2018)
	Bacillus megaterium	Krishnapriya et al. (2015)
	Bacillus pasteurii/ Sporosarcina pasteurii	Gollapudi et al. (1995), Bang et al. (2001), Chahal et al. (2012), Chahal and Siddique (2013)
	Bacillus sphaericus	Van Tittelboom et al. (2010), Achal et al. (2011; 2013), Wang et al. (2012a; 2012b; 2014a; 2014b; 2014c), Erşan et al. (2015a), Seifan et al. (2018)
Denitrifying	*Diaphorobacter nitroreducens*	Erşan et al. (2015a; 2016b; 2016a)
	Pseudomonas denitrificans	Erşan et al. (2016a; 2016b)
Non-ureolytic	*Bacillus alkalinitriculus*	Wiktor and Jonkers (2011)
	Bacillus alkaliphilus	Jang et al. (2020)
	Bacillus cohnii	Sierra-Beltran et al. (2014), Xu et al. (2014), Tziviloglou et al. (2016), Wiktor and Jonkers (2016), Mors and Jonkers (2017a; 2017b), Sharma et al. (2017), Zhang et al. (2017), Chaurasia et al. (2019), Bisht et al. (2020), Tan et al. (2020),
	Bacillus halmapalus	Palin et al. (2017), Sharma et al. (2017)
	Bacillus halodurans	Wiktor and Jonkers (2016)
	Bacillus mucilaginous	Qian et al. (2015), Chen et al. (2016)
	Bacillus pseudofirmus	Stuckrath et al. (2014), Wiktor and Jonkers (2016), Sharma et al. (2017), Alazhari et al. (2018),
	Bacillus subtilis	Gat et al. (2011), Khaliq and Ehsan (2016), Kalhori and Bagherpour (2017), Abdulkareem et al. (2019), Hamza et al. (2020), Schwantes-Cezario et al. (2020),
	Lysinibacillus varieties	Su et al. (2019), Jang et al. (2020)

size also remarkably influence the effectiveness of the self-healing capability of bacteria. For this reason, the references should be consulted for further information on use.

10.4   STRUCTURAL APPLICATIONS

In the technical literature, there are some examples of using the bacteria-based self-healing mortar/concrete for field applications. One of the field applications of self-healing concrete with bacteria was conducted in the Ecuador Andean highlands in 2014. In this region, many canals have a concrete lining used for the irrigation system. It was observed that there was a crack initiation and propagation on the newly constructed concrete lining that causes a great leakage and resulted in a danger to system sustainability. Therefore, the use of self-healing concrete was deemed a feasible solution to build such concrete canals for enhancing the irrigation system's sustainability and performance. To this, concrete linings with a length of 3 m were constructed with the bacteria-based self-healing concrete while for the comparison; another one with a length of 3 m was produced without bacteria. The formwork was removed three days after casting and the water flow inside the canal was allowed after two days. It was reported that no sign of cracking or deterioration on the concrete linings with bacteria was observed even after five months (Sierra-Beltran and Jonkers, 2015). In addition, Sierra-Beltran and Jonkers (2015) have carried out two field applications using the self-healing agent in two different ways: repair mortar and liquid-based system. They have applied the patch repairing mortar form for the purpose of repairing the three existing buildings suffering from the leakages. On the other hand, they have applied the liquid-based system to a structure having multiple micro-cracks suffering from water leakage in rainy weather. They have concluded that both bacteria-based repair materials can be successfully and confidentially employed for repairing purposes to enhance the condition of the building.

Another full-scale application of the bacteria-based self-healing agent was seen as a self-healing repair mortar carried out by Mors and Jonkers (2019). In two projects, the self-healing repair mortar with bacteria was used to show the applicability and functional performance in practice. For that purpose, two different cases were chosen to moderate the applicability for structural rehabilitation and conveyance of cracked concrete basement walls' water tightness. In the first one, the damaged cover of the reinforced concrete columns in a chemical facility located in Limburg, the Netherlands, was repaired. For this, firstly the damaged part up to a depth of 4 cm beyond the reinforcement was taken out and a self-healing repair mortar with bacteria was applied to repair this removed part. It was reported that the applied mortar was categorized as structural repair mortar R3, which could be used for repairing structural concrete having compressive strength up to 45 MPa. Repairing the basement walls of an underground parking garage of the building named Groningen Forum located in Groningen, the Netherlands, was employed in the second project. The walls of this garage were located 20 m below the ground level of the building. First, the cracks with active water leaks were trimmed up to 5 cm to achieve an adequate bonding surface before the self-healing repair mortar with bacteria was manually applied (Sierra-Beltran et al., 2014; Sierra-Beltran and Jonkers, 2015; Mors and Jonkers, 2019). The researchers have semi-annually monitored and determined the self-healing performance of repairing mortar by performing a few non-destructive testing techniques throughout a two-year period. The observations revealed that the repaired patches preserved their water tightness and maintained the sound bond with the underlying concrete.

Apart from using a self-healing agent for repairing purposes, Mors and Jonkers (2019) have carried out two projects in which bacteria-based self-healing concrete was utilized as construction material. In the first project, a wastewater purification tank was constructed using 15 prefabricated concrete members with dimensions of $7.0 \times 2.5 \times 0.15$ m in Limburg, the Netherlands. Three of these 15 prefabricated concrete members were manufactured by a bacteria-based self-healing concrete. It was reported that after operating for more than three years as of September 2019, both reference-type and self-healing members have shown no signs of cracking or other damage or deterioration damage types. On the other hand, in the second project, instead of prefabricated concrete members, in situ cast concrete was preferred to construct a rectangular water reservoir having 47-m length, 5-m height, and 5.5-m width in Hoogvliet, the Netherlands. In order to determine the effectiveness of the self-healing concrete, only half of the concrete used to construct the tank was manufactured as bacteria-based self-healing concrete. Namely, two walls of the tank were made with self-healing concrete while the other two were cast with ordinary concrete. The authors have reported that after being in full operation for more than a year, all walls have not yet shown signs of active leakage. However, they have emphasized for both projects that the mentioned monitoring period is not enough to make a conclusion about the potential advantages of adopting self-healing concrete. As a consequence, it should be noted that the concrete of the first project was manufactured with a 10-kg healing agent dosage while the concrete of the second project had a healing agent dosage of 5 kg.

A large-scale and field application of self-healing concrete used in concrete retaining wall panels of dimensions of 1.8 m × 1 m × 0.15 m was examined by Davies et al. (2018) in the United Kingdom. Four different self-healing agents: being in the first place porous aggregates as bacteria-loaded, microcapsules as sodium silicate-loaded, shape memory polymers, and mineral healing agents supplying networks of vascular flow, and their mechanisms were investigated. The width of crack and permeability reduction through time passed in the mechanical loaded pre-cracked panels were measured to monitor the performance of the four adapted self-healing techniques. Moreover, the bacteria-infused lightweight perlite aggregates were employed in the construction of the bacteria-based self-healing concrete panel. It was pointed out that applied self-healing concrete trials considering different techniques for the panel construction were successful; however, these trials showed that different self-healing methods are suitable for different applications. For this, as reportedly, the damage mechanisms required to prevent should be clearly identified.

10.5   SUMMARY AND CONCLUSIONS

Based on the aforementioned explanations, the following conclusions could be drawn in regard to the healing mechanisms, bacteria species and their effects, and the structural application of the bacteria-based self-healing mortars/concretes.

Choosing the ideal healing species remains challenging since each one has advantages and disadvantages. However, comparing non-ureolytic bacteria species to other species (i.e., ureolytic species), there are a number of significant benefits such as avoiding environmental problems and release of no harmful ammonium emissions in the course of the self-healing process.

The service life of concrete structures is decreased by the micro-cracks developed. Cracking in concrete/reinforced concrete members can be observed due to thermal changes, shrinkage at early ages, loading conditions, freezing and thawing influence, or a combination of them. These cracks permit the entrance of undesirable agents such as water, carbon dioxide, and chloride ions into the structural members, resulting in concrete degradation and reinforcement corrosion. As a result, regular repair and maintenance are required for such structures. In addition to the associated costs of repair and maintenance, the use of additional natural resources for repair could also rise the carbon footprint of concrete/reinforced concrete structures. From this point of view, as reportedly, the usage of bacteria-based self-healing mortar/concrete instead of conventional ones has a great potential to raise the service life of structures owing to healing cracks by the bacterial conversion of incorporated organic compounds into calcium carbonate. However, in order to completely assess whether healed bacterial self-healing concrete members will attain an equivalent or identical life span performance when compared to uncracked ordinary concrete members, long-dated durability testing is necessary. It is necessary to use bacteria-based self-healing concretes more frequently in real engineering projects.

An overview of the usage of these bacteria-based materials indicates that self-healing concrete with bacteria has the potential to be used in the construction of new structures. Besides, the repair materials having bacteria in the form of liquid-based systems or mortar have been successfully applied in the field to repair aged structures. It is worthy to mention that still further studies and investigations are needed to better understand the mechanisms and the efficiency of such healing agents under different weather conditions, chemical attacks, as well as mechanical loading. Besides, more studies are required to commercialize the various methods used to create bacterial self-healing materials.

REFERENCES

Abdulkareem M, Ayeronfe F, Majid MZA, Mohd.Sam AR, Kim JHJ. Evaluation of effects of multi-varied atmospheric curing conditions on compressive strength of bacterial (*Bacillus subtilis*) cement mortar. *Construction and Building Materials*, **218**, 1–7, 2019. https://doi.org/10.1016/j.conbuildmat.2019.05.119

Achal V, Mukherjee A, Basu PC, Reddy MS. Strain improvement of *Sporosarcina pasteurii* for enhanced urease and calcite production. *Journal of Industrial Microbiology and Biotechnology*, **36**(7), 981–988, 2009. https://doi.org/10.1007/s10295-009-0578-z

Achal V, Mukherjee A, Reddy MS. Microbial concrete: Way to enhance the durability of building structures. *Journal of Materials in Civil Engineering*, **23**(6), 730–734, 2011. https://doi.org/10.1061/(ASCE)MT. 1943-5533.0000159

Achal V, Mukherjee A, Sudhakara Reddy M. Biogenic treatment improves the durability and remediates the cracks of concrete structures. *Construction and Building Materials*, **48**, 1–5, 2013. https://doi. org/10.1016/j.conbuildmat.2013.06.061

Alazhari M, Sharma T, Heath A, Cooper R, Paine K. Application of expanded perlite encapsulated bacteria and growth media for self-healing concrete. *Construction and Building Materials*, **160**, 610–619, 2018. https://doi.org/10.1016/j.conbuildmat.2017.11.086

Aldea CM, Song WJ, Popovics JS, Shah SP. Extent of healing of cracked normal strength concrete. *Journal of Materials in Civil Engineering*, **12**(1), 92–96, 2000. https://doi.org/10.1061/(ASCE)0899-1561(2000)12:1(92)

Bang SS, Galinat JK, Ramakrishnan V. Calcite precipitation induced by polyurethane-immobilized *Bacillus pasteurii*. *Enzyme and Microbial Technology*, **28**(4–5), 404–409, 2001. https://doi.org/10.1016/S0141-0229(00)00348-3

Bisht V, Chaurasia L, Singh LP, Gupta S. Bacterially stabilized desert-sand bricks: Sustainable building material. *Journal of Materials in Civil Engineering*, **32**(6), 04020131, 2020. https://doi.org/10.1061/(ASCE)MT. 1943-5533.0003101

Castanier S, Le Metayer-Levrel G, Perthuisot JP. Ca-carbonates precipitation and limestone genesis – The microbiogeologist point of view. *Sedimentary Geology*, **126**(1–4), 9–23, 1999. https://doi.org/10.1016/S0037-0738(99)00028-7

Chahal N, Siddique R. Permeation properties of concrete made with fly ash and silica fume: Influence of ureolytic bacteria. *Construction and Building Materials*, **49**, 161–174, 2013. https://doi.org/10.1016/j. conbuildmat.2013.08.023

Chahal N, Siddique R, Rajor A. Influence of bacteria on the compressive strength, water absorption and rapid chloride permeability of fly ash concrete. *Construction and Building Materials*, **28**(1), 351–356, 2012. https://doi.org/10.1016/j.conbuildmat.2011.07.042

Chaurasia L, Bisht V, Singh L, Gupta S. A novel approach of biomineralization for improving micro and macro-properties of concrete. *Construction and Building Materials*, **195**, 340–351, 2019. https://doi. org/10.1016/j.conbuildmat.2018.11.031

Chen H, Qian C, Huang H. Self-healing cementitious materials based on bacteria and nutrients immobilized respectively. *Construction and Building Materials*, **126**, 297–303, 2016. https://doi.org/10.1016/j. conbuildmat.2016.09.023

Davies R, Teall O, Pilegis M, Kanellopoulos A, Sharma T, Jefferson A, Gardner D, Al-Tabbaa A, Paine K, Lark R. Large scale application of self-healing concrete: Design, construction, and testing. *Frontiers in Materials*, 5, 51, 2018. https://doi.org/10.3389/fmats.2018.00051

De Belie N, Wang J, Bundur ZB, Paine K. Bacteria-based concrete. In: Pacheco-Torgal F, Melchers RE, Shi X, De Belie N, Van Tittelboom K, Saez A. (eds.) Eco-Efficient Repair and Rehabilitation of Concrete Infrastructures. Woodhead Publishing, p. 531–567, 2018. https://doi.org/10.1016/B978-0-08-102181-1.00019-8

De Belie N, De Muynck W. Crack repair in concrete using biodeposition. International Conference on Concrete Repair, Rehabilitation and Retrofitting, Cape Town, South Africa, 24–26 November 2008.

De Muynck W, De Belie N, Verstraete W. Microbial carbonate precipitation improves the durability of cementitious materials: A review. *Ecological Engineering*, **36**(2), 118–136, 2010. https://doi.org/10.1016/j.ecoleng.2009.02.006

De Rooij M, Van Tittelboom K, De Belie N, Schlangen E. Introduction. In: De Rooij M, Van Tittelboom K, De Belie N, Schlangen E. State-of-the-Art Report of RILEM Technical Committee 221-SHC: Self-Healing Phenomena in Cement-Based Materials. Springer Dordrecht, p. 1–17, 2013. https://doi.org/10.1007/978-94-007-6624-2

DeJong JT, Mortensen BM, Martinez BC, Nelson DC. Bio-mediated soil improvement. *Ecological Engineering*, **36**(2), 197–210, 2010. https://doi.org/10.1016/j.ecoleng.2008.12.029

Dry CM. Three designs for the internal release of sealants, adhesives, and waterproofing chemicals into concrete to reduce permeability. *Cement and Concrete Research*, **30**(12), 1969–1977, 2000. https://doi.org/10.1016/S0008-8846(00)00415-4

Edvardsen C. Water permeability and autogenous healing of cracks in concrete. *ACI Materials Journal*, **96**, 448–454, 1999.

Erşan Y, Da Silva FB, Boon N, Verstraete W, De Belie N. Screening of bacteria and concrete compatible protection materials. *Construction and Building Materials*, **88**, 196–203, 2015a. https://doi.org/10.1016/j.conbuildmat.2015.04.027

Erşan Y, De Belie N, Boon N. Microbially induced $CaCO_3$ precipitation through denitrification: An optimization study in minimal nutrient environment. *Biochemical Engineering Journal*, **101**, 108–118, 2015b. https://doi.org/10.1016/j.bej.2015.05.006

Erşan Y, Hernandez-Sanabria E, Boon N, De Belie N. Enhanced crack closure performance of microbial mortar through nitrate reduction. *Cement and Concrete Composites*, **70**, 159–170, 2016a. https://doi.org/10.1016/j.cemconcomp.2016.04.001

Erşan Y, Verbruggen H, De Graeve I, Verstraete W, De Belie N, Boon N. Nitrate reducing $CaCO_3$ precipitating bacteria survive in mortar and inhibit steel corrosion. *Cement and Concrete Research*, **83**, 19–30, 2016b. https://doi.org/10.1016/j.cemconres.2016.01.009

Gat D, Tsesarsky M, Shamir D. Ureolytic calcium carbonate precipitation in the presence of non-ureolytic competing bacteria. *Geo-Frontiers*, 3966–3974, 2011. https://doi.org/10.1061/41165(397)405

Gollapudi UK, Knutson CL, Bang SS, Islam MR. A new method for controlling leaching through permeable channels. *Chemosphere*, **30**(4), 695–705, 1995. https://doi.org/10.1016/0045-6535(94)00435-W

Hamza O, Esaker M, Elliott D, Souid A. The effect of soil incubation on bio self-healing of cementitious mortar. *Materials Today Communications*, **24**, 100988, 2020. https://doi.org/10.1016/j.mtcomm.2020.100988

Hoffmann TD, Reeksting BJ, Gebhard S. Bacteria-induced mineral precipitation: A mechanistic review. *Microbiology*, **167**(4), 001049, 2021. https://doi.org/10.1099/mic.0.001049

Hung CC, Su YF. Medium-term self-healing evaluation of engineered cementitious composites with varying amounts of fly ash and exposure durations. *Construction and Building Materials*, **118**, 194–203, 2016. https://doi.org/10.1016/j.conbuildmat.2016.05.021

Igarashi S, Kunieda M, Nishiwaki T: Research activity of JCI technical committee TC-075B: Autogenous healing in cementitious materials. In: Proceedings of 4th International Conference on Construction Materials: Performance, Innovations and Structural Implications, Nagoya, Japan, p. 91–102, 2009. https://www.jci-net.or.jp/j/publish/research/tcr/tcr2009/TC075B_tcr09.pdf

Jang I, Son D, Kim W, Park W, Yi C. Effects of spray-dried co-cultured bacteria on cement mortar. *Construction and Building Materials*, **243**, 118206, 2020. https://doi.org/10.1016/j.conbuildmat.2020.118206

Jiang Z, Li W, Yuan Z. Influence of mineral additives and environmental conditions on the self-healing capabilities of cementitious materials. *Cement and Concrete Composites*, **57**, 116–127, 2015. https://doi.org/10.1016/j.cemconcomp.2014.11.014

Jonkers HM. Self healing concrete: A biological approach. In: Van Der Zwaag, S. (eds.) Self Healing Materials. Springer Series in Materials Science, vol. 100. Springer, Dordrecht, p. 195–204, 2007. https://doi.org/10.1007/978-1-4020-6250-6_9

Jonkers HM, Thijssen A, Muyzer G, Copuroglu O, Schlangen E. Application of bacteria as self-healing agent for the development of sustainable concrete. *Ecological Engineering*, **36**(2), 230–235, 2010. https://doi.org/10.1016/j.ecoleng.2008.12.036

Jung Y, Kim W, Kim W, Park W. Complete genome and calcium carbonate precipitation of alkaliphilic *Bacillus* sp. AK13 for self-healing concrete. *Journal of Microbiology and Biotechnology*, **30**(3), 404–416, 2020. https://doi.org/10.4014/jmb.1908.08044

Justo-Reinoso I, Heath A, Gebhard S, Paine K. Aerobic non-ureolytic bacteria-based self-healing cementitious composites: A comprehensive review. *Journal of Building Engineering*, **42**, 102834, 2021. https://doi.org/10.1016/j.jobe.2021.102834

Kalhori H, Bagherpour R. Application of carbonate precipitating bacteria for improving properties and repairing cracks of shotcrete. *Construction and Building Materials*, **148**, 249–260, 2017. https://doi.org/10.1016/j.conbuildmat.2017.05.074

Khaliq W, Ehsan MB. Crack healing in concrete using various bio influenced self-healing techniques. *Construction and Building Materials*, **102**, 349–357, 2016. https://doi.org/10.1016/j.conbuildmat.2015.11.006

Krishnapriya S, Venkatesh Babu DL, Arulraj PG Isolation and identification of bacteria to improve the strength of concrete. *Microbiological Research*, **174**, 48–55, 2015. https://doi.org/10.1016/j.micres.2015.03.009

Li VC, Yang EH. Self-healing in concrete materials. In: van der Zwaag S (ed.) Self-Healing Materials: An Alternative Approach to 20 Centuries of Materials Science. The Netherlands: Springer, p. 161–193, 2007.

Mitchell AC, Ferris FG. The coprecipitation of Sr into calcite precipitates induced by bacterial ureolysis in artificial groundwater: Temperature and kinetic dependence. *Geochimica et Cosmochimica Acta*, **69**(17), 4199–4210, 2005. https://doi.org/10.1016/j.gca.2005.03.014

Mors R, Jonkers H. Effect on concrete surface water absorption upon addition of lactate derived agent. *Coatings*, **7**(4), 51, 2017a. https://doi.org/10.3390/coatings7040051

Mors R, Jonkers H. Feasibility of lactate derivative based agent as additive for concrete for regain of crack water tightness by bacterial metabolism. *Industrial Crops and Products*, **106**, 97–104, 2017b. https://doi.org/10.1016/j.indcrop.2016.10.037

Mors RM, Jonkers HM. Bacteria-based self-healing concrete: Evaluation of full scale demonstrator projects. *RILEM Technical Letters*, 4, 138–144, 2019. http://dx.doi.org/10.21809/rilemtechlett.2019.93

Neville AM. Autogenous healing – A concrete miracle? *Concrete International*, **24**(11), 76–82, 2002.

Palin D, Wiktor V, Jonkers H. A bacteria-based self-healing cementitious composite for application in low-temperature marine environments. *Biomimetics*, **2**(3), 13, 2017. https://doi.org/10.3390/biomimetics2030013

Pang B, Zhou Z, Hou P, Du P, Zhang L, Xu H. Autogenous and engineered healing mechanisms of carbonated steel slag aggregate in concrete. *Construction and Building Materials*, **107**, 191–202, 2016. https://doi.org/10.1016/j.conbuildmat.2015.12.191

Parks J, Edwards M, Vikesland P, Dudi A. Effects of bulk water chemistry on autogenous healing of concrete. *Journal of Materials in Civil Engineering*, **22**(5), 515–524, 2010. https://doi.org/10.1061/(ASCE)MT.1943-5533.0000082

Phillips AJ, Cunningham AB, Gerlach R, Hiebert R, Hwang C, Lomans BP, Westrich J, Mantilla C, Kirksey J, Esposito R, Spangler L. Fracture sealing with microbially induced calcium carbonate precipitation: A field study. *Environmental Science and Technology*, **50**(7), 4111–4117, 2016. https://doi.org/10.1021/acs.est.5b05559

Phillips AJ, Gerlach R, Lauchnor E, Mitchell AC, Cunningham AB, Spangler L. Engineered applications of ureolytic biomineralization: A review. *Biofouling*, **29**, 715–733, 2013. https://doi.org/10.1080/08927014.2013.796550

Phillips AJ, Lauchnor E, Eldring JJ, Esposito R, Mitchell AC, Gerlach R, Cunningham AB, Spangler LH. Potential CO_2 leakage reduction through biofilm-induced calcium carbonate precipitation. *Environmental Science and Technology*, **47**(1), 2–9, 2012. https://doi.org/10.1021/es301294q

Qian C, Chen H, Ren L, Luo M. Self-healing of early age cracks in cement-based materials by mineralization of carbonic anhydrase microorganism. *Frontiers in Microbiology*, **6**, 1225, 2015. https://doi.org/10.3389/fmicb.2015.01225

Reeksting BJ, Hoffmann TD, Tan L, Paine K, Gebhard S. In-depth profiling of calcite precipitation by environmental bacteria reveals fundamental mechanistic differences with relevance to application. *Applied and Environmental Microbiology*, **86**(7), e02739–19, 2020. https://doi.org/10.1128/AEM.02739-19

Reinhardt HW, Jooss M. Permeability and self-healing of cracked concrete as a function of temperature and crack width. *Cement and Concrete Research*, **33**(7), 981–985, 2003. https://doi.org/10.1016/S0008-8846(02)01099-2

Rivadeneyra MA, Delgado G, Ramos-Cormenzana A, Delgado R. Biomineralization of carbonates by *Halomonas eurihalina* in solid and liquid media with different salinities: Crystal formation sequence. *Research in Microbiology*, **149**(4), 277–287, 1998. https://doi.org/10.1016/S0923-2508(98)80303-3

Sagripanti JL, Bonifacino A. Comparative sporicidal effects of liquid chemical agents. *Applied and Environmental Microbiology*, **62**(2), 545–551, 1996. https://doi.org/10.1128/aem.62.2.545-551.1996

Schlangen E, Joseph C. Self-healing processes in concrete. In: Gosh SK (ed.) Self-Healing Materials: Fundamentals, Design Strategies, and Applications. Weinheim: Wiley-VCH, p. 141–182, 2009.

Schlegel HG. General Microbiology. 7th edn. Cambridge University Press, Cambridge, UK, 1993.

Schwantes-Cezario N, Camargo GSFN, do Couto AF, Porto MF, Cremasco LV, Andrello AC, Toralles BM. Mortars with the addition of bacterial spores: Evaluation of porosity using different test methods. *Journal of Building Engineering*, **30**, 101235, 2020. https://doi.org/10.1016/j.jobe.2020.101235

Seifan M, Samani AK, Berenjia A. Bioconcrete: Next generation of self-healing concrete. *Applied Microbiology and Biotechnology*, **100**, 2591–2602, 2016. https://doi.org/10.1007/s00253-016-7316-z

Seifan M, Sarmah AK, Ebrahiminezhad A, Ghasemi Y, Samani AK, Berenjian A. Bio-reinforced self-healing concrete using magnetic iron oxide nanoparticles. *Applied Microbiology and Biotechnology*, **102**(5), 2167–2178, 2018. https://doi.org/10.1007/s00253-018-8782-2

Sharma TK, Alazhari M, Heath A, Paine K, Cooper RM. Alkaliphilic *Bacillus* species show potential application in concrete crack repair by virtue of rapid spore production and germination then extracellular calcite formation. *Journal of Applied Microbiology*, **122**(5), 1233–1244, 2017. https://doi.org/10.1111/jam.13421

Sierra-Beltran MG, Jonkers HM, Schlangen E. Characterization of sustainable bio-based mortar for concrete repair. *Construction and Building Materials*, **67**, 344–352, 2014. https://doi.org/10.1016/j.conbuildmat.2014.01.012

Sierra-Beltran MG, Jonkers HM. Crack self-healing technology based on bacteria. *Journal of Ceramic Processing Research*, 16(S.1), 1–7, 2015. http://www.jcpr.or.kr/journal//doi.html?vol=016&no=S1&page=33

Souza LR. Polymeric microcapsules using microfluidics for self-healing in construction materials. PhD thesis, University of Cambridge, Cambridge, UK, 2017. https://doi.org/10.17863/CAM.16673

Stocks-Fischer S, Galinat JK, Bang SS. Microbiological precipitation of $CaCO_3$. *Soil Biology and Biochemistry*, **31**(11), 1563–1571, 1999. https://doi.org/10.1016/S0038-0717(99)00082-6

Stuckrath C, Serpell R, Valenzuela LM, Lopez M. Quantification of chemical and biological calcium carbonate precipitation: Performance of self-healing in reinforced mortar containing chemical admixtures. *Cement and Concrete Composites*, **50**, 10–15, 2014. https://doi.org/10.1016/j.cemconcomp.2014.02.005

Su Y, Feng J, Zhan Q, Zhang Y, Qian C. Non-ureolytic microbial self-repairing concrete for low temperature environment. *Smart Materials and Structures*, **28**(7), 075041, 2019. https://doi.org/10.1088/1361-665X/ab2012

Tan L, Reeksting B, Ferrandiz-Mas V, Heath A, Gebhard S, Paine K. Effect of carbonation on bacteria-based self-healing of cementitious composites. *Construction and Building Materials*, **257**, 119501, 2020. https://doi.org/10.1016/j.conbuildmat.2020.119501

Tziviloglou E, Wiktor V, Jonkers H, Schlangen E. Bacteria-based self-healing concrete to increase liquid tightness of cracks. *Construction and Building Materials*, **122**, 118–125, 2016. https://doi.org/10.1016/j.conbuildmat.2016.06.080

Van Tittelboom K, De Belie N, De Muynck W, Verstraete W. Use of bacteria to repair cracks in concrete. *Cement and Concrete Research*, **40**(1), 157–166, 2010. https://doi.org/10.1016/j.cemconres.2009.08.025

Wang JY, De Belie N, Verstraete W. Diatomaceous earth as a protective vehicle for bacteria applied for self-healing concrete. *Journal of Industrial Microbiology and Biotechnology*, **39**(4), 567–577, 2012b. https://doi.org/10.1007/s10295-011-1037-1

Wang J, Dewanckele J, Cnudde V, Van Vlierberghe S, Verstraete W, De Belie N. X-ray computed tomography proof of bacterial-based self-healing in concrete. *Cement and Concrete Composites*, **53**, 289–304, 2014a. https://doi.org/10.1016/j.cemconcomp.2014.07.014

Wang JY, Snoeck D, Van Vlierberghe S, Verstraete W, De Belie N. Application of hydrogel encapsulated carbonate precipitating bacteria for approaching a realistic self-healing in concrete. *Construction and Building Materials*, **68**, 110–119, 2014b. https://doi.org/10.1016/j.conbuildmat.2014.06.018

Wang JY, Soens H, Verstraete W, De Belie N. Self-healing concrete by use of microencapsulated bacterial spores. *Cement and Concrete Research*, **56**, 139–152, 2014c. https://doi.org/10.1016/j.cemconres.2013.11.009

Wang J, Van Tittelboom K, De Belie N, Verstraete W. Use of silica gel or polyurethane immobilized bacteria for self-healing concrete. *Construction and Building Materials*, **26**(1), 532–540, 2012a. https://doi.org/10.1016/j.conbuildmat.2011.06.054

Wiktor V, Jonkers HM. Bacteria-based concrete: From concept to market, *Smart Materials and Structures*, **25**(8), 084006, 2016. https://doi.org/10.1088/0964-1726/25/8/084006

Wiktor V, Jonkers HM. Quantification of crack-healing in novel bacteria-based self-healing concrete. *Cement and Concrete Composites*, **33**(7), 763–770, 2011. https://doi.org/10.1016/j.cemconcomp.2011.03.012

Wong LS. Microbial cementation of ureolytic bacteria from the genus *Bacillus*: A review of the bacterial application on cement-based materials for cleaner production. *Journal of Cleaner Production*, **93**, 5–17, 2015. https://doi.org/10.1016/j.jclepro.2015.01.019

Wu M, Johannesson B, Geiker M. A review: Self-healing in cementitious materials and engineered cementitious composite as a self-healing material. *Construction and Building Materials*, **28**(1), 571–583. 2012. https://doi.org/10.1016/j.conbuildmat.2011.08.086

Xu J, Yao W. Multiscale mechanical quantification of self-healing concrete incorporating non-ureolytic bacteria-based healing agent. *Cement and Concrete Research*, **64**, 1–10, 2014. https://doi.org/10.1016/j.cemconres.2014.06.003

Xu J, Yao W, Jiang Z. Non-ureolytic bacterial carbonate precipitation as a surface treatment strategy on cementitious materials. *Journal of Materials in Civil Engineering*, **26**(5), 983–991, 2014. https://doi.org/10.1061/(ASCE)MT.1943-5533.0000906

Zamani M, Nikafshar S, Mousa A, Behnia A. Bacteria encapsulation using synthesized polyurea for self-healing of cement paste. *Construction and Building Materials*, **249**, 118556, 2020. https://doi.org/10.1016/j.conbuildmat.2020.118556

Zhang J, Liu Y, Feng T, Zhou M, Zhao L, Zhou A, Li Z. Immobilizing bacteria in expanded perlite for the crack self-healing in concrete. *Construction and Building Materials*, **148**, 610–617, 2017. https://doi.org/10.1016/j.conbuildmat.2017.05.021

Zhang W, Zheng Q, Ashour A, Han B. Self-healing cement concrete composites for resilient infrastructures: A review. *Composites Part B: Engineering*, **189**, 107892, 2020. https://doi.org/10.1016/j.compositesb.2020.107892

Zhong WH, Yao W. Influence of damage degree on self-healing of concrete. *Construction and Building Materials*, **22**(6), 1137–1142, 2008. https://doi.org/10.1016/j.conbuildmat.2007.02.006

11 Structural Health Monitoring as a Tool toward More Resilient Societies

A. Bogdanovic, J. Bojadjieva, Z. Rakicevic, V. Sheshov,
K. Edip, A. Poposka, F. Manojlovski, T. Kitanovski,
I. Gjorgjeska, A. Shoklarovski, I. Markovski, D. Filipovski,
D. Ivanovski, and N. Naumovski

CONTENTS

11.1 INTRODUCTION

Structural health monitoring is a process of health assessment of a structure through an automated monitoring system, which includes important components such as sensor networking, data processing, damage assessment and decision making. The end results of structural health monitoring should be life and safety benefits acquired in the most inexpensive way. Since structural health monitoring in the city of Ohrid is in its development phase, extensive activities have been taken to ensure that the local city municipality benefits from this new technology. Since the structural health monitoring process consists mainly of observation and evaluation of a structure, sample measurements were taken periodically in order to assess the correctness of the instrumentation. First, measurement for the considered structure was done by using different sensors and considering both accelerations and displacements from different triggers, particularly earthquakes occurring in the local region. The soil foundation was monitored precisely in order to follow the shear wave propagation of the travelling waves and their effects on the upper structure. It is important to know that maintaining safe and reliable buildings and infrastructure is important for the society and safety of human lives. As given in the work of Chen [1], despite the initially used necessary design methodology, civil engineering structures deteriorate with time due to various reasons, including failure caused by cyclic loads, environmental factors such as steel corrosion, concrete carbonation and aging of construction materials. Therefore, structural health monitoring is influenced by operational and environmental factors, which involve uncertainties, making it difficult to define structural health in terms of age and usage. According to Brownjohn [2], severe natural disasters such as earthquakes result in demands for quick assessment of conditions of civil structures. Usually, the assessment of existing buildings largely depends on visual inspection that may be inaccurate. It is difficult to accurately evaluate structural conditions from inaccurate visual inspection data even if the inspection is done by experts [3]. For a civil engineering structure, it is important to identify damage to the structure at the earliest possible time so that the needed interventions can be arranged accordingly. To perform identification, a fundamental challenge is to make measurements sensitive to small damages so that multiple damage types can be measured [4]. As given in the work of Acunco [5], the apparent natural frequencies and damping levels vary according to the oscillation amplitude as a result of nonlinear behavior when an RC building is hit by a seismic event. The key point to be considered is to identify the varied parameters by considering a seismic event measured at the base of the building by a physical sensor installed on the ground. This case is compatible with the measurements in our case study since there were also measurements inside the ground, at different depths, to consider the local site effects. Another point which should be considered is explained in the work of [6] in which a new method for estimation of the modal mass ratios of buildings from unscaled model shapes is identified from ambient vibrations. In this approach, the whole mass of the building is concentrated in the centroid of the polygons and the experimental mode shapes are expressed in terms of rigid translations, enabling the mass matrix to be computed easily. It should also be mentioned that, in some countries, structural health monitoring systems are mandatory to be installed in tall buildings [7], which makes our approach in the case study of the Ohrid city even more valuable. For example, the Uniform Building Code (UBC 1997) recommends that, for some buildings in highly seismic regions (e.g., Los Angeles, California), a minimum of three tri-axial accelerometers should be installed. As earthquakes are still regarded as a potential risk for high RC buildings, structural health monitoring systems can help in rapid assessment of the correct conditions of these structures. In this regard, the work of Hoult [8], which includes a rapid estimate of the damage state after an earthquake event is presented. Namely, the methodology proposed is fitted with an extensive seismic monitoring system in which the displacement profile allows a direct calculation of the interstory drift. In the case study of Ohrid, the methodology also includes fitting of the results in order to obtain correct values from the measuring instruments since the main objective is to accurately evaluate a building by using new methodologies. It should be mentioned that the case study of an Ohrid city RC building includes an in-house developed software platform for real-time monitoring of the building from the case study. As given in Limongelli [9], one of the main reasons for using structural health monitoring is its superiority and significance compared to other traditional observational methods, which are mainly costly, time-consuming and dependent on the operator, including uncertainties. Therefore, the maturity of the structural health monitoring, which is evidenced by the development of sensing systems, enables obtaining of data during significant earthquake events. It is also to be mentioned that different types of structures need different phenomena to be monitored. In the case study of Ohrid, considering that a lake and mountains are present in the surrounding location, the monitoring of the selected structure involved the influence of the local site effects for which different measuring instruments were placed at different depths of the soil medium. In recent years, disasters triggered not only by earthquakes have occurred frequently in N. Macedonia, resulting in decline of structural stiffness. Therefore, structural parameter identification under disasters such as earthquakes has obtained more experimental attention [10]. There has been a growing interest in applying health monitoring approaches to assessment of the real-time performance of structures. As given in the work of Giordano [10], sensor-based approaches involve acquisition of data from dynamic measurements without require-ments for undertaking visual inspections. The assessment of the damage level in construction is of a great importance. Another important point is definition of the relationship between the modal properties and the structural integrity of dynamical systems in order to define the vibration-based structural health monitoring [11]. The result of the vibration-based structural health monitoring is to obtain a seismic chart of the structure, which should provide a certain level of seismic damage based on the decrease in vibration frequency experimentally identified in the post-event phase. For this reason, the measuring devices and their communication with the developed computer software should be continuously inspected and updating of the measurements has to be provided. Being on a site consisting of soft layers, the Ohrid region represents a challenging location in which measure-ments on both soil and structures should be done properly. This project shows interesting results regarding the validity of the latest methodologies that can be used to support the process of making decisions about structural usability.

11.2 DESCRIPTION OF THE SELECTED LOCATION

11.2.1 GEOTECHNICAL AND SEISMOLOGICAL CONDITIONS AT THE SITE

The tower location (Figure 11.1) is one of the four instrumented locations within the Ohrid 3D seismic network installed by IZIIS and USGS (United States Geological Surveys) [12]. There is evidence on intensive seismic activity along the investigated location, namely the earthquakes with magnitudes greater than six (M > 6) that happened in the distant past (1906, Ohrid, ML = 6.00; 1911, Ohrid, ML = 6.70). In 2016, an earthquake with a magnitude of 5 according to the European MCS scale was felt in Ohrid. The earthquake epicenter was 12 km northeast from Ohrid. It caused visible damage, particularly to older structures and structures pertaining to cultural heritage, show-ing the gap between the scientific investigations and engineering practice. Previous studies [13–17] performed for Ohrid by UKIM-IZIIS showed that geological conditions in combination with a cer-tain intensity of seismic exposure in some specific regions could give rise to some geotechnically associated hazards that have an unfavorable effect upon engineering structures.

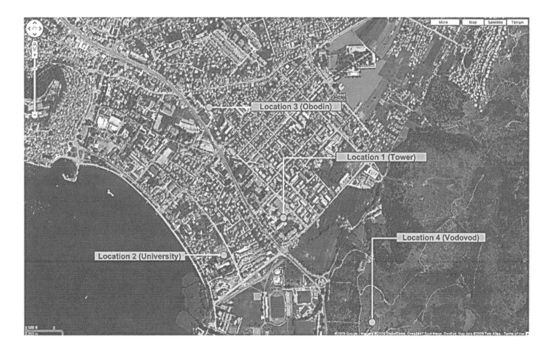

FIGURE 11.1   Locations of the 3D network in Ohrid.

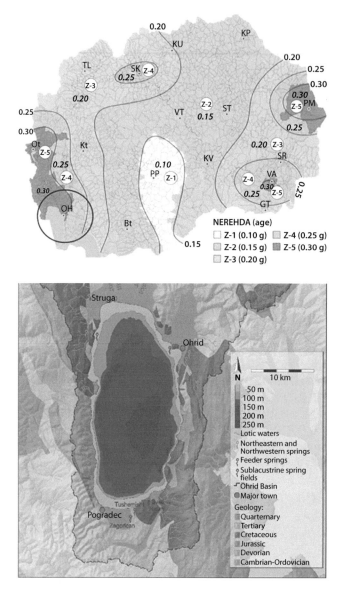

FIGURE 11.2   Seismic hazard map of Macedonia, soil type A, return period of 475 years (top) and geological conditions of the Ohrid region (bottom).

Based on the latest seismic hazard map of Macedonia prepared according to the Eurocodes (PGA) [18], the city of Ohrid is situated in a zone of moderate to high seismicity, with PGA of 0.3 g at bedrock, for a return period of 475 years (Figure 11.2, top) [19].

The Ohrid city lies in the Ohrid Lake watershed area and is characterized by the following geotechnical conditions (Figure 11.2, bottom) [20]:

1. Surface quaternary and deep Pliocene sediments;
2. Surface quaternary sediments consisting of fine gravel and sand as well as organic clays and sand down to depth of 20 m;
3. Heterogeneous nature characterized by unfavorable physical-mechanical characteristics. The underground water level is generally high.

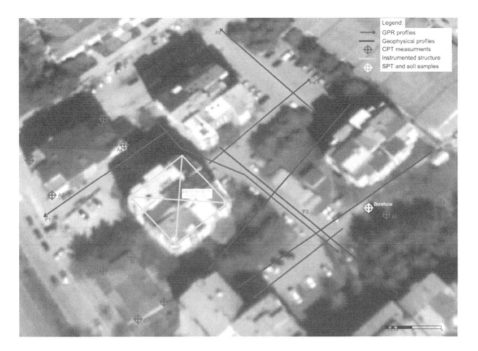

FIGURE 11.3 Layout of the performed investigations near the instrumented building – location – tower.

To define the geotechnical characteristics of the site, data from previous investigations as well as data from additionally performed geophysical and geotechnical investigations and georadar measurements (as presented in Figure 11.3) were used. The results from the geophysical investigations enabled the obtaining of seismic sections down to maximum depth of 150 m whereat local discontinuities and deformations in the terrain structure were defined. The models obtained by analysis of data from the investigations combined with application of seismic refraction, MASW and HVSR, distinguish five lithological media characterized by different physical-mechanical characteristics.

The following lithological media are distinguished:

- A surface layer – dusty, sandy and clayey, with seismic velocity values of $V_s = 150$–200 m/s;
- Subsurface layer of clay, dust and sand with seismic velocity values of $V_s = 200$–400 m/s;
- More compact Quaternary sediments with seismic velocity values in the range of $V_s = 400$–600 m/s;
- Pliocene sediments with seismic velocity values in the range of $V_s = 650$–800 m/s;
- Terrain bedrock, Paleozoic shales with seismic velocity values of $V_s > 1{,}000$ m/s.

From the performed mapping of the soil core and the performed analysis of data obtained from CPT and SPT as well as from the aspect of the lithological composition of the terrain and strength and deformability characteristics, it can be said that the soil on the investigated location is characterized by variable geomechanical characteristics. The records mainly show a lithological structure with alternating occurrence of dusty clays with fine gravel and clayey dust that are moderately plastic and with variable thickness of layers, forming, most commonly, a sandwich structure. The penetration resistance of dusty parts ranges within the limits of $qc = (0{,}5$–$1{,}2)$ MPa and the corrected number of SPT blows is $N60 = 4$, whereas those of the sandy and fine gravel parts are within the limits of $qc = (6{,}0$–$10{,}0)$ MPa with $N60 = 14$. Selected results from geotechnical and geophysical investigations are shown in Figure 11.4.

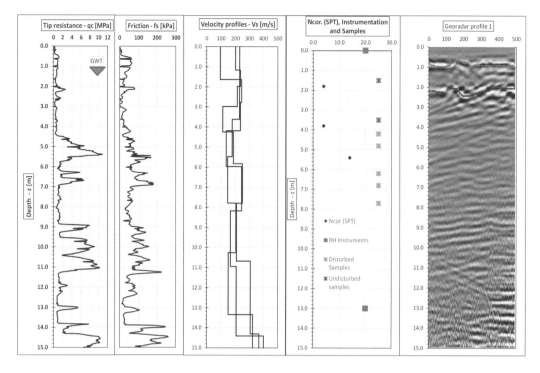

FIGURE 11.4   Extensive in-situ performed investigations versus depth at the location.

11.2.2   DESCRIPTION OF THE BUILDING

A residential-business structure was considered. The ground floor of the structure is intended for business premises, whereas the remaining stories are used for housing. The plan is approximately square, proportioned 24.0 × 24.4 m and is developed into 11 levels: a ground floor, a mezzanine, first to eight story, and an attic. The ground floor has a height of H = 3.84 m, whereas the height of the mezzanine is H = 2.60 m. The remaining stories have an equal height of H = 2.88 m.

The structure represents a mixed reinforced concrete structural system, with square and rectangular columns whose larger dimensions of cross sections at the ground floor are gradually decreased toward the higher stories. In the central part, in the direction of one of the symmetry axes, there run reinforced concrete diaphragms, between which, the staircase space is situated at one end of the building. The foundation is done on piles, through solitary footings that are interconnected by foundation beams – ties in both orthogonal directions. Each solitary footing is placed upon four piles, except for the end ones that are placed on three piles. The diaphragms are also founded on piles, through strip foundation. Figure 11.5 shows the characteristic plan and the characteristic cross section of the building.

11.2.3   INSTRUMENTATION SETUP

Based on previous investigations and taking into account the dynamic behavior of the building, the most favorable locations for installation of the instruments were defined. A total of 10 accelerometers type Kinemetrics FBA-3 (force balance accelerometers) were installed (Figure 11.6a). These were located at four characteristic points of the foundation structure, one on the sixth story and one on the ninth story, three in the profile of the soil deposit and one at the level of the terrain, in the house for data acquisition. They are graphically presented to details in Figure 11.7. The accelerometers installed in the building are protected by an iron box, whereas those at the foundations are

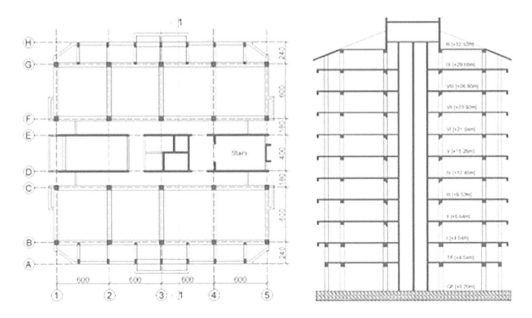

FIGURE 11.5   Characteristic plan and characteristic cross section of the building.

also protected by an iron box that is situated in a specially prepared shaft in the upper level of the foundations (Figures 11.8 and 11.9).

The FBA-3 Force Balance Accelerometer is a high-sensitivity, low-frequency triaxial device suitable for a variety of seismic and structural applications. It is an economical instrument characterized by high reliability, ruggedness and low current drain. The FBA-3 is a spring-mass device that uses variable capacitance transduction. The output is fed back to the parallel combination of capacitor C_0 and the torquer coil, which is an integral part of the mass. From the coil, the feedback

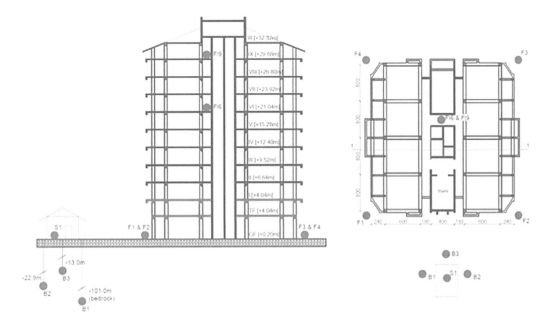

FIGURE 11.6   Kinemetrics FBA-3 (a); acquisition supporting house with three boreholes (b).

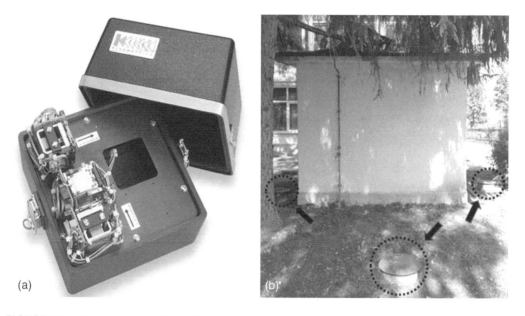

FIGURE 11.7   Graphic presentation of the location of the installed measuring instruments.

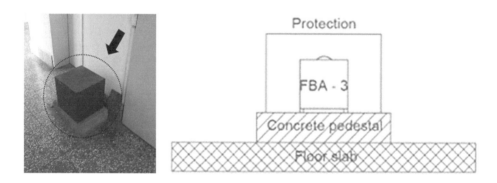

FIGURE 11.8   Kinemetrics FBA-3 placed on the 6th and the 9th floor.

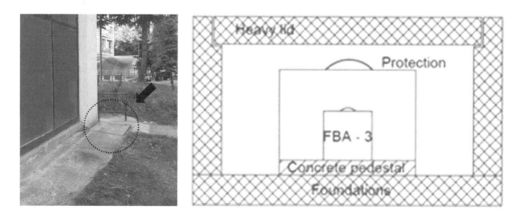

FIGURE 11.9   Kinemetrics FBA-3 placed at the foundations level.

TABLE 11.1
Instrument Identification

Number	Location	Type and Serial Number	Register
1	9th floor	FBA-3 116-1	112-1
2	6th floor	FBA-3 116-2	112-2
3	F1	FBA-3 116-3	112-3
4	S1	FBA-3 117-1	107-1
5	B1	FBA-3 DH 112-1	107-2
6	B2	FBA-3 DH 112-2	107-3
7	B3	FBA-3 DH 112-3	107-4
8	F2	FBA-3 107-2	116-1
9	F3	FBA-3	116-2
10	F4	FBA-3 105-1	116-3

loop is completed through resistors R_0 and R_h. This has the effect of stiffening the system, and thus increasing the natural frequency to 50 Hz; hence, it has a measuring range from DC to 50 Hz. Resistor R_0 (with C_0) controls the damping, which is normally adjusted to 70% of the critical. The acceleration sensitivity is controlled by the gain, K_p, of the post-amplifier. The full-scale range is ± 1.0 g (or optional ¼, ½ and 2 g).

Table 11.1 shows all the measuring instruments with their own serial numbers and location that corresponds with Figure 11.2. Presented in Figure 11.6b is the small house for data acquisition as well as the three bore holes in its immediate vicinity.

11.3 CONCEPT OF LEARNING OBJECTIVES AT THE LOCATION – GEOTECHNICAL ASPECTS

The unique instrumentation setup at the location on one hand and the available extensive geotechnical and geophysical characterization on the other hand, have provided a well-established case study for research on several learning objectives including (Figure 11.10):

- Review of literature on existing relationships to classify the soil profile and define the unit weight of the soil layers, undrained shear strength, friction angle, etc.;
- Definition of the Vs profile from the geophysical investigations and possible establishment of a correlation from the CPT data to predict the Vs profile for the Ohrid region such as in [21–23];
- Investigation of the SPT-CPT, SPT-Vs correlations [24–27];
- Investigation of the liquefaction potential by using the CPT, SPT, Vs methods for the specific location and drawing conclusions on the wider Ohrid region [28–30];
- Investigation of the relationships for definition of the small strain modulus G_{max} obtained on field and from laboratory cyclic triaxial and simple shear tests;
- Performance of GPR measurements to define the soil stratigraphy in complex geologic conditions.

11.3.1 CORRELATIONS IN-SITU AND LAB TESTS

For the investigated location, a large number of field investigations stated in the previous chapters were performed. Among these, there were research boreholes with different depths between 10 to 15 m in the near vicinity of the investigated location. In the Laboratory for Soil Dynamics within

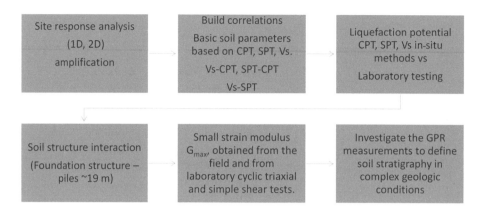

FIGURE 11.10   Concept of learning objectives at the in-situ laboratory.

the Department for Geotechnics and Special Structures, specimens taken from a number of depths of borehole B1 were brought as follows: undisturbed specimens from 1.0 to 1.5 m and from 3.0 to 3.5 m, as well as partially disturbed specimens from depths of 4.0 to 4.3 m, 4.7 to 4.9 m, 6.0 to 6.1 m, 6.7 to 6.9 m and 7.8 to 8.0 (Figure 11.11).

The specimens were stored in a water chamber for the purpose of preservation of their natural humidity until the moment of their testing. In accordance with the established working plan for the amount and type of laboratory tests, the following laboratory tests were performed:

- Grain size distribution curves (ASTM D7928);
- Tests for definition of minimal and maximal compactness, e_{max} and e_{min} (ASTM D7263);
- Tests for definition of natural humidity (ASTM2216);
- Uniaxial tests;
- Triaxial tests (ASTM 3999).

The performed detailed field and laboratory investigations provided an excellent database for further investigations in a larger number of fields, particularly in the field of correlation of field and

FIGURE 11.11   Mapping of borehole B1.

TABLE 11.2
Values of G_{max} Obtained from Different Tests

Depth (m) / Strain (%)	G_{max} DSS (kPa) 0.001	G_{max} TS (kPa) 0.0015	G_{max} Profile 1 (kPa) 0.00001	G_{max} Profile 2 (kPa) 0.00001	G_{max} Profile 3 (kPa) 0.00001
4.0–4.3	14,351	/	51,755	95,613	57,604
4.7–4.9	35,128	/	51,755	29,092	57,604
6.7–6.9	40,825	37,640	33,755	103,469	109,829
7.6–7.8	21,138	23,993	73,091	103,469	69,777

laboratory tests. For example, the maximum value of the shear modulus G_{max} (G_0) within this report is defined in several ways: geophysical measurements through the obtained Vs velocities, uniaxial direct shear tests, triaxial tests through the maximum value of the Young's modulus of elasticity, E_{max}. The obtained values are comparatively shown in Table 11.2. It is worthy to mention that the difference in the values may also be ascribed to the different strain level for which the shear moduli were defined so that, in the case of the triaxial tests, it is the lowest with the value of approximately 0.015%, in the uniaxial tests it is 0.001%, whereas the results from the geophysical measurements are for a strain of less than 0.00001%.

11.3.2 SITE RESPONSE

To obtain an initial general insight into the amplification characteristics of the location of the in-situ geo laboratory, an equivalent linear analysis of the local soil conditions was carried out by use of the Shake 2000 software package. The soil model of the location was obtained from field investigations, through SPT measurements, taking specimens from certain depths, as explained in details under item 3 of this report, performed geophysical measurements and geo-radar measurements, as explained in item 2 of this report. Figure 11.12 shows the model of the soil column that was analyzed by use of the Shake 2000 software package. In accordance with the location of the structure, the soil model was analyzed for input acceleration of 0.30 g, in compliance with the seismic zoning map of the Republic of North Macedonia, for a return period of 475 years, elaborated by UKIM-IZIIS – Skopje.

Used as input excitation in the model during the analysis were the time histories of acceleration of recorded earthquakes that are critical from the aspect of the predominant periods of the location as follows: El Centro, USA, 1940; Petrovac, Oliva – Montenegro, 1979; Ulcinj – Albatros,

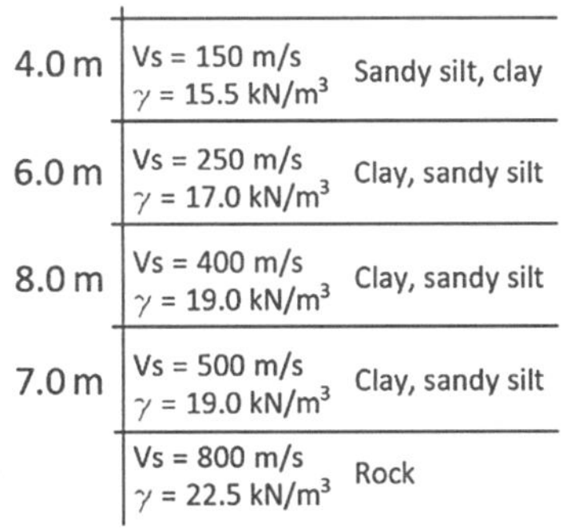

FIGURE 11.12 Mathematical model for analysis of the local site effects.

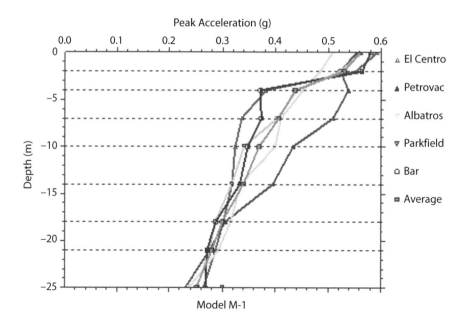

FIGURE 11.13   Peak accelerations along depth for input acceleration $a_{max} = 0.30$ g – Response of the soil column.

Montenegro 1979; Parkfield – California, 1966; Bar – Montenegro, 1979. These time histories of acceleration covered the earthquakes with expected magnitudes and frequency ranges with maximum amplitudes of seismic waves from local and near foci.

In accordance with the results from the regional investigations of the seismic hazard, the accelerograms were scaled to the expected peak acceleration at bedrock, $a_{max} = 0.30$ g, which occurs with a return period of 475 years. Such scaled time histories of acceleration were used as input seismic excitation of the mathematical model of the soil.

The effect of the local medium was evaluated based on the analysis of the dynamic response of the mathematical model. This analysis enabled definition of the peak accelerations along depth of the model as well as the response spectra of the models for the foundation level of the structure. With the analyses of the local soil effects, there were obtained the mean periods of natural vibration of 0.65–0.74 s, for acceleration level of 0.30 g (corresponding to the mean level of deformations). Figure 11.13 and Table 11.3, show the variation of peak accelerations along depths of the models obtained by convolution of selected accelerograms, for input acceleration of $a_{max} = 0.30$ g.

Figure 11.14 shows the spectra for each of the selected time histories of acceleration, for input acceleration of 0.30 g and damping of D = 5%, for foundation level of –2.0 m, along with the mean value from all the analyses. From the obtained spectra, it is clear that the dominant amplitudes occur in the period range of 0.4–1.0 s.

Based on the performed analyses the mean elastic response spectrum was computed for the assumed foundation depth–site spectrum. From the diagram, it can be concluded that the amplitudes of the mean response spectrum of the site exceed the corresponding amplitudes of the spectrum recommended in Eurocode 8 in the range of periods of 0.4–1.8 s. The diagrams show that the surface layers considerably amplify the earthquake effect, which is the result of the low strength characteristics of the soil in these layers.

Still, to get a more accurate insight, one of the next recommended research steps will be to analyze the local soil conditions by a more complex nonlinear analysis of the soil response, using the already recorded earthquake accelerations at the site of the in-situ geo-laboratory and the previously defined seismic hazard.

TABLE 11.3

Peak Accelerations along Depth, Periods of the Soil Column and DAF$_{mean}$

			Peak Ground Acceleration on Rock $a_0 = 0.30$ (g)				
			Peak Acceleration along Depth				
Depth	ACC. El Centro	ACC. Petrovac	ACC. Albatros	ACC. Parkfield	ACC. Bar	Mean Peak ACC. a_{max} (g)	DAF$_{avg}$
0	0.59	0.55	0.51	0.56	0.58	0.56	1.86
−2.0	0.55	0.52	0.49	0.52	0.56	0.53	1.76
−4.0	0.38	0.54	0.45	0.44	0.37	0.44	1.45
−7.0	0.34	0.51	0.41	0.40	0.37	0.41	1.35
−10.0	0.32	0.43	0.40	0.34	0.35	0.37	1.23
−14.0	0.32	0.40	0.33	0.32	0.33	0.34	1.13
−18.0	0.30	0.31	0.32	0.29	0.29	0.30	1.01
−21.0	0.29	0.28	0.30	0.27	0.27	0.28	1.00
−25.0	0.30	0.30	0.30	0.30	0.30	0.30	1.00
Period (s)	0.71	0.71	0.65	0.72	0.74	Mean period: 0.71 s	

11.3.3 SSI – SOIL-STRUCTURE INTERACTION

11.3.3.1 Description of the Structure of the In-Situ Geo-Laboratory

One of the important aspects of the IZIIS in-situ geo-laboratory is that there is a building on the location that represents an excellent example for research of the soil-structure interaction phenomenon. With the recording of real earthquakes on the location, along depth, upon the surface and on the building itself, as well as detailed characterization of the soil conditions with

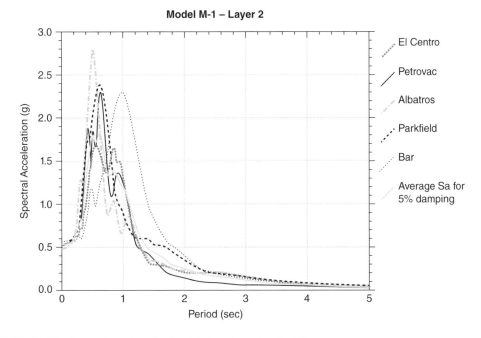

FIGURE 11.14   Spectral acceleration for 5% damping at level −2.0 m.

TABLE 11.4

Parameters for Calculation of the Rule of Thumb for the SSI Importance

Input Parameters:

H	36.1	[m]	– total height of the structure
h	25.2	[m]	– effective height = 0.7H
B	2.3	[m]	– width of the foundation
L	2.3	[m]	– length of the foundation
Bt.o./2	12.3	[m]	– half width of the foundation basis
Lt.o./2	12.3	[m]	– half length of the foundation basis
q	165	[kPa]	– average stress from the structure
Df	1.7	[m]	– depth of foundation
T	0.65	[s]	– period of the first vibration mode

the performed field and laboratory investigations, a lot of opportunities for investigation of this complex phenomenon are opened.

11.3.3.2   Computation of the Relationship Between Structure – Soil Stiffness and Stiffness Coefficients for Interaction

For the considered structure, computation of the soil-structure interaction relationship was performed according to the formula derived by Veletsos and Nair [31] and Bielak [32]. This relationship defines whether the interaction has an important role in the behavior of the structure, i.e., if:

$$h/v_s * T \quad \begin{matrix} > \\ < \end{matrix} \quad \begin{matrix} 0.1 \\ 0.1 \end{matrix} \quad \begin{matrix} \text{SSI is important} \\ \text{SSI is less important} \end{matrix}$$

In the specific case, this relationship amounts to 0.13, showing that the soil-structure interaction has an important role in the behavior of the structure.

In the relation, h – represents the height of the structure up to the center of mass for the first mode shape of the structure, Vs – represents the shear wave velocity at the foundation level and T – represents the first period of the structure.

Based on the calculated value, this ratio is 0.13, which proves the interaction importance for the structural behavior. Table 11.4 shows the input parameters for the calculations of the rule of thumb for the SSI importance.

11.4   CONCEPT OF LEARNING OBJECTIVES AT THE LOCATION – STRUCTURAL ASPECTS

11.4.1   Dynamic Behavior of the Structure – Ambient Vibrations

The method of ambient vibrations enables identification of the natural frequencies, mode shapes and values of the equivalent viscous damping coefficients. This method is based on measurement of the structural response in time domain, i.e., processing of the time-velocity or time-acceleration record taken by a corresponding equipment. The amplitudes of vibration of the structures under the effect of ambient vibrations caused by air flow and other harmonic or random excitations coming from the environment of the considered locations are low and present in each environment, the accelerations being within the limits of 10^{-7} to 10^{-4} cm/s^2. The main advantage of these measurements is the use of equipment, which is light for transport and easy for use as well as the possibility for undisturbed use of the structure that is tested [33–35].

FIGURE 11.15   (a) Accelerometer PCB Piesotronics model 393B12, (b) system for data acquisition module NI cDAQ 9178and (c) module NI 9234.

To record vibrations in both orthogonal directions, accelerometers type PCB Piezotronics model 393B12 (Figure 11.15a), with sensitivity 10,000 mV and range of up to 4.9 m/s^2 or 0.5 g, were used. The measurements were done with a sampling frequency of 2,048 Hz. Used were a total of eight accelerometers with different distributions of measuring points and measuring directions. The recorded signals from the accelerometers were collected by a 12-channel acquisition system – module NI cDAQ-9178 and three modules NI 9234 (Figure 11.15b and c), which transforms the analog signals into digital. The values of the recorded accelerations are expressed into the unit of "ground acceleration – g" (9.81 m/s^2).

The testing procedure consists of recording vibrations in real time and their processing in time and frequency domain. The measurement started with the dynamic calibration test whereat all sensors (accelerometers) were placed in the same position and in the same direction. The resonant frequencies of the structure can be preliminarily defined by use of this test, but still, final definition of natural frequencies is possible upon obtaining the mode shapes of vibration. Following the calibration test, the accelerometers were placed at previously defined points of the structure, at all accessible levels (Figure 11.16). The disposition of the measuring points was conceptualized in such a way that it enabled the obtaining of an optimal number of necessary records for definition of the dynamic characteristics of the structure. At each point, two accelerometers were placed to measure vibrations in transverse, i.e., longitudinal direction. One point (point 1) at the attic level (+29.68 m),

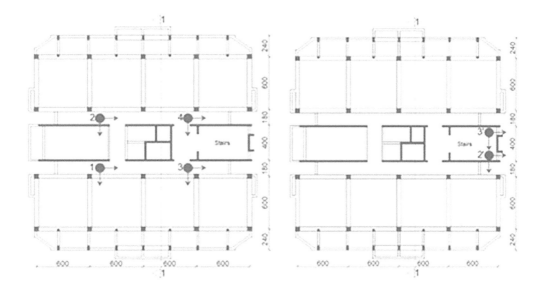

FIGURE 11.16   Position of the accelerometers: (a) mezzanine, story 1 ÷ 8 and attic and (b) at the ground floor.

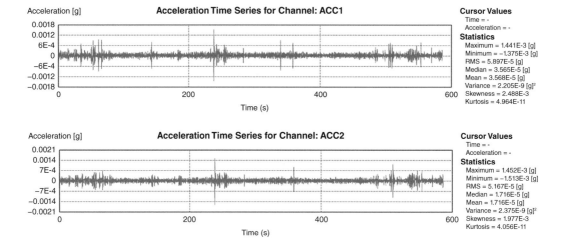

FIGURE 11.17 Records of time histories of acceleration taken by accelerometers 1 and 2 during the first test, attic level.

was selected as referent and the position of the accelerometers at that point was constant during all measurements, whereas the remaining six were displaced from the defined points in pairs of two accelerometers. A total of 42 measuring points were defined, two measuring points in the ground floor and four measuring points at each story. Figure 11.16 shows the position of the accelerometers at the mezzanine, story 1 ÷ 8 and attic (a) and the position of the accelerometers at the ground floor (b). To record time histories of acceleration at all previously defined points, 11 tests were performed, including the calibration test. To eliminate the effect of possible noises (non-stationary excitations) that may occur during a test, the sets of recorded data consisted of time histories of acceleration with time duration of 600 s and sampling frequency of 2,048 Hz.

Once all data from the measurement, i.e., time histories from all measuring points are available, the procedure of decomposition of the response time history (*Frequency Domain Decomposition, FDD*) integrated in the ARTeMIS software package [36] is used to identify the basic dynamic characteristics. The procedure is based on decomposition of the response of the system of a set of independent systems with a single degree of freedom, for each mode separately, and consequently, the fundamental frequencies of vibration of the structure, the corresponding mode shapes and damping are obtained. Figure 11.17 shows two time histories of acceleration recorded by accelerometers placed at the referent point of the first test where one can get a visual insight into the measuring set of 600 s. The results from the transformed records from all tests are shown in Figure 11.18. Using the stated method and applying the identification technique PP (peak peaking), there were defined the natural dynamic characteristics shown in Table 11.5. Figure 11.19 displays the mode shapes for the corresponding natural frequencies of vibration.

Table 11.6 contains information on the dominant frequencies of the structure measured in 1985. If Tables 11.5 and 11.6 are compared, it can be seen that there is not a significant change of frequency and hence no significant change in stiffness of the structure. A slight decline of frequency, i.e., 8% in the y-direction and 6% in the x-direction, is observed.

11.4.2 NUMERICAL SIMULATION AND VERIFICATION OF STRUCTURAL BEHAVIOR

SAP2000 software was used for analysis and analytical modeling of the tested structure (Figure 11.20). Modeling of the structural elements was done in accordance with the existing (available) formwork plans and reinforcement details. The columns and the beams were modeled as "frame" elements, whereas the walls and the slabs were modeled as "shell" elements. The linear

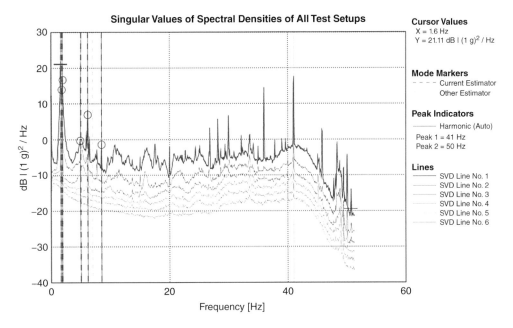

FIGURE 11.18 Definition of frequencies according to the "PP" method from the frequency curve.

TABLE 11.5
Values of Natural Frequencies

Frequency [Hz]	Damping [%]	Complexity [%]	Description
1.59	2.98	0.76	First mode of vibration in transverse direction X
1.80	0.00	16.20	Torsion
1.93	2.48	0.28	First mode of vibration in longitudinal direction Y
5.01	2.93	3.76	Second mode of vibration in transverse direction X
8.45	0.00	23.65	Second mode of vibration in longitudinal direction Y

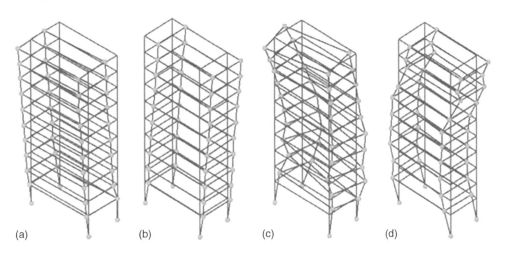

(a)   (b)   (c)   (d)

FIGURE 11.19 Mode shapes for the corresponding natural frequencies of vibration. (a) First mode of vibration in transverse direction X (f = 1.59 Hz). (b) First mode of vibration in longitudinal direction Y (f = 1.93 Hz). (c) Second mode of vibration in transverse direction X (f = 5.01 Hz). (d) Second mode of vibration in longitudinal direction Y (f = 1.59 Hz).

TABLE 11.6

Values of Natural Frequencies, Measurements Done in 1985

Direction	Frequencies [Hz]				Damping [%]
	f1	f2	f3	f4	f1
P1	2.08	2.72	3.52	6.56	3.90
P2	1.76	2.64		6.56	3.70
Torsion	2.00	2.32		6.64	3.60

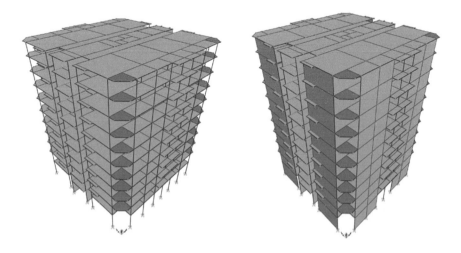

FIGURE 11.20   Analytical model from SAP2000.

structural elements were divided into finite elements by use of the "auto frame mesh – intermediate joints" function, whereas the 2D elements were divided into finite elements of 50 cm by use of the "automatic area mesh" function. The structure is fixed to the base, by which its motion is prevented in all directions.

First of all, to control the model, the structure was loaded with assumed expected loads for such type of structures to the value of 3.0 kN/m² obtaining thus a stable analytical model that behaves as expected. For this combination of loads, the frequencies obtained in the x- and y-direction amount to 1.11 and 2.41 Hz, respectively, Table 11.7. To harmonize the results from the modal analysis of the analytical model with the dynamic characteristics measured on field, a number of analyses were carried out by taking into account the distribution and the effect of the partition walls, as well. For the model with the closest dynamic characteristics, frequencies of 1.59 and 2.31 Hz were obtained

TABLE 11.7

Dynamic Characteristics of the Basic Model (SAP2000 output)

OutputCase Text	StepType Text	StepNum Unitless	Period s	Frequency Cyc/s	CircFreq rad/s	Eigenvalue rad²/s²
MODAL	Mode	1	0.89	1.114	7.00	49.07
MODAL	Mode	2	0.70	1.414	8.88	78.95
MODAL	Mode	3	0.41	2.411	15.15	229.56

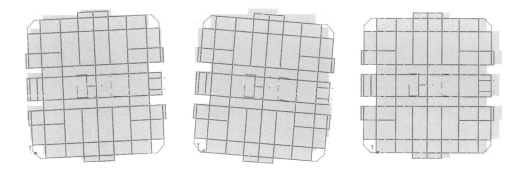

FIGURE 11.21   (a) Translation in x-direction with some rotation (f = 1.59 Hz). (b) Rotation (f = 1.81 Hz). (c) Translation in y-direction (f = 2.31 Hz).

TABLE 11.8
Dynamic Characteristics of the "Fitted" Model (SAP2000 output)

OutputCase Text	StepType Text	StepNum Unitless	Period s	Frequency Cyc/s	CircFreq rad/s	Eigenvalue rad²/s²
MODAL	Mode	1	0.62	1.590	9.99	99.85
MODAL	Mode	2	0.55	1.814	11.40	129.99
MODAL	Mode	3	0.43	2.315	14.55	211.72

in the x- and y-direction, respectively. As to the mode shapes, a clear translation in the y-direction (Figure 11.21c) and translation with some rotation in the x-direction (Figure 11.21a) was obtained.

Tables 11.8 and 11.9 show a comparison of frequencies of the structure in both orthogonal directions measured in 1985, 2021 and obtained analytically from the model in SAP2000.

There is a difference in the frequency in the y-direction obtained from the analytical model that is expected since the loads used for the analytical model correspond to those in the structural project. Taking into account that the building has been undergoing changes in the course of its serviceability period for which there is no document, it has been defined that the analytical model shows well the behavior of the structure in real conditions, meaning that there is a good correlation between the experimental and the analytical results and that the model can be used in further numerical analyses.

11.5   DESCRIPTION OF THE HEALTH MONITORING SYSTEM

The integral system is based on hardware and software components. The hardware includes measuring instruments and a system for acquisition and conversion of analog signals. The software is aimed at organization of the measurements into a corresponding record and enabling the user

TABLE 11.9
Comparison of Dynamic Characteristics

	Y-direction	X-direction
Measurement 1985	2.08	1.76
Measurement 2021	1.90	1.59
SAP2000	2.31	1.59

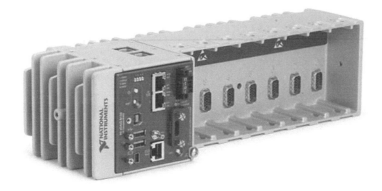

FIGURE 11.22   cDaq-9133 chassis + NI 9205 module.

presentation of the measurements, interaction with the hardware and timely alarming about defined events and conditions of interest [37].

11.5.1 HARDWARE SOLUTION

To define a corresponding system for measurement and acquisition of data, there was a certain period of analysis and definition of necessary characteristics after which measuring equipment was procured from National Instruments. The procured equipment includes a chassis with embedded controller-type cDAQ-9133 and three acquisition modules type 3 NI 9205, each of which enables a 16-channel differential analog-digital conversion of voltage signals from the measuring instruments (Figure 11.22). The chassis controller is based on the Windows 7–embedded operational system that provides simple ways of remote access to the system and simpler configuration, maintenance and upgrading of the software elaborated for this purpose. The selected chassis offers the possibility of attaching eight modules, which will enable upgrading of the system with additional measuring points in the future.

In correlation with the hardware solution, the monitoring software is elaborated in the LabView development platform. This platform offers the possibility for easy integration and putting into operation a wide spectrum of hardware measuring systems produced by National Instruments, facilitating considerably the process of development and upgrading and maintaining of the software solution that is added to this system.

11.5.2 SOFTWARE SOLUTION

From the user point of view, the elaborated software consists of a screen for configuration and a main screen displaying the measurements in real time.

11.5.2.1 Configuration Panel

The configuration panel is used for setting the parameters of the measuring channels as well as definition of the mode of operation of the system. The channels are defined by a corresponding designation and description of the position of the associated measuring instruments in the structure. Set for each channel separately is a calibration coefficient and an activation level (trigger) with whose exceedance the system starts to record. The trigger can involve acceleration or displacement level. The duration of the record is defined via the duration of the pre-event (accumulated record prior to the trigger) and duration of the post-event (record following the trigger). Upon expiry of the defined time following the trigger, the software creates a data file in TDMS format containing the

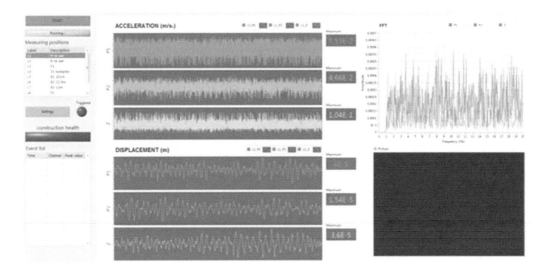

FIGURE 11.23 Presentation of the software for monitoring of the structure in real time.

time histories of accelerations at all measuring points. Further on, the data from this data file may additionally be analyzed and processed in another software for processing and analysis of data.

11.5.2.2 Display Panel

The main user screen is organized into two functional units. A part for display control and alarm and a part for display of measured and computed data in real time.

The software continuously checks and displays the acceleration level at each point, mathematically computing and displaying the displacement at the same time by using the measured acceleration. To find and monitor the characteristic frequencies of the structure, the Fourier spectrum is mathematically computed at the same time and is graphically displayed in a separate part of the display panel. The accelerations and the displacements are displayed individually for each point, along all three axes, as time histories from the last 30 s.

On the basis of the measured and computed data, in the background, the software continuously performs basic analysis of the structure and, based on predefined parameters, it displays the conditions in which the structure is. The conditions of the structure are displayed through an indicator in the form of a traffic light (green – showing that the structure is OK, yellow – showing that caution and additional checks are necessary and red – showing that the structure is not safe).

In this phase, the software is intended for internal use, as means for research, testing and a starting point for further development into a solution that can be used by concerned individuals and institutions. Further development anticipates sending data to a central server that will enable, through a web platform that will be additionally developed, presentation of data in an adapted and generalized form intended for end users (Figure 11.23). With this, the system can be used as means for alarm and basic assessment of possible consequences from inflicted damage, for the purpose of enabling a more efficient response of the concerned services and institutions.

11.5.3 MAINTENANCE AND SUSTAINABILITY OF THE SYSTEM

The possibility of remote access to the system enables a simple way of control of operation, adjustment of parameters of the system as well as a fast way of making software upgrades and other software interventions if needed.

The connection of the system with Internet enables a timely alarm and information to people in charge of the conditions of the system, potential problems or timely information about the condition of the structure of interest. This enables a fast response of competent people in conditions of potential threat for the safety of human lives or need for elimination of any problem within the system for the purpose of enabling its undisturbed operation.

If a need arises in the future, the selected hardware allows a simple increase of the number of measuring points in the structure of interest. The used chassis possesses eight slots for installation of measuring modules. Out of these, three have been used for the present needs, which provides the possibility for extension of the system by additional five modules. If the same type of module is used, the system could be extended with an additional 80 channels in the future. The development platform LabView that is used for software development enables easy integration of additional measuring channels of the system with minimal interventions in the code.

The selection of the hardware and development tools as well as the way in which the entire system is integrated enable a simple and efficient way of maintenance and scalability and contribute to higher reliability of the system.

11.6 CONCLUSIONS

Monitored structures often represent testing sites developed only for research on high-rise buildings where instrumentation is partially either on the structures or in the soil. Damage to buildings, structures and lifelines caused by earthquakes is known to depend on the nature of the arriving seismic energy as well as on the characteristics of the structures and the soil beneath. The absence of high-quality recorded earthquakes in the Balkan region, especially multiple acceleration records at the same stations and different levels of excitation, limits the ability of researchers to study local effects and predict earthquake-triggered damages to buildings and soil failures. The Ohrid 3D seismic network, with its unique instrumentation data, will give a significant scientific contribution by providing real earthquake data referring to both soil and structures in the urban area. With its revitalization, the possibility to study their interaction effects in real time is provided as a sound basis for increased earthquake resilience.

ACKNOWLEDGMENTS

The authors acknowledge the financial support of UKIM-IZIIS.

REFERENCES

[1] Chen, H.-P., Structural Health Monitoring of Large Civil Engineering Structures. 2018.
[2] Brownjohn, J.M., et al., Vibration-Based Monitoring of Civil Infrastructure: Challenges and Successes. Journal of Civil Structural Health Monitoring, 2011, **1**(3), 79–95.
[3] Aktan, A.E., A.J. Helmicki, and V.J. Hunt, Issues in Health Monitoring for Intelligent Infrastructure. Smart Materials and Structures, 1998, **7**(5), 674.
[4] Farrar, C.R., K. Worden, and J. Dulieu-Barton, Principles of Structural Degradation Monitoring. Encyclopedia of Structural Health Monitoring, 2009.
[5] Acunzo, G., et al., Application of Genetic Algorithms for a New Approach for Seismic Building Monitoring: Integrated Measurement Systems with Physical and Virtual Sensors. Bulletin of Earthquake Engineering, 2022, **20**, 1–25.
[6] Acunzo, G., et al., Modal Mass Estimation from Ambient Vibrations Measurement: A Method for Civil Buildings. Mechanical Systems and Signal Processing, 2018, **98**: 580–593.
[7] Aytulun, E. and S. Soyöz, Implementation and Application of a SHM System for Tall Buildings in Turkey. Bulletin of Earthquake Engineering, 2022, **20**(9), 4321–4344.
[8] Hoult, R., A Computationally-effective Method for Rapidly Determining the Seismic Structural Response of High-rise Buildings with a Limited Number of Sensors. Bulletin of Earthquake Engineering, 2022, **20**(9), 4395–4417.

[9] Limongelli, M.P. and M. Çelebi, Seismic Structural Health Monitoring: from Theory to Successful Applications. 2019: Springer.
[10] Giordano, P.F., L.J. Prendergast, and M.P. Limongelli, A Framework for Assessing the Value of Information for Health Monitoring of Scoured Bridges. Journal of Civil Structural Health Monitoring, 2020, **10**(3), 485–496.
[11] Sivori, D., S. Cattari, and M. Lepidi, A Methodological Framework to Relate the Earthquake-induced Frequency Reduction to Structural Damage in Masonry Buildings. Bulletin of Earthquake Engineering, 2022, **201**–36.
[12] Petrovski, J. et al., Characteristics of Earthquake Ground Motions Obtained on the Ohrid Lake, 1995.
[13] Jordanka Chaneva, Application of Field Methods in Liquefaction Hazard Assessment, Master Thesis, 2018: in Macedonian, IZIIS – Skopje.
[14] Jordanka Chaneva, Julijana Bojadjieva, Vlatko Sheshov, Kemal Edip, Toni Kitanovski, Dejan Ivanovski. CPT and SPT In-Situ Methods for Liquefaction Potential Assessment, VERLAG, Weimar, Germany, 5–16 August 2019.
[15] Julijana Bojadjieva, Vlatko Sheshov, Kemal Edip, Jordanka Chaneva, Toni Kitanovski, Dejan Ivanovski. GIS Based Assessment of Liquefaction Potential for Selected Earthquake Scenario. Earthquake Geotechnical Engineering for Protection and Development of Environment and Constructions. Proceedings of the 7th International Conference on Earthquake Geotechnical Engineering. 7th ICEGE, Rome, Italy, 17–20th June, 2019.
[16] Three-dimensional Network of Instruments for Investigation of the Effect of Local Soil Conditions and Behaviour of Structures under the Effect of Earthquakes, Volume IX, IZIIS Report, 1985, 85–154.
[17] Three-Dimensional Strong Motion Array in the Republic of Macedonia. 10th European Conference on Earthquake Engineering, Duma (ed.) © 1995: Balkema, Rotterdam, ISBN 90 5410 528 3.
[18] Milutinovic, Z., R. Shalic, and D. Tomic. Seismic Hazard Map (PGA) for Macedonia, Based on MKC EN 1998-1:2004 – Eurocode 8, Institute of Earthquake Engineering and Engineering Seismology, Ss. Cyril and Methodius University, Skopje, IZIIS Report 2016-26, 2016.
[19] Milutinovic et al. Adopted for National Annex to MKS EN 1998-1:2012 Eurocode 8, 2016.
[20] Hauffe, T., C. Albrecht, K. Schreiber, K. Birkhofer, S. Trajanovski, and T. Wilke. Spatially Explicit Analysis of Gastropod Biodiversity in Ancient Lake Ohrid. Biogeosciences, 2011, 8(1), 175–188.
[21] Mayne, P.W. Interpretation of Geotechnical Parameters from Seismic Piezocone Tests. Proceedings, 3rd International Symposium on Cone Penetration Testing, Las Vegas, 2014, pp. 47–73.
[22] McGann, C.R., B.A. Bradley, M.L. Taylor, L.M. Wotherspoon, and M. Cubrinovski. Development of an Empirical Correlation for Predicting Shear Wave Velocity of Christchurch Soils from Cone Penetration Test Data. Soil Dynamics and Earthquake Engineering, 2015, 75, 66–75.
[23] Ohta, Y., and N. Goto. Empirical Shear Wave Velocity Equations in Terms of Characteristic Soil Indexes. Earthquake Engineering & Structural Dynamics, 1978, 6(2), 167–187.
[24] Robertson, P.K., R.G. Campanella, and A. Wightman. Spt-Cpt Correlations. Journal of Geotechnical Engineering, 1983, 109(11), 1449–1459.
[25] Lunne, T., J.J. Powell, and P.K. Robertson. Cone Penetration Testing in Geotechnical Practice. 2002: CRC Press.
[26] Ahmadi, M.M., and P.K. Robertson. Thin-Layer Effects on the CPT q_c Measurement. Canadian Geotechnical Journal, 2005, 42(5), 1302–1317.
[27] Librić, L., D. Jurić-Kaćunić, and M.S. Kovačević. Application of Cone Penetration Test (CPT) Results for Soil Classification. Građevinar, 2017, 69(1), 11–20.
[28] Robertson, P.K., and C.E. Wride. Evaluating Cyclic Liquefaction Potential Using the Cone Penetration Test. Canadian Geotechnical Journal, 1998, 35(3), 442–459.
[29] Youd, T.L., and I.M. Idriss. Liquefaction Resistance of Soils: Summary Report from the 1996 NCEER and 1998 NCEER/NSF Workshops on Evaluation of Liquefaction Resistance of Soils. Journal of Geotechnical and Geoenvironmental Engineering, 2001, 127(4), 297–313.
[30] Boulanger, R.W., and I.M. Idriss. CPT and SPT Based Liquefaction Triggering Procedures. Report No. UCD/CGM.–14, 1 European Committee for Standardization (2004) Eurocode 7: Geotechnical Design – Part 2: Ground Investigations and Testing. CEN, 2014: Brussels, EN 1997-2.
[31] Veletsos, A.S., and V.D. Nair. Seismic interaction of structures on hysteretic foundations. Journal of the Structural Division, 1975, 101(1), 109–129.
[32] Bielak, J. "Dynamic behavior of structures with embedded foundations." Journal of Earthquake Engineering and Structural Dynamics, 1975, 3(3), 259–274.
[33] Experimental In-Situ Testing of the Structures of Cevahir Sky City Residential Complex In Skopje by Forced and Ambient Vibration Methods, Mihail Garevski, Igor Gjorgjiev, Goran Jekic, Report IZIIS, June 2016.

[34] Rakikevic, Z., A. Bogdanovic, I. Markovski, D. Filipovski, and D. Naumovski. Definition of Dynamic Characteristics of the Structure of the Italian Embassy in Skopje by Experimental Ambient Vibration Method, IZIIS Report no. 46/2016, 2016.

[35] Ehsan Noroozinejad Farsangi, Aleksandra Bogdanovic, Zoran Rakicevic, Angela Poposka and Marta Stojmanovska, Ambient Vibration Testing and Field Investigations of Two Historical Buildings in Europe, Structural Durability & Health Monitoring, Vol. 14, No. 4, 2020, pp. 283–301, Tech Science Press. DOI:10.32604/sdhm.2020.010564

[36] ARTemis Modal Software – for Operational Modal Analysis and Experimental Modal Analysis.

[37] Structural Health Monitoring of Large Civil Engineering Structures, Hua-Peng Chen, Yi-Qing Ni, Wiley, 2018.

12 Toward More Seismic Resilient Construction via Risk-Based Application of Smart Vibration Control Systems

Ali Khansefid and Ali Maghsoudi-Barmi

CONTENTS

12.1 INTRODUCTION

In regions with high seismicity, the construction process of buildings is significantly impacted by the earthquake phenomenon. When buildings are built without the proper seismic-resistant structural systems and/or materials, the resulting damages can be severe. Due to its technical advantages over other conventional structural systems like moment-resisting frames, the application of vibration control systems (VCSs) has grown significantly over the past few decades. As a result, there are several successful applications of vibration control systems that have used various materials and technologies.

Despite vibration control systems' acceptance and usefulness, their application in practical projects is not highly prevalent in many countries, particularly developing ones. The slightly (probably) higher initial costs of the building projects equipped with vibration control systems are one of the primary causes of this unfavorable outcome. However, the fundamental underlying reason is the negligence of the effectiveness of vibration control systems from a risk-based perspective throughout a building's life span.

This chapter tries to address the aforementioned challenges by emphasizing the application of vibration control systems from a risk-based point of view while also taking into account the systems' technical and financial efficacy. In this regard, an attempt is made to review the existing research works in this field, with a focus on new and original ideas. The great performance of these systems from a life-cycle-cost (LCC) viewpoint is another thing that is of interest. In the following,

DOI: 10.1201/9781003325246-12

first a definition of risk and resilience will be explained. Then a brief classification of vibration control systems will be discussed. After that as a main part, the effects of different types of VCSs on the lifetime risk of buildings will be discussed, and finally, a short review of the effects of uncertainties on the seismic risk of buildings with VCSs will be given.

12.2 RISK AND RESILIENCE DEFINITION

In the beginning, it is necessary to deal with the risk and resilience concepts to build a foundation for the next discussions. Literally, the term "risk" is defined as "danger or possible loss or injury" (Cambridge dictionary 2022). But it can be paraphrased technically as "the probability of facing undesired consequences for an event or decision," which is equivalently expressed as "a relationship between degree of undesirable outcome and the expected value of the frequency that the degree of loss is exceeded" by Porter (2015). In the context of natural disasters, the risk is composed of three important components, namely, hazard, exposure to hazard, and vulnerability (Thywissen 2006). Recently, along with the term "risk," another term is used frequently by the decision makers, stakeholders, and natural disaster experts, which is "resilience." The root of this word is a Latin word called *resilio* with the meaning of "to jump back" (Klein et al. 2003). However, again, in the technical context, it means "The ability to recover from (or to resist being affected by) some shock, insult or disturbance" (Cimellaro 2016). For the quantification of a system resilience, an index called "resilience index" is defined as the area below the functionality curve (Figure 12.1) of the system divided by the recovery time.

$$R_I = \frac{1}{(T_R)} \int_{t_0}^{t_0+T_R} Q(t)\, dt \qquad (12.1)$$

where R_I is the resilience index, T_R is the recovery time, t_0 is the onset of hazard occurrence, Q_R is the initial performance capacity of the system, and Q_L is the performance capacity of the system right after the hazard occurrence. Also, as it is shown in Figure 12.1, a significant difference exists between the risk and resilience concepts. The first one deals with the drop in the performance of a system confronting a shock, while the latter focus on keeping the performance level of the system in a specific time frame with the minimum recovery time.

A very important challenge, which is neglected in most of the research works and guidelines (ASCE 2017; European Standard Commission 2005) is the performance evaluation of a building during its life span. As is shown in Figure 12.2, a building may experience several mainshock and aftershock events during its lifetime with different magnitude ranges. Definitely, its performance

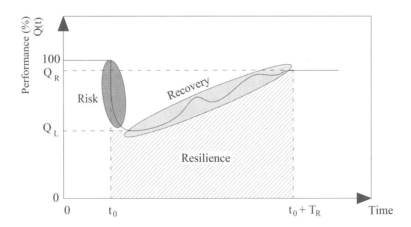

FIGURE 12.1 Functionality curve, resilience index, and risk of a system in confronting an external shock.

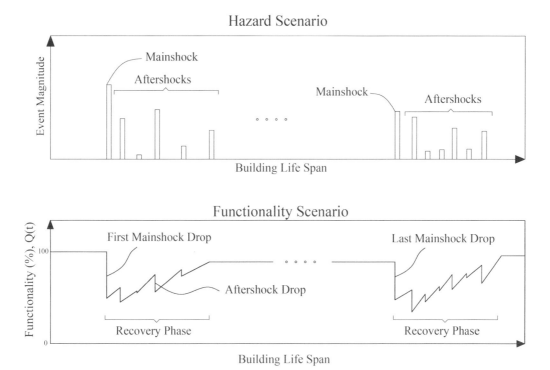

FIGURE 12.2 Schematic of real seismic hazard scenario and functionality curve of a building and/or structure during its life span.

can be affected and decreased after experiencing each of these events. The situation can get worse if it experiences a new aftershock while it is under a repair process from a previous earthquake event, which will increase the required time for the recovery process of the building.

To calculate the seismic risk, the process shown in Figure 12.3 should be followed. First, a modeler should work on the asset (building) model; in other words, the material, geometry, and mechanical characteristics of the structure should be identified. Then, by using the hazard model, the intensity and frequency of occurrence of earthquake events should be simulated, and finally, via the vulnerability models, the response of buildings, and their damage and losses should be estimated probabilistically under the simulated hazard scenario.

A last but not least item is the LCC of buildings, which means the whole cost of the project, including the initial costs, as well as all probable losses in the future. To estimate the LCC, it is very important to first simulate the future probable hazard scenarios appropriately and also have a correct vision of the costs of imposed damages and losses to the building.

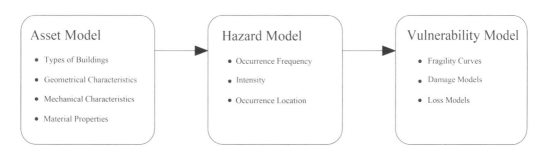

FIGURE 12.3 Steps of seismic risk assessment and its components.

12.3 COMMON TYPES OF STRUCTURAL VIBRATION CONTROL SYSTEMS

VCSs work as an additional seismic protection system to the initial seismic force resisting system of structures. The main role of such systems is to reduce the time history responses of structures exposed to the external loads (earthquakes, wind, impact, blast, etc.). Adding these systems to the structure will lead to several interesting outcomes, including the reduction in structural damage, reduction of non-structural element damage, a decrease of damage to the building content, and the creation of added value to the whole building. Generally, the VCSs are classified into four separate categories, namely, active, semi-active, passive, and hybrid systems (Christopoulos and Filiatrault 2006). This classification is shown in Figure 12.4. Passive systems are a kind of offline device. In other words, the devices will be installed in the building based on their initial design, and their characteristics will be left unchanged. However, on the other side, the active systems can be called

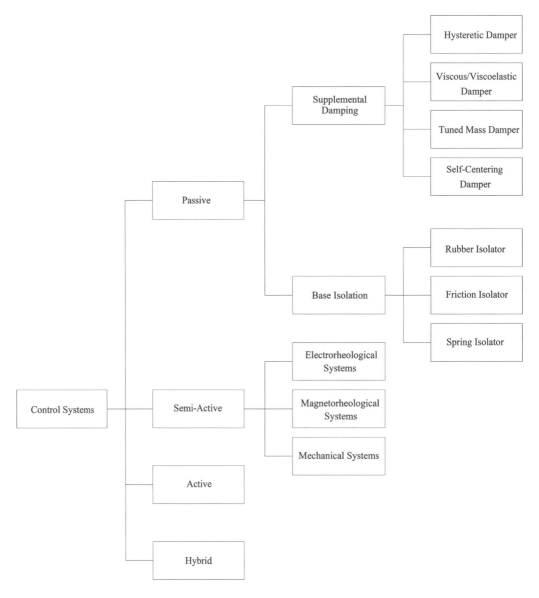

FIGURE 12.4 Classification of vibration control systems.

online systems since their properties/forces are changing continuously under the implementation of external loads by means of a considerable external power source to optimize the building performance. Additionally, the semi-active devices are something in between, i.e., their properties can change while experiencing the external loads without a considerable external energy supply source. Finally, the hybrid system is a combination of the three previous categories. In each of these categories, there are several types of devices, which are also presented in Figure 12.4. It is necessary to keep in mind that these are only selected types of devices. Due to the more extensive application, the prime focus of the chapter will be on the effectiveness of passive systems in seismic risk mitigation, while semi-active and active systems also will be reviewed briefly.

12.4 LIFETIME RISK AND RESILIENCE ASSESSMENT OF VCSs

In this section, the application of VCSs in the improvement of the structural performance will be reviewed from the risk-based perspective. Accordingly, in the next, different types of VCSs will be reviewed one by one.

12.4.1 PASSIVE DAMPERS

12.4.1.1 Hysteretic Dampers

There are different types of hysteretic dampers as VCSs, including but not limited to the ADAS, TADAS (Alehashem et al. 2008), friction dampers (Pall 2004), and rotational friction dampers (Mualla and Belev 2002); however, generally, they can be classified in two types of metallic yielding dampers and friction dampers. These systems are considered displacement-activated devices since their performance relies on their experienced deformations. These systems are among the cheapest types of VCSs. The supremacy of these hysteretic dampers from the technical aspect, compared to conventional buildings, has been proved in several research works (Farzampour and Eatherton 2019; Iwata et al. 2021; Khansefid and Ahmadizadeh 2016). This superiority has not been under the focus of many researchers. However, only a few works dealt with the topic from the risk and resiliency point of view. In the following, some of the innovative studies and their results regarding this challenge will be presented and discussed.

ADAS and TADAS dampers, as widely used metallic devices, are made of tapered steel plates that dissipate the earthquake input energy by their plates yielding. These systems have been used in several practical projects (Bayat and Abdollahzade 2011; Christopoulos and Filiatrault 2006). They are simple and easy to install in the building. However, their application under aftershock sequences is questionable, since they absorb the earthquake energy by experiencing nonlinear large deformation, or in other words, by experiencing the damage. By considering all of these challenges regarding the performance of such buildings, it is shown in the previous works that (Shin and Singh 2017) the application of TADAS dampers can significantly reduce the estimated seismic loss of two 9-story building models in Seattle and Los Angeles as defined samples by FEMA-355C (2000). As shown in Figure 12.5, the application of TADAS dampers reduces the expected collapse cost of these buildings, obtained from the probabilistic risk assessment analysis, even up to 65%.

The slit-friction damper is one of the innovative hysteretic dampers as a seismic protection device (Kim and Shin 2017). As is shown in Figure 12.6(a), the device is made of steel plates and friction pads that work with each other to dissipate the seismic energy. Also, in part (b) of this figure, the preferred installation arrangement is shown, which is called the panel arrangement. This system is well capable of providing a large hysteretic loop. Interestingly, this damper reduces well the loss value of buildings and the repair time under seismic scenarios. Kim and Shin (2017) performed a seismic risk assessment of a 15-story building constructed with concrete material in the 1970s, using the FEMA P-58 (2018) approach, while the building is equipped with different dampers, including (1) slit damper, (2) hybrid damper (slit + friction), and (3) hybrid damper with

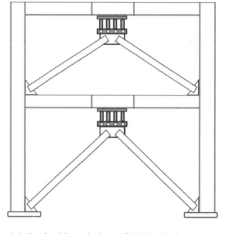 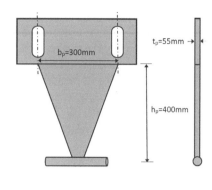

(a) Typical instalation of TADAS damapers.   (b) Plates of TADAS damapers.

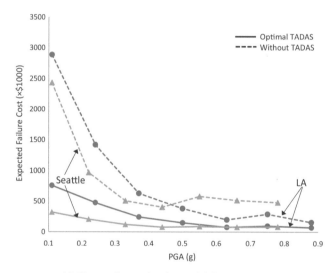

(c) Comparison of estimated failure cost of building
models with and without TADAS dampers.

FIGURE 12.5 TADAS damper and its effects on the estimated failure cost of building models in Seattle and Los Angeles. (From Shin and Singh 2017.)[1]

increased column size. They showed that the repair cost of the building equipped with such a system is much less than the uncontrolled building, as is clearly depicted by the probability density functions (PDFs) of estimated loss value in part (c) of the figure. In addition, the PDF of the estimated repair time in part (d) of the figure depicts that the required time for recovering the building from the damaged condition is at a lower level (on average 15%).

12.4.1.2 Viscous Dampers

One of the other types of VCSs that is widely used in practical construction projects is a viscous fluid damper. This device is made of a cylinder and a piston filled with oil and/or silicon fluid materials (Lago et al. 2018). From a technical aspect, the great performance of this system is proven in several studies (Mansoori and Moghadam 2009; Whittle et al. 2012; Yaghmaei-Sabegh et al.

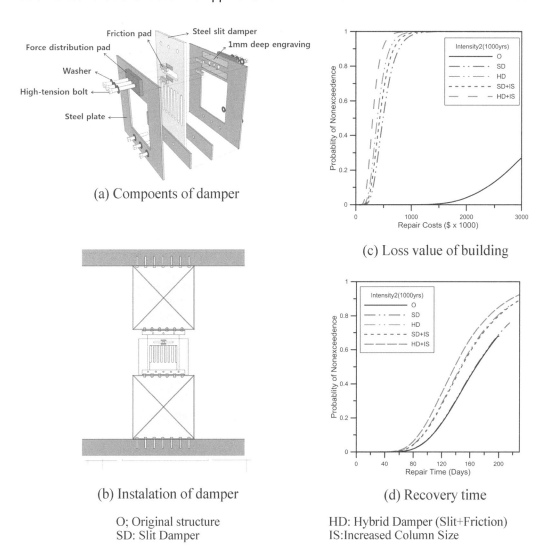

(a) Compoents of damper

(c) Loss value of building

(b) Instalation of damper

(d) Recovery time

O; Original structure   HD: Hybrid Damper (Slit+Friction)
SD: Slit Damper   IS:Increased Column Size

FIGURE 12.6 Innovative slit-friction damper (From Kim and Shim 2017.)[2]

2020; Zhou and Xing 2021). However, it is not enough, and it is necessary to take a step forward in the economic and lifetime performance assessment of such systems. Some limited works in the literature try to deal with this challenge. Their outcomes (Carofilis Gallo et al. 2022; Gidaris and Taflanidis 2012; Gidaris and Taflanidis 2015; Taflanidis and Beck 2009; Taflanidis and Gidaris 2013) emphasize that the application of viscous dampers is completely beneficial in the life span of the building. As an example, the results of a research work by Gidaris and Taflanidis (2015) are presented here in Figure 12.7. It is the result of risk assessment and life-cycle estimation of a three-story building with/without dampers under different seismic hazard scenarios. Accordingly, different site-to-source distances (r_{med}) and different seismicity rates (b_M coefficient of Gutenberg-Richter equation (Kramer 1996)) are considered. The smaller the distance and b_M, the higher the estimated cost. Also, it is thoroughly clear that the LCC of a building is reduced significantly by the usage of a viscous damper, especially in a more severe earthquake hazard scenario (up to 90%).

Another important point is the effect of viscous dampers on the distribution of estimated loss between different damage types. When a viscous damper is added to a building, it is expected that the overall estimated loss will reduce, and the portion related to the structural damage reduces

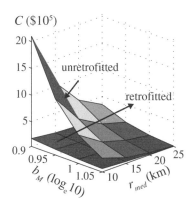

FIGURE 12.7 LCC of a three-story reinforced concrete building with/without viscous dampers. (From Gidaris and Taflanidis 2015.)[3]

more. However, the damage to the acceleration-sensitive parts can be decreased too. This fact is observed by Taflanidis and Gidaris (2013), where they calculated the LCC of a three-story concrete building with and without viscous dampers by considering all components of the building. Interestingly, in addition to a significant reduction in the estimated LCC, the distribution of losses changed considerably. The application of viscous dampers reduces the structural and partition damages more significantly. Also, it is noteworthy to point out that in the building equipped with the dampers, the high portion of LCC is due to the damper's initial prices. This again highlights the fact that viscous dampers will reduce the LCC of buildings tangibly even though the building owner has to pay for the damper's initial cost at the construction phase of the project.

12.4.1.3 Self-Centering Dampers

Among the various vibration control systems are self-centering dampers (SCDs), which provide lateral stiffness and dissipate seismic energy, while re-centering the structure after earthquakes. This will help in the elimination of residual deformation, and consequently, an effective reduction in both downtime and post-earthquake repair costs (Xu et al. 2022). By now, several types of self-centering braced structures have been introduced and examined (Erochko et al. 2013; Fang et al. 2020; Miller et al. 2012; Zhou et al. 2015; Zhu et al. 2020). Some of these systems use smart materials such as shape memory alloys (SMAs), while others use the pre-tensioning of elements to fulfill the self-centering feature.

Following the performance-based studies regarding the novel SCDs, a risk-based lifetime cost study is needed to better justify the application of this system; however, it has received less attention in the literature. Considering four different types of SCBs, namely, 1) single-core SCB with FRP tendons and friction energy dissipation (ED) device (FRP-S-SCB), 2) dual-core SCB with FRP tendons and friction ED device (FRP-D-SCB), 3) single-core SCB with shape memory alloy (SMA) tendons and friction ED device (SMA-F-SCB), and 4) single-core SCB with SMA tendons and viscoelastic ED device (SMA-V-SCB), it is shown that in a system-level analysis and under maximum considered earthquake (MCE), tendon fracture definitely occurs in the FRP tendons, which questions its self-centering capability while regarding the SMA-viscoelastic hybrid braces, the deformation and floor acceleration demands are effectively reduced, which leads to the lowest collapse probability (Fang et al. 2021). Moreover, by investigating the vulnerability functions and corresponding losses to the structures as depicted in Figure 12.8, a single-core SCB arrangement with FRP tendons performs worse than the others since its brace ductility is in a low level, while the arrangement of the dual-core can reduce economic losses through a significantly improvement in the ductility, and lastly, the lowest expected loss belongs to the SMA viscoelastic hybrid brace which is almost 50% less than the other types (Fang et al. 2021).

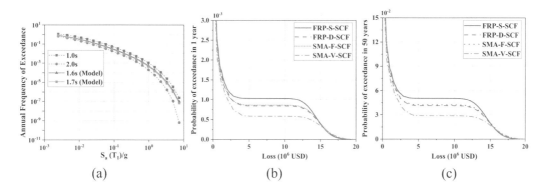

FIGURE 12.8 Time-based analysis of SCBs: (a) hazard, (b) annual probability of exceedance, and (c) probability of exceedance in 50 years. (From Fang et al. 2021.)[4]

Further application of the SCDs can be in retrofitting structures. An innovative hybrid SCD, (Figure 12.9) has been examined as a retrofitting solution against the steel slit plate damper individually, and an enhanced seismic performance was observed, which is due to extra stiffness, energy dissipation, and self-centering capability of SMA bars. More importantly, in the previous research works (NourEldin et al. 2018), it was observed that the LCC of the building frames with such a hybrid damper is much smaller than that of bare frames (about 50%) and the frames with normal slit dampers (about 15%), even though the initial cost of the hybrid damper is in a higher range than that of the bare frames (less than 10%). This is important, since it truly shows the economic distinction of VCSs, along with their accepted performance, as a suitable solution for seismic risk reduction of buildings.

As other types of VCSs, SCDs can be used in both design of new buildings and retrofit of existing civil structures to mitigate corresponding seismic risk through improving their responses. More numerically speaking, Zheng et al. (2018) quantified the advantages of application of a proposed novel SC bearing in terms of resilience and LCC, by taking the direct and indirect costs into account. The schematic of the finite element model and the proposed SC bearing are shown in Figure 12.10. The indirect loss is shown to be related to the seismic hazard intensity and could be in a higher range compared to the direct loss, specifically for the rare scenarios. As shown in Figure 12.11, for the bridges in Los Angeles and Nutbush, the LCC of the bridges with the self-centering bearing is in the lower gamut significantly (on average about 40% less).

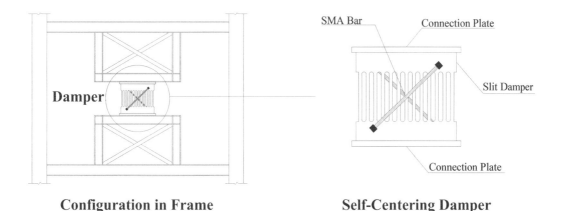

FIGURE 12.9 Steel slit SMA hybrid damper and its installation configuration in building frames.

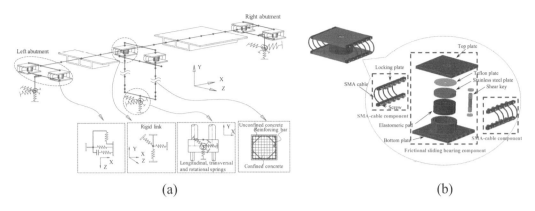

(a)   (b)

FIGURE 12.10 (a) Schematic of a nonlinear FE model of a bridge equipped with SC bearing; (b) configuration of the SC bearing. (From Zheng et al. 2018.)[5]

12.4.1.4 Base Isolation Systems (Rubber, Friction)

Seismic isolation systems (SIS) are a kind of effective VCSs, which substantially reduce the seismic force demands, and in turn, truly can help in providing resiliency in the communities. Since the appearance of this system, several technologies have been used, and various SISs have been emerged, among the most widely used, low and high damping rubber bearings and frictional isolators can be named. Regarding the performance, design, and construction requirements of isolation systems, several studies have been conducted, and specific standards have also been developed. However, a more recent issue is the risk-based performance evaluation and lifetime costs of the aforementioned systems, for which fewer studies have been conducted.

Considering building-type structures, it is shown that depending on the circumstances, seismic isolation can significantly reduce seismic risk and lifetime costs. Regarding the special steel moment or concentrically braced frame (SMF and SCBF) buildings designed according to the standard of ASCE/SEI 7-16 (ASCE 2017), and seismically isolated by frictional bearings, experience expected annual losses beneath half of the values calculated for none-isolated cases (Kitayama and Cilsalar 2022). Five different categories, including structural, non-structural (drift and acceleration based) damage, demolition, and collapse loss are considered in loss estimation. Also, the selection and scaling of earthquake records in a way that the IDA approach set in FEMA P695 can affect the results of estimated total annual losses (EAL), up to twice the one obtained in accordance with the Conditional Spectra approach (Kitayama and Cilsalar 2022; see Figure 12.12). Moreover, as is

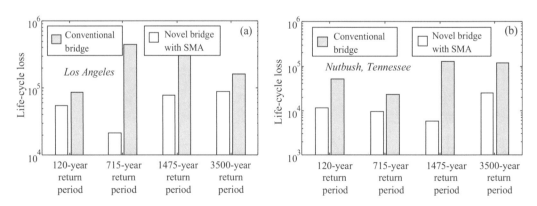

FIGURE 12.11 LCC of the conventional and novel bridges under different earthquake scenarios at (a) Los Angeles, CA, and (b) Nutbush, TN. (From Zheng et al. 2018.)[6]

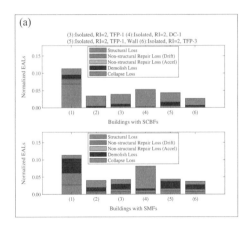
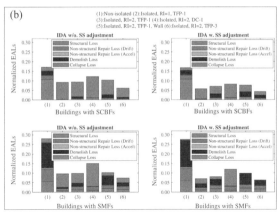

FIGURE 12.12 Normalized EAL for non-isolated and isolated steel frame buildings: TFP and DC stand for triple friction pendulum and double concave friction pendulum bearing, respectively, using (a) conditional spectra approach; (b) IDA with and without spectral shape (SS) effect. (From Kitayama and Cilsalar, 2021.)[7]

shown in Figure 12.12, the contribution of each of the five loss types mentioned above, involved in total losses, has remained the same with and without isolation systems.

Alongside the frictional isolators, rubber base isolators have also shown to be capable of seismic risk mitigation, both for new buildings and/or retrofitted ones. Considering two performance levels, namely Global Collapse (GC) and Usability Preventing Damage (UPD), for code-based new and retrofitted RC buildings and, as it is shown in Figure 12.13, Micozzi et al. (2021) concluded that base-isolated structures limit the damage probability; however, in the highly active seismic regions, the margin of structural collapse probability may be in a low level. In contrast, in the case of fixed support condition, larger damage prevention annual failure rates are experienced by the structure. Further discussion can be made here regarding the isolation system as a retrofitting technique, especially the historical types. Analyzing two different RC buildings in Italy, equipped with isolation systems of (1) high damping rubber bearings; (2) rubber bearings and flat sliding bearings; and (3) curved surface sliders, showed that all SISs effectively limit the onset of damage of non-structural members for higher seismic intensity levels than the ones required to be considered by the current design code; however, they show a little margin toward collapse, beyond the design intensity

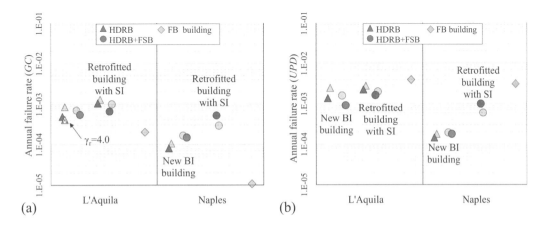

FIGURE 12.13 Normalized risk for L'aquila and Naples sites for (a) global collapse performance level and (b) usability preventing damage performance level. SI, seismically isolated; FB, fixed base; HDRB, high damping rubber bearing; FSB, flat sliding bearing. (From Micozzi et al. 2021.)[8]

level (Cardone et al. 2019). Risk analysis of an old non-ductile RC frame building with/without base isolation retrofit considering mainshock-aftershock sequences discovered that SIS can significantly reduce the seismic risk for higher damage levels; meanwhile, ignorance of aftershock effects can lead to a tangible underestimation of the real existing seismic risk (Han et al. 2014).

Besides the building-type structures, a seismic isolation system is an accepted solution for seismic risk mitigation in infrastructures. Bridges are among the examples, which are equipped with multiple types of SISs all over the world. The high capability of unbounded low damping laminated rubber bearings as an isolation system in mitigating earthquake excitation in bridges, especially in developing countries is proven (Maghsoudi-Barmi et al. 2021), where a significant seismic loss reduction was shown in bridges. Maghsoudi-Barmi et al. (2021) used an optimally designed highway bridge for their investigations. This is a remarkable point; the optimal use of the isolation system will affect the results of seismic risk assessments. This is also confirmed by Asadi et al. (2020). They investigated the LCC of a bridge equipped with lead rubber bearings (LRBs). They considered several combinations of sub-structure and the seismic isolator to reach to cost-effective seismic design. Structural, non-structural, and traffic losses were included in the cost calculations. It was shown that the increase of initial costs regarding the LRBs to a specific level can considerably decrease the damage costs; however, further increase will result in higher costs. For instance, a slight increase of initial costs of LRBs might decrease the lifetime losses by 38%. Moreover, traffic costs were shown to be the most important parameter in the estimated loss.

The implementation of SIS can considerably reduce seismic forces on the structures, and also the components; hence, considering an important structure of nuclear power plant (NPP), it would potentially provide significant benefits regarding the improvement of seismic risk and reduction in construction cost. A seismic risk assessment done on the generic nuclear facility at the sites of the Idaho National Laboratory with moderate seismic hazard, and the Los Alamos National Laboratory with high seismic hazard showed that the application of SISs reduces seismic risk by seven to eight orders of magnitude (Yu et al. 2018). Besides, at all of the considered sites, a cost reduction was observed by the application of SISs, more beneficially at sites of high seismicity rate (Yu et al. 2018). However, as an important note, long-period structures like base-isolated NPPs vibrate with a relatively longer duration, especially when experiencing subduction area earthquakes. As an example, the devastating Tohoku earthquake, followed by the damages and instabilities in the Japanese NPPs, can be named, where the earthquake waves were propagated with relatively long periods of up to 10 s (Ali et al. 2014). To this aim, Ali et al. (2014) reported a procedure on ground motion selection, considering two sets of short periods and long periods for the analysis, and showed that the annual probability of damage limit states is in an unacceptable situation for the isolated NPP. Also, again it was verified that the isolated NPPs are more vulnerable to the long-period seismic motions, which emphasize the appropriate ground motion selection procedure for seismic risk assessment of NPPs.

12.4.1.5   Risk-Based Comparison of Passive VCSs Performance

At the end of this part, it is valuable to have look at the performance of different types of passive VCSs from the risk-based perspective. As common VCSs in the practice, Khansefid (2021) considered hysteretic dampers, viscous dampers, and SISs. The performances of these systems were examined for conventional building types in Tehran's metropolitan area. The earthquake hazard during the lifetime of buildings (50 years) was simulated using an advanced two-step method (Khansefid and Bakhshi 2022), which first generates the probable mainshock-aftershock events of the hazard scenario by simulating their number, occurrence time, and magnitude (see Figure 12.14(a) and (b)). Then, for each of these events, a stochastic acceleration time series would be simulated for a given site condition. Afterward, the building models were analyzed under the seismic scenario during their whole life span using a nonlinear dynamic time history analysis approach, considering the aftershock effects. To estimate the lifetime loss value of building models, cumulative structural damage was considered. Also, the total loss value consists of three main components, including casualty, physical damage, and business interruption were considered.

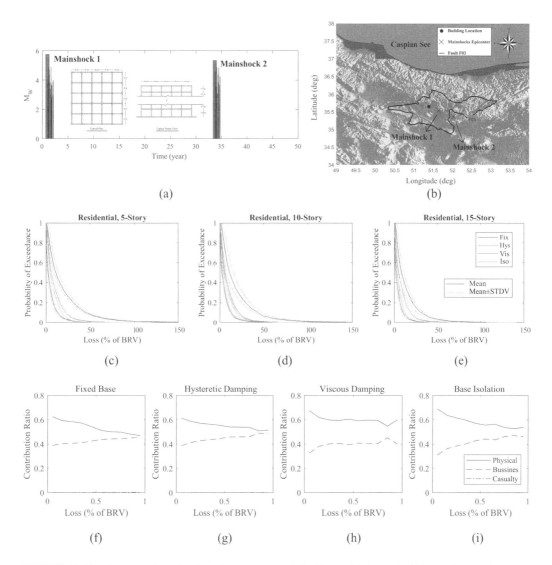

FIGURE 12.14 Comparative seismic risk assessment of 5-, 10-, and 15-story buildings with passive VCSs (hysteretic damper, viscous damper, SIS): (a) randomly generated earthquake event scenario and building models schematic; (b) map of causative fault for randomly generated earthquake scenario; (c) loss curve of 5-story model; (d) loss curve of 10-story model; (e) loss curve of 15-story model; (f) contribution of loss components in the fixed base models; (g) contribution of loss components in the hysteretic damper models; (h) contribution of loss components in the viscous damping models; (i) contribution of loss components in the base-isolated building models. (From Khansefid 2021.)[10]

The performance of building models is impressive. As is shown in Figure 12.14(c)–(e), the PML[9] of all building models with the VCSs, in their 50 years' life span, is reduced significantly. The PML values for the uncontrolled structure, hysteretic damping, SIS, and viscous damping systems, respectively, were equivalent to 47%, 33%, 28%, and 22%, highlighting the superiority of the viscous damping system. Moreover, the contribution of different sources to the estimated lifetime loss value is depicted in Figure 12.14(f)–(i). It is impressive that in all cases, the loss value due to the casualty is near zero, even in an uncontrolled building. In addition, physical damage plays a more important role in total loss. However, by increasing the intensity of the hazard scenario, the contribution of business interruption losses is increased. In other words, in a larger event, the loss of physical damage will be saturated. But the portion related to business interruption still continues to increase.

12.4.2 Semi-Active Systems

Semi-active VCSs are a kind of system that can flexibly change their mechanical characteristic during the time of experiencing an external load without needing a considerable external power source. There are different mechanisms and types of semi-active systems. Some of them only work by changing the stiffness of the device using a control valve in a cylinder and piston device (Figure 12.15(a)) (Kumar et al. 2013), while in other cases, the damping coefficient of viscous dampers is changed by using servo valves (Waghmare et al. 2019). Additionally, there is a more complicated device which is called an electrorheological (ER) damper (Figure 12.15(b)). In this type of damper, a material is used which contains micron-sized dielectric particles sensitive to the electric field like oil. By applying a strong electric field, the particles of this material are aligned, and hence, the strength of the material will increase (Makris et al. 1996), which offers a possibility for a semi-active device. Beside this kind of system, as is shown in Figure 12.15(c), there is another one, which offers a semi-active behavior, called a magnetorheological (MR) damper (Cruze et al. 2021). This system is very similar to the ER dampers. However, the main difference between these two devices is that the MR dampers are sensitive to the magnetic field instead of the electric field. Therefore, the MR material of this system will be polarized and aligned in the presence of a magnetic field, which means a change in the strength properties of the damping device. It is also noteworthy to mention that the usage of ER dampers is less frequent than that of MR dampers due to factors such as the limited range of possible yield stress, sensitivity to pollutants and severe temperatures, and the high voltage requirement.

One of the structural types that enjoys semi-active devices is the bridge. The application of MR dampers in bridges, especially cable bridges, has been studied in several previous works. However, most of them only considered the technical features of this system (Chen et al. 2016; Heo et al. 2014; Ok et al. 2007). On the other side, with the financial perspective, MR dampers can reduce the LCC of bridges through a risk-based approach, as it was shown by Hahm et al. (2013). They compare

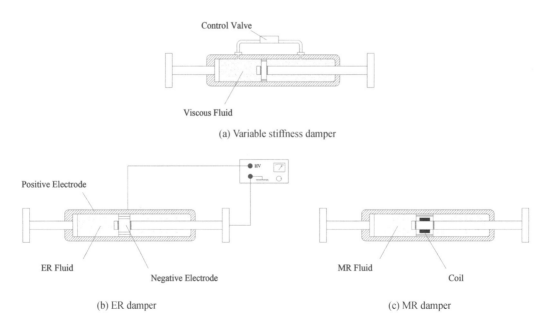

FIGURE 12.15   Schematic of well-known semi-active dampers.

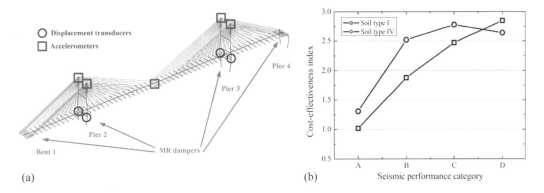

FIGURE 12.16   Application of MR damper in a cable state bridge: (a) arrangement of dampers in bridge; (b) LCC index of a cable state bridge. (From Hahm et al. 2013.)[11]

the LCC (initial building and damper cost plus the lifetime damage cost) of a cable bridge, with and without the presence of MR dampers, by defining a cost factor, which was the ratio of LCC of uncontrolled to the controlled structure (Figure 12.16). Interestingly, the obtained ratio varies from 1 to 2.9 for the severe hazard scenarios on different soil types, which means the absolute economic superiority of MR dampers. Similar studies on a typical frame-type structure (Salajegheh and Asadi 2020) showed that by using the MR dampers in the buildings, there is an optimal MR damper design scheme that can reduce the total LCC of the building by around 16%.

12.4.3   ACTIVE CONTROL SYSTEMS

Active vibration control systems work as online systems. As shown in Figure 12.17, the active control system consists of several components, including sensors, a data logger, a processor, an actuator, and an electrical power source. While an external excitation is experienced by the structure, the sensors start to collect the data regarding the responses of building and/or ground motion vibration.

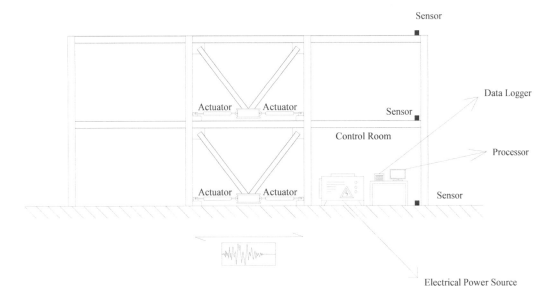

FIGURE 12.17   Schematic of active VCSs in buildings.

Afterward, the data will be transferred to the data logger and then the processor, where the data will be analyzed using complicated optimization algorithms to find the optimal force value that should be exerted on the structure by means of actuators (Soong et al. 1991). As it is clear, the optimization algorithm plays a key role in this system. Several algorithms have been developed for this purpose in past decades, including linear quadratic regulator (LQR), sliding mode control (SMC), optimal polynomial control (OPC), etc. (Khansefid and Ahmadizadeh 2016; Yang et al. 1995, 1996). The performance and outcome of all of these algorithms depend on the location where the sensors are installed. Accordingly, three categories are defined, closed-loop, open-loop, and closed-open loop algorithms. In the first algorithm, the sensors are installed only on the building. In the second one, sensors only record the ground motion, and in the last case, the data regarding both structural response and ground motion are recorded (Soong and Spencer 2000). These active control systems suffer from two major problems that can affect their performance in practice: time delay and spillover. The first one is related to the delay time between the data acquisition of sensors and the implementation of forces by the actuators. While the second problem is about the simplification of building simulation due to the lack of enough sensors in the building. These issues can increase the uncertainty level in the performance of active control systems (Pu 1998; Xue et al. 2018).

The feasibility of using active control systems in the buildings always has been questioned, especially considering the high-level initial cost of these systems besides their technical complication. However, by dealing with this issue from the risk-based perspective, major parts of queries about this system can be answered, as it is done by Khansefid and Bakhshi (2019). In that study, three separate models of 5-, 10-, and 15-story buildings were considered, as typical buildings in Tehran metro city. The buildings were designed in two separate scenarios to look into the efficiency of active VCSs. In the first one, the buildings were designed as uncontrolled traditional structures. While in the second case, buildings with active vibration control systems (LQR algorithm) were designed through an advanced two-step optimization process working on both technical and financial aspects to reach the final optimal design. Afterward, the buildings were analyzed under mainshocks as well as mainshock-aftershock sequences, and the probable loss values of the building were obtained. As is presented in Figure 12.18, the usage of the active VCS reduces the estimated total loss significantly (67% reduction on overage equal to 4% of building replacement value). Additionally, the loss value of casualty is almost zero, which confirms the complete risk mitigation of this loss type. Moreover, the loss of temporal accommodation is very low in both fixed base and active control systems. On the other side, it is observed that the majority of building damage cost (65%) is related to the structural acceleration response (non-structural acceleration sensitive, and content damage), whereas almost 35% of the losses is dependent to the structural drift response (structural, non-structural drift sensitive, and temporal accommodation loss types). In the end, in a summary, the active VCSs not only improve the structural performance of a building but also reduce the economic loss of buildings in their lifespan under probable mainshock-aftershock sequences.

12.5   UNCERTAINTIES IN THE RISK ASSESSMENT OF BUILDINGS WITH SVCs

In the process of seismic risk and resilience assessment, there are major sources of aleatory and epistemic uncertainties. They can be divided into the hazard model, structural model, building damage and loss model, and the repair and recovery model. In Figure 12.19, the sub-models of the resilience assessment are shown. Each of these sub-models has its own feature and is considered to be modeled as a random variable. The probability density function of the resilience index model also is provided in Equation (12.2), where r, t, d, s, and i are resilience index, repair and recovery, damage and loss, structural response, and hazard random variables, respectively. In addition, f(.) is the probability density function.

$$f_{R,T,D,S,I}(r,t,d,s,i) = f_{R,T,D,S,I}(r \mid t,d,s,i).f_{T,D,S,I}(t \mid d,s,i).f_{D,S,I}(d \mid s,i).f_{S,I}(s \mid i).f_I(i) \qquad (12.2)$$

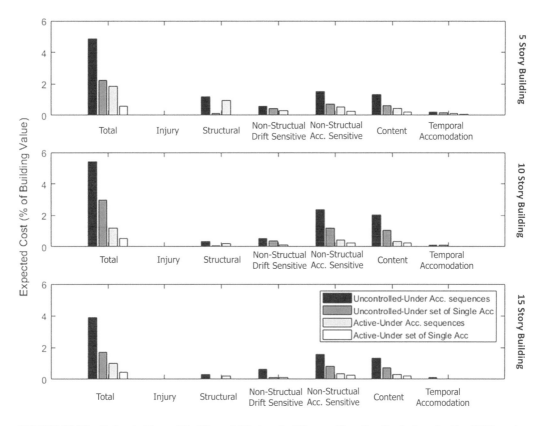

FIGURE 12.18 Estimated loss of 5-, 10-, and 15-story buildings with optimally designed active VCSs under the mainshocks-aftershock sequences. (From Khansefid and Bakhshi 2019.)[12]

In the case of using VCSs in the building, due to the application of very rigorous test protocols (ASCE 2017) in order to be assured about the quality of the vibration control devices, it can be expected that we are confronting low-level uncertainties in the properties of VCSs. Additionally, as it is studied in the previous works, while the supplemental damping system is used (Lavan and Avishur 2013), the sensitivity of structural responses to the uncertainty in the damping properties is negligible. However, in the case of using isolation systems, the uncertainties, especially due to the

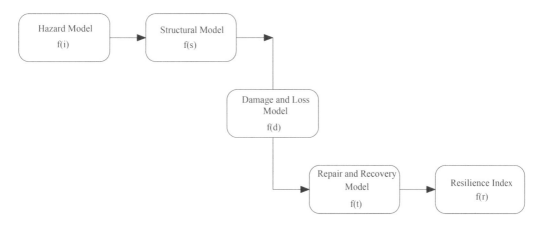

FIGURE 12.19 Sub-models of resilience index assessment and the available uncertainties.

aging phenomenon, can affect the performance of the whole building and isolation systems (Markou et al. 2018, 2019; Roy and Chakraborty 2015). That is why the ASCE7-16 (ASCE 2017) considers the effects of uncertainties in the performance of damping and isolation systems. According to this code, the engineers are required not to use the nominal properties of systems for the design purpose, rather they should first calculate the upper- and lower-bound values of the VCS mechanical characteristics (yield force, stiffness, damping coefficient, etc.) and then perform the analysis and design procedure for two separate scenarios of upper- and lower-bound properties of vibration control devices.

12.6   CONCLUSION

In the end, as was discussed in this chapter, the performance assessment of VCSs from the risk-based perspective is necessary. Since the supremacy of these systems compared to the conventional structural systems, from an economic point of view, only can be proved through risk-based frameworks. However, their technical superiority was already proven in hundreds of academic and practical projects. Additionally, it is shown that for the risk and resilience assessment, it is highly recommended to assess the building performance in its life span, throughout life-cycle-cost calculation, under all probable future seismic hazard scenarios, which can have an effect on the estimated risk and resilience index.

NOTES

1. Reprinted from Engineering Structures, 144, Shin H and Singh MP, Minimum life-cycle cost-based optimal design of yielding metallic devices for seismic loads, Copyright © 2017 with permission from Elsevier
2. Reprinted from Engineering Structures, 130, Kim J and Shin H, Seismic loss assessment of a structure retrofitted with slit-friction hybrid dampers, Copyright © 2017 with permission from Elsevier.
3. Reprinted by permission from Rightslink: Springer, Bulletin of Earthquake Engineering, Performance assessment and optimization of fluid viscous dampers through life-cycle cost criteria and comparison to alternative design approaches. Gidaris I and Taflanidis A, Copyright © 2015.
4. Reprinted from Engineering Structures, 241, Fang C, Ping Y, Zheng Y and Chen Y, Probabilistic economic seismic loss estimation of steel braced frames incorporating emerging self-centering technologies, Copyright © 2021 with permission from Elsevier.
5. Reprinted from Construction and Building Materials, 158, Zheng Y, Dong Y and Li, Y, Resilience and life-cycle performance of smart bridges with shape memory alloy (SMA)-cable-based bearings. Copyright © 2018 with permission from Elsevier.
6. Reprinted from Construction and Building Materials, 158, Zheng Y, Dong Y and Li, Y, Resilience and life-cycle performance of smart bridges with shape memory alloy (SMA)-cable-based bearings. Copyright © 2018 with permission from Elsevier.
7. Reprinted by open access: Springer, Bulletin of Earthquake Engineering, Seismic loss assessment of seismically isolated buildings designed by the procedures of ASCE/SEI 7-16. Kitayama S and Cilsalar H, Copyright © 2022. Open access license: http://creativecommons.org/licenses/by/4.0/. Doi: 10.1016/j.engstruct.2018.06.082.
8. Reprinted from Journal of Earthquake Engineering, Micozzi F et al., Risk Assessment of Reinforced Concrete Buildings with Rubber Isolation Systems Designed by the Italian Seismic Code, Copyright © 2021 with permission from Taylor and Francis. https://www.tandfonline.com/doi/abs/10.1080/13632469.2021.1961937.
9. Probable maximum loss
10. Reprinted from Structures, 34, Khansefid A, Lifetime risk-based seismic performance assessment of buildings equipped with supplemental damping and base isolation systems under probable mainshock-aftershock scenarios. Copyright © 2021 with permission from Elsevier.
11. Reprinted by permission from Rightslink: Springer, KSCE Journal of Civil Engineering, Cost-effectiveness evaluation of an MR damper system based on a life-cycle cost concept. Hahm D, Ok SY, Park W, Koh HM, and Park KS, Copyright © 2013.

12. Reprinted from: Khansefid, A. and Bakhshi, A., 2019. Advanced two-step integrated optimization of actively controlled nonlinear structure under mainshock–aftershock sequences. Journal of Vibration and Control, 25(4), pp. 748–762, Copyright © 2019 SAGE Publishing, doi: 10.1177/1077546318795533.

REFERENCES

Alehashem, S.M.S., Keyhani, A. and Pourmohammad, H., 2008, October. Behavior and performance of structures equipped with ADAS & TADAS dampers (a comparison with conventional structures). In The 14th World Conference on Earthquake Engineering (Vol. 12, p. 17).

Ali, A., Abu Hayah, N., Kim, D. and Cho, S., 2014. Probabilistic seismic assessment of base-isolated NPPs subjected to strong ground motions of Tohoku earthquake, Nuclear Engineering and Technology, 46(5), pp. 699–706.

American Society of Civil Engineers (ASCE), 2017, June. Minimum Design Loads and Associated Criteria for Buildings and Other Structures. American Society of Civil Engineers.

Asadi, P., Nikfar, D. and Hajirasouliha, I., 2020. Life-cycle cost based design of bridge lead-rubber isolators in seismic regions, Structures, 27, pp. 383–395.

Bayat, M. and Abdollahzade, G.R., 2011. Analysis of the steel braced frames equipped with ADAS devices under the far field records, Latin American Journal of Solids and Structures, 8, pp. 163–181.

Cambridge dictionary, 2022. https://dictionary.cambridge.org/dictionary/english-german/risk, accessed in August 2022.

Cardone, D., Conte, N., Dall'Asta, A., Di Cesare, A., Flora, A., Lamarucciola, N., Micozzi, F., Ponzo, F.C. and Ragni, L., 2019. RINTC-E project: The seismic risk of existing Italian RC buildings retrofitted with seismic isolation. 7th International Conference on Computational Methods in Structural Dynamics and Earthquake Engineering Methods in Structural Dynamics and Earthquake Engineering, Crete, Greece.

Carofilis Gallo, W.W., Clemett, N., Gabbianelli, G., O'Reilly, G. and Monteiro, R., 2022. Seismic Resilience Assessment in Optimally Integrated Retrofitting of Existing School Buildings in Italy.

Chen, Z.H., Lam, K.H. and Ni, Y.Q., 2016. Enhanced damping for bridge cables using a self-sensing MR damper, Smart Materials and Structures, 25(8), p. 085019.

Christopoulos, C. and Filiatrault, A., 2006. Principles of Passive Supplemental Damping and Seismic. IUSS Press, Pavia, Italy.

Cimellaro, G.P., 2016. Urban resilience for emergency response and recovery. Fundamental Concepts and Applications.

Cruze, D., Gladston, H., Farsangi, E.N., Banerjee, A., Loganathan, S. and Solomon, S.M., 2021. Seismic performance evaluation of a recently developed magnetorheological damper: Experimental investigation. Practice Periodical on Structural Design and Construction, 26(1), p. 04020061.

Erochko, J., Christopoulos, C., Tremblay, R. and Kim, H., 2013. Shake table testing and numerical simulation of a self-centering energy dissipative braced frame, Earthquake Engineering & Structural Dynamics, 42(11), pp. 1617–1635.

European Standard Commission, 2005. Eurocode 8: Design of structures for earthquake resistance, Part, 1, EN 1998–1.

Fang, C., Ping, Y.W., Chen, Y.Y., Yam, M.C.H., Chem, J. and Wang, W., 2020. Seismic performance of self-centering steel frames with SMA-viscoelastic hybrid braces, Journal of Earthquake Engineering, 26(10), pp. 5004–5031. https://doi.org/10.1080/13632469.2020.1856233

Fang, C., Ping, Y., Zheng, Y. and Chen, Y., 2021. Probabilistic economic seismic loss estimation of steel braced frames incorporating emerging self-centering technologies, Engineering Structures, 241. https://doi.org/10.1016/j.engstruct.2021.112486

Farzampour, A. and Eatherton, M.R., 2019. Parametric computational study on butterfly-shaped hysteretic dampers, Frontiers of Structural and Civil Engineering, 13(5), pp. 1214–1226.

Federal Emergency Management Agency (FEMA), 2018. Seismic Performance Assessment of Buildings Volume 1-Methodology Second Edition, FEMA P-58-1. Applied Technology Council, Washington D.C., USA.

FEMA-355C, 2000. State of the Art Report on System Performance of Steel Moment Frames Subject to Earthquake Ground Shaking. Federal Emergency Management Agency, Washington (DC).

Gidaris, I. and Taflanidis, A., 2015. Performance assessment and optimization of fluid viscous dampers through life-cycle cost criteria and comparison to alternative design approaches, Bulletin of Earthquake Engineering, 13(4), pp. 1003–1028.

Gidaris, I. and Taflanidis, A., 2012, September. Design of fluid viscous dampers for optimal life cycle cost. In Proceedings of 15th World Conference of Earthquake Engineering, Lisbon, Portugal.

Hahm, D., Ok, S.Y., Park, W., Koh, H.M. and Park, K.S., 2013. Cost-effectiveness evaluation of an MR damper system based on a life-cycle cost concept, KSCE Journal of Civil Engineering, 17(1), pp. 145–154.

Han, R., Li, Y. and Lindt, J., 2014. Seismic risk of base isolated non-ductile reinforced concrete buildings considering uncertainties and mainshock–aftershock sequences, Structural Safety, 50, pp. 39–56.

Heo, G., Kim, C. and Lee, C., 2014. Experimental test of asymmetrical cable-stayed bridges using MR-damper for vibration control, Soil Dynamics and Earthquake Engineering, 57, pp. 78–85.

Iwata, M., Kato, T. and Wada, A., 2021. Buckling-restrained braces as hysteretic dampers. In Behaviour of Steel Structures in Seismic Areas (pp. 33–38). CRC Press.

Khansefid, A., 2021. Lifetime risk-based seismic performance assessment of buildings equipped with supplemental damping and base isolation systems under probable mainshock-aftershock scenarios. In Structures (Vol. 34, pp. 3647–3666). Elsevier.

Khansefid, A. and Ahmadizadeh, M., 2016. An investigation of the effects of structural nonlinearity on the seismic performance degradation of active and passive control systems used for supplemental energy dissipation, Journal of Vibration and Control, 22(16), pp. 3544–3554.

Khansefid, A. and Bakhshi, A., 2019. Advanced two-step integrated optimization of actively controlled nonlinear structure under mainshock–aftershock sequences. Journal of Vibration and Control, 25(4), pp. 748–762, doi: 10.1177/1077546318795533.

Khansefid, A. and Bakhshi, A., 2022. New model for simulating random synthetic stochastic earthquake scenarios, Journal of Earthquake Engineering, 26(2), pp. 1072–1089.

Kim, J. and Shin, H., 2017. Seismic loss assessment of a structure retrofitted with slit-friction hybrid dampers, Engineering Structures, 130, pp. 336–350.

Kitayama, S. and Cilsalar, H., 2022. Seismic loss assessment of seismically isolated buildings designed by the procedures of ASCE/SEI 7-16, Bulletin of Earthquake Engineering, 20, pp. 1143–1168.

Klein, R.J., Nicholls, R.J. and Thomalla, F., 2003. Resilience to natural hazards: How useful is this concept? Global Environmental Change Part B: Environmental Hazards, 5(1), pp. 35–45.

Kramer, S.L., 1996. Geotechnical Earthquake Engineering. Pearson Education India.

Kumar, P., Jangid, R.S. and Reddy, G.R., 2013. Response of piping system with semi-active variable stiffness damper under tri-directional seismic excitation, Nuclear Engineering and Design, 258, pp. 130–143.

Lago, A., Trabucco, D. and Wood, A., 2018. Damping technologies for tall buildings: Theory. In Design Guidance and Case Studies (pp. 235–237). Butterworth-Heinemann, UK.

Lavan, O. and Avishur, M., 2013. Seismic behavior of viscously damped yielding frames under structural and damping uncertainties, Bulletin of Earthquake Engineering, 11(6), pp. 2309–2332.

Maghsoudi-Barmi, A., Khansefid, A., Khaloo, A. and Ehteshami Moeini, M., 2021. Probabilistic seismic performance assessment of optimally designed highway bridge isolated by ordinary unbonded elastomeric bearings. Engineering Structures, 247. https://doi.org/10.1016/j.engstruct.2021.113058

Makris, N., Burton, S.A., Hill, D. and Jordan, M., 1996. Analysis and design of ER damper for seismic protection of structures, Journal of Engineering Mechanics, 122(10), pp. 1003–1011.

Mansoori, M.R. and Moghadam, A.S., 2009. Using viscous damper distribution to reduce multiple seismic responses of asymmetric structures, Journal of Constructional Steel Research, 65(12), pp. 2176–2185.

Markou, A.A., Stefanou, G. and Manolis, G.D., 2018. Stochastic response of structures with hybrid base isolation systems, Engineering Structures, 172, pp. 629–643.

Markou, A.A., Stefanou, G. and Manolis, G.D., 2019. Stochastic energy measures for hybrid base isolation systems, Soil Dynamics and Earthquake Engineering, 119, pp. 454–470.

Micozzi, F., Flora, A., Viggiani, L.R.S., Cardone, D., Ragni, L. and Dall'Asta, A., 2021. Risk assessment of reinforced concrete buildings with rubber isolation systems designed by the Italian seismic code. Journal of Earthquake Engineering, 26(14), pp. 7245–7275.

Miller, D.J., Fahnestock, L.A. and Eatherton, M.R., 2012. Development and experimental validation of a nickel-titanium shape memory alloy self-centering buckling-restrained brace, Engineering Structures, 40, pp. 288–298.

Mualla, I.H. and Belev, B., 2002. Performance of steel frames with a new friction damper device under earthquake excitation, Engineering Structures, 24(3), pp. 365–371.

NourEldin, M., Naeem, A. and Kim, J., 2018. Life-cycle cost evaluation of steel structures retrofitted with steel slit damper and shape memory alloy–based hybrid damper, Advances in Structural Engineering, 22(1), pp. 3–16.

Ok, S.Y., Kim, D.S., Park, K.S. and Koh, H.M., 2007. Semi-active fuzzy control of cable-stayed bridges using magneto-rheological dampers, Engineering Structures, 29(5), pp. 776–788.

Pall, A., 2004, August. Performance-based design using pall friction dampers-an economical design solution. In 13th World Conference on Earthquake Engineering, Vancouver, BC, Canada (Vol. 71).

Porter, K., 2015. A beginner's guide to fragility, vulnerability, and risk, Encyclopedia of Earthquake Engineering, 2015, pp. 235–260.

Pu, J.P., 1998. Time-delay compensation in active control of structures, Journal of Engineering Mechanics, 124(9), pp. 1018–1028.

Roy, B.K. and Chakraborty, S., 2015. Robust optimum design of base isolation system in seismic vibration control of structures under random system parameters, Structural Safety, 55, pp. 49–59.

Salajegheh, S. and Asadi, P., 2020. Life-cycle cost optimization of semiactive magnetorheological dampers for the seismic control of steel frames, The Structural Design of Tall and Special Buildings, 29(18), p.e1807.

Shin, H. and Singh, M.P., 2017. Minimum life-cycle cost-based optimal design of yielding metallic devices for seismic loads, Engineering Structures, 144, pp. 174–184.

Soong, T.T., Masri, S.F. and Housner, G.W., 1991. An overview of active structural control under seismic loads, Earthquake Spectra, 7(3), pp. 483–505.

Soong, T.T. and Spencer, B.F., 2000. Active, semi-active and hybrid control of structures, Bulletin of the New Zealand Society for Earthquake Engineering, 33(3), pp. 387–402.

Taflanidis, A.A. and Beck, J.L., 2009. Life-cycle cost optimal design of passive dissipative devices, Structural Safety, 31(6), pp. 508–522. Buildings in Italy. Buildings, 12(6), p. 845.

Taflanidis, A.A. and Gidaris, I., 2013. Life-cycle cost based optimal retrofitting of structures by fluid dampers. In Structures Congress 2013: Bridging Your Passion with Your Profession (pp. 1777–1788).

Thywissen, K., 2006. Components of Risk: a Comparative Glossary. UNU-EHS.

Waghmare, M.V., Madhekar, S.N. and Matsagar, V.A., 2019. Semi-active fluid viscous dampers for seismic mitigation of RC elevated liquid storage tanks, International Journal of Structural Stability and Dynamics, 19(03), p. 1950020.

Whittle, J.K., Williams, M.S., Karavasilis, T.L. and Blakeborough, A., 2012. A comparison of viscous damper placement methods for improving seismic building design, Journal of Earthquake Engineering, 16(4), pp. 540–560.

Xu, L., Xingsi Xie, X. and Li, Z., 2022. Seismic performance and resilience of composite damping self-centering braced frame structures, Fundamental Research. https://doi.org/10.1016/j.fmre.2022.05.009

Xue, K., Igarashi, A. and Kachi, T., 2018. Optimal sensor placement for active control of floor vibration considering spillover effect associated with modal filtering, Engineering Structures, 165, pp. 198–209.

Yaghmaei-Sabegh, S., Jafari-Koucheh, E. and Ebrahimi-Aghabagher, M., 2020, December. Estimating the seismic response of nonlinear structures equipped with nonlinear viscous damper subjected to pulse-like ground records. In Structures (Vol. 28, pp. 1915–1923). Elsevier.

Yang, J.N., Agrawal, A.K. and Chen, S., 1996. Optimal polynomial control for seismically excited non-linear and hysteretic structures, Earthquake Engineering & Structural Dynamics, 25(11), pp. 1211–1230.

Yang, J.N., Wu, J.C. and Agrawal, A.K., 1995. Sliding mode control for nonlinear and hysteretic structures, Journal of Engineering Mechanics, 121(12), pp. 1330–1339.

Yu, C., Bolisetti, C., Coleman, J.L., Kosbab, B. and Whittaker, A.S., 2018. Using seismic isolation to reduce risk and capital cost of safety-related nuclear structures, Nuclear Engineering and Design, 326, pp. 268–284.

Zheng, Y., Dong, Y. and Li, Y., 2018. Resilience and life-cycle performance of smart bridges with shape memory alloy (SMA)-cable-based bearings, Construction and Building Materials, 158, pp. 389–400.

Zhou, Z., Xie, Q., Lei, X.C., He, X.T. and Meng, S.P., 2015. Experimental investigation of the hysteretic performance of dual-tube self-centering buckling-restrained braces with composite tendons, Journal of Composites for Construction, 19(6), p. 04015011.

Zhou, Y. and Xing, L., 2021. Seismic performance evaluation of a viscous damper-outrigger system based on response spectrum analysis, Soil Dynamics and Earthquake Engineering, 142, p. 106553.

Zhu, R.Z., Guo, T. and Mwangilwa, F., 2020. Development and test of a self-centering fluidic viscous damper, Advances in Structural Engineering, 23(13), pp. 2835–2849.

13 Decision Systems to Support Cost-Benefit Analysis Toward Railway Infrastructures Sustainable Rehabilitation

Maria João Falcão Silva, Paula Couto,
Filipa Salvado, and Simona Fontul

CONTENTS

13.1   INTRODUCTION

The construction industry is an economic activity with a great impact on the environment. and natural heritage and naturally has a huge impact on the built environment and architectural heritage. It is also responsible for a significant portion of the negative environmental impacts in terms of final energy consumption (42%), emission of greenhouse gases (50%) and waste production (22%) (Schultmann et al., 2010). In this framework, the European Union has been setting targets and defining policies with a view to preserving environment and resource rationalization, reflected in the Horizon 2020 community program (EC, 2008). The search for solutions to achieve these goals in the construction sector leading necessarily life to a strong apostate in increasing the usefulness of existing infrastructures, through their rehabilitation and maintenance (Mansfield, 2001). The rehabilitation of the major infrastructures of a country, since that ensures the reduction of its vulnerability in the face of risks, thus currently assumes a leading role in increasing the sustainability of the urban environment, constituting an important alternative to the realization of new constructions or to the demolition and reconstruction of the existing ones, as it significantly reduces the consumption of materials and the production of waste (Barbisan et al., 2012). There are issues related to the decision to intervene or not, which need more sustained and consistent validation. The decision to rehabilitate existing infrastructures is complex, as the associated costs require consideration of different levels, given their relevance to all stakeholders in the decision-making process, and are not always easily quantifiable. Furthermore, following recent developments in the European Union, it is essential and urgent to carry out studies of economic basis to support the rehabilitation strategies to be adopted (EC, 2008).

DOI: 10.1201/9781003325246-13

In this context, it seems clear that the use of methodologies based on cost-benefit analysis (CAB) can contribute positively to support decisions related to investment projects in infrastructures. The CAB makes it possible to study the feasibility of projects and assess their impacts based on the comparison of costs and benefits over a given time horizon (Çetinceli, 2005; FEMA 227, 1992; FEMA 228, 1992; Mishan, 1988). Alternatively, in cases where it is difficult to assess in monetary terms the benefits of investing resources or investments, there are methods that use a cost-effectiveness analysis (CEA) (EVALSED, 2013). This kind of analysis allows to identify and select alternative projects with the same objectives and essentially constitutes a tool to support the selection of projects within the scope of a pre-defined program, having little applicability in projects with multiple objectives.

The expansion of light-rail transit (LRT) systems in recent decades is due, among numerous factors, to the fact that they are non-polluting systems, benefiting the urban environment, as they are safe and reliable and, through channel segregation, guarantee fast urban routes, avoiding congestion. Reliability in the travel time and good commercial speed, the high transport capacity, comfort, and accessibility as well as ease of integration in the centers urban factors are factors that translate into strong adherence by those who live and work in the city (Cartier, 2005; Hass-Klau & Cramptom, 2005; Jefferson & Kühn, 1996; Suzuki et al., 2013).

The modern LRT or tramway becomes a source of pride for the city and the population's adherence to these systems are almost always a certainty. A new collective transport culture is present and visible within the public space and is without a doubt, the tramway's most direct contribution to contemporary urban culture. The tramway can be a concrete element of a city densification strategy, considering its corridor of influence and avoiding the location of large equipment outside that perimeter. The LRT can be a fundamental tool in a movement to revalue the centers and may represent a formidable vector of urban regeneration. The restoration of public spaces goes through a significant reduction in the use of the car, speed, and surface assigned to circulation and parking. The use of public surface transport, in its own channel, is the main alternative to the use of cars (Laisney & Grillet-Aubert, 2006).

The purpose of this proposal is to address some of the situations highlighted in the diagnosis on the mobility of parishes located farther south in the municipality of Vila Franca de Xira – Póvoa Santa Iria, Forte da Casa, and Vialonga. The analysis highlights the need to take advantage of the high transport capacity ensured by the railway service essentially in commuting to Lisbon in view of the existing quadruplication of the railway line between Póvoa and Lisbon, based on a cost-benefit analysis (CBA).

Although on a local scale, the objectives of the PET (Strategic Plan Transport) for 2020 were sought to be achieved, namely: (i) Guarantee the right to good accessibility and mobility for all citizens, regardless of the geographic area in which you reside and your socio-economic conditions; (ii) privileging safety, comfort, and speed of travel from each source to the final destination; (iii) transport must promote spatial planning; (iv) transport has to reduce its negative environmental impacts essentially in terms of pollutant emission terms; (v) privileging a global and integrated vision of mobility/accessibility; (vi) promote the modal split of people.

The solution to be presented aims to promote the most sustainable mobility patterns –increasing the cohesion of means of transport, coherence, and social inclusion, more friendly from the point of view of being environmentally friendly, more energy efficient, more comfortable, and safer and with less accident rates. Of the collective means of transport existing in the area under study, the train is evident as it is the means of transport that presents a large capacity offer facilitating the mobility of populations and contributing to the improvement of the quality of life, essentially by decongesting road traffic and reducing pollutant emissions.

13.2 COST-BENEFIT ANALYSIS

The CBA comprise methods to evaluate the net economic impact of an investment project and can be used for a variety of interventions. The CBA is characterized by being an evaluation model that admits monetary unity as the main measure and has been predominantly used in the context of large

FIGURE 13.1   Cost-benefit analysis structure.

public investments during the second half of the twentieth century (Barbisan et al., 2012; Directive 2008/98, 98; Mansfield, 2001; Salvado et al., 2019).

The CBA of investment projects is explicitly required by the EU regulations that govern the Structural Funds (SF), the Cohesion Funds (CF), and the Instrument for Structural Policies for Pre Accession (ISPA) in the case of projects whose budgets exceed, respectively, 50, 10, and 5 million euros. The methodologies based on CBA consist of methods for assessing the net economic impact of an investment project and can be used for a variety of interventions. In these circumstances, the aim of a CBA is to determine whether a project is feasible from the point of view of social welfare through the algebraic sum of the discounted costs and benefits over time.

CBA is based on the monetary value conversion of all costs and benefits, even when they are intangible, through several adjustments (prices distortions; monetary value for the non-monetary impacts; inclusion of indirect effects; and social rates different from financial rates) (Directive 2008/98, 98) and allows for determining if future benefits are enough to justify the current costs of a certain investment project.

For the application of a CBA to different investment projects, to support the decision-making process, the following presented five sequential phases should be considered (Figure 13.1) (EC, 2003; EVALSED, 2013; Falcão Silva et al., 2014; Falcão Silva & Salvado, 2015; Mishan, 1988; Salvado & Falcão Silva, 2016): (i) Context analysis, project objectives, and identification (Phase 1); (ii) feasibility and options analysis (Phase 2); (iii) financial analysis (Phase 3); (iv) economic analysis (Phase 4); (v) risk assessment (Phase 5).

Context Analysis, Project Objectives, and Identification (Phase 1) correspond to the definition of social and economic objectives of the proposed investment project as well as the context of its implementation, design, and materials indicators. The project unit of analysis (individual project, stage of a larger-scale project, group of projects, etc.) must be clearly identified and defined in accordance with the general principles of a CBA (Falcão Silva et al., 2019). In defining the objectives of a certain project, the socioeconomic benefits of its execution and its material indicators must be foreseen. The set of benefits resulting from this project, in terms of well-being, must have an adequate proportion to the costs and the main social and economic effects; both direct and indirect must be considered. If it is not possible to quantify all the direct and indirect social effects of the project in question, due to lack of information; replacement values linked to the objectives of the project must be established (Falcão Silva et al., 2014).

Feasibility and Options Analysis (Phase 2) refers to the development of an analysis to guarantee the project feasibility from a technical point of view, based on the different options initially proposed for the investment project (EVALSED, 2013). The feasibility analysis of a certain project does not only refer to the verification of the possibility of implementing the different specialties of the project, but also implies the consideration of aspects related to marketing, management, analysis of execution, etc., given that different alternatives can be adopted for the project, in view of the socioeconomic objectives identified. For a given intervention, it is necessary to check whether the chosen option corresponds to the best among the possible alternatives and that other options have been duly considered. To this end, an analysis of possible alternatives must have been carried out previously, and it is necessary to consider at least two options: (i) absence of intervention (do-nothing); (ii) existence of intervention (do-something) (Falcão Silva et al., 2014).

Financial Analysis (Phase 3) has three purposes: (i) to gather the necessary information for the cashflow analysis; (ii) evaluate the financial viability of the project (sustainability analysis); and

(iii) assess financial benefits by calculating profitability from the private investor's point of view (financial return on the project and capital) (EVALSED, 2013). The project's cash-flow forecasts are used to calculate the appropriate rates of return in accordance with, the Internal Rate of Financial Return (TIRF), the Internal Rate of Financial Return calculated on the Investment Cost (TIRF/C), and the Internal Rate of Financial Profitability calculated on own funds (TIRF/K), as well as the corresponding Financial Net Present Value (VALF) (Brealey et al., 2007). Financial analysis provides essential information on the relationship between factors of production and the product and their prices and the overall structure of the income and expenditure schedule. It is carried out based on a set of data tables that gather the financial flows of the investment, distributed among the total investment (investment costs and residual value), the operating expenses and revenues (operating costs and eventual revenues), and the sources of financing and the analysis of cash flow for financial viability. The financial analysis should allow the elaboration of tables summarizing the financial flows, namely: (i) return on investment (capacity of net operating income to cover investment costs) that allows the assessment of financial sustainability and the profitability or return of the project. The financial viability of the project is guaranteed if the accumulated net cash flow is positive or null in any of the years considered in the analysis. The financial return of the project is evaluated based on the TIRF and VALF values mentioned above; (ii) return on invested capital, by analyzing investment costs and sources of financing, is the private investor's own funds (when they are actually paid), or public contributions (local, regional, and central), credits at the time they are repaid, in addition to operating costs, interest included, and revenue-related entries (Falcão Silva et al, 2014). It is the starting point for the subsequent economic analysis (EVALSED, 2013) and allows obtaining the necessary information in terms of income and expenses, their relative market prices, and how they are distributed over the expected time for the implementation and exploration of the project.

Economic Analysis (Phase 4) overviews the project suitability in terms of economic values of all costs and benefits (EVALSED, 2013). In the proposed methodology, with the economic analysis, it is intended to evaluate the contribution of a project to the economic well-being of a region or country since, in practice, the economic analysis addresses the evaluation of the opportunity for the society and better use of resources. This analysis makes it possible to move from the perspective of the private investor to that of the public operator in relation to the project in question, when considering the interest of all interested parties in the company and not just that of the private investor or the owners of the project (Brealey et al., 2007). Starting from the financial analysis (study of investment performance, regardless of its financial sources), the economic analysis applies a series of corrections to the financial data, and considers the social benefits and costs not accounted for in the financial analysis. These corrections consist of eliminating tax effects, considering the externalities that lead to social costs and benefits, and converting the market prices used in the financial analysis. This is possible by assigning, for each input and output element, a conversion factor that turns the market prices into fictitious prices. The economic analysis thus consists of three distinct phases: (i) *Tax correction*: Deduction, of the financial analysis flows, of the payments that have no real counterpart in resources, such as subsidies and indirect taxes on factors and products. As for direct public transfers, these are not included in the initial framework to be used for financial analysis, which considers investment costs and not financial resources. (ii) *Correction of externalities*: External costs and benefits for which there is no cash flow need to be included in the exits and inflows. Externalities are the side effects of producing goods or services on other people who are not directly involved with the activity, that is, externalities refer to the impact of a decision on those who did not participate in that decision. They represent positive or negative effects on third parties without the opportunity to prevent them and without the obligation to pay them or the right to be compensated. (iii) *Conversion of market prices into fictitious prices to integrate social costs and benefits (determining conversion factors)*: This correction is necessary since the markets are imperfect and the market prices do not always reflect the opportunity cost of a good. If price distortion is not eliminated, they are not adequate indicators of well-being. Investment projects may favor other socioeconomic agents in addition to the direct recipients of the social benefit generated by the project itself. If so, this should be considered in an appropriate assessment. Externalities should, whenever possible, be quantified in monetary terms, and if not possible, they should be quantified through other indicators. External effects are sometimes difficult to quantify in monetary terms, requiring many approaches that consider artificial market mechanisms capable of measuring the preferences of individuals and society. This is the case, for example, of the costs and benefits associated with the amount of time or of human lives. Regarding the environmental impact, it must be correctly described and assessed. At the end of this analysis, it should be possible to identify the solution that corresponds to the best use of resources, in terms of optimizing the cost-benefit ratio expressed in economic terms; that is, the one that leads to the greatest benefits and the lowest costs (EVALSED, 2013; Falcão Silva et al, 2014).

Risks Assessment (Phase 5) comprises a sensitivity analysis and a risk analysis. The first one performed the selection of variables and critical parameters of the CBA model that have a more pronounced effect (positive or negative) in relation to the value used as the best estimate in the reference case, and have a more pronounced effect on the determined economic parameters. The criteria to be used for the choice of critical variables differ according to the characteristics of the project considered and must be rigorously evaluated in each case. In the context of the proposed methodology, the following steps are proposed to be considered in the sensitivity analysis: (i) identification of all variables used to calculate all parameters of economic and financial analysis, grouping them by categories; (ii) identification of possible interdependencies between variables that could lead to distortions in results and double accounting; iii) quantitative analysis of the impact of variables, in order to select those that are inelastic or that have marginal elasticity. The subsequent quantitative analysis can be limited to the most significant variables, which should be checked in case of doubt; iv) evaluation of the progress of variables over time; v) identification of critical variables. As examples of critical variables related to fictitious costs and benefits prices, we have the market price conversion coefficients, the time value, the cost of avoided deaths, the fictitious prices of goods and services, or the evaluation of externalities. Likewise, critical variables related to quantitative parameters of costs and benefits are, for example, the dimensions of the area used, the incidence of the energy produced, or the secondary raw materials used. After identifying the critical variables, it is necessary, to carry out the risk analysis, to associate each variable with a probability distribution, defined in a range of values around the best estimate used in the reference case. The presentation of the result consists of expressing it in terms of the distribution of probabilities or accumulated probabilities of the IRR or VAL over a given range of values. The cumulative probability curve thus allows a degree of risk to be assigned to the project.

Finally, after the application of all the phases described, it becomes possible to obtain project performance indicators that will help in a decision-making process.

13.3 APPLICATION OF COST-BENEFIT ANALYSIS TO CASE STUDY

The cost-benefit analysis (CBA) is an analytical tool for evaluating an investment decision, allowing to quantify the variation in well-being attributable to it. The primary objective of a CBA is to help a more efficient allocation of resources, demonstrating the suitability for society of a particular intervention by comparison with existing alternatives. A CBA analysis comprises two main components: a financial analysis and an economic analysis. While the financial analysis of the project focuses on the benefits and costs attributed to the infrastructure manager and operator, the economic analysis focuses on the benefits and costs generated for the society.

The present section presents the results of the CBA prepared for the LTR project between Póvoa de Santa Iria and MARL, which contemplates the construction of a new section of line between the Póvoa de Santa Iria railway station and MARL, constituting a system of fast transport. The analysis was based on data on investment and exploration costs and on the results of the study on demand and social and environmental benefits of the section under study, following the methodology of the European Commission for the elaboration of CBA.

As main sources of information, this CBA used: (i) the results of the study on demand and social and environmental benefits of the section under study; (ii) the current European recommendations, namely those of the European Commission (EC) and the European Investment Bank (EIB), referring as a priority source to the "Guide to Cost Benefit Analysis of Investment Projects," European Commission (December 2014). To update prices, the inflation values published in the Banco de Portugal from December 2018, for the period 2018 to 2021, and the Bank's estimates European Central, HICP Inflation forecasts, for the period between 2022 and 2024, have been considered. From 2025 onwards, a constant inflation rate equal to the last forecast value is considered.

13.3.1 CONTEXT ANALYSIS, PROJECT OBJECTIVES, AND IDENTIFICATION

Adopting the light tram system, the aim is to create a fast and quality connection that enhances the articulation of the residence with the railway mode of the residents of the parishes of Vialonga, Forte da Casa, Póvoa de Santa Iria (second phase) and customers and employees of both MARL (western terminus) as well as the Solvay Industrial Complex (eastern end of the route). This system aims to take advantage of the potential installed in the suburban rail transportation of the Azambuja concretely from Póvoa station toward Lisbon. For the success of this system, it will be essential to achieve the best levels of functionality and attraction at the Póvoa Station interface.

Consideration was given to the expected traffic decongestion on the northern line with the implementation of the high-speed line between Lisbon and Porto. The effectiveness and efficiency of transshipment both for the heavy railroad and for the road transport of regional and local service could be reached there with the adoption of transport promotion policies integrated and multimodal public.

Tariff integration policies coordinated with the exploitation of services as well as the promotion of and integration of information to the public are priorities for the success of the proposed system. The creation and/or alteration of the road service in distribution routes from bus stops new system, integrated, and coordinated in terms of tariffs, exploitation, and information to the public could be one of the fundamental elements for the "livelihood" of the various modes of transport collective. This promotes the intermodally and competitiveness of public transport compared to transport individual.

The planned vehicles are those normally used in this type of system throughout Europe, presenting the characteristics given in Table 13.1.

For its circulation, the construction of a new infrastructure, platform, and food is planned electrical by catenary. It is intended to minimize the landscape impacts caused by the system in the region by introducing the various "finishes" available at the platform level. It is proposed to apply

TABLE 13.1
Technical Characteristics of Planned Vehicles

Tara	53 tons
Number of seats	100
Number of seats	248
Maximum speed	100 km/h
Commercial speed	30 km/h
Maximum acceleration	1.25 m/s^2
Maximum deceleration	2.4 m/s^2
Vehicle length	37,000 mm
Vehicle width	2,650 mm
Gauge	1,435 mm
Vehicle height	3,500 mm
Minimum bending radius	25 m

the Ballastless Track system. This system indicated for the rapid mass transport in urban landscape allows the coexistence of people, means of transport, and other machines in crowded spaces both in the construction phase and in the operation phase.

This system can be covered with concrete, bituminous, or other paving material and even grass (green strips), allowing the use by buses, other motor vehicles, bicycles, and people. It offers high comfort and safety and requires minimum levels of care and maintenance.

The route consists of a double track with about 6.20 in width and 6.0 km in length, comprising eight stops. It will be inserted practically in all layouts next to existing streets, totaling a width of 15 m. At stops to insert the central boarding platform, its width increases to 25 m. It starts at the Póvoa Interface Station (point A).

It is proposed that this interface integrates the bus terminal, a car silo, and covered pedestrian connection to the Póvoa Railway Station to be built in the area currently used for parking the train users. It runs between the road and Rio until the access to the existing viaduct over the railroad line. It continues a new viaduct to be built on the CF, in alignment with the roundabout of the EN10 and the EM502. It is then integrated into the current EM 502, with the implementation of Stop B next to the existing warehouses and shopping center. This stop is intended to serve mainly the population of Forte da Casa, but also that of the second phase of Póvoa de Santa Iria. Since this stop is located at the border of the two parishes, the construction of a pedestrian overpass that connects the two slopes would enhance the transfer between parishes.

The compatibility of the route with the EM 502 continues until the underpass to AE1/IP1. There, it is proposed to implement Stop C, close to the existing business niche. From this point, the route is implanted in a cultivation area until it intersects ER19. Stop D will be implemented in the available area between ER 19 and EM 501.

The route is now integrated in the EM 501 until the Mercado Abastecedor de Lisboa – Stop H. In this section, it is foreseen the implantation of 3 intermediate stops – Stops E, F, and G. They are inserted in the urban network and are located next to the conglomerates denser. The termination stop at MARL will integrate the existing car park and the terminus road stop. The aim is to ensure fast, safe, and "environmentally friendly" transport for users of this important infrastructure and for users coming from the eastern (Alpriate and Quintanilho) and western (Mogos, Santa Eulália) conglomerates (Figure 13.2).

FIGURE 13.2 Layout under analysis.

For the calculation of the number of users of the new service, the population served by the collective road transport, is specifically of the routes that end at the railway station of Póvoa and the direct careers to the city of Lisbon. The value of 50% was adopted for the migration of these users to the new service. This value is justified in view of the high occupancy rates that regional and local road services complement, in the case of direct careers, with the inevitable attraction of users by both heavy and light rail. Considering the necessary and urgent reorganization and elimination of barriers created by illegal parking in this area, as well as the search for alternatives to the use of IT, 80% of its users were considered to migrate to the new system.

It can be considered that the new service will have a demand of around 6,290 users/hour in Year 0 and in the morning peak period. It was also intended to reduce the use of IT by employees and customers of the Solvay Industrial Complex and MARL facilities. With the end stops of the new system located next to the facilities of the Solvay Industrial Complex and the supply market of the Lisbon region, this attractiveness can be guaranteed.

Considering the value of 60% for the transition from IT to the new system, there was a daily demand of around 480 users/hour. In view of the values presented for Year 0, based on the hourly capacity/ speed by means of transport presented in the previous figure, it is proposed to implement the fast tram system or light rail tram between the Póvoa Railway Station and the supply market of Lisbon. Based on the data on the characteristics of the planned vehicles and the layout of the system, the offer was calculated. The values obtained for Year 0 in the morning rush hour are shown in Table 13.2.

To capitalize on the attractiveness that the light train system will generate, both for the high standard of service quality, convenience, and speed, as well as for providing the service at affordable costs. In economic and environmental contexts, greater competitiveness will be achieved in relation to individual transport, thus leading to the deterrence of the use of individual transport. Consequently, the operationality between the journeys in public transport, individual transport, and transport linked to logistical and industrial activities that proliferate in the parishes under study will be increased.

It is important to emphasize that the route has characteristics of a cycle path, thus enhancing the practice of sport and leisure with the consequent improvement of the population's quality of life. The proposed system will allow the population to get closer to the Tagus River, reinforcing the mobility of users of Riverside Parks between Póvoa and Vila Franca de Xira, including public holidays and weekends.

13.3.2 Feasibility and Options Analysis

The base scenario, usually called the "do-nothing" scenario, corresponds to the scenario of not pursuing the project. Since, if the project does not come to fruition, more reduced alternative

---

TABLE 13.2

Characteristics of the Planned Vehicles and the Layout of the System for Year 0 in the Morning Rush Hour

Vehicle capacity	248 places
Line extension	6 km
Commercial speed	km/hour
Time of travel	10 minutes
Rotation time	30 minutes
Interval	2 minutes
Frequency	30 veículos/hour
Number of vehicles needed	4
Capacity	992 places/hour/direction

investments are not foreseen to address some of the supply shortfalls that this project aims to resolve. The base scenario will follow the business as usual (BAU) criterion, as indicated by the "Guide to Cost-Benefit Analysis of Investment Projects," European Commission (December 2014). The use of this criterion means that, in a scenario of non-continuation of the project, it is assumed that the network and the service provided by the road operators will remain as they currently exist, generating a level of costs, revenues, and benefits that are in line with the current operating levels.

The assumptions used in the definition of the base scenario include the following: (i) The road network to be considered will be the network in operation on January 1, 2020. (ii) The Lisboa Viva intermodal travel ticket is the only valid ticket in the system. This title is also valid on the Rodoviária Nacional network, coexisting with single-mode tariff, on the railway lines operated by CP Lisboa and in some careers of private bus operators, namely Rodoviária de Lisboa and Rodoviária da Estremadura.

The project scenario, usually referred to as the "do-something" scenario (as opposed to the previous one), corresponds to the scenario for pursuing the LRT project, as defined previously.

The costs and benefits generated by the project are due to changes in the behavior of current "travelers" because of the transport offer provided by the new LRT. Thus, it is important to quantify these changes in demand. The traffic model divides demand according to the following type of passenger: (i) demand captured by public transport (TP): passengers using public transport the local network and the direct road network to Lisbon; (ii) demand captured by individual transport (IT): users "conquered" by IT, that is, those who today use their vehicle and who make their travels, total or partially, using the TP in general (unregulated parking next to the railway station Póvoa); (iii) induced demand: passengers that today do not make a certain trip and that, as a result the increase and improvement in the transport offer provided by the new infrastructure, choose to for making this trip.

The annualization of demand estimates over the project's years of operation was carried out according to the assumptions following. It was assumed to have a five-year ramp-up period and an increase in the demand value in relation to the estimated value for the cruise year during this ramp-up period (Table 13.3). This "ramp-up" effect is only applied to passengers who transfer from individual transport and to the segment of induced travel.

As in the "do-nothing" scenario, the natural growth in demand according to the year is calculated according to the following: (i) In the first three years (2020 to 2022), it is based on the validation forecast model in the collective road transport network, according to the evolution of the explanatory variables "overnight stays" and "fuel sales" referring to the three cities under study (Póvoa) Santa Iria, Forte da Casa, and Vialonga), although weighted with the GDP evolution estimates considering an elasticity of one, according to the formula: $\Delta Procura = \Delta PIB \ per \ capitae$, where e represents the elasticity. (ii) In the following years, the GDP growth estimates presented in the study by the European Commission – The 2018 Ageing Report – Economic and budgetary

TABLE 13.3

Demand in Relation to the Estimated Value for the Cruise Year

Year	Demand Considered/Estimated Demand
1	88.0%
2	92.7%
3	96.3%
4	98.8%
5	100.0%

projections for the 27 EU Member States (2016–2070) are assumed, with an elasticity of 1 against GDP growth figures.

For this study, an analysis period of 30 years was considered, starting in 2020 and ending in 2050. This period is divided as follows: (i) 2 years for studies and construction of the infrastructure: 2020 to 2022; (ii) 28 years of operation: 2023 to 2050. The assumed financial analysis period is based on the reference period adopted by the European Commission for carrying out CBA in the transport sector (between 25 and 30 years). Regarding the social and environmental benefits generated by the initial investment, it was assumed that they will have a 50-year reflection, so the difference (20 years) is considered between the economic and financial analysis through the residual value.

The evolution of demand in the LRT network in the "do-something" and "do-nothing" scenarios is shown in Figures 13.3 and 13.4.

Year	Search Captured to Local TP	Search Captured to Direct TP	Search Captured to Local TI	Search Captured to Terminus TI	Induced Search	Total
2020	-	-	-	-	-	-
2021	-	-	-	-	-	-
2022	-	-	-	-	-	-
2023	10,468	5,825	1,478	3,538	174	21,483
2024	10,631	5,915	1,582	3,785	187	22,100
2025	10,793	6,006	1,668	3,992	197	22,656
2026	10,956	6,097	1,737	4,157	205	23,152
2027	11,120	6,187	1,785	4,270	211	23,573
2028	11,283	6,278	1,811	4,333	214	23,919
2029	11,446	6,369	1,837	4,396	217	24,265
2030	11,608	6,459	1,863	4,458	220	24,608
2031	11,771	6,550	1,889	4,520	223	24,953
2032	11,941	6,645	1,917	4,586	226	25,315
2033	12,120	6,744	1,945	4,655	230	25,694
2034	12,308	6,849	1,975	4,727	233	26,092
2035	12,505	6,958	2,007	4,803	237	26,510
2036	12,712	7,073	2,040	4,882	241	26,948
2037	12,928	7,193	2,075	4,965	245	27,406
2038	13,154	7,319	2,111	5,052	249	27,885
2039	12,291	7,451	2,149	5,143	254	27,288
2040	13,638	7,589	2,189	5,238	258	28,912
2041	13,898	7,733	2,231	5,337	263	29,462
2042	14,162	7,880	2,273	5,439	268	30,022
2043	14,431	8,030	2,316	5,542	273	30,592
2044	14,705	8,182	2,360	5,647	278	31,172
2045	14,984	8,338	2,405	5,755	284	31,766
2046	15,269	8,496	2,451	5,864	289	32,369
2047	15,559	8,658	2,497	5,975	295	32,984
2048	15,855	8,822	2,545	6,089	300	33,611
2049	16,156	8,990	2,593	6,205	306	34,250
2050	16,463	9,161	2,642	6,322	312	34,900

FIGURE 13.3   Evolution of daily demand for scenario do-something.

Year	Local TP	Direct TP	Total TP	Local TI	Terminus TI	Total TI	Total (TP+TI)
2020	20,935	11,649	32,584	2,100	6,700	8,800	41,384
2021	21,291	11,847	33,138	2,136	6,814	8,950	42,088
2022	21,632	12,037	33,669	2,170	6,923	9,093	42,762
2023	21,973	12,227	34,200	2,204	7,032	9,236	43,436
2024	22,316	12,418	34,734	2,239	7,142	9,381	44,115
2025	22,658	12,607	35,265	2,273	7,251	9,524	44,789
2026	23,000	12,798	35,798	2,307	7,361	9,668	45,466
2027	23,342	12,989	36,331	2,341	7,470	9,811	46,142
2028	23,686	13,179	36,865	2,376	7,580	9,956	46,821
2029	24,027	13,369	37,396	2,410	7,689	10,099	47,495
2030	24,368	13,559	37,927	2,444	7,799	10,243	48,170
2031	24,709	13,749	38,458	2,479	7,908	10,387	48,845
2032	25,067	13,948	39,015	2,515	8,022	10,537	49,552
2033	25,443	14,158	39,601	2,552	8,143	10,695	50,296
2034	25,838	14,377	40,215	2,592	8,269	10,861	51,076
2035	26,251	14,607	40,858	2,633	8,401	11,034	51,892
2036	26,684	14,848	41,532	2,677	8,540	11,217	52,749
2037	27,138	15,100	42,238	2,722	8,685	11,407	53,645
2038	27,613	15,365	42,978	2,770	8,837	11,607	54,585
2039	28,110	15,641	43,751	2,820	8,996	11,816	55,567
2040	28,630	15,931	44,561	2,872	9,163	12,035	56,596
2041	29,174	16,233	45,407	2,926	9,337	12,263	57,670
2042	29,728	16,542	46,270	2,982	9,514	12,496	58,766
2043	30,293	16,856	47,149	3,039	9,695	12,734	59,883
2044	30,868	17,176	48,044	3,096	9,879	12,975	61,019
2045	31,455	17,503	48,958	3,155	10,067	13,222	62,180
2046	32,052	17,835	49,887	3,215	10,258	13,473	63,360
2047	32,661	18,174	50,835	3,276	10,463	13,739	64,574
2048	33,282	18,519	51,801	3,339	10,652	13,991	65,792
2049	33,914	18,871	52,785	3,402	10,854	14,256	67,041
2050	34,559	19,230	53,789	3,467	11,060	14,527	68,316

FIGURE 13.4 Evolution of daily demand for scenario do-nothing.

In terms of the evolution of the offer and despite the current deficiencies in the public road passenger service, it was assumed that this would remain constant corresponding to the "do-nothing" scenario. According to data provided by the operator Rodoviária de Lisboa, the careers under study ensure an offer of 21,927 places per day. In the "do-something" scenario, the evolution of the offer corresponds to the offer of the LRT system. As previously mentioned, the LRT system with four vehicles offers 35,712 seats per day with a reserve vehicle.

13.3.3 Financial Analysis

The financial analysis of the LRT project consists of the evaluation of the incremental cash flows of the project; that is, the difference in financial cash flows between scenarios ("do-something" and "do-nothing").

The indicators of the financial return analysis presented are: (i) the sum of the annual financial cash flows for the analysis period (2020–2050); (ii) the Net Financial Updated Value (VALf): the value updated to the year 2020 of financial cash flows; (iii) the Internal Rate of Financial Return

(TIRf): the rate at which cash flows are discounted in order to obtain a VALf equal to zero. For the calculation of the VALf, a financial discount rate of 4% was used, respecting the recommendation of the Guide for the ACB of EC Investment Projects (2014), for the financial discount rate to be applied to countries eligible for the Cohesion Fund European. For the analysis of the financial return from the project's perspective, the annual financial cash flows (excluding VAT) were calculated using the formula:

$$FCF = -IC - OC + RV + OR \qquad (13.1)$$

With FCF corresponding to financial cash flow, IC corresponding to investment costs, OC corresponding to operating costs, RV corresponding to residual value, and OR corresponding to operating revenue.

13.3.3.1 Investment Costs

This section presents the total investment costs that the LRT concessionaire will have to bear in the overall analysis period (2020–2050), valued at constant 2020 prices. Considering that, in the case of the LRT project, it will not be realized, lower alternative investments are not foreseen to fill some of the flaws that the new line intends to solve; the investment costs presented correspond only to the "do-something" scenario. The investment costs considered in this project refer to the direct and indirect costs related to: (i) studies, opinions, projects, and consultancy; (ii) land; (iii) construction work (contract); (iv) machinery and equipment: acquisition of rolling stock, line signaling, integrated security system, the contract, inspection, exploration support systems, radio, voice and data equipment, and ticketing systems equipment; (v) contingencies; (vi) advertising and publicity.

These costs are based on information collected from different entities, and are net of VAT. In the specific case of investment in rolling stock, the acquisition of five vehicles (4 in operation + 1 reserve) was considered, at a base value of 2.8 million euros per vehicle at current prices, which will be delivered in the period 2021 to 2022. It is assumed that all investment costs are incurred between the year of implementation of the project (2020) and the first year of operation of the LRT services on the new constructed section (2023). Figure 13.5 shows the investment costs calendar, net of VAT, at constant 2020 prices, in euros.

	2020	2021	2022	Total	Val 4%
Construction of the 6kn LRT and 8 stops	4,594,085.00	12,877,430.00	10,624,184.00	28,095,699.00	25,768,158.00
Studies, projects, Consulting	395,584.00	395,584.00	0.00	791,168.00	746,108.00
Land	1,643,376.00	1,643,376.00	0.00	3,286,752.00	3,099,562.00
Infrastructure construction					
Viaduct	1,621,283.00	3,242,567.00	0.00	4,863,850.00	4,556,861.00
Stations	177,528.00	177,528.00	177,528.00	532,584.00	492,656.00
Intermodal station - Póvoa	0.00	177,528.00	177,528.00	355,056.00	321,956.00
Terminus Station - Parque MARL	0.00	133,146.00	133,146.00	266,292.00	241,467.00
Platform	391,120.00	782,240.00	0.00	1,173,360.00	1,099,302.00
Superstructure	0.00	360,267.00	360,267.00	720,534.00	653,363.00
Rolling stock	0.00	5,600,000.00	8,400,000.00	14,000,000.00	12,645,084.00
Advertising and publicity	365,195.00	365,195.00	365,195.00	1,095,585.00	1,013,448.00
Trials and hits (exploration)	0.00	0.00	1,010,520.00	1,010,520.00	898,349.00

FIGURE 13.5 Investment costs.

The investment to be made over the reference period of the project has a total value of 28.10 million euros, at constant prices in 2020, which corresponds to an updated value (with discount rate of 4%) of approximately 25.77 million euros.

13.3.3.2 Operating Costs

The operating costs considered report the operation and maintenance costs of the infrastructures under analysis, highlighting that in the "do nothing" scenario these costs are derived from the operation and maintenance of the road transport network while in the "do something" scenario it is an LRT infrastructure in its own channel.

These costs can be broken down into different items, such as (i) infrastructure maintenance costs; (ii) variable operating costs; (iii) energy and catenary costs; (iv) security, inspection, and surveillance items; (v) ticketing costs (Vending Machines – MVAs); (vi) energy and catenary costs; (vii) exploration support systems (ESS), signaling and auxiliary systems.

This section presents the total operating costs that the LRT concessionaire will have to bear in the global analysis period (2020–2050), valued at constant 2020 prices and net of VAT. The estimated values for both scenarios considered the information collected from the operators of this type of transport; in the case of the "do nothing" scenario, Rodoviária de Lisboa was used, and in the "do something" scenario, Metro do Porto and Metro were used south of the Tagus. The annual incremental costs calculated for both scenarios are those indicated in Figures 13.6 and 13.7.

Operating costs incurred during the project's reference period have a total value of approximately 145.80 million euros in the "do-nothing" scenario, which corresponds to an updated value (discount rate of 4%) of approximately 82.81 million euros, and approximately 199.72 million euros in the "do-something" scenario, which corresponds to an updated amount (discount rate of 4%) of approximately 109.96 million euros.

13.3.3.3 Residual Value

To calculate the residual value of investment goods, the method of calculating the net present value of cash flows in the remaining years of the operation's investments was applied. The years of life remaining in the operation are computed by the difference between the reference period and the useful life of the infrastructure and equipment that are part of the investment project. For this purpose, a cruise cash flow of 15.33 million euros (2050 euros) and a period of 20 years of remaining useful life were considered. It is estimated that, in the last year of analysis (2050), the residual value of investment goods is 4.94 million euros (2020 prices) which corresponds to an updated value (rate of 4%) of approximately 2.22 million euros.

13.3.3.4 Operating Revenue

The annual revenue calculation was performed considering that the year consists of 250 working days, 52 Saturdays, and 63 Sundays and holidays. It was also considered for ticketing reference prices for the current year 2020, the value of 1.5 € in the current trips of public transport by road of the local network between parishes and the value of 1.8 € was projected for the LRT ticket. It is noteworthy that for the LRT system, revenue with provision for the lease of commercial spaces was included, the construction of which is considered at all stops, in the total of eight units with a monthly income of 450 € at 2020 prices. Operating revenues calculated for the two systems are shown in Figure 13.8.

The estimated operating revenues over the reference period have a total value of 595.66 million euros for the "nothing" scenario, which corresponds to an updated value of 321.15 million euros, and 465.74 million euros for the "do something" scenario, which corresponds to 266.17 million euros of updated value.

	Short-lived Equipment Replacement Costs	Final Running Costs	Maintenance Costs					Variable Operating Costs			Total
			Personnel Costs	Maintenance and Repair	General Administration	Insurance	Consumption of Raw Materials	Energy	Consumables	Repairs to Keep the Operation Longer	
2020	50,400.00	1,056,240.00	427,680.00	1,271,040.00	264,960.00	80,640.00	154,512.00	4,586.00	493,440.00	57,984.00	3,861,482.00
2021	51,257.00	1,074,196.00	434,951.00	1,292,648.00	269,464.00	82,011.00	157,139.00	4,664.00	501,828.00	58,970.00	3,927,128.00
2022	52,077.00	1,091,383.00	441,910.00	1,313,330.00	273,776.00	83,323.00	159,653.00	4,739.00	509,858.00	59,913.00	3,989,962.00
2023	52,900.00	1,108,627.00	448,892.00	1,334,082.00	278,101.00	84,640.00	162,175.00	4,824.00	517,911.00	60,860.00	4,053,012.00
2024	53,725.00	1,125,922.00	455,895.00	1,354,892.00	282,440.00	85,960.00	164,705.00	4,889.00	525,993.00	61,809.00	4,116,230.00
2025	54,547.00	1,143,148.00	462,870.00	1,375,622.00	286,761.00	87,275.00	167,225.00	4,964.00	534,041.00	62,755.00	4,179,208.00
2026	55,371.00	1,160,430.00	469,850.00	1,396,394.00	291,091.00	88,591.00	169,750.00	5,039.00	542,105.00	63,703.00	4,242,324.00
2027	56,196.00	1,177,700.00	476,860.00	1,417,200.00	295,428.00	89,913.00	172,280.00	5,114.00	550,182.00	64,652.00	4,305,525.00
2028	57,022.00	1,195,012.00	483,870.00	1,438,033.00	299,771.00	91,235.00	174,812.00	5,189.00	558,270.00	65,602.00	4,368,816.00
2029	57,843.00	1,212,220.00	490,838.00	1,458,741.00	304,088.00	92,547.00	177,330.00	5,264.00	566,309.00	66,547.00	4,431,727.00
2030	58,664.00	1,229,434.00	497,808.00	1,479,455.00	308,406.00	93,863.00	179,848.00	5,338.00	574,350.00	67,492.00	4,494,658.00
2031	59,485.00	1,246,646.00	504,777.00	1,500,163.00	312,724.00	95,177.00	182,366.00	5,413.00	582,391.00	68,437.00	4,557,579.00
2032	60,348.00	1,264,722.00	512,096.00	1,521,920.00	317,258.00	96,557.00	185,010.00	5,492.00	590,836.00	69,429.00	4,623,668.00
2033	61,253.00	1,283,693.00	519,778.00	1,544,749.00	322,017.00	98,005.00	187,785.00	5,574.00	599,699.00	70,470.00	4,693,023.00
2034	62,203.00	1,301,590.00	527,834.00	1,568,682.00	327,008.00	99,524.00	190,696.00	5,660.00	608,994.00	71,563.00	4,763,754.00
2035	63,198.00	1,324,448.00	536,279.00	1,593,791.00	332,240.00	101,117.00	193,747.00	5,751.00	618,738.00	72,708.00	4,842,017.00
2036	64,241.00	1,346,301.00	545,128.00	1,620,089.00	337,722.00	102,785.00	196,944.00	5,846.00	628,947.00	73,907.00	4,921,910.00
2037	65,333.00	1,369,188.00	554,395.00	1,647,630.00	343,464.00	104,532.00	200,292.00	5,945.00	639,639.00	75,164.00	5,005,582.00
2038	66,476.00	1,393,149.00	564,097.00	1,676,464.00	349,474.00	106,362.00	203,797.00	6,049.00	650,833.00	76,479.00	5,093,180.00
2039	67,673.00	1,418,225.00	574,251.00	1,706,640.00	355,765.00	108,276.00	207,465.00	6,158.00	662,548.00	77,856.00	5,184,857.00
2040	68,925.00	1,444,463.00	584,875.00	1,738,213.00	362,347.00	110,279.00	211,303.00	6,272.00	674,805.00	79,295.00	5,280,777.00
2041	70,234.00	1,471,908.00	595,987.00	1,771,239.00	369,231.00	112,375.00	215,318.00	6,391.00	687,626.00	80,803.00	5,381,112.00
2042	71,569.00	1,499,874.00	607,311.00	1,804,892.00	376,246.00	114,520.00	219,409.00	6,513.00	700,681.00	82,338.00	5,483,353.00
2043	72,928.00	1,528,372.00	618,850.00	1,839,186.00	383,396.00	116,685.00	223,578.00	6,636.00	714,004.00	83,902.00	5,587,537.00
2044	74,314.00	1,557,411.00	630,608.00	1,874,130.00	390,680.00	118,903.00	227,826.00	6,763.00	727,570.00	85,497.00	5,693,702.00
2045	75,725.00	1,587,001.00	642,590.00	1,909,739.00	398,103.00	121,162.00	232,154.00	6,891.00	741,394.00	87,121.00	5,801,880.00
2046	77,165.00	1,617,154.00	654,799.00	1,946,024.00	405,667.00	123,464.00	236,565.00	7,022.00	755,480.00	88,776.00	5,912,116.00
2047	78,631.00	1,647,880.00	667,240.00	1,982,978.00	413,374.00	125,830.00	241,060.00	7,155.00	769,835.00	90,463.00	6,024,446.00
2048	80,125.00	1,679,190.00	679,917.00	2,020,675.00	421,228.00	128,200.00	245,640.00	7,292.00	784,462.00	92,182.00	6,138,911.00
2049	81,647.00	1,711,095.00	692,836.00	2,059,068.00	429,232.00	130,636.00	250,307.00	7,430.00	799,366.00	93,933.00	6,255,550.00
2050	83,198.00	1,743,606.00	706,000.00	2,098,190.00	437,387.00	133,118.00	255,063.00	7,571.00	814,554.00	95,718.00	6,374,405.00
Total	1,903,015.00	39,881,773.00	16,148,448.00	47,992,245.00	10,004,426.00	3,044,825.00	5,834,103.00	173,174.00	18,631,430.00	2,189,374.00	145,802,813.00
Val 4%	1,080,811.00	22,650,717.00	9,171,456.00	27,257,031.00	5,681,979.00	1,729,298.00	3,313,459.00	98,354.00	10,581,657.00	1,243,448.00	82,808,210.00

FIGURE 13.6   Operating costs – Do-nothing scenario.

	Short-lived Equipment Replacement Costs	Final Running Costs	Maintenance Costs					Variable Operating Costs			Total
			Personnel Costs	Maintenance and Repair	General Administration	Insurance	Consumption of Raw Materials	Energy	Consumables	Repairs to Keep the Operation Longer	
2020	0.00	0.00	0.00	0.00	0.00	0.00	0.00	0.00	0.00	0.00	0.00
2021	0.00	0.00	0.00	0.00	0.00	0.00	0.00	0.00	0.00	0.00	0.00
2022	6,720.00	352,080.00	106,920.00	317,760.00	88,320.00	13,440.00	25,752.00	45,864.00	24,672.00	28,992.00	1,010,520.00
2023	33,600.00	1,760,400.00	534,600.00	1,588,800.00	441,600.00	67,200.00	257,520.00	458,640.00	246,720.00	289,920.00	5,679,000.00
2024	34,124.00	1,787,862.00	542,940.00	1,613,585.00	448,489.00	68,248.00	261,537.00	465,795.00	250,569.00	294,443.00	5,767,592.00
2025	34,646.00	1,815,217.00	551,247.00	1,638,273.00	455,351.00	69,293.00	265,538.00	472,921.00	254,403.00	298,948.00	5,855,837.00
2026	35,169.00	1,842,626.00	559,571.00	1,663,011.00	462,227.00	70,339.00	269,548.00	480,063.00	258,244.00	303,462.00	5,944,260.00
2027	35,698.00	1,870,081.00	567,908.00	1,687,790.00	469,114.00	71,387.00	273,565.00	487,215.00	262,092.00	307,983.00	6,032,833.00
2028	36,218.00	1,897,572.00	576,256.00	1,712,600.00	476,010.00	72,436.00	277,586.00	494,378.00	265,945.00	312,511.00	6,121,512.00
2029	36,740.00	1,924,897.00	584,555.00	1,737,262.00	482,864.00	73,479.00	281,583.00	501,497.00	269,774.00	317,011.00	6,209,662.00
2030	37,261.00	1,952,230.00	592,856.00	1,761,931.00	489,721.00	74,523.00	285,582.00	508,618.00	273,605.00	321,512.00	6,297,839.00
2031	37,783.00	1,979,561.00	601,155.00	1,786,598.00	496,577.00	75,566.00	289,580.00	515,738.00	277,436.00	326,014.00	6,386,008.00
2032	38,331.00	2,008,265.00	609,872.00	1,812,504.00	503,777.00	76,662.00	293,779.00	523,217.00	281,458.00	330,741.00	6,478,606.00
2033	38,906.00	2,038,389.00	619,020.00	1,839,691.00	511,334.00	77,812.00	298,186.00	531,065.00	285,680.00	335,702.00	6,575,785.00
2034	39,509.00	2,069,984.00	628,615.00	1,868,205.00	519,260.00	79,018.00	302,807.00	539,296.00	290,108.00	340,905.00	6,677,707.00
2035	40,141.00	2,103,104.00	638,673.00	1,898,098.00	527,568.00	80,282.00	307,652.00	547,925.00	294,750.00	346,360.00	6,784,553.00
2036	40,803.00	2,137,805.00	649,211.00	1,929,416.00	536,273.00	81,607.00	312,729.00	556,966.00	299,613.00	352,075.00	6,896,498.00
2037	41,497.00	2,174,149.00	660,247.00	1,962,216.00	545,389.00	82,994.00	318,045.00	566,434.00	304,707.00	358,060.00	7,013,738.00
2038	42,223.00	2,212,196.00	671,802.00	1,996,555.00	554,934.00	84,446.00	323,611.00	576,347.00	310,039.00	364,326.00	7,136,479.00
2039	42,983.00	2,252,015.00	683,894.00	2,032,493.00	564,923.00	85,966.00	329,436.00	586,721.00	315,620.00	370,884.00	7,264,935.00
2040	43,778.00	2,293,677.00	696,546.00	2,070,094.00	575,374.00	87,557.00	335,530.00	597,576.00	321,459.00	377,746.00	7,399,337.00
2041	44,610.00	2,337,258.00	709,780.00	2,109,426.00	586,306.00	89,220.00	341,906.00	608,930.00	327,566.00	384,922.00	7,539,924.00
2042	45,458.00	2,381,665.00	723,266.00	2,149,505.00	597,446.00	90,916.00	348,402.00	620,499.00	333,790.00	392,236.00	7,683,183.00
2043	46,322.00	2,426,916.00	737,008.00	2,190,346.00	608,797.00	92,643.00	355,021.00	632,288.00	340,132.00	399,688.00	7,829,161.00
2044	47,202.00	2,473,028.00	751,012.00	2,231,962.00	620,364.00	94,403.00	361,767.00	644,302.00	346,595.00	407,282.00	7,977,917.00
2045	48,099.00	2,520,015.00	765,281.00	2,274,370.00	632,151.00	96,197.00	368,640.00	656,544.00	353,180.00	415,021.00	8,129,498.00
2046	49,012.00	2,567,896.00	779,821.00	2,317,583.00	644,162.00	98,026.00	375,644.00	669,018.00	359,890.00	422,906.00	8,283,958.00
2047	49,944.00	2,616,686.00	794,638.00	2,361,617.00	656,401.00	99,887.00	382,782.00	681,730.00	366,726.00	430,942.00	8,441,353.00
2048	50,892.00	2,666,403.00	809,736.00	2,406,488.00	668,873.00	101,785.00	390,055.00	694,682.00	373,696.00	439,129.00	8,601,739.00
2049	51,859.00	2,717,064.00	825,121.00	2,452,211.00	681,582.00	103,719.00	397,466.00	707,881.00	380,796.00	447,473.00	8,765,172.00
2050	52,845.00	2,768,688.00	840,798.00	2,498,803.00	694,531.00	105,690.00	405,017.00	721,331.00	388,032.00	455,975.00	8,931,710.00
Total	1,182,373.00	61,947,729.00	18,812,349.00	55,909,193.00	15,539,718.00	2,364,741.00	9,036,266.00	############	8,657,297.00	10,173,169.00	199,716,316.00
Val 4%	651,317.00	34,124,376.00	10,362,924.00	30,798,005.00	8,560,171.00	1,302,635.00	4,967,121.00	8,846,382.00	4,758,807.00	5,592,061.00	109,963,799.00

FIGURE 13.7 Operating costs – Do-something scenario.

DO-nothing Scenario		DO-something Scenario		
ANO	Venda de Bilhetes	ANO	Venda de Bilhets	Aluguer de espaços comerciais
2020	14,975,606.00	2020	0.00	0.00
2021	15,230,192.00	2021	0.00	0.00
2022	15,473,875.00	2022	0.00	0.00
2023	15,718,362.00	2023	11,848,001.00	1,314,000.00
2024	15,963,568.00	2024	12,188,106.00	1,317,942.00
2025	16,207,811.00	2025	12,495,339.00	1,321,896.00
2026	16,452,549.00	2026	12,769,142.00	1,325,862.00
2027	16,697,692.00	2027	13,000,870.00	1,329,839.00
2028	16,943,148.00	2028	13,191,983.00	1,333,829.00
2029	17,187,129.00	2029	13,381,948.00	1,337,830.00
2030	17,431,187.00	2030	13,571,971.00	1,341,844.00
2031	17,675,223.00	2031	13,761,979.00	1,345,869.00
2032	17,931,514.00	2032	13,961,528.00	1,349,907.00
2033	18,200,487.00	2033	14,170,950.00	1,353,956.00
2034	18,482,594.00	2034	14,390,600.00	1,358,018.00
2035	18,778,316.00	2035	14,620,850.00	1,362,092.00
2036	19,088,158.00	2036	14,862,094.00	1,366,179.00
2037	19,412,657.00	2037	15,114,749.00	1,370,277.00
2038	19,752,378.00	2038	15,379,258.00	1,374,388.00
2039	20,107,921.00	2039	15,656,084.00	1,378,511.00
2040	20,479,917.00	2040	15,945,722.00	1,382,647.00
2041	20,869,036.00	2041	16,248,690.00	1,386,795.00
2042	21,265,548.00	2042	16,557,416.00	1,390,955.00
2043	21,669,593.00	2043	16,872,006.00	1,395,128.00
2044	22,081,315.00	2044	17,192,575.00	1,399,313.00
2045	22,500,860.00	2045	17,519,233.00	1,403,511.00
2046	22,928,377.00	2046	17,852,099.00	1,407,722.00
2047	23,364,016.00	2047	18,191,289.00	1,411,945.00
2048	23,807,932.00	2048	18,536,923.00	1,416,181.00
2049	24,260,282.00	2049	18,889,125.00	1,420,429.00
2050	24,721,228.00	2050	19,248,018.00	1,424,691.00
Total	595,658,471.00	Total	427,418,548.00	38,321,556.00
Val 4%	321,146,914.00	Val 4%	243,531,931.00	22,634,391.00

FIGURE 13.8 Operating revenue do-nothing scenario or do-something scenario.

13.3.3.5 Financial Results

Table 13.4 shows the result obtained for the indicators of the financial analysis of the project's return, in euros, at constant 2020 prices.

The results obtained demonstrate that the LRT project presents financial profitability from the perspective of the project, given that it has a positive VALf of 43.75 million euros. The project also has a positive IRR of 61%, which confirms its profitability with financing from the concessionaire.

13.3.4   ECONOMIC ANALYSIS

13.3.4.1   Price Correction

The Guide for the Investment Projects ACB (2014), by the European Commission's DG Regional Policy recommends that: (i) prices of factors of production and products to be considered in the

TABLE 13.4
Financial Analysis Results

	Do-nothing	Do-something
Operating revenue	129,918,367.00	84,526,023.00
Exploration costs	−46,124,890.00	−18,859,504.00
Investment costs	−29,106,219.00	−26,666,507.00
Residual value	4,938,023.00	4,748,099.00
Financial cash flow	59,625,281.00	43,748,111.00
TIRf	**61%**	

CBA must be net VAT and other indirect taxes; (ii) the prices of the factors of production to be considered in the CBA must be gross of direct taxes; (iii) payments for pure transfers to people, such as contributions to social security, should be omitted in the calculations; (iv) in certain cases, indirect taxes/subsidies are intended to correct the externalities, so in these situations prices should be gross of those taxes.

Due to the imperfections inherent in most markets, market prices may suffer distortions, not reflecting the true opportunity cost of the goods and services used in the investment and exploration. As such, for calculating the indicators of economic analysis, it was necessary to correct these market prices, replacing them with shadow prices, by applying conversion factors.

For the calculation of the conversion factors, the investment and operating costs were divided in costs with qualified labor, costs with unskilled labor, and other costs investment/exploration. Prices related to investment/operating costs not related to labor were applied by a standard correction factor, calculated as follows:

$$SCF = (TI \ CIF \ prices + TE \ FOB \ prices/(TI \ CIF \ prices + TE \ FOB \ prices + TID) \quad (13.2)$$

With SCF corresponding to Standard correction factor TI CIF process to total imports at CIF prices, TE to total exports at FOB prices, and TID to total import duties. For labor costs, corrections were made to the level of wages considered for skilled and unskilled labor in the calculation of investment costs (and, therefore, residual value), to consider shadow wages when calculating these costs. This methodology consists of applying a correction factor to wages at market prices resulting from the application of the formula $(1 - t)$ for the wages of qualified labor and $(1 - t) \times (1 - u)$ for the wages of unskilled labor, where t is the rate of direct tax and social contributions and i is the unemployment rate in the region. This analysis resulted in the conversion factors (shadow prices) shown in Table 13.5.

A percentage of qualified labor costs of 7% was assumed in total investment costs and 15% of total operating costs, and a percentage of 25% of costs with unskilled labor in total investment costs, and 35% in total operating costs. These percentages reflect values verified in similar projects. As a

TABLE 13.5
Correction Factors (Shadow Prices)

Activity	Correction Factor
Skilled labor	0.77
Unskilled labor	0.69
Remaining investment/operating costs	0.99

TABLE 13.6
Weighted Correction Factors

Description	Weighted Correction Factor
Investment costs	0.9
Operating costs	0.86

result, the weighted correction factors to be applied to the total investment and operating costs are those shown in Table 13.6.

The investment cost correction factor was used to adjust investment expenses and residual value. In turn, the operating cost correction factor was used to calculate the economic operating costs of the LTR. To correct the distortion effect of indirect taxes, a correction factor of 1.23 for VAT to be paid on costs and a correction factor of 6% of VAT to be paid in revenues were also used.

13.3.4.2 Relevant Parameters

The economic analysis of the LRT project consists of the evaluation of the incremental cash flows of the project, that is, the difference in economic cash flows between scenarios ("do-nothing" and "do-something"). The annual economic cash flows were calculated using the following parameters (net of VAT) and formula:

$$ECF = -IC + RV + SEB \tag{13.3}$$

With ECF corresponding to Economic cash-flow, IC to Investment costs, RV to Residual value and SEB to Socioeconomic benefits. The investment costs (in accordance with 4.3.1) and the residual value are corrected to the weighted correction factor of investment costs (according with Table 13.6).

To calculate the residual economic value of the investment project, the method of calculating the net current value of treasury flows was applied in the remaining years of the operation's investments, which include the project's social and environmental benefits. The years of life remaining in the operation are computed by the difference between the reference period and the useful life of the infrastructure and equipment that are part of the investment project. It is assumed that the socioeconomic benefits of the project are felt for 50 years, so they do not end at the end of the last year of analysis. It is estimated that, in the last year of analysis (2050), the economic residual value of 591.59 million euros (2020 prices) that corresponds to an updated value (rate of 5%) of approximately 357.64 million euros.

The socioeconomic benefits are calculated based on the following presented in Section 4.4.3.

13.3.4.3 Benefits and Externalities Quantification

The quantification of the socioeconomic benefits associated with the LRT project, as well as the updating of unit values (time value, values of social and environmental externalities, unit transport costs), following the most current national references or, failing these, the European references, adapting them to the Portuguese reality and taking as reference the documents: (i) Manual for Estimating External Costs in the Transport Sector (IMPACT Deliverable 1), CE Delft, 2008, (commissioned by EC DG TREN); (ii) update of the Manual of External Transport Costs, Final Report, Ricardo-AEA, Report for the European Commission: DG MOVE, January 2014; (iii) HEATCO, Deliverable 5, 2006; (iv) Cost-Benefit Analysis Guide for Investment Projects – Economic Evaluation Tool for Cohesion Policy 2014–2020, December 2014, European Commission.

All coefficients are explained during the presentation of the results for each of the social and environmental benefits considered. As recommended, monetary values were defined and adjusted for 2020, considering the annual information rate and GDP per capita evolution. The source used for this variable was "The 2012 Ageing Report – Economic and budgetary projections for the 27 EU Member States (2010–2060) European Economy 2 | 2012 (provisional version), European Commission."

The same principle of updating and adjusting annual unit values was followed for the other benefits according to the following: (i) in cases where the bibliography suggests the consideration of an elasticity of 1.0 (benefits associated with environmental costs, accidents and noise) this was used; (ii) in cases where the bibliography is silent, a conservative approach is assumed through considering an elasticity of 0.7 (benefits associated with maintenance costs the highway); (iii) in the case of the amount of time associated with the reason for traveling "not at work," as recommended in the EC Investment Projects Guide for ACB (2014), an elasticity of 0.5 compared to the evolution of GDP per capita; (iv) in the case of operating costs for individual transport, with regard to the component of the perceived cost, as recommended in the ACB Guide to Investment Projects of the CE (2014), a constant value was assumed; iv) in the case of the benefits associated with the reduction of costs of road operators, a period of gradual adjustment of three years was assumed.

The EC's Investment Projects ACB Guide recommends dividing the different types of investment socioeconomic benefits in the categories that are following systematized.

The **producer surplus**, which in the context of this analysis can be understood as the gains that fall on the LRT concessionaire and other management entities and operators of transport (which includes users of individual transport as "operators" of their vehicles) as a result of the project under evaluation. The producer surplus includes: (i) the benefits associated with the increase in ticket revenue generated by the LRT, after subtracting the incremental costs associated with their operation (adjusted for their shadow prices as referred to in the economic analysis chapter of this report); (ii) the benefits associated with reducing operating costs of public road transport (adjusted for shadow prices) due to less need for supply, after subtracting revenues that these operators lose to the LRT; (iii) savings in operating costs for individual transport in its component do not perceived, which includes the costs borne by IT passengers as operators of the your individual transport vehicle, such as maintenance costs, overhauls periodic depreciation of the vehicle's value, etc. These differ from operating costs perceived, as the latter are supported by the same IT passengers, but as transport users; (iv) savings that public entities/road managers have in maintenance costs the highway, due to its lower use.

13.3.4.3.1 Benefits Associated with Lower Public Transport Operating Costs by Lower Need for Supply

To assess the decrease in operating costs of public transport operator's road transport, the reduction in transport production resulting from the transfer of users was considered for the LRT. In the case of the LRT project, the reductions in operating costs for public road transport corrected shadow prices accumulated in the 30 years of the analysis period correspond to a benefit of 66.04 million euros, which corresponds to an updated value (rate of 4%) of 35.10 million euros.

13.3.4.3.2 Benefits Associated with Reduced Operating Costs for Individual Transport (Unrecognized Cost Component)

Removing vehicles from circulation translates into important savings in their cost of use. In fact, typically, motorists greatly underestimate the actual costs of using the vehicle, tending to account solely for the cost of fuel. We are facing what is called consumer surplus, for which only the perceived cost of using the vehicle. However, to the extent that the vehicle belongs to them, it is considered that, in this situation, motorists are simultaneously producers of the transport service they use. In this context, unrecognized costs must also be accounted for as producer surplus. For this reason, for the purpose of estimating the social benefits associated with reducing the number of vehicles in circulation, the costs of using them are divided into two parts – perceived costs (relative to the consumer) and unrecognized costs (relative to the producer) – which include the costs maintenance, depreciation of the vehicle value, savings associated with lower need for periodic reviews, etc. Updating the coefficient values for unrecognized operating costs recommended in the bibliography for the reference year (2007) and for the first year of operation (2023) of the LRT project obtained the values 0.2390 and 0.3048, respectively. Differences in operating costs for individual transport (in their non-cost component) perceived accumulated in the 30 years of the LRT project analysis period correspond to a total benefit of 46.74 thousand euros, which corresponds to an updated amount (rate of 4%) of 26.66 thousand euros.

13.3.4.3.3 *Benefits Associated with Reduced Highway Maintenance Costs*

The costs associated with the road include expenses with the maintenance and repair of roads that result from the level of use, and as is natural, the heaviest vehicles present a higher value. In a scenario in which, on the one hand, there is a transfer of passengers from individual transport for public transport (LRT) and, on the other hand, transfer of passengers from the buses to the LRT, these transfers translate into a lower volume of vehicles in circulation on the roads, which reduces maintenance costs. The reference work (Update of the Handbook on External Costs of Transport, Final Report, RicardoAEA, Report for the European Commission: DG MOVE, January 2014) presents values country-specific, which are defined by vehicle class (motorized, light passengers, buses, light goods vehicles, and various types of heavy vehicles) and by type of highway (unspecified, highways, main roads, and other roads). In the present case, the value corresponding to the non-specification of the type of road was used as a reference. Table 13.7 presents the reference values, as well as the value used for the calculation maintenance costs of the highway in the area where the project is being developed, with a weighting of 15% in low-density urban areas and 85% in congested urban areas.

Based on the estimated values of vehicles * km that are no longer performed by passengers who transfer from individual transport to the LRT and reducing the supply of public transport road, it is possible to calculate the benefits resulting from lower costs associated with maintenance of the highway through the difference between the costs to be borne in the project scenario and in the "do-nothing." Reductions in highway maintenance costs accumulated in the 30 years of the analysis period correspond to a benefit of 55.17 thousand euros in the scenario with the project, which corresponds to an updated value (rate of 4%) of 31.47 thousand euros.

The **consumer surplus**, which in the context of the present analysis, can be understood as the gains that fall on transport users because of the project being evaluated. At the consumer surplus include (i) gains related to the reduction of travel time for types of traffic whose mode of transport is changed as a result of the project: traffic captured from public transport and traffic captured for individual transport; (ii) welfare gains generated for passengers who previously did not perform any type of trip and start doing it due to the new network extensions (traffic induced by the project); (iii) benefits associated with reducing the costs of operating individual transport in its perceived component, that is, the costs borne by IT passengers while transport users (such as fuel), after new spending on tickets made by these new users of the metro network; (iv) benefits generated by the reduction of road congestion that the project provides, and which materialize in reductions in the

---

TABLE 13.7
Reference Values for the Calculation of Highway Maintenance Costs

Type of Vehicle	Cost Portugal (€ct/vehicle * km 2010)	Cost Portugal (€ct/vehicle * km 2010)
Passenger car	0.26	0.30
Bus	1.12	1.42

TABLE 13.8

Weighted Average of the Time Values by the Weight of the Trips "In Service" and "Out of Service"

Mode of Transportation	"Out of Service"	In Service	Value (2023)
Public	96.5	3.5	0.1227 €/min
Individual	95.75	4.25	0.1711 €/min

travel time of users of the highway whose mode of transport does not change after the opening of the new network extension.

13.3.4.3.4   Benefits Associated with Reducing the Travel Time of Diverted Traffic

Estimates of benefits associated with travelers' time savings are directly associated with modal shift estimates. Based on these estimates and the distances and average speed levels allowed for the various connections and for the various modes, it was possible to calculate the time savings corresponding to the opening of the LRT system. Analyzed the diverse reference bibliography and considering the difference between real time travel time and the travel time felt by the passenger, as well as the transfer of users from IT to LRT, the value of the assumed time was obtained, updated for the initial year of exploration (2023) of the values recommended by the HEATCO project. The final value adopted corresponds to the weighted average of the time values by the weight of the trips "in service" and "out of service" obtained through the most recent mobility surveys in the cities of Lisbon, Oporto, and Coimbra, as shown (Table 13.8).

The differential of travel time costs in public transport accumulated in the 30 years of the analysis of the LRT project corresponds to a total benefit of 70.52 thousand euros, which corresponds to an updated value (rate of 4%) of 40.23 thousand euros. For individual transport, the difference in accumulated travel time costs over 30 years corresponds to a total benefit of 98.33 thousand euros, which corresponds to an updated amount (rate of 4%) of 56.09 thousand euros.

13.3.4.3.5   Benefits Associated with Reduced Operating Costs for Individual Transport (Perceived Cost Component)

Removing vehicles from circulation translates into important savings in their usage costs, such as the cost of fuel. We are faced with what is called operating costs perceived effects of transport: the costs incurred by individual transport passengers, in their condition of users of that transport (and not of operators of the same). Once again, the formulas proposed by the Department of Transport publication (United Kingdom) were used.to estimate the operating costs of various types of vehicles, including the automobile, in function of the average speed of circulation, updating the values of the coefficients recommended in the bibliography for the reference year (2019), and for the first year of operation of the LRT project. As recommended for the ECB Investment Projects ACB (2014), the relative values the perceived costs were only updated for 2019, assuming to be constant from this year. For 2007, a value of 0.1248 is considered and for 2023 a value of 0.1502. From the perspective of the IT user, the differences in operating costs for individual transport (in their unrealized cost component) accumulated in the 30 years of the LRT project analysis period, correspond to a total benefit of 22.99 thousand euros, which corresponds to an amount updated (rate of 4%) of 12.11 thousand euros.

13.3.4.3.6   Benefits Associated with Reduced Congestion Costs

When capturing for public transport on your own site, you are currently looking for a car and/or this translates, in practice, into the withdrawal of vehicles from circulation, which, in turn, contributes to reducing road congestion. This translates into savings in travel time for road traffic whose mode

TABLE 13.9

Reference Costs Associated with Reduced Congestion Costs

Type of Vehicle	Cost Portugal (€ct/vehicle * km 2010)	Cost Portugal
Public	23	29.26
Individual	57.51	73.16

of transport is not altered by the project. The reference work (Update of the Handbook on External Costs of Transport, Final Report, RicardoAEA, Report for the European Commission: DG MOVE, January 2014) presents values specific for country congestion costs, which are defined by vehicle class (light vehicles, buses, simple and articulated heavy vehicles), by type of road (highways, main roads, and other roads), type of region (metropolitan, urban, and rural), and degree of congestion of the road (free regime, close to capacity, and above capacity). As regards the determination of the reference value to be considered for light vehicles and buses, the following assumptions are considered: (i) metropolitan region; (ii) type of road: 5% of highway, 60% of main roads and 35% of other roads; (iii) degree of congestion: 83% in free regime (that is, 20 hours), 15% in regime close to capacity (3.5 hours), and 2% above capacity (0.5 hours). Table 13.9 shows the reference values applied in metropolitan areas like the study, as well as the amount used to calculate congestion costs road in euros/vehicle * km for the year of entry into service.

The accumulated difference in congestion costs between two scenarios under analysis, in a total of 30 years of the LRT project review period, corresponds to a benefit of 10.36 million euros, which corresponds to an updated value (rate of 4%) of 5.91 million euros.

13.3.4.3.7 Benefits to Society Associated with Positive Externalities

The transfer of demand from road to LRT has positive impacts at the environmental level (reduction in the emission of pollutants and noise pollution) and safety (reduction in accidents). Based on estimated transfer values for individual transport and public transport road and the unit costs that the literature attributes to each of these externalities, it is possible to calculate the benefits associated with the positive externalities created by the project. The individual benefits of each of these benefits are listed below.

13.3.4.3.8 Environmental Benefits, Passengers Transferred from TP, and Passengers Transferred from IT

To estimate the benefit resulting from the reduction of polluting emissions, the range of vehicles * km traveled by car, bus and metro between the do-something scenario, and the do-nothing scenario, using the methodology suggested in the reference work (Update of the Handbook on External Costs of Transport, Final Report, Ricardo-AEA, Report for the European Commission: DG MOVE, January 2014), which presents values of external marginal costs associated with pollution country and mode of transport. For this calculation, the composition of the vehicle fleet in the region in study, the relative weights of the bus fleet by EURO norm and the following assumptions: (i) the area under study is 100% urban; (ii) missions produced by LRT vehicles are equivalent to emissions from a train electric. The unit values per vkm obtained for each mode of transportation are following shown (Table 13.10).

The accumulated differences in the estimated costs of polluting emissions from public transport in the 30 years of analysis of the LRT project correspond to a benefit of 1.83 million euros, which corresponds to an updated value (rate of 4%) of 1.04 million euros. Concerning individual transport, the accumulated differences in the costs of polluting emissions in the 30 years of analysis, correspond to a benefit of 1.51 million euros, which corresponds to an updated value (rate of 4%) of 859 thousand euros.

TABLE 13.10

Passengers Transferred between Mode of Transportation

Year	IT	Bus	Metro
2010	0.0093	0.0454	0.0098
2023	0.0117	0.0571	0.0123

13.3.4.3.9 Benefits from Less Noise Pollution

The methodology for estimating these benefits was based on modal transfers associated with the LRT project: (i) vehicles.km that are no longer covered in individual transport; (ii) vehicles.km that are no longer traveled by bus; (iii) additional vehicles.km covered by the LRT. The reference work (Update of the Handbook on External Costs of Transport, Final Report, RicardoAEA, Report for the European Commission: DG MOVE, January 2014) presents values country noise pollution costs, which are defined by mode of transport vehicle (in this case: light vehicles and buses), by type of area (urban, suburban, and rural) type of traffic (dense or light), and time of day (day and night). As regards the determination of the reference value to be considered for light vehicles and buses, the following assumptions are considered: (i) mode of transport: bus and car; (ii) area: urban (defined as areas with 3,000 inhabitants per km of highway), assumed at 85%; and suburban (area with 700 inhabitants per km of highway) assumed at 15%; (iii) type of traffic: 30% dense traffic and 70% light traffic during the day, and 100% traffic slight at night; (iv) time of day: 97.5% during the day and 2.5% at night. The coefficients for calculating the costs associated with noise considered for the quantification of benefits resulting from noise emission are shown in Table 13.11 in euros/1,000 vkm for the year of commissioning.

 The differences between the costs of noise pollution between the two public transport scenarios of the 30 years of analysis, correspond to a total of benefits associated with the reduction of pollution with a total value of approximately 2.08 million euros of benefits to society, which corresponds to an updated value (rate of 4%) of 1.18 million euros. Regarding individual transport, the differences between the costs of noise pollution between the two scenarios over the 30 years of analysis, correspond to a total of benefits associated with the reduction of noise pollution with a total value of approximately 2.76 million euros of benefits for society, which corresponds to an updated value (rate of 4%) of 1.58 million euros.

13.3.4.3.10 Benefits from Fewer Vehicles in Circulation – Accidents

The methodology for estimating these benefits is based on modal transfers associated with the LRT project, measured in vehicles.km that are no longer covered on the highway. The methodology for calculating these benefits involves estimating impacts in terms of reducing victims of road accidents resulting from the reduction of vehicles in circulation. The most recent available data on

TABLE 13.11

Costs Associated with Noise Considered for the Quantification of Benefits Resulting from Noise Emission

Mode of Transportation	IT	Bus
Car	13.22	16.12
Bus	71.18	86.16
Light rail (2/6 to the surface)	23.73	28.72

TABLE 13.12

Accident Rates per Vehicle.Km, Vehicles.Km Traveled Annually in This Municipality Based on Fuel Sales by Municipality

Type of Victim	Rate
With dead	0.000300
With serious injured	0.001201
With slight injuries	0.044942

road accidents in the municipality under study were used (Vila Franca de Xira). Complementarily, to determine the accident rates per vehicle.km, vehicles.km traveled annually in this municipality based on fuel sales by municipality (Table 13.12).

Based on these accident rates, the average social accident costs were estimated for road with reference to the values presented in the reference work (Update of the Handbook on External Costs of Transport, Final Report, Ricardo-AEA, Report for the European Commission: DG MOVE, January 2014). The difference accumulated in the 30 years of analysis between the costs of road accidents in the two scenarios correspond to a social benefit associated with the decrease in the number of accidents of road transport with a total value of approximately 2.53 million euros, which corresponds to a value updated (rate of 4%) of 1.44 million euros.

Externalities, which in the context of the present analysis, can be understood as the set of net gains generated by the project and which benefit the remaining agents of society that do not are linked to it as producers or consumers. This set of benefits include: (i) benefits resulting from the reduction of pollutant gas emissions that affect the health of the population residing/attending geographic areas close to the transport infrastructure, the agricultural production, biodiversity, and may even cause material damage to buildings; (ii) reducing the contribution to climate change than using more efficient transport energy efficient has the lowest release of greenhouse gases in the workplace energy production; (iii) cost reduction that the noise pollution generated by the use of the infrastructure and that affects the populations that inhabit/frequent geographic areas close to the same, measured in terms of improving the population's comfort or even their health; (iv) security benefits that the new infrastructure will bring, as a result of the decrease in number of accidents.

Figures 13.9–13.11 show the socioeconomic costs and gains produced by the LRT project in total analysis period (2020–2050), for each scenario and in euros (2020 prices). The total socioeconomic benefits generated by the project correspond to the sum of the differences between the scenarios "do-something" and "do-nothing."

It should be noted that the ticket revenues generated for the LRT correspond to a cost for the passenger, so its positive effect on the producer surplus is canceled out with the negative effect on the consumer surplus. The same applies to the revenue that public transport operators lost by transferring passengers to the LRT: this loss of revenue is a cost for these operators, but a gain for passengers. In fact, all types of revenue generated or lost by transport operators is a simple transfer of resources between the operator and the passenger, not increasing the level of well-being in society. As such, and in accordance with the indications in the EC's ACB Guide to investment projects, all income/expenses from the ticket office will not be considered.

In the overall analysis period, all entities that make up the group of producers and consumers of the various modes of transport obtain a reduction in their socioeconomic costs due to the LRT project. The producer surplus recorded a positive change of 231.97 million euros, and the surplus for the consumer a positive variation of 10.55 million euros. The benefits resulting from externalities total 10.71 million euros in the total period under analysis. This makes a total of net socioeconomic

	2020	2021	2022	2023	2024	2025	2026	2027	2028	2029	2030
Producer superplus	0.00	0.00	0.00	6,629,644.00	6,733,066.00	6,836,082.00	6,939,306.00	7,042,703.00	7,146,231.00	7,249,137.00	7,352,074.00
LRT operating costs corrected at shadow prices	0.00	0.00	0.00	4,883,940.00	4,960,129.00	5,036,019.00	5,112,063.00	5,188,233.00	5,264,500.00	5,340,309.00	5,416,141.00
TP operating costs corrected at shadow prices	0.00	0.00	0.00	1,742,791.00	1,769,979.00	1,797,060.00	1,824,195.00	1,851,376.00	1,878,591.00	1,905,643.00	1,932,703.00
TI operating costs not perceived	0.00	0.00	0.00	1,336.00	1,357.00	1,377.00	1,398.00	1,419.00	1,440.00	1,461.00	1,481.00
Highway maintenance costs	0.00	0.00	0.00	1,577.00	1,601.00	1,626.00	1,650.00	1,675.00	1,700.00	1,724.00	1,749.00
Consumer superplus	0.00	0.00	0.00	301,431.00	306,134.00	310,817.00	315,511.00	320,212.00	324,918.00	329,598.00	334,279.00
Time costs - diverted traffic TP	0.00	0.00	0.00	2,015.00	2,047.00	2,078.00	2,109.00	2,141.00	2,172.00	2,204.00	2,235.00
Time costs - diverted traffic TI	0.00	0.00	0.00	2,810.00	2,854.00	2,898.00	2,942.00	2,985.00	3,029.00	3,073.00	3,117.00
Perceived operating costs of TI	0.00	0.00	0.00	657.00	667.00	677.00	688.00	698.00	708.00	718.00	729.00
Congestion costs	0.00	0.00	0.00	295,949.00	300,566.00	305,164.00	309,772.00	314,388.00	319,009.00	323,603.00	328,198.00
Externalities	0.00	0.00	0.00	306,045.00	310,818.00	315,574.00	320,339.00	325,113.00	329,891.00	334,642.00	339,394.00
TP emissions costs	0.00	0.00	0.00	52,361.00	53,177.00	53,991.00	54,806.00	55,623.00	56,440.00	57,253.00	58,066.00
TI emissions costs	0.00	0.00	0.00	43,047.00	43,718.00	44,387.00	45,057.00	45,729.00	46,401.00	47,069.00	47,737.00
TP noise pollution costs	0.00	0.00	0.00	59,309.00	60,234.00	61,155.00	62,079.00	63,004.00	63,930.00	64,851.00	65,772.00
TI noise pollution costs	0.00	0.00	0.00	79,008.00	80,241.00	81,469.00	82,699.00	83,931.00	85,165.00	86,391.00	87,618.00
Accident costs	0.00	0.00	0.00	72,320.00	73,448.00	74,572.00	75,698.00	76,826.00	77,955.00	79,078.00	80,201.00
Total	0.00	0.00	0.00	7,237,120.00	7,350,018.00	7,462,473.00	7,575,156.00	7,688,028.00	7,801,040.00	7,913,377.00	8,025,747.00

FIGURE 13.9 Socioeconomic benefits and externalities 1–3.

	2031	2032	2033	2034	2035	2036	2037	2038	2039	2040	2041
Producer superplus	7,455,003.00	7,563,101.00	7,676,548.00	7,795,534.00	7,920,262.00	8,050,946.00	8,187,812.00	8,331,099.00	8,481,059.00	8,637,958.00	8,802,080.00
LRT operating costs corrected at shadow prices	5,491,967.00	5,571,601.00	5,655,175.00	5,742,830.00	5,834,715.00	5,930,988.00	6,031,815.00	6,137,372.00	6,247,844.00	6,363,429.00	6,484,335.00
TP operating costs corrected at shadow prices	1,959,761.00	1,988,177.00	2,018,000.00	2,049,279.00	2,082,067.00	2,116,421.00	2,152,400.00	2,190,067.00	2,229,489.00	2,270,734.00	2,313,878.00
TI operating costs not perceived	1,502.00	1,524.00	1,547.00	1,571.00	1,596.00	1,622.00	1,650.00	1,679.00	1,709.00	1,741.00	1,774.00
Highway maintenance costs	1,773.00	1,799.00	1,826.00	1,854.00	1,884.00	1,915.00	1,947.00	1,981.00	2,017.00	2,054.00	2,093.00
Consumer superplus	338,958.00	343,874.00	349,032.00	354,442.00	360,112.00	366,054.00	372,277.00	378,793.00	385,609.00	392,744.00	400,206.00
time costs - diverted traffic TP	2,266.00	2,299.00	2,334.00	2,370.00	2,408.00	2,447.00	2,489.00	2,533.00	2,578.00	2,626.00	2,676.00
time costs - diverted traffic TI	3,160.00	3,206.00	3,254.00	3,305.00	3,357.00	3,413.00	3,471.00	3,532.00	3,595.00	3,662.00	3,731.00
perceived operating costs of TI	739.00	750.00	761.00	773.00	785.00	798.00	811.00	826.00	840.00	856.00	872.00
Congestion costs	332,793.00	337,619.00	342,683.00	347,994.00	353,562.00	359,396.00	365,506.00	371,902.00	378,596.00	385,600.00	392,927.00
Externalities	344,146.00	349,136.00	354,372.00	359,956.00	365,624.00	371,657.00	377,975.00	384,589.00	391,512.00	398,754.00	406,329.00
TP emissions costs	58,879.00	59,733.00	60,629.00	61,659.00	62,554.00	63,586.00	64,667.00	65,799.00	66,983.00	68,222.00	69,518.00
TI emissions costs	48,406.00	49,108.00	49,844.00	50,617.00	51,427.00	52,275.00	53,164.00	54,094.00	55,068.00	56,087.00	57,152.00
TP noise pollution costs	66,692.00	67,659.00	68,674.00	69,739.00	70,855.00	72,024.00	73,248.00	74,530.00	75,871.00	77,275.00	78,743.00
TI noise pollution costs	88,845.00	90,133.00	91,485.00	92,903.00	94,389.00	95,947.00	97,578.00	99,285.00	101,073.00	102,942.00	104,898.00
Accident costs	81,324.00	82,503.00	83,740.00	85,038.00	86,399.00	87,825.00	89,318.00	90,881.00	92,517.00	94,228.00	96,018.00
Total	8,138,107.00	8,256,111.00	8,379,952.00	8,509,932.00	8,645,998.00	8,788,657.00	8,938,064.00	9,094,481.00	9,258,180.00	9,429,456.00	9,608,615.00

FIGURE 13.10 Socioeconomic benefits and externalities 2–3.

	2042	2043	2044	2045	2046	2047	2048	2049	2050	Total	Val 5%
Producer superplus	8,969,319.00	9,139,737.00	9,313,392.00	9,490,346.00	9,670,663.00	9,854,406.00	10,041,638.00	10,232,431.00	10,426,846.00	231,968,423.00	117,154,188.08
LRT operating costs corrected at shadow prices	6,607,537.00	6,733,080.00	6,861,009.00	6,991,368.00	7,124,204.00	7,259,564.00	7,397,495.00	7,538,048.00	7,681,271.00	170,886,981.00	86,305,391.29
TP operating costs corrected at shadow prices	2,357,842.00	2,402,641.00	2,448,291.00	2,494,809.00	2,542,210.00	2,590,512.00	2,639,732.00	2,689,887.00	2,740,994.00	60,979,529.00	30,797,326.28
TI operating costs not perceived	1,807.00	1,842.00	1,877.00	1,912.00	1,949.00	1,986.00	2,023.00	2,062.00	2,101.00	46,743.00	23,607.26
Highway maintenance costs	2,133.00	2,174.00	2,215.00	2,257.00	2,300.00	2,344.00	2,388.00	2,434.00	2,480.00	55,170.00	27,863.26
Consumer superplus	407,810.00	415,558.00	423,454.00	431,499.00	439,698.00	448,053.00	456,566.00	380,365.00	474,080.00	10,462,084.00	5,283,809.50
Time costs - diverted traffic TP	2,727.00	2,778.00	2,831.00	2,885.00	2,940.00	2,996.00	3,053.00	3,111.00	3,170.00	70,518.00	35,614.67
Time costs - diverted traffic TI	3,802.00	3,874.00	3,948.00	4,023.00	4,099.00	4,177.00	4,257.00	4,338.00	4,420.00	98,332.00	49,661.96
Perceived operating costs of TI	889.00	906.00	923.00	940.00	958.00	977.00	995.00	1,014.00	1,033.00	22,988.00	11,609.94
Congestion costs	400,392.00	408,000.00	415,752.00	423,651.00	431,701.00	439,903.00	448,261.00	371,902.00	465,457.00	10,270,246.00	5,186,922.93
Externalities	414,050.00	421,918.00	73,557.00	438,104.00	446,428.00	454,908.00	463,552.00	472,360.00	481,334.00	10,352,077.00	5,228,251.16
TP emissions costs	70,839.00	72,185.00	60,472.00	74,954.00	76,378.00	77,829.00	79,308.00	80,815.00	82,351.00	1,819,077.00	918,713.36
TI emissions costs	58,238.00	59,345.00	83,317.00	61,621.00	62,792.00	63,985.00	65,201.00	66,440.00	67,702.00	1,529,028.00	772,225.94
TP noise pollution costs	80,239.00	81,764.00	110,992.00	84,901.00	86,514.00	88,157.00	89,832.00	91,539.00	93,278.00	2,102,860.00	1,062,036.17
TI noise pollution costs	106,891.00	108,922.00	101,596.00	113,101.00	115,250.00	117,439.00	119,671.00	121,944.00	124,261.00	2,755,075.00	1,391,433.24
Accident costs	97,843.00	99,702.00	83,740.00	103,527.00	105,494.00	107,498.00	109,540.00	111,622.00	113,742.00	2,512,597.00	1,268,971.26
Total	9,791,179.00	9,977,213.00	9,810,403.00	10,359,949.00	10,556,789.00	10,757,367.00	10,961,756.00	11,085,156.00	11,382,260.00	252,782,584.00	127,666,248.74

FIGURE 13.11 Socioeconomic benefits and externalities 3–3.

TABLE 13.13

Economic Analysis Indicators

	Total 2020–2050	VALE 5%
Investment costs	−26,195,597.00	−23,495,420.00
Residual value	591,588,426.00	357,636,927.00
Socioeconomic benefits	253,223,749.00	127,666,248.74
Economic cash flow	818,616,578.00	426,030,546.00
TIRE	5.09%	
B/C ratio	5.44	

benefits generated by the project of 253.22 million euros, which corresponds to an updated value (rate of 5%) of approximately 127.89 million euros.

13.3.4.4 Economic Return Indicators

The economic analysis indicators presented are (i) the sum of the annual economic cash flows for the analysis period (2020–2050); (ii) the Economic Net Updated Value (VALE). The value updated to the year 2020 of the project's economic cash flows; (iii) the Internal Rate of Economic Return (IRR). The rate at which cash flows are discounted economic costs of the project to obtain a VALE equal to zero; iv) the benefit-cost ratio (B/C ratio): the percentage of socioeconomic benefits generated for the project in its total costs. For the calculation of the VALE, a social discount rate of 5% was used, respecting the recommendation of the ECB Guide for Investment Projects of the EC (2014), for the social discount rate to be applied to countries eligible for the European Cohesion Fund.

13.3.4.5 Economic Analysis Indicators

The results obtained demonstrate that the LRT project has economic profitability, given that it generates economic benefits that exceed the costs by 818.62 million euros, and has a positive VALUE: 462.03 million euros. The project has a positive TIRE and higher than the social discount rate (5.44%), which confirms its economic viability, and a benefit-cost ratio of 5.44, which means that the project generates socioeconomic benefits more than five times higher than its costs. Table 13.13 presents the result obtained for the economic analysis indicators, at constant prices.

13.3.5 RISKS ASSESSMENT

Considering the sensitivity analysis if the values of the parameters included in the analysis (i.e., CF) vary individually by 10%, the profitability of the project is affected and results in the variations shown in Figure 13.12. The parameters with the greatest influence on the total VAL of the project, and therefore the most elastic variables are (i) the parameters of the analysis model and price dynamics, namely the TA and inflation, and (ii) the individual perception parameter related to the dimension of the aesthetic improvement infrastructure. For none of these variables, considerable changes are foreseen, likely to modify the final profitability of the project.

When analyzing the risk considering optimistic, pessimistic, and probable long-term scenarios, it was observed that the analysis of the optimistic scenario and the analysis of the most probable scenario are practically coincident. The expected verifications for each parameter, according to the current trend, are presented in Table 13.14.

The results are shown in Figure 13.13. The optimistic/most likely scenario presents considerable variation compared to the base scenario.

FIGURE 13.12   Sensitivity analysis of the parameters considered in the CBA.

TABLE 13.14
Project Costs and Benefits Variation

Parameter	Optimistic = Most Probable	Pessimistic
Refresh rate	7.5% (-)	7.5% (+)
Inflation rate	5% (+)	5% (-)
Aesthetic value	7.5% (+)	0%
Customers satisfaction	7.5% (+)	0%
Others	0%	0%

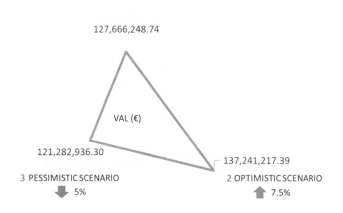

FIGURE 13.13   Scenario analysis under risks assessment.

In this sense, and considering the analysis presented in the previous sections, it is considered that the expansion of light-rail transit (LRT) systems in Vila Franca de Xira are quite advantageous in a near future and, consequently, of greater attractiveness for infrastructure owners as well as for the sustainable development of the railway infrastructures in the metropolitan area of Lisbon.

13.4 CONCLUSIONS

The present paper was enhanced by the need to assist decision-making processes in public rehabilitation investments to maximize benefits while minimizing costs over the life cycle. The results obtained are relevant to introduce in future similar projects (e.g., decision-making experience; information acquired during this evaluation process). The decision on investment projects making should be supported by a CBA, based on economic models cost-benefit, covering several areas, which are particularly relevant: (i) technical, (ii) financial, (iii) environmental, (iv) planning, (v) competitiveness, and (vi) economic and social development. The presented methodology for CBA of investment projects in the rehabilitation of railway tracks has an added value, both technical and scientific, enabling them to (i) express a judgment on the economic and social desirability of these projects; (ii) to establish a comparison between different design alternatives; and (iii) encourage the practice of identifying and accounting costs and economic benefits, even if not immediately convertible into monetary units. However, it requires rigor and methodological consistency. After the implementation of the CBA methodology, a final evaluation that allows a comparative analysis of the results and the initial forecasts, should be carried out.

ACKNOWLEDGMENTS

The authors acknowledge the results from Jose Cunha work developed under the scope of *Support Decision Methods* in the Post-Graduate Course in Railway Rehabilitation.

REFERENCES

Barbisan, A., Spadotto, A., Nora, D., Turella, E., Wergenes, T. *Environmental impacts caused by construction*, 2012
Brealey, R., Allen, F. e Myers, S. *Princípios de Finanças Empresariais*, 8a Edição, McGraw-Hill de Portugal, 2007
Cartier, A. *Le tramway, entre image de la ville et marketing urbain*, Master Recherche "Villes et Territoires", Université de Provence, 2005.
Çetinceli, S. *Cost-benefit analysis for various rehabilitation strategies*. MsC in Civil Engineering. Graduate School of Natural and Applied Sciences, Middle East Technical University. 2005.
EC, *Directive 2008/98/EC of the European Parliament and of the Council of 19 November 2008 on waste and repealing certain directives*, European Commission, 2008.
EC. *Cost Analysis Manual and benefits of investment projects*, DG Regional Policy of European Commission, 2003
EVALSED. *Guide to Evaluating Socio Economic Development*, Technical Manual II, QREN, 2013
Falcão Silva, M.J., Salvado, F. *Cost-benefit analysis: Methodology for decision support in Architecture, Engineering and Construction interventions*, (in PT) Report 288/2015-DED/NEG-LNEC, 2015
Falcão Silva, M.J., Salvado, F., Coelho, E. *Proposed methodology for a cost-benefit analysis of rehabilitation projects* (in PT), in *Proceedings of Jornadas Portuguesas de Engenharia de Estruturas (JPEE2014)*, Lisbon, Portugal, November 26-28, 2014.
Falcão Silva, M.J., Salvado, F., Couto, P. *Cost-benefit analysis applied to waste treatment investment projects*, WASTES: Solutions, Treatments and Opportunities - 5th International Conference. September 4th–6th, 2019.
FEMA 227 *Seismic Rehabilitation of Federal Buildings: A Benefit/Cost Model*. Volume 1: User Manual, FEMA, Washington, D.C., 1992.
FEMA 228 *Seismic Rehabilitation of Federal Buildings: A Benefit/Cost Model*. Volume 2: Supporting Documentation, FEMA, Washington, D.C., 1992.

Hass-Klau, C., Cramptom G. *Economic Impact of Light Rail Investments: Summary of theResults of 15 Urban Areas in France, Germany, UK and North America*, Urban TransportDevelopment, Section 4., pp. 245–255, 2005.

Jefferson, C., Kühn, A. Multimodal LRT vehicles: a development for the future? In: Recio, J., Sucharov, L. (Eds.), *Urban Transport and the Environment II*. WIT Press, Barcelona, pp. 433–442, 1996.

Laisney, F., Grillet-Aubert, A. *Tramway, espaces publics et mobilités*, Rapport de rechercheIPRAUS "Architectures du transport", 301 p, 2006.

Mansfield, J. *Refurbishment: some difficulties with a full definition*, 7th International Conference Insp. Appr. Repairs & Maintenance, Nottingham-UK, 2001.

Mishan, E. *Cost-Benefit Analysis*, 4th Edition, Unwin Hyman, Boston, 1988.

Salvado, F., Falcão Silva, M.J. *Applicability of cost-benefit analysis methodology. Externalities identification and evaluation in buildings rehabilitation* (in PT), Report 84/2016-DED/NEG-LNEC, 2016.

Salvado, F., Falcão Silva, M.J., Couto, P., Baião, M. *Performance indicators for cost-benefit analysis applied to investment projects*, IABSE Symposium Towards a Resilient Built Environment – Risk and Asset Management, Guimarães, Portugal, March 27-29, 2019.

Schultmann et al., *Collection of background information for the development of EMAS pilot reference sectoral documents*: The Construction sector, 2010.

Suzuki, H., Cervero, R., Iuchi, K. *Transforming cities with transit: transit and landuse integration for sustainable urban development*. Washington, DC: World Bank, 2013.

14 Adaptive System Supervising the Response Control for Smart Structures

Zubair Rashid Wani, Manzoor Tantray, and Ehsan Noroozinejad Farsangi

CONTENTS

14.1 INTRODUCTION

This chapter proposes a response-based adaptive (RBA) controller that uses the stability of model-based controller and adaptability of intelligent control strategies. The theory of the proposed controller is followed by mathematical representation (Wani, 2023). The parameters of the control algorithm are obtained from the optimization of structural response/s and the properties of the magnetorheological (MR) damper employed. Next, a five-story multi-degree freedom frame is modeled and the efficiency of the proposed RBA controllers are examined and compared with the H2/LQG algorithm. Next, the experimental validation of the proposed control strategy with an MR damper placed at the ground story of a multi-story structure subjected to diverse sets of ground motions is performed to verify the effectiveness of structural control. The experimental verification of the control strategy starts with a detailed description of the complete experimental setup, followed by the structural identification of the model. Then a detailed description of the non-contact digital image correlation (DIC) technique employed in this study has been presented. Afterward, the instruments and transducers designed and involved in implementing the control system and recording the structural response are discussed. The novel power amplifier circuit designed and fabricated to deliver the required input current based on the control signal is discussed. Input seismic excitations are discussed before experimental results are compared and deliberated. The experimental results are also compared with corresponding analytical results to obtain a better understanding of structural control and validating the numerical results. Finally, the chapter wraps up with the highlights of the proposed control strategy in mitigating the structural response on a shake table.

14.2   BASIC PRINCIPLE

The principle of RBA control is founded on the model-based control with an adaptive part based on the structural response. First, the structure to be controlled is modeled numerically with an MR damper as the control device subjected to seismic excitation (Elenas & Meskouris, 2001; Lupoi et al., 2006; Snieder & Şafak, 2006; Spencer & Nagarajaiah, 2003; Tubaldi & Kougioumtzoglou, 2015; Wani, Tantray, Noroozinejad Farsangi, et al., 2022; Wani et al., 2021). As obtained from the previous chapter, the velocity of the MR damper is directly proportional to the damping force; also, the phase angle between the two is zero. Therefore, the velocity of the installation layer in a multi-storied structure is taken as the varying parameter of the control strategy (Wani & Tantray, 2020). Next, particular response quantity/ies of the structure i.e., displacement, acceleration, drift, base shear, etc., are selected by the designer as objective functions. In the numerical model, variable gain (G) is provided to the velocity feedback signal, which is directly supplied to the MR damper as the input current. The value of G is linearly changed (in small increments) from zero to max saturation value of the MR damper. The whole control strategy is simulated in time, for each seismic excitation considered (where n is the number of increments). Finally, for all considered seismic excitations, the performance curves corresponding to the structural response to be mitigated (selected by the designer) and the linear gain are plotted to acquire the optimal gain (G_{opt}), from performance curves. The optimal control gain and other parameters as attained by testing of the damper and applied to the control strategy for maximum reduction in structural response. If the designer wishes to select multiple objective functions (responses) then the same procedure is followed. The multiple optimal control gains obtained corresponding to each response quantity are averaged and a trade-off between the objective responses is performed to achieve the required global minima (Wani, Tantray, & Farsangi, 2021b; Wani et al., 2021; Wani & Tantray, 2021). The whole principle of RBA control is shown in Figure 14.1.

14.3   MATHEMATICAL MODELLING OF THE CONTROL STRATEGY

Several mathematical models have been considered to demonstrate the dynamic behavior of civil engineering structures e.g., framed building model, cantilever beam model, shear story model, and torsional coupled model (Al-Bedoor, 2001; Cai et al., 2014; Kanninen, 1973; Lukic et al., 2018; Moghaddam & Hajirasouliha, 2006; Rabe et al., 1996; Wani & Tantray, 2020). The following assumptions are considered before mathematical modeling of the structure using a shear beam model.

- The stiffness of the floor and beams of each floor is infinite.
- The rigid-floor-decks are sustained by massless flexible columns and columns are axially inextensible.
- The structural response is only considered in one principal path and response in other orthogonal directions is taken as insignificant.
- All points at the ground level of the structure experience the same level of ground acceleration.

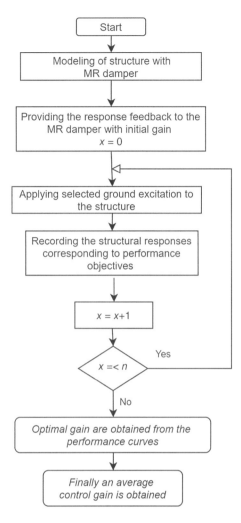

FIGURE 14.1 Flowchart indicating principle of RBA control strategy.

A pictorial description of the shear beam model of a five-story building is shown in Figure 14.2, where m_r, k_r, and c_r are the floor mass, stiffness, and damping, respectively. x_g is the ground acceleration and x_r is the floor displacement with reference to the original position of the frame.

The governing equation of the controlled system is

$$[M]\{\ddot{x}(t)\}+[C]\{\dot{x}(t)\}+[K]\{x(t)\}=\{P(t)\}+[B]\{f(t)\} \tag{14.1}$$

where $\{\ddot{x}(t)\}$, $\{\dot{x}(t)\}$ and $\{x(t)\}$ are acceleration, velocity, and displacement response vectors respectively. $\{P(t)\}$ can be any external excitation i.e., earthquake or wind load. $\{f(t)\}$ is the control forces generated by the control device installed in the structure and [B] is its position matrix. The lumped mass matrix for a structure with "n" degrees of freedom is given by

$$M = \begin{bmatrix} m_1 & 0 & \cdots & \cdots & 0 \\ 0 & m_2 & \cdots & \cdots & 0 \\ \cdots & \cdots & \ddots & \cdots & \cdots \\ \cdots & \cdots & \cdots & \ddots & \cdots \\ 0 & 0 & \cdots & \cdots & m_n \end{bmatrix}$$

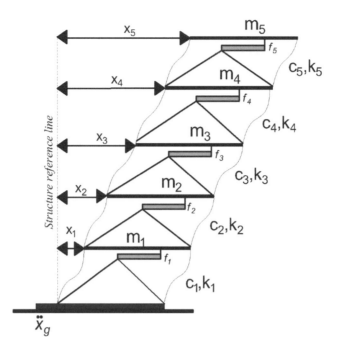

FIGURE 14.2 Pictorial description of the shear beam model of a five-story building.

The structural stiffness matrix [K] is given by

$$K = \begin{bmatrix} k_{11} & k_{12} & k_{13} & \cdots & k_{1n} \\ k_{21} & k_{22} & k_{23} & \cdots & k_{2n} \\ k_{31} & k_{32} & k_{33} & \cdots & k_{3n} \\ \cdots & \cdots & \cdots & \ddots & \cdots \\ k_{n1} & k_{n2} & k_{n3} & \cdots & k_{nn} \end{bmatrix}$$

And the damping matrix [C] is calculated by the most popular Rayleigh damping hypothesis.

$$[C] = \varepsilon_1[M] + \varepsilon_2[K]$$

where ε_1 and ε_2 are the coefficients that determine the contribution of mass and stiffness to the damping of the structure.

$$\varepsilon_1 = \frac{2\omega_1\omega_2\left(\xi_1\omega_2 - \xi_2\omega_1\right)}{\omega_2^2 - \omega_1^2}$$

and

$$\varepsilon_2 = \frac{2\omega_1\omega_2\left(\xi_2\omega_2 - \xi_1\omega_1\right)}{\omega_2^2 - \omega_1^2}$$

where ω_1 and ω_2 are first- and second-order frequencies and ξ_1, ξ_2 are corresponding damping coefficients. If the structure is a building and excitation load P(t) is earthquake excitation, then Equation (14.1) can be written as

$$[M]\{\ddot{x}(t)\} + [C]\{\dot{x}(t)\} + [K]\{x(t)\} = -[M]\ddot{x}_g(t) + [B]\{f(t)\} \tag{14.2}$$

state space version of the above equation can be written as

$$\{\dot{Z}(t)\} = [A]\{Z(t)\} - [D_0]P(t) + \{L\}[G]\{f(t)\}$$

$$\{Y\} = [E_0][Z(t)] \tag{14.3}$$

where $\{Y\}$ is the output vector, $\{Z(t)\}$ and $\{\dot{Z}(t)\}$ is the state vector of displacement and velocity, and $[A] = \begin{bmatrix} [0] & [I] \\ -[M]^{-1}[K] & -[M]^{-1}[C] \end{bmatrix}$ is the structural system matrix. $[D_0] = \begin{bmatrix} [0] \\ [M]^{-1}[B_0] \end{bmatrix}$;

$[G] = \begin{bmatrix} [0] \\ [M]^{-1}B \end{bmatrix}$ is the input matrix; $\{L\}$ is the vector corresponding to the position of the damper

and B_0 represents the place of external excitation.

For RBA control strategy, the velocity $\{\dot{Z}(t)\}$ of the installation layer is taken as the varying parameter for the control current. The gain (G_{optimal}) applied to the control signal is attained from the local minima of the performance graphs obtained after iterative simulations are performed, for different sets of seismic excitations. The performance objective/s (response) of the structure is/are selected by the designer from the direct and derived model outputs.

The damping force is directly related to the current supplied (shown in equation) and damping force is in phase with the corresponding velocity. Therefore,

$$f(t) \propto I \quad \text{and} \quad f(t) \propto \dot{Z}(t)$$

$$I \propto \dot{Z}(t) \tag{14.4}$$

As the current to the MR damper cannot be negative, an absolute value of the velocity makes better sense.

$$i = G_{\text{optimal}} * abs\{\dot{Z}(t)\} + i_0 \tag{14.5}$$

where G_{optimal} = control gain and i_0 = initial value of the current. Next, the optimal control gain is to be obtained from one or multiple performance curves shown in the equations.

P-curve for acceleration:

$$f_1[Z(t)] = \int_0^t \left\{ \sqrt{\sum_{i=1}^r \left[\frac{\partial^2\{Z_r(t)\}}{\partial t^2} \right]^2} \right\} dt \quad \forall \quad G = [0, n] \tag{14.6}$$

P-curve for inter-story displacement:

$$f_2[Z(t)] = \int_0^t \left\{ \sqrt{\sum_{i=1}^r [Z_r - Z_{r-1}]^2} \right\} dt \quad \forall \quad G = [0, n] \tag{14.7}$$

P-curve for absolute displacement:

$$f_3[Z(t)] = \int_0^t \left\{ \sqrt{\sum_{i=1}^r [Z_r]^2} \right\} dt \quad \forall \quad G = [0, n] \tag{14.8}$$

And, P-curve for base shear:

$$f_4[Z(t)] = \int_0^t \left\{ \sqrt{\sum_{i=1}^r \left[[K_r]*Z_r\right]^2} \right\} dt \qquad \forall \quad G = [0,n] \tag{14.9}$$

where r = floors of the structure, n = increments for each seismic excitation, and $[K_r]$ is the stiffness of the corresponding floor. The lowest point curve for each P-curve is attained by:

$$\frac{\partial\{f_1[Z(t)]\}}{\partial G} = 0; \quad \frac{\partial\{f_2[Z(t)]\}}{\partial G} = 0; \quad \frac{\partial\{f_3[Z(t)]\}}{\partial G} = 0; \quad \text{and} \quad \frac{\partial\{f_4[Z(t)]\}}{\partial G} = 0 \tag{14.10}$$

The parameters, i.e., the optimal control gain and the initial value of current as obtained, are used in Equation (14.5) to obtain a maximum reduction in overall structural response against seismic excitation based on the objective function/s (Wani, Tantray, & Farsangi, 2021a; Wani et al., 2021; Zubair et al., 2022).

14.4   INPUT SEISMIC EXCITATIONS

The particulars of the designated ground motions are organized in Table 14.1. The records are scaled as per ASCE-7 scaling of the earthquake ground motion matching the target spectrum.

14.5   NUMERICAL STUDY

The simulation is carried on a five-story structure with only one installed, as shown in Figure 14.3. The lumped mass matrix, stiffness matrix, and damping matrix of the structure are obtained, as explained in the previous section. Table 14.2 shows the structural matrices and details of the structure in consideration. Table 14.3 shows the mode shapes and natural frequencies, as obtained using the signal processing toolbar module of MATLAB. Figure 14.4 shows the pictorial representation of normalized mode shapes.

The structure is subjected to ground acceleration of recorded seismic events shown in the previous section. The MR damper in the structure is controlled and compared for three different strategies, namely (1) RBA control with single response objective, (2) RBA control with multiple response objectives, and (3) benchmark H2/LQG strategy.

TABLE 14.1
Scaled Excitations as per ASCE-7

S No.	Earthquake Record	Year	Recording Station	PGA (m/sec²)
1	Kobe (Japan)	1995	Nishi	4.530
2	Imperial valley	1940	El-Centro array #9	3.104
3	Northern Calif	1954	Ferndale City Hall	1.632
4	Northridge	1994	Lake Hughes #12A	0.943
5	Friuli (Italy)	1976	Cordeiro	0.687
6	S Fern	1971	Hawaii, USA	0.981
7	Chi Chi	1999	Taiwan	0.197
8	Big Bear	1992	California, USA	0.343
9	Dinar	1995	Turkey	0.392
10	Kozani	1995	Greece	0.235

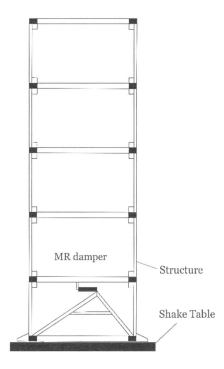

FIGURE 14.3   Five-floor framed structure with damper at ground floor placed on shake table.

TABLE 14.2
Structural Matrices and Details of Structure Considered in This Study

Parameters	DOF				
	1	**2**	**3**	**4**	**5**
Mass matrix,	310	0	0	0	0
M	0	310	0	0	0
(kg)	0	0	310	0	0
	0	0	0	310	0
	0	0	0	0	310
Stiffness matrix,	130688	−65344	0	0	0
K	−65344	130688	−65344	0	0
(N/m)	0	−65344	130688	−65344	0
	0	0	−65344	130688	−65344
	0	0	0	−65344	65344
Damping matrix,	842.16	−376.56	0	0	0
C	−376.56	842.16	−376.56	0	0
[N/(m/sec)]	0	−376.56	842.16	−376.56	0
	0	0	−376.56	842.16	−376.56
	0	0	0	−376.56	465.61

Total height of structure	6 m
Floor area of Structure	2 m × 2 m

TABLE 14.3

Natural Frequencies and Mode Shape of the Structure

Model matrix	1	1	1	1	1
	1.919	1.3097	0.2846	−0.8308	−1.6825
	2.6825	0.7154	−0.919	−0.3097	1.8308
	3.2287	−0.3728	−0.5462	1.0882	−1.3979
	3.5133	−1.2036	0.7635	−0.5944	0.5211
Natural frequencies	$\omega_1 = 0.7047$ Hz	$\omega_2 = 2.057$ Hz	$\omega_3 = 3.242$ Hz	$\omega_4 = 4.165$ Hz	$\omega_5 = 4.751$ Hz

For this study, the objective responses selected by the designer are mentioned as under:

- *For RBA with single response objective (RBA1)*: Overall reduction in inter-story displacement of each floor is taken as the objective function
- *For RBA control with multiple response objectives (RBA2)*: Overall reduction in absolute acceleration and inter-story displacement responses are selected as objective functions.

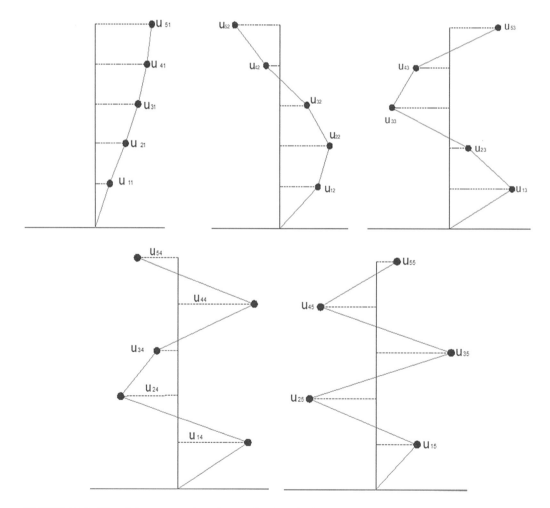

FIGURE 14.4 Pictorial representation of normalized mode shapes.

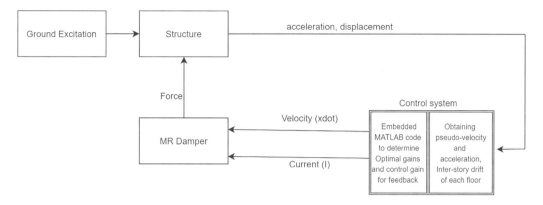

FIGURE 14.5 Control process used to implement RBA with single response objective (RBA1)

The proposed control strategy gives the designer the flexibility to choose any one, two, or multiple structural response as the objective parameters based on the design requirements. However, the selection of multiple performance objectives will involve a trade-off between those structural responses.

14.5.1 RBA Control with Single Response Objective

The pictorial depiction of a complete closed-loop system employing an RBA control algorithm with a single response objective can be visualized in Figure 14.5. Table 14.4 lists the simulation configuration parameters considered in the MATLAB study. The parameters of the control systems are obtained from performance graphs, which are attained after sequential iteration using increments gains for different sets of ground motions. Equation (14.6) is used to acquire the ordinate portion of the curves and the incremental gain forms the abscissa. Table 14.5 shows the MATLAB code written in combination with the Simulink model to generate the required curves. Figure 14.6 shows the performance graphs corresponding to the summation of inter-story displacement and incremental gains, for all sets of ground motion. The turning point of each curve is obtained to determine an average control gain for the control strategy.

Figure 14.7 shows a magnified view of one of the performance graphs. The optimal control gain ($G_{optimal}$) as obtained from the performance graphs for the RBA control system with a single response objective is $G_{optimal} = 10$ A sec m^{-1} and $I_0 = 0.2$ A.

14.5.2 RBA Control with Multiple Response Objective

Compared to the RBA control with a single response objective, RBA control with multiple response objective functions determine the control parameters based on the trade-off between the overall

TABLE 14.4

Model and Simulation Configuration Parameters in MATLAB

Parameter	Value	Description
Simulation time	Variable	Equal to the time history of excitation
Solver type	Fixed-step	Auto (Automatic solver selection)
Timestep	0.001	Fixed time step
Sampling time	0.001	Taken equal to the time step

TABLE 14.5
MATLAB Code for Generation of Performance Curve for Each Seismic Excitation Employing RBA1 Control Strategy

```
isd1=[ ]; % null Vector of ∑ inter-story drift for each simulation %        1
for G =0:1:200                                                              2
% Gain fed to the feedback loop after each iteration % varying from 0 to 200 %
sim('StructurewithMRdamper');                                               3
% the model with damper is created in Simulink with appropriate feedback loop
is run %
isd1=[isd1 isd];                                                            4
% isd is ∑ inter-story drift of all 5 floors obtained from each simulation %
end                                                                         5
plot(isd1)    % curve for gain (G) and isd1 is obtained %                   6
Gopt = islocalmin(isd1)                                                     7
% local mimina of the performnace curve i.e. optimal gain for each seismic
excitation%

%%The optimal control gain for the control algorithm is obtained have structure
is simulated for all considered excitations.%%
```

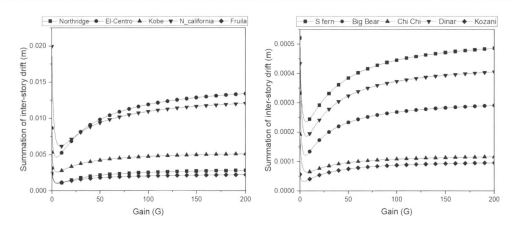

FIGURE 14.6   P-curves for RBA1 control strategy.

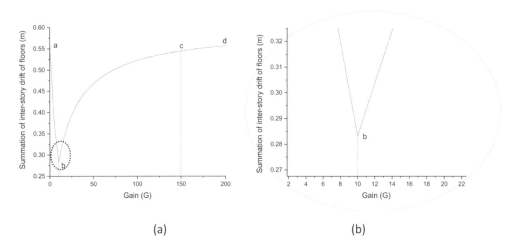

(a)   (b)

FIGURE 14.7   a) One of the P-curves with b) zoomed in to the area of interest.

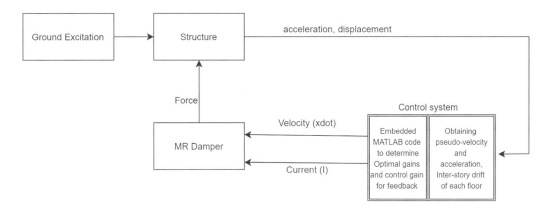

FIGURE 14.8 Control process used to implement RBA with multiple response objectives (RBA2).

reduction of selected structural responses. Figure 14.8 shows a pictorial description of the closed-loop system employing an RBA control algorithm with multiple response objectives. The overall inter-story displacement and absolute acceleration responses are designated for objective functions to be minimized, for the control strategy. The simulation process is carried with the same initial parameters discussed in the previous section. Table 14.6 shows the MATLAB code written to obtain the required performance envelopes. Figures 14.9 and 14.10 show the performance graphs corresponding to the summation of inter-story displacement and summation of absolute acceleration with incremental gains, for all sets of ground motion. Separate optimal gains are obtained from the two

TABLE 14.6
MATLAB Code for Generation of Performance Curve for Each Seismic Excitation Employing RBA2 Control Strategy

```
isd1=[ ];  % null Vector of ∑ intersory drift for each simulation %        1
aar1=[ ];  % null Vector of ∑ absolute acceleration response for each
simulation %                                                               2
for G =0:1:200                                                             3
% Gain fed to the feedback loop after each interation % varying from 0 to 200 %
sim('StructurewithMRdamper');                                              4
% the numerical model of structure with MR damper created in simulink with
appropriate feedback loop is run %
isd1=[isd1 isd];                                                           5
aar1=[aar1 aar ];                                                          6
% isd is ∑ intersory drift of all 5 floors obtained from each simulation %
% arr is ∑ absolute acceleration of all 5 floors obtained from each simulation %
end                                                                        7
plot(isd1)    % curve for gain (G) and isd1 is obtained %                  8
plot(aar1)    % curve for gain (G) and aar1 is obtained %                  9
G1 = islocalmin(isd1)                                                      10
G2 = islocalmin(aar1)                                                      11
Gopt = (G1+G2)/2                                                          12
% local mimina of the performnace curveS i.e. optimal gain for each seismic
excitation%
```

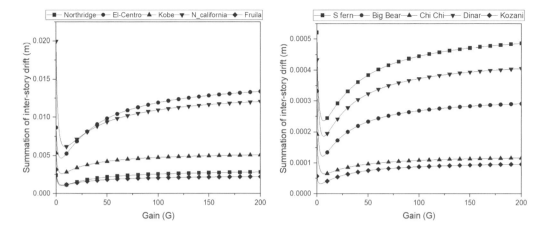

FIGURE 14.9 P-curves for RBA2 control strategy.

objective functions corresponding to the turning points of the graphs. Finally, the control gain for the closed-loop control is obtained by taking an average of two optimal gains.

The average optimal control gain ($G_{optimal}$) as obtained from the performance graphs for RBA control system with multiple response objective is $G_{optimal} = 7$ A sec m^{-1} and $I_0 = 0.2$ A.

14.5.3 Simulation Results

The performance is assessed by observing the peak inter-story displacement, absolute acceleration, base shear time history, and control current time history responses. Each simulation is carried for the time period equal to the time history of a particular seismic ground motion. The RBA control with single and multiple response objectives are designated as RBA(1) and RBA(2), respectively, and H2/LQG strategy.

14.5.3.1 Maximum Structural Response

Figure 14.11 depicts each floor's maximum drift and acceleration response. It can be shown that the RBA(1) and RBA(2) control techniques surpassed the standard COC approach in reducing the structure's inter-story response. Furthermore, RBA(1) outperformed RBA(2) in decreasing inter-story

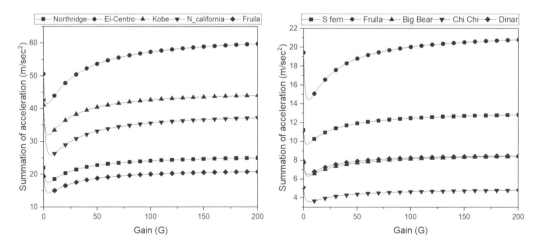

FIGURE 14.10 P-curves for RBA2 control strategy corresponding to minimization of floor acceleration.

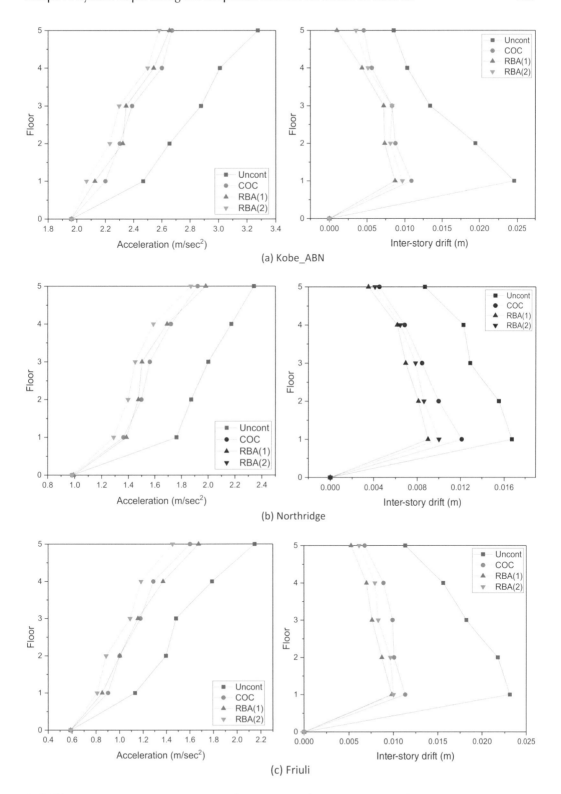

(a) Kobe_ABN

(b) Northridge

(c) Friuli

FIGURE 14.11 Comparison of maximum floor responses for all control strategies.

drift reaction. RBA(1) control values for acceleration responses, on the other hand, were equivalent to COC for high-intensity recordings and somewhat higher for low-intensity ground movements. However, RBA(2) controller outclassed other strategies in response mitigation. The reason for better performance of RBA strategies is their rapid adaptability and flexibility of the control current that generates the required control force at each particular instant and the objective function/s of RBAs play a critical role in mitigating a particular structural response. The yielding of the structure is directly related to its peak response.

14.5.3.2   Time History Response

The time history for a top floor response and base shear of the structure employing the proposed control strategies are shown in Figures 14.12 and 14.13. It is clear that the proposed RBA control techniques outperformed the matching COC approach, which is indicative of the fact that RBA control strategies are not only able to mitigate the peak responses but the overall time history response. Also, the reduction in base shear corresponds to the smooth transition of the control current. Figure 14.14 shows the control force and input current time history comparison of control strategies. Compared to the COC strategy, the input current for RBA controllers never crosses the saturation mark of 2A; hence, the power requirement is even lower.

14.6   EXPERIMENTAL SETUP

The details of the shake table and the structure to be controlled using the MR damper are discussed in this section. The seismic experimentations were conducted on a simulation assembly comprising of a 3 meter × 3 meter shake table. The peak cargo capacity is 1.5×10^4 kg, with a variable spectrum of 0.1–100 Hz. Figure 14.15 shows the horizontal and vertical actuators used to control the movement of the shake table. The structure five-storied framed structure to be tested on the shake table is first connected to a base plate through columns. Figure 14.16 shows the position of trial-axial accelerometers placed on each floor and one on the base plate. Also, the position of the strain gauges to determine the column base strain is displayed.

Figure 14.17 shows the closed loop of the control strategy. The measured velocity (acceleration response) from the installation layer of the MR damper is fed to *National Instruments* USB-6002 through a GRANITE-12 multi-channel Daq. The feedback signal is fed LabView (software platform), where a decision regarding the amount of control current to be fed to the MR damper is taken. The output signal (voltage) is received through the output channel of USB-6002, which is supplied to a power amplifier circuit to provide a required control current based on the output voltage. The required control current is fed to the MR damper in real time for optimal response control against external seismic excitation.

The structural responses are also measured using DIC technique and wired transducers. All the hardware and software platforms mentioned above are discussed in detail in subsequent sections. For now, the details of the structure and the MR damper are discussed here.

14.6.1   Details of the Structure

This experiment's basic structure is a five-story steel frame. The structure's overall height is 6 m, a plan size of 2 meters × 2 meters. The steel sections 40 mm × 40 mm are used as columns, steel C channel sections of 100 mm × 40 mm are used as beams. Figure 14.18 shows the XY and ZY faces.

Each level of the construction has reinforced concrete (RCC) slabs that operate as lumped floor weights. Every RCC slab is 2.2 meters × 0.67 meters × 0.1 meters and is reinforced in both orientations with 10 mm steel bars. As indicated in Figure 14.19, a steel bar is forged with the slab to serve as one of the endpoints of the MR damper. The specification of the considered structure is designed based on the theory that maximum damping force is more than 20% of the total mass of the structure (inertial force) at 5 g ground acceleration, taking into consideration the payload capacity of the shake table.

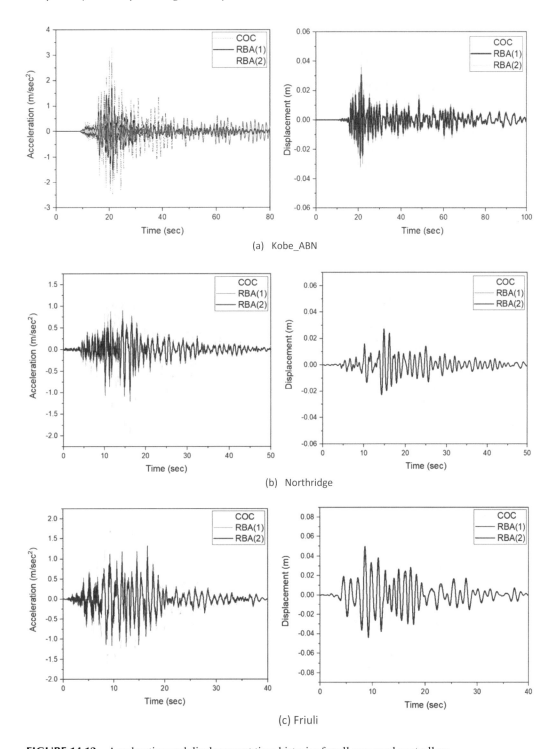

(a) Kobe_ABN

(b) Northridge

(c) Friuli

FIGURE 14.12 Acceleration and displacement time histories for all proposed controllers.

14.6.2   MR DAMPER CONFIGURATION

To determine and experimentally validate the vibration control of the structure using semi-active control strategies, an MR damper is placed in a Chevron brace arrangement. As illustrated in Figure 14.20, one end of the MR damper is attached to the A-frame built in the center of the frame and the other end is fixed to the embedded section in the slab. The MR damper's axis is kept in the horizontal alignment.

14.7   SYSTEM IDENTIFICATION

For the five-story shear model of the structure in consideration, the structural mass matrix $[M]$ and stiffness matrix $[K]$ as obtained from structural and material characteristics are as follows.

$$M = \begin{bmatrix} 310 & 0 & 0 & 0 & 0 \\ 0 & 310 & 0 & 0 & 0 \\ 0 & 0 & 310 & 0 & 0 \\ 0 & 0 & 0 & 310 & 0 \\ 0 & 0 & 0 & 0 & 290 \end{bmatrix} \text{ kg;}$$

$$K = \begin{bmatrix} 1.3069 & -0.6534 & 0 & 0 & 0 \\ -0.6534 & 1.3069 & -0.6534 & 0 & 0 \\ 0 & -0.6534 & 1.3069 & -0.6534 & 0 \\ 0 & 0 & -0.6534 & 1.3069 & -0.6534 \\ 0 & 0 & 0 & -0.6534 & 0.6534 \end{bmatrix} \times 10^5 \text{ N/m}$$

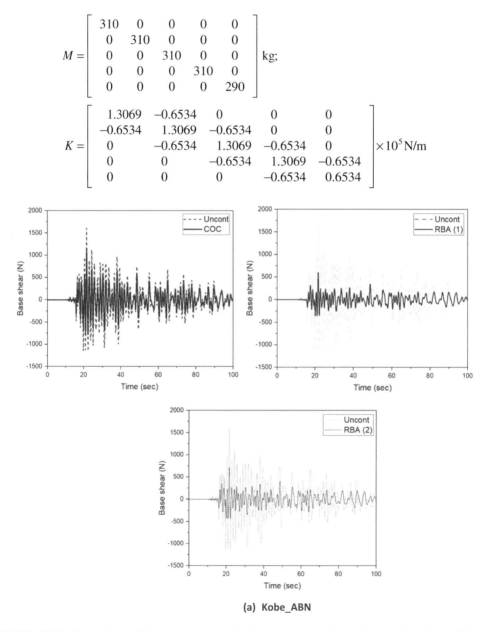

(a) Kobe_ABN

FIGURE 14.13   Base shear of the structure employing the proposed control strategies for a) Kobe, b) Northridge, and c) Friuli ground motion. (*Continued*)

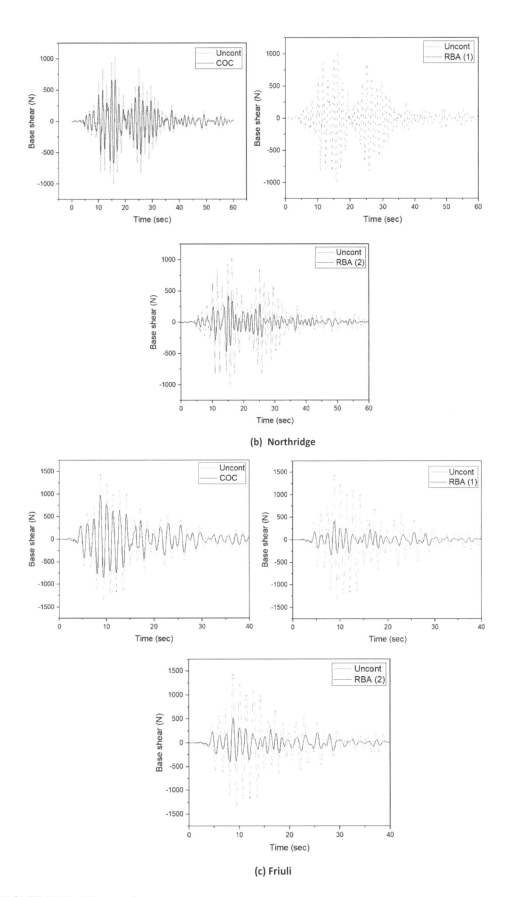

(b) Northridge

(c) Friuli

FIGURE 14.13 (*Continued*)

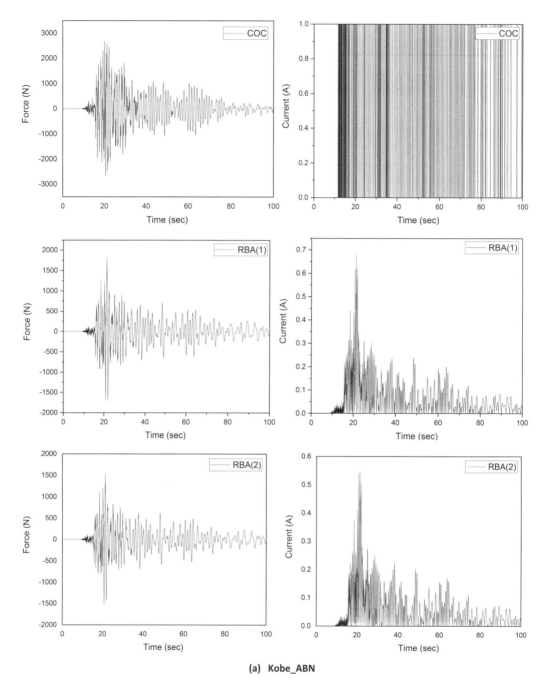

(a) Kobe_ABN

FIGURE 14.14   Control force and current evaluation of control strategies. (*Continued*)

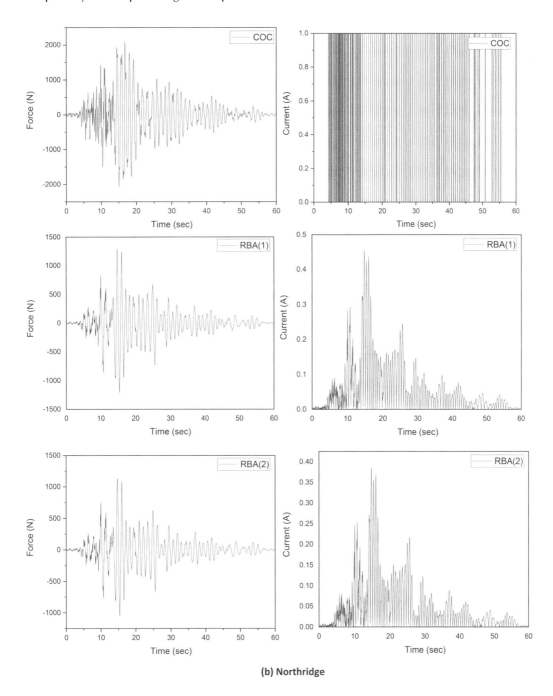

(b) Northridge

FIGURE 14.14   (*Continued*)

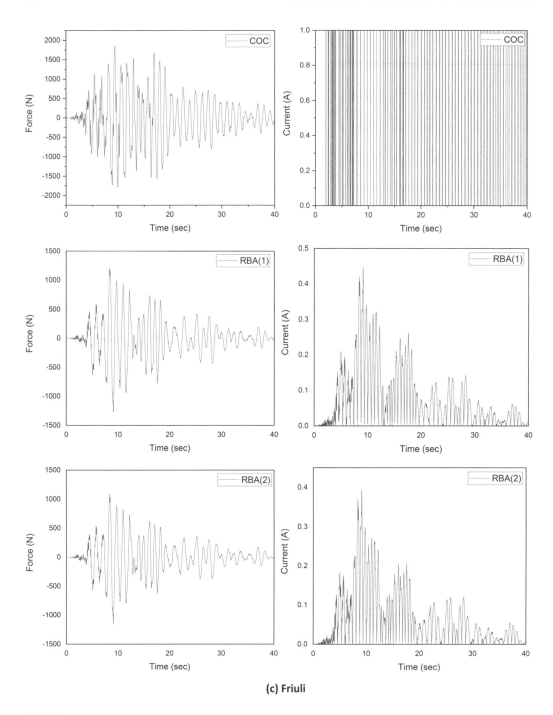

(c) Friuli

FIGURE 14.14   (*Continued*)

FIGURE 14.15   Horizontal and vertical actuators used to control the movement of the shake table.

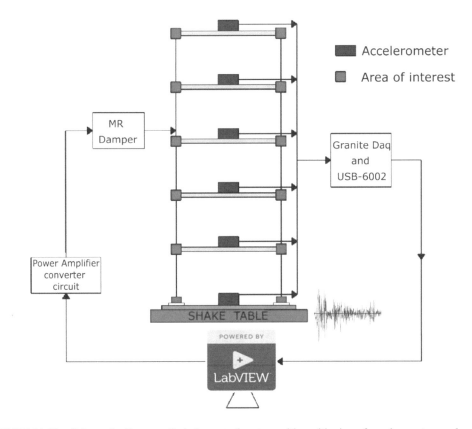

FIGURE 14.16   Schematic diagram of whole control system with positioning of accelerometers and point of interests of the framed structure on shake table.

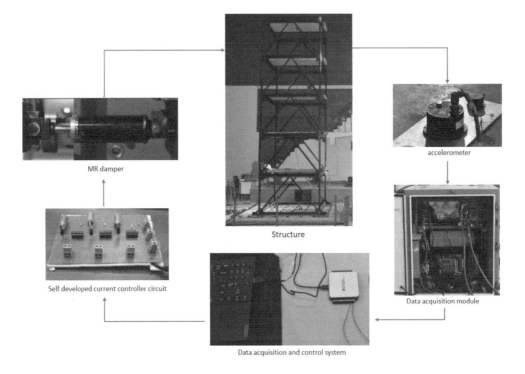

FIGURE 14.17   Pictorial representation of structure of the control strategy.

FIGURE 14.18   The steel frame with lumped mass: a) XY face and b) ZY face.

| (a) | (b) |

FIGURE 14.19   Shows a) an iron probe embedded in the slab and b) fasteners for fixating the RCC slab.

The structure on the shake table is subjected to pink noise. Pink noise signal has an inverse relationship between power spectral density and signal frequency, which is different from white noise having equal power distribution per frequency. The five natural frequencies of 2.5783, 7.5387, 13.7653, 18.6055, and 22.59821 as obtained from the Fourier transform of the acceleration response. The damping matrix [C] is also obtained from subsequent structural analysis. Also, the predicted and identified natural frequencies are shown in Table 14.7.

$$
C = \begin{bmatrix}
842.16 & -376.56 & 0 & 0 & 0 \\
-376.56 & 842.16 & -376.56 & 0 & 0 \\
0 & -376.56 & 842.16 & -376.56 & 0 \\
0 & 0 & -376.56 & 842.16 & -376.56 \\
0 & 0 & 0 & -376.56 & 465.61
\end{bmatrix} \text{Ns/m}
$$

FIGURE 14.20   Chevron brace arrangement of MR damper.

TABLE 14.7
Numerical and Experimental Natural
Frequencies of First Five Mode Shapes

Mode	Natural Frequency	
	Numerical	Experimental
1st	2.5893	2.5783
2nd	7.5534	7.5387
3rd	13.8932	13.7653
4th	18.7023	18.6055
5th	22.6301	22.59821

14.8   SOFTWARE, SENSORS, AND INSTRUMENTATIONS

The implementation structural control system discussed previously involves multiple hardware and software platforms that are adopted and fabricated. The transducers and DIC technique used to determine the dynamic structural responses are discussed in this section. Also, the fabrication and design of the novel power amplifier circuit is discussed.

14.8.1   LABVIEW

LabView is a software platform created by *National Instruments* as a visual programming language that interacts well with their hardware product. LabView uses virtual instruments (VIs) to perform several applications, i.e., control system, data measurement, hardware configuration, etc. VI allows a non-programmer to just drag and drop VIs and operations to create a graphical program.

The control system developed using MATLAB is created in LabView for the hardware implementation and experimental verification of the same. The real-time data from the accelerometers is directly fed to the LabView environment through the configured hardware device (USB-6002) where it is converted into corresponding velocity data using numerical techniques. The signal from the accelerometers and the control parameters obtained from the numerical study are used to generate the required control signal, which is to be fed to the MR damper as a control current. LabView is also used to record the input and the output signal of the controller.

14.8.2   DIGITAL IMAGE CORRELATION SETUP

The following factors are addressed in order to produce an assessment of the structure on the shaking table using DIC: Prepping the region of interest lighting conditions, calibration, hardware setup, and stream capture settings (Wani, Tantray, & Noroozinejad Farsangi, 2022). The flowchart representing the whole process of implementing DIC with a pattern matching algorithm has been shown in Figure 14.21. The main components are discussed below.

14.8.2.1   Cameras

The acA1920-25 series cameras were used to monitor the full-field response of the structure. The features of the camera are illustrated in Table 14.8. Each pixel of the camera is equal to 2.2 mm on the structure. Cameras were installed on an adjustable tripod, which aids in camera placement and stability. The position of the cameras is shown in Figure 14.22. All cameras are connected to a trigger to enable simultaneous activation.

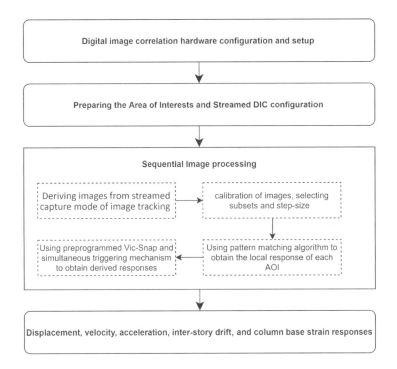

FIGURE 14.21 Flowchart involved in implementation of DIC with pattern matching algorithm.

TABLE 14.8
Camera Specification Used in the Study

Parameter	Value
PoW requirement	2.4 W/2.3 W
Triggering	BNC hardware trigger
Interface	USB 3.0
Sensor size	1/3.4″
Sensor	On Semiconductor
Pixel size (mm^2)	2×2
Frame rate	160 frames per second
Resolution	$1,900 \times 1,000$
Resolution	2.4 MP

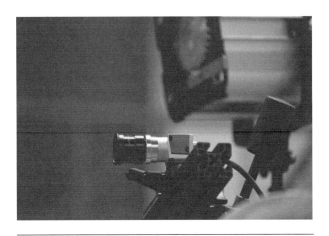

FIGURE 14.22 Placement of cameras related to the structure.

14.8.2.2 AOI (Area-of-Interest)

For the structure under examination, 12 AOIs were created. The AOI is formed by the beam-column connection at each floor, as shown in Figure 14.23. Furthermore, the speckles are dispersed randomly along the AOI to avoid tessellation, which can create mistakes during picture post-processing.

14.8.2.3 Lighting, Positioning, and Focusing

A pair of LED lights for each camera is provided to eliminate dark regions during a shaking test. After adjusting the illumination, the cameras are placed orthogonal to the AOI design. Before performing the shaking table test, Vic-Snap data capture software permits the employment of a digital crosshair (Figure 14.24) for correct placement and centering.

14.8.2.4 Streamed Capture

Because this study includes the shaking test of the steel frame, the standard fps technique of capture is inadequate to overcome the complexity induced in system dynamics with the addition of MR dampers. Furthermore, most methods for determining frame rate are dependent on trial and error. To address this issue, the DIC streaming capture technique is used, where frequency of capture is computed as a Nyquist factor multiplied by resonant frequency (Yadav, 2009). The system identification is performed and natural frequencies are obtained as shown in Figure 14.25.

14.8.2.5 Calibration and Post-Processing of Sequential Images

The features that control the accuracy and reliability of response measurement from image tracking are as follows:

- **Selecting a proper subset.**
  The size and shape of a subset (sub-image) selected for a particular AOI depend on the dimension of each speckle and the number of pixels requires to monitor the motion of

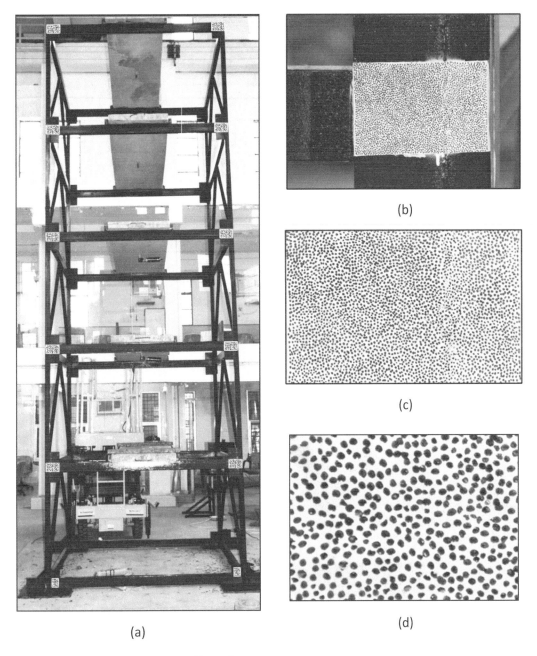

(b)

(c)

(d)

(a)

FIGURE 14.23 AOI to be monitored by DIC.

structure effectively. Depending on the geometry of each AOI, the geometry of a subset selected for postprocessing is indicated by the red box in Figure 14.26. The subset size must be sufficiently large to ensure that the region used for correlation includes a sufficiently distinctive pattern to differentiate itself from another subset (if selected). However, larger subsets are susceptible to approximation errors for the underlying displacements. Therefore, to guarantee a reliable full-field measurement, an optimal subset size of 30×30 pixels is selected by the trial-and-error method.

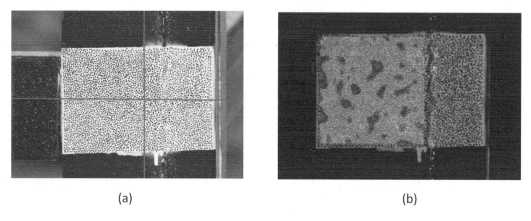

(a)   (b)

FIGURE 14.24 Centering and focusing of AOIs.

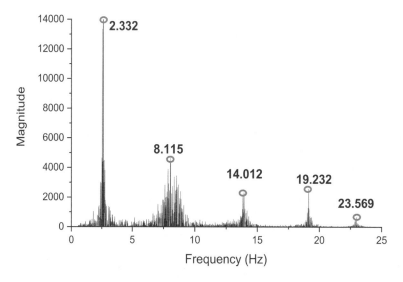

FIGURE 14.25 Frequency plot of structure showing resonant frequencies.

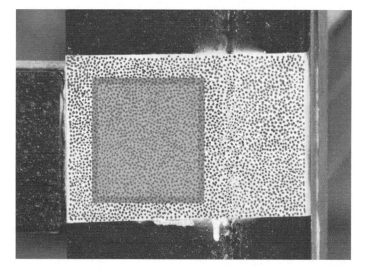

FIGURE 14.26 Subset for response monitoring.

- **Selecting a step size.**
  The step size has a significant effect on spatial resolution than the size of a subset. More DIC data points and thus higher spatial resolution are provided by smaller step sizes. Therefore, a step size equal to 1/8th of the subset size is selected for this study. Also, step size determines the number of data points tracked in a subset.
- **Pattern matching algorithm.**
  The field of DIC has been mostly concentrated on small movements and deformations. Also, previous DIC-monitored shake table tests conducted were limited to structure corresponding to low-frequency bandwidth. For this purpose, the differential method of image matching which was able to determine the motion between the two images performed adequately (Ngeljaratan & Moustafa, 2020).
  The reference image is denoted by R, and the displaced image is denoted by Q. Therefore, we aim at minimalizing the squared change in gray values, also called the sum of squares deviation. The optimal value of the Disp vector is obtained by

$$\bar{\mathbf{d}}_{\text{opt}} = argmin \sum |Q(\mathbf{x} + \bar{\mathbf{d}}) - R(\mathbf{x})|^2 \tag{14.11}$$

First-degree Taylor series for the cost fiction can be obtained from iterative simulation of above equation.

$$\chi^2\left(\bar{d}_x + \Delta_x, \bar{d}_y + \Delta_y\right) = \sum \left| Q(\mathbf{x} + \bar{\mathbf{d}}) - \frac{\partial Q}{\partial x}\Delta_x - \frac{\partial Q}{\partial y}\Delta_y - R(\mathbf{x}) \right|^2 \tag{14.12}$$

where $\bar{d}_x$ and $\bar{d}_y$ = reference estimates for the subset, Δ_x and Δ_y are motion update changes sought after each reiteration. Equating previous equations,

$$\begin{bmatrix} \Delta_x \\ \Delta_y \end{bmatrix} = \begin{bmatrix} \sum\left(\frac{\partial Q}{\partial x}\right)^2 & \sum\frac{\partial Q}{\partial x}\frac{\partial Q}{\partial y} \\ \sum\frac{\partial Q}{\partial x}\frac{\partial Q}{\partial y} & \sum\left(\frac{\partial Q}{\partial y}\right)^2 \end{bmatrix}^{-1} \begin{bmatrix} \sum\frac{\partial Q}{\partial x}(R-Q) \\ \sum\frac{\partial Q}{\partial y}(R-Q) \end{bmatrix} \tag{14.13}$$

The mean estimate of motion for the kth iteration using $\bar{\mathbf{d}}^{k+1} = \bar{\mathbf{d}}^k + \Delta$ until convergence is obtained.

14.8.2.6 Derivation of Responses

The reaction of AOI is monitored in relation to the associated reference picture, which is captured at the beginning of each test.

$$\begin{bmatrix} d_{Xn} = X_n - X_r \\ d_{Zn} = Z_n - Z_r \end{bmatrix} \tag{14.14}$$

where d_{Zn} and d_{Xn} present the displacement of AOI in vertical and horizontal directions for the *n*th image.

The AOI velocity agrees to an average velocity obtained using the numerical difference method for two consecutive images (Liu et al., 2015). Equation (14.15) represents the average 2D velocity.

$$d'_{Zr}(t) = \frac{d_{Zr}(t + \Delta t) - d_{Zr}(t)}{\Delta t} \quad \text{and} \quad d'_{Xr}(t) = \frac{d_{Xr}(t + \Delta t) - d_{Xr}(t)}{\Delta t} \tag{14.15}$$

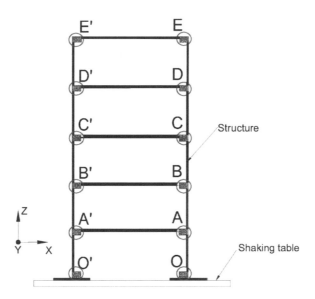

FIGURE 14.27 AOIs demarcated on each floor of the structure.

where $d'_{Zr}(t)$ and $d'_{Xr}(t)$ represent the vertical and horizontal velocities of an AOI during dynamic testing. Also, O to E and O' to E' represent the position of different AOIs are shown in Figure 14.27. The random noise associated with streamed data is minimized using a second-order Wiener filter, whilst retaining the min and max (peak) values.

The acceleration response is computed to signify the dynamic behavior of a structure. The equation signifies the numerical method employed by DIC to compute the acceleration values.

$$d''_{Zr}(t) = \frac{d'_{Zr}(t + \Delta t) - d'_{Zr}(t)}{\Delta t} \quad \text{and} \quad d''_{Xr}(t) = \frac{d'_{Xr}(t + \Delta t) - d'_{Xr}(t)}{\Delta t} \, for \; r = \text{A' to E' and A to E}$$

$$(14.16)$$

where $d''_{Zr}(t)$ and $d''_{Xr}(t)$ are the acceleration values of an AOI in global Z and X directions. Also, the acceleration values obtained have high-frequency noise associated with them, which can be eliminated using the filtering option in frequency domain.

Vic Snap-2D was also programmed to determine drift in both directions'. $\Delta d^X_{RR'}(t)$ and $\Delta d^Y_{RR'}(t)$ in Equation (14.17) depict the drift corresponding with nominated sites on the x- and z-axis.

$$\Delta d^X_{AB}(t) = d^X_B(t) - d^X_A(t)$$
$$\Delta d^Z_{AB}(t) = d^Z_B(t) - d^Z_A(t)$$

$$(14.17)$$

where $d^X_B(t)$, $d^X_A(t)$, and $d^Z_B(t)$, $d^Z_A(t)$ are the displacement of points B and A in horizontal and vertical. Also, the engineering strains are obtained by:

$$\varepsilon_{AB}(t) = \left[\tilde{J}_{AB}(t) - \tilde{J}_{AB}(0) \right] / \tilde{J}_{AB}(0)$$

$$(14.18)$$

$$\text{and} \; \tilde{J}_{AB}(t) = \sqrt{\left(\tilde{d}^X_{AB}(t) \right)^2 + \left(\tilde{d}^Z_{AB}(t) \right)^2} = \sqrt{\left[\tilde{d}^X_B(t) - \tilde{d}^X_A(t) \right]^2 + \left[\tilde{d}^Z_B(t) - \tilde{d}^Z_A(t) \right]^2}$$

where $\tilde{J}_{AB}(t)$ is the length from point A to B, $\tilde{d}^X_{AB}(t)$ and $\tilde{d}^Z_{AB}(t)$ are the deformation in the X and Z directions, respectively.

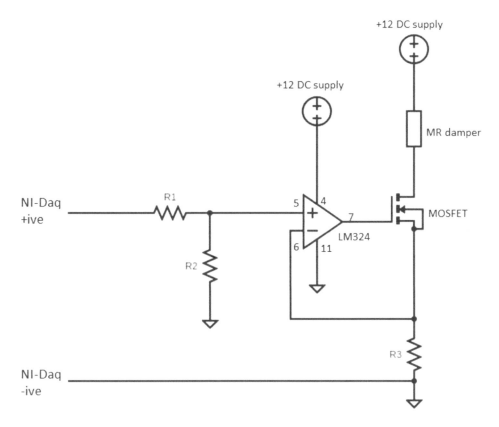

FIGURE 14.28   Power amplifier circuit to deliver the required current to the MR damper.

14.8.3   NOVEL POWER AMPLIFIER AND CONVERTER CIRCUIT

Most of the controllers used to implement the control strategy have voltage driven output that do not mesh well with the current driven MR damper and other semi-active control devices. Therefore, a device or a circuit is introduced between the controller and the MR damper to act as a converter and power amplifier. Most of these devices are either rigid in design, discontinuous in conversion, and/or expensive. Therefore, a novel power amplifier and converter circuit is **proposed and fabricated** that would overcome the drawbacks of traditional circuits.

The NI-DAQ voltage is sent into the self-developed power amplifier circuit, which transforms it to the appropriate current. Figure 14.28 illustrates the circuit diagram. The range of output current can be changed and designed as per the requirements of the particular MR damper by changing R3. Figure 14.29 shows the pictorial representation of circuit fabricated with three input and output channels. The conversion of voltage to corresponding current values is shown in Table 14.9.

14.8.4   NI-DAQ-6002 AND 6009

The input voltage from the transducers to the control unit and the output control voltage to the power amplifier circuit provided by the control unit were executed using USB-DAQ-6002 and 6009. These are the hardware components manufactured by National Instruments (NI) that perform amicability with already mentioned LabView software by NI. The Analog Input (AI) channels of these DAQ systems are used to convert the analog signal from the transducers to corresponding digital values that is used as the feedback to the control system. The AO channels provide the low-power control

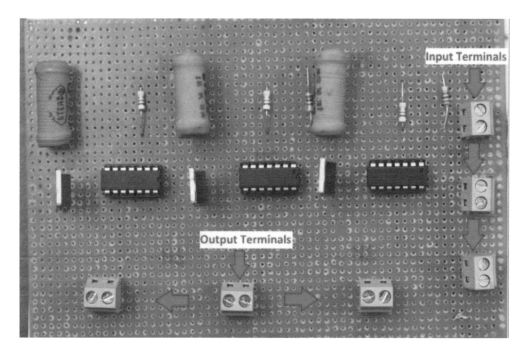

FIGURE 14.29 Actual circuit fabricated on a PCB with three channels.

voltage from the control system to the power amplifier circuit before it is supplied to the MR damper as a control current.

14.9   INPUT EXCITATIONS AND TESTING CONDITIONS

Figure 14.30 depicts the accelerograms under examination, which were retrieved from the PEER GMB by the University of Berkeley. The detail of each ground motion is presented in Table 14.10. The ground motions are scaled as per ASCE-7 specification of designing the target spectrum.

14.10   RESULTS AND DISCUSSIONS

14.10.1 Peak Floor Responses

Figure 14.31 depicts the peak column base strain, floor displacement, acceleration, and inter-story drift of controlled structures for all specified seismic excitations. For all ground motions, there is

TABLE 14.9
Conversion Chart for Circuit

Voltage Input to Circuit (V)	Output Current (A)
0.0	0
2.0	0.2
4.0	0.4
6.0	0.6
8.0	0.8
10.0	1.0

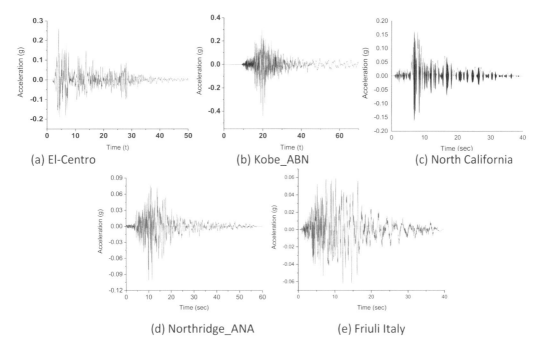

(a) El-Centro   (b) Kobe_ABN   (c) North California

(d) Northridge_ANA   (e) Friuli Italy

FIGURE 14.30   Time history records of selected seismic ground excitations.

a significant decrease in seismic reactions for the control scheme when compared to uncontrolled cases.

For all earthquake events, the drop in top floor acceleration ranges from 6.3 to 31.4%. Furthermore, drift and base strain mitigation ranges from 26.22% to 35.8% and 16.73% to 41.12%, respectively. This shows appropriate structural protection in the event of a power or control outage during a seismic event. But the maximum drift and acceleration increase for low-intensity excitation compared to passive-off control. In addition, relative to the uncontrolled condition, there is a considerable reduction in base strain of 16.67%, 22.32%, and 38%, for El-Centro, Kobe, and Northridge, respectively. Figures 14.31(d) and 31(e) show a significant increase in base strain of 10% to 48.34%. As a result, it is possible to conclude that the passive technique outperforms for strong ground movements, but it is detrimental for structures subjected to low-impact events.

RBA2 control, like passive-off control, reduced base strain by 21.5% to 62.4% as compared to the passive technique. For all seismic data evaluated, the RBA2 technique was credible in lowering the highest drift response by 24.6% to 55.1% compared to passive-on control. For RBA2, maximum acceleration responses are equivalent and, in a few situations, slightly lower. In comparison to the uncontrolled situation, the RBA2 technique was able to lower maximum acceleration values by 20%

TABLE 14.10
Earthquake Records Considered for the Study

S. No.	Earthquake Record	Year	Recording Station	PGA (m/sec²)
1	Imperial valley (California)	1940	El-Centro array #9	3.104
2	Kobe (Japan)	1995	Nishi	2.053
3	Friuli (Italy)	1976	Cordeiro	0.687
4	Northern Calif (USA)	1954	Ferndale City Hall	1.632
5	Northridge (Los Angeles)	1994	Lake Hughes #12A	0.943

to 35.6%. In particular, RBA2 control reduced the maximum acceleration response for El-Centro and Kobe ABN by 13.6% and 7.65%, respectively. However, 4.45%, 8.51%, and 13.32% increase in acceleration values were observed for different ground motions. As a result, the RBA2 method was capable of dramatically lowering structural reactions.

14.10.2 ROOT MEAN SQUARE (RMS) RESPONSES

Table 14.11 depicts the structure's normalized RMS response, which is regarded as a critical performance measure for achieving the requisite control. When compared to the uncontrolled example, the devised control approach was able to minimize the RMS displacement by 42.2% to 59.8% for all ground recordings. The RBA2 control was able to accomplish 56.8%, 62.8%, and 59.8% mitigation in first-, third-, and fifth-floor RMS displacement respectively. When compared to passive-on values, the RBA2 reduced 24.7%, 23%, and 7.7% reductions for the identical circumstance. Furthermore, the RBA2 technique outperformed passive-on in terms of reducing the RMS acceleration response. The highest reduction in RMS acceleration is 37.9%, 47.5%, and 40.9% for the first, third, and fifth floors, respectively. Meanwhile, as compared to the uncontrolled instance, the RMS acceleration for the RBA2 method was lowered by 61.3%, 56.3% for the third floor, and 55% for the top level (Figure 14.32).

In overall, RMS acceleration and RMS displacement response of top floor are reduced by 31.4% and 51.8%, respectively, for all ground recordings evaluated. The average RMS acceleration and RMS displacement decrease for the first, third, and fifth floors is 35.6% and 58.4%, respectively,

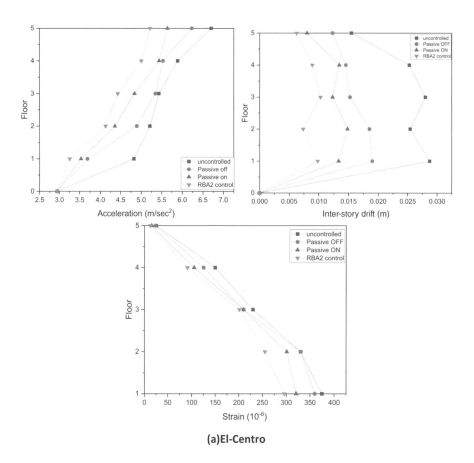

(a)El-Centro

FIGURE 14.31 Maximum structural responses obtained for selected seismic excitations.

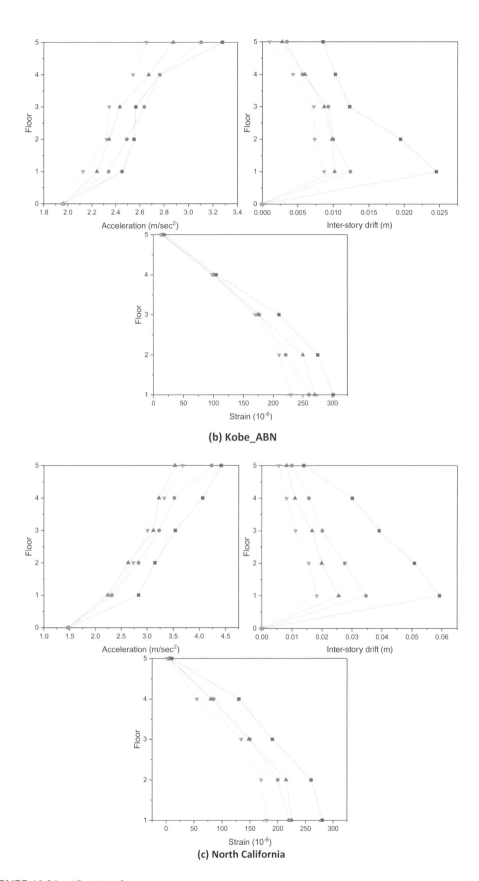

(b) Kobe_ABN

(c) North California

FIGURE 14.31 *(Continued)*

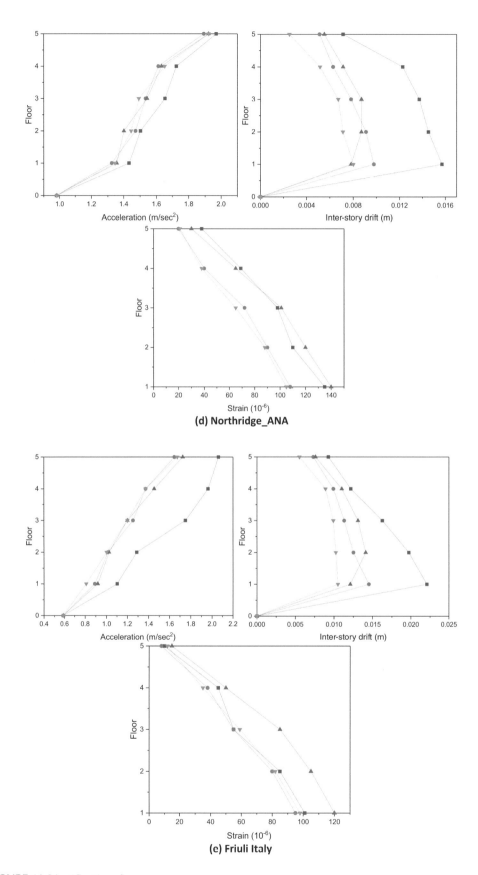

(d) Northridge_ANA

(e) Friuli Italy

FIGURE 14.31 (*Continued*)

TABLE 14.11

Normalized Root-Mean-Square Responses of the Structure

		Acceleration					Displacement				
Controller	Floor	El-Centro	Kobe	North Calif	Northridge	Friuli	El-Centro	Kobe	North Calif	Northridge	Friuli
Uncontd	1st	1.000	1.000	1.000	1.000	1.000	1.000	1.000	1.000	1.000	1.000
P-off		0.696	0.642	0.832	0.781	0.793	0.712	0.781	0.801	0.762	0.771
P-on		0.611	0.641	0.755	1.114	1.231	0.689	0.721	0.671	0.781	0.791
RBA2		0.532	0.541	0.722	0.704	0.892	0.442	0.506	0.521	0.673	0.742
Uncontd	3rd	1.000	1.000	1.000	1.000	1.000	1.000	1.000	1.000	1.000	1.000
P-off		0.725	0.677	0.822	0.786	0.779	0.747	0.801	0.801	0.727	0.737
P-on		0.593	0.535	0.723	0.926	1.177	0.579	0.612	0.615	0.772	0.771
RBA2		0.556	0.489	0.705	0.772	0.811	0.427	0.382	0.537	0.582	0.692
Uncontd	5th	1.000	1.000	1.000	1.000	1.000	1.000	1.000	1.000	1.000	1.000
P-off		0.753	0.711	0.812	0.761	0.764	0.781	0.821	0.801	0.691	0.702
P-on		0.612	0.601	0.691	1.231	1.123	0.489	0.522	0.579	0.782	0.771
RBA2		0.554	0.561	0.688	0.794	0.872	0.412	0.426	0.552	0.512	0.600

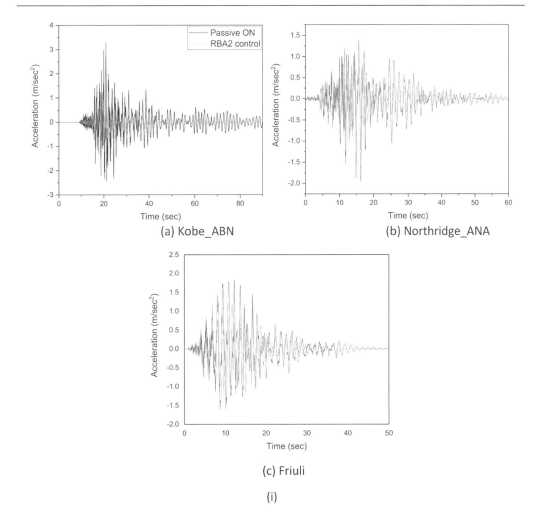

(a) Kobe_ABN

(b) Northridge_ANA

(c) Friuli

(i)

FIGURE 14.32 i) Acceleration Response, ii) Displacement Response.

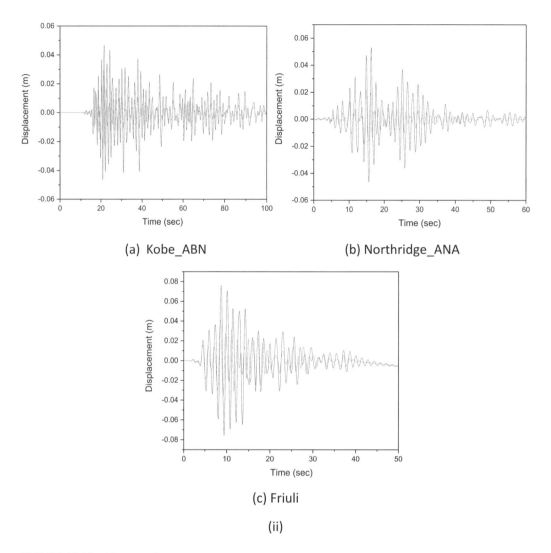

(a) Kobe_ABN   (b) Northridge_ANA

(c) Friuli

(ii)

FIGURE 14.32 *(Continued)*

when compared to uncontrolled, and 12% and 32.4%, respectively, when compared to passive-on control. As a result of the RMS values, it is possible to conclude that the RBA2 technique successfully minimized the acceleration responses while greatly attenuating the structure's displacement reaction.

14.11 CONCLUSIONS

This chapter begins with an introduction to the theory of RBA control and its response optimization technique using incremental gain simulation. This control strategy employs the optimization technique to attain an optimal control gain for a velocity feedback control loop, based on minimization of response/s objectives. The study develops as an easy, adaptable and computationally less intense technique that requires fewer system feedback and variable to perform the optimization process. Next, a mathematical description of the shear beam model of a structure with an MR damper. The state-space equation and the corresponding system matrices are attained to be used as a structural

block in the model. Also, the input seismic excitations used in the simulation are selected to characterize a wide range of ground motions.

Then, a controlled (with known system matrices) is mathematically modeled using a MATLAB/
Simulink environment. The structure is simulated to obtain the required control parameters for
RBA control strategies with single and multiple response objectives before implementing the same.
Next is experimental validation of proposed control strategies of the five-story steel frame on the
shake table. An MR damper installed at the ground floor employing four different control strategies
was able to attenuate acceleration, inter-story drift, and column base strain response of the structure
compared to the uncontrolled case. After discussing the physical specification of the structure and
the shake table, required system identification is carried out to determine the model parameters of
the structure used in this study. The structural parameters of the numerical model show close relation with the corresponding experimental values and therefore further experimental investigation
can be performed. Next, the hardware involved in controlling and recording the structural response
is discussed. The unique streaming DIC with a pattern matching algorithm designed to efficiently
capture the global response of the structures has been extensively deliberated. This was followed by
the design and fabrication of a novel power amplifier circuit, required to convert the control voltage
to the corresponding current values for the MR damper. Finally, the input seismic excitations to
which the structure is subjected on the shake table and the control strategies employed are explained
before the results are discussed.

The results of the numerical study demonstrated the superiority of the proposed RBA control
algorithms against the traditional H_2/LQG strategy using clipped optimal control. However, the
degree of reduction in structural responses using RBA control strategies depends on the target function selected. The main motivation behind the deployment and implementation of the RBA controller is to develop an algorithm that makes complete utilization of current range. Also, the RBA
controller provides a smooth and gradual transition of current, thus increasing the overall stability
of the system. The experimental findings revealed that the proposed controller outperforms other
existing controllers in terms of computation time efficiency and adaptability in obtaining the global
optimum. As can be concluded from the results, the RBA2 was able to significantly attenuate both
story drift and absolute acceleration compared to passive control strategies. However, the reduction
in response quantities is achieved as per the trade-off between the performance objectives selected
by the designer.

REFERENCES

Al-Bedoor, B. O. (2001). Modeling the coupled torsional and lateral vibrations of unbalanced rotors. *Computer Methods in Applied Mechanics and Engineering.* https://doi.org/10.1016/S0045-7825(01)00209-2

Cai, J., Bu, G., Yang, C., Chen, Q., & Zuo, Z. (2014). Calculation methods for inter-story drifts of building structures. *Advances in Structural Engineering.* https://doi.org/10.1260/1369-4332.17.5.735

Elenas, A., & Meskouris, K. (2001). Correlation study between seismic acceleration parameters and damage indices of structures. *Engineering Structures.* https://doi.org/10.1016/S0141-0296(00)00074-2

Kanninen, M. F. (1973). An augmented double cantilever beam model for studying crack propagation and arrest. *International Journal of Fracture.* https://doi.org/10.1007/BF00035958

Liu, X., Tong, X., Yin, X., Gu, X., & Ye, Z. (2015). Videogrammetric technique for three-dimensional struc-
tural progressive collapse measurement. *Measurement: Journal of the International Measurement Confederation.* https://doi.org/10.1016/j.measurement.2014.11.023

Lukic, R., Poletti, E., Rodrigues, H., & Vasconcelos, G. (2018). Numerical modelling of the cyclic behavior of
timber-framed structures. *Engineering Structures.* https://doi.org/10.1016/j.engstruct.2018.03.039

Lupoi, G., Franchin, P., Lupoi, A., & Pinto, P. E. (2006). Seismic fragility analysis of structural systems.
Journal of Engineering Mechanics. https://doi.org/10.1061/(asce)0733-9399(2006)132:4(385)

Moghaddam, H., & Hajirasouliha, I. (2006). An investigation on the accuracy of pushover analysis for estimating the seismic deformation of braced steel frames. *Journal of Constructional Steel Research.* https://
doi.org/10.1016/j.jcsr.2005.07.009

Ngeljaratan, L., & Moustafa, M. A. (2020). Structural health monitoring and seismic response assessment of bridge structures using target-tracking digital image correlation. *Engineering Structures*. https://doi.org/10.1016/j.engstruct.2020.110551

Rabe, U., Janser, K., & Arnold, W. (1996). Vibrations of free and surface-coupled atomic force microscope cantilevers: Theory and experiment. *Review of Scientific Instruments*. https://doi.org/10.1063/1.1147409

Snieder, R., & Şafak, E. (2006). Extracting the building response using seismic interferometry: Theory and application to the Millikan Library in Pasadena, California. *Bulletin of the Seismological Society of America*. https://doi.org/10.1785/0120050109

Spencer, B. F., & Nagarajaiah, S. (2003). State of the Art of Structural Control. *Journal of Structural Engineering*. https://doi.org/10.1061/(asce)0733-9445(2003)129:7(845)

Tubaldi, E., & Kougioumtzoglou, I. A. (2015). Nonstationary stochastic response of structural systems equipped with nonlinear viscous dampers under seismic excitation. *Earthquake Engineering and Structural Dynamics*. https://doi.org/10.1002/eqe.2462

Wani, Z. R., & Tantray, M. (2021). Study on integrated response-based adaptive strategies for control and placement optimization of multiple magneto-rheological dampers-controlled structure under seismic excitations. *JVC/Journal of Vibration and Control*. https://doi.org/10.1177/10775463211000483

Wani, Z. R., & Tantray, M. A. (2020). Parametric Study of Damping Characteristics of Magneto-Rheological Damper: Mathematical and Experimental Approach. *Pollack Periodica Pollack*, *15*(3), 37–48. https://doi.org/10.1556/606.2020.15.3.4

Wani, Z. R., Tantray, M., & Farsangi, E. N. (2021a). Shaking Table Tests and Numerical Investigations of a Novel Response-Based Adaptive Control Strategy for Multi-Story Structures with Magnetorheological Dampers. *Journal of Building Engineering*, 102685. https://doi.org/10.1016/j.jobe.2021.102685

Wani, Z. R., Tantray, M., & Farsangi, E. N. (2021b). Investigation of proposed integrated control strategies based on performance and positioning of MR dampers on shaking table. *Smart Materials and Structures*, *30*(11), 115009. https://doi.org/10.1088/1361-665x/ac26e6

Wani, Z. R., Tantray, M. A., Iqbal, J., Farsangi, E. N., Wani, Z. R., Tantray, M. A., Iqbal, J., & Farsangi, E. N. (2021). Configuration assessment of MR dampers for structural control using performance-based passive control strategies. *Structural Monitoring and Maintenance*, *8*(4), 329. https://doi.org/10.12989/SMM.2021.8.4.329

Wani, Z. R., Tantray, M., & Noroozinejad Farsangi, E. (2022). In-Plane measurements using a novel streamed digital image correlation for shake table test of steel structures controlled with MR dampers. *Engineering Structures*, *256*, 113998. https://doi.org/10.1016/J.ENGSTRUCT.2022.113998

Wani, Z. R., Tantray, M., Noroozinejad Farsangi, E., Nikitas, N., Noori, M., Samali, B., & Yang, T. Y. (2022). A Critical Review on Control Strategies for Structural Vibration Control. *Annual Reviews in Control*, *54*, 103–124. https://doi.org/10.1016/J.ARCONTROL.2022.09.002

Wani, Z. R., Tantray, M., & Sheikh, J. I. (2021). Experimental and numerical studies on multiple response optimization-based control using iterative techniques for magnetorheological damper-controlled structure. *Structural Design of Tall and Special Buildings*, *30*(13). https://doi.org/10.1002/tal.1884

Yadav, A. (2009). Nyquist-Shannon Sampling Theorem. In Digital Communication.

Zubair, R., Manzoor, T., & Ehsan, N. F. (2022). Acceleration Response-Based Adaptive Strategy for Vibration Control and Location Optimization of Magnetorheological Dampers in Multistoried Structures. *Practice Periodical on Structural Design and Construction*, *27*(1), 04021065. https://doi.org/10.1061/(ASCE)SC.1943-5576.0000648

15 SiDMACIB
A Digital Tool for an Automated Structurally Informed Design of Interlocking Masonry Assemblages

Elham Mousavian and Claudia Casapulla

CONTENTS

15.1 INTRODUCTION

Form-finding methods that maximize structural efficiency through optimal distribution of materials have been a long-standing design effort throughout history to construct, e.g., arches, vaults and domes. New digital tools emerged in the recent decades have brought form-finding methods to a new level and now we can not only design but also construct expressive structural forms with complex geometries [1–3]. While most of the form-finding digital tools solely concentrate on optimizing the overall geometry, the structural performance of a real-life complex structure that needs to be segmented into smaller parts, not only depends on the overall geometry of the assembly, but also on the shape of each individual segment of the assembly and how the segments are bonded together.

How to segment a structure can considerably change the physical and structural properties of an architectural model, e.g., the load that a segmental structure bears may reduce compared to the unsegmented model since a continuum is broken up to some segments. To remedy this situation, the constituent segments of a discrete assembly can be kept together by fasteners, mortar, etc. These fillers however can be eroded over time. The position of the joints shared between the segments of an assembly and their shape can also significantly affect the structural performance of a discrete

DOI: 10.1201/9781003325246-15

assembly. The influence of the joint shape and location on the structural behavior of discrete assemblies have already been taken into account in the literature.

Referring to the historic masonry structures, it was observed that different bond patterns [4] or stereotomy methods [5] determining the joint shape and layout may perform considerably different load-bearing capacity. Thanks to the new digital fabrication and construction advancements, it is now possible to fabricate segments with complex shapes made of masonry and other quasi-brittle materials like glass using, e.g., 3D concrete printing [6], and CNC stone cutting [7] techniques. Therefore, joints between the segments can now take complex shapes that perfectly interlock the segments together and considerably improve their structural performance. These advancements have resulted in the recent exponential attention to design discrete assemblages composed of convex and concave interlocking blocks with complex geometries.

Recently, several studies were carried out on the relation between different joint shapes and the structural or physical properties of assemblages like energy dissipation and sound absorption [8, 9]. Regarding load bearing capacity, the influence of interlocking joint shapes on the in-plane or out-of-plane behaviors of flat structures [10–12], or simple geometries like cylindrical structures [13] were addressed. However, these studies are yet limited to a few numbers of overall geometries and interlocking joint geometries. On the other hand, they do not develop form-finding methods that automatically find the optimal geometry of joints maximizing the structural performances.

Quite a few digital tools have been developed in the literature to improve the structural performance of discrete assemblages through optimizing the geometry of the segments of an assembly. The first generation of such tools was in fact some simple rules of thumbs applicable to the very limited structural typologies; e.g., Rippmann [3] showed that the bearing capacity of a compression-only structure composed of rigid units does not change after being segmented by joints normal to the flow of axial forces. Later, Yao et al. [14] developed a tool to analyze the stability of interlocking models composed of rigid units and provided some suggestions on removal of the possible instabilities.

A novel CAD plugin was developed by the authors of this paper within the MSCA_IF project SiDMACIB No. 791235 (Horizon 2020) to design structurally feasible masonry interlocking assemblages [15]. This digital tool presents a flexible data structure to model interlocking single layer assemblages with diverse complex geometries. Also, in this study the segments are considered breakable. SiDMACIB first extended the limit analysis to interlocking joints and developed a shape optimization procedure to automatically tune the geometric parameters of the interlocking joints to remove the infeasibility. This method remains limited to corrugated interlocking faces within the European project. Later, other shape and/or topology optimization procedures were implemented to model structurally optimal interlocking joints when the segments can take a wide range of complex convex or concave (interlocking) geometries. The authors of this paper developed an optimization algorithm to find the best layout of joints for a free-form structure composed of convex and concave interlocking blocks with complex shapes [16]. Given the overall structure and a super set of potential joints, this optimization procedure finds the best subset of the joints that corresponds to the maximum load the system can sustain.

Adopting the continuum topology optimization, Aharoni et al. [17] developed a density-based topology optimization procedure to segment a continuum to several components. Given the number of segments, *the same number of* interconnected continuum domains are developed, while each of them tries to find the best shape and position of one of the interlocking rigid segments of the assembly. The last tool developed by Liu et al. [18] is a topological stereotomy design process that first models the overall structure by several stacked units and then, using the graph theory, unifies possible adjacent stacked units to model larger interlocking segments with complex shapes corresponding to a stable solution. Such an exponential growth in developing computational form-finding and optimization methods for interlocking assemblages demonstrates the importance of the structurally informed design of interlocking joints in the architectural design of the digital age.

This manuscript details how the SiDMACIB tool has been developed and its subsequent derivations. After a brief review on limit analysis of masonry systems, Section 15.2 presents how this method is updated to analyze the interlocking assemblages with corrugated joints. To validate the proposed method, an experimental investigation and a discrete element (DE)–based numerical analysis is then presented. The validation procedure, particularly, registers the torsion-shear capacity of an interlocking joint obtained by the proposed static problem, experimental test and DE analysis, and compares them to each other. Section 15.3 presents the computational setup developed to model the single-layer interlocking assemblage and, implementing the proposed static problem of Section 15.2, to analyze the structural feasibility of the interlocking assemblages. Section 15.4 then demonstrates a novel joint shape adjustment method to remove the infeasibility of an interlocking assemblage. To this aim, first a method is developed to measure the amount of the sliding infeasibility of an assemblage. Measuring the infeasibility guides the designer on how to adjust the shape (or other parameters) of the joints to reduce the infeasibility. In the second step, a shape optimization procedure is presented that minimizes the measured infeasibility though an automated adjustment of the shapes of the joints. In Section 15.5, the developed plugin is used to analyze and optimize an interlocking model (a skew arch). In the end, the conclusions and future works are provided in Section 15.6.

15.2 STRUCTURAL ANALYSIS OF INTERLOCKING BLOCKS (STATIC PROBLEM)

According to the limit analysis theorems developed by Kooharian [19] and Heyman [20], a masonry assemblage can be modeled as a number of fully rigid blocks so that the plastic failures of the assemblage can only occur at the interfaces shared between the blocks. Based on this assumption, the structural feasibility of a masonry assemblage can be assessed through finding the stress state or strain rate at the interfaces between the rigid blocks when static or kinematic approaches are implemented, respectively [21, 22].

Adopting the static problem of limit analysis, Livesley [22] proposed a model in which the stress state of an interface is abstracted to the internal forces at one or several contact points distributed on the interface. These internal forces are found through solving an equilibrium problem aimed at equilibrating the external forces subjected to the centroid of the rigid blocks and the internal forces at the contact points. In such an equilibrium problem, the internal forces are constrained by several yield conditions and when no solution exists for the constrained equilibrium problem, the assemblage is considered structurally infeasible.

Two common abstraction models based on Livesley's method are: the convex contact model, in which a flat interface is represented by a single contact point at the centroid of the face; and the concave contact model that abstracts the planar interface to several contact points usually distributed on the edges of the face. In the former model, three internal forces and three internal moments at the single contact point demonstrates the stress state of the face. In the latter model, on the contrary, three internal forces (normal to each other) at every contact point represent the face stress state.

The extension of the convex contact model to an interlocking interface seems a challenging issue since such an interface has a non-isotropic behavior and the yield functions covering all the possible plastic failures of the face would be highly complex. On the other hand, and to the best of the authors' knowledge, there are only two works in the literature that have extended the concave contact model to the interlocking non-planar interfaces. The first effort has been carried out by the authors of this manuscript [23] in which the concave contact model has been extended to the non-planar interfaces with corrugated shapes, as elaborated in detail in this section. Later, Kao et al., [24] in order to analyze the stability of discrete assemblies made of the blocks with concave geometries (i.e., concave polyhedrons [25]) distributed a large number of contact points on the edges of a non-planar interface with arbitrary shapes.

While the latter work enables us to model interlocking assemblages with diverse shapes, a large number of points on each interface can increases the computational expenses and therefore the proposed method can be hardly extended to assemblages with a large number of blocks. SiDMACIB, instead, explores the minimum number of contact points required to represent a non-planar

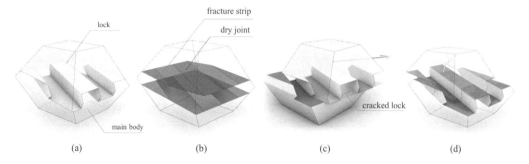

FIGURE 15.1   (a) Two stacked interlocking blocks with corrugated joints shared in between; (b) abstraction of the interlocking joint into two types of failure strips; (c) shear; and (d) sliding failures of the interlocking joint.

corrugated interface with one or a number of locks (projections attached to the main body of a block) having rectangular cross sections (Figure 15.1(a)).

As mentioned above, a convex masonry block (i.e., a convex polyhedron [25]) has been usually treated to be fully rigid in the literature, so that the only potential failure planes within a masonry assemblage are the joints between the rigid entities. However, several works considered the blocks to be breakable through introducing a number of failure planes within a block from which the block can crack [26, 27]. Unlike a block with convex geometry, the fracture possibility of an interlocking block modeled as a concave polyhedron should be seriously taken into account. This work, therefore, particularly considers the possibility of fracture and separation of the locks from the main body of the block. Other possible fractures of the interlocking block have been excluded in this manuscript meaning that the locks and the block main body of the block are considered rigid entities. Such an abstracted model is demonstrated in Figure 15.1(b), where the fracture plane from which a lock can be detached from the main body of the block is depicted in strips. Apparently, for a discrete assemblage composed of such interlocking blocks, the other type of the failure planes are the joints between the blocks (Figure 15.1(b)).

A lock can tangentially crack, and then slide and/or twist with respect to the main body when the lateral forces subjected to the lock are greater than the torsion-shear resistance of the fracture plane connecting the lock to the main body. A pure shear crack of the locks of an interlocking block is shown in Figure 15.1(c). Furthermore, an entire interlocking block can simply slide with respect to another block, parallel to the lock direction, assuming that the shear resistance of the locks is always greater than the frictional resistance of the interlocking faces (Figure 15.1(d)).

On the other hand and considering the tensile resistances of both the fracture plane and dry joints to be zero, each lock can be separated and displaced normally from the main body of the block when the fracture plane connecting that lock to the main body is fully in tension. Also, when this plane is partially in tension, the lock can rock, i.e., rotate with respect to the main body. Similarly, an interlocking block can be entirely separated or rock with respect to another block when the dry joints shared between two blocks are in tension. As mentioned above, assuming that the internal forces can only be distributed on the failure planes, Livesley [22] simplified each plane to a number of contact points (concave contact model) and, solving the equilibrium equation, found the internal forces at these points. In the next section, how the concave contact model can be extended to the interlocking joints is elaborated.

15.2.1   CONCAVE CONTACT MODEL FOR INTERLOCKING BLOCKS

According to the classical linearized concave contact model developed by Livesley [22] for the flat interfaces, first a 2D face is abstracted to a number of contact points distributed on the face; then the internal force at each point equilibrating the external forces subjected to every single block is

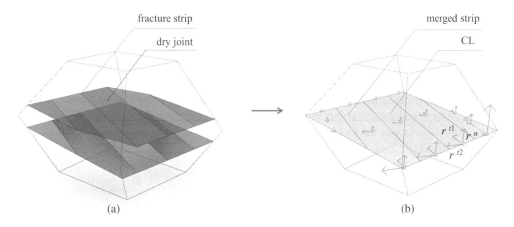

fracture strip   merged strip

dry joint   CL

(a)   (b)

FIGURE 15.2   (a) Fracture strip and dry joint above each other; (b) contact points distributed on the CL of the merged strips.

found. Each internal force is decomposed into three components that are separately constrained by linearized yield conditions: one component normal to the face, and two other components tangential to the face and normal to each other.

To extend this model to the corrugated interfaces, each pair of fracture and dry joint strips located one above the other is considered merged (Figure 15.2(a)), and contact points are distributed over the centerline (CL) of each merged face (Figure 15.2(b)).

The three components of an internal force belonging to each contact point are the normal force r^n, the tangential force r^{t1} parallel to the CL, and the other tangential force r^{t2} normal to the CL. The best location of the contact points and their number on each CL are investigated in the next section. Here, it is only assumed that a number of contact points are distributed on every CL of an interlocking joint.

According to the failure scenarios presented in the previous section, at contact point i, r_i^n should always be in compression: $y_1 = r_i^n \leq 0$; r^{t1} should meet the Coulomb's friction law: $y_2 = |r_i^{t1}| \leq (\mu r_i^n)$, where μ is the friction coefficient at the dry joint; and, r_i^{t2}, instead, must be less than the shear resistance of the contact point: $y_3 = |r_i^{t2}| \leq (p_i T_0)$. The latter resistance is a portion p_i of the overall pure shear resistance of the lock T_0, which in this work is set to be $(\tau_k a b)$, where a and b are length and thickness of the lock, respectively, and τ_k is the material shear strength. It should be considered that the summation of portions p_1 to p_m on a CL with m contact points is equal to one.

For each block, the external forces applied to its centroid must equilibrate the constrained internal forces at all the contact points distributed on the block interfaces as follows:

$$
\begin{cases}
C_{eq} \cdot \vec{r} + \vec{E} = 0 & \textit{Equilibrium equation} \\
\textit{subjected to:} \quad y_1 = r^n \leq 0 & \textit{compression constraint} \\
\quad\quad\quad\quad\quad y_2 = |r^{t1}| \leq \mu |r^n| & \textit{friction constraint} \\
\quad\quad\quad\quad\quad y_3 = |r^{t2}| \leq p \cdot T_0 & \textit{shear constraint}
\end{cases}
\tag{15.1}
$$

where $\vec{E}$ is the vector of the external forces and torques applied to the centroid of the block, $\vec{r}$ is the vector of the internal forces and C_{eq} is the equilibrium coefficient matrix, elaborated in detail in Section 15.3.

Rocking, sliding, and twisting of a block with respect to another block usually are combined with each other. Therefore, the constraints of the equilibrium equation problem must be developed in such a way that all the possible combined failures are perfectly covered. In particular, when a lock is subjected to non-uniformly distributed lateral forces, it can twist and slide with respect to the main body of the block, and sliding constrains should be able to represent the mixed torsion-shear failure at the interlocking joint, similar to what is implemented for planar joints [28]. In the proposed concave contact model, the three parameters determining the mixed torsion-shear resistance at a fracture plane are the number of the contact points and their location on the CL of the lock, along with the portion p of the overall lock pure shear resistance T_0 allocated to each contact point. The next section explores the best combination of these three parameters that can represent the realistic torsion-shear behavior of the lock of an interlocking block.

15.2.1.1 Torsion-Shear Interaction at the Lock Interface

As explained in the previous section, this work is developed with the assumption that the main body of a block and its attached locks are rigid entities. According to the proposed model, external forces are considered as pointed forces applied to the centroids of these rigid bodies. For example, to reproduce the torsion-shear behavior of their interface, a lateral force V normal to CL and a torque M can be applied to the lock centroid simultaneously (Figure 15.3(a)).

This section particularly investigates how to formulate the linear yield function y_3 for each contact point on a CL. As briefly mentioned in the previous section, this paper investigates how the three parameters of contact point number and location, and shear resistance portion, may affect the torsion-shear resistance of the interface, using the function y_3 for each contact point on its CL (Equation (15.1)).

Generally speaking, the stress state of a face can ideally be represented by the internal forces at a very large (close to infinite) number of contact points distributed uniformly on it; an approach that apparently is highly computational expensive. In this work, instead, seven different options are presented with a different number and location of contact points (Figure 15.3(b)). These options are divided in two sets consisting of two and three contact points per lock, as follows.

For the first set with two contact points per CL, the shear resistance of both contact points 1 and 2 at a lock interface is considered half of its total shear resistance T_0, i.e., p_1 and p_2 are 0.5. Four different locations are considered for these contact points with a distance of a (Option 1), 2/3 a (Option 2), 1/2 a (Option 3) and 1/3 a (Option 4) from each other, respectively. The second set consists of three contact points per CL, located at the end points and the middle point of a CL, while these points can have different shear resistances. This means that different portions of T_0 are allocated to the different contact points of a CL, under the following condition: $p_1 + p_2 + p_3 = 1$. Herein, three options for the shear resistance allocation at contact points 1 to 3 on a CL are considered: $p_1 = p_2 = p_3 = 0.333$ (Option 5); $p_1 = p_2 = 0.25$ and $p_3 = 0.5$ (Option 6); $p_1 = p_2 = 0.1665$ and $p_3 = 0.667$ (Option 7).

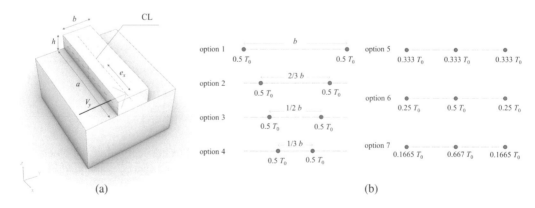

(a)   (b)

FIGURE 15.3 (a) Loading condition to analyze the torsion-shear capacity of the cohesive contact between the lock and the main body; (b) seven options for the contact point distribution on a CL.

To measure the torsion-shear resistance using the seven concave contact model options presented above, experimental and numerical tests are used for comparison, as described in the following subsection.

15.2.2 VALIDATION OF THE PROPOSED CONTACT MODEL

In order to find the most realistic option for the contact points among those in Figure 15.3(b), the numerical and experimental results obtained in previous works ([29, 30]) are herein used. These consist of analytical contact formulations in limiting conditions (using limit analysis), an ad hoc experimental investigation and numerical tests based on DE model, considering various eccentricities of V with respect to the lock centroid (Figure 15.3(a)). In the following, only the experimental and numerical modeling are briefly reported in order to compare the related results with the proposed options.

15.2.2.1 Experimental Investigation Setup

An ad hoc test setup was realized to measure the torsion-shear resistances at the cohesive interface connecting the rigid lock and the main body of the specimens that were casted using 3D printed molds. The designed specimen comprises a main body, with dimensions $100 \times 90 \times 50$ mm^3 and a $100 \times 30 \times 15$ mm^3 cuboidal lock located on its upper face ($a \times b \times h$ in Figure 15.3(a)).

Four test setups, S1, S2, S3 and S4, simulate the shear and torsion-shear failures, considering one of the two customized mortars detailed in [29]. S1 represents the pure shear loading (four specimens tested), while the three sets S2, S3 and S4 measure the torsion-shear capacity (three specimens tested per set), providing in total 13 experimental results.

15.2.2.2 Specimen Characterization

The specimen geometric properties are similar to the interlocking block presented in Section 15.2.1.1 (Figure 15.3(a)). Several mortar mixtures were explored in the first phase of the experimental program to find the mixture compatible to the limit load of 500 N (due to the employed instrument), and also to obtain a failure mode in the form of clear cutting of the lock from the main block (cohesive crack). After several investigations, two non-standard mortars were finally accepted, of which only one is considered in this manuscript. The composition and curing time of this mortar is reported in Table 15.1.

15.2.2.3 Test Setup

The test setup was particularly designed to register the pure shear and the combined torsion-shear resistances of the lock interface, with the main goal of both reducing the effect of bending [31] and neglecting the pre-compression force, considered as the main challenges of the classical experimental setup for shear tests (UNI EN 1052-3 2007) [32]. To this aim, the horizontal load was applied on the lock front face by the static gravity load using a pulley system and an electric hydraulic jack, with monotonically increasing value until failure. Figure 15.4 presents the overall test setup in which the main body of the block was kept fixed to a based board to avoid sliding and to the upper

TABLE 15.1
Composition and Mechanical Properties of the Tested Mortar

Mortar	Pozzolana	Sand	Cement	Lime	Curing Time	Tensile Strength [MPa]	Compressive Strength [MPa]	Density [kN/mc]
M	47%	38%	15%	-	28 days	0.37	0.75	14.0

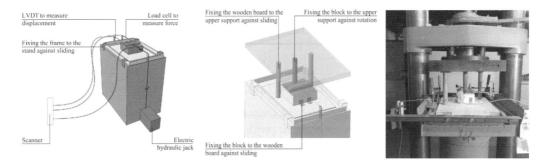

FIGURE 15.4 Experimental test setup.

support to avoid rotation. Applying an iron cup on the lock and a 3D printed cup-shaped box containing the main body of the block allowed to keep the lock and the main body as rigid.

The load was applied under displacement control at a constant rate of 3 mm/min. Forces were measured by a load cell with 500 N maximum capacity and 10 Hz acquisition frequency, while implementing three linear variable displacement transducers (LVDTs) with a displacement range of ± 50 mm (positioned to the side and back faces of the lock) the displacements were registered. The transducers' data was then acquired by a digital scanner, distributed by Vishay Measurements Group.

15.2.2.4 Testing Program and Main Results

To reproduce the pure shear resistance and torsion-shear combinations, different positions and orientations for the pulley system were employed, as reported in Table 15.2. The table presents all the four sets, S1 (pure shear) and S2, S3 and S4 (torsion-shear), in which the horizontal force V is applied with various eccentricities and directions with respect to the vertical midplane of the lock.

Table 15.3 presents the results related to each set, considering that four tests were carried out for S1, and three tests were executed for each of the sets S2, S3 and S4. It is remarkable that the coefficients of variation (CV) mostly present a low dispersion of the frequency distributions, showing a good test setup reliability.

TABLE 15.2

Experimental Set Configurations and Dimensions (in mm)

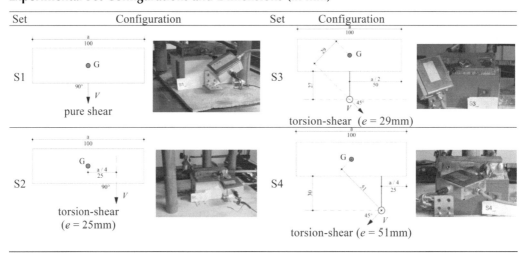

TABLE 15.3
Torsion-shear Capacities of the 26 Specimens

Set	ecc. [mm]	Shear Force V [N]				CV [%]	Average V [N]
		T1	T2	T3	T4		
1	0	250	242	226	243	3.7	240
2	25	213	203	181	-	6.7	199
3	29	181	193	192	-	2.9	189
4	51	118	133	106	-	9.3	119

15.2.2.5 DE Model Setup

Discrete element (DE) method is one the most well-developed numerical modeling approaches for masonry structures, in which units of a discrete assembly can move independently of each other. These units can be rigid or deformable when consist of finite deformable elements. Solving the equation of motion, the DE method allows computing large displacements and rotations of the units with respect to each other, while it can automatically update the contacts between the moving bodies during the analysis [33].

Using 3DEC software, two different models were developed in [29] and improved in [30] to simulate the interlocking block behavior presented in Section 15.2.1.1 (see Figure 15.3(a)):

1. In the first model (rigid model), an interlocking block was decomposed into two rigid hexahedral entities representing the block main body and the lock on top of that. These entities were connected by a cohesive contact, considering that in 3DEC, contacts (unlike the units) are always deformable interfaces consisting of 2D finite elements;
2. In the second approach (deformable model), an interlocking block was modeled as a concave deformable polyhedron that consists of finite tetrahedral elements.

The torsion-shear resistance of the lock interface was presented for these two models through recording the maximum lateral loads at different eccentricities. In the following, first, the load and boundary conditions implemented in both rigid and deformable models are presented. Then, the designed rigid and deformable interlocking models are demonstrated in detail.

Loading condition: As shown in Figure 15.3(a), the torsion-shear capacity of the contact shared between the lock and the main body of the block is measured by fully fixing the main body and make the upper lock free to displace only in the X-Y plane, while the lock is subjected to a combination of lateral force V_y and torque $M_y = V_y \cdot e_x$, where e_x is the eccentricity with respect to the lock centroid. To model this loading condition in 3DEC, the combined lateral force and torque can be

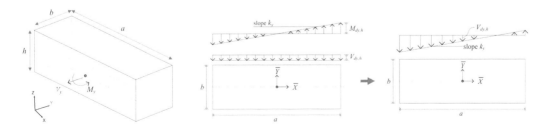

FIGURE 15.5   Linearly distributed forces representing the combined lateral force and torque.

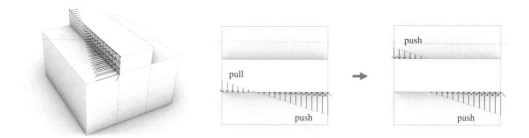

FIGURE 15.6 Substitution of the tensile triangular distributed forces pulling the lock face by compressive triangular distributed forces subjected to the other vertical face of the lock.

represented by a linearly distributed force on one of the vertical faces of the lock (Figure 15.5). This is expressed by the formulations:

$$V_{dy,h} = \frac{V_y}{ah}; \quad M_{dy,h} = \frac{6M_y}{a^2h} \quad k_y = \frac{12M_y}{a^3h} \tag{15.2}$$

where k_y is the slope of the distribution and $(a \cdot h)$ is the face area.

Depending on the value of e_x, the linear force can either form a fully compressive-trapezoidal distributed force or consist of two compressive and tensile triangular distributed forces. For the latter case, if the tensile stress of the lock is not enough, the lock can fail before reaching its maximum torsion-shear resistance. Instead, as shown in Figure 15.6, the tensile triangular distributed force can be substituted by a compressive triangular distributed force applied to the other vertical face of the lock. It should be observed that the measured torsion-shear capacity of this proposed amended model is the same as the initial loading model.

Rigid model: A rigid model consists of two rigid cuboids with different width, stacked over each other. These two rigid blocks are connected by a cohesive contact, considering that the contacts in 3DEC can only be deformable. The interface shared between the blocks in contact is automatically identified by the software and it was observed that, due to the difference in width of the lock and main body of the block, the generated contact produced several computational errors. To remedy this situation, the main body of the block was divided into three parts, as shown in Figure 15.7(a), and the generated deformable contact was discretized differently, namely 'normal', 'radial' and 'rad8' (Figure 15.7(b)). The torsion shear capacity of the cohesive contact was then calculated for each of these three discretization models.

Deformable model: 3DEC software can model concave blocks only if they are modeled as deformable polyhedrons discretized into finite elements, while the software implements uniform-strain internal meshing for these discrete elements [34, 35]. A concave polyhedron was modeled by unifying two cuboids representing the main body of the block and the lock. It was observed that discretizing such a model presents several computational errors. To amend this drawback, the main body of the block was first discretized to three sections and then these three parts together with the lock were unified to generate the concave interlocking block.

3DEC provides different discretization methods including 'quad' and 'edge'. Many of these methods are inapplicable to the concave interlocking block due to several computational errors, and it was observed that the 'edge' keyword by which the blocks are discretized into the tetrahedral zones, is the only reliable discretization method applicable to the different mesh sizes.

The effect of mesh size on the accuracy of the FE analyses of quasi-brittle materials has been widely acknowledged and studied [36, 37]. Apparently, smaller mesh sizes increase the result accuracy and yet the computational expenses of the analysis. However, the dependence of results on the

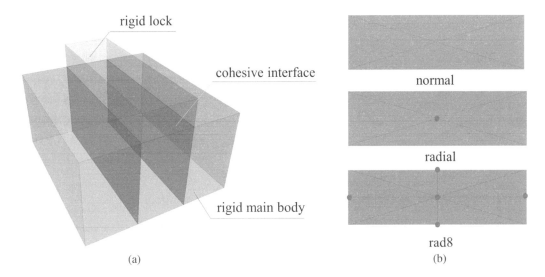

(a)   (b)

FIGURE 15.7   (a) DE rigid model including rigid lock and main body of the block and the deformable cohesive contact in between; (b) different discretization models for the contact interface.

mesh could be different in various parts of the model; e.g., the torsion-shear resistance of the lock interface is mostly dependent on the mesh size of the zone connecting the lock and the main body together, while the mesh size in the other locations of the lock and the main body has relatively no effect on the results, though it can considerably change the computational time.

Similarly, the size of the finite element in the X, Y and Z directions can have different impact on the accuracy of the results. For the analysis of the lock torsion-shear capacity, for example, the results are much more affected by changes in the mesh size in Z directions compared with the mesh sizes in the X and Y directions. Based on what explained, a new deformable model was proposed in which the lock was first divided in two vertically stacked parts, with a very thin lower part connected to the block main body (Figure 15.8(a)). These two parts together with the main body of the block were then unified to model a concave polyhedron (Figure 15.8(b)). Now, using any mesh size (determined by the edge keyword), the size of the finite elements inside the lower/thinner part of the lock in the Z direction is equal to the height of the same part, which is a very small value (Figure 15.8(c)). It was observed that, using this model, the measured pure shear capacity of the lock becomes completely mesh independent and the influence of the mesh size on the combined torsion-shear resistance of the lock considerably reduces [30].

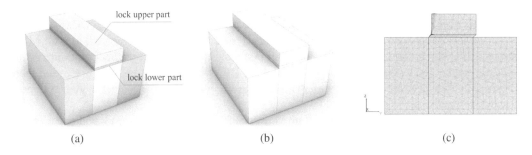

(a)   (b)   (c)

FIGURE 15.8   (a) DE revised deformable model, in which the lock is discretized into two stacked segments; (b) unified parts; (c) plastic deformation of the lock interface.

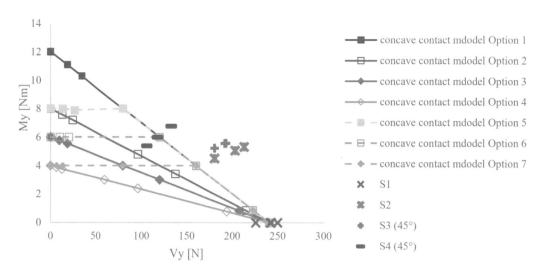

FIGURE 15.9 Torsion-shear curves of the seven concave contact model options and the experimental tests.

15.2.2.6 Comparison of the Results

In this subsection, the torsion-shear capacity of the cohesive contact between the lock and main body of the block is presented for different eccentricities, using the different numerical and experimental models reported in the previous subsections.

Figure 15.9 shows the torsion-shear curves belonging to the seven concave contact model options categorized into two sets, including two and three contact points per CL, respectively (see Section 15.2.1.1). The figure also presents the results obtained by the experimental investigation. The material properties implemented to the concave contact models were driven from the experimental tests, according to which the material density is set to be 1,400 kg/m^3 and the shear strength at the cohesive interface is considered to be equal to the average experimental shear strength 8×10^4 N/m^2 [29].

The first observation in Figure 15.9 is that all the torsion-shear curves belonging to the concave contact model options of the first set (two contact points per CL) are the linear in one variable, while the other belonging to the second set of options (three contact points per CL) is divided into two linear segments. On the other hand, it is evident that the regression curve of the experimental results is not linear and it mostly follows the curves of Options 5 to 7 (second set). Particularly, it can be observed that Option 5 has the best agreement with the results of the experimental test. While, among the options of the first set with two contact points per CL, Option 2 is closer to the results of the experimental tests on the safe side.

On the other hand, Figure 15.10 shows the torsion-shear results obtained by the DE rigid model when three discretization models of 'normal', 'radial' and 'rad8' are implemented, as well as the results of the seven options. The main reason of depicting these results together is the interesting observation that the torsion-shear curves of the three DE rigid models are almost linear and more importantly they perfectly coincide with the linear curves of the concave contact model Options 1, 2 and 3, respectively.

To demonstrate which DE rigid model provides results closer to the actual torsion-shear capacity, the torsion-shear curves obtained by the DE rigid 'normal', 'radial' and 'rad8' models, are compared to the concave contact models Options 5 and 2 (that present the best agreement with the experimental results), together with the results of the experimental tests (Figure 15.11). Table 15.4 shows the material properties considered for the rigid polyhedrons and the cohesive contact between them used in DE analysis.

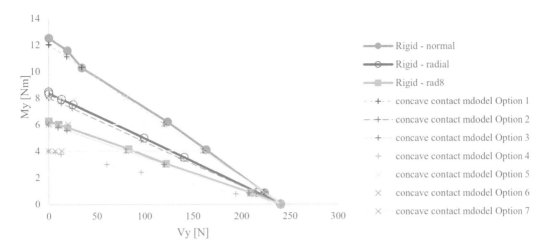

FIGURE 15.10 Torsion-shear curves of the seven concave contact model options and the of DE rigid model using three different contact discretization approaches.

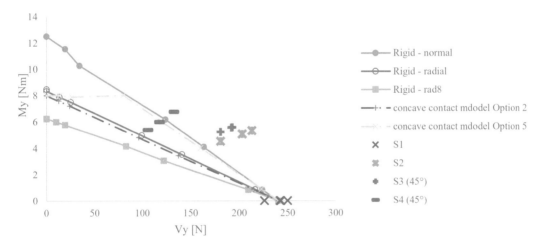

FIGURE 15.11 Torsion-shear curves of the concave contact model Options 2 and 5, of the DE rigid model using three different contact discretization methods, and of the experimental tests.

TABLE 15.4
Material Properties Implemented in the DE Rigid Model

Block material	Density	1,400 kg/m³	Cohesive joint material	Normal stiffness	10^{10} N/m
	Young's modulus	10^{10} N/m²		Shear stiffness	10^{10} N/m
	Poisson's ratio	10^{-10}		Cohesion	8×10^4 N/m²
				Tensile strength	3.7×10^5
				Friction angle	2 deg

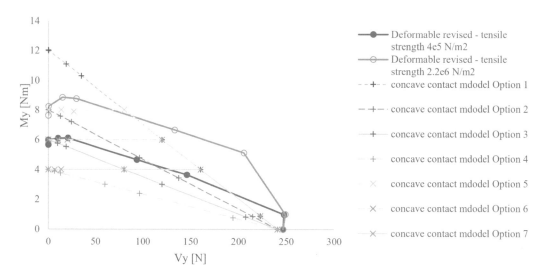

FIGURE 15.12 Torsion-shear curves belonging to the seven concave contact model options and DE deformable model using two different tensile strengths.

The results presented in Figure 15.11 show that the DE rigid 'radial' model can provide more realistic and conservative results when compared to the experimental ones. This linear curve very close to Option 2 is below the regression curve of the experimental results, while the corresponding pure torsion value relatively matches that of Option 5.

On the other hand, Figure 15.12 shows the torsion-shear results obtained by the DE revised deformable model when the lock is discretized into two parts and the height of the lower part is 0.0005 m. The mesh size (average edge length of tetrahedral zones) of this concave block is set to be 0.004 m, while two tensile strengths, i.e., σ_M (which is equal to the mortar average tensile strength obtained by the experimental test) and σ_{Mmax} (which is the maximum possible tensile strength according to Mohr-Coulomb's model [30]), are used. All the implemented material properties provided according to the experimental tests are presented in Table 15.5.

The first observation is that these torsion-shear curves are convex and non-linear, and the curvature of the graphs increases when the tensile strength rises. Comparing the torsion-shear curves obtained by the DE revised deformable model and those obtained by the seven concave contact model options, it can be seen that the curves belonging to DE model with tensile strength 4e5 N/m² and Option 3 are in better agreement with each other, and their pure shear and pure torsion coincide. Similarly, the curves belonging to DE model with tensile strength 2.2e6 N/m² and Option 5 are close to each other, and their pure shear and pure torsion are to a great extent similar.

TABLE 15.5

Material Properties Implemented in the DE Deformable Model

Density		1,400 kg/m³
Shear modulus (N/m²)		10^{10}
Bulk modulus (N/m²)		5×10^9
Cohesion τ_M (N/m²)		8×10^4
Friction angle ϕ (deg)		2
Tensile strength	Experiment σ_M	3.7×10^5
	$\sigma_M{}^{max}$	2.2×10^6

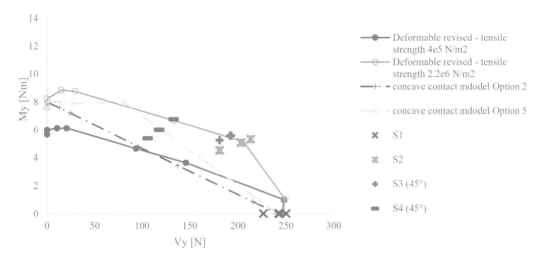

FIGURE 15.13 Torsion-shear curves belonging to the concave contact model Options 2 and 5, DE deformable model using two different tensile strengths.

On the other hand, Figure 15.13 presents the torsion-shear results obtained by the DE deformable model with tensile strengths σ_M and σ_M^{max}, as well as the results of Options 2 and 5, together
with the results of the experimental tests. This figure shows that the DE deformable model with
the maximum tensile strength 2.2e6 N/m^2 can provide more realistic results when compared to
the experimental results. Similarly, the torsion-shear curve of this model is closer to Option 2 and
mostly to Option 5.

All the carried-out comparisons among the proposed concave contact model and experimental
and numerical tests show that the concave contact model Option 5 with three contact points per CL
can reproduce more realistic results. However, since the number of contact points can impact the
computational time, the designer might choose to model interlocking joints with two contact points
per CL, using Option 2 for conservative results.

In the following, these two options are used to analyze interlocking assemblages. However, all
seven options are available in the SiDMACIB plugin presented in the next sections.

15.3 COMPUTATIONAL SETUP FOR SINGLE-LAYER
ASSEMBLAGES OF INTERLOCKING BLOCKS

Adopting the concave contact model presented in Section 15.2, this section presents a computational setup by which interlocking single-layer assemblages can be modeled and analyzed. Using
this setup, single-layer structures with arbitrary shapes including flat wall panels, vaults, domes
and freeform geometries can be modeled and analyzed. The setup also allows us to model different
bond patterns including stacked or running bonds, while openings can also be considered within the
assemblage. In the following, the developed modeling and analytical setups are presented, together
with their implementation in the SiDMACIB plugin [38].

15.3.1 Modeling Setup

This subsection presents data structures by which the information related to the contact points
involved in solving the static problem can be arranged. These data structures are developed for
single layer discrete assemblies composed of hexahedral blocks, while the faces of a block can be
either conventional (flat) or interlocking.

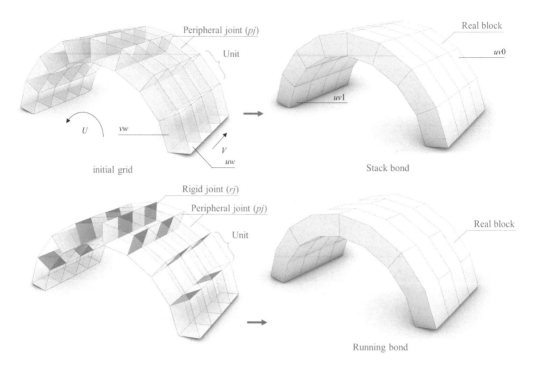

FIGURE 15.14 Modeling interlocking single layer assemblages with stack or running bond patterns in SiDMACIB.

A discrete assembly with stacked bond are herein represented by two grids on the extrados and intrados of the single-layer assemblage, while both grids have the same number of rows (*u* direction) and columns (*v* direction). The block faces are arranged in four data trees collecting the faces on the extrados (*uv*0), intrados (*uv*1), as well as faces in the *u* (*vw*) and *v* (*uw*) directions (Figure 15.14). The single-layer shell can be open or closed, either in *u* or *v* directions. If the shell is closed in the *u* direction, the *vw* faces of the first and last columns match each other and, likewise, if the shell is closed in the *v* direction, the *uw* faces of the first and last rows coincide.

Every *uv*0 and *uv*1 face can be chosen as a support that transfer the loads to the ground. Similarly, every *vw* face at the first and last rows, and every *uw* face at the first and last columns of the block grids can be chosen as a support. The internal forces can only be distributed on the selected supports as well as on all the *uw* and *vw* faces that are not located at the first and last rows or columns. These faces are herein referred to as 'active' while the other faces are named 'inactive' since they are not involved in the static problem solving. As mentioned above, openings can also be considered inside the shell through the removal of one or several blocks of the assembly. In this case, all the faces belonging to the removed blocks are set to be inactive and excluded from the static problem computational setup.

On the other hand, each active face can be a conventional flat or an interlocking face. The two parameters required to model an interlocking face are the number and orientation of the locks. To simplify the shape exploration, however, the number of the involved parameters is considered limited to eight, so that each pair of lock number and orientation determines the interlocking shape of the *uv*0, *uv*1, *vw* and *uw* faces, respectively.

As mentioned above, using the SiDMACIB digital tool, the designer can model the running bond pattern to layout the blocks on the single layer assemblage, while the running bond courses can be developed in either *u* or *v* directions (Figure 15.14). In this case, every two adjacent so-called units of an assemblage grid composes one block, so that the face shared between the two units (herein called rigid joint (*rj*)) acts as a part of the block, and the faces on the boundary of the block act as dry joints between the blocks (herein named peripheral joints (pj)). Forces distributed on the rigid joints are unconstrained and can take any value, while the peripheral joints between the blocks are constrained by the yield functions y_2 and y_3 in Equation (15.1).

15.3.2 Analytical Setup

Generally, the static problem of an interlocking assemblage with corrugated joints represented in Equation (15.1) can be upgraded as follows to include both stacked and running bond patterns:

$$\begin{cases} C_{eq} \cdot \vec{r} + \vec{E} = 0 & \textit{Equilibrium equation} \\ \textit{subjected to:} \quad y_1 = r^{pj,n} \leq 0 & \textit{compression constraint} \\ \quad y_2 = |r^{pj,t1}| \leq \mu |r^{pj,n}| & \textit{friction constraint} \\ \quad y_3 = |r^{pj,t2}| \leq 0.5\,T_0 & \textit{shear constraint} \end{cases} \tag{15.3}$$

where $\vec{E}$ is the vector of the external forces and torques applied to the centroids of the units, (considering that in the case of running bond assemblages, each block consists of two units), and $\vec{r}$ is the vector of the internal forces at the active rigid rjs and peripheral pjs joints. C_{eq} is the equilibrium coefficient matrix that is arranged as follows:

$$C_{eq} = \left[C_{uv0}^{sub} \mid C_{uv1}^{sub} \mid C_{uw}^{sub} \mid C_{vw}^{sub} \right] \tag{15.4}$$

where submatrices $C_{uv0}{}^{sub}$, $C_{uv1}{}^{sub}$, $C_{vw}{}^{sub}$ and $C_{uw}{}^{sub}$ contain the coefficients related to the internal forces on the $uv0$, $uv1$, vw and uw faces, respectively. An example of submatrix $C_{uv0}{}^{sub}$ is:

$$uv0_{1;1}\, uv0_{1;2} \ldots uv0_{q;p}$$

$$\begin{matrix} b_{1;1} \\ b_{1;2} \\ \vdots \\ b_{q;p} \end{matrix} \begin{bmatrix} C_{11} & & & \\ & C_{22} & & \\ & & \ddots & \\ & & & C_{pq\,pq} \end{bmatrix} = C_{uv0}^{sub} \tag{15.5}$$

For an assemblage with p rows and q blocks in each row, $C_{uv0}{}^{sub}$ is a sparse matrix that consists of submatrices C_{11} to $C_{pq\,pq}$, where C_{ij} is a coefficient submatrix for the ith interface acting on the jth block. Using C_{11} for Interface 1 with m contact points acting on a block, a part of an equilibrium equation for the block can be written as follows:

$$\left[\qquad\qquad C_{11} \qquad\qquad \ldots \right] \cdot \vec{r} \;=\; \vec{E}$$

$$\begin{bmatrix} e_1^{nx} & e_1^{t1x} & e_1^{t2x} & \cdots & e_1^{nx} & e_1^{t1x} & e_1^{t2x} & \cdots \\ e_1^{ny} & e_1^{t1y} & e_1^{t2y} & & e_1^{ny} & e_1^{t1y} & e_1^{t2y} & \cdots \\ e_1^{nz} & e_1^{t1z} & e_1^{t2z} & & e_1^{nz} & e_1^{t1z} & e_1^{t2z} & \cdots \\ (\vec{e_1^n}\times\vec{d_{1\_1}})^x & (\vec{e_1^{t1}}\times\vec{d_{1\_1}})^x & (\vec{e_1^{t2}}\times\vec{d_{1\_1}})^x & (\vec{e_1^n}\times\vec{d_{m\_1}})^x & (\vec{e_1^{t1}}\times\vec{d_{m\_1}})^x & (\vec{e_1^{t2}}\times\vec{d_{m\_1}})^x & \cdots \\ (\vec{e_1^n}\times\vec{d_{1\_1}})^y & (\vec{e_1^{t1}}\times\vec{d_{1\_1}})^y & (\vec{e_1^{t2}}\times\vec{d_{1\_1}})^z & (\vec{e_1^n}\times\vec{d_{m\_1}})^y & (\vec{e_1^{t1}}\times\vec{d_{m\_1}})^y & (\vec{e_1^{t2}}\times\vec{d_{m\_1}})^y & \cdots \\ (\vec{e_1^n}\times\vec{d_{1\_1}})^z & (\vec{e_1^{t1}}\times\vec{d_{1\_1}})^z & (\vec{e_1^{t2}}\times\vec{d_{1\_1}})^z & (\vec{e_1^n}\times\vec{d_{m\_1}})^z & (\vec{e_1^{t1}}\times\vec{d_{m\_1}})^z & (\vec{e_1^{t2}}\times\vec{d_{m\_1}})^z & \cdots \end{bmatrix} \cdot \begin{bmatrix} r_{1\_1}^n \\ r_{1\_1}^{t1} \\ r_{1\_1}^{t2} \\ \vdots \\ r_{m\_1}^n \\ r_{m\_1}^{t1} \\ r_{m\_1}^{t2} \\ \vdots \end{bmatrix} = \begin{bmatrix} EF^x \\ EF^y \\ EF^z \\ ET^x \\ ET^y \\ ET^z \end{bmatrix}$$

$$\tag{15.6}$$

where $\overrightarrow{e_k^n}$, $\overrightarrow{e_k^{t1}}$ and $\overrightarrow{e_k^{t2}}$ are the unit vectors of the local coordinates on the interface, and $\overrightarrow{d_{1\_1}}$ and $\overrightarrow{d_{m\_1}}$ are the vectors connecting the block centroid to contact points 1 and m, respectively. Also, *EF* and *ET* represent the external forces and torques applied to the block centroid. Apparently, to complete the equilibrium equation for the block, the coefficient submatrices related to the other interfaces acting on this block are required to be added to Equation (15.6).

15.3.3 IMPLEMENTATION AND VALIDATION

The data structures designed to model single-layer interlocking assemblages and the computational setup elaborated to analyze the proposed static problem, were implemented in the Rhinoceros Grasshopper GH plugin SiDMACIB [38]. Grasshopper is a visual programming environment within the Rhinoceros 3D software. The plugin is developed using C# programming language within the Visual Studio environment. All the modeling tasks, by which an interlocking assemblage can be generated, are carried out directly using C#, together with the coefficient matrices of the equilibrium problem and its inequality constraints, and the vectors of internal and external forces. These data are then used to solve the static problem by the least square optimization method, using lsqlin in MATLAB. Linear equality and non-equality constraints are involved, while MATLAB is used as a backend in the Visual Studio environment.

Figure 15.15 shows the GH component Static Problem developed to analyze the structural feasibility of the assemblage. The required inputs are listed in the left side of the component while the outputs are demonstrated in its right side. In this component, l0P and l1P collect the coordinates of the points at the corners of the unit cells, located on the extrados (layer 0) and intrados (layer 1). Structural supports can then be chosen among all *uv*0 and *uv*1 faces as well as *vw* and *uw* faces on the edges of the assemblage, using UVS0, UVS1, UWS0 and VWS0 input parameters, respectively. Furthermore, excluding *vw* and *uw* faces from the analysis, using UWE0 and VWE0 input parameters, openings can be generated inside the assemblage. It is worth noting that *uv*0 and *uv*1 faces are inactive and excluded from the analysis by default, unless they are selected as supports.

A bond pattern of the assemblage can be determined by the designer between two options of running and stacked bonds using a BP input parameter. To model the interlocking joints, four separated GH components were developed in the SiDMACIB plugin for the lock centerlines (CL), through determining the lock number and orientation for the *uv*0, *uv*1, *vw* and *uw* faces. The endpoints of these CLs can be imported to the GH component Static Problem using UV0LP, UV1LP, UW0LP and VW0LP input parameters. If no CL endpoint is imported as an input, the faces are considered conventional and flat by default.

External forces and torques applied to the centroid of each unit can then be inserted using EF and ET input parameters. Besides, the block weight can be separately assigned by determining the material density using D input parameter. Friction coefficient at the joint and the material shear resistance are also assigned using FC and ShSt input parameters. Finally, the contact point distribution among the seven options presented in Section 15.2.1.1 (see Figure 15.3(b)) can be chosen by the designer through the OP input parameter.

On the other side, the main output of the GH component Static Problem is the problem solution residual. In fact, MATLAB lsqlin always returns an output for both solvable and unsolvable problems, providing the closest possible answer that the solver can find. Instead, when residual is a vector of all zeros, it can be assured that the problem has at least one valid solution. Res output parameter in the GH component Static Problem is the summation of all elements of the residual vector. Besides that, the internal forces at each contact point are provided as outputs of this GH component. Four output parameters of UV0IF, UV1IF, UW0IF and VW0IF show the internal forces on the *uv*0, *uv*1, *uw* and *vw* faces, respectively. Also, UV0Points, UV1Points, UW0Points and VW0Points output parameters show the coordinates of the contact points on the *uv*0, *uv*1, *uw* and *vw* faces at which the internal forces are placed.

The implemented computational setup has been successfully validated by the authors, as comprehensively reported in previous manuscripts [23, 39, 40]. These papers compared the limit

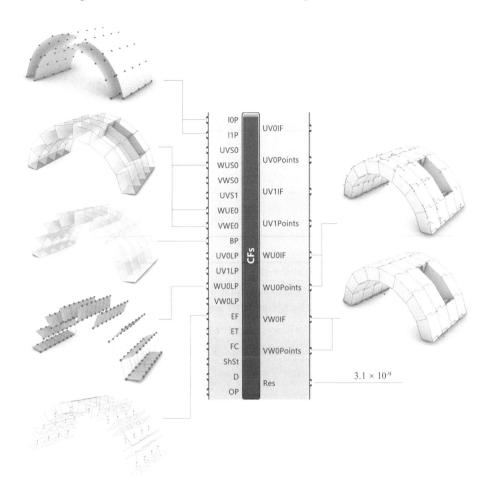

FIGURE 15.15   GH component static problem.

conditions of various assemblages like the circular arch [23], the hemispherical dome and the pavilion vault [39], as well as different wall panels with running bond and openings, and skew arch [40], with results existing in the literature. Some few but key examples that briefly illustrate this comparison are provided in Tables 15.6–15.10.

On the other hand, there is a big limitation to validate the applicability of the plugin to the interlocking assemblages, using the literature. In fact, to the authors' knowledge, there is no comprehensive study existing on the interlocking curved structures presented above when joints have corrugated shapes.

TABLE 15.6

Ratio of Minimum Thickness to Radius (t/R) for a Semi-Circular Arch with 10 m Centerline Radius and 20 Blocks, under Their Own Weight, Obtained by SiDMACIB and Gilbert et al. [41]

	Friction Coefficient: 0.4		Friction Coefficient: 0.35		Friction Coefficient: 0.3147	
	t/R	Diff	t/R	Diff	t/R	Diff
SiDMACIB	0.1062	0.014	0.157	0.003	0.192	0.026
Gilbert et al. [41]	0.10778		0.15646		0.19726	

TABLE 15.7

Ratio of Minimum Thickness to Radius (t/R) for a Hemispherical Dome with 10 m Centerline Radius, 20 Rows and 10 Blocks at Each Row, under Their Own Weight, Obtained by SiDMACIB and Heyman [20]

	Friction Coefficient: 0.27		Friction Coefficient: 0.23		Friction Coefficient: 0.2	
	t/R	Diff	t/R	Diff	t/R	Diff
SiDMACIB	0.044	0.035	0.075	0.136	0.097	0.03
Heyman [20]	0.0425		0.066		0.1	

TABLE 15.8

Minimum Thickness of a Pavilion Vault with 3 m Generatrix Radius, Spanning on a Square of 6 m by 6 m, with 6 Rows and 32 Blocks at Each Row, under Their Own Weight, with 1 N/m³ Specific Weight, Obtained by SiDMACIB and D'Ayala and Tomasoni [42]

	Friction Coefficient: 0.5		Friction Coefficient: 0.4		Friction Coefficient: 0.3	
	t [m]	Diff	t [m]	Diff	t [m]	Diff
SiDMACIB	0.16	0.059	0.24	0	0.56	0.125
D'Ayala and Tomasoni [42]	0.17		0.24		0.64	

TABLE 15.9

Ratio of Minimum Thickness to Radius (t/R) for a Helicoidal Skew Arch with Skew Angle 45°, Radius 3 m and Length 5 m, with 50 Rows and 4 Blocks at Each Row, under Their Own Weight, Obtained by SiDMACIB and the DE Analysis Carried Out in Mousavian et al. [40]

	Friction Coefficient: 1.73		Friction Coefficient: 0.84		Friction Coefficient: 0.58	
	t/R	Diff	t/R	Diff	t/R	Diff
SiDMACIB	0.043	0	0.043	0.188	0.047	0.573
Mousavian et al. [40]	0.043		0.053		0.11	

TABLE 15.10

Load Factor for Wall Panels 1 to 3 of Gilbert et al. [41], with b = Number of Blocks and c = Number of Contacts, Obtained by Gilbert et al. [41] and SiDAMCIB

Example No.	b × c	Associative λ [41]	Associative λ – SiDMACIB	Diff
1	33 × 83	0.64286	0.64995	0.011%
2	55 × 141	0.58000	0.59279	0.022%
3	46 × 102	0.40369	0.40368	2.47×10^{-5}%

15.4 JOINT SHAPE ADJUSTMENT

The static problem elaborated on in the previous section only shows if an interlocking assemblage is feasible or not. When an assemblage is structurally infeasible, implementing a checking approach may allow the designer to change parameters like number and orientation of the locks, aimed at exploring the parameter values representing a feasible model. Such a shape exploration is however non-directional because the binary information of being structurally feasible or infeasible cannot direct the designer on how to adjust the parameters to remove an infeasibility. Instead, if the infeasibility can be measured, e.g., using natural numbers, then the designer can simply observe if through the parametric changes, the structural infeasibility is decreased or increased. Joint shape adjustment to reduce and finally remove the infeasibility not only can be carried out manually by the designer but also automatically through a shape optimization problem, whose variables are the geometric parameters of the joints and its objective function is the minimization of the measured structural infeasibility. In the following, first, the static problem is reformulated to measure the infeasibility. Then, adopting this novel infeasibility measurement method, a shape optimization problem is developed to automatically tune the joint shapes.

15.4.1 SLIDING INFEASIBILITY MEASUREMENT

Measuring the structural feasibility of a discrete assembly was introduced for the first time by Whiting et al. [43]. As explained above, a discrete assembly with dry joints between blocks cannot be in the tension and therefore Whiting et al. [43] developed a method to measure the value of tensile forces at the joints and introduced it as the infeasibility measure. In that manuscript, this value is called normal infeasibility measure since it only measures the tensile forces exhibited in a model. Instead, this work presents a method to calculate the so-called sliding infeasibility, showing the tangential internal forces within an assemblage that violate the friction or shear constraints.

In the following, how the sliding infeasibility measure (SIM) can be formulated is explained. Removing the friction and shear constraints from Equation (15.3), the tangential internal forces can take any arbitrary value. So, r^{t1} and r^{t2} that violate the friction and shear constraints, respectively, can be collected and their sum can be considered as the SIM. Apparently, since r^{t1}s and r^{t2}s can be negative or positive, the summation of their absolute values shows the SIM. In this case, SIM is a natural value that is zero when the model is structurally feasible.

However, for the static problem, usually more than one solution (i.e., several vectors of internal forces) can exist for an assemblage, leading to the question: which of these solutions are more suitable to be used to measure the SIM? As explained above, the main purpose of the SIM is to measure the distance of an infeasible model from becoming feasible. Therefore, if more than one solution exists for an assemblage, we are looking for the solution (the vector of internal forces) in which the summation of r^{t1}s and r^{t2}s violating the sliding constraints is smaller than that for the other possible solutions. This in fact shows the shortest distance between the current status of the model and a structurally feasible model, among all the others. To obtain this solution, each tangential internal force can be decomposed into two components, one bounded by the sliding constraint and one unbounded. This means that for contact point i, r_i^{t1} is decomposed to r_i^{t1a} (limited by the friction constraint) and r_i^{t1b} (unconstrained), and similarly r_i^{t2} is decomposed to r_i^{t2a} (bounded by the shear constraint) and r_i^{t2b} (unbounded), as follows:

$$\begin{cases} r_i^{t1} = r_i^{t1a} + r_i^{t1b} \\ r_i^{t2} = r_i^{t2a} + r_i^{t2b} \\ \left| r_i^{t1a} \right| \leq \mu \left| r_i^{n} \right| \quad \text{friction constraint} \\ \left| r_i^{t2a} \right| \leq p\, T_0 \quad \text{shear constraint} \end{cases} \tag{15.7}$$

Then, SIM can be found using the following equation, which is in fact an optimization problem that minimizes the value of unbounded components of tangential forces r^{t1b} and r^{t2b}. This problem is subjected to the equality constraint of the equilibrium equation, and three inequality constraints: the first one limits the normal forces to be compressive; the second one limits the bounded tangential forces parallel to the lock CL, r^{t1a}, according to the friction Coulomb's law; and the third one keeps the bounded tangential forces normal to the CL, r^{t2a}, within the shear constraint. The formulation is:

$$
\begin{cases}
SIM_\theta = \min \sum_{\eta=1}^{L} \left[\left(r_\eta^{pj,t1b} \right)^2 + \left(r_\eta^{pj,t2b} \right)^2 \right] & \textit{Objective function} \\[2ex]
\textit{subjected to:} \quad C_{eq,\theta} \cdot \vec{r} + \vec{E} = 0 & \textit{Equilibrium equation} \\[1ex]
\qquad\qquad r^{pj,n} \leq 0 & \textit{compression constraint} \\[1ex]
\qquad\qquad \left| r^{pj,t1a} \right| \leq \mu \left| r^{pj,n} \right| & \textit{friction constraint} \\[1ex]
\qquad\qquad \left| r^{pj,t2a} \right| \leq 0.5\, T_0 & \textit{shear constraint}
\end{cases} \tag{15.8}
$$

where $r_\eta^{pj,t1b}$ and $r_\eta^{pj,t2b}$ are the unbounded tangential components parallel and normal to the lock at the generic contact point η, respectively, L is the total number of the contact points, and $\vec{r}$ contains r^{t1a}, r^{t1b}, r^{t2a}, r^{t2b} and r^n at the contact points.

15.4.1.1 Implementation and Validation

Implemented in Equation (15.8), the GH component SIM has been developed in the GH plugin SiDMACIB, to measure the SIM. Same as the GH component Static Problem, the related matrices and vectors to solve Equation (15.8) are built using C# language and the fmincon in MATLAB, which is a non-linear programming solver used as a backend.

Figure 15.16 demonstrates the SIM component and its input and output parameters. The input parameters of this component are identical to those of the GH component Static Problem. The output parameters instead classify the valid (VUV0IF, VUV1IF, VVW0IF, VUW0IF) and invalid (IUV0IF, IUV1IF, IVW0IF, IUW0IF) internal forces, separately. Once the proposed static problem is validated, it is implemented to verify the SIM measurement method. To do so, the structural feasibility of various interlocking structures can be analyzed by changing the friction values, and their SIMs can also be measured. Apparently, it is expected that the SIM of the structurally feasible model be very close to zero.

This infeasibility method was comprehensively applied to different assemblages including the circular arch, the hemispheric dome and the pavilion vault [39]. Different lock orientations were explored to show how the infeasibility of the model can be reduced by changing this geometric parameter. Table 15.11 summarizes the related results through the selection of only two key lock orientations for the assemblages presented in Section 15.3.3.

15.4.2 Shape Optimization

In this section, a novel shape optimization algorithm is developed that minimizes the SIM of a given interlocking assemblage, through the automated adjustment of its joint shape. As discussed above, two parameters of the lock number and orientation determine the joint shape. To date, however, the optimization has only involved the lock orientation.

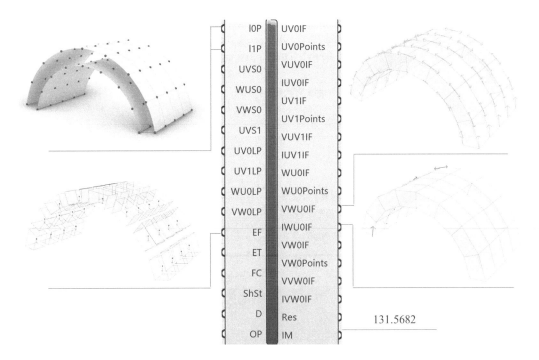

131.5682

FIGURE 15.16   GH component SIM.

TABLE 15.11
Structural Feasibility Analysis (Using Static Problem) and Measurement (Using SIM Method) for Different Interlocking Assemblages Presented in Tables 15.6–15.8 [39]

	Circular Arch			
	Lock Orientation 0		Lock Orientation $\pi/2$	
Friction Coefficient	Feasibility	SIM	Feasibility	SIM
0.33	yes	0.026	no	0.157
10	yes	4×10^{-4}	yes	0.011
	Hemispheric Dome			
	Lock Orientation 0		Lock Orientation $\pi/2$	
Friction Coefficient	Feasibility	SIM	Feasibility	SIM
0.2	yes	0.002	No	0.071
10	yes	3×10^{-5}	yes	0.007
	Pavilion Vault			
	Lock Orientation 0		Lock Orientation $\pi/2$	
Friction Coefficient	Feasibility	SIM	Feasibility	SIM
0	No	0.138	No	0.230
10	yes	21×10^{-5}	yes	30×10^{-5}

In fact, this shape optimization is an iterative procedure allowing to calculate $SIM_{\theta,i}$ at each iteration i, for a set of lock orientations $\{\theta_{uv0,i}, \theta_{uv1,i}, \theta_{uw,i}, \theta_{vw,i}\}$. Once computed $SIM_{\theta,i}$, apparently, the related internal forces $\vec{r_{\theta,i}}$ are also found. This optimization is formulated as follows:

$$\begin{cases} \min S(\theta, r) \\ subjected\ to: 0 \leq \theta \leq \pi \end{cases} \tag{15.9}$$

where:

$$S(\theta,r) = \begin{cases} SIM_\theta = \min \sum_{\eta=1}^{L} \left[\left(r_\eta^{pj,t1b} \right)^2 + \left(r_\eta^{pj,t2b} \right)^2 \right] \\ subjected\ to:\ C_{eq,\theta} \cdot \vec{r} + \vec{E} = 0 \\ \qquad\qquad r^{pj,n} \leq 0 \\ \qquad\qquad \left| r^{pj,t1a} \right| \leq \mu \left| r^{pj,n} \right| \\ \qquad\qquad \left| r^{pj,t2a} \right| \leq p\ T_0 \end{cases} \tag{15.10}$$

The variables of this shape optimization are in fact the set of lock orientations $\theta = \{\theta_{uv0}, \theta_{uv1}, \theta_{uw}, \theta_{vw}\}$, which are continuous variables, limited between zero and π rad, as well as $\vec{r}$, which in this type of shape optimization is known as a state variable [44], showing the state of the internal forces at θ.

15.4.2.1 Implementation

Implemented in Equations (15.9) and (15.10), the GH component Shape Optimization has been developed using C# language for modeling and MATLAB's fmincon for the shape optimization. fmincon is a MATLAB function for constrained non-linear and multivariable optimizations. The most challenging part of the application of an external solver like fmincon in the Visual Studio environment is the presence of four geometric variables (lock orientations) in the shape optimization. This means that when the value of the lock orientation is updated during the shape optimization in MATLAB, all the related geometric parameters like the coordinate of the contact points must be updated.

Using C# language through Visual Studio environment, we can simply use the geometric functions provided by RhinoCommon library developed for .NET programming to find the contact point coordinates for the given lock orientations. On the contrary, RhinoCommon is not applicable to MATLAB, and instead Rhino3dmIO library developed also by Rhino can be used. To this aim, a MATLAB function has been developed that accepts the lock orientations as input and returns coefficient matrices and vectors developed based on the coordinates of the contact points and local coordinates of the faces. The actions providing these outputs have been wrapped as a Dynamic Link Library (DLL) file developed using the Rhino3dmIO library.

Figure 15.17 demonstrates the GH component Shape Optimization. All the input parameters of this component are similar to those of the GH components Static Problem and SIM, except the input parameters assigning the lock geometric parameters. In other words, instead of importing the contact points located on the CLs (see Sections 15.3.3 and 15.4.1.1), in this GH component, the lock number (UV0N, UV1N, UW0N, VW0N) and orientation (UV0O, UV1O, UW0O, VW0O) are assigned directly to the component and used as the initial points of the shape optimization. The output of this GH component is thus the set of optimal lock orientations $\{\theta_{uv0,opt}, \theta_{uv1,opt}, \theta_{uw,opt}, \theta_{vw,opt}\}$.

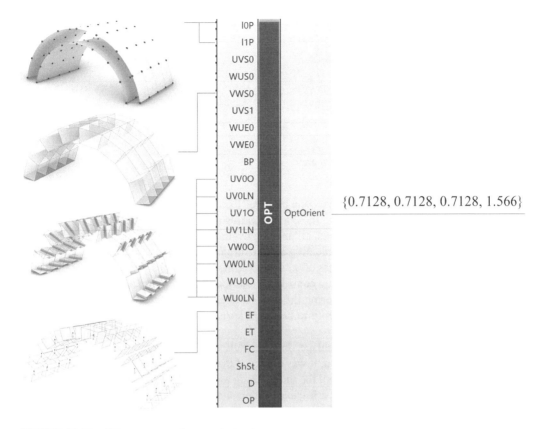

$\{0.7128, 0.7128, 0.7128, 1.566\}$

FIGURE 15.17   GH component shape optimization.

15.5   APPLICATION OF THE SiDMACIB PLUGIN

The application of SiDMACIB to parametrically model and analyze the feasibility of an inter-locking assemblage and optimize its possible infeasibility by automated interlocking joint shape adjustment has been widely presented in [23, 39] and [40]. In this book chapter, how an interlocking skew arch can be analyzed and optimized using SiDMACIB, involving all the three analytical GH components (Static Problem, SIM and Shape Optimization), is presented. The selection of this benchmark is particularly interesting firstly because its interlocking joints have complex shapes and secondly because its structural stability is highly dependent on the sliding resistance of the joints.

15.5.1   MODELING

Any arbitrary arch can be generated through extruding a profile curve along an extrusion line. When the extrusion line is not perpendicular to the profile curve, the modeled arch is called skew arch, while this angle is known as the skew angle. Skew arch particularly became a popular struc-tural system for bridges in 19th century in Europe. Three types of skew arches including helicoidal, logarithmic and false skew arches, are categorized based on their construction methods (bond pat-terns). This manuscript specifically focuses on the helicoidal arch.

    To model a helicoidal skew arch with skew angle θ, first its development surface, which is in fact the flattened skew arch, must be modeled. This surface is a rectangle rotated through the angle θ with respect to the Y-axis. To model an arch with m rows and n blocks in each row, the development surface is divided into m segments in the u direction and n segments in the v direction, considering

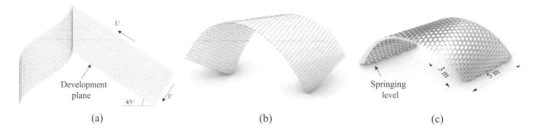

FIGURE 15.18 Modeling a helicoidal skew arch.

that each two units in one row represents one block of the assemblage with running bond pattern (Figure 15.18(a)). Modeling the flat grid, the corresponding grid on the circular surface of the arch can be found (Figure 15.18(b)). In the end, all units below the springing level are removed to obtain the final model (Figure 15.18(c)), representing a skew arch with circular profile, skew angle 45°, radius and length of 3 and 5 m, thickness of 0.24 m, 20 rows and 10 units (five blocks) in each row. While the vw faces of the assemblage are considered interlocking using three locks, other faces are set to be conventional and flat. Considering the friction angle to be 70 degrees, the material shear strength to be 5×10^4 N/m^2 and the density to be 2,400 kg/m^3, the assemblage is only subjected to its own weight.

To analyze this interlocking assemblage, first the GH component Static Problem was used. Through a parametric shape exploration, it was observed that for the lock orientations θ_{vw} between 0 and 0.45 rad with 0.05 rad intervals, all the interlocking assemblages are infeasible. The lock orientation 0.46 is the first assemblage as a feasible model, shown in Figure 15.19(a) with its internal forces.

To validate the proposed static problem, the same assemblage modeled in Rhino Grasshopper was imported to 3DEC and analyzed using DE method, as reported in detail in [40]. The interlocking blocks were simplified to rigid main bodies and locks, connected by cohesive interfaces in between. To validate the proposed static problem, two assemblages with lock orientations 0.1 rad (infeasible) and 0.46 rad (feasible) were analyzed in 3DEC. As a result, the former assemblage starts

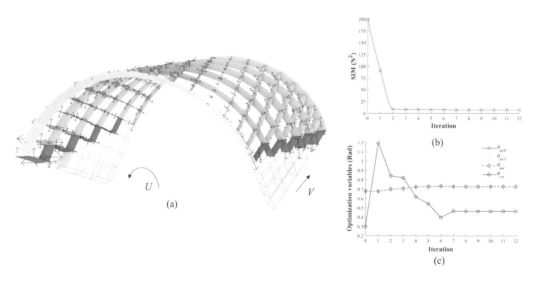

FIGURE 15.19 (a) Internal forces at the interlocking skew arch with the lock orientation $\theta vw = 0.46$ rad found by solving the static problem; (b) SIMs and (c) optimization variables at the shape optimization iterations.

to collapse after enough computation cycles, while the latter, reaches the equilibrium problem with the solve ratio 10^{-5}.

In the second step, the SIM related to each of the interlocking assemblages with different lock orientations between 0.3 and 0.46 rad was calculated using the GH SIM component. The SIM measured for the infeasible interlocking assemblage with the lock orientation 0.3 rad is a large value of 199.8 N^2, while the SIM for the feasible interlocking assemblage with lock orientation 0.46 is suddenly reduced to SIM = 7.050711 N^2. Since internal forces distributed on this assemblage are even greater than 10^3 N, such SIM can perfectly correspond to a feasible model.

Lastly, given the initial lock orientation 0.3 rad to the *vw* interlocking faces, the GH component Shape Optimization was implemented to minimize SIM through finding the optimal lock orientation. To do so, the initial set of lock orientations θ_0 = {0, 0, 0, 0.3} was considered as the optimization starting point, while, through considering the lock numbers to be N = {0, 0, 0, 2}, only *vw* faces were modeled as interlocking joints and *uv*0, *uv*1 and *uw* faces were set to be conventional flat faces. Before the execution of the shape optimization, fmincon in MATLAB automatically reset the first three initial lock orientations of θ_{uv0}, θ_{uv1} and θ_{uw} to 0.6775, since 0 is on the lower bound of the optimization variables. Apparently, the optimization is independent of these values because the corresponding faces are conventional and not involved in the shape optimization. Instead, the important initial value of θ_{vw} was kept to be 0.3 (as primarily assigned) by fmincon before initiation of the optimization.

Figures 15.19(b) and (c) show the SIM and the lock orientation changes during the optimization, respectively. The first observation is that the SIM measured manually (reported above) and automatically is identical. On the other hand, the values of θ_{uv0}, θ_{uv1} and θ_{uw} were kept almost the same during the optimization since they have no impact on the SIM value, as expected.

The presented optimization, however, has two major limitations. First, it is highly computational expensive. Running the shape optimization problem on an Intel Core i7 based PC (at 1.4 GHz) with 16.0 GB of RAM, under Microsoft Windows 10, the CPU time was about 90 minutes. Using MATLAB solvers as a backend and application of a DLL file in the MATLAB optimization function could be one of the main reasons for such a drawback. In the future, other optimization packages developed for the C# environment will be tested to observe how the computational efficiency of the optimization can be improved. On the other hand, fmincon is a constrained non-linear function provided by MATLAB that can adopt any non-linear objective function and constraints. Although this solver provides us with a great flexibility in formulating the objective and constraint functions, reaching the global optimal solution is very difficult and challenging. Instead, if other optimization methods, e.g., like second-order cone programming, can be developed using the proper formulation of the objective functions and constraints, then the accuracy of the solver in finding the global optima may highly increase. This will be considered more in detail in future works.

15.6 CONCLUSIONS AND FUTURE WORKS

This book chapter presented a novel framework called SiDMACIB, developed to model interlocking assemblages with corrugated joints, analyze their structural feasibility and remove the possible infeasibility through automated adjustment of the shape of the interlocking joints. This digital tool is in fact a GH plugin containing several GH components to model, analyze and optimize the interlocking assemblages.

To develop this tool, first the concave contact model of limit analysis was extended to interlocking joints through a novel contact point distribution. Also, new sliding constraints were introduced to the static problem, developed according to the orthotropic sliding behavior of the corrugated joints. Different options for the contact number and location, and for the related contact point shear resistance, were provided and the resulting torsion-shear resistance of the lock interface of a simple cubic interlocking block (with a single lock) was registered and compared to the results obtained by the novel experimental investigation and DE-based numerical tests.

The experimental test was developed based on a new setup to analyze the torsion-shear resistance at the cohesive interface between the main body of an interlocking block and its lock. A customized mixture was explored to find the appropriate mortar. Casting this mortar using a 3D printed mold, the interlocking block was fabricated. Four test setups, S1, S2, S3 and S4, simulated the shear and torsion-shear failures through the application of a horizontal load on the lock front face by the static gravity load using a pulley system and application of an electric hydraulic jack. The torsion-shear displacements were registered using three LVDTs.

Apart from the experimental investigation, a DE-based numerical analysis was developed using 3DEC software. Two different modeling approaches were designed, i.e., the rigid model, where the rigid main body is connected with the rigid lock by a cohesive interface, and the deformable model, represented by a concave deformable polyhedron with finite material shear strength. Also, in order to reduce the computational time when small mesh sizes are implemented, a new deformable model was proposed. The mesh sizes at the location connecting the lock and the main body were considered to be small, while in the other locations where the mesh sizes have no effect on the results, these were set to be large. The torsion-shear of the lock interface was investigated for both models and compared to the results obtained by the proposed concave contact model.

Comparing the results, it was observed that the concave contact model Options 2 and 5, DE rigid model with radial contact discretization, and DE deformable model with maximum tensile strength provide more realistic results.

The proposed static problem was implemented to develop the GH component Static Problem within the GH plugin SiDMACIB. Developing a flexible data structures and computational setups, free-form single-layer assemblages with stacked and running bond patterns and openings within the assemblage can be modeled.

To quantify the sliding infeasibility in an interlocking assemblage, a new variable SIM was introduced and formulated using the proposed static problem. This method was implemented to develop a GH component SIM, through which the designer can change the geometric parameters like lock number and rotation to reduce the infeasibility. The SIM minimization through joint orientation adjustment can also be carried out automatically using a new shape optimization procedure that was implemented to develop the GH component Shape optimization. In the end, all the analytical GH components of the SiDMACIB plugin were successfully applied to analyze and optimize a helicoidal skew arch.

In the future, the proposed static problem will be extended to other interlocking joint shapes like corrugated $\vee$ - $\wedge$ shaped or sinusoidal joints. On the other hand, due to the high computational expenses of the shape optimization, formulating new convex optimization problems will be taken into account. The shape optimization will also be expanded through introducing different geometric parameters to the problem. One high level goal is to model each joint using several control points, so that the ideal shape of the joints corresponding to the structurally feasible model can be found through automated adjustment of the control points. Using this approach, interlocking joints with diverse shapes can be modeled.

ACKNOWLEDGMENTS

This project has received funding from the European Union's Horizon 2020 research and innovation program under the Marie Skłodowska-Curie Grant Agreement No. 791235. It reflects only the authors' view, and the Agency is not responsible for any use that may be made of the information it contains.

REFERENCES

1. Lewis, W. J. (2003). Tension Structures: Form and Behaviour. Thomas Telford.
2. Veenendaal, D., Block, P. (2012). An overview and comparison of structural form finding methods for general networks. International Journal of Solids and Structures, 49(26), 3741–3753.

3. Rippmann, M. Funicular Shell Design: Geometric approaches to form finding and fabrication of discrete funicular structures. Doctoral dissertation, ETH Zurich, 2016.
4. Boni, C., Ferretti, D., Lenticchia, E. (2022). Effects of brick pattern on the static behavior of masonry vaults. International Journal of Architectural Heritage, 16(8), 1199–1219.
5. Gáspár, O., Sajtos, I., Sipos, A. A. (2021). Friction as a geometric constraint on stereotomy in the minimum thickness analysis of circular and elliptical masonry arches. International Journal of Solids and Structures, 225, 111056.
6. BRG. (2021) Striatus-3D concrete printed masonry bridge. https://www.striatusbridge.com/
7. Weir, S., Moult, D., Fernando, S. (2016). Stereotomy of wave jointed blocks. In: Reinhardt, D., Burry, J., Saunders, R. (Eds.), Robotic Fabrication in Architecture, Art and Design 2016, Springer, Switzerland, 284–293.
8. Dyskin, A. V., Estrin, Y., Pasternak, E. (2019). Topological interlocking materials. In: Estrin, Y., Bréchet, Y., Dunlop, J., Fratzl, P. (Eds.), Architectured Materials in Nature and Engineering, Springer, Berlin, Germany, 23–49.
9. Cui, S., Yang, Z., Lu, Z. (2020). An analytical model for the bio-inspired nacreous composites with interlocked 'brick-and-mortar' structures. Composites Science and Technology, 193, art. no. 108131.
10. Mirkhalaf, M., Zhou, T., Barthelat, F. (2018). Simultaneous improvements of strength and toughness in topologically interlocked ceramics. Proceedings of the National Academy of Sciences, 115(37), 9128–9133.
11. Weizmann, M., Amir, O., Grobman, Y. J. (2016). Topological interlocking in buildings: A case for the design and construction of floors. Automation in Construction, 72, 18–25.
12. Javan, A. R., Seifi, H., Lin, X., Xie, Y. M. (2020). Mechanical behaviour of composite structures made of topologically interlocking concrete bricks with soft interfaces. Materials & Design, 186, art. no. 108347.
13. Xu, W., Lin, X., Xie, Y. M. (2020). A novel non-planar interlocking element for tubular structures. Tunnelling and Underground Space Technology, 103, art. no. 103503.
14. Yao, J., Kaufman, D. M., Gingold, Y., Agrawala, M. (2017). Interactive design and stability analysis of decorative joinery for furniture. In: Proceedings of the International Conference on Computer Graphics and Interactive Techniques (Proc. SIGGRAPH), ACM Transactions on Graphics (TOG), 36(2), art. no. 20.
15. Mousavian, E., Casapulla, C. (2018–2020), SiDMACIB Project. Marie Skłodowska-Curie Individual Fellowship, Grant Agreement No. 791235, https://cordis.europa.eu/project/id/791235
16. Mousavian, E., Casapulla, C. (2022). Joint layout design: Finding the strongest connections within segmental masonry arched forms. Infrastructures, 7(1), 9.
17. Aharoni, L., Bachelet, I., Carstensen, J. V. (2021). Topology optimization of rigid interlocking assemblies. Computers & Structures, 250, 106521.
18. Liu, B. (2022). Topological stereotomic design of systems of interlocking stackable modular blocks for constructing multi-storey funicular masonry buildings. Master Thesis, Delft University of Technology.
19. Kooharian, A. (1952). Limit analysis of voussoir (segmental) and concrete archs. Journal Proceedings, 49(12), 317–328.
20. Heyman, J. (1966). The stone skeleton. International Journal of Solids and Structures, 2(2), 249–279.
21. Livesley, R. K. (1978). Limit analysis of structures formed from rigid blocks. International Journal for Numerical Methods in Engineering, 12(12), 1853–1871.
22. Livesley, R. K. (1992). A computational model for the limit analysis of three-dimensional masonry structures. Meccanica, 27(3), 161–172.
23. Mousavian, E., Casapulla, C. (2020). Structurally informed design of interlocking block assemblages using limit analysis. Journal of Computational Design and Engineering, 7(4), 448–468.
24. Kao, G. T. C., Iannuzzo, A., Thomaszewski, B., Coros, S., Van Mele, T., Block, P. (2022). Coupled rigid-block analysis: Stability-aware design of complex discrete-element assemblies. Computer-Aided Design, 146, 103216.
25. Mortenson, M. E. (1999). Mathematics for Computer Graphics Applications. Industrial Press Inc., 214
26. Portioli, F., Cascini, L., Casapulla, C., D'Aniello, M. (2013). Limit analysis of masonry walls by rigid block modelling with cracking units and cohesive joints using linear programming. Engineering Structures, 57, 232–247.
27. Lourenço, P. B., Rots, J. G. (1997). Multisurface interface model for analysis of masonry structures. Journal of Engineering Mechanics, 123(7), 660–668.
28. Casapulla, C., Maione, A. (2018). Modelling the dry-contact interface of rigid blocks under torsion and combined loadings: Concavity vs. convexity formulation. International Journal of Non-Linear Mechanics, 99, 86–96.
29. Casapulla, C., Mousavian, E., Argiento, L., Ceraldi, C., Bagi, K. (2021). Torsion-shear behaviour at the interfaces of rigid interlocking blocks in masonry assemblages: Experimental investigation and analytical approaches. Materials and Structures, 54(3), 1–20.

30. Mousavian, E., Bagi, K., Casapulla, C. (2022). Torsion-shear behaviour at interlocking joints: Calibration of discrete element-deformable models using experimental and numerical analyses. International Journal of Architectural Heritage, 17(1), 212–229. doi: 10.1080/15583058.2022.2101034

31. Montazerolghaem, M., Jaeger, W. (2014). A comparative numerical evaluation of masonry initial shear test methods and modifications proposed for EN 1052-3. Proceedings of the 9th International Masonry Conference, Guimarães, Portugal, 7–9 July 2014.

32. UNI EN 1052-3 (2007). Methods of test for masonry – Part 3: Determination of initial shear strength.

33. Cundall, P. A. (1971). A computer model for simulating progressive, large-scale movement in blocky rock system. Proceedings of the International Symposium on Rock Mechanics, Rubrecht, 1971.

34. Jean, M. (1999). The non-smooth contact dynamics method. Computer Methods in Applied Mechanics and Engineering, 177(3–4), 235–257.

35. Munjiza, A. A. (2004). The Combined Finite-Discrete Element Method. John Wiley & Sons.

36. Bažant, Z. P., Planas, J. (2019). Fracture and Size Effect in Concrete and Other Quasi-Brittle Materials. Routledge.

37. Hicks, M. A. (1998). Adaptive mesh simulation of passive earth pressure failure. In: Application of Numerical Methods to Geotechnical Problems, Springer, Vienna, 493–502.

38. Mousavian, E. (2022) SiDMACIB plugin for Rhino-Grasshopper, https://github.com/ElhamMousavian

39. Mousavian, E., Casapulla, C. (2020). Quantifiable feasibility check of masonry assemblages composed of interlocking blocks. Advances in Engineering Software, 149, 102898.

40. Mousavian, E., Bagi, K., Casapulla, C. (2022). Interlocking joint shape optimization for structurally informed design of block assemblages. Journal of Computational Design and Engineering, 9(4), 1279–1297.

41. Gilbert, M., Casapulla, C., Ahmed, H. M. (2006). Limit analysis of masonry block structures with non-associative frictional joints using linear programming. Computers and Structures, 84(13–14), 873–887.

42. D'Ayala, D. F., Tomasoni, E. (2011). Three-dimensional analysis of masonry vaults using limit state analysis with finite friction. International Journal of Architectural Heritage, 5(2), 140–171.

43. Whiting, E., Ochsendorf, J., Durand, F. (2009). Procedural modeling of structurally-sound masonry buildings. In: Proceedings of International Conference on Computer Graphics and Interactive Techniques (SIGGRAPH Asia 2018), Yokohama (Japan), ACM Transactions on Graphics (TOG), 28(5), art. no. 112, pp. 1–9.

44. Whiting, E. J. W. (2012). Design of structurally-sound masonry buildings using 3D static analysis. Doctoral dissertation, Massachusetts Institute of Technology.

16 Parametric Study of Low-Rise Buildings Using a Smart Hybrid Rocking Structural System with Vertical Dampers

Mehrdad Piri, Ali Akbari, Vahidreza Gharehbaghi, Ehsan Noroozinejad Farsangi, and Mohammad Noori

CONTENTS

16.1 INTRODUCTION

Currently, conventional building systems are being replaced by systems in which the possible damage during an extreme event, such as strong ground motions, is removed or mitigated in the worse scenario. Such systems, i.e., Damage Control Systems (DCS), save the time needed to reconstruct a building from scratch after demolishing the building because of being uninhabitable and are more cost-effective. Furthermore, DCS benefit from various elements that absorb the asserted energy and sacrifice themselves to prevent the main structural elements from harm by excessive forces or moments. These elements, including fuses, dampers, elastic footings, post-tensioning tendons (PTs), etc., are most of the time replaceable and repairable in case they are damaged. By deploying DCS worldwide and subsequently building resilient cities, both human lives and financial resources are saved during a disastrous incident. One range of the earlier mentioned systems is a Rocking Shear Wall (RW), and as it appears from the name, the primary mechanism in RWs is the rocking motion. In RWs, the edges of the walls are allowed to uplift, and as a result of the vertical difference between the two footings (slip displacement mechanism), energy-dissipating devices can control the lateral displacements. This mechanism can result from either a stepping rocking wall or a pinned one. When the case is a stepping RW, controlling the lateral displacement can be easier because the heavy weight of the RW is effective in returning the wall to its original position. However, the edges on which the stepping wall is rocking about undergo damage and sometimes can be difficult to be fixed in this type of RWs. Since a pinned RW cannot rock around the edges and only rocks on a pinned support, it does not benefit from the effectiveness of the weight. Rather, it could be detrimental and make the entire structure experience larger displacements. Here, several energy-dissipating devices are needed to work together to restrain the RW.

DOI: 10.1201/9781003325246-16

347

Rocking motion was first noticed in the early ages when the Egyptian, Chinese, Persian, and Greek empires used to build their palaces and structures using free-standing columns. Since then, researchers have developed various methods to improve the seismic behavior of free-standing or controlled rocking columns [1]. More recently, in 2019, Makris and Aghagholizadeh equipped free-standing columns with energy-dissipating devices. They concluded that, although in some cases, using energy-dissipating devices made the seismic response worse, in most cases, the seismic response of the damped structures was lower than that of the undamped ones [2]. A year later, Ríos-García and Benavent-Climent carried out a study to enhance the use of controlled rocking columns. They used steel bars as anchors in rocking columns and allowed the columns to uplift in a controlled range. The results showed that, by using dampers and the gap, the inter-story displacements and story accelerations in upper floors decreased by 3 and 1.5 times, subsequently [3]. The concept of rocking columns was soon transferred to bridge engineering, where Beck and Skinner in 1974 applied the rocking pier technique to enhance the performance of bridges when exposed to seismic excitations. In their study, bridge piers could take steps and separate from the foundation. It was found that not only by using energy-absorbing elements between the piers and the foundation, the number of steps and their amplitude decrease sharply, but also it is a more cost-effective method of providing resistance against earthquakes [4]. More recently, Xu et al. in 2019 proposed a double-column rocking pier to enhance the seismic resiliency of bridges. In their model, columns were connected through repairable shear fuses to absorb the seismic energy and unbounded pre-stressed steel strands to control the lateral displacements [5].

In 1991, Priestly developed the application of rocking concrete buildings, considering the cost-effectiveness of precast concrete structures and being easy to build and suitable for mass construction, to enhance their seismic performance in the regions with high seismicity under the PRESS program [6]. Lately, there has been more research on reinforced concrete frames using rocking walls. Guo et al. upgraded an existing RC frame with rocking shear walls and observed that by using the rocking concept and friction dampers in RC frames, not only do the structures tend to be more resilient but also the deformation pattern would be refined, which subsequently prevents the soft story failure [7]. Xu et al. also tried to improve the seismic performance of these structures by deploying disc springs on the rocking toes to hinder them from possible damage. This method could also improve the self-centering and deformation ability of the structures as well as control the residual displacements induced by earthquakes [8]. Mottier et al. used a different rocking bracing frame in which the braced rocking frame could rock about the side columns. Friction dampers were also deployed to control the upper-story drifts and steel bars to yield under tension and buckle under compression. The results revealed that because of the impacts the columns had to undergo, the forces inside the beams increased markedly [9]. To solve this problem and to prevent the fixed connections in moment frames from forming plastic hinges and subsequently failing, Piri and Massumi proposed a rocking concrete shear wall that was attached to a gravity frame. In this hybrid system, not only do the connections in the frames not take any damage during severe ground motions, but the rocking shear wall remains elastic, and if damaged, the harmed elements can be replaced or repaired. This also reduces construction costs, as the frames are no longer designed to carry lateral loads [10]. To reduce the costs even more, Massumi et al. remodeled and modified a previously proposed rocking wall that was rectangular-shaped. In the new model, the border elements method was deployed, and the rocking wall was dumbbell-shaped. This study concluded that by using this strategy, the new structures used significantly less concrete and had more capacity than the conventional ones [11].

The application of rocking walls is not limited solely to RC and steel frames, but is also of great importance in masonry and timber structures. In 2017, Niu and Zhang conducted experimental research on the seismic behavior of masonry walls. It was observed in this study that by separating the piers from the spandrels, not only the failure mode changes from shear to rocking but also provides the structure with self-centering and energy-dissipating abilities [12]. More recently, a new steel-timber rocking wall was used by Li et al., and friction dampers were utilized to connect a rocking wall to a timber frame. The results revealed that by using this type of connection, residual deformations could be controlled, and the connections would not take any damage by forming plastic hinges. Rather, the seismic energy would disappear through the slip displacement made in the friction dampers [13]. Moreover, the use of shape memory alloys [14], springs [15], post-tensioning tendons [16], and pressurized fluid devices [17] were investigated by different researchers, all pointing out the efficiency of such energy-dissipating devices in reducing the displacements and providing better self-centering properties to resist earthquakes. Some experiments investigated diaphragm connections [18], the design [19], cost-benefit efficiency [20], and higher mode effects [21] of rocking structures in general.

The current study investigates the influence of different parameters on the seismic response of a 3D low-rise building in which the lateral resistant system is a rocking wall equipped with energy-dissipating devices. The connections in the frames are assumed to be hinged to mitigate the possibility of being damaged by the connections and subsequently forming plastic hinges. In the end, an efficient model is proposed based on the study, considering various components. Then the proposed model is evaluated to observe whether the method is accurate.

16.2 CASE STUDY

In this paper, a 3-story model from the SAC (Structural Engineers Association of California) [22] steel project is considered to propose and build a resilient model. This 3-story building is in Los Angeles, on stiff soil (type S2), and designed to be an office building. The plan and elevation view of the building can be seen in Figure 16.1. The case model comprises a combination of moment and gravity frames. As a result, the forces are only accumulated in the areas with moment frames, and the gravity frames that only stand the gravity loads would remain immune to the possible damage through lateral loads. By looking at Figure 16.1, the results show that the highlighted lines show the moment-resisting frames that are on the outside frames, while internal dashed lines indicate the gravity frames. The shaded demonstrates where the penthouse is placed on the rooftop, and it is not modeled as a separate story. Rather, only extra loads apply to the marked areas.

The load distribution used for the analysis is as follows; Penthouse dead load: 116 psf, roof dead load (except for the penthouse areas): 83 psf, floor dead load: 96 psf, reduced live load: 20 psf. Applying all these loads would eventually lead to the following seismic mass for each story; First and second floors: 65.53 kips-sec^2/ft and roof: 70.90 kips-sec^2/ft. Additionally, columns in the perimeter moment frames shown in bold in Figure 16.1 bend about their strong axis, while in the gravity frames, columns bend about their strong axis in North-South (NS) direction. Since the study building is a steel structure, A36 steel and dual grade A36 Gr. 50 steel were used to represent the behavior of beams and columns, respectively. Table 16.1 shows the designed sections of the LA-3 story building presented by the SAC joint venture based on the load distribution, material properties, plan layout, elevation view, and column specifications that were told earlier.

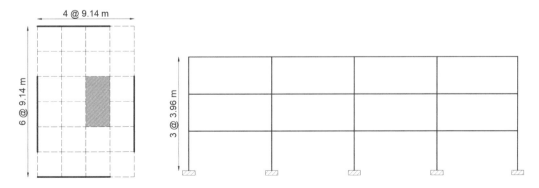

FIGURE 16.1 SAC project's plan and elevation views [22].

TABLE 16.1
Designed Sections According to the SAC Project [22]

LA-3 Story	Moment Frames			Gravity Frames		
	Columns			Columns		
Story/Floor	Exterior	Interior	Beams	Below Penthouse	Others	Beams
1	W14X257	W14X311	W33X118	W14X82	W14X68	W18X35
2	W14X257	W14X311	W30X116	W14X82	W14X68	W18X35
3	W14X257	W14X311	W24X68	W14X82	W14X68	W16X26

The models are built according to the information and specifications given above. To reach reliable results, verification was done in this study by conducting both modal and push-over analyses. Table 16.2 provides the modal results compared to the original ones that were extracted from FEMA355C. As seen in the table, the results are in accordance with the original study, having errors less than 10%, so the models are soundly modeled.

Furthermore, push-over analysis was also done to evaluate the nonlinear behavior of the models. To this end, various push-over lateral load patterns, including response spectrum, inverted triangle, and seismic coefficient, were used to pick the most critical response. Then the results were extracted and compared to the SAC curves, as seen in Figure 16.2. The figure indicates a slight difference between the results. However, these are negligible since the difference is less than 10% throughout the curve.

16.3 METHODOLOGY

After the verified models are built, it is time to model the self-centering concrete shear wall and release the fixed connections. The self-centering ability in this study is provided by a pinned base, allowing the wall to rock about it. Since it is challenging to control the upper-story drifts in pinned rocking walls, especially when the connections are hinged, a double hybrid rocking wall was presented. By deploying this method, there is a space between the doubled walls in which friction dampers could be implemented and easily become replaced if damaged. These dampers start to work using the slip displacement once the lateral load is applied. Additionally, post-tensioning cables are utilized to provide more self-centering capability in the rocking shear wall. To dissipate more seismic energy that was asserted to the structure, energy-dissipating devices were added at the base corner of the walls to limit the upper floor's movements and prevent the edges from bearing damage. It is worth mentioning that the connections between the walls and the surrounding columns are through link beams that are released at both ends, preventing the moments from being transferred to the main columns. As was previously noted, the main beam-column connections and column-foundation connections are pinned and not fixed so that they may rotate as the rocking shear wall rocks back and forth. Figure 16.3 indicates a schematic view of the above-described hybrid rocking

TABLE 16.2
LA 3-Story Model Verification Using Modal Analysis

Modes	The Models			SAC Models		
	First	Second	Third	First	Second	Third
Fundamental Period (Second)	1.04	0.33	0.17	1.03	0.33	0.17
Modal Mass (Percent)	81	14	4.2	82.8	13.5	3.7

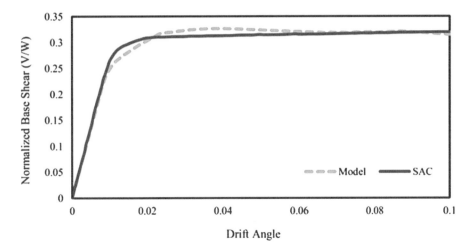

FIGURE 16.2 LA 3-story push-over verification of the model compared to SAC.

concrete shear wall and other elements. It also should be noted that both column-beam and column-foundation connections are pinned to let the rocking mechanism dominate. Otherwise, the heavy rocking wall would harm the fixed connection by doing a rocking motion several times until plastic hinges form all over the frames.

Following the modeling stage that was done in SAP2000, now it is time for the material properties to be assigned. All the specifications used in this study are selected to meet the general considerations related to energy-dissipating devices. Regarding the PTs, four tendons were used in each wall with elasticity modulus, yield strength, and cross-section of 198, 1.528, and 25.8 cm², respectively. Furthermore, the friction dampers that were deployed between the two walls, according to Figure 16.3, had a stiffness of 200 kN/mm, and only one damper in each story was used. Turning to the energy-dissipating bearings, with the ability to endure both tension and compression, they were designed to withstand a vertical stiffness of 110 kN/mm and negligible horizontal stiffness. Finally, to represent the behavior of the concrete in the rocking concrete shear wall, concrete type C30 was applied.

It is worth mentioning that, since the rocking wall in this study is pinned from the base, which allows the free rocking motion to happen, the design of the wall lacks significance. Because the wall's

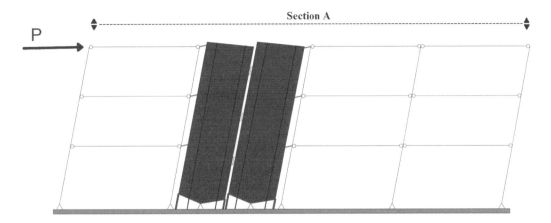

FIGURE 16.3 Schematic deformed elevation view of the model used in this study while exposed to lateral loads.

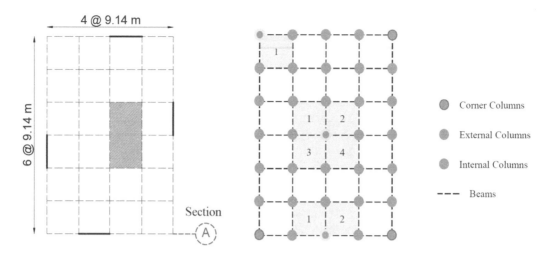

FIGURE 16.4 Plan view of the new 3-story building, rocking walls' location, and design layout of the plan.

only function is to rock until energy-dissipating devices activate and subsequently absorb and dissipate the asserted energy. In this case, considering the fact that in a low-rise building, the pinned concrete rocking wall remains rigid due to being short and thick, and according to research by Grigorian and Grigirian [23], the design of the pinned rocking wall has no impact on the final responses. Therefore, in this study, the rocking shear wall was assumed to be rigid, irrespective of the design.

The described rocking concrete shear wall was implemented in the external frames, as seen in Figure 16.4, with bold lines while dashed lines show released connections. By this layout, the center of the mass and center of the stiffness would locate at one point, which subsequently results in preventing torsion from happening. When this happens, the sway mechanism would be dominant, which is ideal for rocking motion. Additionally, as Figure 16.4 depicts, using this method has made it possible to achieve a uniform system when it comes to designing the steel column sections. According to the figure, they are categorized into three, namely corner, external, and internal columns. This uniformity also reduces execution costs.

The obvious reason behind the change in the cross-section is that the hinged frames only stand the gravity loads (the rocking wall is the only lateral load-carrying element), so the columns with more loading areas are designed differently. In this case (according to Figure 16.4), corner, external, and internal columns each carry one, two, and four areas, respectively. Table 16.3 summarizes the designed sections using the Iranian Seismic Code [24], AISC 360-10 [25], and the earlier-mentioned properties.

The primary model is now built and designed. This parametric study includes altering some of the rocking shear wall's main properties; therefore, the designed frame remained unchanged in all the models. This single design is because of the function of the gravity frames that are solely designed to withstand gravity loads (as their connections are not fixed). Since the changes in the main model only concern the lateral load-bearing capacity, hence the design of the gravity elements

TABLE 16.3
Designed Sections of LA 3-Story

Story	Corner Columns	External Columns	Internal Columns	Beams
	●	●	●	- - -
1st, 2nd, and 3rd	W14X30	W14X43	W14X61	W10X12

TABLE 16.4

Study Models, Their Characteristics, and the Change in Each Model

Model Number	Characteristics				
	PT Distance from the Centerline	PT Relative Length	Friction Damper Stiffness (kN/mm)	Foot Bearing	Wall Width (1st, 2nd, and 3rd Floor) (cm)
1 (Primary)	0.5d	97%	200	Yes	25-25-25
2	0d	97%	200	Yes	25-25-25
3	1d	97%	200	Yes	25-25-25
4	0.5d	98%	200	Yes	25-25-25
5	0.5d	99%	200	Yes	25-25-25
6	0.5d	97%	0	Yes	25-25-25
7	0.5d	97%	60	Yes	25-25-25
8	0.5d	97%	200	No	25-25-25
9	0.5d	97%	200	Yes	25-20-20

would not vary. The general approach to this paper is to change one element of the rocking wall one at a time, while other components remain constant.

Different setups in this investigation comprised the following changes:

1. The distance of the PT tendons from the centerline of the wall
2. Relative length of the PTs
3. Stiffness of the friction dampers
4. The existence of the foot bearings
5. Reduced width of the wall in height

Table 16.4 summarizes each model's properties and indicates whether the targeted property remained unchanged (marked with light gray) or it changed in the study model (marked with dark gray) compared to the primary model (mentioned with Model No. 1 values).

In order to extract the dynamic characteristics of earlier described models, modal analysis was done in the first place. By comparing Tables 16.4 and 16.5, the influence of the changed parameters on the fundamental period and mass participation ratios can be witnessed.

In order to reach the study goals, nonlinear time-history analysis was done with different hazard levels, including 50% in 50, 10% in 50, and 2% in 50 years. The ground motions were presented and modified by SAC according to the project's properties and included two main horizontal directions. In this study, 14 pairs of several ground motions that SAC formerly used were selected and applied to the models. Furthermore, nonlinear direct integration type with the Hilber-Hughes-Taylor method, 5% of damping, and P-delta were other parameters that were considered throughout the analysis. Table 16.6 presents the ground motions used in this study.

TABLE 16.5

Dynamic Characteristics of All Study Models Related to the First Mode

1st mode	Models								
	1 (Primary)	2	3	4	5	6	7	8	9
Fundamental Period (Second)	0.70	0.71	0.67	0.74	0.79	0.97	0.79	0.75	0.84
Mass Participation Ratio (Percent)	84	84	85	84	84	84	84	84	84

TABLE 16.6
Ground Motion Records Used in the Nonlinear Time-History Analysis in This Study

EQ	Description	Earthquake Magnitude	Distance (km)	Scale Factor	Time Step (sec)	PGA (g's)
		10% in 50 years				
1	Imperial Valley, El Centro, 1940	6.9	10.0	2.01	0.020	0.68
2	Imperial Valley, Array 06, 1979	6.5	1.2	0.84	0.010	0.30
3	Northridge, Newhall, 1994	6.7	6.7	1.03	0.020	0.68
4	Northridge, Rinaldi, 1994	6.7	7.5	0.79	0.005	0.58
5	Northridge, Sylmar, 1994	6.7	6.4	0.99	0.020	0.82
		2% in 50 years				
6	Loma Prieta, 1989	7.0	3.5	0.82	0.010	0.47
7	Northridge, 1994	6.7	7.5	1.29	0.005	0.94
8	Northridge, #1, 1994	6.7	6.4	1.61	0.020	1.33
9	Tabas, 1974	7.4	1.2	1.08	0.020	0.99
		50% in 50 years				
10	Coyote Lake, 1979	5.7	8.8	2.28	0.010	0.59
11	Landers, 1992	7.3	64.0	2.63	0.020	0.34
12	Morgan Hill, 1984	6.2	15.0	2.35	0.020	0.55
13	Parkfield, Cholame 8W, 1966	6.1	8.0	2.92	0.020	0.79
14	North Palm Springs, 1986	6.0	9.6	2.75	0.020	0.52

16.4   RESULTS AND DISCUSSIONS

Nonlinear time-history dynamic analysis was performed on Table 16.4's study models using ground motions earlier described in Table 16.6. In order to investigate the impact of altering various specifications in the primarily built model, it was extracted the base shear chart, inter-story displacement curves, and energy absorbed by the structure during each earthquake event. Then the primary model was compared to the models in which the same element was changed (Model 1 was compared to models 2 and 3, 1 to 4 and 5, 1 to 6 and 7, 1 to 8, and 1 to 9).

As can be seen in Figure 16.5, and by comparing the base reaction figures of the primary model (M1) to the ones with changed parameters, placing the post-tensioning tendons at the centerline (M2) would reduce the amount of base reaction while locating them at the edges (M3) would significantly increase this amount. This is logical since, in the latter state, the rocking motion is harder when it is restricted at both ends. By reducing the pre-stressed ratio of the PTs (M4 and M5), although the structure tended to be softer, the amount of base reaction rose gently. Turning to the models with no friction and softer dampers (M6 and M7), it is logical to conclude that the softer the friction damper, the lower the base reaction. This result also complied with the rest of the models where elastic bearings were removed, and the wall's height was reduced in height (M8 and M9), meaning that in both of them, the structures became softer and subsequently recorded less base reaction figures. Reduction in the wall's thickness in height had a higher impact on the model's loss of base reaction than removing the footings due to the reduction in concrete usage, resulting in less weight.

To further assess and compare the performance of the models, the average inter-story displacements in all models have been extracted and shown in Figure 16.6. It can be witnessed in Figure 16.6 (a) that while placing the PT tendons at the center of the wall would make the inter-story displacements more uniform (M2), locating them at the edges would decrease the upper floor's drifts (M3). Regarding changing the PTs' pre-stressed ratios, the results show that reducing the forces in the PTs would lead to more uniform responses (M4 and M5 in Figure 16.6 (b)). Turning to the change in the friction dampers' stiffness, it was evident from the beginning that reducing the stiffness of the

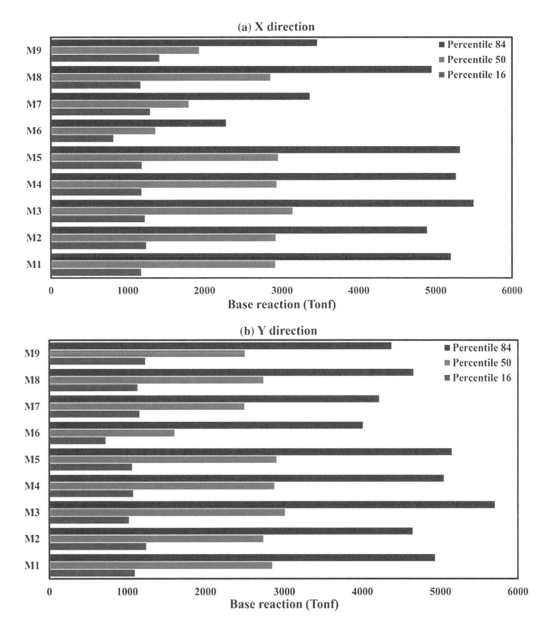

FIGURE 16.5 Base shear comparison of the models in three different percentiles in (a) x- and (b) y-directions.

friction dampers would cause higher inter-story displacements as the structures with such properties would become softer (M7 in Figure 16.6 (c)). However, completely removing the dampers and
using only link beams between the walls significantly increased inter-story displacements (M6 in
Figure 16.6 (c)). Therefore, in this state, the drifts are in borders with the limitation of 5% in regulations, which is unacceptable. Eventually, the models without wall footings (M8) and reduced wall
thickness in height (M9) both positively made the drifts more uniformly distributed. Notably, reducing the walls' thickness made the drifts less when the stories went up (see Figure 16.6 (d)). This is
because not only there is less amount of concrete used in this model (13%), but also the center of the
mass is closer to the ground level, which is seismically more ideal compared to the state in which
there is a massive weight in height and hard to restrain. To draw a general conclusion, it is logical by
looking at the figure that the softer the structures, the more uniform the responses become.

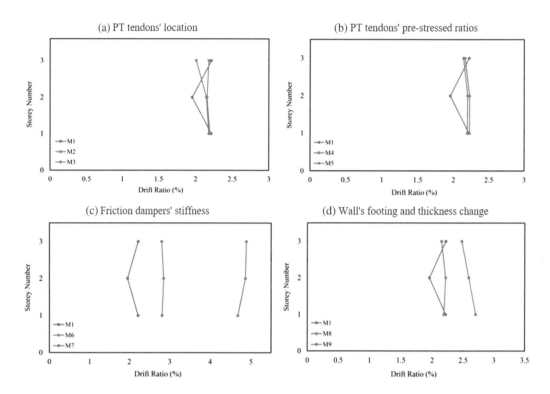

FIGURE 16.6 Average inter-story displacement of the models compared to the primary one (M1) in all the records.

Table 16.7 summarizes the amount of energy absorbed by the models during each ground motion. Furthermore, it can be seen in the last row of the table that some models, i.e., M2, M4, M5, and M6, absorbed almost the same amount of recorded average energy as the primary model with only a slight difference. Compared to M1, the models with PTs at the edges, reduced dampers' stiffness and thickness in height accounted for 3, 6, and 7% less energy absorption, respectively. On the contrary, the model in which the dampers were removed from the space between the walls took, on average, 22% more energy from the seismic energy than the main model.

Now that the results are drawn from the parametric study, it is time to define the indices to find the optimized model. Three indices are taken into account in this study: the index of base reaction, the upper floor's joint displacement, and energy absorbed by the structures. To this end, each data is scaled to the maximum one in the first place, forming a range of numbers from the least to 1. The lowest one represents the best performance in each of the defined indices. Then the three indices are combined using the SRSS method; thus, the lowest combined result is the optimized one. The reason behind choosing the lowest scaling method is that it is assumed in this study that the structures with the lowest upper floors' displacements and lowest energy while being resilient and softer (lowest base shear) are the most optimized ones. This was also because, in the previous sections, it was witnessed that softer structures had more uniform responses to the ground motion records.

$$I_x = \sqrt{I_{Bx}^2 + I_{Dx}^2 + I_{Ex}^2} \qquad (16.1)$$

$$I_y = \sqrt{I_{By}^2 + I_{Dy}^2 + I_{Ey}^2} \qquad (16.2)$$

TABLE 16.7

The Amount of Energy Absorbed by the Models during Each Ground Motion Record (Tons.m²/sec²)

EQ/Model	M1	M2	M3	M4	M5	M6	M7	M8	M9
1	14,587.3	12,721.6	19,342.2	14,925.5	15,271.7	41,076.4	10,247.1	13,711.8	8,949.7
2	36,833.0	39,385.8	33,442.2	37,447.0	37,571.4	26,652.8	39,575.2	38,135.5	40,715.3
3	14,901.3	15,049.9	13,034.8	15,052.3	14,953.3	25,797.7	11,071.2	15,159.7	10,836.3
4	12,991.4	11,381.3	12,989.9	13,213.7	13,334.5	10,474.7	5,111.9	12,143.9	5,712.3
5	4,897.8	5,123.4	4,488.2	48,73.1	4,865.8	3,872.5	5,831.7	4,903.5	6,422.3
6	2,259.2	2,311.1	1,911.1	2,249.1	2,236.2	1,903.5	1,500.2	2,262.8	1,530.8
7	6,748.6	6,670.0	7,503.0	6,791.4	6,834.0	7,722.7	9,812.3	6,684.0	8,748.3
8	14,150.7	14,012.4	1,2625.8	14,187.0	14,207.8	10,564.1	15,014.0	14,281.9	15,382.9
9	5,678.5	5,688.4	4,945.8	5,666.4	5,634.1	10,308.3	4,185.1	5,719.4	4,092.5
10	3,402.2	3,639.4	2,870.3	3,331.4	3,296.0	4,205.0	4,346.3	3,516.9	3,949.2
11	713.3	915.1	480.0	686.7	656.5	866.3	540.9	781.5	727.8
12	792.2	981.9	753.4	775.0	772.8	1,156.2	1,760.2	872.1	1,641.8
13	2,232.3	2,396.2	2,130.9	2,199.0	2,132.5	2,279.9	2,913.2	2,299.8	2,684.6
14	1,328.9	1,465.1	1,554.5	1,320.6	1,335.8	1,953.6	2,321.3	1,308.7	2,094.5
Average	8,679.8	8,695.8	8,433.7	8,765.6	8,793.0	10,631.0	8,159.3	8,698.7	8,106.3
$\left(\dfrac{M_1 - M_n}{M_1}\right)$	0.0%	+0.1%	−3%	+1%	+1.3%	+22%	−6%	+0.2%	−7%

Where I_B, I_D, and I_E stand for the defined indices for base reaction, joint displacement, and energy absorbed by the structure, respectively. I_x and I_y also represent the combined SRSS response of all the mentioned indices in two primary directions. In order to prevent repetition, only the results for the x-direction are presented in this study. Tables 16.8–16.10 show the scaled numbers of base reaction, joint displacement, and energy absorbed by the models in the x-direction in every ground motion record.

TABLE 16.8

Scaled Base Reaction Indices of All Models in Each Ground Motion Record in X-Direction

EQ / I_{Bx}	M1	M2	M3	M4	M5	M6	M7	M8	M9
1	0.838	0.842	1.000	0.845	0.847	0.506	0.724	0.801	0.688
2	0.962	0.964	1.000	0.961	0.960	0.700	0.797	0.940	0.892
3	0.964	0.942	1.000	0.968	0.971	0.496	0.835	0.937	0.822
4	0.983	1.000	0.978	0.980	0.974	0.379	0.894	0.973	0.936
5	0.985	0.926	0.966	0.993	1.000	0.397	0.445	0.946	0.493
6	0.867	0.777	1.000	0.884	0.899	0.586	0.330	0.814	0.376
7	0.973	1.000	0.956	0.959	0.956	0.346	0.859	0.974	0.917
8	0.985	0.927	0.965	0.993	1.000	0.374	0.444	0.946	0.493
9	0.971	1.000	0.971	0.981	0.986	0.378	0.384	0.953	0.477
10	0.882	0.911	1.000	0.891	0.901	0.462	0.869	0.872	0.883
11	0.893	1.000	0.739	0.868	0.848	0.558	0.537	0.920	0.721
12	0.785	0.821	0.848	0.776	0.765	0.444	0.935	0.788	1.000
13	0.993	0.946	0.970	0.994	1.000	0.480	0.615	0.959	0.694
14	0.751	0.689	0.877	0.772	0.789	0.571	0.953	0.681	1.000

TABLE 16.9
Scaled Joint Displacement Indices of All Models in Each Ground Motion Record in X-Direction

EQ / I_{Dx}	M1	M2	M3	M4	M5	M6	M7	M8	M9
1	0.577	0.598	0.602	0.575	0.572	1.000	0.761	0.580	0.640
2	0.441	0.459	0.413	0.438	0.435	1.000	0.563	0.449	0.548
3	0.616	0.644	0.547	0.611	0.606	1.000	0.800	0.627	0.765
4	0.745	0.784	0.639	0.738	0.730	1.000	0.959	0.762	0.954
5	0.727	0.745	0.599	0.719	0.714	1.000	0.467	0.737	0.487
6	0.436	0.416	0.441	0.439	0.441	1.000	0.253	0.428	0.255
7	0.741	0.775	0.637	0.735	0.727	1.000	0.948	0.758	0.947
8	0.711	0.729	0.586	0.704	0.699	1.000	0.456	0.721	0.476
9	0.783	0.812	0.693	0.781	0.777	1.000	0.495	0.776	0.580
10	0.617	0.634	0.588	0.615	0.614	1.000	0.790	0.620	0.724
11	0.516	0.597	0.393	0.497	0.477	1.000	0.469	0.558	0.564
12	0.516	0.566	0.505	0.505	0.493	1.000	0.937	0.542	0.888
13	0.572	0.596	0.516	0.567	0.560	1.000	0.573	0.583	0.589
14	0.545	0.528	0.539	0.553	0.560	0.875	1.000	0.523	0.968

Now that all the data is scaled to the max amount in each earthquake record, indices in both directions can be calculated using Equations (16.1) and (16.2). Tables 16.11 and 16.12 summarize the combined indices in both directions using the above-mentioned equations. As seen in the figures, shaded areas represent the lowest index in each ground motion record. Most of the shaded areas are recorded to be in the last four models, especially the ones with changed friction dampers' stiffness.

By taking both directions into account using the following equation, the final index can be drawn, as shown in Table 16.13.

$$I_T = \sqrt{I_x^2 + I_y^2} \tag{16.3}$$

TABLE 16.10
Scaled Energy Absorption Indices of All Models in Each Ground Motion Record in X-Direction

EQ / I_{Ex}	M1	M2	M3	M4	M5	M6	M7	M8	M9
1	0.824	0.836	1.000	0.840	0.807	0.779	0.915	0.824	0.738
2	0.662	0.606	0.819	0.676	0.690	1.000	0.852	0.630	0.699
3	0.839	0.871	0.761	0.835	0.830	0.837	1.000	0.853	0.959
4	0.812	0.839	0.689	0.811	0.811	0.464	0.933	0.841	1.000
5	0.992	0.992	0.717	0.986	0.974	0.671	0.340	1.000	0.401
6	0.576	0.493	0.705	0.591	0.608	1.000	0.141	0.537	0.161
7	0.797	0.985	0.694	0.810	0.810	0.429	0.913	0.855	1.000
8	0.955	0.978	0.747	0.978	0.965	0.634	0.342	1.000	0.405
9	0.999	0.981	0.849	1.000	0.995	0.559	0.288	0.976	0.411
10	0.882	0.931	0.786	0.867	0.861	0.802	1.000	0.906	0.917
11	0.739	1.000	0.458	0.701	0.663	0.871	0.493	0.819	0.763
12	0.439	0.514	0.465	0.432	0.432	0.570	0.908	0.472	1.000
13	0.978	1.000	0.922	0.966	0.960	0.990	0.756	0.990	0.855
14	0.415	0.362	0.400	0.422	0.428	0.601	1.000	0.404	0.832

TABLE 16.11
Combined Indices in x-direction Using Equation (16.1)

EQ / I_x	M1	M2	M3	M4	M5	M6	M7	M8	M9
1	1.309	1.329	1.537	1.323	1.302	1.364	1.393	1.287	1.195
2	1.248	1.228	1.357	1.254	1.260	1.578	1.296	1.218	1.259
3	1.419	1.435	1.370	1.417	1.414	1.395	1.529	1.414	1.476
4	1.477	1.522	1.356	1.471	1.463	1.166	1.609	1.495	1.669
5	1.576	1.549	1.344	1.574	1.568	1.268	0.729	1.562	0.801
6	1.129	1.010	1.301	1.150	1.171	1.531	0.439	1.065	0.482
7	1.460	1.603	1.342	1.455	1.449	1.142	1.571	1.502	1.655
8	1.545	1.532	1.354	1.562	1.555	1.242	0.722	1.554	0.796
9	1.598	1.619	1.464	1.604	1.602	1.206	0.690	1.569	0.855
10	1.391	1.448	1.401	1.387	1.390	1.363	1.543	1.402	1.465
11	1.269	1.535	0.955	1.222	1.177	1.439	0.866	1.352	1.191
12	1.037	1.122	1.091	1.022	1.007	1.234	1.605	1.066	1.670
13	1.507	1.500	1.434	1.498	1.495	1.486	1.130	1.497	1.249
14	1.016	0.940	1.104	1.039	1.058	1.205	1.705	0.949	1.622

Based on what was seen in this study and Table 16.13, the model with reduced dampers' stiffness (M7), despite being softer than the primary model, had the least final index. This means that M7 had relatively lower joint displacements, less energy absorption, and was more resilient than the others, thus is the final optimized model.

16.5 FURTHER DISCUSSIONS

To complete the study, a more optimized model in which all the efficient characters are combined into a single one will be proposed based on the results from the parametric study. Then the given model will be evaluated to see whether the method is accurate. The OptRockWall model is proposed

TABLE 16.12
Combined Indices in y-direction Using Equation (16.2)

EQ / I_y	M1	M2	M3	M4	M5	M6	M7	M8	M9
1	1.388	1.371	1.215	1.388	1.396	1.199	1.542	1.356	1.691
2	1.549	1.644	1.164	1.526	1.502	1.181	0.811	1.579	0.974
3	1.096	1.040	1.358	1.111	1.126	1.341	1.527	1.055	1.358
4	1.479	1.458	1.472	1.497	1.512	1.402	1.473	1.450	1.489
5	0.998	0.986	1.097	1.002	1.004	1.655	1.017	0.976	1.000
6	0.923	0.813	1.111	0.942	0.960	1.630	0.840	0.864	0.796
7	1.465	1.446	1.470	1.491	1.509	1.378	1.479	1.455	1.491
8	1.010	0.997	1.100	1.012	1.015	1.647	1.027	0.986	1.010
9	1.336	1.050	1.545	1.377	1.410	1.439	0.809	1.205	0.752
10	0.966	1.043	0.750	0.947	0.931	1.602	1.299	0.981	1.270
11	1.385	1.391	1.263	1.390	1.387	1.545	1.236	1.393	1.313
12	0.961	1.177	0.937	0.941	0.931	1.340	1.629	1.037	1.382
13	1.008	1.123	0.927	0.995	0.974	1.064	1.732	1.035	1.480
14	1.350	1.423	1.417	1.346	1.351	1.517	1.386	1.315	1.362

TABLE 16.13

The Final index, Showing the SRSS Answers of All Responses in Both Directions in Each Earthquake Record

EQ / I_T	M1	M2	M3	M4	M5	M6	M7	M8	M9
1	1.908	1.910	1.959	1.917	1.909	1.816	2.068	1.870	2.081
2	1.990	2.052	1.788	1.975	1.960	1.971	1.519	1.994	1.592
3	1.793	1.772	1.929	1.800	1.807	1.935	2.151	1.764	2.016
4	2.091	2.108	2.001	2.098	2.104	1.823	2.171	2.082	2.247
5	1.865	1.836	1.734	1.866	1.862	2.085	1.242	1.842	1.291
6	1.026	0.912	1.206	1.046	1.065	1.580	0.630	0.965	0.639
7	2.068	2.159	1.991	2.083	2.092	1.790	2.148	2.091	2.238
8	1.846	1.828	1.745	1.861	1.857	2.063	1.246	1.840	1.296
9	2.083	1.930	2.128	2.114	2.134	1.878	1.053	1.979	1.159
10	1.694	1.784	1.589	1.680	1.673	2.103	2.007	1.711	1.949
11	1.878	2.072	1.583	1.851	1.819	2.111	1.509	1.941	1.773
12	1.414	1.626	1.438	1.389	1.371	1.821	2.277	1.488	2.167
13	1.813	1.874	1.708	1.798	1.785	1.828	2.058	1.820	1.936
14	1.690	1.705	1.796	1.700	1.716	1.938	2.187	1.622	2.138
Average	1.797	1.826	1.757	1.799	1.797	1.910	1.733	1.786	1.754

with the characteristics given in Table 16.14, and the reason each feature is chosen to be implemented on the proposed model is also mentioned.

The model is built using the information in Table 16.14, Sections 16.2 and 16.3, with precisely the same setting for plans, elevations, modeling details, design, analysis, ground motions, assumptions, etc. Then the results were extracted and compared to the others in terms of base reaction, joint displacements, and energy absorption.

In Figures 16.7 and 16.8 and Table 16.15, the results of the nonlinear time-history analysis regarding OptRockWall's model and its comparison to the former optimized one can be seen. The proposed OptRockWall recorded, on average, the same amounts of inter-story displacements but with lower- and upper-story displacements, which means this model had good control over vibration and did not pass it to the upper floors. This was also true about the average amount of base reaction. However, according to Table 16.15, the OptRockWall benefits from 13% less concrete usage, which makes this model lighter than the M7. As a result, the OptRockWall had a higher performance level than the M7. Turning to the energy absorbed by the structures, although the M7 showed it could absorb 6% less energy from seismic excitations, the OptRockWall recorded even less energy

TABLE 16.14

Proposed Model's Specifications Based on the Parametric Study

Model	Characteristics				
	PT distance from the centerline	PT relative length	Friction damper stiffness (kN/mm)	Foot bearing	Wall width (1st, 2nd, and 3rd floor) (cm)
OptRockWall	1d	97%	100	No	25-20-20
Reason	Highest base shear, upper floors control	Drift control (to stay within the code's range)	Less energy absorption, Drift uniformity	Fewer costs, Easier to execute, drift uniformity	13% less concrete, better seismic performance, less energy absorption

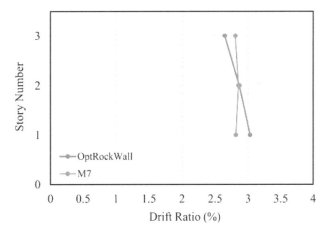

FIGURE 16.7   Inter-story displacement comparison between the OptRockWall and M7.

absorption by 10% less than the primary model. It is logical to conclude that the optimized characteristics are well projected in the OptRockWall because of this study.

The three defined indices (i.e., I_B, I_D, and I_E) used in the previous section were calculated once more for the new OptRockWall and can be seen and compared in Table 16.16, where the final index regarding the OptRockWall stood lower than every model's I_T. This means that the parametric study and the method in finding an optimized model are accurate.

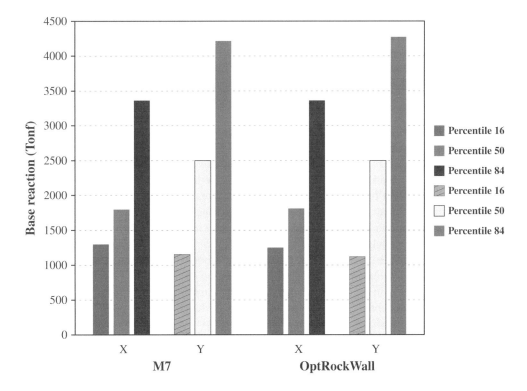

FIGURE 16.8   Base reaction comparison between the OptRockWall and M7.

TABLE 16.15

Comparison of Different Components among the Primary, Previously Optimized, and OptRockWall Models

Model/Term	Fundamental Period	Mass Participation Ratio	Amount of Concrete Used	Energy Absorption by the Model	Energy Absorption Difference $\left(\dfrac{M_1 - M_n}{M_1} \right)$
M1	0.70 sec	First mode 84%	235 tons	8679.8 tons.m²/sec²	0.0%
M7	0.79 sec	First mode 84%	235 tons	8159.3 tons.m²/sec²	−6%
OptRockWall	0.80 sec	First mode 85%	200 tons	7853.3 tons.m²/sec²	−10%

TABLE 16.16

Calculated Indices Using Equations (16.1)–(16.3) and Compared Optimized Models with the Primary One

Model/Index	Average I_B		Average I_D		Average I_E		I_x	I_y	I_T
	x	y	x	y	x	y			
M1	0.917	0.886	0.610	0.506	0.779	0.600	1.345	1.180	1.801
M7	0.687	0.805	0.677	0.658	0.720	0.714	1.203	1.261	1.742
OptRockWall	0.680	0.809	0.673	0.665	0.628	0.645	1.144	1.224	1.675

16.6   CONCLUSION

This chapter is allocated to a parametric study of the seismic behavior of a low-rise building in which a hybrid self-centering wall with different properties is connected to a gravity frame. Nonlinear time-history analysis was done to reach the study goals, and parameters, including PT forces, their location in the rocking wall, dampers' stiffness, the absence or presence of foot bearings, and thickness of the wall in height, were investigated. The results are as follows:

- The influence of changing the location of the PTs in the self-centering wall was investigated, and it was seen that although placing the PTs at the centerline did not affect the amount of base reaction, locating them at the edges increased this amount by 7%. Furthermore, by locating the PTs at the centerline, the inter-story displacements became more uniformly distributed at almost the same drift rates as the primary model and M3. It is worth mentioning that putting the PTs at the edges decreased the upper floor's displacement compared to the 1st story; hence it is seismically more ideal. Turning to the energy absorbed by the models, M3 recorded 3% less energy absorption, while the change in the M2 slightly raised the share of energy received by earthquakes. In the end, M3, among the other models, accounted for the best performance based on the results.
- Changing the forces in the PTs also were investigated in this chapter, and it was witnessed that decreasing the forces slightly (from 97% of relative length to 98% and 99%) influenced little on the amount of base reaction as well as the energy absorption. However, this change was positive when it came to the inter-story displacements, where M4 and M5 showed more uniform responses but with more drift ratios (3–4%).
- Removing the dampers between the coupled self-centering wall and using link beams instead halved the amount of base reaction while the inter-story displacements became twice. This model absorbed 22% more seismic energy than the base model. Decreasing

the stiffness of the dampers (to 60 kN/mm) also had almost the same effect, with 25% less base reaction and 30% more drifts but with 6% less energy absorption. It was concluded from this investigation that the presence of the damper with a reasonable amount of stiffness would be crucial to control the displacements and absorbing less energy.

- Removing the walls' foot bearings did not influence the responses much. It is advisable in the pinned rocking walls not to use such footings since they are challenging to execute, costly and unnecessary (foot edges are immune to possible damage).
- Turning to the change in the thickness of the wall, 30 less base reactions and 20% more drifts were observed. However, this model showed 7% less energy absorption and 13% less concrete usage. Moreover, reducing the thickness in height in the self-centering wall made the center of the mass become closer to the ground level, and as a result, the rooftop's displacements became less compared to the 1st two stories. Less concrete usage also means less load from lateral loads and less cost for materials and execution.
- With regard to the optimization method given in this chapter, it should be noted that three indices (I_B, I_D, and I_E) were used to represent the behavior in each model, and then the SRSS method was deployed to extract unique answers for every single model. The results indicated that the M7 indicated the best performance when subjected to the same series of ground motion records since it not only became softer and resilient but, more importantly, recorded less energy absorption and more uniform inter-story displacements.
- It was tried in this chapter to evaluate the optimization method by combining the results from the previous models into one final model, as the OptRockWall. Most efficient elements were picked from each parameter, and then the newly generated model was analyzed with the same specifications as the previous ones. The results showed that the new OptRockWall model recorded the minimum amount of final index, recording better performance even than the previously announced optimized model, i.e., M7. The OptRockWall preserved all the desirable behavior of each parameter, taking 10% less energy from seismic excitations, using 13% less concrete, displaying better inter-story displacements distributions, and finally, having less rooftop displacement.

REFERENCES

[1] C. Zhong and C. Christopoulos, "Self-centering seismic-resistant structures: Historical overview and state-of-the-art," Earthq. Spectra, vol. 38, no. 2, pp. 1321–1356, 2022.

[2] N. Makris and M. Aghagholizadeh, "Effect of supplemental hysteretic and viscous damping on rocking response of free-standing columns," J. Eng. Mech., vol. 145, no. 5, p. 4019028, 2019, doi: 10.1061/(asce)em.1943-7889.0001596.

[3] G. Ríos-García and A. Benavent-Climent, "New rocking column with control of negative stiffness displacement range and its application to RC frames," Eng. Struct., vol. 206, p. 110133, 2020, doi: 10.1016/j.engstruct.2019.110133.

[4] J. L. Beck and R. I. Skinner, "The seismic response of a reinforced concrete bridge pier designed to step," Earthq. Eng. Struct. Dyn., vol. 2, no. 4, pp. 343–358, 1973, doi: 10.1002/eqe.4290020405.

[5] L. Xu, X. Lu, Q. Zou, L. Ye, and J. Di, "Mechanical behavior of a double-column self-centering pier fused with shear links," Appl. Sci., vol. 9, no. 12, p. 2497, 2019, doi: 10.3390/app9122497.

[6] M. J. N. Priestley, "Overview of PRESSS research program," PCI J., vol. 36, no. 4, pp. 50–57, 1991, doi: 10.15554/pcij.07011991.50.57.

[7] T. Guo, Z. Xu, L. Song, L. Wang, and Z. Zhang, "Seismic resilience upgrade of RC frame building using self-centering concrete walls with distributed friction devices," J. Struct. Eng., vol. 143, no. 12, p. 4017160, 2017, doi: 10.1061/(asce)st.1943-541x.0001901.

[8] L. Xu, S. Xiao, and Z. Li, "Hysteretic behavior and parametric studies of a self-centering RC wall with disc spring devices," Soil Dyn. Earthq. Eng., vol. 115, pp. 476–488, 2018.

[9] P. Mottier, R. Tremblay, and C. Rogers, "Shake table test of a two-story steel building seismically retrofitted using gravity-controlled rocking braced frame system," Earthq. Eng. Struct. Dyn., vol. 50, no. 6, pp. 1576–1594, 2021.

[10] M. Piri and A. Massumi, "Seismic performance of steel moment and hinged frames with rocking shear walls," J. Build. Eng., 2022, doi: 10.1016/j.jobe.2022.104121.

[11] A. Massumi, M. Piri, and M. Nematnezhad, "Enhancing the performance of moment frames using a dumbbell-shaped rocking shear wall with additional devices," 2022.

[12] N. Lijun and W. Zhang, "Experimental Study on a Self-Centering Earthquake-Resistant Masonry Pier with a Structural Concrete Column," Adv. Mater. Sci. Eng., vol. 2017, pp. 1–15, 2017, doi: 10.1155/2017/6379168.

[13] Z. Li et al., "Lateral Performance of Self-Centering Steel–Timber Hybrid Shear Walls with Slip-Friction Dampers: Experimental Investigation and Numerical Simulation," J. Struct. Eng., vol. 147, no. 1, p. 4020291, 2021, doi: 10.1061/(asce)st.1943-541x.0002850.

[14] Y. Pang, W. He, and J. Zhong, "Risk-based design and optimization of shape memory alloy restrained sliding bearings for highway bridges under near-fault ground motions," Eng. Struct., vol. 241, p. 112421, 2021, doi: 10.1016/j.engstruct.2021.112421.

[15] A. Hashemi, P. Zarnani, R. Masoudnia, and P. Quenneville, "Experimental testing of rocking cross-laminated timber walls with resilient slip friction joints," J. Struct. Eng., vol. 144, no. 1, p. 4017180, 2018, doi: 10.1061/(asce)st.1943-541x.0001931.

[16] P. Chi, T. Guo, Y. Peng, D. Cao, and J. Dong, "Development of a self-centering tension-only brace for seismic protection of frame structures," Steel Compos. Struct., vol. 26, no. 5, pp. 573–582, 2018, doi: 10.12989/scs.2018.26.5.573.

[17] S. Kitayama and M. C. Constantinou, "Fluidic Self-Centering Devices as Elements of Seismically Resistant Structures: Description, Testing, Modeling, and Model Validation," J. Struct. Eng., vol. 143, no. 7, p. 4017050, 2017, doi: 10.1061/(asce)st.1943-541x.0001787.

[18] T. C. Steele and L. D. A. Wiebe, "Large-Scale Experimental Testing and Numerical Modeling of Floor-to-Frame Connections for Controlled Rocking Steel Braced Frames," J. Struct. Eng., vol. 146, no. 8, p. 4020163, 2020, doi: 10.1061/(asce)st.1943-541x.0002722.

[19] S. Polyakov, Design of earthquake resistant structures. Mir Publishers, 1974.

[20] M. NourEldin, A. Naeem, and J. Kim, "Life-cycle cost evaluation of steel structures retrofitted with steel slit damper and shape memory alloy-based hybrid damper," Adv. Struct. Eng., vol. 22, no. 1, pp. 3–16, 2019, doi: 10.1177/1369433218773487.

[21] C. Zhong and C. Christopoulos, "Shear-controlling rocking-isolation podium system for enhanced resilience of high-rise buildings," Earthq. Eng. Struct. Dyn., vol. 51, no. 6, pp. 1363–1382, 2022, doi: 10.1002/eqe.3619.

[22] Fedral Emergency Management agency (FEMA), "FEMA 355C - State of the Art Report on Systems Performance of Steel Moment Frames Subject to Earthquake Ground Shaking," Rep. No. FEMA-355C, Federal Emergency Management Agency, Washington, DC. Taylor & Francis Washington, DC, p. 344, 2000.

[23] C. E. Grigorian and M. Grigorian, "Performance control and efficient design of rocking-wall moment frames," J. Struct. Eng., vol. 142, no. 2, p. 4015139, 2016, doi: 10.1061/(asce)st.1943-541x.0001411.

[24] "Building and Housing Research Center, Standard No. 2800. Iranian Code of Practice for Seismic Resistant Design of Buildings, Third Revision, Building and Housing Research Center, Tehran.," Build. Hous. Res. Cent. Tehran, Iran, 2005.

[25] A. I. of S. Construction, "Specification for Structural Steel Buildings," Am. Inst. Steel Constr. Chicago-Illinois, pp. 1–612, 2010.

17 Review of the Use of Waste Resources in Sustainable Concrete

Aggregate Types, Recycling Procedures, Fresh and Hardened Properties, Structural Characteristics, and Numerical Formulation

Arash Karimi Pour and Ehsan Noroozinejad Farsangi

CONTENTS

17.1 INTRODUCTION

Large amounts of trash are dumped into the environment daily, harming both the ecosystem and human health [1, 2]. Global solid waste generation is accelerating as business and technology advance and the human population rises [3–5]. Yearly, about 5 billion tons of carbon dioxide are emitted from factories to produce construction materials. Additionally, the claim for waste aggregates to produce concrete aggregate is about 4,300 million worldwide [6–8]. In addition, the destruction and upkeep of old buildings as well as the building of new ones produce a lot of garbage after the construction chain [9, 10]. For example, the dumped waste materials increased yearly by 14 million, and 15 million tons in the United States and China, respectively [11, 12]. Therefore, inappropriate waste materials resource disposal is a major issue that many nations face. All buildings and devastation leftover must be transferred to designated landfills [13, 14]. Even while recycling and reuse are excellent substitutes to waste resources, only 30% of the entire waste aggregate volume is assigned to recycling, for instance [15]. However, waste aggregate recycling is only about 80% in nations like Japan that lack natural resources and land. Some nations are charging for landfill dumping.

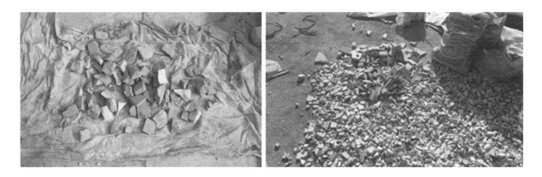

FIGURE 17.1   Example from used recycled aggregates. (Provided by current authors.)

New Zealand implemented a $10/tons fee in 2008 as landfill sites became increasingly limited [16]. Another motivation for recycling is when non-hazardous construction and devastating building materials are recycled or repurposed, buildings that are under construction can receive one to two points toward their certification [17]. Despite the modest amount of waste materials recycling, this procedure can produce a substance referred to as recycled particles [18, 19]. This substance could be viewed as a sustainable response to the depletion of ordinary sources and the rising need for land for waste resources. Therefore, a variety of subjects are covered in the current literature review, including the characterization of waste aggregates, the recycling procedure, the impact of coarse waste aggregates in concrete, and several strategies for enhancing the characteristics of concrete produced by waste [20, 21]. The structural characteristics of recycled concrete are also offered, along with the current standards for concrete produced by recycled aggregates. This study might help engineering, standards, researchers, and agencies understand recycled aggregate concrete better, promoting its custom and bringing about significant environmental assistance. Several elements, including architectural methods and waste-dumped materials, can have an impact on waste aggregate composition. However, most of the weight of waste aggregates is made up of concrete, mortar, and ceramics [22]. An example of common waste granite aggregates widely used in the Middle East is presented in Figure 17.1, which was provided by the authors.

The Brazilian National Council for the Environment is categorizing waste aggregates into four main groups, as shown in Figure 17.2. The waste aggregate arrangement in the United States covers the types mentioned in Figure 17.2 as well as recovered construction materials such as wood and

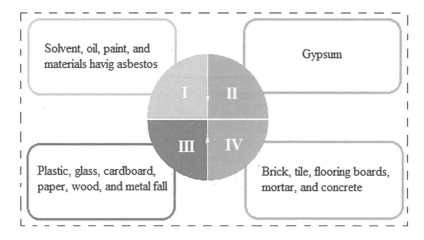

FIGURE 17.2   Classification of waste aggregates for concrete construction. (Provided by current authors.)

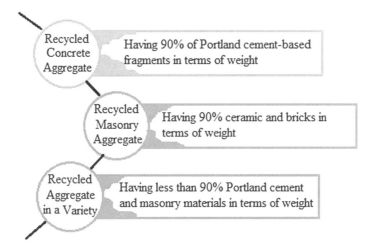

FIGURE 17.3 Main sources of production of recycled aggregates from waste materials resources. (Provided by current authors.)

glass [23]. But the main waste materials resources converted into various forms of recycled aggregates are crushed concrete and masonry materials [24], as shown in Figure 17.3.

Therefore, glass and plastic can be used as substitutes for natural coarse and fine aggregates The concrete's compressive strength as well as several long-term qualities like permeability and resistance to chloride ion penetration could be enhanced with glass and plastic recycled coarse aggregates incorporation [25–27]. Additionally, the pozzolanic activity is anticipated to be significantly increased if the glass could be ground even greater [28–30]. While because of the poor link among plastic particles and concrete matrix as well as the low strength and low elastic moduli, the mechanical characteristics of concrete containing plastic coarse aggregates decrease when the substitution ratio rises [31, 32]. Recycled fine aggregates could also be utilized in for rigid concrete runways where good water drainage is required because of the increase in water absorption [33]. While concrete with inadequate characteristics may result from the short bonding between these mechanisms and the paste matrix [34].

17.2 REUSING WASTE AGGREGATES PROCEDURE

The quantity of recycling phases and the method of crushing have an impact on the properties of coarse and fine recycled aggregates [35, 36]. The recycling process also enhances aggregate shape, resulting in rounder and less acute particles, due to impact and peeling-off effects [24, 37, 38]. The adhering mortar is ground into powder using a mechanical crushing procedure that involves a rapid revolving eccentric gear [39]. According to the highest particle size and the anticipated value of the finished product, various recycled processes can be used. The regular recycling procedure includes various steps of crushing, screening, and categorization to get rid of impurities like glass and plastic. This approach uses a jaw crusher as the primary crusher, which can handle huge chunks of concrete and any remaining rebars. The residual material is then sieved after the iron leftovers are detached utilizing a magnetic sieve. In a subordinate crusher, like an impression or rotary crusher, aggregate bigger than 20 mm were thus once more crumpled. If required, the subordinate crushing might be repeated. Figure 17.4 shows a schematic illustration of the regular recycling procedure [16, 40–42]. However, other mechanical techniques enable recycled aggregates of superior quality with properties comparable to those of conventional ones.

An eccentric rotor, a screw rod, or a better jaw crusher are used in place of the impact or rotary crusher in these techniques [43]. The input material is better processed using this more complicated

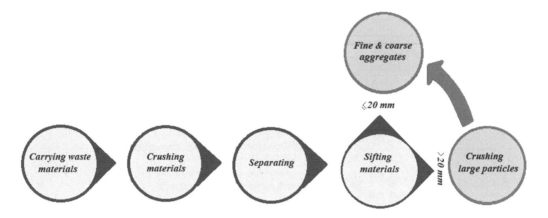

FIGURE 17.4   Illustration of the regular recycling procedure. (Provided by current authors.)

mechanical method, which also removes attached mortar and reduces aggregate size through constant friction. In addition to mechanical grinding, heating the recycled aggregate is another option for weakening the mortar that has been attached. The attached mortar develops cracks because of the large temperature throughout the thermal procedure, losing mechanical strength as a result and being simply detachable from the unique aggregate surface [44–46]. After thermal action up to 700°C in a rotary oven, the heating and rotting process yields waste aggregates with just 2% of attached concrete paste [47]. A technique called heating and rubbing was introduced in 2005. This process involves heating rough-crushed concrete smithereens slighter than 40 mm to around 300°C in a fuel heater for 40 to 60 minutes, followed by crushing and rubbing in a tube mill [48]. In 2011, an impression pounder and specialized microwave tandoor were used in place of a kerosene furnace to optimize this procedure [49]. This innovative method uses less energy because the material is heated for only two minutes. An illustration of these heat-based recycling techniques is demonstrated in Figure 17.5.

Regardless of the type of aggregates used in concrete, microwave flagging pre-treatment is operative [50]. It also works when the solid is only uncovered to brief microwave boiler energy. In general, a greater temperature makes it easier to remove the attached mortar; however, when the temperature exceeds 500°C, the qualities of recycled aggregates may be compromised. Thermal grinding produces excellent results, but it also uses a lot of energy and emits more carbon dioxide. Thermal expansion is a sophisticated wet recycling method that uses water. In this technique, waste materials are submerged in water for 2 hours to completely soak the mortar that has been attached [51–53]. Specimens are then desiccated at roughly 500°C for 2 hours before being immersed in cold water. While immediate cooling generates stress and, as a result, cracks in the adhered paste,

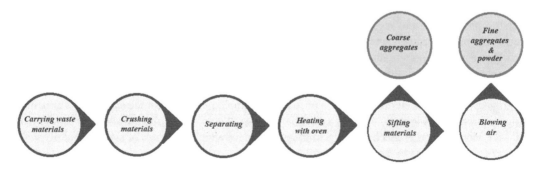

FIGURE 17.5   Illustration of the heating recycling procedure. (Provided by current authors.)

which may then be simply repaired, heating stimulates the creation of water vapor in the soaked attached mortar [54]. Autogenous cleaning, which involves recycled aggregates crashing into one another in a revolving mill drum, is another wet approach. The remaining contaminants and fines are then removed from the material by cleaning it with water and drying it [55]. The technique is more expensive even though these wet recycling procedures successfully eliminate the mortar that has stuck to it.

17.3 PRIOR EXPERIMENTAL RESULTS

17.3.1 FRESH-STATE CHARACTERISTICS

The workability is directly impacted by the shape of the coarse aggregates. Concrete created by smoothed particles uses fewer mixing water than concrete prepared by angular pebbles to achieve the same workability [56, 57]. Since the W/C fraction significantly influences the compressive resistance, spherical aggregates are therefore more favorable. Additionally, particles that are elongated, uneven, and lamellar have bigger specific surfaces. They may become trapped between the reinforced rebars, causing the concrete to fill in irregularly and form concreting gaps, a condition that is extremely detrimental to the functionality of structural concrete components [58, 59]. As opposed to spherical aggregates, however, angular particles provide greater mortar adherence. Thus, using angular aggregates results in more durable concretes for the same water-to-cement fraction while ensuring correct casting and compaction. The shape of coarse aggregates consequently has two opposing characteristics: the rounded shape, which improves workability but decreases adhesion, and the angular shape, which has inferior flowability but strengthens the binding among aggregates and paste. In general, recycled aggregates have a higher form index and are flatter and longer than natural aggregates [60]. The fact that recycled aggregates might absorb more water from cement paste and, as a result, produce concrete with poor workability and less control over the effective water-to-cement fraction in the cement paste is another crucial factor to consider [61]. This greater water absorption capacity of recycled concrete aggregate can be attributed to the material's adhesion to a mortar that has pores and micro-cracks [62, 63]. Recycled masonry aggregate and varied recycled aggregate also absorb more water than regular aggregate because ceramic particles are inherently extra absorbent than natural aggregates [64]. As a result, before mixing, recycled aggregates are saturated to surge both the workability and compressive resistance of hardened concrete [65]. The moisture situations, form, and substitution fraction have an impact on the flowability of concretes constructed with waste aggregates [62]. Concrete loses cohesiveness as the replacement ratio rises, which has an impact on the similarity of the new concrete throughout molding.

17.3.2 HARDENED-STATE CHARACTERISTICS

In conventional concrete, the interfacial transition zone among the concrete paste and aggregate is a two-phase compound. The interfacial transition zone acts as a link between the matrix and the aggregate [66, 67]. As a result of the voids and micro-cracks in the interfacial transition zone, which prevent stress from being transferred, the toughness of the concrete might be small even though the individual components are stiff. Additionally, the mechanical resistance of the aggregate itself is hampered by this pore arrangement. Recycled aggregate has a larger aggregate devastating index, larger Los Angeles abrasion index, and minor resistance to crushing and degradation [38, 62, 63, 68]. For illustration, Figure 17.6 contrasts a few characteristics between natural granitic coarse aggregate and recycled concrete aggregates [69].

At higher ages, the hydration process continues and produces an increasing amount of calcium silicate hydrates, making this impact on porosity less significant. According to Kwan et al. [70], capillary gaps get smaller and microstructure gets denser as age. These findings indicated that the structural integrity of recycled aggregate concretes is not directly impacted by the presence of

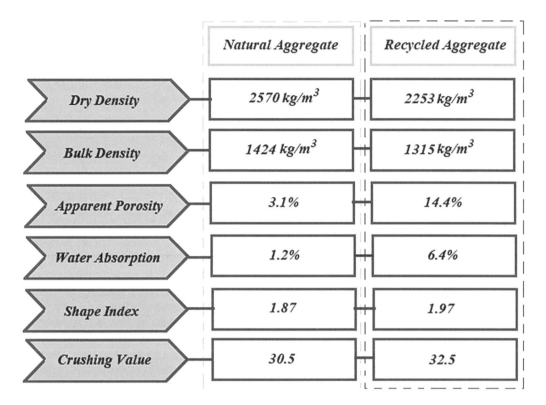

FIGURE 17.6 Comparison between a few characteristics of natural and recycled aggregate. (Provided by current authors.)

significant voids or cracks. It has been well-researched how employing recycled aggregate in place of natural coarse aggregate influences the mechanical characteristics of concrete. The compressive resistance of recycled aggregate concrete rises with age, like conventional concrete [71–73]. However, the compressive resistance at a given age often declines as the recycled aggregate substitution fraction rises. Nevertheless, the findings of the numerous research that have already been published concerning the compressive resistance of concrete made with recycled particles vary greatly. This substantial variety can be explained by two key factors. First, depending on the value of the mechanical qualities of the particle. Additionally, various investigations typically use diverse approaches, such as using concretes by the same W/C fractions or concretes with the same workability. The deductions also depend on the form, magnitude, content, and mechanical characteristics of the recycled aggregates [74].

According to numerous research, as the substitution fraction of recycled aggregate rises, so does the tensile resistance. For larger substitution fractions, this drop can reach 13% [75–81]. The utilization of recycled aggregates in structural parts is directly impacted by this decrease in tensile strength. In comparison to beams built with normal concrete, recycled concrete beams have lower shear resistance and flexural cracking loads. When Bravo et al. [43] examined the mechanical characteristics created from recycled aggregate from various sites in Portugal, they found that particles with a larger percentage of clay produced the lowest results. The clay surrounds the grains of recycled aggregate and inhibits the bond among the aggregate and matrix. Based on findings from multiple authors [38, 55, 60, 62, 70–130], Figure 17.7 shows the relationship between compressive resistance and recycled substitution fraction for various concrete classes. Larger concrete grades and particularly higher recycled aggregate percentages show a significant difference in the outcomes. The results start to show more variation for C35 to C60 over 50% recycled aggregate. For C25 to C30

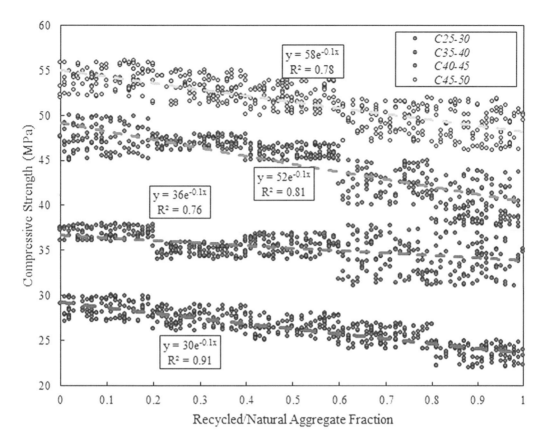

FIGURE 17.7   Relation among the compressive resistance and recycled/natural aggregate fraction as per previous experimental reports. (Provided by current authors.)

concrete grade, this expressive variation does not exist; the results vary roughly continuously independent of the recycled aggregate percentage. It is feasible to see a decline in compressive resistance for all varieties of concrete as the recycled aggregate ratio rises. But for higher concrete grades, this decrease is much more expressive (C45 to C60). The failure of the planes might be the cause of this. To conclude that the interfacial transition zone was the restrictive strength component, Butler et al. [100] found that failure surfaces for a concrete C30 happened close to the aggregate. However, the failure plane for concrete C50 was mostly via the aggregates, representing that the resistance of the coarse aggregate was the restrictive resistance component. Therefore, it is reasonable to expect that for advanced concrete grades, where the interfacial transition zone is greater, the concrete resistance will decrease more noticeably as the recycled aggregate fraction rises. Alternatively, for lower concrete grades, the interfacial transition zone may be more limiting for resistance as per the characteristics of the recycled aggregate, and a surge in the recycled aggregate fraction may not have a momentous impact on the resistance of the concrete.

According to Figure 17.7, the following equation is recommended to foresee the compressive resistance of concrete based on the recycled/natural aggregate ratio:

$$f_c = \alpha e^{-1R/N} \begin{cases} \alpha = 30 & for\ C25-30 \\ \alpha = 36 & for\ C35-40 \\ \alpha = 52 & for\ C40-45 \\ \alpha = 58 & for\ C45-50 \end{cases} \qquad (17.1)$$

TABLE 17.1

Previous Models Proposed for Elastic Moduli-Compressive Resistance Prediction

Study	Model
Cabral et al [22]	$E_c = 2.58 \times f_c^{0.63}$
Lovato [131]	$E_c = 5.74 \times f_c^{0.5} - 13.39$
Ravindrarajah and Tam [132]	$E_c = 4.63 \times f_c^{0.5}$

Where, f_c and R/N indicate the compressive resistance of concrete and recycled-to-natural aggregates ratio, respectively. As the amount of recycled aggregate rises, several studies have also demonstrated that the concrete's elastic moduli decrease [38, 55, 60, 62, 70–130]. However, this decline becomes more apparent when recycled aggregate contains more mortar that has been bonded to it [38]. The behavior of reinforced concrete structural parts is directly impacted by the elastic moduli, making it a crucial parameter to monitor. The relationships between the recycled aggregate concrete's compressive resistance and elastic modulus have been the subject of several different equations proposed by previous investigations. Table 17.1 provides several of these equations. Each of these authors developed their equation based solely on an analysis of their experimental findings as a result, in this work, the experimental results for the elastic moduli of multiple studies were combined into a single graphic curve, as illustrated in Figure 17.8 [38, 55, 60, 62, 70–130]. This was done to increase sampling. Then, using the three equations in Table 17.1 and

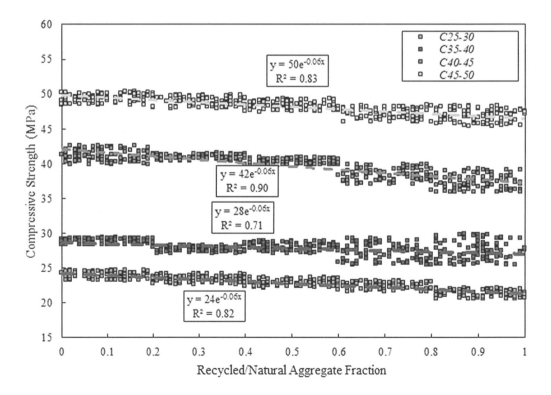

FIGURE 17.8 Relationship between the elastic moduli and recycled/natural aggregate fraction as per previous experimental reports. (Provided by current authors.)

the experimentally determined compressive strength, the theoretically corresponding modulus of elasticity was derived. In this figure, the comparison with the experimental results and projected technique by the previous investigation are shown, as well.

According to Figure 17.8, the following equation is recommended to foresee the elastic moduli of concrete based on the recycled/natural aggregate ratio. This formula has an acceptable correlation with experimental results ($R^2 > 0.71$) which could be used as a good prediction tool.

$$E_c = \beta e^{-0.06R/N} \begin{cases} \beta = 24 & for\ C25-30 \\ \beta = 28 & for\ C35-40 \\ \beta = 42 & for\ C40-45 \\ \beta = 50 & for\ C45-50 \end{cases} \tag{17.2}$$

Where E_c and R/N indicate the elastic moduli of concrete and recycled-to-natural aggregates ratio, respectively. As the percentage of recycled particles in concrete rises, Figure 17.8 demonstrates how the elastic moduli decrease. Due to the attendance of two distinct interfacial transition zones, this impact was observed. Recycled concretes are more deformable than conventional concretes because this second interfacial transition zone inclines to be frailer than the concrete paste of traditional concretes, which lowers the concrete resistance [133–136]. Additionally, it is evident from Figure 17.8 that there is a significant difference between experimental findings and those attained using the theoretical curves put out by previous investigations. The fundamental reason for this difference is that the theoretical curves only take the compressive resistance into account, even though the recycled aggregates are highly heterogeneous. Therefore, extra characteristics, such as density, water absorption, or level cement configuration, should probably be taken into account for better estimation. Due to the momentous variety of recycled aggregates, various criteria should be considered while designing mixes. Considering the unique properties of recycled aggregates, several authors offered several methods for the mixture design of concretes using recycled aggregates and the forecast of their compressive resistance. By adjusting the aggregate/cement fraction determined by the ACI technique, Bairigi et al. [137] completed it likely to design recycled aggregate concretes with a needed resistance from 15 to 30 MPa in 1990. Fathifazl et al. [138] presented a novel method in 2009. To attain the same mortar capacity as a mixture formed completely of natural aggregates, this technique, known as corresponding mortar capacity, studies the amounts of individual stages and modifies the coarse aggregate and concrete matrix. Pepe [139] proposed a similar year that the large permeability of recycled aggregates affects the amount of free water that is accessible for the combination and, as a result, the temporal evolution of compressive resistance. The findings demonstrated that this theoretical technique accurately predicted the temporal development of compressive resistance seeing the water absorption capability, an essential characteristic of recycled aggregates. Additionally, a new relationship is advanced to foresee the elastic moduli of concrete having various recycled aggregate contents with different compressive strengths, as presented in Equation (17.3) and Figure 17.9, as well. Regarding this figure, there is a high relationship among the proposed model and experimental results ($R^2 > 0.99$). Therefore, Equation. (17.3) could be employed as a very precise technique to forecast the relationship among elastic moduli and compressive resistance of concrete with different recycled aggregate fractions and compressive strengths.

$$\begin{cases} E_c = 2.56 f_c^{0.7} & for\ C25-40 \\ E_c = 3.46 f_c^{0.7} & for\ C40-45 \\ E_c = 6.36 f_c^{0.7} & for\ C45-50 \end{cases} \tag{17.3}$$

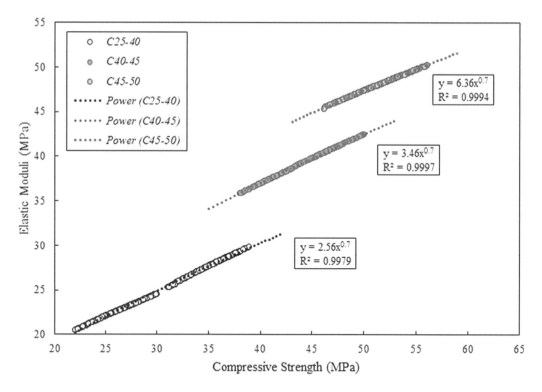

FIGURE 17.9 Relation among the elastic moduli and compressive resistance as per previous experimental reports. (Provided by current authors.)

17.4 IMPROVEMENT IN THE PERFORMANCE OF CONCRETE MADE BY WASTE AGGREGATES

Utilizing the pozzolanic micro-powders, employing polymer emulsion, bacterial carbonate bio-deposition, carbonation, and various types of fiber (steel, polypropylene) are a few techniques used in addition to mixing design methods to enhance the physical characteristics of waste particles and reduce the performance break among natural and recycled aggregate concrete paste [140, 141]. Other techniques include pre-emerging in acid, magnetic water, and a two-stage mixing approach. When employing recycled aggregates, changing the order in which the components are mixed is sufficient to alter the qualities of the concrete. This procedure involves dividing the necessary water into two equivalent portions and adding each component to the mixing at two distinct times.

This caused a coating of the cement slurry to appear on top of the recycled aggregate according to previous investigations [142–144]. Additionally, the new interfacial transition zone's overall volume of voids and CH crystals shrank [122]. As the hydration results of cement in the solid paste are melted, pre-emerging in acid, for instance, could be utilized to eliminate the adhering mortar and improve the superiority of recycled aggregate while also dropping permeability and water absorption capability [86, 145–147]. Though, using acids with greater attention puts the safety of the workforce in jeopardy and adds harmful ions, including hydrogen chloride and sulfuric acid, to the recycled aggregates, which can affect recycled concrete's durability. Therefore, washing processed recycled aggregates requires a significant amount of water. In addition, the way this washing water is disposed of and the leftover effects of the treatment cause new environmental issues. Because acid acetic is more cost-effective, safer, and cleaner than hydrogen chloride or sulfuric acid as a treatment solution, Wang et al. [148] utilized it in their tests to make up for these limitations. Alternatively, certain techniques for sealing the pores of recycled aggregates have been discovered.

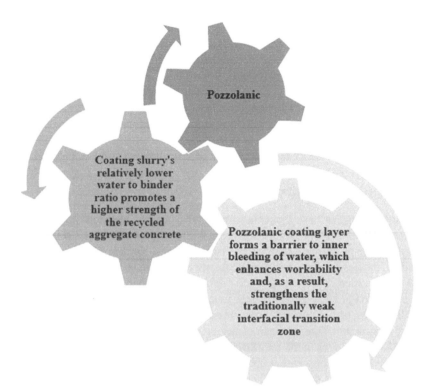

FIGURE 17.10 Mechanisms affect the interfacial transition zone and better workability. (Provided by current authors.)

Polymer emulsions, which are water-repellent, lower the capacity of porous materials, like recycled aggregates, to absorb water [128, 149–153]. Typically, there are two ways to apply this treatment: covering the concrete surface with polymer emulsion or dispersing the polymer emulsion into freshly formed concrete during the production process. Concrete made with recycled aggregate can be made more durable using either technique. The coating procedure, however, is more efficient since it may reach a deeper level of impregnation [154]. Modifying the interfacial transition zone microstructure with the addition of pozzolanic micro-powders can be done through either physical or chemical means. The ability of mineral admixtures to fill interfacial transition zone micro-pores in the first is what distinguishes them. Meanwhile, the pozzolanic reaction in the presence of water is connected to the chemical mechanism [114]. Because of the following factors, a denser interfacial transition zone and better workability and mechanical properties are seen when an inorganic mixture is incorporated into concrete, as shown in Figure 17.10. Pozzolanic material causes the calcium hydroxide that has built up in the pores and on the exterior of the adhered paste to be consumed, resulting in the formation of novel hydration results and a more solid calcium silicate hydrate erection and tricalcium aluminate, which can chemically bind chloride ions and strengthen the bonding among recycled aggregate and cement matrix [155].

It has already been extensively researched how bacterial carbonate bio-deposition can strengthen and protect the surface of stones and concrete. It is because some bacteria can cause calcium carbonate crystals to form by engaging in a variety of physiological processes. These physiological processes create an uninterrupted impermeable coating on the surface and seal the holes, acting as a block to prevent water and other corrosive chemicals from penetrating [156–159]. The concrete's mechanical properties and durability are enhanced by its dense surface and reduced permeability. Another novel technique is carbonation, in which the carbon dioxide combines with the principal cement hydration, resulting in the old cement paste, calcium oxide and hydrated calcium silicate, to produce calcium carbonate and silica gel, which fills the pores of the adhering cement mortar. In general, this technique reduces permeability, consolidation the porous cement mortar's porous surface and reduces its water absorption, that is typically attended by a consolidation impact [160]. Because carbonation action may absorb carbon dioxide released by industrial processes, it is effective and environmentally benign. However, compared to other ways, the time required is substantially more. Efficiency is further influenced by humidity and carbon dioxide concentration. The technique is useful for non-reinforced structures, but because of carbonation, which lowers the cement paste's PH, steel rebars lose their passivity and become more susceptible to corrosion [160–164].

17.5 STRUCTURAL CHARACTERISTICS

Though recycled aggregates have been the subject of extensive research, their current uses are limited to low-utility areas like pavements and landscaping [165]. This state could be clarified by the patchy availability of recycled sources, the lack of applicable designing codes, the absence of professional data, the dearth of monetary motivations and administration provisions, and the widespread belief that recycled aggregate concrete performs worse than traditional concrete [166, 167]. Usually, the utilization of recycled aggregates in structural concretes is restricted by international regulations. Depending on the intended concrete resistance grade and the properties of the recycled aggregate, such as structure, density, water engagement, and the amount of impurities, the allowable amount of recycled aggregates in the mixture varies. The aggregates must, however, possess specified qualities, such as water absorption up to 7%, sulfate ion and chlorine content less than 0.1%, and contaminant content less than 1%. The standards that control the usage of recycled aggregates in various nations are summarized in Table 17.2.

As in the German requirements [198, 199], the growth in the permitted recycled aggregate amount is typically attended by a decrease in the permitted water engagement capability and proportion of pollutants. When using recycled aggregate made from concrete waste, up to 90% of the recycled aggregate may be used if the water absorption capacity is at a maximum of 10% and the contaminants are at most 0.2%; however, when using recycled aggregate made from demolition waste, the recycled aggregate content is only allowed to be 70% and the water absorption capacity and contaminants are only allowed to be 15% and 0.5%, respectively [200–211]. Many researchers have looked at the structural characteristics of RC beams with varying proportions of recycled coarse aggregate in its place of natural aggregates to determine the pertinency of recycled aggregates on a structural member. The literature review [212–219] shows, for instance, that even with advanced proportions of recycled coarse aggregate, flexural moments and ductility at serviceableness are not meaningfully influenced when beams are manufactured to exhibit ductile performance. This is consistent with the small impact of concrete characteristics on the bending performance behavior of beams. In a similar vein, recycled aggregate exhibits yielding and final behavior like conventional concretes due to the ductile design of the steel reinforcing.

The inferior sharpness of recycled aggregates, the earlier reduction, and the inferior tensile resistance of recycled aggregate concretes, according to the authors, result in more significant and sooner cracking than conventional concrete. Due to the reduced interfacial bonding and friction among the recycled aggregate and matrix [220, 221], which could impair the bonding resistance among concrete and rebars, beams had a noticeably lower ductility ratio as the recycled aggregate substitution fraction increased [216–218]. Regarding crack patterns, Bai and Sun [218] found similarities among recycled aggregate and regular concrete. Both recycled and concrete beams exhibit the same performance during four-point bending tests: crack development started with the emergence of bending cracks in the highest moment zone, and then extra bending cracks appeared between the load and support regions. Other writers [212–219] have noted similar phenomena. Following a four-point flexural test, Figure 17.11 compares a traditional concrete beam with a 100% waste aggregate specimen. Ignjatovic et al. [217] and Kang et al. [216] confirmed that even for recycled aggregate

TABLE 17.2

The Main Improvement of Various Techniques

Technique	Consequences
Fibers incorporation [9, 168–185]	• Bridging role of fibers reduces the cracks width and transfers stresses over the cracks. • Increasing the tensile, compressive, and flexural strengths due to higher tensile strength of fibers relative to concrete matrix. • The elastic moduli raise with an increase in the fibers content. • Increasing the fiber content reduces the workability as a result of higher water absorption of fibers.
Two-stage mixing method [120, 142–144]	• The necessary water is divided into two equivalent portions and extra to the mixture twice. • Fills the pores and crevices in the recycled aggregate to create a more solid interfacial transition zone. • Recycled aggregate concrete's mechanical characteristics and toughness behavior are enhanced.
Pre-soak in acid [86, 145–148]	• The attached mortar's hydration byproducts are dissolved in an acidic solution in recycled aggregates. • Water absorption declines when the interfacial transition zone gets less porous and broken, and specific density rises. • Raises the ratio of chloride to sulfate in recycled aggregate concrete. • Using acid acetic is less expensive, safer, and cleaner than using hydrogen chloride or sulfuric acid.
Carbonation [160–164]	• Calcium carbonate and silica gel are produced when carbon dioxide combines with calcium and hydrated calcium silicate, filling recycled aggregate pores. • Increases density while reducing porosity and water absorption. • It is a method that is good for the environment because it can capture carbon dioxide released by industrial activities. • Concrete constructed with treated recycled aggregates has better mechanical qualities and less drying shrinkage. • Carbonation enhances both the old and new interfacial transition zone.
Including pozzolanic micro powders [88, 117, 124, 158, 191–198]	• Calcium hydroxide consumption makes the calcium silicate hydrate structure denser, which lowers porosity. • Enhances the compactness and mechanical performance of recycled aggregate concrete. • Efficiency is influenced by pozzolanic materials' reactivity, calcium hydroxide content, and particle size in the adhering mortar. • Fly ash addition enhances concrete's flowability and reduces water absorption. • When compared to Portland cement, its delayed binder feature greatly reduces concrete permeability in older construction and boosts compressive strength.
Polymer emulsion usage [128, 149–154]	• Recycled aggregate's surface is sealed when it is submerged in a polymer mix, which fills the adhering paste pores. • Treatment increases crushing index and density while considerably reducing recycled aggregate water absorption. • Reduced drying shrinkage and increased resistance to chloride ion diffusion and carbonation are two characteristics of treated recycled aggregate concrete.
Bio-deposition of carbonate by bacteria [194–197]	• Bacterial physiological activity results in calcium carbonate crystal precipitation, which continuously forms an anti-water coating on the surface of the recycled aggregate. • Concrete made with recycled aggregate has less permeability, which improves its mechanical qualities and durability. • Efficiency is influenced by PH, temperature, the amount of calcium, and the number of bacteria. • The bacteria solution can be used by immersion and spraying. Immersion is more effective even if spraying is more practical.

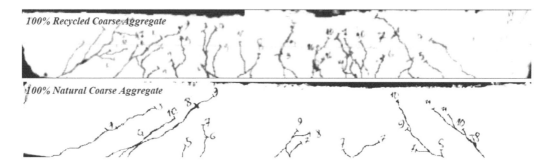

FIGURE 17.11 Comparison between the conventional concrete beam and 100% incorporation recycled aggregate concrete beam.

substation fractions over 50%, the cracking form of waste aggregate concrete beams is identical to the control specimen. In contrast, Seara-Paz et al. [212] and Arezoumandi et al. [215] found that beams with a larger proportion of waste aggregate had a decrease in cracking arrangement and an increase in crack diameter. The different results for cracking spacing and cracking breadth may be due to the different reinforcing rates used by each author. The authors noted that although the fracture pattern is similar to that of conventional concrete concerning the shear performance of waste aggregate concrete beams, shear resistance is lower due to the recycled concrete's lower tensile resistance [96–98, 127, 222–224]. Additionally, the macroscopic study demonstrated that shear failures in recycled concrete beams occurred through the recycled aggregates rather than at the mortar and aggregate contact, as is typically the case for conventional concrete [225].

Tosic et al. [226] collected 217 laboratory findings, built a record, and assessed the relevance of the Eurocode 2 standards for the bending and shear resistances of recycled aggregate RC beams. They concluded that whereas shear resistance forecasts for beams with stirrups in Eurocode 2 are accurate, flexural resistance predictions are incorrect. However, most investigations consider recycled aggregate concrete beams' minimal transverse reinforcement ratios. Consequently, the authors suggested doing extra studies on recycled aggregate RC beams with stirrup fractions greater than the minimum. Pacheco et al. [227] also achieved comparable outcomes. Despite certain ambiguities, references to recycled aggregate concrete parts may be found in actual applications. As a result, the request for recycled aggregate to produce structural concrete members such as beams, columns, and slabs is highly recommended for evaluation.

17.6 CONCLUSION

One of the eco-friendly keys to the rising waste dumping disaster and the reduction of natural aggregate resources brought on by the creative industry is the recycling of construction and demolition debris, which results in recycled aggregate. Recycled aggregate has, however, primarily been utilized in low-value claims up until now, such as the base of the pavement. Recycled aggregates are further permeable and have a further complicated microstructure than natural aggregates. The endurance of the material is impacted by the greater porosity, which serves as a possible channel for both hostile chemicals like chloride ions and water as well. Recycled coarse aggregates often require further water than conservative concrete to achieve a similar slump due to their higher water absorption. This influences the homogeneousness of the concrete throughout molding, which lowers the mechanical resistance of the concrete. Despite this, there is no precise debate on curing techniques for waste aggregate concretes in the literature.

Therefore, the drop in compressive resistance grows as the amount of recycled aggregate fraction increases. Additionally, according to the literature, there are two separate interfacial transition zones present in concrete, which is why the elastic moduli decrease as the percentage of recycled particles in concrete rises interfacial transition zones. The second interfacial transition zones of recycled concretes tend to be more deformable than the paste aggregate matrix of simple concretes because the strength of the concrete decreases because of the second interfacial transition zones' weakness. The results of various literature analyses used in the current study also demonstrate that there is a significant disparity between experimental results and theoretical curves that only take compressive strength into account. More variables could likely be taken into account for improved estimation, such as density, water engagement, or even cement configuration. The research of the works also reveals that numerous researchers have experimented with various methods to alter the mechanical behavior and long-term qualities of recycled aggregate concrete, enabling a larger replacement ratio. The count of pozzolanic micro powders and polymer fibers, as well as pre-soaking in acid, bacterial carbonate bio-emerging, and carbonation, are the techniques that are most frequently used in the literature. The findings from these techniques are pertinent, and the characteristics of the recycled concrete were frequently extremely like those discovered for normal concrete.

There are still certain obstacles in the way of using waste aggregate in structural concrete. There is disagreement, for instance, about the mixed design approach that finest accommodates the momentous variety of recycled aggregates and their unique properties. Additionally, though there are encouraging findings in the previous works about the current formulations for strength estimate, there are still some concerns regarding the performance of recycled aggregate concrete in bending and shear loads as well as the structural parts' long-term integrity. As a result, most global standards and specifications provide a 30% substitution limit for structural parts for recycled aggregate. Recycled aggregate has a lot of potential for the environment, despite the current uncertainties. Therefore, it is crucial to progress investigations about the viability of employing increasing proportions of recycled aggregate on structural concretes to enhance energy savings and environmental preservations. Additionally, it could be practical to create methodological provisions and recommendations for the manufacture and worth assurance of structural recycled aggregate concrete, which would enhance how users and contractors generally see this substance.

Finally, the proposed models in this review to foresee the compressive resistance and elastic moduli of concrete with various compressive strengths based on recycled coarse aggregate fraction could be utilized as beneficial tools since they are developed using a wide range of experimental results.

REFERENCES

1. Şimşek O, Sefidehkhan HP, Gökçe HS. Performance of fly ash-blended Portland cement concrete developed by using fine or coarse recycled concrete aggregate. *Construct. Build. Mater.* 2022; 357: 129431.
2. Shmlls M, Abed M, Horvath T, Bozsaky D. Multicriteria based optimization of second generation recycled aggregate concrete. *Case Stud. Construct. Mater.* 2022; 17: e01447.
3. Huang L, Krigsvoll G, Johansen F, Liu Y, Zhang X. Carbon emission of global construction sector. *Renew. Sustain. Energy Rev.* 2018; 81: 1906–1916.
4. European Aggregates Association. A Sustainable Industry for a Sustainable Europe. *Annual Review.* 2018.
5. U.S. Department of the Interior. Minerals Yearbook - Sand and Gravel, Construction. 2018.
6. ISWA. ISWA Report at a Glance. International Solid Waste Association, AUT, Viena, 2015.
7. USEPA, Advancing Sustainable Materials Management: Fact Sheet Assessing Trends in Material Generation. *Recycling, Composting, Combustion.* 2014.
8. Huang B, Wang X, Kua H, Geng Y, Bleischwitz R, Ren J. Construction and demolition waste management in China through the 3R principle. *Resour. Conserv. Recycl.* 2018; 129: 36–44.
9. Karimi Pour A, Ghalehnovi M, Golmohammadi M, Brito J. Experimental investigation on the shear behaviour of stud-bolt connectors of steel-concrete-steel fibre-reinforced recycled aggregates sandwich panels. *Materials.* 2021; 14(18): 5185.
10. Ouyang K, Liu J, Liu S, Song B, Guo H, Li G, Shi C. Influence of pre-treatment methods for recycled concrete aggregate on the performance of recycled concrete: A review. *Resour Conserv. Recycl.* 2023; 188: 106717.
11. European Environment Agency. EU as a Recycling Society – Present Recycling Levels of Municipal Waste and Construction & Demolition Waste in the EU. 2009.

12. BRELPE. Panorama Dos Resíduos Sólidos No Brasil 2016, 2016.
13. USEPA. Industrial and construction and demolition (C&D) landflls. 2018.
14. European Parliament and Council. Commission Decision of 3 May 2000 Replacing Decision 94/3/EC Establishing a List of Wastes Pursuant to Article 1(a) of Council Directive 75/442/EEC on Waste, 2000.
15. Bohmer S. Aggregates Case Study: Final Report Referring to Contract. N. 150787-2007 F1SC-AT, 2008.
16. Ccanz. Best Practice Guide for the Use of Recycled Aggregates in New Concrete, 2011.
17. U.S. Green Building Council. LEED V4 for building design and construction, 2018.
18. Karimi Pour A, Edalati M, Brito J. Influence of magnetized water and water/cement ratio on the properties of untreated coal fine aggregates concrete. *Cem. Conc. Com.* 2021; 122: 104121.
19. Rezaiee-Pajand M, Mohebi Najm Abad J, Karimi Pour A, Rezaiee-Pajand A. Propose new implement models to determine the compressive, tensile and flexural strengths of recycled coarse aggregate concrete via imperialist competitive algorithm. *J. Build. Eng.* 2021; 40: 102337.
20. Karimi Pour A, Edalati M. Retrofitting of the corroded reinforced concrete columns with CFRP and GFRP fabrics under different corrosion levels. *Eng. Str.* 2021; 228: 111523.
21. Ghalehnovi M, Yousefi M, Karimipour A, Brito J, Norooziyan M. Investigation of the behaviour of steel-concrete-steel sandwich slabs with bi-directional corrugated-strip connectors. *App. Sci.* 2020; 10: 8647.
22. Cabral AEB, Schalch V, Molin DCCD, Ribeiro JLD. Mechanical properties modeling of recycled aggregate concrete. *Construct. Build. Mater.* 2010; 24(4): 421–430.
23. United States Environmental Protection Agency. Sustainable management of construction and demolition materials. 2018.
24. Silva V, Brito J, Dhir RK. Properties and composition of recycled aggregates from construction and demolition waste suitable for concrete production. *Construct. Build. Mater.* 2014; 65: 201–217.
25. Katerusha D. Investigation of the optimal price for recycled aggregate concrete — An experimental approach. *J. Clean. Prod.* 2022; 365: 132857.
26. Xiao J, Zhang H, Tang Y, Deng Q, Wang D, Poon CS. Fully utilizing carbonated recycled aggregates in concrete: Strength, drying shrinkage and carbon emissions analysis. *J. Clean. Prod.* 2022; 377: 134520.
27. Yuan WB, Mao L, Li LY. A two-step approach for calculating chloride diffusion coefficient in concrete with both natural and recycled concrete aggregates. *Sci. Total Environ.* 2023; 856(2): 159197.
28. Park SB., Lee BC, Kim JH. Studies on mechanical properties of concrete containing waste glass aggregate. *Cement Concr. Res.* 2004; 34(12): 2181–2189.
29. Ismail ZZ, AL-Hashmi EA. Recycling of waste glass as a partial replacement for fine aggregate in concrete. *Waste Manag.* 2009; 29(2): 655–659.
30. Kou SC, Poon CS. Properties of self-compacting concrete prepared with recycled glass aggregate. *Cement Concr. Compos.* 2009; 31(2): 107–113.
31. Choi YW, Moon DJ, Chung JS, Cho SK. Effects of waste PET bottles aggregate on the properties of concrete. *Cement Concr. Res.* 2005; 35(4): 776–781.
32. Siddique R, Khatib J, Kaur I. Use of recycled plastic in concrete: A review. *Waste Manag.* 2008; 28(10): 1835–1852.
33. Albano C, Camacho N, Hernández M, Matheus A., Gutiérrez A. Influence of content and particle size of waste pet bottles on concrete behavior at different w/c ratios. *Waste Manag.* 2009; 29(10): 2707–2716.
34. Sáez del Bosque IF, Zhu W, Howind T, Matías A, Sánchez de Rojas MI, Medina C. Properties of interfacial transition zones (ITZs) in concrete containing recycled mixed aggregate. *Cement Concr. Compos.* 2017; 81: 25–34.
35. Khitab A, Kırgız MS, Nehdi ML, Mirza J, de Sousa Galdino AG. Karimi Pour A. Mechanical, thermal, durability and microstructural behavior of hybrid waste-modified green reactive powder concrete. *Cons. Build. Mat.* 2022; 344: 128184.
36. Fasihihour N, Mohebbi Najm Abad J, Karimi Pour A, Mohebbi MR. Experimental and numerical model for mechanical properties of concrete containing fly ash: Systematic review. *Measurement.* 2022; 188: 110547.
37. Duan ZH, Poon CS. Properties of recycled aggregate concrete made with recycled aggregates with different amounts of old adhered mortars. *Mater. Des.* 2014; 58: 19–29.
38. Nagataki S, Gokce A, Saeki T, Hisada M. Assessment of recycling process induced damage sensitivity of recycled concrete aggregates. *Cement Concr. Res.* 2004; 34(6): 965–971.
39. Shi C, Li Y, Zhang J, Li W, Chong L, Xie Z. Performance enhancement of recycled concrete aggregate – A review. *J. Clean. Prod.* 2016; 112: 466–472.
40. Movassaghi R. *Durability of Reinforced Concrete Incorporating Recycled Concrete as Aggregate (RCA).* University of Waterloo. 2006.

41. Behera M, Bhattacharyya SK, Minocha AK, Deoliya R, Maiti S. Recycled aggregate from C&D waste & its use in concrete – a breakthrough towards sustainability in construction sector: A review. *Construct. Build. Mater.* 2014; 68: 501–516.

42. Matias D, Brito J, Rosa A, Pedro D. Mechanical properties of concrete produced with recycled coarse aggregates – influence of the use of superplasticizers. *Construct. Build. Mater.* 2013; 44: 101–109.

43. Quattrone M, Angulo SC, John VM. Energy and CO_2 from high performance recycled aggregate production. *Resour. Conserv. Recycl.* 2014; 90: 21–33.

44. Karimi Pour A, Ghalehnovi M, Brito J. New model for the lap-splice length of tensile reinforcement in concrete elements. *J. Str. Eng.* 2021; 147: 12.

45. Xing W, Tam VWY, Le KN, Butera A, Hao JL, Wang J. Effects of mix design and functional unit on life cycle assessment of recycled aggregate concrete: Evidence from CO2 concrete. *Cons. Build. Mat.* 2022; 348: 128712.

46. Ulucan M, Alyamac KE. A holistic assessment of the use of emerging recycled concrete aggregates after a destructive earthquake: Mechanical, economic and environmental. *Waste Manag.* 2022; 146: 53–65.

47. Mulder E, de Jong TPR, Feenstra L. Closed cycle construction: An integrated process for the separation and reuse of C&D waste. *Waste Manag.* 2007; 27(10): 1408–1415.

48. Shima H, Tateyashiki H, Matsuhashi R, Yoshida Y. An advanced concrete recycling technology and its applicability assessment through input-output analysis. *J. Adv. Concr. Technol.* 2005; 3(1): 53–67.

49. Akbarnezhad A, Ong KCG, Zhang MH, Tam CT, Foo TWJ. Microwave-assisted beneficiation of recycled concrete aggregates. *Construct. Build. Mater.* 2011; 25: 3469–3479.

50. Bru K, Touzé S, Bourgeois F, Lippiatt N, Ménard Y. Assessment of a microwave-assisted recycling process for the recovery of high-quality aggregates from concrete waste. *Int. J. Miner. Process.* 2014; 126: 90–98.

51. Rezaiee-Pajand M, Rezaiee-Pajand A, Karimi Pour A, Mohebi Najm Abbadi J. A particle swarm optimization algorithm to suggest formulas for the behaviour of the recycled materials reinforced concrete beams. *Int. J. Optim. Civil Eng.* 2020; 10(3): 451–479.

52. Karimi Pour A, Chaboki HR, Ghalehnovi M. Investigation of flexural behavior of concrete beams made of recycled aggregate. *Concr. Res. Quart. J.* 2019.

53. Chaboki HR, Ghalehnovi M, Karimi Pour A, Khatibinia M. Investigation the shear behaviour of recycled aggregate concrete beams. *J. Struct. Constr. Eng.* 2021; 7(4): 82–99.

54. de Juan MS, Gutiérrez PA. Study on the influence of attached mortar content on the properties of recycled concrete aggregate. *Construct. Build. Mater.* 2009; 23(2): 872–877.

55. Pepe M, Toledo Filho RD, Koenders EAB, Martinelli E. Alternative processing procedures for recycled aggregates in structural concrete. *Construct. Build. Mater.* 2014; 69: 124–132.

56. Yan P, Wu J, Lin D, Liu X. Uniaxial compressive stress–strain relationship of mixed recycled aggregate concrete. *Constr. Build. Mat.* 2022; 350: 128663.

57. Kim J. Influence of quality of recycled aggregates on the mechanical properties of recycled aggregate concretes: An overview. *Constr. Build. Mat.* 2022; 328: 127071.

58. Chen Y, Zhang S, Ye P, Liang X. Mechanical properties and damage constitutive of recycled aggregate concrete with polyvinyl alcohol fiber under compression and shear. *Case Stud. Construct. Mater.* 2022; 17: e01466.

59. Bao J, Zheng R, Yu Z, Zhang P, Song Q, Xu J., Gao S. Freeze-thaw resistance of recycled aggregate concrete incorporating ferronickel slag as fine aggregate. *Construct. Build. Mater.* 2022; 356: 129178.

60. Bravo M, Brito J, Pontes J, Evangelista L. Mechanical performance of concrete made with aggregates from construction and demolition waste recycling plants. *J. Clean. Prod.* 2015; 99: 59–74.

61. Padmini AK, Ramamurthy K, Mathews MS. Influence of parent concrete on the properties of recycled aggregate concrete. *Construct. Build. Mater.* 2009; 23(2): 829–836.

62. Poon CS, Shui ZH, Lam L, Fok H, Kou SC. Influence of moisture states of natural and recycled aggregates on the slump and compressive strength of concrete. *Cement Concr. Res.* 2004; 34(1): 31–36.

63. Sagoe-Crentsil KK, Brown T, Taylor AH. Performance of concrete made with commercially produced coarse recycled concrete aggregate. *Cement Concr. Res.* 2001; 31(5): 707–712.

64. Cavalline TL, Weggel DC. Recycled brick masonry aggregate concrete: Use of brick masonry from construction and demolition waste as recycled aggregate in concrete. *Struct. Surv.* 2013; 31(3): 160–180.

65. Brand AS, Roesler JR, Salas A. Initial moisture and mixing effects on higher quality recycled coarse aggregate concrete. *Construct. Build. Mater.* 2015; 79: 83–89.

66. Chaboki HR, Ghalehnovi M, Karimi Pour A. Study of the flexural behaviour of recycled aggregate concrete beams. *J. Concr. Res.* 2019; 12: 45–60.

67. Chaboki HR, Ghalehnovi M, Karimi Pour A, Khatibinia M. Study on the Flexural Behavior of Recycled Concrete Beams. *5th Nati Conf on App Res in Civ Eng, Arch and Urb Man.* 2018.

68. Andreu G, Miren E. Experimental analysis of properties of high performance recycled aggregate concrete. *Construct. Build. Mater.* 2014; 52: 227–235.

69. Salgado FDA, Silva FDA. Properties of recycled aggregates from different composition and its influence on concrete strength, *Rev. IBRACON de Estruturas e Mater.* 2021; 14(6).

70. Kwan WH, Ramli M, Kam KJ, Sulieman MZ. Influence of the amount of recycled coarse aggregate in concrete design and durability properties. *Construct. Build. Mater.* 2012; 26(1): 565–573.

71. Quan X, Wang S, Li J, Luo J, Liu K, Xu J, Zhao N, Liu Y. Utilization of molybdenum tailings as fine aggregate in recycled aggregate concrete. *J. Clean. Prod.* 2022; 372: 133649.

72. Chen Y, Li P, Ye P, Li H, Liang X. Experimental investigation on the mechanical behavior of polyvinyl alcohol fiber recycled aggregate concrete under triaxial compression. *Construct. Build. Mater.* 2022; 350: 128825.

73. Xiao J, Tang Y, Chen H, Zhang H, Xia B. Effects of recycled aggregate combinations and recycled powder contents on fracture behavior of fully recycled aggregate concrete. *J. Clean. Prod.* 2022; 366: 132895.

74. Le HB, Bui QB. Recycled aggregate concretes – A state-of-the-art from the microstructure to the structural performance. *Construct. Build. Mater.* 2020; 257.

75. Gomez-Soberon JMV. Porosity of recycled concrete with substitution of recycled concrete aggregate: An experimental study. *Cement Concr. Res.* 2002; 32: 1301–1311.

76. Mefteh H, Kebaïli O, Oucief H, Berredjem L, Arabi N. Influence of moisture conditioning of recycled aggregates on the properties of fresh and hardened concrete. *J. Clean. Prod.* 2013; 54: 282–288.

77. Yang J, Du Q, Bao Y. Concrete with recycled concrete aggregate and crushed clay bricks. *Construct. Build. Mater.* 2011; 25(4): 1935–1945.

78. Brito J, Ferreira J, Pacheco J, Soares D, Guerreiro M. Structural, material, mechanical and durability properties and behaviour of recycled aggregates concrete. *J. Build. Eng.* 2016; 6: 1–16.

79. Arezoumandi M, Steele AR, Volz JS. Evaluation of the bond strengths between concrete and reinforcement as a function of recycled concrete aggregate replacement level. *Structures.* 2018; 16: 73–81.

80. Etxeberria M, Vazquez E, Marí A. Microstructure analysis of hardened recycled aggregate concrete. *Mag. Concr. Res.* 2006; 58: 683–690.

81. Hamad BS, Dawi AH, Daou A, Chehab GR. Studies of the effect of recycled aggregates on flexural, shear, and bond splitting beam structural behavior. *Case Stud. Construct. Mater.* 2018; 9: e00186.

82. Poon CS, Shui ZH, Lam L. Effect of microstructure of ITZ on compressive strength of concrete prepared with recycled aggregates. *Construct. Build. Mater.* 2004; 18: 461–468.

83. Tabsh WS, Abdelfatah AS. Influence of recycled concrete aggregates on strength properties of concrete. *Construct. Build. Mater.* 2009; 23(2): 1163–1167.

84. De Oliveira MB, Vazquez E. The influence of retained moisture in aggregates from recycling on the properties of new hardened concrete. *Waste Manag.* 1996; 16(1–3): 113–117.

85. Dilbas H, Simsek M, Çakir O. An investigation on mechanical and physical properties of recycled aggregate concrete (RAC) with and without silica fume. *Construct. Build. Mater.* 2014; 61: 50–59.

86. Ismail S, Ramli M. Mechanical strength and drying shrinkage properties of concrete containing treated coarse recycled concrete aggregates. *Construct. Build. Mater.* 2014; 68: 726–739.

87. Kou SC, Poon CS. Long-term mechanical and durability properties of recycled aggregate concrete prepared with the incorporation of fly ash. *Cement Concr. Compos.* 2013; 37(1): 12–19.

88. Xiao J, Li J, Zhang C. Mechanical properties of recycled aggregate concrete under uniaxial loading. *Cement Concr. Res.* 2005; 35(6): 1187–1194.

89. Bairagi NK, Ravande K, Pareek VK. Behaviour of concrete with different proportions of natural and recycled aggregates. *Resour. Conserv. Recycl.* 1993; 9(1–2): 109–126.

90. Xuan D, Zhan B, Poon CS. Assessment of mechanical properties of concrete incorporating carbonated recycled concrete aggregates. *Cement Concr. Compos.* 2016; 65: 67–74.

91. Kazmi SMS, Munir MJ, Wu YF, Patnaikuni I, Zhou Y, Xing F. Effect of recycled aggregate treatment techniques on the durability of concrete: A comparative evaluation. *Construct. Build. Mater.* 2020; 264: 120284.

92. Bui NK, Satomi T, Takahashi H. Improvement of mechanical properties of recycled aggregate concrete basing on a new combination method between recycled aggregate and natural aggregate. *Construct. Build. Mater.* 2017; 148: 376–385.

93. Kou S, Poon C. Mechanical properties of 5-year-old concrete prepared with recycled aggregates obtained from three different sources. *Mag. Concr. Res.* 2008; 60(1): 57–64.

94. Leite MB, Monteiro PJM. Microstructural analysis of recycled concrete using X-ray microtomography. *Cement Concr. Res.* 2016; 81: 38–48.

95. Martínez-Lage I, Vazquez-Burgo P, Velay-Lizancos M. Sustainability evaluation of concretes with mixed recycled aggregate based on holistic approach: Technical, economic and environmental analysis. *Waste Manag.* 2020; 104: 9–19.

96. Rahal KN, Alrefaei YT. Shear strength of recycled aggregate concrete beams containing stirrups. *Construct. Build. Mater.* 2018; 191: 866–876.

97. Pradhan S, Kumar S, Barai SV. Shear performance of recycled aggregate concrete beams: An insight for design aspects. *Construct. Build. Mater.* 2018; 178: 593–611.

98. Al-Mahmoud F, Boissiere R, Mercier C, Khelil A. Shear behavior of reinforced concrete beams made from recycled coarse and fine aggregates. *Structures.* 2020; 25: 660–669.

99. Poon CS, Chan D. The use of recycled aggregate in concrete in Hong Kong. *Resour. Conserv. Recycl.* 2007; 50(3): 293–305.

100. Butler L, West JS, Tighe SL. The effect of recycled concrete aggregate properties on the bond strength between RCA concrete and steel reinforcement. *Cement Concr. Res.* 2011; 41(10): 1037–1049.

101. Talamona D, Hai Tan K. Properties of recycled aggregate concrete for sustainable urban built environment. *J. Sustain. Cement-Based Mater.* 2012; 1(4): 202–210.

102. Khan MT, Jahan I, Amanat KM. Splitting tensile strength of natural aggregates, recycled aggregates and brick chips concrete. *Proc. Inst. Civ. Eng.: Construct. Mater.* 2020; 173(2): 79–88.

103. Bui NK, Satomi T, Takahashi H. Recycling woven plastic sack waste And PET bottle waste as fiber in recycled aggregate concrete: An experimental study. *Waste Manag.* 2018; 78: 79–93.

104. Limbachiya M, Meddah MS, Ouchagour Y. Use of recycled concrete aggregate in fly-ash concrete. *Construct. Build. Mater.* 2012; 27(1): 439–449.

105. Medina C, Zhu W, Howind T, Sanchez De Rojas MI, Frías M. Influence of mixed recycled aggregate on the physical-mechanical properties of recycled concrete. *J. Clean. Prod.* 2014; 68: 216–225.

106. Kou SC, Poon CS, Etxeberria M. Influence of recycled aggregates on long term mechanical properties and pore size distribution of concrete. *Cement Concr. Compos.* 2011; 33(2): 286–291.

107. Ahmadi M., Farzin S, Hassani A, Motamedi M. Mechanical properties of the concrete containing recycled fibers and aggregates. *Construct. Build. Mater.* 2017; 144: 392–398.

108. Moallemi Pour S, Shahria Alam M. Investigation of compressive bond behavior of steel rebar embedded in concrete with partial recycled aggregate replacement. *Structures.* 2016; 7: 153–164.

109. Fonseca N, Brito J., Evangelista L. The influence of curing conditions on the mechanical performance of concrete made with recycled concrete waste. *Cement Concr. Compos.* 2011; 33: 637–643.

110. Cantero B, Saezdel Bosque IF, Matías A, Medina C. Statistically significant effects of mixed recycled aggregate on the physical-mechanical properties of structural concretes. Construct. *Build. Mater.* 2018; 185: 93–101.

111. Pepe M, Toledo Filho RD, Koenders EAB, Martinelli A. A novel mix design methodology for recycled aggregate concrete. *Construct. Build. Mater.* 2016; 122: 362–372.

112. Chaboki HR, Ghalehnovi M, Karimi Pour A. Investigation of shear behavior of concrete beams made of recycled aggregate. Arch and Urban Man. 2018.

113. Mas B, Cladera A, Del Olmo T, Pitarch F. Influence of the amount of mixed recycled aggregates on the properties of concrete for non-structural use. *Construct. Build. Mater.* 2012; 27(1): 612–622.

114. Younis KH, Pilakoutas K. Strength prediction model and methods for improving recycled aggregate concrete. *Construct. Build. Mater.* 2013; 49: 688–701.

115. Malesev M, Radonjanin V, Marinkovic S. Recycled concrete as aggregate for structural concrete production. *Sustainability.* 2010; 2(5): 1204–1225.

116. Wagih AM, El-Karmoty HZ, Ebid M, Okba SH. Recycled construction and demolition concrete waste as aggregate for structural concrete. *HBRC J.* 2013; 9(3): 193–200.

117. Kou SC, Poon CS. Enhancing the durability properties of concrete prepared with coarse recycled aggregate. *Construct. Build. Mater.* 2012; 35: 69–76.

118. Islam K, Rahman J, Sakil K, Rezaul M, Muntasir AHM. Flexural response of fiber reinforced concrete beams with waste tires rubber and recycled aggregate. *J. Clean. Prod.* 2021; 278: 123842.

119. Ali R, Hamid R. Workability and compressive strength of recycled concrete waste aggregate concrete. *Appl. Mech. Mater.* 2015; 754(755): 417–420.

120. Tam VVWY, Gao XFX, Tam CCM. Microstructural analysis of recycled aggregate concrete produced from two-stage mixing approach. *Cement Concr. Res.* 2005; 35(6): 1195–1203.

121. Kim K, Shin M, Cha S. Combined effects of recycled aggregate and fly ash towards concrete sustainability. *Construct. Build. Mater.* 2013; 48: 499–507.

122. Tam VWY, Butera A, Le KN. Carbon-conditioned recycled aggregate in concrete production. *J. Clean. Prod.* 2016; 133: 672–680.

123. Katz A. Treatments for the improvement of recycled aggregate. *J. Mater. Civ. Eng.* 2004; 16(6): 597–603.

124. Abd Elhakam A, Mohamed AE, Awad E. Influence of self-healing, mixing method and adding silica fume on mechanical properties of recycled aggregates concrete. *Construct. Build. Mater.* 2012; 35: 421–427.

125. Pepe M, Koenders EAB, Faella C, Martinelli E. Structural concrete made with recycled aggregates: Hydration process and compressive strength models. *Mech. Res. Commun.* 2014; 58: 139–145.

126. Corinaldesi V. Structural concrete prepared with coarse recycled concrete aggregate: From investigation to design. *Adv. Civ. Eng.* 2011: 1–7.

127. Etman EE, Afefy HM, Baraghith AT, Khedr SA. Improving the shear performance of reinforced concrete beams made of recycled coarse aggregate. *Construct. Build. Mater.* 2018; 185: 310–324.

128. Kou SC, Poon CS. Properties of concrete prepared with PVA-impregnated recycled concrete aggregates. *Cement Concr. Compos.* 2010; 32(8): 649–654.

129. Amorim J, Roberto P, Lima L, Batista M, Dias R, Filho T. Cement & concrete composites compressive stress – Strain behavior of steel fiber reinforced recycled aggregate concrete. *Cement Concr. Compos.* 2014; 46: 65–72.

130. Corinaldesi V. Mechanical and elastic behaviour of concretes made of recycled-concrete coarse aggregates. *Construct. Build. Mater.* 2010; 24: 1616–1620.

131. Lovato PS, Verifcaçaodos parametros de controle de agregados reciclados de resíduos de construçao e demoliçao para utilizaçao em concreto, *Dissertaçaode Mestrado, Universidade Federal do Rio Grande do Sul* (UFRGS), 2007.

132. Ravindrarajah RS, Tam C. Properties of concrete made with crushed concrete as coarse aggregate. *Mag. Concr. Res.* 1985; 37(130): 29–38.

133. Zhu L, Wen T, Tian L. Size effects in compressive and splitting tensile strengths of polypropylene fiber recycled aggregate concrete. *Construct. Build. Mater.* 2022; 341: 127878.

134. Zhang H, Xiao J, Tang Y, Duan Z, Poon CS. Long-term shrinkage and mechanical properties of fully recycled aggregate concrete: Testing and modelling. *Cem. Con Com.* 2022; 130: 104527.

135. Zhu L, Ning Q, Han W, Bai L. Compressive strength and microstructural analysis of recycled coarse aggregate concrete treated with silica fume. *Construct. Build. Mater.* 2022; 334: 127453.

136. Tošić N, Martínez DP, Hafez H, Reynvart I, Ahmad M, Liu G, de la Fuente A. Multi-recycling of polypropylene fibre reinforced concrete: Influence of recycled aggregate properties on new concrete. *Construct. Build. Mater.* 2022; 346: 128458.

137. Bairagi NK, Vidyadhara HS, Tech M, Ravande K. Mix Design Procedure for Recycled Aggregate Concrete, 1990.

138. Fathifazl G, Abbas A, Razaqpur AG, Isgor OB, Fournier B, Foo S. New Mixture Proportioning Method for Concrete Made with Coarse Recycled Concrete Aggregate. 2010.

139. Pepe M. *A Conceptual Model for Designing Recycled Aggregate Concrete for Structural Applications.* Springer International Publishing Switzerland, 2015; 7–16.

140. Sutcu M, Gencel O, Erdogmus E, Kizinievic O, Kizinievic V, Karimi Pour A, Velasco PM. Low cost and eco-friendly building materials derived from wastes: Combined effects of bottom ash and water treatment sludge. *Construct. Build. Mater.* 2022; 324: 126669.

141. Katman HYB, Khai WJ, Kırgız MS, Nehdi ML, Benjeddou O, Thomas BS, Papatzani S, Rambhad K, Kumbhalkar MA, Karimi Pour A. Transforming Conventional Construction Binders and Grouts into High-Performance Nanocarbon Binders and Grouts for Today's Constructions. *Buildings.* 2022

142. Tam VWY, Tam CM. Diversifying two-stage mixing approach (TSMA) for recycled aggregate concrete: TSMAs and TSMAsc. *Construct. Build. Mater.* 2008; 22(10): 2068–2077.

143. Tam VWY, Gao XF, Tam CM. Comparing performance of modified two-stage mixing approach for producing recycled aggregate concrete. *Mag. Concr. Res.* 2006; 58(7): 477–484.

144. Tam VWY, Tam CM. Assessment of durability of recycled aggregate concrete produced by two-stage mixing approach. *J. Mater. Sci.* 2007; 42(10): 3592–3602.

145. Ismail S, Ramli M. Engineering properties of treated recycled concrete aggregate (RCA) for structural applications. *Construct. Build. Mater.* 2013; 44: 464–476.

146. Tam VWY, Tam CM, Le KN. Removal of cement mortar remains from recycled aggregate using presoaking approaches. *Resour. Conserv. Recycl.* 2007; 50(1): 82–101.

147. Saravanakumar P, Abhiram K, Manoj B. Properties of treated recycled aggregates and its influence on concrete strength characteristics. *Construct. Build. Mater.* 2016; 111: 611–617.

148. Wang L, Wang J, Qian X, Chen P, Xu Y, Guo J. An environmentally friendly method to improve the quality of recycled concrete aggregates. *Construct. Build. Mater.* 2017; 144: 432–441.
149. Spaeth V, Djerbi Tegguer A. Improvement of recycled concrete aggregate properties by polymer treatments. *Int. J. Sustain. Built Environ.* 2013; 2(2): 143–152.
150. Spaeth V, Lecomte JP. Hydration process and microstructure development of integral water repellent cement based materials. *5th Int. Conf. Water Repellent Treat. Build. Mater.* 2008; 254: 245–254.
151. Wittmann FH, Xian Y, Zhao T, Beltzung F, Giessler S. Drying and Shrinkage of Integral Water Repellent Concrete. 2006; 229–242.
152. Tsujino M, Noguchi T, Tamura M, Kanematsu M, Maruyama I. Application of conventionally recycled coarse aggregate to concrete structure by surface modification treatment. *J. Adv. Concr. Technol.* 2007; 5(1): 13–25.
153. Zhao W, Wittmann TFH, Jiang R, Application of Silane-Based Compounds for the Production of Application of Silane-Based Compounds for the Production of Integral Water Repellent Concrete. *HVI, 6th International Conference on WR Treatment of Building Materials,* 2011; 137e144.
154. Zhu YG, Kou SC, Poon CS, Dai JG, Li QY. Influence of silane-based water repellent on the durability properties of recycled aggregate concrete. *Cement Concr. Compos.* 2013; 35(1): 32–38.
155. Silva RV, Brito J, Neves R, Dhir R. Prediction of chloride ion penetration of recycled aggregate concrete. *Mater. Res.* 2015; 18(2): 427–440.
156. Chang YC, Wang YY, Zhang H, Chen J, Geng Y. Different influence of replacement ratio of recycled aggregate on uniaxial stress-strain relationship for recycled concrete with different concrete strengths. *Structures.* 2022; 42: 284–308.
157. Zhan P, Xu J, Wang J, Zuo J, He Z. A review of recycled aggregate concrete modified by nanosilica and graphene oxide: Materials, performances and mechanism. *J. Clean. Prod.* 2022; 375: 134116.
158. Chen S, Wang H, Guan J, Yao X, Li L. Determination method and prediction model of fracture and strength of recycled aggregate concrete at different curing ages. *Construct. Build. Mater.* 2022; 343: 128070.
159. Algourdin N, Pliy P, Beaucour PL, Noumowé A, di Coste D. Effect of fine and coarse recycled aggregates on high-temperature behaviour and residual properties of concrete. *Construct. Build. Mater.* 2022; 341: 127847.
160. Thiery M, Dangla P, Belin P, Habert G, Roussel N. Carbonation kinetics of a bed of recycled concrete aggregates: A laboratory study on model materials. *Cement Concr. Res.* 2013; 46: 50–65.
161. Zhang J. Performance Enhancement of Recycled Concrete Aggregates through Carbonation. 2015; 27.
162. Zhang J, Shi C, Li Y, Pan X, Poon CS, Xie Z. Influence of carbonated recycled concrete aggregate on properties of cement mortar. *Construct. Build. Mater.* 2015; 98: 1–7.
163. Kou SC, Zhan BJ, Poon CS. Use of a CO2 curing step to improve the properties of concrete prepared with recycled aggregates. *Cement Concr. Compos.* 2014; 45: 22–28.
164. Zhan B, Poon CS, Liu Q, Kou S, Shi C. Experimental study on CO2 curing for enhancement of recycled aggregate properties. *Construct. Build. Mater.* 2014; 67: 3–7.
165. Winter MG. A conceptual framework for the recycling of aggregates and other wastes. *Municip. Eng.* 2002; 151(3): 177–187.
166. RILEM. Final Report of the RILEM Technical Committee 217-PRE. *Springer Netherlands.* 2013.
167. Tam VWY, Soomro M, Evangelista ACJ. A review of recycled aggregate in concrete applications. *Construct. Build. Mater.* 2018; 172: 272–292.
168. Karimi Pour A, Brito J, Ghalehnvoi M, Gencel O. Torsional behaviour of rectangular high-performance fibre-reinforced concrete beams. *Structures.* 2022; 35: 511–519.
169. Karimi Pour A, Ghalehnovi M, Edalati M, Brito J. Properties of fibre-reinforced high-strength concrete with nano-silica and silica fume. *Appli. Sci.* 2021; 11: 9696.
170. Karimi Pour A, Jahangir H, Rezazadeh Eidgahee D. A thorough study on the effect of red mud, granite, limestone and marble slurry powder on the strengths of steel fibres-reinforced self-consolidation concrete: Experimental and numerical prediction. *J. Build. Eng.* 2021; 44: 103398.
171. Gencel O, Kazmi SMS, Munir MJ, Kaplan G, Bayraktar OY, Ozturk Yarar D, Karimi Pour A, Ahmad MA. Influence of bottom ash and polypropylene fibers on the physico-mechanical, durability and thermal performance of foam concrete: An experimental investigation. *Construct. Build. Mater.* 2021; 306: 124887.
172. Ghalehnovi M, Karimi Pour A, Anvari A, Brito J. Flexural strength enhancement of recycled aggregate concrete beams with steel fibre-reinforced concrete jacket. *Eng. Struct.* 2021; 240: 112325.
173. Ghalehnovi M, Karimi Pour A, Brito J, Chaboki HR. Crack width and propagation in recycled coarse aggregate concrete beams reinforced with steel fibres. *Appl. Sci.* 2020; 10.

174. Karimi Pour A. Effect of untreated coal waste as fine and coarse aggregates replacement on the properties of steel and polypropylene fibres reinforced concrete. *Mech Mat.* 2020; 150: 103592.

175. Rezaiee-Pajand M, Karimi Pour A, Mohebbi Najm Abad J. Crack spacing prediction of fibre-reinforced concrete beams with lap-spliced bars by machine learning models. *Iran. J. Sci. Technol. Trans Civ. Eng.* 2020; 45: 833–850.

176. Anvari A, Ghalehnovi M, Brito J, Karimi Pour A. Improved bending behaviour of steel fibres recycled aggregate concrete beams with a concrete jacket. *Mag. Con Res.* 2019; 1–19.

177. Farokhpour M, Ghalehnovi M, Karimi Pour A, Amanian M. Effect of polypropylene fibers on the behavior of recycled aggregate concrete. *5th Nat Conf on Rec Achin Civ Eng, Arch and Urba.* 2019.

178. Farokhpour Tabrizi M, Ghalehnovi M, Karimi Pour A. The effect of polypropylene fibers on the recycled aggregate concrete. *6th Nat Conf on Appl Res in Civ Eng, Arch, and Urb Man.* 2019.

179. Ghalehnovi M, Farokhpour Tabrizi M, Karimi Pour A. Investigation of the effect of steel fibers on failure extension of recycled aggregate concrete beams with lap-spliced bars. *Sharif J. Civ. Eng.* 2019; 18(12).

180. Farokhpour M, Ghalehnovi M., Karimi Pour A. Structural performances of concrete beams with hybrid, fiber-reinforced polymer-steel reinforcements. *7th Nat and 3th Int Conf in Civ Eng.* 2018.

181. Karimi Pour A, Ghalehnovi M, farokhpour M. The effect of polypropylene fiber on recovered aggregate concrete. *5th Nati Conf on Rec Ach in Civ Eng, Arch and Urb.* 2018.

182. Karimi Pour A. *Investigation of Lap-Spliced Reinforcing Bars in Steel Fibres' Concrete (SFC) Under Static and Cyclic Loading.* Ferdowsi University of Mashhad. 2017.

183. Karimi Pour A, Esfahani MR. The effect of steel fibers on flexural cracking of fiber in reinforced concrete beams with lap-spliced bars. *J. Ferdowsi Civil Eng.* 2017; 31.

184. Karimi Pour A, Shirkhani A, Kırgız MS, Noroozinejad Farsangi E. Experimental investigation of GFRP-reinforced concrete columns made with waste aggregates under concentric and eccentric loads. *Stru. Conc.* 2023; 24(1).

185. Karimi Pour A. Experimental and numerical evaluation of steel fibres RC patterns influence on the seismic behaviour of the exterior concrete beam-column connections. *Eng. Str.* 2022; 263: 114358.

186. Limbachiya M, Meddah M, Ouchagour Y. Performance of portland silica fume cement concrete produced with recycled concrete aggregate. *ACI Mater. J.* 2012; 109(1): 91–100.

187. Kou SC, Poon CS, Agrela F. Comparisons of natural and recycled aggregate concretes prepared with the addition of different mineral admixtures. *Cement Concr. Compos.* 2011; 33(8): 788–795.

188. Singh LP, Karade SR, Bhattacharyya SK, Yousuf MM, Ahalawat S. Beneficial role of nanosilica in cement based materials – A review. *Construct. Build. Mater.* 2013; 47: 1069–1077.

189. Lima C, Caggiano A, Faella C, Martinelli E, Pepe M, Realfonzo R. Physical properties and mechanical behaviour of concrete made with recycled aggregates and fly ash. *Construct. Build. Mater.* 2013; 47: 547–559.

190. Somna R, Jaturapitakkul C, Rattanachu P, Chalee W. Effect of ground bagasse ash on mechanical and durability properties of recycled aggregate concrete. *Mater. Des.* 2012; 36: 597–603.

191. Somna R, Jaturapitakkul C, Made AM. Effect of ground fly ash and ground bagasse ash on the durability of recycled aggregate concrete. *Cement Concr. Compos.* 2012; 34(7): 848–854.

192. Kurda R, de Brito J, Silvestre JD. Influence of recycled aggregates and high contents of fly ash on concrete fresh properties. *Cement Concr. Compos.* 2017; 84: 198–213.

193. Tangchirapat W, Buranasing R, Jaturapitakkul C. Use of high fineness of fly ash to improve properties of recycled aggregate concrete. *J. Mater. Civ. Eng.* 2010; 22(6): 565–571.

194. Wang J, Vandevyvere B, Vanhessche S, Schoon J, Boon N, Belie ND. Microbial carbonate precipitation for the improvement of quality of recycled aggregates. *J. Clean. Prod.* 2017; 156: 355–366.

195. Hammes F, Verstraete W. Key roles of pH and calcium metabolism in microbial carbonate precipitation. *Rev. Environ. Sci. Biotechnol.* 2002; 1(1): 3–7.

196. Grabiec AM, Klama J, Zawal D, Krupa D. Modification of recycled concrete aggregate by calcium carbonate biodeposition. *Construct. Build. Mater.* 2012; 34: 145–150.

197. Qiu D, Tng QS, Yang EH. Surface treatment of recycled concrete aggregates through microbial carbonate precipitation. *Construct. Build. Mater.* 2014; 57: 144–150.

198. D.I.N. DIN. 4226-100 Aggregates for Concrete and Mortar - Part 100: Recycled Aggregates. Deutsches Institut Fur Normung. 2002.

199. E.N. DIN. 12620 - Aggregates for Concrete. Deutsches Institut Fur Normung. 2003.

200. LNEC, E 471-2009. Guia para utilizaçao de agregados reciclados grossos em betoes de ligantes hidraulicos, Laboratorio Nacional de Engenharia Civil. Ministerio das Obras Públicas, Transportes e Comunicaçoes, Lisboa, PRT, 2009; 8.

201. CSIRO HB. Guide to the Use of Recycled Concrete and Masonry Materials. *Commonwealth Scientifc and Industrial Research Organisation (CSIRO). Australia.* 2002; 80.
202. Ghoniem A. Deep learning shear capacity prediction of fibrous recycled aggregate concrete beams strengthened by side carbon fiber-reinforced polymer sheets. *Comp. Stru.* 2022; 300: 116137.
203. Ke XJ, Tang ZK, Yang CH. Shear bearing capacity of steel-reinforced recycled aggregate concrete short beams based on modified compression field theory. *Structures.* 2022; 45: 645–658.
204. Abushanab A, Alnahhal W. Flexural behavior of reinforced concrete beams prepared with treated wastewater, recycled concrete aggregates, and fly ash. *Structures.* 2022; 45: 2067–2079.
205. Yuan CM, Cai J, Chen QJ, Liu X, Huang H, Zuo Z, He A. Experimental study on seismic behaviour of precast recycled fine aggregate concrete beam-column joints with pressed sleeve connections. *J. Build. Eng.* 2022; 58: 104988.
206. Gao D, Zhu W, Fang D, Tang J, Zhu H. Shear behavior analysis and capacity prediction for the steel fiber reinforced concrete beam with recycled fine aggregate and recycled coarse aggregate. *Structures.* 2022; 37: 44–55.
207. Kumar M, Ekbote AG, Singh PK, Rajhans P. Determination of shear strength of self-compacting treated recycled aggregate concrete beam elements. *Mater. Today.* 2022; 65(2): 715–722.
208. Visintin P, Dadd L, Alam MU, Xie T, Bennetta B. Flexural performance and life-cycle assessment of multi-generation recycled aggregate concrete beams. *J. Clean. Prod.* 2022; 360: 132214.
209. Anike EE, Saidani M, Olubanwo AO, Anya UC. Flexural performance of reinforced concrete beams with recycled aggregates and steel fibres. *Structures.* 2022; 39: 1264–1278.
210. Seara-Paz S, González-Fonteboa B, Martínez-Abella F, Eiras-López J. Deformation recovery of reinforced concrete beams made with recycled coarse aggregates. *Eng. Struct.* 2022; 51: 113482.
211. Pan Z, Zheng W, Xiao J, Chen Z, Chen Y, Xu J. Shear behavior of steel reinforced recycled aggregate concrete beams after exposure to elevated temperatures. *J. Build. Eng.* 2022; 48: 103953.
212. Seara-Paz S, Gonzalez-Fonteboa B, Martínez-Abella F, Eiras-Lopez J. Flexural performance of reinforced concrete beams made with recycled concrete coarse aggregate. *Eng. Struct.* 2018; 156: 32–45.
213. Choi WC, Yun H, Kim SW. Flexural performance of reinforced recycled aggregate concrete beams. *Mag. Concr. Res.* 2012; 64(9): 837–848.
214. Zhao S, Sun C. Experimental study of the recycled aggregate concrete beam flexural performance. *Appl. Mech. Mater.* 2013; 368–370.
215. Arezoumandi M, Smith A, Volz JS, Khayat KH. An experimental study on flexural strength of reinforced concrete beams with 100% recycled concrete aggregate. *Eng. Struct.* 2015; 88: 154–162.
216. Kang THK, Kim W, Kwak YK, Hong SG. Flexural testing of reinforced concrete beams with recycled concrete aggregates. *ACI Struct. J.* 2014; 111(3): 607–616.
217. Ignjatović IS, Marinković SB, Mišković ZM, Savić AR. Flexural behavior of reinforced recycled aggregate concrete beams under short-term loading. *Mater. Struct.* 2013; 46(6): 1045–1059.
218. Bai W, Sun B. Experimental study on flexural behavior of recycled coarse aggregate concrete beam. *Appl. Mech. Mater.* 2010; 29(32): 543–548.
219. Pradhan S, Kumar S, Barai SV. Performance of reinforced recycled aggregate concrete beams in flexure: Experimental and critical comparative analysis. *Mater. Struct.* 2018; 51(3): 1–17.
220. Butler LJ, West LS, Tighe SL. Bond of reinforcement in concrete incorporating recycled concrete aggregates. *J. Struct. Eng.* 2015; 141(3).
221. Seara-Paz S, Gonzalez-Fonteboa B, Eiras-Lopez J, Herrador MF. Bond behavior between steel reinforcement and recycled concrete. *Mater. Struct.* 2014; 47(1–2): 323–334.
222. Hamad BS, Dawi AH, Daou A, Chehab GR. Studies of the effect of recycled aggregates on flexural, shear, and bond splitting beam structural behavior. *Case Stud. Construct. Mater.* 2018; 9.
223. Mohammed TU, Shikdar KH, Awal MA. Shear strength of RC beam made with recycled brick aggregate. *Eng. Struct.* 2019; 189: 497–508.
224. Arezoumandi M, Smith A, Volz JS, Khayat KH. An experimental study on shear strength of reinforced concrete beams with 100% recycled concrete aggregate. *Construct. Build. Mater.* 2014; 53: 612–620.
225. Wardeh G, Ghorbel E, Fouré B. Poutres soumises à l'effort tranchant, in: F. Larrard, H. Colina (Eds.), *Le Beton Recycle, Marne-la-Vallee*: Ifsttar. 2018; 459–473.
226. Tosic N, Marinkovic S, Ignjatovic I. A database on flexural and shear strength of reinforced recycled aggregate concrete beams and comparison to Eurocode 2 predictions. *Construct. Build. Mater.* 2016; 127: 932–944.
227. Pacheco JN, de Brito J, Chastre C, Evangelista L. Uncertainty of shear resistance models: Influence of recycled concrete aggregate on beams with and without shear reinforcement. *Eng. Struct.* 2020; 204.

18 Seismic Multi-Hazard Risk and Resilience Modeling of Networked Infrastructure Systems

Milad Roohi, Saeideh Farahani, Ali Shojaeian, and Behrouz Behnam

CONTENTS

18.1 INTRODUCTION

Communities worldwide increasingly rely on civil infrastructure to provide essential services that promote economic growth, social inclusion, governance, and quality of life (Roohi et al., 2020). Severe earthquake-induced hazards manifest themselves in various physical, social, and economic losses to the infrastructure and subsequently disrupt the functionality of communities. In recent decades, substantial lifeline damages have been observed during various earthquakes, including Loma Prieta, 1989; Northridge, 1994; Kobe, 1995; and Tohoku, 2011, which has highlighted a vital need to adopt proper mitigation, recovery, and adaptation strategies to help communities worldwide to improve the integrity, sustainability, and resilience of their civil infrastructure and lifeline systems (Roohi et al., 2020).

Researchers and engineers have expended significant efforts to accelerate the development of knowledge and tools for seismic risk and resilience modeling to enable stakeholders, insurers, and officials better understand and manage their risk. The modeling aims to quantify probabilistic metrics

DOI: 10.1201/9781003325246-18

that can assist in evaluating the current condition of infrastructure and identifying its vulnerability. This requires estimating the seismic damage and consequent losses of lifeline systems that are crucial for emergency planners, government, and financial organizations. In recent years, many studies have addressed seismic vulnerability and resilience of lifeline systems. Furthermore, in the United States and other parts of the world, extensive investments have been made to develop advanced regional risk and community resilience platforms. However, the use of these platforms (especially in underdeveloped and developing countries) would not be feasible due to the lack of access to these platforms, and limited resources allocated for data collection, resilience planning, and adaptation.

This chapter is an effort to review recent advancements in multi-hazard seismic risk and resilience of civil infrastructure and to adopt and advance the knowledge and tools for developing seismic risk and resilience models for civil infrastructure with a focus on underdeveloped and developing countries. First, this chapter elaborates on recent advancements in seismic risk and resilience modeling. This is followed by presenting a general method of approach for modeling seismic risk and resilience of networked infrastructure to introduce readers to various aspects of modeling including data collection, hazard modeling, vulnerability and recovery modeling, and risk/resilience quantification. This chapter continues by presenting a real-world case study to illustrate the application of the method of approach on civil infrastructure located in Iran. The case study case presents fire-following earthquake impact modeling of urban gas network infrastructure subject to two major geological seismic hazards, including ground shaking and ground failure simultaneously.

18.2   RECENT ADVANCEMENTS IN SEISMIC RISK AND RESILIENCE OF NETWORKED INFRASTRUCTURE

The birth of seismic catastrophe risk modeling goes back to the 1980s when initial efforts began to develop computer-based models to measure natural hazard loss potential by linking actuarial science, engineering, meteorology, and seismology with advances in information technology and geographic information systems (GIS). These models usually consist of four main components, including 1) the exposure component that provides portfolio characteristics in terms of attributes such as geography, construction, occupancy, and year built; 2) the hazard component that simulates the intensity and distribution of hazard events within the region of study; 3) vulnerability component that assesses physical damage and loss ratio for structures and their content; and 3) financial component that translates all physical loss ratios into economic metrics.

Following the 1994 Northridge and 1995 Kobe earthquakes, researchers and engineers realized the need to develop seismic design and assessment methods, which can improve the seismic vulnerability of structures and mitigate earthquake losses. These efforts resulted in the development of an important engineering concept known as performance-based earthquake engineering (PBEE). PBEE includes concepts and techniques related to the design, construction, and maintenance of structures to assure predictable performance objectives are met under earthquakes (SEAOC, 1995). The Pacific Earthquake Engineering Research Center (PEER) developed performance-based engineering (PBE) to ensure (as much as possible) predictable performance objectives are met subject to natural hazards (Moehle and Deierlein, 2004). This framework provides a robust and probabilistic methodology based on three logical steps, including 1) hazard analysis, 2) damage analysis, and 3) loss analysis. The outcome of every step is characterized by one of three generalized variables: Intensity Measure (IM), Damage Measure (DM), and Decision Variable (DV). Using the Total Probability Theorem, the PEER PBEE framework equation can be expressed:

$$\lambda[DV] = \iiint p[DV \mid DM]\, p[DM \mid EDP]\, p[EDP \mid IM]\, p[IM \mid D]\, dIM\,.\,dEDP\,.\,dDM \quad (18.1)$$

where the expression $p[X \mid Y]$ refers to the probability density of X conditioned on knowledge of Y; D denotes facility location, structural, non-structural, and other features; $p[IM \mid D]$ is the probability of experiencing a given level of intensity; $p[EDP \mid IM]$ is the conditional probability of experiencing a level of response, given a level of ground motion intensity; $p[DM \mid EDP]$ is the conditional probability of experiencing the damage state, given a level of structural response; $p[DV \mid DM]$ is the conditional probability of experiencing a loss of certain size, given a level of damage. The expected loss or value of the decision variable $p[DV]$ is calculated as the sum of these quantities' overall levels of intensity, response, damage, and loss. An alternative expression for the PEER framework in Equation (18.1) can be written as

$$\lambda[DV] = \iiint p[DV \mid DM]\, p[DM \mid IM]\, p[IM \mid D]\, dIM \,.\, dEDP \,.\, dDM \qquad (18.2)$$

where $p[DM \mid IM]$ is the conditional probability of experiencing the damage state, given an intensity level, which can be calculated using fragility functions. A fragility function of a structural system, F_r, relates a given hazard intensity measure to the conditional probability of exceeding a particular damage state k, given by DS_k. Damage states consist of insignificant (DS_1), moderate (DS_2), extensive (DS_3), and complete (DS_4). The fragility functions are often described by a lognormal cumulative distribution function (Kennedy and Ravindra, 1984), given by

$$F_r[IM] = p[DM \mid IM] = p[DS > DS_k \mid IM] = \Phi\left[\frac{1}{\beta}.ln\left(\frac{IM}{IM_{med}}\right)\right] \qquad (18.3)$$

where IM is the intensity measure at the location of the structure. Φ is the standard cumulative probability function. β and IM_{med} denote the parameters of the lognormal cumulative distribution function corresponding to a particular damage state k.

The Federal Emergency Management Agency (FEMA) in the United States developed the HAZUS methodology (FEMA, 2020) which "is a nationally applicable standardized methodology that contains models for estimating potential losses due to earthquakes, floods, and hurricanes. HAZUS uses geographic information systems (GIS) technology to estimate the physical, economic and social impacts of disasters. It graphically illustrates the limits of identified high-risk locations due to earthquakes, hurricanes, and floods. Users can visualize the spatial relationships between populations and other fixed geographic assets or resources for the specific hazard being modeled – a crucial function in the pre-disaster planning process."

Furthermore, FEMA P-58 (FEMA, 2012a) provides a guideline to quantify the performance measures of a structure in terms of casualties, repair cost, repair time, and unsafe placards. These measures can be compared with performance objectives to determine if a structural system meets the requirements defined by decision-makers. The P-58 procedure is a flexible and not a prescriptive procedure and it also provides an application called Performance Assessment Calculation Tool (PACT) for users.

The main drawback of the methodologies and guidelines as HAZUS and FEMA P8 is that they assess the vulnerability of an infrastructure system without considering the complex interdependencies embedded (Rinaldi et al., 2001). This highlighted the need to develop resilience modeling methodologies to account for the interaction of an individual structural system or components with other systems within a community including physical, social, and economic systems. The resilience modeling of a community provides the analyst with the ability to estimate the probability of expected damages and losses in component- and system-level scales subjected to a natural hazard and, subsequently, make informed decisions (Roohi et al., 2020). Accordingly, community resilience modeling has been the focus of numerous research studies. Bruneau et al. (2003) developed a conceptual framework to define and quantitatively assess the seismic resilience of communities. The framework was further extended and illustrated by Renschler et al. (2010) and led to the development of the PEOPLES framework for evaluating resilience based on seven identified dimensions (Renschler et al., 2010; Cimellaro et al., 2016). The functionality of a system over the post-event recovery time range, $Q(t)$, is often used to quantify the resilience of the system, R, as follows (Bruneau et al., 2003):

$$R = \frac{1}{t_h - t_0} \int_{t_0}^{t_h} Q(t)\, dt \qquad (18.4)$$

where t_0 = the occurrence time of an event and t_h = the investigated time point after the recovery process. The functionality of a system can be narrowly classified into five discrete states consisting of restricted entry, restricted use, pre-occupancy, baseline functionality, and full functionality. In a broader classification, functionality states can be categorized into nonfunctional and functional.

In the literature, there are only a few frameworks capable of performing multi-disciplinary and multi-scale community resilience modeling. The NIST-funded Center for Risk-Based Community Resilience Planning (NIST-CoE) has developed the IN-CORE that allows users to develop physics-based models of interdependent physical systems combined with socio-economic systems and optimize community resilience planning and post-disaster recovery strategies (van de Lindt et al., 2018). Loggins et al. (2019) introduced the Civil Restoration with Interdependent Social Infrastructure Systems (CRISIS) model, a computer-aided decision-support model to find the optimized scheduled repair of damaged civil infrastructures for the restoration of social infrastructure services that depend on the damaged civil systems.

18.3   SEISMIC MULTI-HAZARD RISK AND RESILIENCE MODELING OF NETWORKED INFRASTRUCTURE SYSTEMS

This section presents an integrated probabilistic seismic multi-hazard risk and restoration modeling method of approach for resilience-informed decision-making in infrastructure networks. In particular, the main objective is to account for the multi-hazard effects of earthquakes consisting of seismic waves propagation, liquefaction, and landslide. The individual and combined effects of these three seismic-induced hazards can significantly disrupt the integrity of distributed networked infrastructures. The methodology subsequently estimates the probabilistic measure of the combined damage state of network components, quantifies component- and system-level functionality and restoration, and performs an economic loss analysis. The methodology consists of the following four steps: 1) infrastructure inventory data collection, 2) seismic multi-hazard analysis, 3) physical damage and restoration analysis, and 4) loss estimation. The methodology can be implemented in a manner consistent with the HAZUS guideline. In the following, the main steps of the methodology are discussed in more detail.

18.3.1   INFRASTRUCTURE INVENTORY DATA COLLECTION

The methodology begins by collecting infrastructure inventory data which includes the geographical location of various components of infrastructure systems and the classification of these components according to FEMA standards (FEMA, 2012b). The location data and shapefile map of the network components are the primary data needed for GIS-based analyses. The attribute values of inventory data are then used as input to the following steps to evaluate the physical damage and loss to network components.

18.3.2   SEISMIC MULTI-HAZARD ANALYSIS

The second step is to perform a seismic multi-hazard analysis. This requires first investigating site conditions for the entire distributed network, which consists of the fault and seismic condition of the study region of interest, liquefaction and landslide potential map, and dynamic properties of soil.

In addition, this step requires quantifying potential earth science hazards (PESH) parameters such as permanent ground deformation (PGD) and transient ground displacement (TGD). The first earthquake effect, which can damage lifelines and infrastructure networks, is the TGD, which is caused by seismic wave propagation. The second one is the PGD, which may result in liquefaction, landslide, and ground failure. For seismic risk and resilience assessment of lifelines and infrastructure networks which are widespread throughout the countries, investigating the TGD and PGD is of vital importance.

18.3.2.1 Transient Ground Deformation (TGD)

TGD, which is caused by seismic waves propagation, can be determined using different methods that are defined in the following. Seismic hazard analysis for evaluating PGA can be performed through a deterministic or probabilistic approach. In a deterministic seismic hazard analysis (DSHA) a particular earthquake scenario is assumed based on which the hazard is then estimated (Reiter, 1991). A probabilistic seismic hazard analysis (PSHA) considers the uncertainties of the earthquake in an explicit manner. Some of the most critical uncertainties are the earthquake size, location, and time of occurrence. Some of the attenuation relationships can explicitly calculate peak ground velocity (PGV) by employing PSHA. PSHA begins by describing the magnitude, location, and timing of earthquakes that cause structural damage. A source model composed of N earthquake scenarios given by (E_n) is then developed, where each scenario has magnitude (m_n), location (L_n), and rate (r_n). Each scenario can be written as a function of its magnitude, location, and rate as follows:

$$E_n = E\left(m_n, L_n, r_n\right) \tag{18.5}$$

From the scenario location, distance to the site, D_n, can be determined. With m_n and D_n, the distribution of possible ground-motion levels for each scenario can be obtained using an attenuation relationship as follows:

$$P_n\left(\ln PGA\right) = \frac{1}{\sigma_n \sqrt{2\pi}} e^{-\left(\ln\ PGA - g(m_n, D_n)\right)^2 / 2\sigma_n} \tag{18.6}$$

PSHA can be performed by combining ground motion prediction models through logic trees and quantifying seismic intensity measures consisting of PGA and spectral acceleration (e.g., 1.0-sec spectral acceleration). For estimating the TGD caused by seismic waves propagation (ground shaking), PGV is needed. As HAZUS proposed, for obtaining PGV, the first step is to calculate the spectral acceleration by having a soil classification of a region in terms of dynamic properties. HAZUS technical manual recommends an empirical relationship to calculate PGV as a function of spectral acceleration at T = 1.0 s based on the following equation:

$$PGV = \left(\frac{386.4}{2\pi} \times S_{A1}\right) \Big/ 1.65 \tag{18.7}$$

where S_{A1} is the spectral acceleration at 1.0-s period. It is evident that the prediction of PGV from spectral acceleration may lead to higher degrees of uncertainty compared with a direct calculation procedure.

18.3.2.2 Permanent Ground Deformation (PGD)

Principal forms of PGD are landsliding, seismic settlement, and lateral spreading due to soil liquefaction. Although PGD hazards are usually limited to small regions within the infrastructure networks, their potential for damage is considerable as they can impose large deformation on infrastructure network components. Many models, which consider each of the geotechnical hazards to evaluate the PGD, were developed over the last decade. The main barrier to the implementation of models for lifeline systems is the requirement for very detailed geotechnical data. Therefore, the

HAZUS FEMA (2012b), which is a relatively more uncomplicated and practical model, is a preferable choice for the study plan here.

18.3.2.2.1   Liquefaction

Liquefaction is a soil behavior phenomenon in which saturated soil loses a substantial amount of strength due to excessive pore-water pressure generated by and accumulated during strong earthquake ground shaking. Permanent ground displacements due to lateral spreading and differential settlement are commonly considered significant potential hazards associated with liquefaction. Soil liquefaction has caused significant damage to infrastructure networks in past earthquakes. Zoning regions that might experience liquefaction hazard is hence of particular importance. Once it has been evaluated, the next stage is to determine the likelihood of liquefaction. Defining criteria are required to determine liquefaction susceptibility, and the conditions required to trigger liquefaction are complex and beyond the scopes planned in this section. In HAZUS methodology, liquefaction susceptibility classes are categorized based on deposit type, age, and general distribution of cohesion-less sediments. A conditional probability of liquefaction can be calculated using Equation (18.8) for each of the susceptibility classes, and a given value of PGA. This probability of liquefaction has been developed for $M = 7.5$ earthquake moment magnitude and 5 feet of groundwater depth. The liquefaction possibility is highly influenced by the duration of an earthquake as reflected by earthquake magnitude, M, as well as the groundwater depth:

$$P[Liquefaction] = \frac{P\,[Liquefaction \mid PGA = pga]}{K_M K_W} P_{ml} \qquad (18.8)$$

where $P\,[Liquefaction \mid PGA = pga]$ is the conditional liquefaction probability for a given susceptibility category at a specified level of PGA, P_{ml} is the proportion of the map unit susceptible, K_W is the groundwater correction factor, and K_M is the moment magnitude correction factor. The expected value of liquefaction-induced PGD can be stated here as a function of PGA (Sadigh et al., 1986). K_M is the correction factor which is determined using Equation (18.9), and K_W is the correction factor for groundwater depths other than 5 feet, determined using Equation (18.10):

$$K_M = 0.0027M^3 - 0.0267M^2 - 0.2055M + 2.9188 \qquad (18.9)$$

$$K_W = 0.022d_w + 0.93 \qquad (18.10)$$

where d_w denotes the depth of the groundwater in feet, and P_{ml} is a correction factor that is assumed to be 0.10 for moderately susceptible soils. The conditional expected value of the PGD, which is dependent on the occurrence of liquefaction, can be stated as a function of PGA, as shown in Equation (18.8).

18.3.2.2.2   Landslide

There are three types of landslides, as proposed by Meyersohn (1991). The first type includes rockfall and rock topple, which can cause damage to aboveground infrastructure network components by a direct impact on falling rocks. The second type includes flow and debris flow in which the transported material behaves as a viscous fluid. For loss estimation purposes, this type of landslide is part of the liquefaction potential assessment rather than the landslide potential. The third type includes earth slump and earth slide, in which the earth moves as a block. It is usually developed alongside natural slopes, river channels, and embankments. Earthquake-induced landslide of a hillside slope occurs when the static plus inertia forces within the slide mass cause the factor of safety to drop below 1.0 temporarily. The value of PGA within the slide mass required to cause the factor of safety to drop to 1.0 is denoted by the critical or yield acceleration (ac). This acceleration value is determined based on pseudo-static slope stability analyses and/or empirically based on observations of slope behavior during past earthquakes. The landslide hazard evaluation requires the characterization of a region's landslide susceptibility or sub-regions soil/geologic conditions. The landslide susceptibility is characterized by the geologic group, which is described by slope angle, and critical acceleration. The acceleration required to initiate slope movement is a complex function of slope geology, steepness, groundwater conditions, type of landslide, and history of previous slope performance. The landslide-induced permanent ground displacements are determined using Equation (18.11):

$$E[PGD] = E\left[\frac{d}{a_{is}}\right] a_{is} n \tag{18.11}$$

where $E\left[\dfrac{d}{a_{is}}\right]$ is the expected displacement factor, a_{is} is the induced acceleration (in a decimal fraction of g's), and n is the number of cycles. A relation derived from the results of Makdisi et al. (1978) is used to calculate downslope displacements. Based on HAZUS, at any given location, there is a specified probability of having a landslide susceptible deposit, and that landsliding either occurs or does not occur within susceptible deposits depending on whether the induced peak ground acceleration exceeds the critical acceleration.

18.3.3 PHYSICAL DAMAGE AND RESTORATION ANALYSIS

The third step is to perform damage and restoration analysis by i) estimating the exceeding probability for each damage state given PESH parameters (i.e., PGD and TGD) for each network component including nodes and links, and ii) using the exceeding probability of damage estimates to quantify functionality of system components days following an event. The following two subsections discuss these two steps in more detail.

18.3.3.1 Physical Damage Analysis

Damage analysis can be performed using fragility functions; assuming that infrastructure components having similar characteristics would perform identical subject to a given hazard event. Thus, fragility functions associated with various damage states are assigned to each component given its structural characteristics. A fragility function relates a given hazard intensity measure to the conditional probability of reaching or exceeding a particular damage state i, given by $DS(i)$. Damage states consist of insignificant (or none), $DS(1)$; moderate, $DS(2)$; extensive, $DS(3)$; and complete, $DS(4)$. The fragility functions are often described by a lognormal cumulative distribution function given by

$$P\left[ds \geq DS(i) \mid PESH\right] = \phi\left[\frac{1}{\beta_{DS(i)}} \ln\left(\frac{PESH}{\overline{PESH}_{DS(i)}}\right)\right] \tag{18.12}$$

where, $\overline{PESH}_{DS}$ is the median value of $PESH$ parameter at which the infrastructure component reaches the threshold of damage state, DS. β_{DS} is the standard deviation of the natural logarithm of $PESH$ parameter for DS. ϕ is the standard normal cumulative distribution function. The damage analysis process should account for infrastructure components' vulnerability to various $PESH$ parameters as infrastructure are vulnerable due to PGDs only, but bridges are vulnerable subject to both PGDs and TGDs. Therefore, the probability of exceeding each damage state for bridges must be combined to accurately account for the combined effects of $PESH$ parameters and obtain the combined probability of exceeding a specific damage state $DS(i)$ given by $P_{COMB}\left[ds \geq DS(i)\right]$ and calculated using the following equation:

$$P_{COMB}\left[ds \geq DS(i)\right] = 1 - \prod_{j=1}^{m}\left(1 - P\left(PESH_j\right) P[ds \geq DS(i) \mid PESH_j]\right) \tag{18.13}$$

where j is the *PESH* parameter index, and m is the number of *PESH* parameters considered in damage analysis. The combined probabilities for exceeding various damage states are then used to calculate discrete damage state probabilities, given by $P_{COMB}[ds = DS(i)]$, as follows:

$$P_{COMB}[ds = DS(1)] = 1 - P_{COMB}[ds \geq DS(2)] \tag{18.14}$$

$$P_{COMB}[ds = DS(i)] = P_{COMB}[ds \geq DS(i)] - P_{COMB}[ds \geq DS(i+1)] \tag{18.15}$$

$$P_{COMB}[ds = DS(n)] = P_{COMB}[ds \geq DS(n)] \tag{18.16}$$

18.3.3.2 Restoration Analysis

Many studies have provided various mathematical models and algorithms to analyze and optimize the disrupted components' restoration and recovery process to maximize the system's resilience. Two types of general trends are realized by analyzing these studies, which are (i) top-to-bottom (centralized) (González et al., 2016), and (ii) bottom-to-top (decentralized) (Rangrazjeddi et al., 2022) approaches. Based on the HAZUS methodology, the discrete damage state probabilities, $P_{COMB}[ds = DS(i)]$ are used to estimate the functionality of network components and determine the functionality of each component. Functionality is defined as a component's capacity to serve its intended objectives consisting of structural integrity, safety, and utilities. The pre and post even functionality of component i, is denoted by $Q(0)$ and $Q(t)$, respectively, in which functionality states can be categorized into nonfunctional and functional states (i.e., in a broader classification) using a damage-to-functionality model or can be estimated using restoration curves to estimate the functionality of each component give the exceeding probability of damage states days following an event. If the damage-to-functionality models or restoration curves are not available for infrastructure components, the restoration curves can be employed to quantify the functionality percentage given by

$$Q(t) = \sum_{i=1}^{n} \omega_i(t) P_{COMB}[ds = DS(i)] \tag{18.17}$$

where $\omega_i(t)$ is restoration percent for damage state i at time t after an event.

18.3.4 Loss Estimation

The final step is to employ the most probable damage state obtained from the combined discrete damage states probability calculation of Step 3 to perform an economic loss analysis. Direct economic loss estimation of the infrastructure systems depends on the components' locations and associated seismic hazard intensity measures. The economic loss of each segment, represented by *EL*, is calculated based on probabilities of being in a particular damage state $\left(P_{COMB}[ds = DS(i)]\right)$, the damage ratios ($DR_i$) for each damage state, $DS(i)$, and the replacement value of the component. *EL*s are computed by multiplying the compounded damage ratio (DR_C) by the replacement value (RV), for k^{th} segment, as follows:

$$EL_k = DR_{Ck} RV \tag{18.18}$$

where the compounded damage ratio (DR_{Ck}) is evaluated as the probabilistic combination of damage ratios as follows:

$$DR_{Ck} = \sum_{i=0}^{n} DR_i \times P[ds = DS(i)]_k \tag{18.19}$$

where $P[ds = DS(i)]$ is the probability of being in the damage state i.

18.4 REAL-WORLD ILLUSTRATIVE CASE STUDY OF CIVIL INFRASTRUCTURE SYSTEM

Iran is located on the Alpide earthquake belt, in the active collision zone between the Eurasian and Arabian plates. This issue makes Iran a country that suffers from geotechnical seismic hazards associated with frequent destructive earthquakes. The geographical and geological variety of the country, the existence of different mountain ranges, and coastal areas with sandy soils, among other reasons, are the main factors that make Iran a country where landslides and liquefaction can occur during or after an earthquake (Farahani et al., 2020a; Shojaeian et al., 2021). Iran is selected as a developing country in one of the most seismic-prone areas in the world to show the illustrative case study that can demonstrate the recent advancements in multi-hazard seismic risk and resilience of civil infrastructure and to adopt and advance the knowledge and tools for developing seismic risk and resilience models.

Urban gas pipeline network has a vital role in the severity of earthquake damage as gas pipelines are prone to various hazards, mainly because they may pass through regions with different geographical characteristics where earthquake response may be different (Wijaya et al., 2019). A gas distribution system is composed of pipelines, natural gas pressure regulating stations, valves, and demand nodes. The seismic-induced damages to the pipelines have direct effects on the integrity of components and performance of the system (Esposito, 2011; Behnam, 2017; Farahani et al., 2020b). During an earthquake, damage to high-pressure natural gas distribution networks, buried fuel pipelines, and aerial power distribution lines damages can cause ignitions and generate widespread fire. This section presents a comprehensive seismic and post-seismic risk assessment of the gas pipeline network of Asaluyeh city in Iran.

18.4.1 Gas Network Inventory Data Collection

In the current section, the earthquake and post-earthquake vulnerability of gas pipelines of Asaluyeh, a city in the south of Iran, is investigated. As shown in Figure 18.1, Asaluyeh is a sleepy coastal city, near the Persian Gulf. Seismologically, the city is located in the Zagros earthquake state, which is one of the seismic regions of Iran that is a part of the Himalayan Alpine belt and one of the youngest continental encounter zones on the Earth. The gas pipeline network in the city is also shown in Figure 18.1. As documented by the Bushehr Province gas organization, there are three main lines with a diameter of 8 in. About 40% of the secondary pipes have a diameter of 6 in., and the rest lines have a 4 in. diameter. The pipe diameter inside buildings is 2 in., but they are not involved here for simplicity.

18.4.2 Hazard Analysis

18.4.2.1 Probabilistic Seismic Hazard Analysis

Here, the historical and instrumental records covering the period from the eighth century to the present to a radius of 200 km are used. The Kijko method is used to estimate seismic parameters (Kijko et al., 2016). To determine the maximum acceleration on the bedrock, three attenuation relationships are used with the help of a logic tree. Due to the small size of the area (less than one geographical degree), as well as the seismicity impact of neighboring regions in the case studied, earthquake occurrences in a radius of approximately 200 km from the city center have been taken into account. The dispersion of earthquakes with magnitudes of more than 3 Richter (both historical and instrumental) is shown in Figure 18.2. Also, Figure 18.3 shows the range of seismic sources in the city.

To determine the spectral acceleration, six different attenuation relationships are employed. Finally, the uniform hazard spectrum, which is determined using the logic tree method and probabilistic approach, is presented in Figure 18.4. The spectral accelerations (SA) in different return

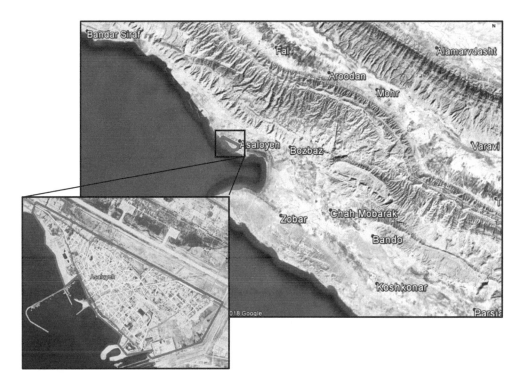

FIGURE 18.1   The borders and the gas pipeline network in Asaluyeh City.

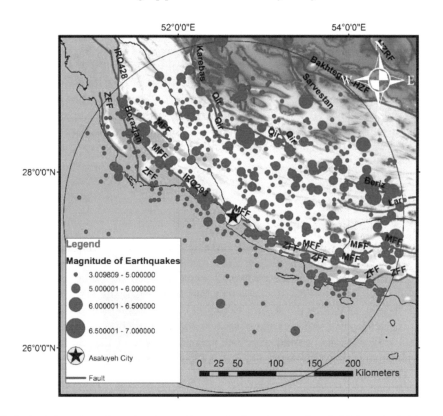

FIGURE 18.2   Dispersion of earthquakes with magnitudes of more than 3 Richter (both historical and instrumental).

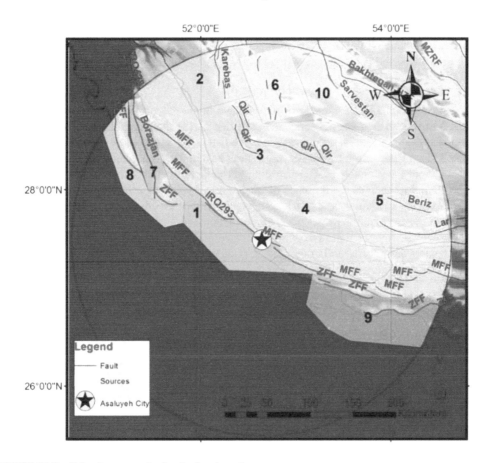

FIGURE 18.3 Seismic sources in the Asaluyeh region.

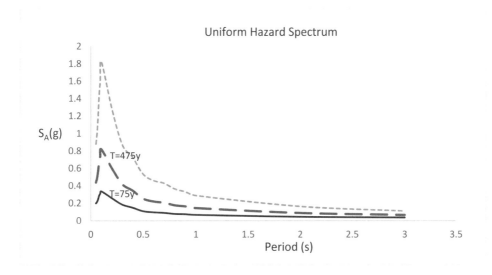

FIGURE 18.4 Uniform hazard spectrum (using the probabilistic seismic hazard analysis).

periods can be obtained using Figure 18.4. Therefore, using Equation (18.7) makes it possible to calculate PGV throughout the length of the pipes. For a 475-year return period, the spectral acceleration for the 1-s period is 0.14 g; hence, the PGV values can be determined.

18.4.2.2 Permanent Ground Deformation (PGD)

The geo-seismic potential for damage is considerable as they can impose large deformation on pipelines. Where the urban gas networks were affected by the mentioned hazards, relatively high pipeline damage rates are often observed in localized areas. The geological investigations indicate that the soil has a slope angle below 10° in all parts of the city. Thus, liquefaction is assumed as the only source of the probable PGD. According to the geotechnical and geological conditions of the Asaluyeh region, such as the presence of sandy and silty soils and high levels of groundwater on the coast during earthquakes, the risk of liquefaction will be considered as a probable factor that can affect the gas pipeline by permanent deformations. As the Asaluyeh sediments are related to the Holocene, the liquefaction susceptibility is categorized in a range with moderate sensitivity. Interested readers are referred to Farahani et al. (2020c) for further information about the hazard analysis for this case study.

18.4.3 Vulnerability Modeling and Physical Damage Analysis

For pipelines, two damage states, which are leak and break, are considered. Generally, when a pipe sustains a degree of damage due to ground failure, the type of damage is likely to be a break, while when a pipe experiences seismic wave propagation, the type of damage is likely to be local leaking. According to the HAZUS, the seismic pipeline damage is independent of the pipes' size, pipes' class, and mechanical specifications. The only available criterion for damage evaluation of the pipelines is either brittle or ductile. The brittle pipelines are old and connected using gas-welded joints, while the welded steel pipes with arc-welded joints are classified as ductile pipelines. Given the fact that the Asaluyeh gas pipelines are ductile, damage due to seismic waves consists of 80% leaks and 20% breaks, while damage due to ground failure consists of 20% leaks and 80% breaks. Equations (18.20) and (18.21) express the relationship between repair rate and PGV, and PGD, respectively.

$$R.R = 0.00003 \times PGV^{2.25} \qquad (18.20)$$

$$R.R = 0.3 \times Prob[liq] \times PGD^{0.56} \qquad (18.21)$$

The quantity and the probability of leaks and breaks because of PGD and PGV are shown in Table 18.1. The pipelines' damage probability is calculated based on the following experimental relationship (Equation (18.22)) by regarding the fact that the number of pipe failures follows Poisson's distribution (O'Rourke and Jeon, 1999):

$$P_f = 1 - exp(-R.R \times L) \qquad (18.22)$$

TABLE 18.1

The Number and the Probability of Leaks and Breaks because of PGD and PGV

	Number of Failures	Failure Probability	Leak	Leakage Probability	Break	Break Probability
PGV	1.75	0.826	1.40	0.753	0.35	0.295
PGD	1.25	0.713	0.25	0.221	1.00	0.632
SUM	3.00	-	1.65	-	1.35	-

18.4.4 Fire Following Earthquake Analysis

A fault tree analysis (FTA) is used here to calculate the ignition probability of damaged pipelines as schematically shown in Figure 18.5. Two factors, which are considered as main factors of the post-earthquake ignition, are the spark source and the failure of the gas network. The damage to the aerial power distribution lines is supposed as an only factor a causing spark. Furthermore, it is assumed that the probability of a spark is equal to the probability of a spark caused due to the aerial power lines' damages. The mentioned probability is assumed to be 0.6 (Crespellani et al., 2006). The vulnerability of the electric power networks at different failure levels is calculated based on the fragility curves presented by HAZUS. Given the PGA of 0.31 g and the fragility curves, the probability of electric power lines, which is at a minor failure level, is equal to 64% and 11% at a moderate failure level. According to the HAZUS definition of the different failure levels, the minor level refers to the states that 4% of all circuits will be failed and 12% for the moderate level. Now, using the calculated probabilities of various damage states and employing Equations (18.28) and (18.29), the probability of ignition due to leakage is 0.35, and 0.32 due to breaking:

$$P_{IL} = 1 - (1 - P_{ILS}) \times (1 - P_{ILM}) \tag{18.23}$$

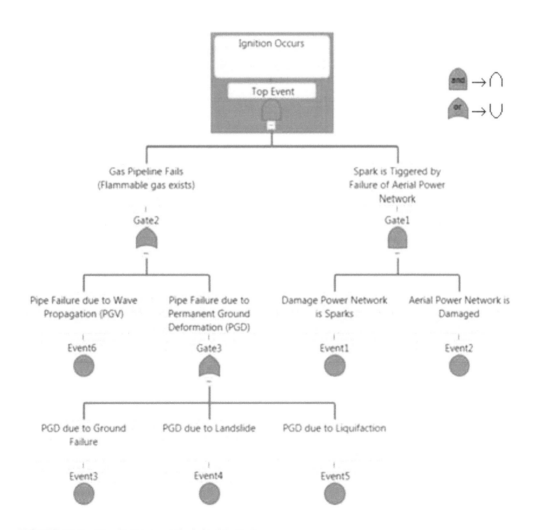

FIGURE 18.5 The fault tree analysis in this study.

$$P_{IB} = 1 - (1 - P_{IBS}) \times (1 - P_{IBM}) \qquad (18.24)$$

For evaluating the consequences of gas pipelines' damages, not only required information on the pipelines is collected, but also climate conditions, which are obtained using the official website of the Iranian meteorological organization, are determined. The main purpose of materials distribution modeling is to estimate the concentration of released material in the environment in three directions over time. In this section, the process hazard analysis software tool (PHAST), which can model material release and calculate the radius of the post-earthquake fire spread, is used.

Since a majority portion of natural gas consists of methane, the target substance in this section is hence methane gas. Although toxic gases are also produced and that can be a damaging factor in some scenarios, it is believed that here heat radiation is a dominant factor. Here, the analysis is performed under a sudden and eruptive fire scenario. As shown in Figure 18.6(a), it can be inferred that the fire jet is propagated up to 45 m in the direction of the wind. A jet fire would occur when a pressurized release is ignited by any source. The jet fire model is based on the radiant fraction of total combustion energy, which is assumed to arise from a point slowly along the jet fame path. Figure 18.6(b) shows the gas concentration in the wind direction. The failure causes the dispersion of gases to the surrounding, and it depends on the wind stability class, wind speed, solar intensity, and ground conditions the concentration may vary. The maximum dispersion reached a distance of 43 m. Figure 18.6(c) presents the flash fire characterized by high temperature, short duration, and a rapidly moving flame front. It occurs when a vapor cloud of flammable material burns. The cloud is typically ignited on the edge and burns toward the release point. The duration of the flash fire is very short (seconds), but it may continue as a jet fire if the release continues. The flash fire reached a maximum distance of 58 m. Finally, Figure 18.6(d) presents the cloud footprint, which is the limited area of gas cloud propagation. According to the analysis results, the cloud reached a maximum downwind distance of 43 m. Human damage is considered equal to the number of people affected by the consequences. Regarding Asaluyeh's population, which is 31,319 people, and the area of the affected radiation zone, the estimated number of damaged people due to the fire following the earthquake is equal to 30 people.

18.5 DISCUSSION AND FUTURE RESEARCH NEEDED

The previous sections presented the recent advancements in seismic risk and resilience modeling of civil infrastructure. Also, a general method of approach was presented to model the impacts of seismic multi-hazards on civil infrastructure. A real-world case study illustrated the implementation of the methodology on urban gas networked infrastructure systems. The results show that the presented methodology can perform component- and system-level damage, functionality, and restoration analysis by accounting for the combined effects of geo-seismic hazards consisting of shake, liquefaction, and landslide, as three major ones. The physical and economic metrics quantified can assist officials with resilience-informed decision-making by identifying vulnerable components/ segments of a distributed infrastructure network along with quantifying the component- and system-level restoration days following a major event. This information can help policymakers and other stakeholders to develop mitigation plans to minimize the impacts of seismic multi-hazards on the most critical and vulnerable segments of the civil infrastructure by i) developing and implementing retrofit plans to reduce immediate loss and improve the resilience of infrastructure components or ii) developing recovery plans that could identify resources (e.g., construction material, labor, and accelerated construction equipment) needed for replacement of infrastructure component that are predicted to be the most vulnerable components of the infrastructure network. The physical models can be chained with advanced socio-economic models to simulate changes in a series of economic stability and sociological metrics.

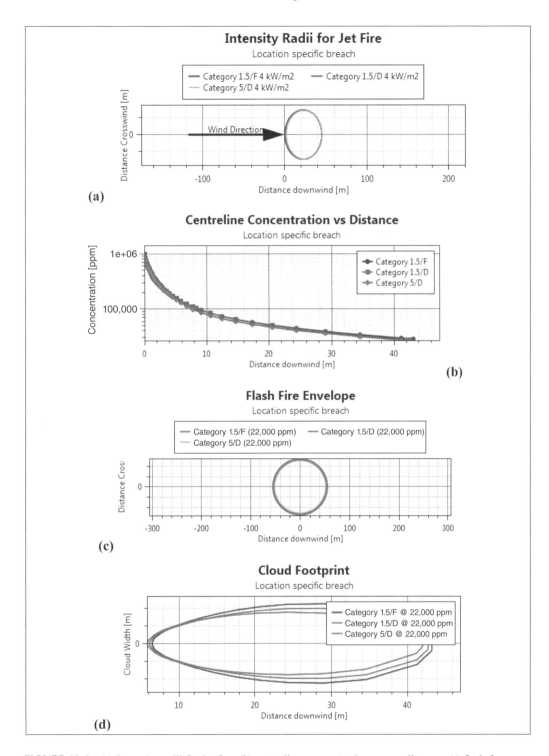

FIGURE 18.6 (a) Intensity radii for jet fire; (b) centerline concentration versus distance; (c) flash fire envelope; (d) cloud footprint.

Despite the recent advancements in community resilience science and tools, there are still several practical and knowledge gaps that can be identified to be a focus of future research. In the following, three major issues and gaps are discussed in detail.

1. Resilience modeling and planning in underdeveloped and developing countries would not be feasible due to limited resources allocated for data collection, resilience planning, and adaptation. One of the main challenges in such countries would be the lack of economic loss, retrofit and repair cost estimation methodologies and datasets to calibrate the accuracy of the economic loss estimation. Effective communication of the seismic risk and resilience studies with policymakers worldwide can facilitate the development and implementation of effective policies that can lead to low-cost solutions to improve the resilience of underdeveloped countries.

2. The research community currently understands very little about how the accuracy of physical damage analysis and post-hazard decision-making can be affected by interdependent infrastructure models of a community, which can be developed with various levels of resolution and uncertainty depending on community data availability, the scale and hazard of interest, and computational resources. This major challenge underscores the need to develop data-driven and model-data fusion methods that can account for data availability, resolution, and uncertainty in risk and resilience modeling. Addressing this knowledge gap will lead to the development of a robust resilience measurement framework that integrates multi-resolution computational models with heterogeneous data to quantify resilience metrics and make informed decisions.

3. There has been significant progress in the civil infrastructure sensing and health monitoring paradigm in which various data-driven, model-driven, and model-data fusion frameworks have been developed for the integrity assessment of civil infrastructure. Still, the linkage between structural health monitoring and infrastructure resilience has not been clearly understood, which highlights the need to develop comprehensive monitoring and control approach to address various challenges toward the development/advancement of smart resilient infrastructure systems. To rapidly and reliably assess the performance and functionality of civil infrastructure systems, it is necessary to develop and implement systematic and autonomous methods. This will involve leveraging recent advancements in structural health monitoring to detect, localize, and quantify structural damage and potentially identify the cause of damage at earlier stages. The data and information obtained from this process can then be incorporated into infrastructure resilience models to enhance their accuracy.

18.6 CONCLUSIONS

This chapter introduced readers to recent advancements in seismic risk and resilience modeling. A general method of approach was presented for multi-hazard seismic risk and resilience modeling civil infrastructure with a focus on underdeveloped and developing countries. A real-world case study was presented to illustrate the application of the method of approach on a civil infrastructure located in Iran. The case presented fire-following earthquake impact modeling of urban gas network infrastructure and demonstrated the metrics that can be quantity and adopted by policymakers and other stakeholders to develop mitigation plans to minimize the impacts of seismic multi-hazards on the most critical and vulnerable segments of the civil infrastructure. Also, an in-depth discussion and future research needs were presented with a focus on 1) advancing the knowledge and tools for resilience modeling and planning in underdeveloped and developing countries, 2) addressing the knowledge gap concerning factors that can affect the accuracy of infrastructure resilience models, and 3) the need to integrate smart sensing and monitoring technologies and data in the process of infrastructure risk and resilience modeling.

REFERENCES

B. Behnam. *Post-Earthquake Fire Analysis in Urban Structures: Risk Management Strategies.* CRC Press, Boca Raton, FL, 2017.

M. Bruneau, S. E. Chang, R. T. Eguchi, G. C. Lee, T. D. O'Rourke, A. M. Reinhorn, M. Shinozuka, K. Tierney, W. A. Wallace, and D. Von Winterfeldt. A framework to quantitatively assess and enhance the seismic resilience of communities. *Earthquake Spectra*, 19(4):733–752, 2003.

G. P. Cimellaro, C. Renschler, A. M. Reinhorn, and L. Arendt. Peoples: a framework for evaluating resilience. *Journal of Structural Engineering*, 142(10):04016063, 2016.

T. Crespellani, J. Facciorusso, and S. Renzi. Seismic risk analysis of a lifeline system subjected to permanent ground deformations. In *First European Conference on Earthquake Engineering and Seismology*, Geneva, Switzerland, 2006.

S. Esposito. Systemic seismic risk analysis of gas distribution networks. 2011.

S. Farahani, B. Behnam, and A. Tahershamsi. Macrozonation of seismic transient and permanent ground deformation of Iran. *Natural Hazards and Earth System Sciences*, 20(11):2889–2903, 2020a.

S. Farahani, B. Behnam, and A. Tahershamsi. Probabilistic seismic multi-hazard loss estimation of Iran gas trunklines. *Journal of Loss Prevention in the Process Industries*, 66:104176, 2020b.

S. Farahani, A. Tahershamsi, and B. Behnam. Earthquake and post-earthquake vulnerability assessment of urban gas pipelines network. *Natural Hazards*, 101(2):327–347, 2020c.

FEMA. *Hazus Earthquake Model Technical Manual (Hazus 4.2 sp3).* Department of Homeland Security, FEMA, Washington, DC, 2020.

FEMA. *Seismic Performance Assessment of Buildings, Volume III*, FEMA p-58. Department of Homeland Security, FEMA, Washington, DC, 2012a.

FEMA. *Multi-Hazard Loss Estimation Methodology: Earthquake Model.* Department of Home Land Security, FEMA, Washington, DC, pages 235–260, 2012b.

A. D. González, L. Dueñas-Osorio, M. Sánchez-Silva, and A. L. Medaglia. The interdependent network design problem for optimal infrastructure system restoration. *Computer-Aided Civil and Infrastructure Engineering*, 31(5):334–350, 2016.

Kennedy, R. P., & Ravindra, M. K. (1984). Seismic fragilities for nuclear power plant risk studies. *Nuclear Engineering and Design*, 79(1), 47–68. doi:10.1016/0029-5493(84)90188-2

A. Kijko, A. Smit, and M. A. Sellevoll. Estimation of earthquake hazard parameters from incomplete data files. Part III. Incorporation of uncertainty of earthquake-occurrence model. *Bulletin of the Seismological Society of America*, 106(3):1210–1222, 2016.

R. Loggins, R. G. Little, J. Mitchell, T. Sharkey, and W. A. Wallace. Crisis: Modeling the restoration of interdependent civil and social infrastructure systems following an extreme event. *Natural Hazards Review*, 20(3):04019004, 2019.

F. Makdisi, H. Seed, and A. Norouzi. Attenuation relations for peak horizontal and vertical accelerations of earthquake ground motion in Iran: a preliminary analysis. 104, 1978.

W. D. Meyersohn. *Analytical and Design Considerations for the Seismic Response of Buried Pipelines.* Cornell University, 1991.

J. Moehle and G. G. Deierlein. A framework methodology for performance-based earthquake engineering. In *13th World Conference on Earthquake Engineering*, volume 679. WCEE Vancouver, 2004.

T. O'Rourke and S.-S. Jeon. Factors affecting the earthquake damage of water distribution systems. In *Optimizing Post-Earthquake Lifeline System Reliability*, pages 379–388. ASCE, 1999.

A. Rangrazjeddi, A. D. González, and K. Barker. Adaptive algorithm for dependent infrastructure network restoration in an imperfect information sharing environment. *PLoS One*, 17(8):e0270407, 2022.

L. Reiter. *Earthquake Hazard Analysis: Issues and Insights.* Columbia University Press, 1991.

C. S. Renschler, A. E. Frazier, L. A. Arendt, G. P. Cimellaro, A. M. Reinhorn, and M. Bruneau. *A Framework for Defining and Measuring Resilience at the Community Scale: The PEOPLES Resilience Framework.* MCEER Buffalo, 2010.

S. M. Rinaldi, J. P. Peerenboom, and T. K. Kelly. Identifying, understanding, and analyzing critical infrastructure interdependencies. *IEEE Control Systems Magazine*, 21(6):11–25, 2001.

M. Roohi, J. W. van de Lindt, N. Rosenheim, Y. Hu, and H. Cutler. Implication of building inventory accuracy on physical and socio-economic resilience metrics for informed decision making in natural hazards. *Structure and Infrastructure Engineering*, 17(4):534–554, 2020.

K. Sadigh, J. Egan, and R. Youngs. *Specification of Ground Motion for Seismic Design of Long Period Structures, Earthquake Notes.* 57:13, 1986.

SEAOC. *Vision 2000: Performance Based Seismic Engineering of Buildings.* Structural Engineers Association of California, 1995.

A. Shojaeian, S. Farahani, B. Behnam, and M. Mashayekhi. Seismic resilience assessment of Tehran's southern water transmission pipeline using GIS-based analyses. *Journal of Numerical Methods in Civil Engineering*, 6(2):93–106, 2021.

J. W. van de Lindt, B. Ellingwood, T. P. McAllister, P. Gardoni, D. Cox, W. G. Peacock, H. Cutler, M. Dillard, J. Lee, L. Peek, et al. Modeling community resilience: Update on the center for risk-based community resilience planning and the computational environment in-core. 2018.

H. Wijaya, P. Rajeev, and E. Gad. Effect of seismic and soil parameter uncertainties on seismic damage of buried segmented pipeline. *Transportation Geotechnics*, 21:100274, 2019.

19 A Novel Design Method of Single TMD Exposed to Seismic Effects

Huseyin Cetin, Ersin Aydin, and Baki Ozturk

CONTENTS

19.1 INTRODUCTION

Classical design in buildings subjected to dynamic phenomena like earthquakes and wind loads attempts to absorb energy via internal force-deformation behavior. After a building material hits its yield point in response to earthquake excitation or other dynamic external stimuli, the energy is attenuated by nonlinear behavior and local damage. In both the classical and contemporary phases of earthquake-resistant structural design, the structure consumes energy by deforming in response to large external forces. Active, passive, semi-active, and mixed control systems were introduced to building systems in the contemporary era, and they are energy-consuming technical components. A properly constructed passive control component called tuned mass damper (TMD) is known to dampen structural vibrations caused by dynamic loads. The efficiency of these dampers in averting the destruction or collapse of vibration-affected buildings have been remarkable. Typically, passive TMDs are tailored to a certain mode, particularly the first mode of structure (Den Hartog, 1956; Falcon et al., 1967; Ioi and Ikeda, 1978; Warburton and Ayorinde, 1980). TMDs are often more effective when situated on the top story of a building (Villaverde and Koyama, 1993; Connor and Klink, 1996; Sato-Brito and Ruiz, 1999). Adjustable mass dampers decrease the system's hysterically dampened energy. The quantity of this dissipated energy in the system is directly proportional to the structural damage. Consequently, TMDs will effectively safeguard nonlinear structures from dynamic impacts (Pinkaew et al., 2003; Christopoulos and Filiatrault, 2006; Lee et al., 2006; Sgobba and Marano, 2010). Moreover, if TMDs are appropriately constructed, they may minimize seismic damage which are structural and nonstructural (Almazan et al., 2012; Daniel and Lavan, 2014), allowing them to be employed efficiently to mitigate both earthquake and wind impacts. Numerous studies (Chen and Wu, 2001; Moon 2010; Frans and Arfiadi, 2015; Cetin et al., 2017; Hussan et al., 2018; Kim and Lee, 2018; Erdogan and Ada, 2021) have been examined both TMDs and multiple TMDs (MTMDs) in terms of their impact on structures.

DOI: 10.1201/9781003325246-19

In this chapter, a passive TMD design method that makes use of transfer functions and an algorithm known as differential evolution (DE) is implemented to reduce the regular and irregular shear building structure of the seismic response. In order to create the governing equations for the frequency domain, the theory of random vibration and excitation with critical probabilities (Takewaki, 2009) are utilized as tools. The approach optimizes TMD parameters, which are coefficients of stiffness, mass, and damping. As objective functions that need to be decreased in relation to the maximum and minimum values of the TMD parameters, the mean-square of the top floor's absolute acceleration and displacement is used. The single TMD shear building model examined using earthquake acceleration data to understand the suggested technique. The results are compared to the scientific literature.

19.2 BUILDING MODEL WITH TMDs

Figure 19.1 depicts a passive TMD-building structure configuration for a multi-story shear building structure, m_{di}, c_{di}, k_{di} denoting TMD mass, damping, and stiffness, respectively. Let's say $\ddot{x}_{gr}(t)$ is the zero-mean stationary random ground acceleration and for a building with a single TMD or MTMD, the displacement vector is denoted as $x(t)$. The following are some ways to express an under control building structure's equation of motion:

$$(M_B + M_T)\ddot{x}(t) + (C_B + C_T)\dot{x}(t) + (K_B + K_T)x(t) = -(M_B + M_T)r\ddot{x}_{gr}(t) \qquad (19.1)$$

where the system's acceleration and velocity are denoted by $\ddot{x}(t)$ and $\dot{x}(t)$, respectively. r is the vector of influence coefficients consisting of 1's. While the stiffness, mass, and damping matrices of building structure are K_B, M_B, and C_B, C_T, K_T, and M_T represent TMD's damping, stiffness, and mass matrices, respectively.

The Fourier Transform of Equation (19.1) can be written as

$$\left(K_{sys} + i\omega C_{sys} - \omega^2 M_{sys}\right)X(\omega) = -M_{sys}r\ddot{X}_{gr}(\omega) \qquad (19.2)$$

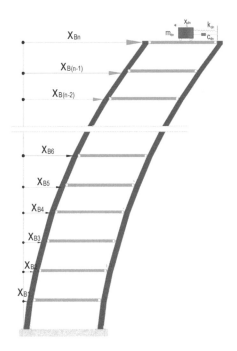

FIGURE 19.1   Representation of a shear building model incorporating TMD.

where M_{sys}, K_{sys}, and C_{sys} are the building structure with a single TMD or MTMD's mass, stiffness, and damping matrices. i expresses the imaginary part of the complex component, $X(\omega)$ is the Fourier Transformed displacement vector, and $\ddot{X}_{gr}(\omega)$ is the zero mean ground acceleration showing stationary character. The shortening of the equation above could be expressed as

$$A_{sys}X(\omega) = -M_{sys}r\ddot{X}_{gr}(\omega) \qquad (19.3)$$

Matrix A_{sys} is defined as follows by this equation:

$$A_{sys} = \left(K_{sys} + i\omega C_{sys} - \omega^2 M_{sys}\right) \qquad (19.4)$$

The Fourier Transform of $X(\omega)$ is obtained by multiplying both sides by A_{sys}^{-1} in Equation (19.3):

$$X(\omega) = -A_{sys}^{-1}M_{sys}r\ddot{X}_{gr}(\omega) \qquad (19.5)$$

If this equation is expressed with regard to displacement transfer function $H_D(\omega)$, it has the following form:

$$X(\omega) = H_D(\omega)\ddot{X}_{gr}(\omega) \qquad (19.6)$$

where $H_D(\omega)$ is defined as follows according to A_{sys}^{-1},

$$H_D(\omega) = -A_{sys}^{-1}M_{sys}r \qquad (19.7)$$

These formulas were utilized by Takewaki (2009) to design the ideal viscous dampers. The inter-story drift vector is the same way Fourier transform is represented as

$$\delta(\omega) = -T_{sys}A_{sys}^{-1}M_{sys}r\ddot{X}_{gr}(\omega) \qquad (19.8)$$

where T_{sys} is the transformation matrix stated as follows:

$$T_{sys} = \begin{bmatrix} 1 & 0 & 0 & & & \cdots & & & 0 \\ -1 & 1 & 0 & 0 & \cdots & & & & \cdots \\ 0 & -1 & 1 & 0 & & & & & 0 \\ 0 & 0 & -1 & 1 & \cdots & & & & 0 \\ \cdots & & \cdots & \cdots & \cdots & & & & \\ & & & & & \cdots & \cdots & & \cdots \\ 0 & 0 & \cdots & & & 0 & -1 & 1 & 0 \\ 0 & 0 & \cdots & & & 0 & 0 & -1 & 1 \end{bmatrix}_{2nx2n} \qquad (19.9)$$

The revised form of Equation (19.8) is

$$\delta(\omega) = H_\delta(\omega)\ddot{X}_{gr}(\omega) \qquad (19.10)$$

where $H_\delta(\omega)$ is the inter-story drift transfer function with regard to input acceleration. $H_\delta(\omega)$ is denoted as

$$H_\delta(\omega) = -T_{sys}A_{sys}^{-1}M_{sys}r \qquad (19.11)$$

The absolute acceleration's Fourier transform is expressed as

$$\ddot{X}_{AA}(\omega) = H_{AA}(\omega)\ddot{X}_{gr}(\omega) \tag{19.12}$$

The following equation represents absolute acceleration in frequency domain, which is indicated by the symbol $H_{AA}(\omega)$:

$$H_{AA}(\omega) = \left(1 + \omega^2 A_{sys}^{-1} M_{sys} r\right) \tag{19.13}$$

The following is the mean square response that the random vibration theory predicts for the ith story's acceleration and displacement, as well as for the inter-story drift respectively:

$$\sigma_{AAi}^2 = \int_{-\infty}^{\infty} |H_{AAi}(\omega)|^2 S_{gr}(\omega) d\omega = \int_{-\infty}^{\infty} H_{AAi} H_{AAi}^* S_{gr}(\omega) d\omega \tag{19.14}$$

$$\sigma_{Di}^2 = \int_{-\infty}^{\infty} |H_{Di}(\omega)|^2 S_{gr}(\omega) d\omega = \int_{-\infty}^{\infty} H_{Di} H_{Di}^* S_{gr}(\omega) d\omega \tag{19.15}$$

$$\sigma_{\delta i}^2 = \int_{-\infty}^{\infty} |H_{\delta i}(\omega)|^2 S_{gr}(\omega) d\omega = \int_{-\infty}^{\infty} H_{\delta i} H_{\delta i}^* S_{gr}(\omega) d\omega \tag{19.16}$$

In this case, $()^*$ denotes the conjugated complex form. When referring to the power spectral density (PSD) of ground acceleration, the symbol $S_{gr}(\omega)$ is used to signify the function $\ddot{x}_{gr}(t)$. The absolute values of the ith story's acceleration, inter-story drift, and displacement transfer functions are denoted by the notations $|H_{AAi}(\omega)| H_{\delta i}$ and $|H_{Di}(\omega)|$, respectively.

19.3   OPTIMIZATION PROBLEM AND METHOD

Global optimization may be broken down into deterministic and stochastic sub-problems. The deterministic optimization method known as the tunneling algorithm (Levy and Montalvo, 1985) is based on the use of gradients. To put things another way, genetic algorithms are a kind of stochastic global optimization that relies on a population-based approach (Goldberg, 1989). Natural occurrences that can be classified as either physical, biological, and chemical serve as inspiration for a kind of metaheuristic evolutionary algorithms. These numerical algorithms apply algorithmic processes all through the optimization process (Yang et al., 2016).

Nonlinearly constrained optimizations may be solved using either a gradient-based or a direct search approach in the numerical realm. In gradient-based methods, gradients and Hessians are used. Augmented Lagrange method, nonlinear interior point method, and sequential quadratic programming point method could be some examples of nonlinear restricted gradient methods. The difference between direct search methods and gradient-based algorithms is that the latter makes use of derivative data. Nelder-Mead, the genetic algorithm, differential evolution, and simulated annealing are all instances of direct search algorithms. Convergence takes longer with direct search, but they're more tolerant to noise (Champion and Strzebonski, 2008).

Several factors for optimizing TMD design may be found in the literature. Minimizing structural displacement for example was a goal of Den Hartog (1956), Thompson (1981), Jacquot and Hoppe (1973), and Fujino and Abe (1993). Improved dynamic structural stiffness was a contribution by Falcon et al. (1967). Increased structure-TMD damping was achieved by Luft (1979), who also investigated some mixed criteria that integrated frequency tuning with displacement and TMD damping determination to provide maximum effective damping. Additionally, he calculated the bare minimum of the TMD's mass. Minimizing structural acceleration was a topic of research for Ioi and Ikeda (1978).

There are a wide number of critical purposes that may be required of various types of structures. Inelastic structures may benefit more from focusing on reducing displacements than elastic ones do on decreasing accelerations. The acceleration or stress response may be more important for nonstructural elements and nonstructural components in the structures, even if the displacement (or deformation) response of an inelastic structure is often more important when considering damage to the structure. Therefore, engineers consider their end result while deciding which objective functions to use. In chapter, we consider using the DE optimization approach to get the minimum values for three objectives. Therefore, in this setting, the mean-square values of absolute acceleration (f_1) and displacement (f_2) at the top floor, as well as the total of the mean square values of inter-story drifts are lowered separately. These aims consist of,

$$f_1\left(m_{di}, c_{di}, k_{di}\right) = \sigma_{AAN}^2 \tag{19.17}$$

$$f_2\left(m_{di}, c_{di}, k_{di}\right) = \sigma_{DN}^2 \tag{19.18}$$

$$f_2\left(m_{di}, c_{di}, k_{di}\right) = \sigma_{\delta i}^2 \tag{19.19}$$

Upper and lower limit design characteristics for the ith story may be stated as passive restrictions.

$$0 \le m_{di} \le \bar{m}_d \tag{19.20}$$

$$0 \le c_{di} \le \bar{c}_d \tag{19.21}$$

$$0 \le k_{di} \le \bar{k}_d \tag{19.22}$$

where $\bar{m}_{di} \, \bar{c}_{di}, \, \bar{k}_{di}$ are the upper bounds of mass, damping, and stiffness coefficients of ith TMD, respectively. $0 \le m_{di} \le \bar{m}_d$ limitations provide facilities for TMD mass optimization of quantities.

19.3.1 Differential Evolution Optimization Method

Extensively used stochastic global optimization technique differential evolution optimization algorithm was created by Storn and Price (1997). This algorithm is broadly powerful. Figure 19.2 depicts the flowchart for the DE algorithm. With this flowchart in mind, the DE approach may be described as follows (Penunuri et al., 2011; Wu et al., 2018; Biswas et al., 2019):

As g indicates the generation, the ith target vector for this generation is provided by

$$\vec{X}_i^g = \{x_{1,i}^g, x_{2,i}^g, \ldots, x_{D,i}^g\}, \quad i = 1, \ldots, NP \tag{19.23}$$

in which (D) is the dimension of the problem and the elements of the goal vector rely on design parameters as $x_{1,i}^g = m_{d,i}^g$, $x_{2,i}^g = c_{d,i}^g$, and $x_{3,i}^g = k_{d,i}^g$.

The number of populations, denoted by NP, may be calculated based on the prevailing conditions. Typically, NP is restricted between 2D to 40D. (Ronkkonen et al., 2005). The conventional DE technique consists of the following steps: (1) upper and lower limitations are used to define the initial population matrices; (2) during mutation, the donor vector is identified; (3) a test vector is

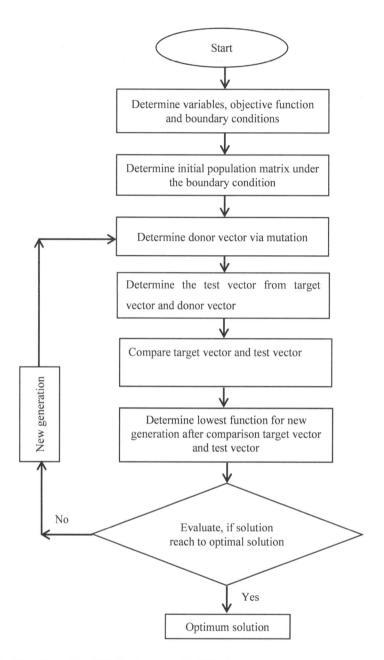

FIGURE 19.2 Flowchart of the DE. (Cetin and Aydin 2019a.)

constructed of a target vector and a donor vector by the execution of a crossover; (4) after comparing the outcomes of the target vector to those of the donor vector, the function with the lowest value
is chosen for the next generation. Following is an explanation of these expressions:

Considering the Initialization, initial generation is determined as $g = 0$ when generating the
population matrix considered. The jth component of the ith vector may be written as follows as an
example of initialization,

$$x_{ij}^0 = x_{j,Low} + rand_{i,j}(0,1).\left(x_{j,Up} - x_{j,Low}\right) \tag{19.24}$$

where the statement $rand_{i,j}(0,1)$ is shown instantaneously for i and j as a real number uniformly distributed between 0 and 1. Herein, $i = 1, 2, 3, ..., NP$ and $j = 1, 2, 3, ..., D$. $x_{j,Low}$ is the lowest limit and $x_{j,up}$ is the upper limit of the jth component of dimension.

Mutational determination of the donor vector can be explained as follows.

Firstly, random target vectors $\vec{X}_{R_1^i}^g$, $\vec{X}_{R_2^i}^g$ and $\vec{X}_{R_3^i}^g$ are selected as random in terms of each target vector $\vec{X}_i^g$ as R_1^i, R_2^i ve $R_3^i \in [1, NP]$. The donor vector $\vec{V}_i^g$ that is constructed of these random vectors by mutation is denoted by

$$\vec{V}_i^g = \vec{X}_{R_1^i}^g + \mathrm{F}.\left(\vec{X}_{R_2^i}^g - \vec{X}_{R_3^i}^g\right) \tag{19.25}$$

In Equation (19.25), F is a factor of scaling ranging between 0 and 2. However, this factor is often set between 0 and 1. Considering Crossover, donor vector consisting of target vector and mutation is subjected to crossover such that test vector $\vec{U}_i^g = \{u_{1,i}^g, u_{2,i}^g, u_{D-1,i}^g, u_{D,i}^g,\}$ may be constructed. This decision is as described below:

$$u_{j,i}^g = \begin{cases} v_{j,i}^g, \text{ If } rand_{i,j}(0,1) \le C_r \text{ or } j = j_{rand} \\ x_{j,i}^g, \text{ otherwise} \end{cases} \tag{19.26}$$

where C_r represents the rate of crossover between 0 and j_{rand}, where rand is a random integer number in between 1 and D.

Taking into account selection, the comparison of the target vector $\vec{X}_i^g$ and the test vector $\vec{U}_i^g$, the vector with the lower score (the superior one) is the one that is passed on to the following generation. The expression of the target vector, which is then passed on to the next generation, is as follows:

$$\vec{X}_i^{g+1} = \begin{cases} \vec{U}_i^g, \text{ If } f\left(\vec{U}_i^g\right) \le f\left(\vec{X}_i^g\right) \\ \vec{X}_i^g, \text{ If } f\left(\vec{U}_i^g\right) > f\left(\vec{X}_i^g\right) \end{cases} \tag{19.27}$$

The suggested procedure employs the DE algorithm for optimization in this section. The mentioned algorithm is a stochastic direct search technique that efficiently locates the global optimum. The DE algorithm is often used in optimization, and as it is included in a large number of coding systems and it is simple to implement. Additionally, the approach searches directly stochastic to reach global optimum and gradient or Hessian are not needed for computations. The suggested parameters of TMD optimization technique rely on transfer functions and differential development in shear building is very dependable and straightforward to implement.

19.4 NUMERICAL EXAMPLES FOR SINGLE PASSIVE TMD

19.4.1 MODEL OF A SEVEN-STORY BUILDING

The suggested approach is used on a seven-story shear frame model shown in Figure 19.3 in order to determine the ideal settings of TMD installed on the top floor. The optimal design of these characteristics is evaluated using El Centro (NS) ground movements.

To identify the ideal TMD parameters, objective functions f_1, f_2, and f_3 are all minimized. Using transfer functions and the DE optimization approach, the mass, damping, and stiffness coefficients of TMD are derived.

The PSD function with white noise excitation is developed based on earthquake frequency content. Furthermore, the PSD function of the input ground acceleration is assessed as a constant value

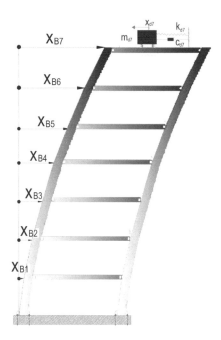

FIGURE 19.3 Model of a seven-story shear building frame with TMD.

for band-limited white noise excitation. The studies cover the frequency spectrum of the first three natural frequencies of the building model. In addition, because it is commonly accepted that a single TMD can effectively regulate the first mode, it is sufficient to choose a frequency range that encompasses the first three modes. This range may be expanded if necessary in systems where higher mode effects predominate. For the PSD function, $S_{gr}(\omega) = 0.015$ m^2/s^3 of input acceleration $\ddot{u}_{gr}(\omega)$ as a constant value, the frequency range is selected as -60 rad/s $\leq \omega \leq -0.2$ rad/s and 0.2 rad/s $\leq \omega \leq 60$ rad/s. Except for this region, it is assumed that $S_{gr}(\omega)$ is zero. Without control, the first and second mode natural damping ratios of the shear frame model are specified as $\xi_1 = \xi_2 = 0.02$. Each storey's stiffness coefficient is $k_i = 5 \times 10^7$ N/m, and the story masses of the model structure are $m_i = 6 \times 10^4$ kg for $i = 1, \ldots, 7$. The natural cyclic frequency of the model building structure is thus computed to $\omega_{Bi} = \{6.03, 17.84, 28.87, 38.63, 46.71, 52.75, 56.47\}$ rad/s. Without a damper, the normalized first mode shape vector of the model structure is $\phi_1 = \{0.21, 0.41, 0.59, 0.75, 0.87, 0.96, 1.\}$. The highest limits of damping, stiffness coefficients, and the mass of TMD are determined to be $\bar{c}_d = 8 \times 10^4$ Ns/m, $\bar{k}_d = 1.5 \times 10^6$ N/m, $\bar{m}_d = 2.1 \times 10^4$ kg (5% of the total mass of the main structure), respectively. Due to the size constraint of TMD, the mass ratio is typically equal or less than 5% for the majority of applications (Connor and Klink, 1996). To discover the ideal values of m_d, c_d, and k_d, the mean-square of top absolute acceleration σ_{AA7}^2, the mean-square of top displacement σ_{D7}^2, and the sum of the mean-square of inter-story drifts $\Sigma_{i=1}^{N} \sigma_{\delta i}^2$ are minimized independently. After determining the optimal TMD settings, the structure equipped with a TMD system is subjected to the seismic load using time-history analysis. The findings of the suggested approach were compared to those of existing methods described in the scientific literature (Den Hartog, 1956; Warburton, 1982; Sadek et al., 1997). Below is a quick summary of these approaches from the literature and the outcomes of these methods for the chosen model. In the suggested technique, the ideal upper limit mass of TMD is determined as 5% of the overall mass of the structure to comparison.

Den Hartog (1956) minimized the displacement of the main structure exposed to harmonic stimulation in order to determine the optimal tuning parameters, which are the optimum frequency ratio $f_{opt} = \omega_d / \omega_B$ and the optimum damping ratio of TMD ξ_{dopt}, as shown in the following expressions:

$$f_{opt} = \frac{1}{1+\mu} \tag{19.28}$$

$$\xi_{dopt} = \sqrt{\frac{3\mu}{8(1+\mu)}} \tag{19.29}$$

Here, $\mu = \dfrac{m_d}{m_B}$ is the mass ratio, ω_d and ω_B are the TMD and structure frequencies, and m_d and m_B are the TMD and primary structure masses. The mass ratio of the first mode is calculated as $\mu_1 = 0.092$; $f_{opt} = 0.915$, which is the optimal frequency ratio for the first mode. The TMD tuning frequency is calculated as $\omega_d = 5.52$ rad/s. The damper's stiffness coefficient is computed as $k_d = 641,021$ N/m. The optimal damping ratio and damping coefficient are estimated as $\xi_{dopt} = 0.178$ and $c_d = 41,309.6$ Ns/m.

According to Warburton's (1982) approach, which is based on white-noise excitation, the optimal tuning frequency and damping ratio are expressed as follows:

$$f_{opt} = \frac{1}{1+\mu} \sqrt{1 - \mu/2} \tag{19.30}$$

$$\xi_{dopt} = \sqrt{\frac{\mu(1 - \mu/4)}{4(1+\mu)(1 - \mu/2)}} \tag{19.31}$$

Therefore, the optimal structural model parameters are $f_{opt} = 0.89$, $\xi_{dopt} = 0.155$, $\omega_d = 4.42$ rad/s, $k_d = 611,437$ N/m, and $c_d = 33,337.6$ Ns/m, respectively.

Sadek et al. (1997) established a unique optimization technique for TMD design in MDOF systems. Their tuning parameters are formulated as follows:

$$f_{opt} = \frac{1}{1 + \mu_i \phi_{ij}} \left(1 - \xi_i \sqrt{\frac{\mu_i \phi_{ij}}{1 + \mu_i \phi_{ij}}} \right) \tag{19.32}$$

$$\xi_{opt} = \phi_{ij} \left(\frac{\xi_i}{1 + \mu_i} + \sqrt{\frac{\mu_i}{1 + \mu_i}} \right) \tag{19.33}$$

Here, ϕ_{ij} is the amplitude of the ith vibration mode at the jth floor and TMD location for a unit participation factor. ξ_i represents the ith mode structural damping ratio. The mass ratio μ_1 for the first mode is computed as follows:

$$\mu_1 = \frac{m_d}{\phi_1^T M_B \phi_1} \tag{19.34}$$

The first mode shape vector without normalization is ϕ_1. It is determined using the formulas $\phi_1 = \{0.21, 0.41, 0.59, 0.75, 0.87, 0.96, 1\}$ and $\phi_{ij} = \phi_{17} = 1$. The mass ratio of the first mode is $\mu_1 = 0.092$. The tuning frequency and damping ratio are $f_{opt} = 0.91$ and $\xi_{opt} = 0.31$, respectively. According to the approach of Sadek et al. (1997), the stiffness coefficient and damping coefficient of TMD are estimated as $k_d = 633,589$ N/m and $c_d = 71,290$ Ns/m.

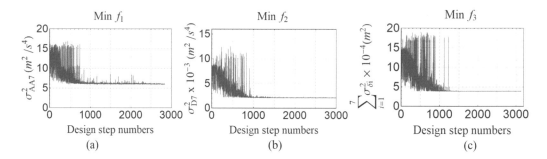

FIGURE 19.4 Variation of objective functions (a) f_1, (b) f_2, and (c) f_3 based on the number of design steps.

19.4.1.1 Outcomes of the Method Proposed

The mean-square of top floor absolute acceleration (f_1), top floor displacement (f_2), and the sum of the mean-square of inter-story drifts (f_3) are reduced using the DE method in accordance with Section 19.3. Optimal design parameters are found according to three objective functions. Figure 19.4 illustrates the change of objective functions based on design processes and the convergence of optimization. Figure 19.5 depicts the evolution of TMD's design parameters during its optimization history. As demonstrated in Figure 19.5(a), TMD mass tends toward the upper limit of mass amount, but TMD parameters k_d and c_d converge to optimal values in Figures 19.5(b) and (c) for three distinct objective functions. In Table 19.1, the optimal TMD parameters based on three distinct objective functions and the optimal values according to published methodologies are compared. Table 19.1 demonstrates that, despite the fact that the mass of each design is the same, the stiffness and damping coefficients vary. The TMD mass is often constrained by building requirements.

Common practice in literature express that the mass ratio restriction should not exceed 5%. The suggested approach just takes the maximum and minimum mass limitations and does not set the mass. The technique will account for the variance in non-fixed mass owing to limitations in the coefficients of stiffness and damping, which may fluctuate. The best mass ratio value approaches the top limit of mass ratio. The mass ratio determined as a consequence of the suggested optimization is utilized to determine the ideal TMD using the literature-based approaches in order to create a comparison. While the design according to f_1 yields the greatest value for k_d, the computed damping coefficient c_d is greater than all other methods with the exception of Sadek et al. (1997). Since the minimization of f_1 and f_2 focuses on displacements, their conclusions are compatible with Den Hartog and Warburton's approaches (1982). Given that Den Hartog (1956) and Warburton (1982) both use displacement-based methodologies, it is preferable to comprehend the presence of this consistency. Moreover, because f_2 and f_3 are displacement-based objective functions, the results of f_1 differ from those of the other objective functions because f_1 is an acceleration-based objective function that yields slightly different results from the other two objective functions and the methods in the literature, which are also displacement-based.

Figure 19.6 depicts the frequency responses of absolute values of top-floor absolute acceleration transfer functions $|H_{AA7}|$, top displacement transfer function $|H_{D7}|$, and first-floor inter-story drift (ID) transfer function $|H_{\delta 1}|$ with respect to the results obtained from the objective functions f_1, f_2, and f_3. It is evident from Figure 19.6 that TMD efficiently minimizes the behavior in terms of the transfer functions of the first floor's top absolute acceleration, top displacement, and inter-story drift. Although the ideal designs for the three distinct objective functions are distinct, they are highly compatible in lowering the structure's reaction. Specifically, the behavior of optimum designs corresponding to objective functions f_2 and f_3 is similar; however, the behavior of the design corresponding to objective function f_1 is somewhat different from the other two.

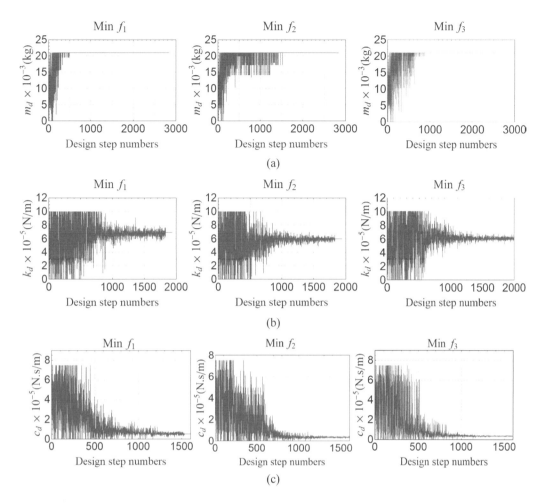

FIGURE 19.5 The fluctuation of TMD mass m_d (a), TMD stiffness coefficient k_d (b), and TMD damping coefficient c_d (c) as a function of design step numbers for f_1, f_2, and f_3.

The peak displacements, absolute accelerations, inter-story drift ratios (IDRs), root mean squares (RMSs) of displacements, and root mean squares (RMSs) of accelerations are displayed in comparison with other method in literature. According to the El Centro earthquake, Figure 19.7 demonstrates that the ideally constructed TMD for the three objective functions in the current investigation yields near and successful response decrements.

The acceleration and displacement values at the seventh level and the IDR of the first floor are analyzed in terms of both their response values and their percentage decrease due to ground

TABLE 19.1

Optimal TMD Parameters for a Model of a Seven-Story Structure

TMD Parameters	Den Hartog (1956)	Warburton (1982)	Sadek et al. (1997)	Min f_1	Min f_2	Min f_3
c_{d7} (kg)	641,021	611,437	633,589	701,886	606,125	632,573
k_{d7} (N/m)	41,309.60	33,337.60	71,290	54,185	34,095	36,344
m_{d7} (Ns/m)	21,000	21,000	21,000	20,998.60	20,964.80	20,949.80

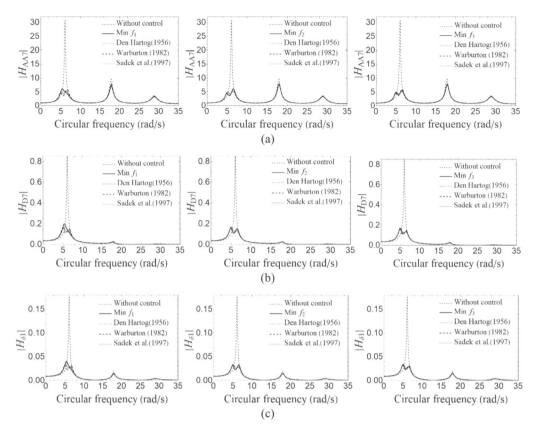

FIGURE 19.6 Frequency responses of absolute value of seventh-floor absolute acceleration transfer functions $|H_{AA7}|$ (a), displacement transfer function $|H_{D7}|$ (b), and first-floor inter-story drift transfer function $|H_{\delta 1}|$ (c) relative to objective functions f_1, f_2, and f_3.

movements. Examining the percentages of reductions in further detail reveals that each objective function reduces its goal more effectively, and that the approaches described here are highly consistent with well-known methodologies in the literature.

Examining the variations of the transfer functions in relation to the excitation frequency and the mass of TMD reveals that the behavior is minimal for all purpose functions at a mass which converges to 21,000 kg. There are similarities between c_d and k_d modifications. When c_d and k_d reach the optimal values shown in Table 19.1, all function values reach a minimum. This indicates that the optimal designs developed are able to minimize the functions that represent the structural behavior outlined in this work.

19.4.2 Comparison of Single TMD Performance Between Model of a Four-Story Regular and Irregular Buildings

In this chapter, a four-story building model shown in Figure 19.8 having distributed regular (Cetin and Aydin, 2019b) and irregular stiffness shear frame (Cetin and Aydin, 2019b) is chosen as an example for comparison. The characteristic of these models are depicted in Table 19.2. For both models, the highest limits of damping, stiffness coefficients, and the mass of TMD are determined to be $c_d = 7.5 \times 10^4$ Ns/m, $\bar{k}_d = 1.0 \times 10^6$ N/m, $\bar{m}_d = 2.88 \times 10^4$ kg (6% of the total mass of the main structure), respectively. The spectral power density function of the input seismic acceleration is

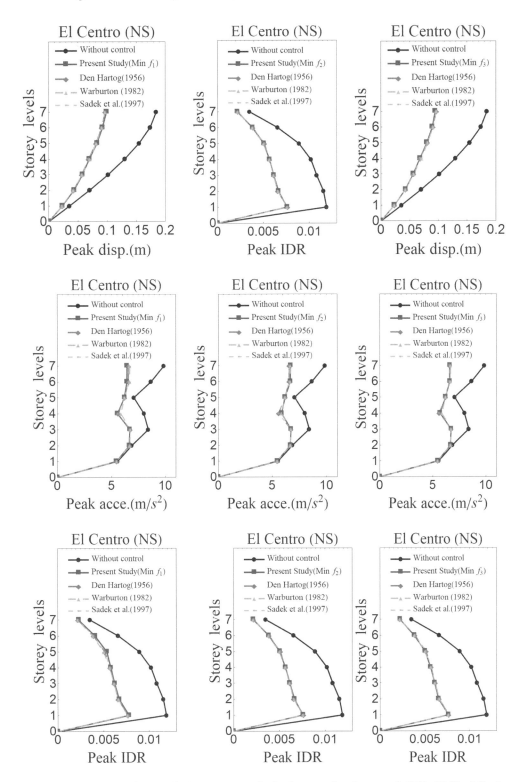

FIGURE 19.7 Graphs of peak displacements, peak absolute accelerations, peak IDR, RMS of displacements, and RMS of accelerations based on f_1, f_2, f_3 and their comparisons. (*Continued*)

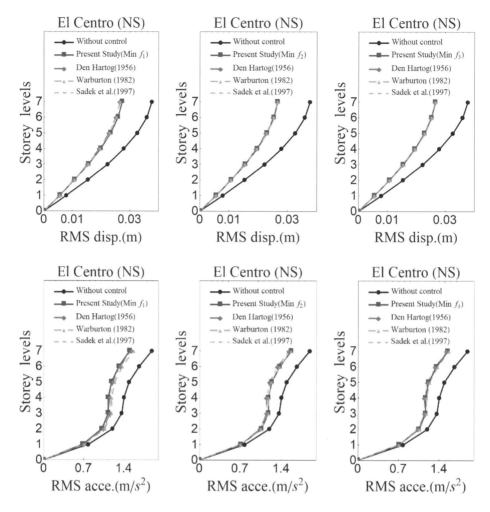

FIGURE 19.7 (*Continued*)

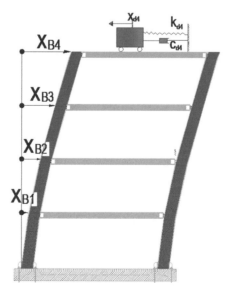

FIGURE 19.8 Model of a four-story shear building frame with TMD.

TABLE 19.2
Characteristics of Regular and Irregular Shear Building Structures

	ith Story Number	k_i	m_i
Regular building model	1	2.5×10^7 N/m	12×10^7 N/m
(Cetin and Aydin,	2	2.5×10^7 N/m	12×10^7 N/m
2019a)	3	2.5×10^7 N/m	12×10^7 N/m
	4	2.5×10^7 N/m	12×10^7 N/m
Regular building model	1	2.5×10^7 N/m	12×10^7 N/m
(Cetin and Aydin,	2	2.75×10^7 N/m	12×10^7 N/m
2019b)	3	3×10^7 N/m	12×10^7 N/m
	4	3.25×10^7 N/m	12×10^7 N/m

modeled as band-limited white noise as a constant value in the ranges -30 rad/s $\leq \omega \leq -0.2$ rad/s and 0.2 rad/s $\leq \omega \leq 30$ rad/s. $S_{gr}(\omega) = 0.015$ m²/s³ is taken, out of in this range it is evaluated as zero. The first and second mode damping ratios of the sliding frame are taken as $\xi_1 = \xi_2 = 0{,}02$. Under these conditions, optimum values of m_d, c_d, and k_d are calculated individually minimizing of the top-floor accelerations $\sigma^2_{AA4}(f_1)$, the square mean of the displacements of the top floor $\sigma^2_{D4}(f_2)$, and the sum of the square mean of the relative displacements of the floors' $\Sigma_1^N \sigma^2_{\delta i}(f_3)$ DE algorithm. After finding the optimum TMD parameters, the TMD-structure system is tested under the El Centro (NS) earthquake. The results found in the example presented in the literature Den Hartog (1956), Warburton (1982), and Sadek et al. (1997). Optimal design parameters are determined according to the objective functions (f_1), (f_2), (f_3). The variation of the objective functions has been according to the design steps convergence to the optimal values individually. Table 19.3 shows calculated optimum TMD parameters for this regular and irregular build structures. In Table 19.4, a comparison of the shear building response reduction (%) of regular shear building and irregular shear building for the El Centro (NS) earthquake is shown. As seen in the table, TMD is much more effective for response reduction of the irregular building.

The suggested approach may be modified to build nume TMDs, placement of TMDs, and regulate wide- and narrow-band frequencies. In addition, the inclusion of TMD mass as a design parameter in the optimization problem provides the foundation for optimizing numerous TMDs. It is unclear which option to choose for the multitudes of numerous TMDs. This chapter, although it only addresses the optimization of a single TMD, may also serve as a reference for the optimization of several TMDs.

TABLE 19.3
Optimum TMD Parameters for Regular and Irregular Shear Building Structures

	TMD Parameters	Den Hartog (1956)	Warburton (1982)	Sadek et al. (1997)	Min f_1	Min f_2	Min f_3
Regular building	m_d (kg)	28,800	28,800	28,800	28,800	28,800	28,800
model (Cetin and	k_d (N/m)	594,355	563,612	532,139	617,234.89	582,546	562,682
Aydin, 2019a)	c_d (Ns/m)	49,063	39,538.40	74,136.30	60,501.01	42,031.10	45,986.20
Regular building	m_d (kg)	28,800	28,800	28,800	28,800	28,800	28,800
model (Cetin and	k_d (N/m)	646,591	614,821	577,562	665,037.68	582,546	562,682
Aydin, 2019a)	c_d (Ns/m)	40,313.50	40,313.50	75,274	43,891.34	42,031.10	45,986.20

TABLE 19.4
Comparison of Regular and Irregular Shear Building's Response Reduction (%) for the El Centro (NS) Earthquake

		Present study Min (f_1) Min (f_2) Min (f_3)		Den Hartog (1956)		Warburton (1982)		Sadek et al. (1997)	
Response	Story Number	Decrement for Regular Shear Building (%)	Decrement for Irregular Shear Building (%)	Decrement for Regular Shear Building (%)	Decrement for Irregular Shear Building (%)	Decrement for Regular Shear Building (%)	Decrement for Irregular Shear Building (%)	Decrement for Regular Shear Building (%)	Decrement for Irregular Shear Building (%)
Peak displacement (m)	4	25.530 26.950 25.532	41.019 38.217 42.038	26.241	40.701	26.241	41.465	22.695	34.331
Peak absolute acceleration (m/s²)	4	4.022 3.925 3.883	5.568 5.470 5.610	3.883	5.610	3.883	5.316	3.745	5.386
Peak IDR	1	31.034 36.782 31.609	32.984 30.890 32.984	32.184	32.984	32.759	39.211	27.586	25.333
RMS displacement (m)	4	38.890 39.530 38.675	60.000 59.403 61.194	39.530	60.448	38.675	60.597	35.470	56.418
RMS acceleration (m/s²)	4	22.810 21.462 21.053	40.781 39.226 40.025	22.222	40.415	21.053	39.043	21.637	39.638

EL Centro (NS)

Source: Cetin and Aydin (2019a, 2019b).

19.5   CONCLUSION OF THE REMARKS

Using the DE approach, this chapter develops a passive TMD design method for shear frames based on transfer functions. The suggested technique determines the optimal TMD mass, stiffness, and damping coefficients, accordingly. In addition, three objective functions under seismic ground movement are employed to evaluate the method's dependability. As objective functions to be reduced, the mean-square top floor absolute acceleration, mean-square top floor displacement, and the sum of mean-square inter-story drifts are selected, taking higher and lower limitations into account. Throughout the optimization process, the fluctuation of TMD mass, stiffness, and damping coefficients in relation to design step numbers are monitored. The frequency response based on transfer functions is discovered in order to comprehend the efficiency of optimum TMD settings. Considering ground motions, the top displacement, the top acceleration, the top inter-story drift ratio, the root mean square of displacements, and the root mean square of absolute acceleration behaviors are calculated in order to compare the proposed method with well-known methods in the literature. In addition, seismic performance of TMD in a regular shear building and a building having stiffness irregularity is compared. This chapter's findings may be expressed as follows:

- While the majority of published approaches optimize just the stiffness and damping parameters, the suggested method optimizes the mass parameter as well. Due to the space limitations of TMD, the mass ratio is typically less than 5% for the majority of applications. In this research, therefore the top limit of TMD is set at 5% and 6% in Examples 1 and 2. When TMD mass is included as a design parameter in optimization, the mass of all designs converges to the upper limit. This may be instructive and practical for the design of various TMD in future research.
- The majority of optimum TMD approaches regulate a certain mode, which is often the first mode. The frequency range of the suggested approach is broad. The designer may broaden this range if the impact of higher modes is addressed.
- If objective functions based on displacement and acceleration are taken into account, the computed TMD parameters are comparable, notwithstanding modest variances.
- It is shown that optimal TMDs are extremely effective seismic movements on the structural behavior.
- For the El Centro (NS) earthquake, a comparison is executed between the shear building response decrease (%) of a regular and irregular shear building. According to the chart, TMD is considerably more effective for reducing the reaction of irregular building than the regular building.
- The suggested TMD optimization approach based on transfer functions and the differential evolution algorithm is very dependable and straightforward to implement in shear buildings. DE algorithm is often used in optimization, as it is included in many programming languages and is simple to use and access. In addition, the approach is a direct stochastic search and requires no gradient or Hessian computations.

REFERENCES

Almazan, J. L., Espinoza, G. and Aguirre, J. J. (2012). "Torsional balance of asymmetric structures by means of tuned mass dampers." *Engineering Structures*, 42, 308–328, DOI: 10.1016/j.engstruct.2012.04.034.

Biswas, P. P., Suganthan, P. N., Wu, G. and Amaratunga, G. A. J. (2019). "Parameter estimation of solar cells using datasheet information with the application of an adaptive differential evolution algorithm." *Renewable Energy*, 132, 425–438.

Cetin, H. and Aydin, E. (2019a). "A new tuned mass damper design method based on transfer functions." *KSCE Journal of Civil Engineering*, 23(10), 4463–4480.

Cetin, H. and Aydin, E. (2019b). "Optimal Design of Adjustable Mass Dampers (TMDs) in Irregular Structures Under Seismic Excitation." *21st National Mechanics congress*, 02–06 September, Nigde Omer Halisdemir University, Nigde, Turkiye

Cetin, H., Aydin, E. and Ozturk, B. (2017). "Optimal damper allocation in shear buildings with tuned mass dampers and viscous dampers." *International Journal of Earthquake and Impact Engineering*, 2(2), pp. 89–120.

Champion, B. and Strzebonski, A. (2008). *Constrained optimization*, Wolfram Mathematica Tutorial Collection, USA, pp. 41–54.

Chen, G. and Wu, J. (2001). "Optimal placement of multiple tuned mass dampers for seismic structures." *Journal of Structural Engineering, American Society of Civil Engineering*, 127(9), 1054–1062.

Christopoulos, C. and Filiatrault, A. (2006). *Principles of Passive Supplemental Damping and Seismic Isolation*, University of Pavia, IUSS Press, Pavia, Italy.

Connor, J. J. and Klink, B. S. A. (1996). *Introduction to Motion Based Design*, Computational Mechanics Publication, Boston, USA, pp. 145–196.

Daniel, Y. and Lavan, O. (2014). Gradient based optimal seismic retrofitting of 3D irregular buildings using multiple tuned mass dampers. *Computers and Structures*, 139, 84–97.

Den Hartog, J. P. (1956). *Mechanical Vibrations*, 4th ed. McGraw-Hill, New York, pp. 87–117.

Erdogan, Y. S. and Ada, M. (2021). "A computationally efficient method for optimum tuning of single-sided pounding tuned mass dampers for structural vibration control. *International Journal of Structural Stability and Dynamics*, 21(05), 2150066.

Falcon, K. C., Stone, B. J., Simcock, W. D. and Andrew, C. (1967). "Optimization of vibration absorbers: a graphical method for use on idealized systems with restricted damping." *Journal of Mechanical Engineering Science*, 9(5), 374–381, DOI:10.1243/JMES_JOUR_1967_009_058_02.

Frans, R. and Arfiadi, Y. (2015). "Designing optimum locations and properties of MTMD systems." *Procedia Engineering*, 125, 892–898.

Fujino, Y. and Abe, M. (1993). "Design formulas for tuned mass dampers based on a perturbation technique." *Earthquake Engineering and Structural Dynamics*, 22(10), 833–854,

Goldberg, D. E. (1989). *Genetic Algorithm in Search, Optimization, and Machine Learning*. Addison-Wesley Longman Publishing, Boston, USA, pp. 1–25.

Hussan, M., Rahman, M. S., Sharmin, F., Kim, D. and Do, J. (2018). "Multiple tuned mass damper for multimode vibration reduction of offshore wind turbine under seismic excitation." *Ocean Engineering*, 160, 449–460.

Ioi, T. and Ikeda, K. (1978). "On the dynamic vibration damped absorber of the vibration system." *Bulletin of the Japan Society Mechanical Engineering*, 21(151), 64–71, DOI: 10.1299/jsme1958.21.64.

Jacquot, R. G. and Hoppe, D. L. (1973). "Optimal Random Vibration Absorber." *Journal of Engineering Mechanics*, 99(3), 612–616.

Kim, S. Y. and Lee, C. H. (2018). "Optimum design of linear multiple tuned mass dampers subjected to white-noise base acceleration considering practical configurations." *Engineering Structures*, 171, 516–528.

Lee, C. L., Chen, Y. T., Chung, L. L. and Wang, Y. P. (2006). "Optimal design theories and applications of tuned mass dampers." *Engineering Structures*, 28(1), 43–53, DOI: 10.1016/j.engstruct.2005.06.023.

Levy, A. V. and Montalvo, A. (1985). "The tunneling algorithm for the global minimization functions." *SIAM Journal on Scientific and Statistical Computing*, 6(1), 15–29.

Luft, R. W. (1979). "Optimal Tuned Mass Dampers for building." *Journal of the Structural Division*, ASCE, 105(12), 2766–2772.

Moon, K. S. (2010). Vertically distributed multiple tuned mass dampers in tall buildings: Performance analysis and preliminary design. *The Structural Design of Tall and Special Buildings*, 19(3), 347–366.

Penunuri, F., Escalante, R. P., Villanueva, C. and Oy, D. P. (2011). "Synthesis of mechanisms for single and hybrid tasks using differential evolution." *Mechanism and Machine Theory*, 46(10), 1335–1349, DOI: 10.1016/j.mechmachtheory.2011.05.013.

Pinkaew, T., Lukkunaprasit, P. and Chatupote, P. (2003). "Seismic effectiveness of tuned mass dampers for damage reduction of structures." *Engineering Structures*, 25(1), 39–46.

Ronkkonen, J., Kukkonen, S. and Price, K. V. (2005). "Real-parameter optimization with differential evolution", *IEEE Congress on Evolutionary Computation*, pp. 506–513.

Sadek, F., Mohraz, B., Taylor, A. W. and Chung, R. M. (1997). "A method of estimating the parameters of tuned mass dampers for seismic applications." *Earthquake Engineering and Structural Dynamics*, 26(6), 617–635, DOI: 10.1002/(SICI)1096-9845(199706)26:6<617::AID-EQE664>3.0.CO;2-Z.

Sato-Brito, R. and Ruiz, S. E. (1999). "Influence of ground motion intensity on the effectiveness of tuned mass dampers." *Earthquake Engineering and Structural Dynamics*, 28(11), 1255–1271.

Sgobba, S. and Marano, G. C. (2010). "Optimum design of linear tuned mass dampers for structures with nonlinear behavior." *Mechanical System and Signal Processing*, 24(6), 1739–1755, DOI: 10.1016/j.ymssp.2010.01.009.

Storn, R. and Price, K. V. (1997). "Differential evolution-a simple and efficient heuristic for global optimization over continuous spaces." *Journal of Global Optimization*, 11(4), 341–359.

Takewaki, I. (2009). *Building Control with Passive Dampers: Optimal Performance-Based Design for Earthquakes*. John Wiley & Sons Ltd. (Asia), Singapore, pp. 51–75.

Thompson, A. G. (1981). "Optimum tuning and damping of a dynamic vibration absorber applied to a force excited and damped primary system." *Journal of Sound and Vibration*, 77(3), 403–415,

Villaverde, R. and Koyama, L. A. (1993). "Damped resonant appendages to increase inherent damping in buildings." *Earthquake Engineering and Structural Dynamics*, Vol. 22, No. 6, pp. 491–507.

Warburton, G. B. (1982). "Optimal absorber parameters for various combinations of response and excitation parameters." *Earthquake Engineering and Structural Dynamics*, 10(3), 381–401,

Warburton, G. B. and Ayorinde, E. O. (1980). "Optimum absorber parameters for simple systems." *Earthquake Engineering and Structural Dynamics*, 8(3), 197–217.

Wu, G., Shen, X., Li, H., Chen, H., Lin, A. and Suganthan, P. N. (2018). "Ensemble of differential evolution variants." *Information Science*, 423, 172–186.

Yang, X. S., Bekdaş, G. and Nigdeli, S. M. (2016). Metaheuristics and optimization in civil engineering. *Modeling and Optimization in Science and Technologies*, Springer, Switzerland, pp. 1–42.

20 Intelligent Controller Optimizing Structural Performance and Control Devices

Zubair Rashid Wani, Manzoor Tantray,
and Ehsan Noroozinejad Farsangi

CONTENTS

20.1 INTRODUCTION

Because placing a control device on each level is impractical, the location of several devices in a building is critical for achieving effective stabilization. Researchers' recent efforts to find the appropriate location of devices have mostly focused on passive devices. Lindberg et al. employed a model controller to establish the best location and quantity of control devices (Lindberg & Longman, 1984). To identify the position, Ibidapo employed the mode forms of a building as an optimal criterion. More often than not, these techniques are issue specific, and location does not correspond to the overall response.

Among the other methods used for optimizing control and location of devices are: adaptive control systems have been investigated in order to decrease structural reaction while dealing with change in system parameters (Al-Fahdawi, Barroso, & Soares, 2019). A gradient and incremental algorithm were established by Takewaki (1997) and Xu and Teng (2002). Chronological and reiterative approaches were also investigated (Wu et al., 1997). Heuristic search methods were used for deciding the best actuator configuration (Lopez Garcia & Soong, 2002) (Zhang & Soong, 1992). Despite the fact that these techniques are efficient and robust, a max is attained. The usage of genetic algorithm (GA) is an optimization policy with a discontinuous function to determine the optimal solution with finite subspace. Simpson et al. considered GAs to determine the ideal location of control devices for efficiently control (Simpson & Hansen, 1996). Sing et al. used GA to regulate the positioning of a passive controller in a tall building (Singh & Moreschi, 2002). GA and a gradient-based approach were employed to determine the ideal spatial location of an active controller (Abdullah et al., 2001).

The design of the placement mechanism should be simple and intuitive, with minimal computing cost. The control design methodology should include the operational aim, response mitigation, and the quantity and positioning of devices. To accommodate for all of these factors, the positioning method should be linked to the control approach (Wani, 2023). As a result, positions and control method parameters are acquired concurrently using this approach. However, the majority of earlier researchers evaluated and identified the control technique and device setting as a distinct objective function, which doesn't always produce the utmost economical and reliable control when many control devices are used.

This chapter starts with developing and integrating the device location algorithm with the already discussed response-based adaptive (RBA) control strategy. The numerical results are then associated with the corresponding GA. The optimal number, control, and placement parameters of dampers obtained in the numerical simulation are used in the experimental validation. For the brevity of the study, the device setting optimization algorithm is only unified with RBA control strategies using single performance parameters. The results of the experimental verification are contrasted with traditional controllers, i.e., passive-OFF and passive-ON.

20.2 INTEGRATED CONTROLLER: PRINCIPLE

The suggested technique takes the control device placement algorithm into account while designing the controller. By integrating innovative and resilient model-based techniques with location optimization strategy, both parameters are accomplished at the same time. This technique involves minimizing an objective function by tackling global objective functions and obtaining an accurate solution. The suggested integrated RBA control technique is oriented on optimizing the response and location of devices. A gain is computed for each device based on objective selected by the designer, and its exact location is derived concurrently using performance curves. Based on the structural matrices, the developer assigns base shear, drift, acceleration, displacement, or other response characteristics as the target function of the control system.

The process starts with optimizing the controller and location of one damper, which involves placing the damper at different floors and obtaining the corresponding performance curves for each seismic ground motion. Comparing the structural performance curves for magnetorheological (MR) damper at different floors, the location and defining constraints are attained. Next, keeping the parameters and configuration of the first MR damper fixed, the structure is simulated for different locations and configurations of the second MR damper with respect to the first. The comparative analysis of performance curves obtained after the second round of simulations helps to fix the control parameters and the location of the second MR damper. Similarly, the rest of the MR dampers are optimized sequentially. The dampers required for a structure depends on several factors: (i) number of floors, (ii) structural importance, (iii) desired %age of control required for a particular structure, and (iv) cost of the project.

20.3 CONTROLLER FORMULATION

20.3.1 Integrated Adaptive Control Strategy

The following steps are associated with the realization of the suggested controller linked with a device allocation method.

20.3.1.1   Step 1: System State Modeling

The equation governing the system state with dissipation devices encountering dynamic ground motions can be represented as shown below (Peng et al., 2018):

$$[\mathbf{K}]\{\mathbf{x}(t)\}[+[\mathbf{C}]\{\dot{\mathbf{x}}(t)\} + \mathbf{M}\{\ddot{\mathbf{x}}(t)\} = [\mathbf{B}]\{\mathbf{f}(t)\} - [\mathbf{M}]\{\Gamma\}\ddot{\mathbf{x}}_g(t) \tag{20.1}$$

where mass, damping, and stiffness matrices are [M], [C], and [K]. $\ddot{\mathbf{x}}_g(t)$ = external force vector and [**B**] device distribution matrices. In state-space form:

$$\{\mathbf{Y}\} = [\mathbf{E_0}][\mathbf{Z}(t)]$$
$$\{\dot{\mathbf{Z}}(t)\} = [\mathbf{A}]\{\mathbf{Z}(t)\} - [\mathbf{D_0}]\mathbf{P}(t) + \{\mathbf{L}\}[\mathbf{G}]\{\mathbf{f}(t)\} \tag{20.2}$$

$$[\mathbf{A}] = \begin{bmatrix} [\mathbf{0}] & [\mathbf{I}] \\ -[\mathbf{M}]^{-1}[\mathbf{K}] & -[\mathbf{M}]^{-1}[\mathbf{C}] \end{bmatrix}; \quad [\mathbf{G}] = \begin{bmatrix} [\mathbf{0}] \\ [\mathbf{M}]^{-1}\mathbf{B} \end{bmatrix}; \quad \text{and} \quad [\mathbf{D_0}] = \begin{bmatrix} [\mathbf{0}] \\ [\mathbf{M}]^{-1} \end{bmatrix}$$

20.3.1.2   Step 2: Varying Element of the Controller

The velocity was chosen as the shifting component of the suggested adaptive control techniques because it is proportionate to the force as established by testing process (Wani, Tantray, Noroozinejad Farsangi, et al., 2022). The velocity is attained from integrating the acceleration responses for experimental verification using signal processing (Wani, Tantray, & Farsangi, 2021b; Wani, Tantray, & Sheikh, 2021b).

$$\text{Force exerted by damper:} \quad f(t) \propto I \quad \text{and} \quad f(t) \propto \dot{Z}(t) \tag{20.3}$$

$$I \propto \dot{Z}(t) \tag{20.4}$$

$$I \propto absolute\,[\dot{Z}(t)] \tag{20.5}$$

$$I = \mathbf{G}_{\text{optimal}} * absolute\left\{\dot{Z}(t)\right\} + I_0 \tag{20.6}$$

where I_0 = value of current at t_0 and Gc = control gain.

20.3.1.3   Step 3: Target Function for Controller

The adaptive control process demands an optimizing variable that must be reduced in order for the structure to respond optimally. When determining the target requirements for response adaptive techniques, the serviceability and safety restrictions are taken into account (Wani, Tantray, & Farsangi, 2021a; 2021b; Wani, Tantray, Iqbal, et al., 2021a; Wani, Tantray, & Noroozinejad Farsangi, 2022; Wani, Tantray, & Sheikh, 2021c). Some of the objective functions (P-curves) that can be constructed to evaluate the system performance and safety are discussed following.

P-curves for acceleration attenuation:

$$f_1[Z(t)] = \int_0^t \left\{ \sqrt{\sum_{i=1}^r \left[\frac{d^2\{Z_r(t)\}}{dt^2} \right]^2} \right\} dt \quad \forall \quad G = [0, n] \tag{20.7}$$

P-curves for inter-story displacement attenuation:

$$f_2[Z(t)] = \int_0^t \left\{ \sqrt{\sum_{i=1}^r [Z_r - Z_{r-1}]^2} \right\} dt \qquad \forall \quad G = [0, n] \tag{20.8}$$

P-curve for absolute displacement mitigation:

$$f_3[Z(t)] = \int_0^t \left\{ \sqrt{\sum_{i=1}^r [Z_r]^2} \right\} dt \qquad \forall \quad G = [0, n] \tag{20.9}$$

And, P-curve for base shear response:

$$f_4[Z(t)] = \int_0^t \left\{ \sqrt{\sum_{i=1}^r [[K_r] * Z_r]^2} \right\} dt \qquad \forall \quad G = [0, n] \tag{20.10}$$

20.3.1.4   Step 4: Spectral Matching of Ground Excitations

The ground movements needed to train the proposed controller to get the requisite design variables are carefully chosen from the PEER Ground Database. Each collection of excitations is chosen based on three key aspects identified, which are listed below.

   a. The collection must account for spectral variation.
   b. The collection need to incorporate ground movements with wide velocity range, and
   c. ground movements with frequency spectra that matched the goal response were chosen.

20.3.1.5   Step 5: Controller Training

The adaptive control approach incorporates the device localization vector {L}. The incremental simulations approach is used to determine the ideal number, comparative placements, and design parameters for the controllers (Zubair et al., 2022).

   a. The procedure begins with the installation of a damper on base level and the assignment of gain Gc = 0 to the variable constraint, forwarded as input current.
   b. The simulated excitations are specified and adjusted for controller structure, corresponding to the target spectrum.
   c. The target response is then calculated for Gc = 1 to p, and the cycle is continued for all conceivable seismic excitations, where the value of p is determined by the extreme value of current that a damper is designed for.
   d. The dampers are optimized as have been illustrated in the publications (Wani, Tantray, & Farsangi, 2021b). The control gains are then selected by observing the P-curves and using the following mathematical equations:

$$\frac{\partial \{f_1[Z(t)]\}}{\partial G} = 0; \quad \frac{\partial \{f_2[Z(t)]\}}{\partial G} = 0; \quad \frac{\partial \{f_3[Z(t)]\}}{\partial G} = 0; \quad and \quad \frac{\partial \{f_4[Z(t)]\}}{\partial G} = 0 \tag{20.11}$$

   e. The same process is repeated to determine until the design parameters for all control devices are determined.

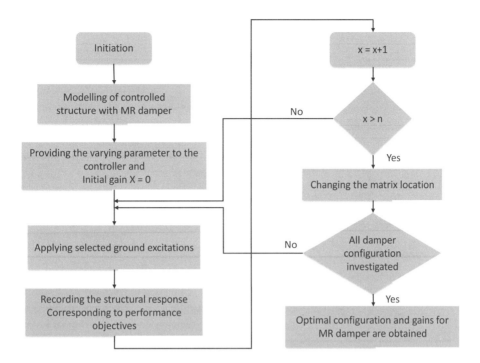

FIGURE 20.1 Flowchart indicating the control system.

Figure 20.1 depicts the entire procedure in a flowchart. The positioning and optimization for each controller are evaluated and determined based on the design requirements selected.

20.4 NUMERICAL STUDY

The structure considered in this study has been adopted from the articles (Wani, Tantray, & Farsangi, 2021a). The model for the control device (MR damper) is also obtained from literature (Wani & Tantray, 2020)

20.4.1 CONTROL STRATEGIES

The designing of the controller entails implementing the processes outlined in Section 20.3.1, detailed below:

a. The controlled structure is modeled in Simulink, and the velocity is selected as a feedback parameter.
b. The seismic excitations are selected and scaled as per ASCE-7 standards and design specifications involved in spectral matching of the seismic excitations.
c. The target objective is selected and the controller is named as the same:
   1. **Inter-story drift response-based adaptive (IDRBA) control:** entrails combining the drift response attenuation with the optimization algorithm.
   2. **Acceleration response-based adaptive (ARBA) control:** entrails combining the acceleration response attenuation with the optimization algorithm.

Figure 20.2 depicts the control process in a closed loop. The key distinction is whether the developer considers serviceability or safety as priority or both. Figures 20.3 and 20.4 illustrate the

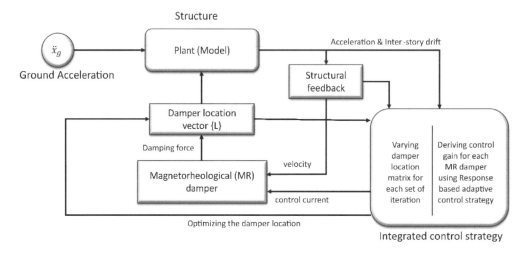

FIGURE 20.2 Closed loop control process.

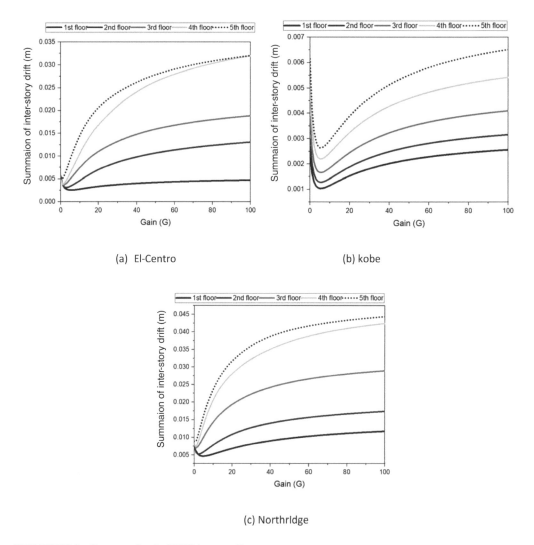

(a) El-Centro

(b) kobe

(c) Northridge

FIGURE 20.3 P-curves for the IDRBA controller.

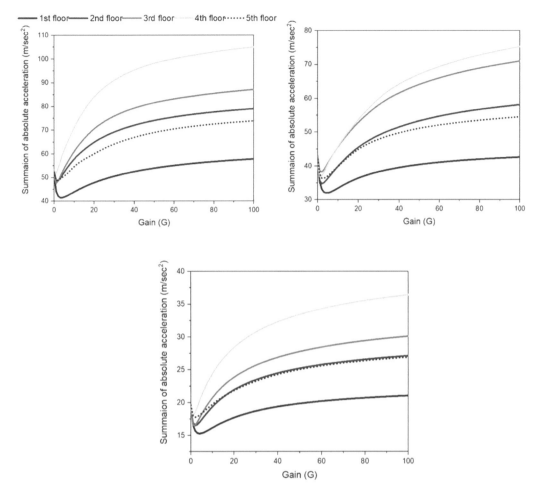

FIGURE 20.4   P-curves for the ARBA controller.

P-curves for the proposed controllers developed by implementing the methods in Section 20.3.1. Only three MR dampers were employed in the study as the increase beyond that did not have much effect on the performance (Wani & Tantray, 2021a). Table 20.1 shows the settings and full realization of the MR damper that were obtained following careful investigation of P-curves.

20.4.2   QUANTUM OF DAMPERS

The two controllers presented are contrasted for a varied number of dampers; peak displacement (Td), max base shear (Bs), and peak acceleration (Ta) are evaluated. Figure 20.5 depicts an assessment of the attenuation of the specified estimated values by the two suggested controllers. The statistics show that as the quantity of control devices grows, so does the %age mitigation in all response values. However, the decrease in %age response after three is minimal. As a result, it is determined that the quantum of dampers should be limited to three and installed in line-predicted configurations.

20.4.3   SIMULATION RESULTS

The responses are examined after numerical modeling of two proposed controllers and a GA strategy. The obtained results are examined for the appropriate number of control devices and ground

TABLE 20.1

Control Parameters for Each Controller

Control Strategy	Configuration	Control Parameters

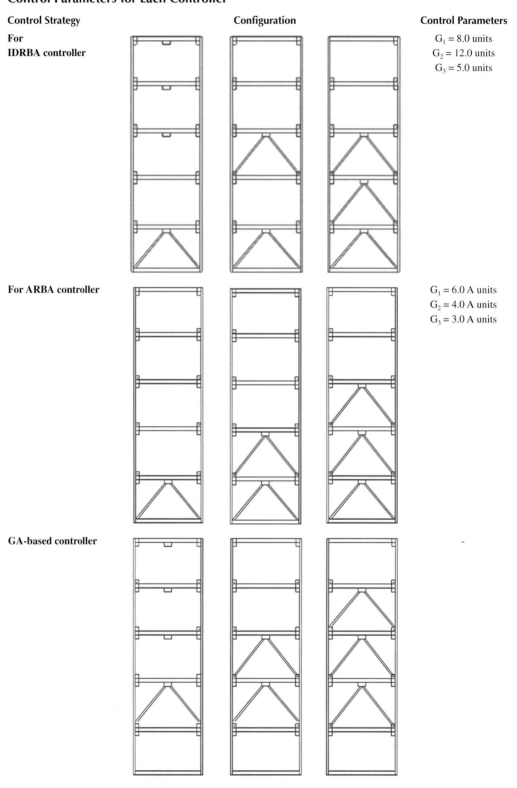

For
IDRBA controller

$G_1 = 8.0$ units
$G_2 = 12.0$ units
$G_3 = 5.0$ units

For ARBA controller

$G_1 = 6.0$ A units
$G_2 = 4.0$ A units
$G_3 = 3.0$ A units

GA-based controller

–

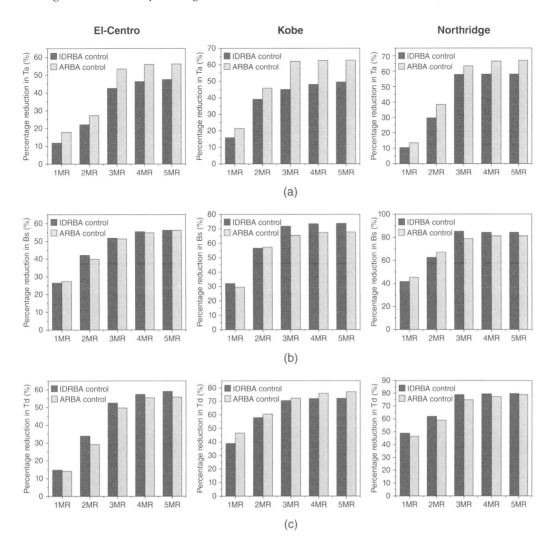

FIGURE 20.5   Reduction of performance quantities to determine the damper number.

motions. For ideal installations, Figures 20.6 and 20.7 demonstrate the peak drift and acceleration response. The IDRBA controller beat ARBA and GA in attenuation of overall drift response.
Furthermore, the ARBA outperformed IDRBA and GA in terms of dampening peak acceleration
reactions. The decrease in response, on the other hand, was not continuous and was directly related
to the severity of excitation (pga). Because the COC algorithm considers control devices to be separate entities, thus the resistance of the controller intensifies floor lockup. The mean peak decrease in
drift by the IDRBA control method is 24.53%, 44.48%, and 53.28% compared to COC, respectively.
Furthermore, the mean peak acceleration reduction by the ARBA controller is 3.24% for one, 8.96%
for two, and 16.43% for three dampers, respectively.

Figure 20.8 depicts the base shear time history for proposed controller compared to the COC
strategy. Because of the controller flexibility and adjustability to ground acceleration, the IDRBA
controller was effective in decreasing the base shear quite substantially more than the comparable
COC approach, as seen in the figure. Furthermore, the IDRBA controller reduces base shear better
than ARBA as it is precisely related to drift, which is the target function of IDRBA.

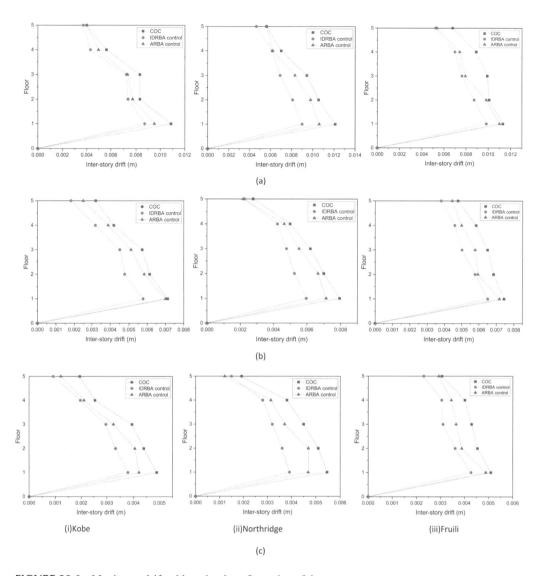

FIGURE 20.6 Maximum drift with optimal configuration of dampers.

20.5 EXPERIMENTAL VERIFICATION OF INTEGRATED CONTROL STRATEGIES

A total of 45 shaking table studies were carried out to verify the controlled structure using an integrated controller and passive strategies. The framed structure, with one, two, and three MR dampers installed at different floors, has been illustrated in Figure 20.9. Figure 20.9(**iv**) shows the lumped mass model of the structure.

20.5.1 ADDITIONAL DEVICES AND SENSORS

Compared to the experiments conducted in the previous chapter, the validation of integrated control strategy employing multiple MR dampers in the structure simultaneously requires additional equipment and recording channels which are mentioned under (Wani, Tantray, & Noroozinejad Farsangi, 2022).

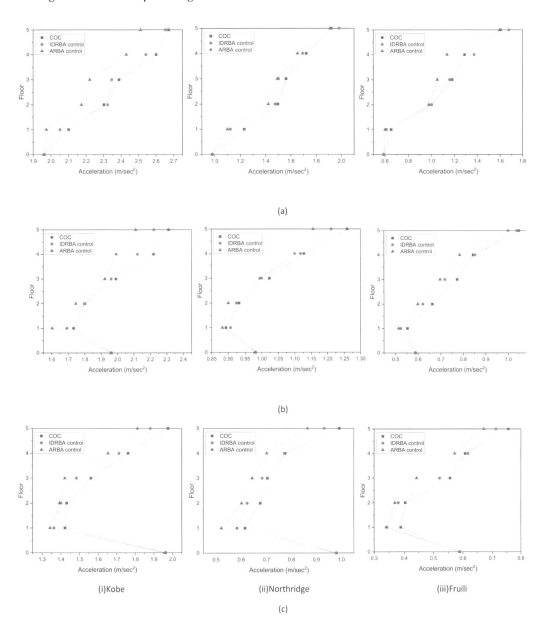

FIGURE 20.7 Maximum acceleration response with optimal configuration of dampers.

20.5.1.1 USB 6009

The hardware components manufactured by National Instruments (NI) perform amicability with the already mentioned LabVIEW software by NI. The difference between USB-6009 and the already mentioned USB-6002 hardware systems is that USB-6009 has three analog output (AO) ports compared to two AOs for USB-Daq-6002. The extra AO channel comes into the picture when three controllers are employed in a controller using passive and proposed strategies.

20.5.1.2 Multi-Channel Power Amplifier Circuit

In the previous chapter, only a single line of input supply voltage is fed through the amplifier circuit to the MR damper as a control current. However, a max of three dampers is employed at the same

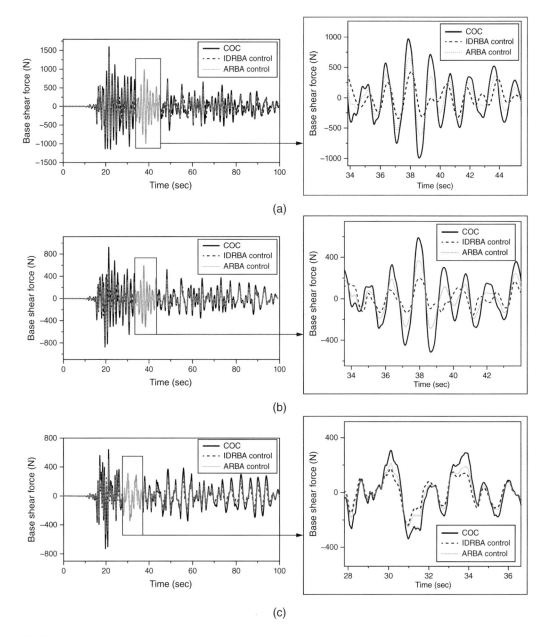

FIGURE 20.8 Base shear record with optimal configuration of dampers.

time to control the structural response; therefore, the multiple channel power amplifier circuit is used to simultaneously provide the required control current to all MR dampers.

In addition to this hardware, a few force transducers to obtain the damping force of MR dampers are installed at different floors and response feedback channels from the damper installation layers are connected to the control system.

20.5.2 Input Excitation and Testing Condition

The ground motion has been adopted from literature (Wani & Tantray, 2021b). Following five cases of namely uncontrolled (without MR dampers), passive off-and-on control, and two proposed

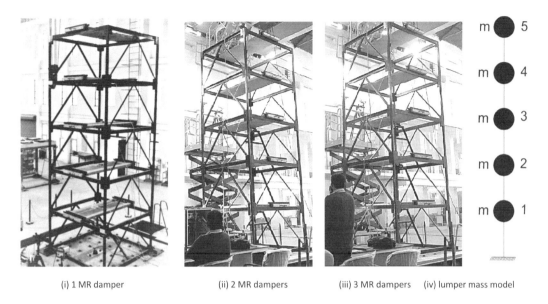

(i) 1 MR damper   (ii) 2 MR dampers   (iii) 3 MR dampers   (iv) lumper mass model

FIGURE 20.9 Positioning of controller at different floors with lumped masses.

strategies are well thought out to understand the performance of algorithms. For the passive controller, the optimal placement of multiple MR dampers as obtained from literature is the successive assignment of dampers (Singh & Moreschi, 2002). Also, the optimal placement of MR dampers and control parameters of proposed controllers is taken from the same as discussed in Section 20.3.2. These strategies are defined as below.

- **Uncontrolled:** This is the condition when the structure is bare, i.e., no MR damper is installed in the structure
- **Passive-off:** the current supply is fixed at a constant zero ampere (0 A) throughout the shake table tests.
- **Passive-on:** The current supply is fixed at a constant of 0.8 A throughout the shake table tests.
- **IDRBA control:** The control current is obtained from optimizing the overall floor drift integrated with the damper location algorithm.
- **ARBA control:** The control current is obtained from optimizing the overall acceleration response of the structure integrated with the damper location algorithm.

20.5.3 Results and Discussions

20.5.3.1 Max Floor Response

The column base strain, absolute acceleration, and max drift for discussed control strategies have been illustrated in Figures 20.10–20.12.

For the passive off situation, it is deemed as the control system's fail-safe state. The passive controller solely makes use structure's inherent reactivity. When a damper, the average peak acceleration, drift, and strain were lowered by 6.30% to 13.45%, 24.45% to 40.67%, and 14.23% to 22.45%, respectively. The mean max decrease in response quantities for two MR dampers is 17.27% to 22.62%, 33.33% to 42.66%, and 27.70% to 35.11%, respectively, when compared to the uncontrolled scenario.

When a continuous supply of current is given to dampers, it works as a backup plan in the event of controller let-down. Passive-on control dissipates reduces the reaction by exploiting the viscosity

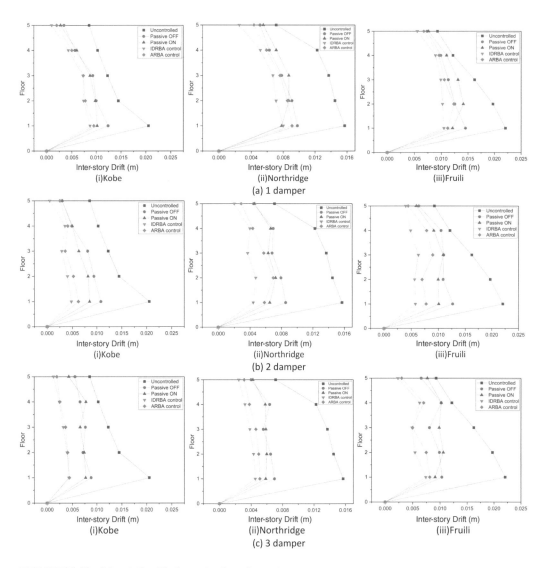

FIGURE 20.10   Max drift with the optimal configuration.

of the MR fluid. Peak reduction in response is 8.45% to 14.22%, 23.76% to 25.12%, and 9.76% to 16.65%, for acceleration, drift, and base strain, respectively. T, the maximum peak diminution in for deployment of two dampers, is 16.77% to 26.43%, 35.35% to 40.45%, and 24.11% to 29.80%, respectively. Finally, the three MR dampers' response reduction is 32.22% to 35.45%, 39.66% to 49.23%, and 37.32% to 40.76%, for acceleration, drift, and strain, respectively.

For IDRBA, the mean attenuation in peak responses with a single MR damper placed strategically is 19.97% to 25.45% in acceleration, 47.98% to 54.06% in floor drift, and 27.44% to 34.75% in base strain, respectively. Likewise, for ideally located two dampers, the mean drop in peak responses is 26.55% to 39.65% for acceleration, 52.44% to 69.47% for drift, and 46.44% to 52.67% for base strain, respectively. Finally, the three dampers' average mitigation is 47.52% to 56.91% in acceleration response, 63.78% to 79.44% in drift response, and 54.44% to 62.94% in base strain response, respectively.

The second proposed ARBA controller, employing one MR damper, mitigated the response by 24.46% to 32.81% for acceleration, 41.31% to 49.44% for floor drift, and 23.22% to 28.31% for base

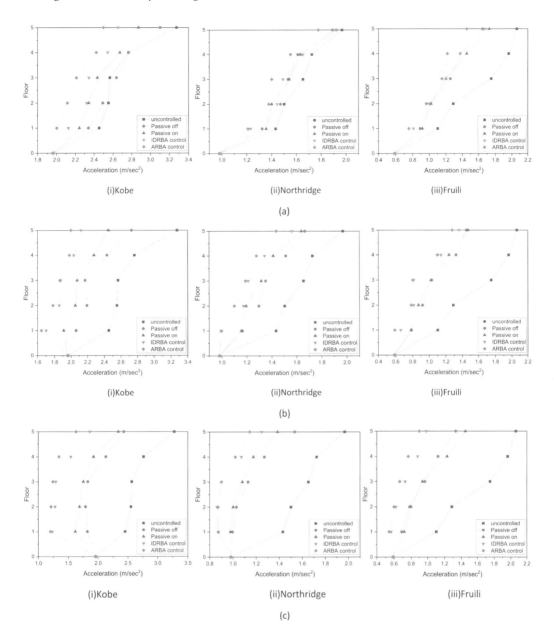

FIGURE 20.11 Max acceleration for optimal configuration.

strain, respectively. The two dampers had 37.66% to 47.55% reduction in acceleration, 45.23% to 52.72% in story drift, and 38.48% to 48.43% base strain, respectively. Finally, the three dampers' peak response reduction is 52.32% to 67.71% for acceleration, 58.32% to 68.45% for floor drift, and 51.35% to 54.85% for base strain, respectively. A detailed examination of the findings revealed that ARBA controller outclassed the IDRBA controller in moderating the max acceleration by 10% to 15%.

20.5.3.2 Hysteresis and FFT Curves

The fluctuation in restoring force with ground movement when exposed to Kobe is shown in Figure 20.13 (for the ARBA controller only). The analysis indicates that as the dampers increased,

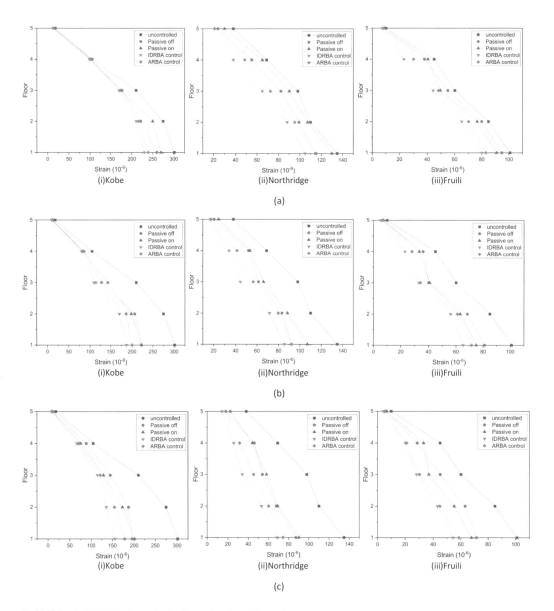

FIGURE 20.12 Max base strain for optimal configuration.

the force dissipation requirement of the system was successfully dispersed among controllers put at different levels. Furthermore, the forces generated IDRBA is greater than that analogous ARBA due to greater control parameters attained during simulations.

Figures 20.14 and 20.15 show the FFT curve responses for proposed controllers with varying damper numbers and compared with a bare frame. The maximum response significantly mitigated and spread across different modes of vibration. For determined configuration of dampers, the mean maximum acceleration pertaining to the fundamental MOV was reduced by 33.50% for one damper, 69.61% for two dampers, and 93.63% three dampers using the ARBA strategy, and 27.31% for one damper, 61.05% for two dampers, and 82.58% for three dampers using the IDRBA control strategy. However, when the reaction is dispersed across different MOVs, the maximum absorption for ARBA is 33.49%, 52.87%, and 87.00%, whereas the IDRBA control technique is 27.31%, 43.25%, and 76.44%. Similarly, for displacement, a considerable drop in fundamental mode is 41.12%,

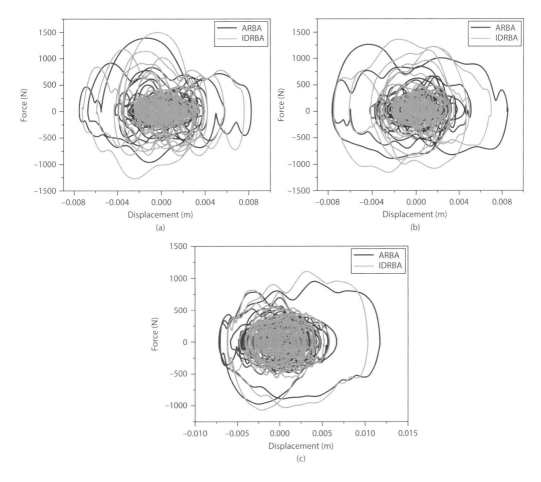

FIGURE 20.13   Hysteresis graphs indicating the energy dissipation involving proposed ARBA controllers.

75.34%, and 8.644% using ARBA and 51.12%, 84.42%, and 94.12% using IDRBA. Furthermore, the maximum deflection considering all modes was reduced by 25.54%, 66.11%, and 79.34% when ARBA was used, and 35.42%, 78.63%, and 89.83% when IDRBA was used.

As a result, the suggested integrated controllers were able to greatly attenuate all reactions when compared to the passive and bare frame because of the continuous adaptability of the control current and dependence of control parameters on the location of MR dampers.

20.6   CONCLUSION

This chapter dealt with both the mathematical and experimentation of two integrated controllers with multiple optimizing objectives. The control algorithms not only supplied optimal control current to the MR damper, but also took care of the location/configuration optimization for multiple MR dampers. In the numerical simulation, the control strategies discussed in the previous chapter were combined with location optimization system to determine the position/conformation of dampers taking into deliberation the target function of the control system. IDRBA and ARBA control strategies based on minimizing overall drift and acceleration were formulated and compared with the benchmark GA controller. The control parameters, i.e., gain and location of dampers are gained for each controller and extensive simulations were carried out for seismic excitations of varying intensities. The outcomes specified the dominance of integrated controllers in reducing the structural response compared to GA strategy.

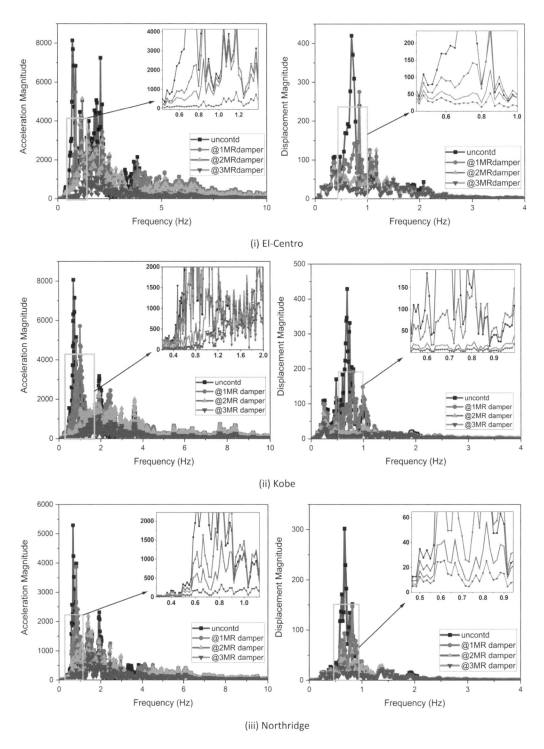

FIGURE 20.14   FFT curves for the IDRBA controller.

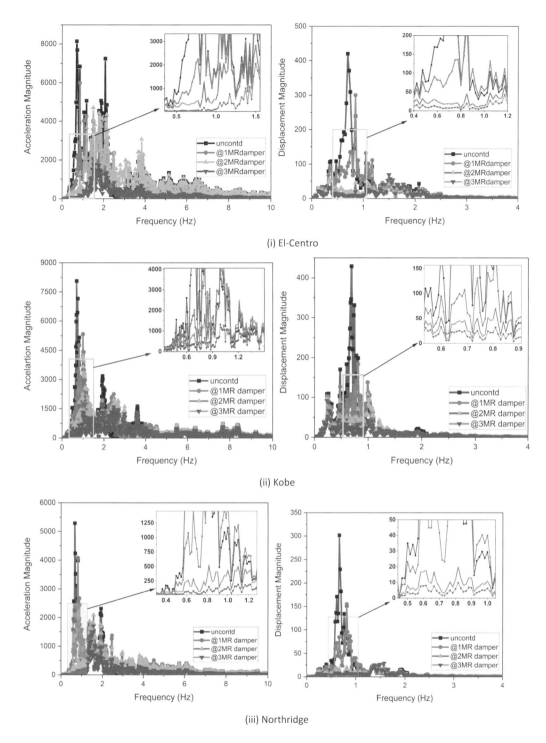

FIGURE 20.15 FFT curves for the ARBA controller.

Next, the experimental verification of the numerically designed integrated control strategies was implemented on a five-story steel structure. All the optimal configurations of MR dampers with respect to the corresponding integrated and passive strategies are considered. For each control strategy, the response reduction increased with a surge in the quantity of control devices from one to three. It was observed that proposed controllers implicitly mitigated responses compared to the bare frame configuration and passive controllers, because of the continuous adaptability of the control current and dependence of control parameters on the location of the MR dampers. Also, the designer can select the target function of the control system based on the importance of the structure and desired response to be controlled. Also, the observed experimental results remained in decent agreement with the corresponding numerical values.

REFERENCES

Abdullah, M. M., Richardson, A., & Hanif, J. (2001). Placement of sensors/actuators on civil structures using genetic algorithms. *Earthquake Engineering and Structural Dynamics*. https://doi.org/10.1002/eqe.57

Lindberg, R. E., & Longman, R. W. (1984). On the number and placement of actuators for independent modal space control. *Journal of Guidance, Control, and Dynamics*. https://doi.org/10.2514/3.56366

Lopez Garcia, D., & Soong, T. T. (2002). Efficiency of a simple approach to damper allocation in MDOF structures. *Journal of Structural Control*. https://doi.org/10.1002/stc.3

Peng, Y., Yang, J., & Li, J. (2018). Parameter identification of modified Bouc–Wen model and analysis of size effect of magnetorheological dampers. *Journal of Intelligent Material Systems and Structures*. https://doi.org/10.1177/1045389X17740963

Simpson, M. T., & Hansen, C. H. (1996). Use of genetic algorithms to optimize vibration actuator placement for active control of harmonic interior noise in a cylinder with floor structure. *Noise Control Engineering Journal*. https://doi.org/10.3397/1.2828399

Singh, M. P., & Moreschi, L. M. (2002). Optimal placement of dampers for passive response control. *Earthquake Engineering and Structural Dynamics*. https://doi.org/10.1002/eqe.132

Takewaki, I. (1997). Optimal damper placement for minimum transfer functions. *Earthquake Engineering and Structural Dynamics*. https://doi.org/10.1002/(SICI)1096-9845(199711)26:11<1113::AID-EQE696>3.0.CO;2-X

Wani, Z. R., & Tantray, M. (2021a). Study on integrated response-based adaptive strategies for control and placement optimization of multiple magneto-rheological dampers-controlled structure under seismic excitations. *Journal of Vibration and Control*, 28(13–14), 10775463211000484. https://doi.org/10.1177/10775463211000483

Wani, Z. R., & Tantray, M. (2021b). Study on integrated response-based adaptive strategies for control and placement optimization of multiple magneto-rheological dampers-controlled structure under seismic excitations. *JVC/Journal of Vibration and Control*. https://doi.org/10.1177/10775463211000483

Wani, Z. R., Tantray, M., & Farsangi, E. N. (2021a). Shaking table tests and numerical investigations of a novel response-based adaptive control strategy for multi-story structures with magnetorheological dampers. *Journal of Building Engineering*, 102685. https://doi.org/10.1016/j.jobe.2021.102685

Wani, Z. R., Tantray, M., & Farsangi, E. N. (2021b). Investigation of proposed integrated control strategies based on performance and positioning of MR dampers on shaking table. *Smart Materials and Structures*, 30(11), 115009. https://doi.org/10.1088/1361-665x/ac26e6

Wani, Z. R., Tantray, M. A., Iqbal, J., Farsangi, E. N., Wani, Z. R., Tantray, M. A., Iqbal, J., & Farsangi, E. N. (2021a). Configuration assessment of MR dampers for structural control using performance-based passive control strategies. *Structural Monitoring and Maintenance*, 8(4), 329. https://doi.org/10.12989/SMM.2021.8.4.329

Wani, Z. R., Tantray, M., & Noroozinejad Farsangi, E. (2022). In-Plane measurements using a novel streamed digital image correlation for shake table test of steel structures controlled with MR dampers. *Engineering Structures*, 256, 113998. https://doi.org/10.1016/J.ENGSTRUCT.2022.113998

Wani, Z. R., Tantray, M., Noroozinejad Farsangi, E., Nikitas, N., Noori, M., Samali, B., & Yang, T. Y. (2022). A critical review on control strategies for structural vibration control. *Annual Reviews in Control*, 54, 103–124. https://doi.org/10.1016/J.ARCONTROL.2022.09.002

Wani, Z. R., Tantray, M., & Sheikh, J. I. (2021b). Experimental and numerical studies on multiple response optimization-based control using iterative techniques for magnetorheological damper-controlled structure. *Structural Design of Tall and Special Buildings*, 30(13). https://doi.org/10.1002/tal.1884

Wani, Z. R., Tantray, M., & Sheikh, J. I. (2021c). Experimental and numerical studies on multiple response optimization-based control using iterative techniques for magnetorheological damper-controlled structure. *The Structural Design of Tall and Special Buildings*, *30*(13), e1884. https://doi.org/10.1002/TAL.1884

Wani, Z. R., & Tantray, M. A. (2020). Parametric study of damping characteristics of magnetorheological damper: Mathematical and experimental approach. *Pollack Periodica*, *15*(3). https://doi.org/10.1556/606.2020.15.3.4

Wu, B., Ou, J. P., & Soong, T. T. (1997). Optimal placement of energy dissipation devices for three-dimensional structures. *Engineering Structures*. https://doi.org/10.1016/S0141-0296(96)00034-X

Xu, Y. L., & Teng, J. (2002). Optimum design of active/passive control devices for tall buildings under earthquake excitation. *Structural Design of Tall Buildings*. https://doi.org/10.1002/tal.193

Zhang, R. H., & Soong, T. T. (1992). Seismic design of viscoelastic dampers for structural applications. *Journal of Structural Engineering (United States)*. https://doi.org/10.1061/(ASCE)0733-9445(1992)118:5(1375)

Zubair, R., Manzoor, T., & Ehsan, N. F. (2022). Acceleration response-based adaptive strategy for vibration control and location optimization of magnetorheological dampers in multistoried structures. *Practice Periodical on Structural Design and Construction*, *27*(1), 04021065. https://doi.org/10.1061/(ASCE)SC.1943-5576.0000648

21 Rapid Discrete Optimal Design of High-Damping Rubber Dampers for Elastic-Plastic Moment Frames under Critical Double Impulse

Koshin Iguchi, Kohei Fujita, and Izuru Takewaki

CONTENTS

21.1 INTRODUCTION

Uncertain natural hazard environments bring a lot of risks to structures and infrastructures. To respond to such risks, structural control and worst-case analysis play an important role. These two approaches are thought to be effective for upgrading the resilience of structures and infrastructures (Bruneau et al. 2003, Cimellaro et al. 2010). Although active structural control and passive structural control have been introduced for earthquake and wind loadings (De Domenico et al. 2019, Lagaros et al. 2013, Soong and Dargush 1997, Takewaki 2009, Takewaki and Akehashi 2021, Uetani et al. 2003), passive structural control is acknowledged as a reliable method especially for earthquake loading. On the other hand, worst-case analysis is believed to be the most effective method for taking into account various uncertainties encountered in structural properties and earthquake ground motion properties.

In passive structural control, discrete optimal design methods are desired because some passive dampers are used as multiple numbers of machine products. In the field of discrete optimal design, the branch and bound method and the genetic algorithm (GA) (Apostolakis and Dargush 2010, Idels and Lavan 2021, Uemura et al. 2021) are two principal methods. As other methods of discrete optimal design of passive dampers, SSA (sequential search algorithm) was proposed by Zhang and Soong (1992). In SSA, a response index, for example inter-story drift, is selected and a damper corresponding to the maximum response index is added sequentially. This procedure is repeated until the response index satisfies the target or the total quantity of added dampers reaches the specified value. Although this procedure is simple, the computational load is quite large.

Recently, Kanno (2013) used a second-order cone model based on a 0-1 mixed-integer programming approach for solving a discrete optimal design problem of viscous dampers in shear building models. Furthermore, Idels and Lavan proposed a gradient-based optimization for moment frames and nonlinear viscous dampers. They compared their method with GA.

In this chapter, the MDOF moment frames including high-damping rubber dampers are subjected to the double impulse and the corresponding one-cycle sine wave as a representative of long-period, pulse-type ground motions and a relaxed sequential search algorithm (RSSA) for discrete optimal damper design is proposed. It should be remarked that the treatment of the critical (resonant) double impulse and the corresponding one-cycle sine wave enables a simple but realistic overcoming of the difficulty in dealing with uncertainties of input ground motions. Furthermore, the relaxation of SSA provides structural engineers with a powerful and rapid measure for finding a realistic discrete optimal damper design.

21.2 DOUBLE IMPULSE REPRESENTING PULSE-TYPE GROUND MOTION

The double impulse (DI) of two single impulses with opposite signs was introduced to represent the main part of a near-fault pulse-type ground motion simulated as a one-cycle sine wave (OCSW) (Kojima and Takewaki 2015). The motivation is that, while the combination of free and forced-vibration components is inevitable in the analysis of response to the forced input, the double impulse induces only a free-vibration component. This enables to avoid the transcendental equation for resonance curves and use a simple energy balance law for obtaining the maximum response without time-history response analysis. The main part is first simulated by OCSW $\ddot{u}_{g\sin}(t)$, as shown in Equation. (21.1) (see Figure 21.1(a)) and then transformed into a double impulse $\ddot{u}_g(t)$, expressed by Equation. (21.2) (see Figure 21.1(b)).

$$\ddot{u}_{g\sin}(t) = A_p \sin \omega_p t \qquad (21.1)$$

$$\ddot{u}_g(t) = V\delta(t) - V\delta(t - t_0) \quad (\delta(t) : \text{Dirac delta function}) \qquad (21.2)$$

where A_p, ω_p, V, and t_0 indicate the acceleration amplitude and circular frequency of OCSW, the velocity amplitude of the double impulse, and the time interval of the two impulses, respectively. Kojima and Takewaki (2015) used a criterion on the same maximum Fourier amplitude in this transformation.

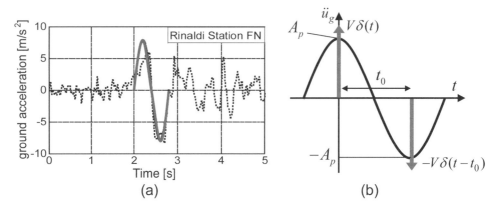

FIGURE 21.1 Simulation of main part of ground motion by double impulse: (a) transformation of main part of Rinaldi station FN motion (Northridge 1994) into OCSW; (b) re-transformation into double impulse.

The ratio a of A_p to V in this transformation is introduced by

$$A_p = aV \qquad (21.3)$$

The parameter a as a function of $t_0 = \pi / \omega_p$ can be derived as

$$a(t_0) = \frac{A_p}{V} = \frac{\max \left| \sqrt{2 - 2\cos(\omega t_0)} \right|}{\max \left| \dfrac{2\pi t_0}{\pi^2 - (\omega t_0)^2} \sin(\omega t_0) \right|} \qquad (21.4)$$

The maximum velocity V_p of OCSW can be expressed by

$$V_p = 2A_p / \omega_p \qquad (21.5)$$

21.3 HIGH-DAMPING RUBBER DAMPERS

The present visco-elastic high-damping rubber dampers show peculiar characteristics compared with ordinary other visco-elastic dampers (see Figure 21.2). The explanation in this section comes from the reference (Tani et al. 2009).

The proposed damper model consists of three elements: (1) elastic-plastic element (element 1), (2) elastic element due to dynamic effect (element 2), and (3) viscous element (element 3). The elastic-plastic element indicates the strain dependency. The stiffness in the frequency range 0.2–2.0 Hz shows a tendency different from the static stiffness. The elastic element due to dynamic effect expresses this property. The viscous element indicates the viscosity of the damping material. The detailed relations of shear stress τ with shear strain γ (or shear strain velocity $\dot{\gamma}$) in these three elements can be found in the reference (Tani et al. 2009).

The high-damping rubber dampers exhibit special characteristics for non-stationary loading. The essence of the model is the employment of reaction modification factors, applied to stationary-loop properties, for gradually decreasing loops. The experiment demonstrated that such modification factors are unnecessary for gradually increasing loops because the maximum strain γ_{max} is updated successively in the gradually increasing loops.

The reaction shear stress in a stationary loop can be obtained by

$$\tau = \tau_1 + \tau_2 + \tau_3 \qquad (21.6)$$

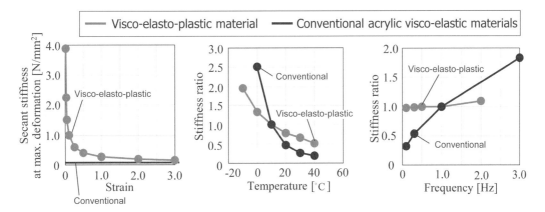

FIGURE 21.2 Comparison of present high-damping rubber damper (visco-elasto-plastic material) with conventional visco-elastic damper.

where τ_i denotes the reaction shear stress of element i in the stationary loop corresponding to γ_{max}. On the other hand, the reaction shear stress for a non-stationary condition may be described by

$$\tau = \beta_2(\beta_1\alpha_1\tau_1 + \alpha_2\tau_2 + \alpha_3\tau_3) \tag{21.7}$$

where α_i = reaction modification factor of element i in a non-stationary loop, β_1 = reaction magnification factor for the virgin loop, and β_2 = material randomness factor. α_1 presents the reaction modification factor of element 1 defined for gradually decreasing loops. α_2 and α_3 express the reaction modification factors of elements 2 and 3, respectively. A more detailed explanation can be obtained from the reference (Tani et al. 2009). This damper element has been incorporated into the general-purpose structural analysis program (SNAP 2015).

21.4 OPTIMIZATION ALGORITHM

21.4.1 Optimal Damper Placement Problem and Optimization Algorithm

While many investigations on damper optimization have been conducted mainly for elastic frames, some studies have been done for elastic-plastic building structures (Akehashi and Takewaki 2019, Attard 2007, Aydin et al. 2007, Idels and Lavan 2021, Uetani et al. 2003). In this chapter, the maximum inter-story drift and the maximum floor acceleration are the response targets for optimization with discrete sets of high-damping rubber dampers. This multi-objective optimization is tackled by introducing the following objective function.

$$\gamma_i = \beta\frac{\delta_i}{\delta_d} + (1-\beta)\frac{\ddot{U}_i}{\ddot{U}_d} \tag{21.8}$$

where δ_i, $\ddot{U}_i$ are the maximum inter-story drift in the ith story and the maximum horizontal acceleration at the ith floor and δ_d, $\ddot{U}_d$ are the target values of the maximum inter-story drift and the target value of the maximum horizontal floor acceleration. In addition, β is a weighting coefficient between the maximum inter-story drift and the maximum horizontal floor acceleration. For a larger value of β, the maximum inter-story drift is to be focused.

Optimization algorithm:

1. Compute the maximum inter-story drifts and the maximum horizontal floor acceleration in every story and at every floor under the COCSW (critical one-cycle sine wave) corresponding to the critical double impulse obtained by the zero base shear force criterion after the first impulse (Akehashi and Takewaki 2019). Obtain γ_{max} among γ_i.
2. Compute the index γ_i for all stories. Set $\gamma_i = 0$ for stories where dampers have already been allocated to the upper limit.
3. Add a predetermined amount of dampers (discrete quantity) to stories where the corresponding indices γ_i satisfy $\gamma_i \geq \gamma_{max}(1-0.01x)$ for a prescribed value of x (x %). This value x is called a relaxation level. Repeat the procedures (1)–(3) until the following conditions are satisfied.
4. Stop the installation of added dampers to the specific story (story i) if both $(\delta_i / \delta_d) < 1$ and $(\ddot{U}_i / \ddot{U}_d) < 1$ are reached. Then, the index γ_i is changed to 0.
5. Stop the damper allocation procedure when one of the following conditions is satisfied: 1) the dampers are not updated, 2) the total amount of dampers reached the upper bound.

Figure 21.3 shows the comparison of the proposed algorithm with SSA. Since the general-purpose structural analysis program (SNAP 2015) can be called in the MATLAB environment, the proposed algorithm can be implemented smoothly and rapidly.

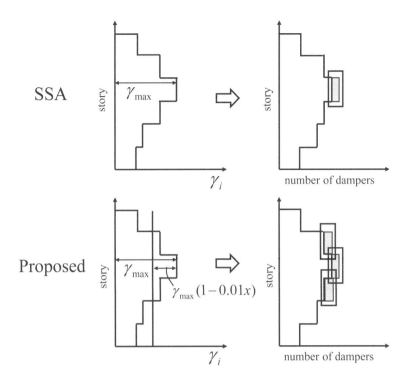

FIGURE 21.3 Comparison of the proposed algorithm with SSA.

21.5   DESIGN EXAMPLE FOR SHEAR BUILDING MODEL

Consider first a 40-story shear building model. Each floor mass is $1,280 \times 10^3$(kg) and the common story height is 4(m). The story stiffness distribution and the lowest mode are shown in Figure 21.4. The fundamental natural period of the model without damper is 4.0(s) and the structural damping ratio is 0.02 (stiffness-proportional). The common yield inter-story drift is 0.03(m) and the post-yield stiffness ratio are 0.05.

High-damping rubber dampers of a thickness of 25(mm) are used as a double sheet and the temperature of 25(°C) and the frequency of 0.25(Hz) are assumed for determining the material properties. The upper bound of the number of dampers in each story is 16 (increment of added damper is 1) and the upper bound of the total number of dampers is 320.

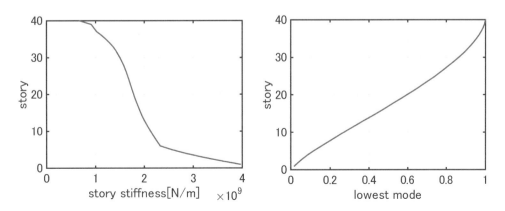

FIGURE 21.4 Story stiffness of shear building model and lowest mode.

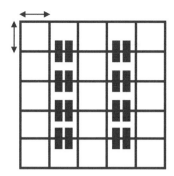

FIGURE 21.5 Building plan and possible damper location.

The critical double impulse with the velocity amplitude of 1.0(m/s) was treated and the corresponding equivalent sine wave with the resonant period was used as an input base motion. It should be remarked that the resonant period of the critical double impulse and the corresponding equivalent sine wave is determined by the criterion of the zero base shear force after the first impulse (Akehashi and Takewaki 2019, Kojima and Takewaki 2015).

Figure 21.5 shows the building plan and possible damper locations. While the influence of dampers on the inter-story drifts is relatively clear, the sensitivity of damper allocation for the top acceleration is not so simple. To investigate this sensitivity of damper allocation for the top acceleration, an analysis was made. Figure 21.6 illustrates the damper locations with high sensitivity for the top acceleration with respect to step number. It can be observed that the dampers installed to the top several stories are sensitive to the suppression of the top acceleration.

In the following examples, a weighting coefficient $\beta = 0.5$ in Equation. (21.8) was employed. This means equal weighting on the inter-story drift and the top acceleration. Four combinations $\delta_d = 50$ mm, $\ddot{U}_d = 2000$ mm/s²; $\delta_d = 70$ mm, $\ddot{U}_d = 2000$ mm/s²; $\delta_d = 50$ mm, $\ddot{U}_d = 3000$ mm/s²; and $\delta_d = 70$ mm, $\ddot{U}_d = 3000$ mm/s² are considered for four control ranges $x = 0\%$, 2%, 5%, and 10%. The case of $x = 0\%$ corresponds to SSA by Zhang and Soong (1992).

Figure 21.7 shows the optimal damper placement, the maximum floor acceleration, and the maximum inter-story drift for $x = 0\%$, 2%, 5%, and 10% in the case of $\delta_d = 50$ mm, $\ddot{U}_d = 2000$ mm/s². It was found that, while the dampers in the lower middle stories are effective for reducing the inter-story drifts, the dampers in the top several stories play an important role for suppressing the top acceleration. However, the suppression rate of the top acceleration is low compared to the reduction rate of the inter-story drifts.

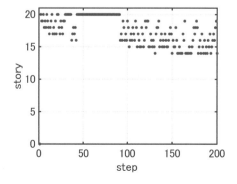

FIGURE 21.6 Sensitive damper allocation for top acceleration.

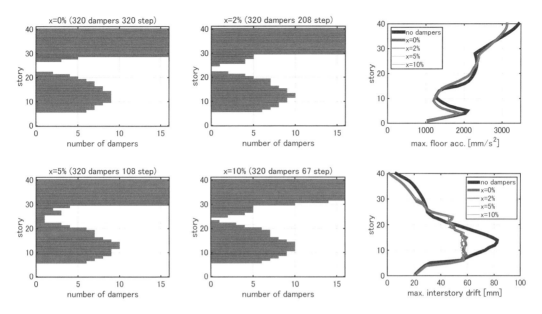

FIGURE 21.7 Optimal damper placement, maximum floor acceleration, and maximum inter-story drift for $x = 0\%$, 2%, 5%, and 10% ($\delta_d = 50$ mm, $\ddot{U}_d = 2000$ mm/s²).

Figure 21.8 presents those distributions in the case of $\delta_d = 70$ mm, $\ddot{U}_d = 2000$ mm/s². Since the weighting rate on the inter-story drifts becomes small, the allocation quantities to the lower middle stories decrease.

Figure 21.9 illustrates those distributions in the case of $\delta_d = 50$ mm, $\ddot{U}_d = 3000$ mm/s². Since the weighting rate on the top acceleration becomes small, the allocation quantities to the top several stories decrease.

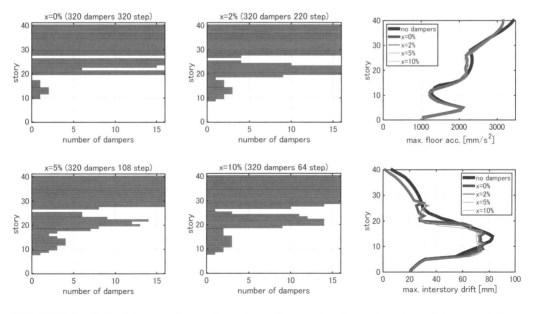

FIGURE 21.8 Optimal damper placement, maximum floor acceleration, and maximum inter-story drift for $x = 0\%$, 2%, 5%, and 10% ($\delta_d = 70$ mm, $\ddot{U}_d = 2000$ mm/s²).

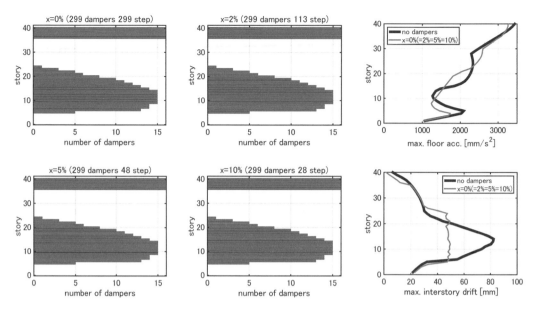

FIGURE 21.9 Optimal damper placement, maximum floor acceleration, and maximum inter-story drift for $x = 0\%$, 2%, 5%, and 10% ($\delta_d = 50$ mm, $\ddot{U}_d = 3000$ mm/s^2).

Figure 21.10 shows those distributions in the case of $\delta_d = 70$ mm, $\ddot{U}_d = 3000$ mm/s^2. Since both constraints on the inter-story drifts and on the top acceleration are relaxed in this case, the allocated damper quantities decrease both in the lower middle stories and the top several stories.

It can be observed from Figures 21.7–21.10 that the proposed algorithm provides a rapid procedure (small number of steps) for a larger value of x with a nearly exact result for the optimal damper location by SSA corresponding to $x = 0\%$.

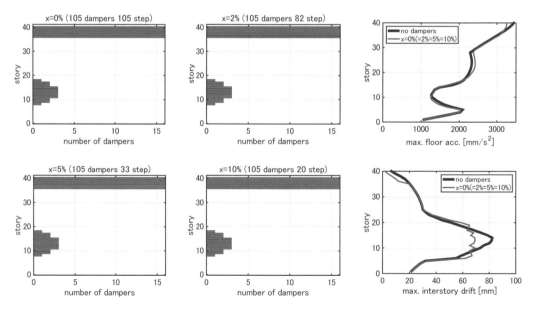

FIGURE 21.10 Optimal damper placement, maximum floor acceleration, and maximum inter-story drift for $x = 0\%$, 2%, 5%, and 10% ($\delta_d = 70$ mm, $\ddot{U}_d = 3000$ mm/s^2).

21.6 DESIGN EXAMPLE FOR MOMENT RESISTING FRAME

Consider next a 20-story plane steel moment resisting frame. Figure 21.11 shows the frame section and the types of member sections together with the first four natural modes. The member cross sections are shown in Table 21.1. The steel used is called SS400 in Japan and has a tensile strength larger than $400(N/mm^2)$. The composite beam-slab action was considered and the stiffness of beams was magnified by a factor of 2.

The story heights are 6(m) in the first story and 4(m) in other stories. The span lengths are 9.6(m) in the central span and 12.8(m) in the outer spans. The fundamental natural period of the model without a damper is 2.655(s) and the structural damping ratio is 0.02 (stiffness-proportional).

The upper bound of the number of dampers in each story is 4 (increment of added damper is 1/4) and the upper bound of the total number of dampers is 90. These parameters are set so as to reflect the relation of a plane frame with the corresponding three-dimensional frame structure. Since the three-dimensional frame structure has four plane frames, the increment 1/4 of added damper in a plane frame indicates one damper in the three-dimensional frame structure.

The critical double impulse with the velocity amplitude of 1.0(m/s) and the corresponding equivalent sine wave with the resonant period was used as an input base motion. As in the shear building model, the resonant period of the critical double impulse and the corresponding equivalent sine wave is determined by the criterion of the zero base shear force after the first impulse (Akehashi and Takewaki 2019, Kojima and Takewaki 2015).

In the following examples, a weighting coefficient $\beta = 0.5$ in Equation. (21.8) was employed.

Figure 21.12 shows the optimal damper placement, the maximum floor acceleration, and the maximum inter-story drift for $x = 0, 10, 99\%$ in the case of $\delta_d = 50$ mm, $\ddot{U}_d = 4000$ mm/s^2. The case of $x = 0\%$ corresponds to SSA by Zhang and Soong (1992). The range $x = 99\%$ means that the dampers except the stories with $\gamma_i = 0$ are increased (added). It can be found that the suppression effect of the top acceleration is remarkable compared to the shear building model. The story stiffness distribution may affect this result.

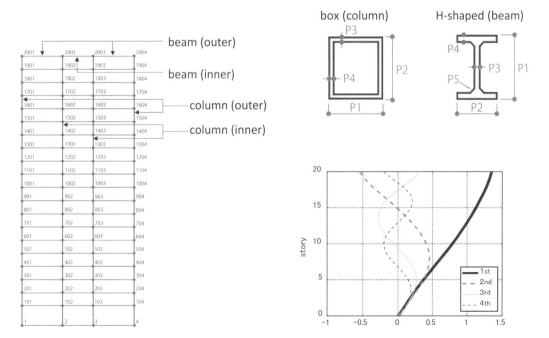

FIGURE 21.11 Moment-resisting frame, member sections, and first four natural modes.

TABLE 21.1
Member Cross Section

		Beam (H-Shaped)							Column (Box)							
		Inner Span [mm]				Outer Span [mm]				Inner Span [mm]				Outer Span [mm]			
		P1	P2	P3	P4	P1	P2	P3	P4	P1	P2	P3	P4	P1	P2	P3	P4
story	20	600	300	14	25	600	300	14	32	550	550	25	25	500	500	22	22
	19	600	300	14	25	600	300	14	32	550	550	25	25	500	500	22	22
	18	700	300	14	28	700	350	16	32	550	550	25	25	500	500	22	22
	17	700	300	14	28	700	350	16	32	550	550	25	25	500	500	22	22
	16	750	350	16	28	750	350	16	36	600	600	28	28	550	550	25	25
	15	750	350	16	28	750	350	16	36	600	600	28	28	550	550	25	25
	14	750	350	16	28	750	350	16	36	600	600	28	28	550	550	25	25
	13	750	350	16	28	750	350	16	36	600	600	28	28	550	550	25	25
	12	750	350	16	28	750	350	16	36	650	650	28	28	600	600	28	28
	11	750	350	16	28	750	350	16	36	650	650	28	28	600	600	28	28
	10	800	300	16	32	800	300	16	32	650	650	28	28	600	600	28	28
	9	800	300	16	32	800	300	16	32	650	650	28	28	600	600	28	28
	8	800	300	16	32	800	300	16	32	650	650	32	32	600	600	28	28
	7	850	250	16	32	850	300	16	32	650	650	32	32	600	600	28	28
	6	850	250	16	32	850	300	16	32	650	650	32	32	600	600	28	28
	5	850	250	16	32	850	300	16	32	700	700	32	32	650	650	28	28
	4	850	300	16	32	850	300	16	32	700	700	32	32	650	650	28	28
	3	850	300	16	32	850	300	16	32	750	750	36	36	700	700	36	36
	2	850	300	16	32	850	300	16	32	750	750	36	36	700	700	36	36
	1	900	300	19	32	900	300	19	32	800	800	36	36	750	750	36	36

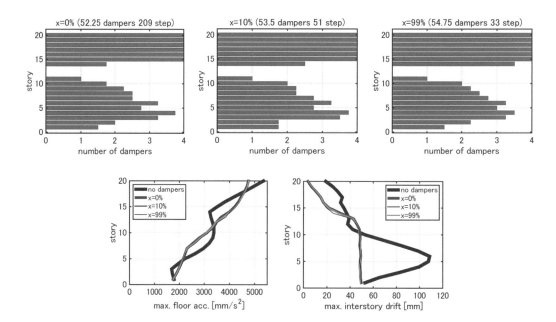

FIGURE 21.12 Optimal damper placement, maximum floor acceleration, and maximum inter-story drift for $x = 0\%$, 10%, and 99% ($\delta_d = 50$ mm, $\ddot{U}_d = 4000$ mm/s^2).

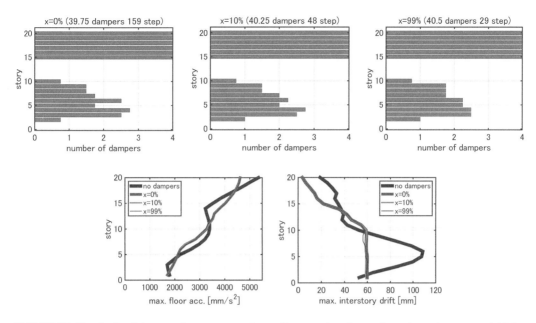

FIGURE 21.13 Optimal damper placement, maximum floor acceleration, and maximum inter-story drift for $x = 0\%$, 10%, and 99% ($\delta_d = 60$ mm, $\ddot{U}_d = 4000$ mm/s^2).

Figure 21.13 presents those distributions in the case of $\delta_d = 60$ mm, $\ddot{U}_d = 4000$ mm/s^2 and Figure 21.14 illustrates those distribution in the case of $\delta_d = 70$ mm, $\ddot{U}_d = 4000$ mm/s^2.

Figures 21.15–21.17 show those distributions in the case of $\delta_d = 50$ mm, $\ddot{U}_d = 5000$ mm/s^2, $\delta_d = 60$ mm, $\ddot{U}_d = 5000$ mm/s^2, and $\delta_d = 70$ mm, $\ddot{U}_d = 5000$ mm/s^2, respectively.

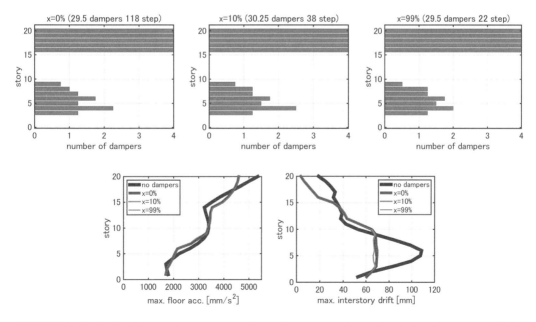

FIGURE 21.14 Optimal damper placement, maximum floor acceleration, and maximum inter-story drift for $x = 0\%$, 10%, and 99% ($\delta_d = 70$ mm, $\ddot{U}_d = 4000$ mm/s^2).

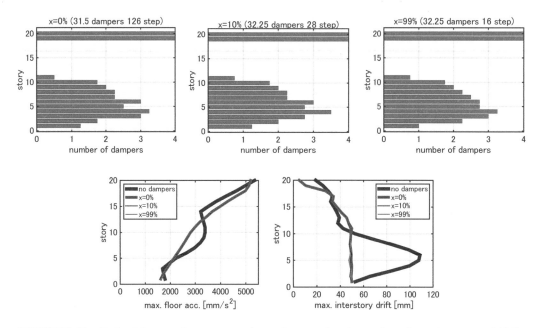

FIGURE 21.15   Optimal damper placement, maximum floor acceleration, and maximum inter-story drift for $x = 0\%$, 10%, and 99% ($\delta_d = 50$ mm, $\ddot{U}_d = 5000$ mm/s^2).

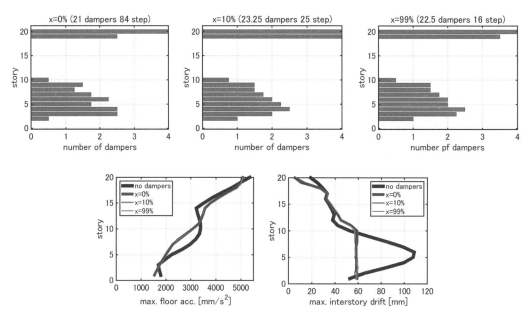

FIGURE 21.16   Optimal damper placement, maximum floor acceleration, and maximum inter-story drift for $x = 0\%$, 10%, and 99% ($\delta_d = 60$ mm, $\ddot{U}_d = 5000$ mm/s^2).

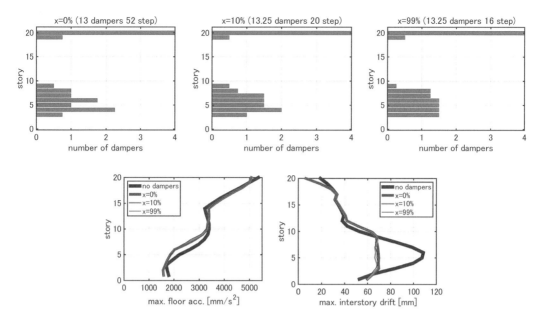

FIGURE 21.17   Optimal damper placement, maximum floor acceleration, and maximum inter-story drift for $x = 0\%$, 10%, and 99% ($\delta_d = 70$ mm, $\ddot{U}_d = 5000$ mm/s²).

As in the case of the shear building model, it can be observed that the proposed algorithm provides a rapid procedure (small number of steps) for a larger value of x with a nearly exact result for the optimal damper location by SSA.

21.7   CONCLUSIONS

An innovative rapid method was proposed for finding the discrete optimal design of high-damping rubber dampers for elastic-plastic moment frames under the critical double impulse representing near-fault pulse-type ground motions. The high-damping rubber damper is a special kind of visco-elastic dampers. These high-damping rubber dampers were installed into elastic-plastic steel moment frames and a rapid discrete optimal design algorithm was proposed based on a conventional SSA (sequential search algorithm) by Zhang and Soong (1992). The concept of SSA was relaxed by extending the effective upgrading range. A double impulse was used as a substitute of pulse-type ground motions and only the critical double impulse resonant to the elastic-plastic response was treated for overcoming a difficulty in modeling an input uncertainty by introducing the concept of the worst-case scenario. The proposed algorithm has the following characteristics.

1. The proposed algorithm is simple even for elastic-plastic main structures and ordinary structural analysis programs can be used as simulators of responses. Since the ordinary structural analysis programs can be called in the controlling program, e.g., MATLAB, all the procedures can be conducted automatically.
2. The proposed algorithm provides a rapid procedure (quite small number of steps) with a nearly exact result for the optimal damper location by SSA. The structural designer can make a decision in the conflicting multi-objective environment between the assurance of accuracy and the computational efficiency (small step number).

ACKNOWLEDGMENTS

Part of the present work is supported by the Grant-in-Aid for Scientific Research (KAKENHI) of Japan Society for the Promotion of Science (No.18H01584) and Sumitomo Rubber Industry Co. These supports are greatly appreciated.

REFERENCES

Akehashi, H. and Takewaki, I. (2019). Optimal viscous damper placement for elastic-plastic MDOF structures under critical double impulse, Frontiers in Built Environment, 5, 20.

Apostolakis, G. and Dargush, G.F. (2010). Optimal seismic design of moment-resisting steel frames with hysteretic passive devices, Earthq. Eng. Struct. Dyn., 39(4), 355–376.

Attard, T.L. (2007). Controlling all inter-story displacements in highly nonlinear steel buildings using optimal viscous damping, J. Struct. Eng., ASCE, 133(9), 1331–1340.

Aydin, E., Boduroglub, M.H. and Guney, D. (2007). Optimal damper distribution for seismic rehabilitation of planar building structures, Eng. Struct., 29, 176–185.

Bruneau, M., Chang, S.E., Eguchi, R.T., Lee, G.C., O'Rourke, T.D., Reinhorn, A.M., Shinozuka, M., Tierney, K., Wallace, W.A. and von Winterfeldt, D. (2003). A framework to quantitatively assess and enhance the seismic resilience of communities, Earthq. Spectra, 19(4), 733–752.

Cimellaro, G., Reinhorn, A. and Bruneau, M. (2010). Framework for analytical quantification of disaster resilience, Eng. Struct., 32(11), 3639–3649.

De Domenico, D., Ricciardi, G. and Takewaki, I. (2019). Design strategies of viscous dampers for seismic protection of building structures: A review, Soil Dyn. Earthq. Eng., 118, 144–165.

Idels, O. and Lavan, O. (2021). Optimization Based Seismic Design of Steel Moment Resisting Frames with Nonlinear Viscous Dampers, Struct. Control Health Monit., e2655.

Kanno, Y. (2013). Damper placement optimization in a shear building model with discrete design variables: a mixed-integer second-order cone programming approach, Earthq. Eng. Struct. Dyn., 42, 1657–1676.

Kojima, K. and Takewaki, I. (2015). Critical earthquake response of elastic-plastic structures under near-fault ground motions (Part 1: Fling-step input), Frontiers in Built Environment, 1, 12.

Lagaros, N., Plevris, V. and Mitropoulou, C.C. (Eds.) (2013). Design optimization of active and passive structural control systems, Information Science Reference, Hershey, PA, USA.

SNAP. (2015). An elastic-plastic analysis program for arbitrary-shape three-dimensional frame structures, Ver.6.1. Tokyo: Kozo System, Inc.

Soong, T.T. and Dargush, G.F. (1997). Passive energy dissipation systems in structural engineering, John Wiley & Sons, Chichester.

Takewaki, I. (2009). Building control with passive dampers: Optimal performance-based design for earthquakes, John Wiley & Sons, Asia, Singapore.

Takewaki, I. and Akehashi, H. (2021). Comprehensive review of optimal and smart design of nonlinear building structures with and without passive dampers subjected to earthquake loading, Frontiers in Built Environment, 7, 631114.

Tani, T., Yoshitomi, S., Tsuji, M. and Takewaki, I. (2009). High-performance control of wind-induced vibration of high-rise building via innovative high-hardness rubber damper, J. of The Structural Design of Tall and Special Buildings, 18(7), 705–728.

Uemura, R., Akehashi, H., Fujita, K. and Takewaki, I. (2021). Global simultaneous optimization of oil, hysteretic and inertial dampers using real-valued genetic algorithm and local search, Frontiers in Built Environment, 7, 795577.

Uetani, K., Tsuji, M. and Takewaki, I. (2003). Application of optimum design method to practical building frames with viscous dampers and hysteretic dampers, Eng. Struct., 25, 579–592.

Zhang, R.H. and Soong, T.T. (1992). Seismic design of viscoelastic dampers for structural applications, J. Struct. Eng., ASCE, 118(5), 1375–1392.

22 Damage Detection in Reinforced Concrete Structures Using Advanced Automatic Systems
An Overview

*Denise-Penelope N. Kontoni, Aman Kumar,
Harish Chandra Arora, Hashem Jahangir,
and Nishant Raj Kapoor*

CONTENTS

Structural health monitoring has the potential to attain sustainability.

Authors

22.1 INTRODUCTION

Civil engineering infrastructures are the hugely invested assets and national interests of any nation. Concrete is the most used manufactured building material in the world due to its accessibility and relative affordability. Furthermore, civil engineering constructions have a longer service life and safety depending on the conditions of the structures, and it is very expensive to replace them rather than to maintain or repair them [1]. Nuclear power plants, bridges, high-rise buildings, electricity utilities, and dams are the essential constructions used by all nations. The requirement for lower life cycle costs, safety, security, and post-disaster condition assessments has given rise to civil structural health monitoring (SHM) [2]. SHM is a critical element for enhancing the security and upkeep of crucial infrastructures like bridges, buildings, dams, etc. For many years, SHM has developed into a standard procedure in the aeronautical industry. This method is still being taken into consideration for normal civil infrastructure system applications. The first step in creating a SHM design is to specify the monitoring goals [3]. The structural health state is accurately and in actual time reported by SHM. Non-destructive examinations are used to determine the severity of the damage, estimate residual life, and forecast impending accidents. The complexity of structures, as well as the growth in their development and construction, has made SHM a difficult work. The need to protect the safety of the structures and the lives involved with them has risen, which has also raised the demand for SHM. Early damage detection is possible using SHM, making it possible to take action before any loss happens [4]. Civil structures must be equipped with SHM to maintain structural integrity and safety. In order to monitor, examine, and detect structural problems continuously and with low labor inputs, SHM plans to develop automated solutions [5].

SHM is used on new and old structures; the major goal is to gather information on the significant intrinsic pressures and distortions that are frequently caused by construction stresses during fabrication and erection processes. In most nations, civil infrastructures constitute a major investment, but they are degrading quickly as a result of a variety of factors, including environmental erosion, earthquake damage, impact loading, inadequate building methods and materials, and so on. It is no longer acceptable to utilize damaging techniques to learn about materials, the state of structures, or the behavior of structures under load. Periodic maintenance can reduce the damages as well as repair and rehabilitation costs of the structures. The time duration required for periodic maintenance is subjected to the type and present condition of the structure.

The primary causes of unexpected collapses for historical and current structures are protracted periods of neglect and catastrophic disasters. To confirm the long-term endurance of major structures, SHM techniques were used in the early 1970s to monitor components of industrial plants and massive structures [6]. SHM systems weren't used to keep an eye on civil constructions in general until much later. Over time, SHMs have demonstrated their value in lowering monitoring costs and streamlining maintenance procedures [7]. In reality, the use of SHM techniques enables the storage of crucial information on how the behavior of materials changes as a result of seismic, environmental, aging, and accidental occurrences. The following parameters are typically monitored by SHM systems: accelerations, displacement evolution, compressive and tensile stresses, deterioration of building materials, humidity, and temperature [8]. It is required for SHM designers to use sensors to ensure continuous monitoring. Such sensors must be installed in the structure at key locations without endangering it. The adoption of SHM founded on non-invasive observing techniques is the final trend, particularly for existing structures. In order to produce mathematical/numerical finite element (FE) simulations to forecast the damage of an object, the gathered data are then expanded, relocated, and used as beginning parameters.

22.2 DAMAGES IN RC STRUCTURES

Different types of damage in reinforced concrete (RC) structures fall under various categories. It is possible to categorize damages according to their categories, causes, modes of attack, frequency of flaws, types of structural weaknesses, quantity and scope of corrective actions, etc. The deterioration and damages in the structures reduce their performance as well as the overall safety of the structures. The various types of damages in RC structures are shown in Figure 22.1.

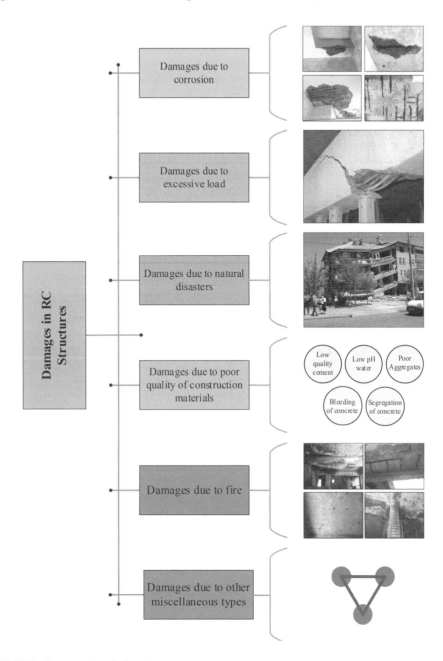

FIGURE 22.1   Damages in reinforced concrete structures.

22.2.1   Damages Due to Corrosion

Today, the most often utilized structural material for construction and infrastructures is reinforced concrete (RC). There are many benefits of using steel and concrete together. Steel enhances the tensile strength of concrete; steel is guarded against corrosion by a passive film (coating) that develops in the alkaline concrete pore solution environment. The overall performance of the RC structures is good, but once the passive film has been damaged due to chloride and/or carbonation infiltration, this leads to the origination of the corrosion process [9, 10]. The durability of the RC structures is dependent on the resistance of concrete against adverse environmental factors and its capacity to safeguard the embedded steel. However, for various reasons, the embedded reinforcement in the concrete becomes corroded, and concrete experiences extraordinarily high tensile forces due to volumetric expansion caused by the by-products of steel corrosion [11]. Concrete is prone to developing cracks between the bars (inclined cracking) or from the bar to the surface because its tensile strength is relatively low compared to its compressive strength (delamination). The extent of rust pressure damage to the concrete varies greatly depending on the thickness of the concrete cover and the reciprocal distance between the steel bars and their diameter.

Corrosion in steel not only decreases the strength of the structural member, but at the same time, it also reduces ductility. The volume of the rust is two to six times the actual diameter of the steel, and then this volumetric expansion puts pressure on the concrete cover [12]. A passive layer forms on the steel due to the high alkalinity of the concrete (pH over 12.5), which lowers the corrosion attack to insignificant levels. Corrosion cannot take place as long as this passive layer is maintained. This protective layer may be destroyed by two processes: carbonation of concrete and chloride attack [13].

22.2.1.1 Corrosion Mechanisms

The potential of hydrogen is not only the single factor that determines the corrosive or passive state of the steel. An additional important factor controlling the development of various chemical substances of a metal is electrochemical potential. Concrete reinforcement corrosion is an electrochemical activity. The corrosion reaction is made up of two half-cell processes. As seen in Equation (22.1), iron discharged electrons and dissolves in a ferrous form at the anode. As seen in Equation (22.2), these electrons travel to the cathode, where the breaking of oxygen structure occurs in the absence of oxygen.

$$Fe \rightarrow Fe^{2+} + 2e^- \tag{22.1}$$

$$O_2 + 2H_2O + 4e^- \rightarrow 4OH^- \tag{22.2}$$

According to Equation (22.3), the cathodic process can also result in water reduction when oxygen is not present.

$$2e^- + 2H_2O \rightarrow 2OH^- + H_2 \tag{22.3}$$

22.2.1.1.1 Chloride-Based Corrosion

Construction exposed to marine environments frequently experiences chloride-based corrosion of RC structures. Equation (22.4) shows the anodic reaction and role of chloride ions in the corrosion process.

$$Fe^{2+} + 2Cl^- + 4H_2O \rightarrow FeCl_2.4H_2O \tag{22.4}$$

$$FeCl_2 + 2H_2O \rightarrow Fe(OH)_2 + 2HCl \tag{22.5}$$

In concrete, chlorides can be in "free" or "bound" form. Additionally, it is well known that while bound chlorides are regarded safe, only free chloride ions are responsible for the corrosion process. According to this theory, the chloride content of concrete mix must adhere to rigorous guidelines. The main causes of chloride-induced rebar corrosion are exposure to maritime environments and/or the use of deicing solutions in cold areas. Water that includes chlorides can infiltrate through concrete with repeated soaking and drying cycles, causing the concrete to deteriorate, and is known as a chloride attack [14].

For concrete structures exposed to chloride content, the maximum concentration of the chloride content given by "ACI 318 [15]" is less than 0.15% by cement mass, but "Chinese code GB 50666" [16] specifies a limit of 0.06% by cement mass. Depending on whatever standard is used, namely the cement grade, the highest possible overall chloride ions percentage in "British Standard BS 5328-1" [17] ranges from 0.1% to 0.4% by cement mass. For prestressed RC constructions, the value is significantly lower. As per Indian standards, the chloride limit should not be more than 2,000 mg/l for plain concrete and 500 mg/l for RC concrete [18].

22.2.1.1.2 Carbonation-Based Corrosion

Steel corrosion in concrete is also a concern from atmospheric CO_2. The available carbon dioxide in the air is also responsible for steel corrosion in steel. Due to the concentration of both industry and population, the majority of constructions are exposed to CO_2. The alkaline components of cement paste, particularly $Ca(OH)_2$, can chemically react with CO_2 to neutralize the pore solution. As shown in Equation (22.6), this procedure is known as carbonation. Carbonation can reduce the pH of concrete lower than 8.5. Below the 8.5 pH value, the passive coating of steel reinforcement is no longer active, and the process of corrosion begins [9, 19].

$$Ca(OH)_2 + H_2O + CO_2 \rightarrow CaCO_3 + 2H_2O \qquad (22.6)$$

Carbonation-induced corrosion is typically characterized by a homogeneous attack on the steel surface rather than pitting [20]. In carbonation-induced corrosion, the anode and cathode are consistently distributed across the steel surface. Their area ratio is typically measured to be 1.0, making this kind of corrosion known as general corrosion.

22.2.2 Damages Due to Excessive Load

The excessive load above the permissible limits can also be responsible for damaging the structures. The damages due to blast loadings are also considered under this type of damage. These damages can occur due to the following reasons: mistakes in the design calculations, an extension of the stories without strengthening the existing stories, installation of heavy machinery, change in the use of the building, etc.

22.2.3 Damages Due to Poor Quality of Construction Materials

The basic ingredients of concrete are cement, fine aggregate, coarse aggregate, and water. Admixtures can be added in the concrete to enhance its workability, and it also depends on the design mix of the concrete. Utilizing recycled aggregates can sometimes result in poor concrete. Although some construction companies might use them to show off their commitment to sustainability, incorporating a lot of recycled elements into a concrete mix might degrade the substance's quality and possibly affect its strength and finish. Some of the flaws that show up immediately after construction include honeycombing, surface cracks, and cold joints.

22.2.4 Damages Due to Fire

Fire disaster is one of the most common and devastating natural disasters in the world. Many sources in buildings increase the risk of fire, and the initial stages of fire pose a serious threat to human life, material loss, structural integrity, and the environment. High temperatures may not necessarily cause damage to the RC structures because these structures have high natural fire resistance. Buildings that have been burned down have concrete that does not have uniform cross-sectional characteristics. The concrete layer closest to the surface suffered the most significant degradation. Compared to wood and stainless-steel structures, reinforced concrete buildings have superior natural fire resistance. The reinforcement is protected from heat as the concrete covering seeps deeper into the cross section of the member. This enables a building's load-bearing capacity to remain roughly high throughout and after a fire. Despite this, concrete is rarely totally destroyed in a fire, in contrast to steel and wood [21]. When concrete is exposed to elevated temperatures, its mechanical properties are weakened by thermal, physical, and chemical processes. RC beams, as well as slabs close to the ceiling, are more prone to damage and are immediately exposed to escalating convective currents in the case of fire. Due to concrete's higher fire resistance, RC structures that have been exposed to fire maintenance have a significant amount of residual capability, and it is frequently possible to reuse the structure in the future. On the other side, thermal deterioration may result in a permanent loss of strength and usability. To make precise decisions concerning future usage and the necessity of repair of fire-damaged structures, a thorough assessment of the residual capacity is required [22].

22.2.5 DAMAGES DUE TO NATURAL DISASTERS

Natural catastrophes occur when a threat overwhelms an extremely vulnerable community, frequently causing mortality and morbidity [23]. In the last decade, there have been over 300 natural disasters worldwide that have cost billions of dollars and affected millions of people annually. Natural disasters can be earthquakes, tsunamis, cyclones, etc. One of the most dangerous natural disasters is earthquakes, which result in significant damage to property as well as life. An earthquake is an unforeseen natural occurrence that causes significant harm to populations and the environment. In RC structures, structural damage during an earthquake may be caused by severe deformations, cumulative damage from repetitive load reversals, or a combination of the two. Up to 10,000 people per year die as a result of these hazards. Additionally, there are annual economic losses in the billions of dollars. In the past 25 years, major earthquakes have occurred all over the world, including Kobe (1995), Afghanistan (1998), Kocaeli (1999), Turkey (2001), India (2003), Haiti (2010), Chile (2010), Van (2011), etc. These earthquakes forced the construction industry to take drastic action to prevent collapse and lessen structural damage. For instance, the 1995 Kobe earthquake in Japan claimed the lives of more than 6,434 people, nearly 4,600 of whom were from Kobe. About 300,000 people lost their homes as a result of the 1999 Kocaeli earthquakes, which resulted in more than 17,000 fatalities, 40,000 injuries, and more [24].

22.3 STRUCTURAL HEALTH MONITORING

To maintain integrity and safety and estimate the deterioration/degradation of civil infrastructures, SHM is a useful tool [25]. Different authors define SHM in different ways; for example, Bassoli et al. [26] define the SHM as "It is the process of characterization of existing civil structures for structural identification and damage detection purposes". Saisi et al. [27] define SHM as "It is the continuous interrogation of sensors installed in the structure aimed at extracting features which are representative of the current state of structural health". Guidorzi et al. [28] define the SHM as "Aims to give a continuous diagnosis of the 'state' of the different parts and of the full assembly of these parts constituting the structure as a whole". The process of determining and monitoring structural integrity and evaluating the type of damage to a structure is known as structural health monitoring (SHM) [29].

The fundamental concept behind SHM is that it includes the characterization of a structural system through its reaction under loading conditions. Over the past few decades, the need for structural health monitoring has grown rapidly, fueling innovation in a variety of sensing technologies [30]. Damage detection is first done through visual inspection, and then other methods like X-rays, magnetic fields, acoustic techniques, and thermal measures are used in the field of SHM. Destructive and non-destructive techniques are also used to detect the damages in the structures. The details of these non-destructive methods are available in previous studies [31–34]. Statistical process control (SPC), non-destructive evaluation (NDE), semi-destructive testing (SDE), condition monitoring (CM), and damage prognosis (DP) are the six primary components that make up the damage detection process. Fiber-optic sensors, an accelerometer for SHM, vibrating wire transducers, a linear variable differential transformer (LVDT), load cells, tilt meters, inclinometers (slope indicators), noise emission sensors, temperature sensors, and others are all included in SHM [35].

"Damage definition", "damage identification", "damage localization", and "damage quantification" are the four stages that typically make up the SHM process. Since the 1970s, SHM has mostly been researched in relation to contact-type sensors. These touch sensors include strain gauges, fiber-optic sensors, accelerometers, and acoustic emission transducers. Most of the mounted sensors would be damaged during the earthquake, and there wouldn't be adequate power sources to run these sensor systems, making it difficult to use them for post-disaster monitoring. It would be appropriate to characterize a structure's health as a divergence from "a sound condition" as a consequence of harm and degradation that would necessitate structural strengthening, retrofitting, or repair.

Nowadays, novel structural health monitoring techniques such as acoustic emissions, optic fiber sensing, laser scanning, X-ray micro-computed tomography, unmanned aerial vehicle, digital image correlation, shearography tests, and artificial intelligence has been more rapidly used in the field of damage detection of RC structures. Further, these techniques are described below. The different types of novel existing and novel structural health monitoring techniques are shown in Figure 22.2.

22.3.1 Acoustic Emission

An Arabian alchemist, Jabir ibn Hayyan, is thought to have made the first recorded observations of AE in the eighth century [36]. But, observing the defects and damages in the stressed material by

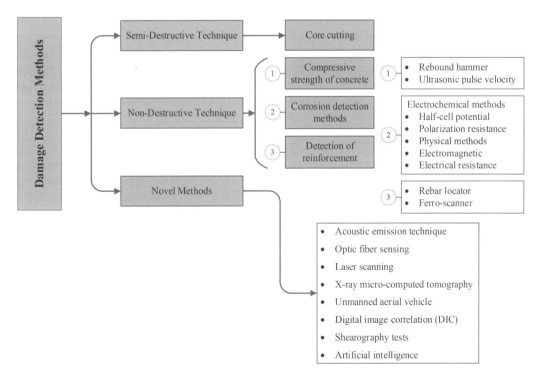

FIGURE 22.2   Types of SHM techniques.

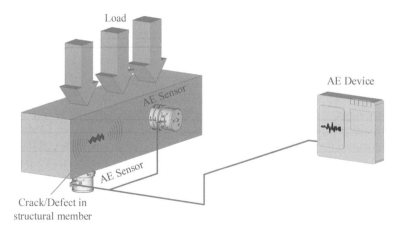

FIGURE 22.3 Acoustic emission.

using AE has been used for more than 60 years. This technology has been used for structural health monitoring, non-destructive assessment (NDE), and materials research. Early in the 1980s, AE was used as an NDE method for polymer matrix composites (PMCs), and since then, its use has grown along with PMCs' expanding range of applications, which in the past ten years have expanded to include major structures like the airframes of commercial and military aircraft [37]. An abrupt release of energy from a local source within a material under stress results in a transient elastic wave known as an AE. The causes of AE are the mechanisms of fracture or deformation phenomena inside materials, such as microstructural separation, crack growth, and movement between material phases. The strain energy generated from newly formed crack surfaces often relates to the energy of sonic emission. A properly configured AE sensor array can detect and locate AE waves, which spread according to the material's sound velocity.

The foundation of the AE NDT method is detecting and converting high-frequency elastic waves to electrical signals. To do this, piezoelectric transducers are directly coupled to the surface of the structure being tested, as shown in Figure 22.3, and then the structure is loaded. During structure loading, the output of each piezoelectric sensor is amplified by a low-noise preamplifier, filtered to remove any extraneous noise, and then further processed by a suitable embedded controller [38].

22.3.2 Optical Fiber Sensor

It has become possible to measure strain and temperature with great accuracy and reliability because of the development of distributed technologies like Raman and Brillouin optical fiber sensing systems [39]. The study of optical measurement techniques dates back more than a century, but in the 1970s, the telecommunications industry's invention of low-loss, high-quality optical fiber waveguides gave them a boost [40]. Various distributed fiber-optical sensors are currently being modified and optimized for installation in a variety of circumstances, such as embedding in concrete, surface mounting in existing surfaces, etc.

The operation of fiber-optic sensors (OFSs) is based on the principle of sending a light signal through a fiber and evaluating the quality of the received or sent signal. The required numbers are then translated from the altered signal characteristics, enabling the measurement of a variety of mechanical, physical, and chemical factors.

In SHM of civil engineering structures, fiber-optic sensors are an excellent choice. Fiber-optic sensors are durable, so they can endure hostile environments and are immune to electromagnetic interference, unlike conventional measuring devices like electrical sensors. Furthermore, because

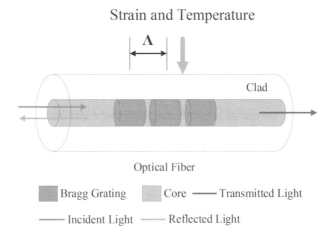

FIGURE 22.4 Optical fiber sensor.

optical glass fibers have low light attenuation, they can be multiplexed and interrogated over a distance of several kilometers. It is a powerful emerging device in the area of SHM with various advantages, such as high precision, compact size, and noise. An optical fiber sensor is made up of a fiber core, cladding layer (CL), and, if desired, an exterior coating or jacket to safeguard the fiber from mechanical and environmental forces, as shown in Figure 22.4 [41], and various types of optical fiber sensors are shown in Figure 22.5 [42]. Since the CL has a lower refractive index than the fiber core, it is possible to achieve 100% internal reflection of the light waves to direct their propagation down the fiber. Optical fiber sensors may track the parameters of external changes by detecting and examining the changes in light characteristics brought on by changes in the environment or object being measured.

The innovative OFSs were created primarily by depositing a unique coating material with a specified thickness on the fiber. This coating material can deteriorate in an outdoor environment and modify the light's characteristics for sensing purposes. The corrosion initiation time can be

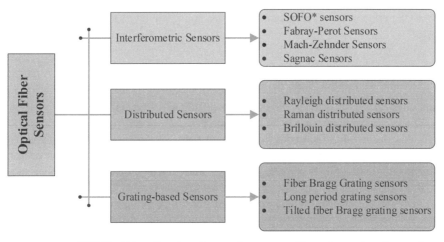

*SOFO = Fabry-Perot interferometric sensors and low coherent interferometric

FIGURE 22.5 Various types of optical fiber sensors.

estimated by designing an OFS with an iron film formed at the cleaved end of the fiber [43]. The iron coating thickness has decreased due to the corrosive atmosphere, which has reduced the intensity of light reflection.

22.3.3 LASER SCANNING

Non-contact laser sensing is presented as an alternative method for SHM of civil structures. The two most popular types of laser sensing devices are laser Doppler vibrometers (LDVs) and light detection and ranging (LiDAR).

LiDAR can work in two ways, (1) TOF measurement and (2) phase-shift measurement. The first one, which is most frequently utilized for commercial LiDAR, is straightforward to comprehend. The LiDAR emits a strong laser pulse, and the target object then reflects back to the LiDAR. The target's distance is calculated when the LiDAR detects the reflected laser pulse.

Applications of LiDAR for SHM often involve employing a dense 3D data set with greater resolution and precision to analyze 3D objects, structural geometry, deformations, fracture information, and visualization, as shown in Figure 22.6 [44, 45].

22.3.4 X-RAY MICRO-COMPUTED TOMOGRAPHY

For many years, X-ray computed tomography (CT) and micro-computed tomography (micro-CT) have been used to analyze non-medical materials. In the last ten years, the approach has advanced from a qualitative imaging tool to a quantitative analysis method, particularly in the field of materials science [46].

22.3.5 SHEAROGRAPHY TEST

Shearography, also known as speckle pattern shearing interferometry, is a non-destructive testing (NDT) method that evaluates the quality of materials in a variety of situations, including vibration analysis, strain measurement, and non-destructive testing. The shearography techniques have a number of benefits over conventional NDT methods, including the ability to test large areas of a structure quickly (up to 1 m^2 per minute [47]), the availability of contactless techniques, the relative lack of sensitivity to the effects of environmental variations, and the ability to perform well on honeycomb materials [48].

22.3.6 UNMANNED AERIAL VEHICLE

Visual and physical examination is the most dependable and frequent inspection of the structural elements. The impact of human errors in inspection has been closely examined, assessed, and constrained for decades in other industries (such as aerospace, automotive, and civil engineering). The standard inspection technique in the aviation industry is currently automated inspection devices with software programs [49]. In high-tech sectors, unmanned/automated inspection and maintenance methods are the greatest options to achieve the lowest possible failure rate and highest possible maintenance level [50]. The civil engineering sector is also going to adopt automated inspection technology for the SHM of the structures. The unmanned or automated techniques have the ability to enhance and automate SHM practices in the civil engineering sector.

In the past, conventional techniques have been used on a small scale using either ground vehicles or aerial. Ground vehicles were the first robotic inspection vehicles commonly used in transportation engineering. Climbing robots are now being used to inspect reinforced concrete constructions. Using electron bombardment, these robots can spot corrosion in its early stages [51]. Moveable

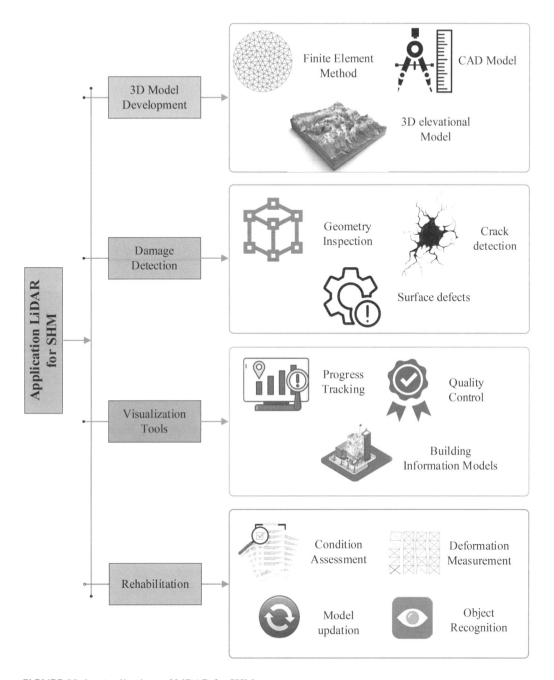

FIGURE 22.6 Applications of LiDAR for SHM.

suction cups help the robot to move and enable inspection in challenging to access areas. The application of UAV in SHM is shown in Figure 22.7.

As per the Unmanned Aerial Vehicle System Association (UAVSA), "UAS is a combination of an Unmanned Aerial Vehicle (UAV), either a fixed-wing or multi-copter aircraft, the payload (what it is carrying), and the ground control system [52]". Any aircraft or aerial device that has the ability to fly without an onboard human pilot is considered a UAS. They are also referred to as drones, remote-controlled helicopters, and remotely piloted and remotely operated aircrafts. The

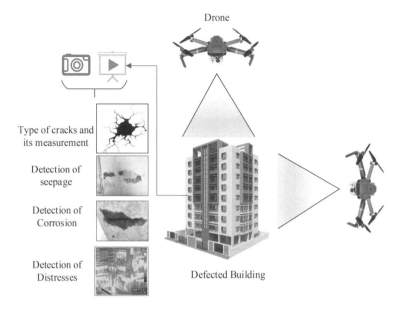

FIGURE 22.7 SHM with unmanned aerial vehicles/drones.

characteristics of the UAS vary according to their intended use, including maximum flight time, weight of the mounted sensors, maximum flight altitude, etc. UAVs can have fixed wings or platforms that can take off and land vertically.

UASs made their initial appearance in the United States a century ago. As the United States entered WWI, UAVs were developed to bomb enemy targets not long after the first successful construction of man-operated aircraft. However, due to engine trouble and a string of mishaps, this operation was abandoned. German engineers created an autonomous aircraft that could fly one-way missions at a top speed of 650 km/h at an altitude of 300 m during World War I. Beginning in 1959 and continuing to the present, the sole purpose of UASs was entirely military.

UASs have been crucial to American successes and air supremacy in a variety of missions and threats [53].

Depending on the needs of the task, a UAS may have one or more sensors can be installed. Most often, a UAS's sensors must be non-contact, which severely restricts the NDE methods that may be used. Thermal and visible cameras are the most often used sensors for assessing the building. Additionally, a variety of sensors are available that are required for autopilot operations. The most popular sensors fit on UASs are described in this section, along with their uses.

22.3.6.1 Visual (Video/Image) Cameras

The most popular and commonly utilized sensors on UASs for remote sensing are cameras. These sensors have a visible spectral range, or a wavelength range, ranging between 390 and 700 nanometers. The method of data collection may be impacted by unfavorable temperatures, poor lighting, high-frequency motors, considerable vibrations, and abrupt turning of the UAS [54].

22.3.6.2 Thermal Infrared Sensor

Thermal sensors are able to gauge a surface's radiated energy and interpret it into temperature. Infrared thermography employs both passive and active methods. The passive method relies solely on the thermal characteristics of material and structures, which are different from the specimen's ambient temperature. In active thermography, the material surface is excited by an external heat or cooling source, which enables the TIR sensors to detect differences in the thermal signature of specimens at various locations.

22.3.7 Digital Image Correlation

For evaluating the structural soundness and state of engineering systems, photogrammetry approaches have been studied. They have shown to be useful techniques that may be applied to assess structural health and spot damage before in-service breakdowns take place.

Sutton initially described the process for performing digital image correlation (DIC) in 1983 [55]. DIC is the most trending method to measure strain and displacements. The DIC technique's fundamental principle is "to take into account a set of points in the reference image and find the corresponding set in the deformed image". The numerical (digital) representation of an image is a huge matrix, where each element denotes the condition of a single element, known as a pixel, in the image. The processing of the images is necessary to calculate the displacement and movement of all existing units. The computational cost of DIC is very high, and the accuracy of DIC can be enhanced by using higher-order interpolation. Commonly, 2D DIC images are popular, but for volumetric digital image correlation 3D DIC images are being used. The number of pixels and the numerical range of a pixel (gray-level resolution) are two factors that affect how accurately images are captured with the DIC technique. A pixel's greater numerical range would improve the ability to distinguish between various light intensities and colors, improving image accuracy. Additionally, an image's accuracy improves with more pixels in it [56].

It is widely used to examine full-field displacement and strain, the progression of damage in structural systems, the spread of cracks, the identification of material deflections, and the emergence of damage. DIC can be used for periodic or ongoing surveillance, taking pictures of the targeted system at different intervals. Software may be used to visually inspect images from various time periods, and the data can be used to calculate the deformations [57]. There have been numerous proposed vision-based algorithms, such as edge detector, segmentation, thresholding, filter-based algorithms, etc. These methods have been used to evaluate the damage to samples of wood, steel, concrete, bridge surfaces, and pavement.

Additional uses for DIC include biomechanical applications, biomaterials, biological tissues, aerospace, computer vision, and industrial and automotive usage. The fundamental idea behind digital image correlation (DIC) is to compare photos that have been acquired of a target system at various degrees of deformation and then analyze those images using correlation-based matching algorithms. By tracking a neighborhood of pixels' gray levels and producing deformations in two as well as three-dimensional strain and vector fields, this method can be used to determine material deformation. A wide variety of sources, including portable commercial cameras, and high-speed video recorders, are available for taking pictures. Unexpected alterations that could cause irregularities would be easily discovered because any structural variations may be precisely contrasted with the pictures that were taken. The calculation of crack width and crack detection serve as examples of DIC's use. As in the RC beam in Figure 22.8, the location of the principal "vertical crack" would be promptly determined if a targeted component underwent load crack inspections. Even so, there can be a few tiny cracks that are difficult to see with the human eye.

22.3.8 Artificial Intelligence

In the computer industry, artificial intelligence (AI) is one of the most popular tools for tackling complex problems in the fields of engineering, medicine, and agriculture. Three primary types of artificial intelligence (AI) techniques are "machine learning (ML)", "deep learning (DL)", and "reinforcement learning (RL)". A data analytics technique called machine learning (ML)

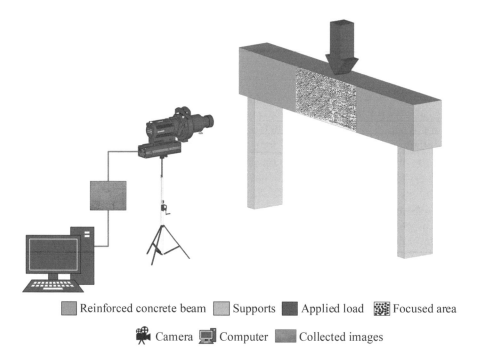

FIGURE 22.8 DIC step for crack and displacement measurement.

teaches computers to learn from experience in a similar way to how people and animals do. ML algorithms don't rely on a model based on predetermined equations to "learn" knowledge; instead, they use computer techniques directly from the data. DL is a highly specific form of learning [58].

The majority of the time, deep learning is used to conduct classification tasks on images, sounds, and texts. A computer agent learns to complete a task using RL, a type of machine learning approach, by interacting with a dynamic environment and learning by doing. This learning technology enables the system to make a number of decisions that raise a task's reward measure without requiring human engagement or being explicitly programmed to accomplish the aim [25, 59]. In ML, the SHM work is divided into four categories: detection, localization, assessment, and prediction [60]. The application of AI in SHM is shown in Figure 22.9.

In numerous fields, including robotics, data mining and pattern recognition, knowledge representation, and agent systems, numerous AI techniques derived from conventional AI, as well as hybrid methods, have already been effectively applied. Additionally, the academic discipline of structural health monitoring (SHM), which is novel and fast expanding, gains more and more from AI developments. Identifying a structure's state and actual behavior accurately is the main objective of SHM research nowadays. Additionally, it aims to create the proper tools to support and help the participating human professionals carry out their particular monitoring activities. The most commonly used AI techniques in SHM are the artificial neural network, GPR, EL, SVM, recurrent, conventional neural networks, etc. The summary of AI techniques used in SHM is shown in Table 22.1.

22.4 CONCLUSIONS

This chapter has covered various types of damages occurring in RC structures, as well as conventional and novel methods of damage detection using automatic technology. There have been many

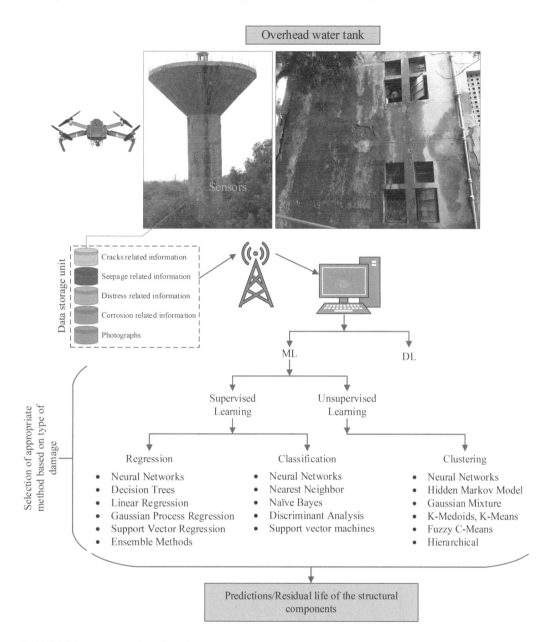

FIGURE 22.9 Application of AI in SHM.

forms and causes of deterioration in concrete structures. The practicing engineer is advised to use a straightforward and practical classification of damages by sources, which incorporates chemical attack, fire, overloading by static and dynamic loads, and malicious damages. This chapter has provided explicit insight into the different types of damages in RC concrete structures and novel techniques to detect damages. The novel SHM methods like acoustic emission, optic fiber sensing, laser scanning, X-ray micro-computed tomography, shearography, unmanned aerial vehicle, digital image correlation, and artificial intelligence–based techniques have been described with their basic principles.

TABLE 22.1
Summary of AI Studies in SHM

Reference	Year	Type of Structure	AI Method	Input Parameters	Findings
González and Zapico [61]	2008	Building	ANN	Frequencies and mode shapes	Degradation of mass and stiffness for the identification of damage index
Chang et al. [62]	2018	Building	ANN	Structure's modal characteristics when subjected to external vibrations	After crucial occurrences, damage patterns in terms of stiffness reduction
Soyoz and Feng [63]	2008	Highway	ANN	Frequencies, mode shapes	Long-term structural parameters, and age
Salazar et al. [64]	2015	Dam	ANN	Time, H_up, Precip, lag (Precip), T_air, ∂(H_up), season, precip	Tan_Disp, Rad_Disp, leakage flow
Nourani and Babakhani [65]	2013	Dam	ANN	H_up, H_dn	Pore pressure
Simon et al. [66]	2013	Dam	ANN	H_up, season, T_amb, T_air	Mass, stiffness, Rad_Disp
Peng et al. [67]	2017	Bridge	ANN	Temperature, defection after stretching, height of stretched section	Defection variation
Yang et al. [68]	2008	Bridge	ANN	Deflection of points	Normality of points
Wang et al. [69]	2017	Building	CNN	Building images	Existence of cracks
Maeda et al. [70]	2018	Road	SSD using Inception V2, and Mobile Net	Road images	Road cracks
Hüthwohl et al. [71]	2019	Bridges	Inception V3	Exposed reinforcement, cracks, rust, spalling	Deck of RC bridges
Murao et al. [72]	2019	Building	YOLO-v2	Existence of cracks	Concrete crack image
Ni et al. [73]	2018	Building	Dual-scale convolutional neural networks	Crack localization, crack width	Laboratory and outdoor
Li et al. [74]	2017	Bridge	SVM	Impact echo (IE) signal collection	Damage classification of the deck
Kerh and Ting [75]	2005	Bridge	SVM	Young modulus of soil, pile length, natural frequency	Scour depth nearby a bridge pier
Rafei and Adeli [76]	2018	Building	Unsupervised deep Boltzman machine	Sensing of the structure's ambient vibration response	Condition assessment, Structural Health Index
Cheng and Zheng [77]	2013	Dam	SVM	T_air, H_up, precip, T_conc	Rad_Disp, uplift pressure

ABBREVIATIONS

ACI	American concrete institute
AE	Acoustic emission
ANN	Artificial neural network
CM	Condition monitoring
CNN	Conventional neural network
DIC	Digital image correlation
DL	Deep learning
DP	Damage prognosis
FE	Finite element
LDV	Laser Doppler vibrometers
LiDAR	Light detection and ranging
LVDT	Linear variable differential transformer
ML	Machine learning
NDE	Non-destructive evaluation
OFS	Optic fiber sensors
PMC	Polymer matrix composites
RC	Reinforced concrete
RL	Reinforcement learning
SDE	Semi-destructive testing
SHM	Structural health monitoring
SPC	Statistical process control
SVM	Support vector machines
ToF	Time of flight
UAS	Unmanned aircraft system
UAV	Unmanned aerial vehicle
UAVSA	Unmanned Aerial Vehicle System Association

REFERENCES

1. Ken P. Chong, "Health monitoring of civil structures." *Journal of Intelligent Material Systems and Structures* 9, no. 11 (1998): 892–898. 10.1177/1045389X9800901104
2. Farhad Ansari, "Practical implementation of optical fiber sensors in civil structural health monitoring." *Journal of Intelligent Material Systems and Structures* 18, no. 8 (2007): 879–889. 10.1177/1045389X06075760
3. F.N. Catbas. "Structural health monitoring: Applications and data analysis." In *Structural Health Monitoring of Civil Infrastructure Systems*, pp. 1–39. Woodhead Publishing, 2009. https://doi.org/10.1533/9781845696825.1
4. Xiaoya Hu, Bingwen Wang, Han Ji. "A wireless sensor network-based structural health monitoring system for highway bridges." *Computer-Aided Civil and Infrastructure Engineering* 28, no. 3 (2013): 193–209. https://doi.org/10.1111/j.1467-8667.2012.00781.x
5. F. K. Chang. "Structural health monitoring: a summary report on the first international workshop on structural health monitoring." In *Proceedings of the 2nd International Workshop on Structural Health Monitoring*, pp. 3–11. Technomic Publishing Company, 1999.
6. Shi-Wei Lin, Ting-Hua Yi, Hong-Nan Li, Liang Ren. "Damage detection in the cable structures of a bridge using the virtual distortion method." *Journal of Bridge Engineering* 22, no. 8 (2017): 04017039. https://doi.org/10.1061/(ASCE)BE.1943-5592.0001072
7. Gangbing Song, Chuji Wang, Bo Wang. "Structural health monitoring (SHM) of civil structures." *Applied Sciences* 7, no. 8 (2017): 789. https://doi.org/10.3390/app7080789
8. Md Anam Mahmud, Kyle Bates, Trent Wood, Ahmed Abdelgawad, Kumar Yelamarthi. "A complete internet of things (IoT) platform for structural health monitoring (SHM)." In *2018 IEEE 4th World Forum on Internet of Things (WF-IoT)*, pp. 275–279. IEEE, 2018. 10.1109/WF-IoT.2018.8355094
9. Shamsad Ahmad. "Reinforcement corrosion in concrete structures, its monitoring and service life prediction—a review." *Cement and Concrete Composites* 25, no. 4–5 (2003): 459–471. https://doi.org/10.1016/S0958-9465(02)00086-0

10. Luca Bertolini, Bernhard Elsener, Pietro Pedeferri, Elena Redaelli, Rob B. Polder. *Corrosion of Steel in Concrete: Prevention, Diagnosis, Repair.* John Wiley & Sons, 2013. https://doi.org/10.1002/9783527651696

11. Roberto Capozucca. "Damage to reinforced concrete due to reinforcement corrosion." *Construction and Building Materials* 9, no. 5 (1995): 295–303. https://doi.org/10.1016/0950-0618(95)00033-C

12. Tracy Dawn Marcotte. "Characterization of chloride-induced corrosion products that form in steel-reinforced cementitious materials." (2001).

13. F. Hunkeler. "Corrosion in reinforced concrete: Processes and mechanisms." *Corrosion in Reinforced Concrete Structures* (2005): 1–45. https://doi.org/10.1533/9781845690434.1

14. S. Jianxia. "Durability design of concrete hydropower structures." (2012): 377–403. https://doi.org/10.1016/B978-0-08-087872-0.00619-3

15. ACI Committee. *"Building Code Requirements for Structural Concrete (ACI 318-08) and Commentary."* American Concrete Institute, 2008.

16. GB 50666-2011, *Code for Construction of Concrete Structures.* Beijing: China Architecture & Construction Press, 2011.

17. British standard 8110: Part 1, *Structural Use of Concrete – Code of Practice for Design and Construction.* London UK: British Standards Institute; 1997.

18. IS 456 : 2000, Plain and Reinforced Concrete – Code of Practice. Bureau of Indian Standards (BIS), New Delhi, India, 2000.

19. Amala James, Ehsan Bazarchi, Alireza A. Chiniforush, Parinaz Panjebashi Aghdam, M. Reza Hosseini, Ali Akbarnezhad, Igor Martek, Farzad Ghodoosi. "Rebar corrosion detection, protection, and rehabilitation of reinforced concrete structures in coastal environments: A review." *Construction and Building Materials* 224 (2019): 1026–1039. https://doi.org/10.1016/j.conbuildmat.2019.07.250

20. L. J. Parrott. "A study of carbonation-induced corrosion." *Magazine of Concrete Research* 46, no. 166 (1994): 23–28. https://doi.org/10.1680/macr.1994.46.166.23

21. Di Qin, PengKun Gao, Fahid Aslam, Muhammad Sufian, Hisham Alabduljabbar. "A comprehensive review on fire damage assessment of reinforced concrete structures." *Case Studies in Construction Materials* (2021): e00843. https://doi.org/10.1016/j.cscm.2021.e00843

22. Ankit Agrawal, V. K. R. Kodur. "A novel experimental approach for evaluating residual capacity of fire damaged concrete members." *Fire Technology* 56, no. 2 (2020): 715–735. https://doi.org/10.1007/s10694-019-00900-1

23. Abhaya S. Prasad, Louis Hugo Francescutti. "Natural disasters." *International Encyclopedia of Public Health* (2017): 215. https://doi.org/10.1016/B978-0-12-803678-5.00519-1

24. Burak Yön, Erkut Sayın, Onur Onat. "Earthquakes and structural damages." *Earthquakes-Tectonics, Hazard and Risk Mitigation. InTech, Rijeka* (2017): 319–339. 10.5772/65425

25. Nishant Raj Kapoor, Aman Kumar, Harish Chandra Arora, Ashok Kumar. "Structural health monitoring of existing building structures for creating green smart cities using deep learning." In *Recurrent Neural Networks*, pp. 203–232. CRC Press, 2023. 10.1201/9781003307822-15

26. Elisa Bassoli, Loris Vincenzi, Marco Bovo, Claudio Mazzotti. "Dynamic identification of an ancient masonry bell tower using a MEMS-based acquisition system." In *2015 IEEE Workshop on Environmental, Energy, and Structural Monitoring Systems (EESMS) Proceedings*, pp. 226–231. IEEE, 2015. 10.1109/EESMS.2015.7175882

27. Antonella Saisi, Carmelo Gentile, Antonello Ruccolo. "Continuous monitoring of a challenging heritage tower in Monza, Italy." *Journal of Civil Structural Health Monitoring* 8, no. 1 (2018): 77–90. https://doi.org/10.1007/s13349-017-0260-5

28. Roberto Guidorzi, Roberto Diversi, Loris Vincenzi, Claudio Mazzotti, Vittorio Simioli. "Structural monitoring of a tower by means of MEMS-based sensing and enhanced autoregressive models." *European Journal of Control* 20, no. 1 (2014): 4–13. https://doi.org/10.1016/j.ejcon.2013.06.004

29. Peter C. Chang, Alison Flatau, S. C. Liu. "Health monitoring of civil infrastructure." *Structural Health Monitoring* 2, no. 3 (2003): 257–267.

30. D. Inaudi. "Sensing solutions for assessing the stability of levees, sinkholes and landslides." In *Sensor Technologies for Civil Infrastructures*, pp. 396–421. Woodhead Publishing, 2014. https://doi.org/10.1533/9781782422433.2.396

31. Ali Hafiza, Thomas Schumacher, Anis Raad. "A self-referencing non-destructive test method to detect damage in reinforced concrete bridge decks using nonlinear vibration response characteristics." *Construction and Building Materials* 318 (2022): 125924. https://doi.org/10.1016/j.conbuildmat.2021.125924

32. Jie Wang, Tuo Xu, Li Zhang, Tianying Chang, Jin Zhang, Shihan Yan, Hong-Liang Cui. "Nondestructive damage evaluation of composites based on terahertz and X-ray image fusion." *NDT & E International* 127 (2022): 102616. https://doi.org/10.1016/j.ndteint.2022.102616

33. S. L. Toh, H. M. Shang, F. S. Chau, C. J. Tay, T. E. Tay. "Damage detection using holography and shearography." In *Fracture of Engineering Materials and Structures*, pp. 489–496. Springer, Dordrecht, 1991. https://doi.org/10.1007/978-94-011-3650-1_71

34. Aman Kumar, Jasvir Singh Rattan, Nishant Raj Kapoor, Ajay Kumar, Rahul Kumar. "Structural health monitoring of existing reinforced cement concrete buildings and bridge using nondestructive evaluation with repair methodology." *Advances and Technologies in Building Construction and Structural Analysis* (2021): 87. 10.5772/intechopen.101473

35. Aman Kumar, Navdeep Mor, "Importance and Methods of Non-Destructive Testing (NDT) Techniques for Building Maintenance", *AkiNik Publications*, (2019). 10.22271/ed.book.432

36. Introduction to Acoustic Emission Testing. 2022. Accessed 25 July 2022. https://www.nde-ed.org/NDETechniques/AcousticEmission/AE_History.xhtml

37. J. Q. Huang. "Non-destructive evaluation (NDE) of composites: Acoustic emission (AE)." In *Non-Destructive Evaluation (NDE) of Polymer Matrix Composites*, pp. 12–32. Woodhead Publishing, 2013. https://doi.org/10.1533/9780857093554.1.12

38. F. Sarasini, C. Santulli. "Non-destructive testing (NDT) of natural fibre composites: Acoustic emission technique." In *Natural Fibre Composites*, pp. 273–302. Woodhead Publishing, 2014. https://doi.org/10.1533/9780857099228.3.273

39. Riccardo Belli, Branko Glisic, Daniele Inaudi, Berhane Gebreselassie. "Smart textiles for SHM of geostructures and buildings." In *4th International Conference on Structural Health Monitoring on Intelligent Infrastructure (SHMII-4)*, pp. 200922–24. 2009.

40. K. T. V. Grattan, T. Sun. "Optical-fiber sensors: Temperature and pressure sensors." *MRS Bulletin* 27, no. 5 (2002): 389–395.

41. Manjusha Ramakrishnan, Ginu Rajan, Yuliya Semenova, Gerald Farrell. "Overview of fiber optic sensor technologies for strain/temperature sensing applications in composite materials." *Sensors* 16, no. 1 (2016): 99. https://doi.org/10.3390/s16010099

42. Honglei Guo, Gaozhi Xiao, Nezih Mrad, Jianping Yao. "Fiber optic sensors for structural health monitoring of air platforms." *Sensors* 11, no. 4 (2011): 3687–3705. https://doi.org/10.3390/s110403687

43. Changxu Li, Wenlong Yang, Min Wang, Xiaoyang Yu, Jianying Fan, Yanling Xiong, Yuqiang Yang, Linjun Li. "A review of coating materials used to improve the performance of optical fiber sensors." *Sensors* 20, no. 15 (2020): 4215. https://doi.org/10.3390/s20154215

44. Elise Kaartinen, Kyle Dunphy, Ayan Sadhu. "LiDAR-based structural health monitoring: Applications in civil infrastructure systems." *Sensors* 22, no. 12 (2022): 4610. https://doi.org/10.3390/s22124610

45. Wallace Mukupa, Gethin Wyn Roberts, Craig M. Hancock, Khalil Al-Manasir. "A review of the use of terrestrial laser scanning application for change detection and deformation monitoring of structures." *Survey Review* 49, no. 353 (2017): 99–116. https://doi.org/10.1080/00396265.2015.1133039

46. Eric Maire, Philip John Withers. "Quantitative X-ray tomography." *International Materials Reviews* 59, no. 1 (2014): 1–43. https://doi.org/10.1179/1743280413Y.0000000023

47. Kadir Güçlüer. "An investigation of the effect of different aggregate types on concrete properties with thin section and nondestructive methods." *J. Eng. Res* 9 (2021): 15–24.

48. Peiyu Wang, Licheng Zhou, Guo Liu, Yongmao Pei. "In situ near-field microwave characterization and quantitative evaluation of phase change inclusion in honeycomb composites." *NDT & E International* 121 (2021): 102469.

49. Kara A. Latorella, Prasad V. Prabhu. "A review of human error in aviation maintenance and inspection." *Human Error in Aviation* (2017): 521–549. https://doi.org/10.1016/S0169-8141(99)00063-3

50. Flavio Prieto, Tanneguy Redarce, Richard Lepage, Pierre Boulanger. "An automated inspection system." *The International Journal of Advanced Manufacturing Technology* 19, no. 12 (2002): 917–925. https://doi.org/10.1007/s001700200104

51. Alexis Leibbrandt, Gilles Caprari, Ueli Angst, Roland Y. Siegwart, Robert J. Flatt, Bernhard Elsener. "Climbing robot for corrosion monitoring of reinforced concrete structures." In *2012 2nd International Conference on Applied Robotics for the Power Industry (CARPI)*, pp. 10–15. IEEE, 2012. 10.1109/CARPI.2012.6473365

52. Francesco Nex, Costas Armenakis, Michael Cramer, Davide A. Cucci, Markus Gerke, Eija Honkavaara, Antero Kukko, Claudio Persello, Jan Skaloud. "UAV in the advent of the twenties: Where we stand and what is next." *ISPRS Journal of Photogrammetry and Remote Sensing* 184 (2022): 215–242. https://doi.org/10.1016/j.isprsjprs.2021.12.006

53. Kendra L.B. Cook. "The silent force multiplier: The history and role of UAVs in warfare." In *2007 IEEE Aerospace Conference*, pp. 1–7. IEEE, 2007. 10.1109/AERO.2007.352737

54. Bharat Sharma Acharya, Mahendra Bhandari, Filippo Bandini, Alonso Pizarro, Matthew Perks, Deepak Raj Joshi, Sheng Wang. "Unmanned aerial vehicles in hydrology and water management: Applications, challenges, and perspectives." *Water Resources Research* 57, no. 11 (2021): e2021WR029925. https://doi.org/10.1029/2021WR029925

55. Michael A. Sutton, W. J. Wolters, W. H. Peters, W. F. Ranson, S. R. McNeill. "Determination of displacements using an improved digital correlation method." *Image and Vision Computing* 1, no. 3 (1983): 133–139. https://doi.org/10.1016/0262-8856(83)90064-1

56. M. A. Rastak, Mahmood M. Shokrieh, L. Barrallier, R. Kubler, S. D. Salehi. "Estimation of residual stresses in polymer-matrix composites using digital image correlation." In *Residual Stresses in Composite Materials*, pp. 455–486. Woodhead Publishing, 2021. https://doi.org/10.1016/B978-0-12-818817-0.00001-9

57. Daniel Reagan, Alessandro Sabato, Christopher Niezrecki. "Unmanned aerial vehicle acquisition of three-dimensional digital image correlation measurements for structural health monitoring of bridges." In *Nondestructive Characterization and Monitoring of Advanced Materials, Aerospace, and Civil Infrastructure 2017*, vol. 10169, pp. 68–77. SPIE, 2017.

58. Machine Learning. 2021. Accessed 20 July 2021. https://in.mathworks.com/discovery/machine-learning.html

59. Deep Learning. 2021. Accessed 28 July 2021. https://in.mathworks.com/discovery/deep-learning.html

60. Keith Worden, Graeme Manson. "The application of machine learning to structural health monitoring." *Philosophical Transactions of the Royal Society A: Mathematical, Physical and Engineering Sciences* 365, no. 1851 (2007): 515–537. https://doi.org/10.1098/rsta.2006.1938

61. María P. González, José L. Zapico. "Seismic damage identification in buildings using neural networks and modal data." *Computers & Structures* 86, no. 3–5 (2008): 416–426. https://doi.org/10.1016/j.compstruc.2007.02.021

62. Chia-Ming Chang, Tzu-Kang Lin, Chih-Wei Chang. "Applications of neural network models for structural health monitoring based on derived modal properties." *Measurement* 129 (2018): 457–470. https://doi.org/10.1016/j.measurement.2018.07.051

63. Serdar Soyoz, Maria Q. Feng. "Long-term monitoring and identification of bridge structural parameters." *Computer-Aided Civil and Infrastructure Engineering* 24, no. 2 (2009): 82–92. https://doi.org/10.1111/j.1467-8667.2008.00572.x

64. Fernando Salazar, M. A. Toledo, E. Oñate, R. Morán. "An empirical comparison of machine learning techniques for dam behaviour modelling." *Structural Safety* 56 (2015): 9–17. https://doi.org/10.1016/j.strusafe.2015.05.001

65. Vahid Nourani, Ali Babakhani. "Integration of artificial neural networks with radial basis function interpolation in earthfill dam seepage modeling." *Journal of Computing in Civil Engineering* 27, no. 2 (2013): 183–195. https://doi.org/10.1061/(ASCE)CP.1943-5487.0000200

66. A. Simon, M. Royer, F. Mauris, J. Fabre. "Analysis and interpretation of dam measurements using artificial neural networks." In *Proceedings of the 9th ICOLD European Club Symposium, Venice, Italy*. 2013.

67. Jiafan Peng, Shunong Zhang, Dongmu Peng, Kan Liang. "Application of machine learning method in bridge health monitoring." In *2017 Second International Conference on Reliability Systems Engineering (ICRSE)*, pp. 1–7. IEEE, 2017. 10.1109/ICRSE.2017.8030793

68. Jianxi Yang, Jianting Zhou, Fan Wang. "A study on the application of GA-Bp neural network in the bridge reliability assessment." In *2008 International Conference on Computational Intelligence and Security*, vol. 1, pp. 540–545. IEEE, 2008. 10.1109/CIS.2008.29

69. Kelvin CP Wang, Allen Zhang, Joshua Qiang Li, Yue Fei, Cheng Chen, Baoxian Li. "Deep learning for asphalt pavement cracking recognition using convolutional neural network." In *Airfield and Highway Pavements 2017*, pp. 166–177. 2017. https://doi.org/10.1061/9780784480922.015

70. Hiroya Maeda, Yoshihide Sekimoto, Toshikazu Seto, Takehiro Kashiyama, Hiroshi Omata. "Road damage detection and classification using deep neural networks with smartphone images." *Computer-Aided Civil and Infrastructure Engineering* 33, no. 12 (2018): 1127–1141. https://doi.org/10.48550/arXiv.1801.09454

71. Philipp Hüthwohl, Ruodan Lu, Ioannis Brilakis. "Multi-classifier for reinforced concrete bridge defects." *Automation in Construction* 105 (2019): 102824. https://doi.org/10.1016/j.autcon.2019.04.019

72. Saki Murao, Yasutoshi Nomura, Hitoshi Furuta, Chul-Woo Kim. "Concrete crack detection using UAV and deep learning." (2019).

73. FuTao Ni, Jian Zhang, Zhi Qiang Chen. "Zernike-moment measurement of thin-crack width in images enabled by dual-scale deep learning." *Computer-Aided Civil and Infrastructure Engineering* 34, no. 5 (2019): 367–384. https://doi.org/10.1111/mice.12421

74. Kenshin Li, Liang Ushiroda, Qiang Yang, Jizhong Song, Xiao. "Wall-climbing robot for non-destructive evaluation using impact-echo and metric learning SVM." *International Journal of Intelligent Robotics and Applications* 1, no. 3 (2017): 255–270.

75. Tienfuan Kerh, S. B. Ting. "Neural network estimation of ground peak acceleration at stations along Taiwan high-speed rail system." *Engineering Applications of Artificial Intelligence* 18, no. 7 (2005): 857–866. https://doi.org/10.1016/j.engappai.2005.02.003

76. Mohammad Hossein Rafiei, Hojjat Adeli. "A novel unsupervised deep learning model for global and local health condition assessment of structures." *Engineering Structures* 156 (2018): 598–607. https://doi.org/10.1016/j.engstruct.2017.10.070

77. Lin Cheng, Dongjian Zheng. "Two online dam safety monitoring models based on the process of extracting environmental effect." *Advances in Engineering Software* 57 (2013): 48–56. https://doi.org/10.1016/j.advengsoft.2012.11.015

23 Efficient Representation of Random Fields for Training the Kriging Predictor in Adaptive Kriging Reliability Assessment of Civil Structures

Koosha Khorramian, Abdalla Elhadi Alhashmi, and Fadi Oudah

CONTENTS

23.1   INTRODUCTION

Active learning Kriging (AK) has been used to assess the reliability of civil structures modeled using random finite element (FE) simulation. Random FE refers to modeling the spatial variability in the structural response using random fields, while AK is a form of artificial intelligence (AI) used to optimize the number of required FE simulations by using learning functions. The use of random FE-AK-based reliability estimation facilitates assessing complex structural limit states that otherwise are difficult, if feasible, to accurately assess. Examples include assessing the reliability of structures with considerable spatial variation in the resistance due to various degradation mechanisms like assessing an aged steel-reinforced concrete bridge deck experiencing pitting corrosion, concrete damage due to freeze-thaw effects, corroded steel trusses, or modeling soil applications where the soil material properties vary spatially in the considered volume [1–4]. For these structures, random FE is employed to assess the limit state while AK is used to optimize the number of required simulations to accurately predict the reliability of the considered limit state.

The practical application of random FE-AK-based techniques is generally limited in civil engineering because of the need for further optimization to improve the efficiency of the method without compromising the accuracy of the estimated reliability. In this chapter, a major challenge that hinders the efficiency of using random FE-AK-based methods to assess the reliability of civil structures experiencing high spatial variation in the material properties is addressed. The challenge essentially relates to the number of variables used in training the Kriging predictor when random fields are utilized to conduct random FE. In current practice, the number of variables used to represent the random field in training the Kriging predictor is equal to the number of standard normal variables used to generate the random field when random field generation techniques like expansion optimal linear estimation (EOLE) are used [1–3]. Should the volatility of the random field be high (e.g., high spatial variation in the considered structure), the number of standard normal variables needed to model the field will be high and so the number of variables used to train the Kriging predictor. The latter adversely impacts the number of required random FE simulations needed in AK analysis to arrive at a reasonably accurate estimate of the reliability of the considered limit state.

The objective of this chapter is to provide practical recommendations for the considered number of variables representing the random field in random FE modeling when training the Kriging predictor in AK reliability. An alternative approach is proposed to represent random fields in training the Kriging predictor in AK reliability analysis by defining alternative measures to represent the random fields based on the overall effect of the random field in the resistance model. The application of the proposed method, upon validation, will allow utilizing AK reliability analysis to assess complex problems that otherwise may not be feasible to analyze using the current AK analysis frameworks. The idea is to select a region with the most influence in the resistance model (named the influence region herein), where the random field is applied, and use a combination of statistical parameters of the field values for the influence region such as mean, standard deviation, minimum, and maximum as representative variables of the random field.

The chapter includes an overview of random fields for structural applications followed by a brief overview of the AK reliability analysis with a focus on the use of random fields. The alternative approach for representing the effect of random fields in the AK training process is then explained in detail. The chapter includes two examples to validate the accuracy of the alternative representative random field variables. The first example is a reliability assessment of a pile group foundation modeled with nonlinear FE analysis using three-dimensional (3D) random fields for soil modulus of elasticity and friction angle. The second example is a reliability assessment of reinforced concrete (RC) girder modeled using a one-dimensional (1D) random field to represent the spatial variation in the moment of resistance determined using sectional analysis.

23.2   RANDOM FIELD REALIZATIONS

A random field can be considered as the joint probability distribution of a set of dependent random variables in a continuous media such as soil. Soil materials are known to be notoriously variable. Their properties vary with depth (i.e., variation of the soil stiffness with depth due to the overburden pressure) and they can vary from one location of a pile to the other pile in a pile group structure. For example, if an engineer was required to obtain the soil properties using boreholes, the borehole data would be dissimilar at different locations in a three-dimensional (3D) space. To account for this variability, random fields are used. Several methods exist to generate the realizations of random fields such as the orthogonal series expansion (OSE) [1], Karhunen–Loeve (KL) expansion [1, 2], the expansion optimal linear estimation (EOLE) [3, 4], and local average subdivision [5]. In this study, EOLE is used as the method for the realization of random fields.

To generate random fields using EOLE, the first step is to produce a mesh (called RF mesh herein) based on the specified geometry. The RF mesh density may be different than the FE mesh density. Let M_{RFE} represent an n-dimensional array containing all vectors corresponding to RF mesh. $M_{RFE} = \left\{ RFE^{(1)}, RFE^{(2)}, \ldots, RFE^{(N_{mesh})} \right\}$, where N_{mesh} is the number of discretized mesh points, $RFE^{(i)}$ is the i_{th} vector representing a point in 3D space, $RFE^{(i)} = \left(X_{RFE_i}, Y_{RFE_i}, Z_{RFE_i} \right)$. The mean, standard deviation, and correlation matrix will be evaluated at the centroid of each point in M_{REF}. After determining the mean value for each point in the mesh, the standard deviation will be evaluated as the mean multiplied by the coefficient of variation of the random field. A lognormal distribution is typically utilized for geotechnical and structural engineering applications to model the correlation between the points in space using a Gaussian correlation function. The equation for the covariance matrix in a 3D random field is presented in Equation (23.1):

$$C_{ZZ}(i,j) = \sigma_i \sigma_j \rho_{ij}; \; \rho_{ij} = \exp\left\{-\frac{\left(X_{RFEi} - X_{RFEj}\right)^2}{l_x} - \frac{\left(Y_{RFEi} - Y_{RFEj}\right)^2}{l_y} -- \frac{\left(Z_{RFEi} - Z_{RFEj}\right)^2}{l_z}\right\}$$

(23.1)

where $\boldsymbol{C_{ZZ}}$ is the correlation matrix with $C_{ZZ}(i,j)$ as its component for row i and column j; σ_i and σ_j are the standard deviation of the points $\boldsymbol{RFE}^{(i)}$ and $\boldsymbol{RFE}^{(j)}$; ρ_{ij} is the correlation function evaluated between every two points in space, $\boldsymbol{RFE}^{(i)}$ and $\boldsymbol{RFE}^{(j)}$; and l_x, l_y, and l_z are the correlation length in the x, y, and z directions. Since EOLE requires a Gaussian process for the realization of random fields [3], a Nataf transformation [6] is used for the EOLE method as presented in Equation (23.2) to transform the lognormal-lognormal correlation matrix to a Gaussian-Gaussian correlation matrix.

$$\rho'_{ij} = x\rho_{ij} = \frac{\ln\left(1 + \rho_{ij} v_i v_j\right)}{\rho_{ij}\sqrt{\ln\left(1 + v_i^2\right)\ln\left(1 + v_j^2\right)}}\rho_{ij}$$

(23.2)

where ρ'_{ij} is the correlation for the standard normal field between two points $\boldsymbol{RFE}^{(i)}$ and $\boldsymbol{RFE}^{(j)}$ in the RFE mesh, x is a factor converting ρ_{ij} to ρ'_{ij} (i.e., converting from lognormal to gaussian), and v_i and v_j are the coefficient of variation of i_{th} and j_{th} point in the mesh for $\boldsymbol{RFE}^{(i)}$ and $\boldsymbol{RFE}^{(j)}$, respectively. The eigenvectors and eigenvalues of the correlation matrix can be calculated for random field realization using Equations (23.3) and 23.4:

$$\hat{H}(\boldsymbol{Z},\theta) = \mu_{lny} + \sigma_{lny}\sum_{i=1}^{M}\frac{\xi_i(\theta)}{\sqrt{\lambda_i}}\boldsymbol{\psi}_i^T \boldsymbol{C}_{Z,Z_i}$$

(23.3)

$$\mu_{lny} = \ln(\mu_y) - \frac{1}{2}\sigma_{lny}^2; \quad \sigma_{lny} = \sqrt{\ln\left(1 + \frac{\sigma_y^2}{\mu_y^2}\right)}$$

(23.4)

where $\hat{H}(\boldsymbol{Z},\theta)$ is the Gaussian field realizations, $\boldsymbol{Z}$ is the desired point in space for which the field value is evaluated, μ_{lny} and σ_{lny} are the normalized mean and standard deviation of a lognormally distributed field, M is the number of participating eigenmodes (i.e., considered number of eigenvalues of the covariance matrix), θ is an independent identifier corresponding to a set of M standard normal variables ($\xi_i(\theta)$), and λ_i and $\boldsymbol{\psi}_i^T$ are i_{th} greatest eigenvalue and eigenvector of the standard normal field, respectively; $\boldsymbol{C}_{Z,Z_i}$ is the i_{th} vector of correlation between all points in the RF mesh and the desired point $\boldsymbol{Z}$. The realizations of the field can be transformed from normal to lognormal field values by applying Equation (23.5):

$$\hat{H}_{LN}(\boldsymbol{Z},\theta) = \exp\left\{\hat{H}(\boldsymbol{Z},\theta)\right\}$$

(23.5)

where $\hat{H}_{LN}(\boldsymbol{Z},\theta)$ is the lognormal field value.

The realizations of random fields can be used to generate different trials for simulation and to conduct reliability analysis. As an example, to consider randomness in soil modulus of elasticity, for each point $\boldsymbol{Z}$ in the FE mesh, the field value can be calculated for a specific realization of the field to be fed to the FE model. The corresponding response is attributed to the specifically generated realization of the random field. Also, the response corresponds to the specific set of ξ_i used in EOLE formulation which distinguishes each trial from other trials and generates a unique random

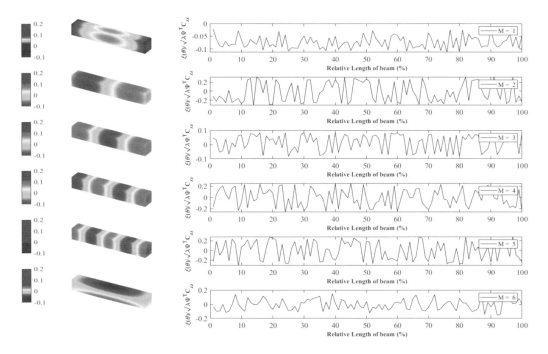

FIGURE 23.1   Greatest six eigenmodes of a one-dimensional random field example.

field realization. This set of unique ξ_i is called representative random variables, that is used both for the generation of field values for FE mesh and to represent the random field in AK reliability. The number of these representative random variables plays a key role in dimensionality and, in turn, the efficiency of AK reliability.

The number of required representative random variables (i.e., M) is determined based on field characteristics such as correlation length, field coefficient of variation, and a prescribed target variance threshold. The idea for the selection of M is to consider enough eigenvectors to reach the target variance threshold. In other words, the target variance threshold controls the required number of sentences including eigenvectors inside Equation (23.3) (i.e., $\sum_{i=1}^{M} \frac{\xi_i(\theta)}{\sqrt{\lambda_i}} \psi_i^T C_{Z,Z_i}$). The values are sorted such that the largest participating eigenmode (M out of N_{mesh}) can be taken for the field generation.

To illustrate the concept of eigenmodes of the random field, let's consider a sample rectangular beam with a one-dimensional (1D) random field along the beam length. Figure 23.1 illustrates the variation of the greatest six eigenmode shapes along the beam (i.e., the terms inside summation in Equation (23.3)). Figure 23.2 shows the cumulative effect of the eigenmodes by showing the summation of sentences in Equation (23.3) for up to six eigenmodes. The variance in the cumulate effect of the eigenmodes diminishes with the increase in the number of modes considered for the field realization. The number of modes is typically determined based on a prescribed target variance. For example, a target variance of 5% is an acceptable limit for the M value truncation for the simulation of a random field in soil [7, 8]. The variance of the lognormal random field that should be compared to the target variance can be calculated using Equation (23.6) [3].

$$Var\left[H(Z) - \hat{H}(Z)\right] = \sigma_{lny}^2 \left[1 - \sum_{i=1}^{M} \frac{1}{\sqrt{\lambda_i}} \psi_i^T C_{Z,Z_i}\right] \tag{23.6}$$

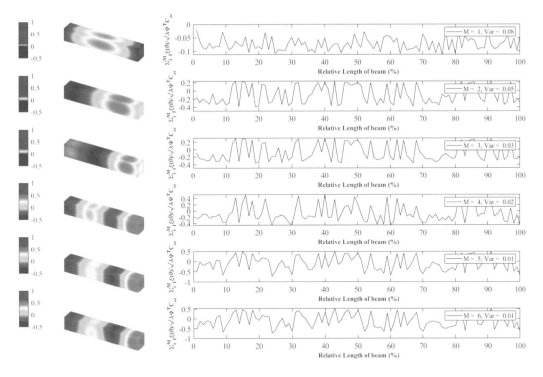

FIGURE 23.2   Cumulative effect of summation of eigenmodes of a one-dimensional random field example.

where $H(Z)$ and $\hat{H}(Z)$ are the actual and approximated values for the random field realization for point $\mathbf{Z}$. The number of required eigenmodes is the same as the number of representative random variables that are used to train the predictor in AK reliability, which is reviewed in the next section.

23.3   RELIABILITY ANALYSIS

Monte Carlo simulation (MCS) and active learning Kriging Monte Carlo simulation (AK-MCS) are utilized in this study to perform the reliability analysis, where MCS is used to verify the accuracy of the AK-MCS results. Figure 23.3(a) illustrates the general framework for MCS and AK-MCS. MCS is conducted using the original performance function, which includes the resistance obtained from the FE model and the load models. The AK-MCS method provides an efficient surrogate model to the performance function, where the original model is replaced with a Kriging predictor. The Kriging predictor is a surrogate mathematical function composed of a regression function and a gaussian process, while more information about Kriging can be found in the literature [9].

The reliability analysis starts with the random generation of n number of trials (Figure 23.3(a)). Each trial includes a random vector X, which contains a realization of random variables and random fields. For MCS and AK-MCS, the original performance function including the FE analysis and the Kriging performance function would be evaluated for each trial, respectively. Figure 23.3(b) shows a trial in which the EOLE realization of the random field with M terms was built to be fed to the FE model while only the values of $\xi_i(\theta)$ in Equation (23.3) were used to be fed to the Kriging performance function. In other words, for MCS, the full realization of the random field for the whole FE mesh is required, while for AK-MCS, M random variables (i.e., $\xi_i(\theta)$) represent the random field instead. Once the performance function is evaluated for all n trials, the number of failed trials (i.e., n_f) can be assessed, where failure corresponds to a negative realization of the original performance function ($G(X) < 0$) for MCS or a negative realization of the surrogate performance

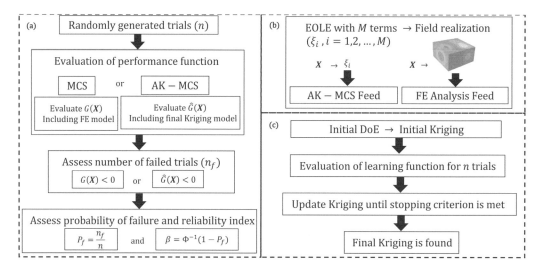

FIGURE 23.3 Algorithms: (a) reliability analysis for MCS and AK-MCS; (b) EOLE field realization; (c) brief review of the active learning process.

function ($\hat{G}(X) < 0$) for AK-MCS. The probability of failure (i.e., P_f) is defined as the number of failed trials to the total number of trials, and the reliability index (i.e., β) is defined as the inverse of cumulative distribution function (CDF) of the standard normal distribution (i.e., Φ^{-1}) of the complement of the probability of failure (i.e., $\beta = \Phi^{-1}(1-P_f)$) if the performance function follows a normal distribution.

The Kriging predictor in AK-MCS is initially trained using a limited set of design of experiments (DoE) or referred to as design points (i.e., a set of inputs and outputs for the original performance function), as illustrated in Figure 23.3(c). The size of DoE is selectively increased in a stepwise manner using active learning to achieve a desired accuracy for the Kriging predictor. The idea of active learning is based on selecting a training candidate for updating the DoE using a learning function so that the trained Kriging predictor becomes sufficiently accurate in finding the sign of the performance function that is required to determine the failure domain. The learning process continues until a stopping criterion is reached, where the Kriging predictor at the last step of training is referred to as the final Kriging predictor. The stopping criterion dictates the size of DoE.

23.4 ALTERNATIVE METHOD FOR REPRESENTING RANDOM FIELD VARIABLES IN AK RELIABILITY

As discussed in the previous section, a set of standard normal random variables representing the random field are utilized in training the Kriging predictor in AK-MCS analysis. The number of standard normal random variables in EOLE (referred to as EOLE RVs herein) can vary based on the characteristics of the random field such as the correlation length, the coefficient of variation of the random field, and the target variance for the random field. AK-MCS becomes less effective, from a computational standpoint, when estimating the reliability of systems with large numbers of EOLE RVs. As a result, an alternative approach to represent the random field variation in training the Kriging predictor is desired for efficient AK-MCS analysis.

The idea of the alternative approach is to build a distribution of random field values of a specific region of the model where the response is most sensitive to variations in the random field. This region is named as the influence region and is typically smaller in volume than the whole modeled system, as shown in Figure 23.4(a). The boundaries of the influence region can be determined based

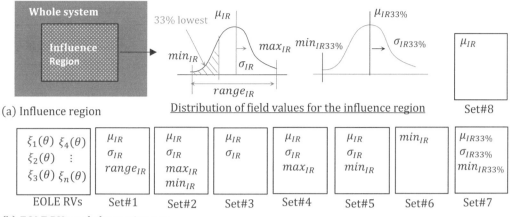

(a) Influence region

(b) EOLE RVs and alternative sets

FIGURE 23.4 Schematic illustration for determining alternative representations of the random field to be used in AK analysis: (a) influence region and (b) EOLE RVs and alternative sets.

on engineering judgment, past performance, or a sensitivity analysis depending on the complexity of the problem of interest. For example, the soil influence region for a soil-pile group interaction problem is expected to be a soil volume close to the piles since engineering judgment and prior engineering knowledge indicate that pile group resistance is more sensitive to the soil properties in the vicinity of the pile than the soil properties far from the piles. For the distribution of random field values in the influence region, main characteristics such as mean (μ_{IR}), standard deviation (σ_{IR}), minimum (min_{IR}), maximum (max_{IR}), and range ($range_{IR}$) of random field values can be distinctly calculated for different realizations of random fields, as shown in Figure 23.4(a). Also, a sub-distribution of the random field values within the influence region can be determined. For instance, a distribution of the lowest one-third values of the influence region can be utilized as an alternative in situations where the resistance model is governed by the lowest group of random field realizations. For this case, the mean ($\mu_{IR33\%}$), standard deviation ($\sigma_{IR33\%}$), and minimum ($min_{IR33\%}$) of the 33% lowest values within the influence region can be determined.

Multiple combinations of the mentioned characteristics of the distribution of the field values would be considered as different alternative representation sets of the random field. Set #1 to Set #8 in Figure 23.4(b) were evaluated in this research in training the Kriging predictor in AK-MCS. The evaluation was conducted using two practical examples as outlined in the following section. The evaluation was presented in terms of the accuracy of AK-MCS in predicting the reliability index and the number of training points for the Kriging predictor.

23.5 VALIDATION OF ALTERNATIVE SOLUTION

23.5.1 Example 1: Soil-Pile Group Interaction

23.5.1.1 Model Description

The finite element (FE) program ABAQUS/Standard [10] was employed to examine the behavior of free-headed laterally loaded pile group foundations installed in sandy soil, as shown in Figure 23.5. All parts (i.e., soil, piles, and pile caps) were modeled in three-dimensional (3D) space using four nodes of brick elements (C3D8R). The pile group was comprised of four circular piles arranged in a square (2×2) configuration with a 0.61 m pile diameter (D) and a pile spacing of three times D. The pile-pile cap interface was assumed to be rigid. To eliminate the effect of boundaries, the soil continuum sides were set to be as ten times D [11, 12] (i.e., 17×17 m). The transitional degrees of

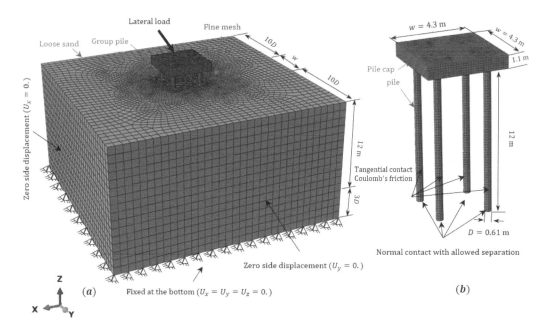

FIGURE 23.5 FE model description: (a) pile group in loose sand and (b) pile group.

freedom of the soil base were restrained in 3D, while the elements along the soil perimeter were allowed to move vertically. The friction (i.e., tangential contact) and bearing behaviors (i.e., normal contact) at the pile-soil interface were simulated using surface-to-surface contact enforced with the penalty and hard contact methods, respectively. The hard contact was assumed to allow separation between the bottom of the pile and the soil. Tangential contact was defined between the piles and the neighboring soil using Coulomb's frictional model.

The steel piles and the concrete pile cap were modeled using linear elastic-perfectly plastic and linear elastic material models, respectively. The Young's modulus of the piles and the pile cap were 200 GPa and 24.6 GPa, respectively. The linear elastic portion of the model is defined based on Hook's law while the plastic part is defined based on the Mohr-Coulomb failure criterion. The controlling parameters for the elastic branch are the soil Young's modulus, E, and Poisson's ratio, ν. The plastic branch (yielding) is defined by the friction angle and the cohesion in Equation (23.7). Two random fields were considered for soil Young's modulus, E, and the soil friction angle, ϕ.

$$\tau = c + \sigma_v \tan(\phi) \tag{23.7}$$

where τ is the yield stress, c is the cohesion, σ_v is the overburden pressure, and ϕ is the friction angle. The analysis consisted of three steps. In the first step, the initial soil stresses and own weight were defined to capture the effect of in-situ geostatic stresses, where the in-situ stresses were stabilized using the soil's weight. In the following step, the pile was introduced to the model, and its weight was applied to capture initial settlements and contact forces. Finally, the load was applied in a controlled manner until failure (i.e., force-controlled).

23.5.1.2 Parametric Study

A parametric analysis was conducted to evaluate the accuracy and number of training points required to train the Kriging predictor in AK-MCS to estimate the reliability of the pile group under lateral wind load using the 8 alternative sets of random field representations shown in Figure 23.4. A total of 162 AK-MCS reliability analyses were performed consisting of nine sets of random field

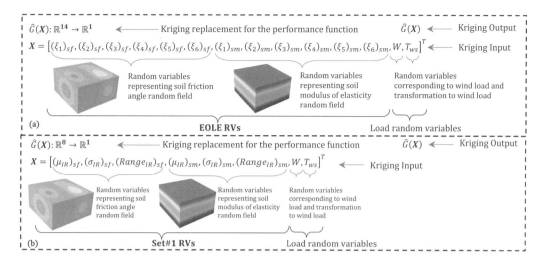

FIGURE 23.6 Kriging input and outputs: (a) EOLE RVs and (b) Set #1.

representations (EOLE RVs and Set #1 to #8 in Figure 23.4) and six configurations of the influence region, where each of the 54 analyses (9 sets × 6 configurations) was conducted three times to consider the effect of initial sampling on the accuracy and efficiency of the analysis.

Two random fields were considered for soil Young's modulus, E, and the soil friction angle, ϕ with six representative random variables for each field (i.e., standard normal variables in EOLE), and two random variables were considered for the lateral wind load to form a total of 14 random variables as input for the Kriging predictor for the EOLE RVs analysis, as shown in Figure 23.6(a). The input vector of random variables for Kriging using Set #1 (see Figure 23.4) consisted of three variables for each random field in addition to the wind load random variables to form a total of eight variables, as shown in Figure 23.6(b). Analysis for Sets #2 to #8 was conducted similarly to Set#1 but with different representations of the random field, as shown in Figure 23.4.

The six configurations of the considered influence regions are shown in Figure 23.7. The influence regions include "Full" with 34 m × 34 m × 16 m, "Middle" with 8 m × 8 m × 12 m, "Wide" with 16 m × 16 m × 12 m, "Tight" with 4.88 m × 4.88 m × 12 m, "Middle WH" with 8 m × 8 m × 12 m without the pile wholes considered as four 0.61 m × 0.61 m × 12 m holes at the center of piles, and "Around Pile" with four 2.5 m × 2.5 m × 12 m soil at the center of each pile and removing four 0.61 m × 0.61 m × 12 m holes at the center of piles. All considered influence regions are at the center of the soil block near piles of different sizes.

Sample distributions of field values for soil modulus of elasticity and soil friction angle are shown in Figures 23.8 and 23.9, respectively, for the considered configurations of the influence regions. Different distribution shapes were observed since the domain under which the random field values are collected to build the distribution is different for each influence region. As a result, for MCS trials, the values of mean, standard deviation, minimum, and maximum for the distribution of the field values in various influence regions have different shapes, as shown in Figures 23.8 and 23.9.

As a reference for the accuracy of compared cases, the example was solved using MCS with 173 trials [13, 14], which showed a reliability index of 1.2007. Using AK-MCS including a Gaussian correlation function, linear regression function, and KO learning function [15, 16], a reliability index of 1.1972 was determined for all AK-MCS analyses, while the required number of added points varied by varying the representative variables of the random field and influence region. DACE MATLAB toolbox [17, 18] was used as the Kriging predictor. Table 23.1 presents the number of added training points in the AK-MCS analysis for the considered cases, where the three values per cell in the table

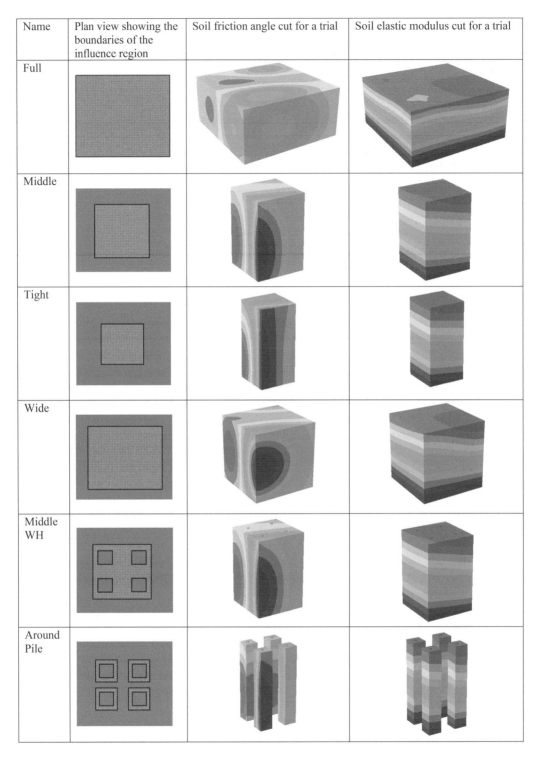

FIGURE 23.7   Random field realization contours of the modulus of elasticity and soil friction angle within the influence region.

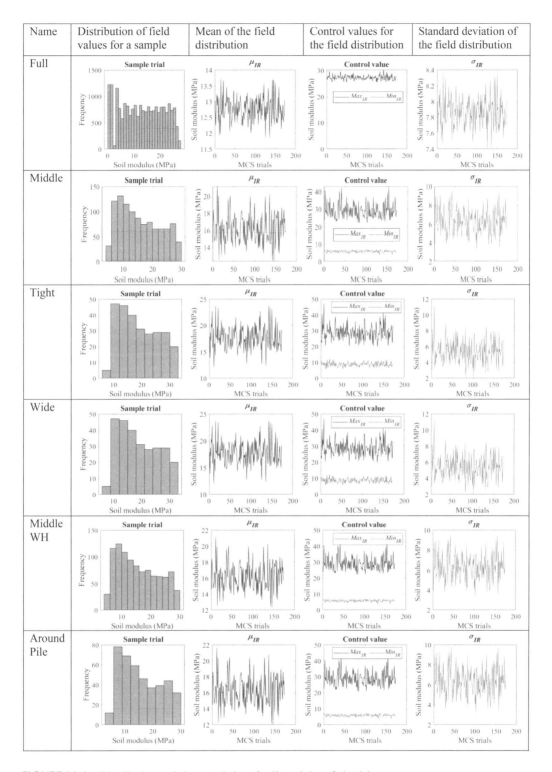

FIGURE 23.8   Distribution and characteristics of soil modulus of elasticity.

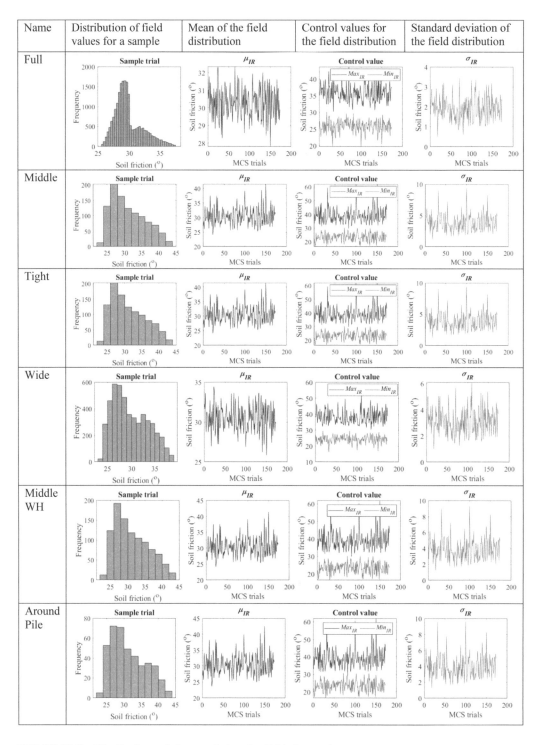

FIGURE 23.9 Distribution and characteristics of soil friction angle.

TABLE 23.1

Number of Added Training Points for AK-MCS in Example 1

Influence Region	EOLE RVs	Set #1	Set #2	Set #3	Set #4	Set #5	Set #6	Set #7	Set #8	Average
Full	15, 25, 24	60, 61, 65	60, 57, 65	60, 59, 57	62, 53, 51	60, 56, 66	57, 57, 59	59, 56, 60	60, 59, 63	55.0
Middle	26, 18, 24	29, 26, 35	28, 26, 20	36, 34, 31	25, 27, 23	22, 23, 19	43, 37, 38	25, 35, 32	50, 50, 51	30.9
Tight	20, 30, 27	36, 40, 29	23, 33, 34	18, 29, 35	33, 34, 32	33, 28, 31	53, 47, 46	23, 37, 33	47, 63, 61	35.4
Wide	22, 21, 26	54, 39, 38	51, 51, 40	55, 52, 58	53, 53, 52	52, 53, 53	52, 52, 49	53, 56, 50	49, 54, 53	47.8
Middle WH	29, 21, 28	31, 34, 34	23, 32, 27	37, 24, 37	18, 26, 26	35, 37, 28	38, 39, 42	23, 34, 32	56, 53, 54	33.3
Around piles	16, 21, 25	25, 28, 24	26, 24, 18	35, 34, 31	23, 35, 27	32, 26, 33	39, 47, 40	35, 33, 28	49, 52, 51	31.7
Average	23.2	38.2	35.4	40.1	36.3	38.2	46.4	39.1	54.2	-

correspond to the number of training points per each of the three repetitive analyses per analysis case (unique set and influence region). The number of points to train the initial predictor for all analyses is 19.

Results indicate that introducing the concept of influence region enhanced the efficiency of the AK-MCS analysis by comparing the "Full" influence region to all others. The most efficient influence region was "Middle" and then "Around Pile". Also, the analysis showed that EOLE RVs results in the lowest number of training points, while all other alternative sets of random field representation required a larger number of training points. Set #2 was the best alternative representation of the soil random field because it yielded a less average number of added training points as compared with other alternative sets. Although the alternative sets (Set #1 to Set #8) did not reduce the number of added training points for this example, the concept of alternative sets to represent the random field in training the Kriging predictor is validated. However, utilizing Set #1 to Set #8 yield less utilization of the computer RAM when the predictor is trained, which enhances the efficiency of the AK-MCS analysis.

It is concluded that all considered sets yield an accurate representation of the reliability index when utilized to assess the reliability using AK-MCS. However, the required number of training points for the Kriging predictor varies based on the set parameters and is higher for the alternative sets compared to the conventional approach of considering the realizations of the standard normal variables in training the Kriging predictor. The utility of the proposed alternative sets would be most relevant when conducting the AK-MCS would not be feasible using EOLE RV representations due to the high number of standard normal variables required to realize the random field. Future work should be conducted to utilize the alternative representations of the random field for high dimensionality problems (i.e., problems requiring a high number of standard normal variables to represent the random field).

23.5.2 EXAMPLE 2: REINFORCED CONCRETE GIRDER

A concrete girder subjected to four-point bending is considered in this example, as shown in Figure 23.10, as presented in the referenced study [19]. The 15 m long concrete girder is subjected to two concentrated loads at 2 m from the ends, considering only dead and live loads with a dead-to-live ratio of 4 [20, 21]. To simplify the problem, the original 15 random variables in the referenced study [19, 22] were reduced to only five random variables including concrete strength, the yield stress of steel, dead load, live load, and live load transformations. The use of AK-MCS in the reliability assessment of bridge pier [23] and girders [24] without random fields and bridge girders with random fields [25] have been validated in literature. The moment of resistance is calculated using a sectional analysis where a one-dimensional (1D) random field is applied to the calculated resistance to consider the spatial variation along the beam resulting from degradation actions such as corrosion and freeze-thaw damage. The considered 1D random field was generated with a mesh size of 100 mm, a correlation length of 1,000 mm, a coefficient of variation of 0.05, a mean value of 1, and a Gaussian correlation function for the random field. By satisfying a target variance of 1% for EOLE realization in Equation (23.6), the required M value should be 22. Therefore, for AK-MCS using EOLE RVs, a total of 27 random variables are required inclusive of the load and resistance effect random variables.

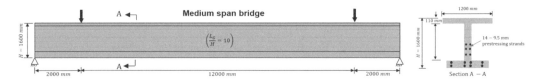

FIGURE 23.10 Details of the considered reinforced concrete girder.

TABLE 23.2
Sensitivity Analysis Results for Example 2

Examination	Crude MCS	EOLE RVs	Set #1 (Entire Beam)	Set #1 RVs (Constant Moment Region)	Set #6 RVs (Constant Moment Region)
Reliability index	3.37	3.36	3.36	3.37	3.36, 3.34, 3.33
Number of training points	3×10^7	>301	>301	>301	69, 70, 62

As a reference point, crude MCS analysis with 30 million trials was conducted, which showed a reliability index of 3.372. EOLE RV, Set #1, and Set #6 of Figure 23.4(b) were considered in the analysis. Two influence regions including the entire beam length and the constant moment region were considered.

The results of the AK-MCS analysis with 200,000 trials, Gaussian correlation function, constant regression function, and KO learning function are shown in Table 23.2. The reliability indexes of the AK-MCS analyses are within a 1% margin of error as compared with the crude MCS although the analysis with the EOLE RV and Set #1 did not converge for up to 300 training points (i.e., was terminated at 300 points). Set #6 converged at an average of 67 points but yielded a larger margin of error as compared with Set #1. This result implies that the stopping criterion can be efficiently met by only considering the minimum field value in the influence region (Set #6 with constant moment region set as the influence region). This occurred because only the minimum field value in the mid-section is enough to determine the minimum resistance of the girder in four-point bending when a sectional analysis is utilized to determine the girder moment of resistance.

23.6 SUMMARY AND CONCLUSION

In this chapter, an alternative method for considering the effect of random fields to train the Kriging predictor in active learning Kriging Monte Carlo Simulation (AK-MCS) reliability analysis was recommended and validated. The alternative method considers a specific region within the modeled structure (the influence region) and considers the characteristics of the distribution of random field values within the influence region as the alternative representative variables for training the Kriging predictor, as opposed to the traditional way of training the kriging predictor using the standard normal variables used in generating the realizations of the random field in methods like the expansion optimal linear estimation (EOLE).

To validate the accuracy and efficiency of the proposed alternative method, two examples were considered including the reliability assessment of a laterally loaded pile group where the soil properties were modeled using a three-dimensional (3D) random field and the resistance was determined using nonlinear finite element (FE) analysis, and the reliability assessment of a reinforced concrete girder where the spatial variation of the moment of resistance along the girder was modeled using a one-dimensional (1D) random field and the resistance was modeled using sectional analysis.

A parametric study was conducted by considering multiple sizes and shapes of the influence region, and by considering various sets of alternative variables to represent the random field in training the Kriging predictor. The analysis results validated the accuracy of the Kriging predictor in AK-MCS for predicting the reliability index of the considered problems when compared to crude MCS. The results showed that the selection of the influence region and alternative variables representing the random field in training the Kriging predictor affect the number of required training points of the AK-MCS analysis. More investigation is required to assess the influence region for various reliability problems and to provide more data on the validity of the alternative representative random variables for AK-MCS. Also, a 3D random field with a larger number of representative random variables should be investigated to further examine the efficiency of the alternative method.

ACKNOWLEDGMENT

The authors would like to acknowledge Norlander Oudah Engineering Limited (NOEL) and Dalhousie University for their financial contribution and support of this project.

DATA AVAILABILITY

Some or all data, models, or codes generated or used during the study are available from the corresponding author by request.

REFERENCES

[1] A.-K. El Haj and A.-H. Soubra, "Efficient estimation of the failure probability of a monopile foundation using a Kriging-based approach with multi-point enrichment," *Computers and Geotechnics*, vol. 121, p. 103451, 2020.

[2] A.-K. El Haj, A.-H. Soubra and J. Fajoui, "Probabilistic analysis of an offshore monopile foundation taking into account the soil spatial variability," *Computers and Geotechnics*, vol. 106, pp. 205–216, 2019.

[3] C. Petrie, "*Reliability Analysis of Externally Bonded FRP Strengthened Beams Considering Existing Conditions: Application of Stochastic FE and Conditional Probability*," MASc Thesis, Dalhousie University, Halifax, Nova Scotia, Canada, 2022.

[4] F. Oudah and A. Alhashmi, "Time-Dependent Reliability Analysis of Degrading Structural Elements using Stochastic FE and LSTM Learning," in *CSCE Annual Conference*, Whistler, BC, Canada, 2022.

[5] K. Khorramian, A. Alhashmi and F. Oudah, "Optimized Active Learning Kriging Reliability Based Assessment of Laterally Loaded Pile Groups Modeled Using Random Finite Element Analysis," *Computers and Geotechnics*, vol. 154, p. 105135, 2023.

[6] Zhang, Jun and B. Ellingwood, "Orthogonal series expansions of random fields in reliability analysis," *Journal of Engineering Mechanics*, vol. 120, no. 12, pp. 2660–2677, 1994.

[7] P. D. Spanos and R. Ghanem, "Stochastic finite element expansion for random media," *Journal of Engineering Mechanics*, vol. 115, no. 5, pp. 1035–1053, 1989.

[8] B. Sudret and A. Der-Kiureghian, "Stochastic finite element methods and reliability," Dept. of Civil and Environmental Engineering, Univ. of California, Berkeley, CA, 2000.

[9] C.-C. Li and A. Der Kiureghian, "Optimal discretization of random fields," *Journal of Engineering Mechanics*, vol. 119, no. 6, pp. 1136–1154, 1993.

[10] G. A. Fenton and E. H. Vanmarcke, "Simulation of random fields via local average subdivision," *Journal of Engineering Mechanics*, vol. 116, no. 8, pp. 1733–1749, 1990.

[11] A. Nataf, "Determination des distribution don t les marges sont donnees," *Comptes Rendus de l Academie des Sciences*, vol. 225, pp. 42–43, 1962.

[12] K. Khorramian and F. Oudah, "Active Learning Kriging-Based Reliability for Assessing the Safety of Structures: Theory and Application," in *Leveraging Artificial Intelligence in Engineering, Management, and Safety of Infrastructure*, Taylor and Francis (CRC), 2022, pp. 184–231.

[13] S. ABAQUS CAE, "Dassault Systemes Corporation," Providence, Rhode Island, 2013.

[14] M. S. Fayyazi, T. Mahdi and W. L. Finn, "Group reduction factors for analysis of laterally loaded pile groups," *Canadian geotechnical journal*, vol. 51, no. 7, pp. 758–769, 2014.

[15] J. Dong, F. Chen, M. Zhou and X. Zhou, "Numerical analysis of the boundary effect in model tests for single pile under lateral load," *Bulletin of Engineering Geology and the Environment*, vol. 77, no. 3, pp. 1057–1068, 2018.

[16] K. Khorramian, A. Alhashmi and F. Oudah, "Effect of random fields in stochastic FEM on the structural reliability assessment of pile groups in soil," in *Canadian Society of Civil Engineering Annual Structural Specialty Conference (CSCE)*, Whistler, BC, Canada, 2022.

[17] K. Khorramian and F. Oudah, "New Learning Functions for Active Learning Kriging Reliability Analysis Using a Probabilistic Approach: KO and WKO Functions," *Structural and Multidisciplinary Optimization*, 2022.

[18] K. Khorramian and F. Oudah, "Reliability Analysis of Structural Elements with Active Learning Kriging Using A New Learning Function: KO Function," in *CSCE 2022 Annual Conference*, Whistler, BC, Canada, 2022.

[19] S. N. Lophaven, H. B. Nielsen and J. Søndergaard, "A Matlab Kriging Toolbox," Technical University of Denmark, Kongens Lyngby, Technical Report No. IMM-TR-2002-12, 2002.

[20] F. Oudah, M. H. El Naggar and G. Norlander, "Unified System Reliability Approach for Single and Group Pile Foundations – Theory and Resistance Factor Calibration," *Computers and Geotechnics*, vol. 108, pp. 173–182, 2019.

[21] A. E. Alhashmi, F. Oudah and M. H. El-Naggar, "Application of Binomial System-Based Reliability in Optimizing Resistance Factor Calibration of Redundant Pile Groups," *Computers and Geotechnics*, vol. 129, p. 103870, 2021.

[22] S. N. Lophaven, H. B. Nielsen and J. Søndergaard, "DACE: A Matlab Kriging Toolbox." *Vol. 2. IMM Informatics and Mathematical Modelling*, The Technical University of Denmark, pp. 1–34, 2002.

[23] E. Buckley, K. Khorramian and F. Oudah, "Optimal Active Learning Kriging Predictor Configuration for Calculating the Reliability Index of Bridge Piers," in *11th International Conference on Short and Medium Span Bridgers*, Toronto, Ontario, Canada, 2022.

[24] E. Buckley, K. Khorramian and F. Oudah, "Application of Adaptive Kriging Method in Bridge Girder Reliability Analysis," in *CSCE Annual Conference*, Virtual, 2021.

[25] K. Khorramian, S. Salili and F. Oudah, "Spatial Variability-Based Reliability Analysis Framework for Concrete Bridge Girders," in *11th International Conference on Short and Medium Span Bridges (SMSB)*, Toronto, Ontario, Canada, 2022.

24 Intelligent Systems in Construction
Applications, Opportunities and Challenges in AR and VR

Pravin R. Minde, Abhaysinha G. Shelake, and Dipak Patil

CONTENTS

24.1 AR AND VR IN THE CONSTRUCTION INDUSTRY

Nowadays, in construction industry people are using software like MSP and BIM for getting accurate data for construction activity execution. But this is not giving the construction industry a hold on issues like wastage, delay in execution of work, cost escalation, etc. Now, time comes when the construction industry should start thinking about the use of visual imaging to cope with its problems. The code term for construction experts to enter the experimental domains of installation and operation is "virtuality" (Brandon P. et al. 2005). Earlier, the construction industry was far more dependent on physical prototyping. Imagine a client comes to purchase a flat. In order to assure him work quality and specifications, you have two options in front of you. The first is to construct a sample flat/bungalow and the second is to prepare virtual model of the same flat/bungalow by using VR. Choosing VR will actually help you to deal with cost, convenience, risk and safety issues associated with flat/bungalow construction (Ahmed 2019). Also, customer satisfaction is far better compared to the physical prototype option. Hence, VR proves to be a better solution as compared to a physical prototype. As a new concept, some researchers still question the applicability of VR over physical prototypes (Avhad & Hinge 2017). Some researchers believe that confusion between AR and VR is an enduring problem. However, one can say the end application in both is same and VR is doing it in better way as compared to physical protypes. Researchers drafted many different names to VR (virtual reality). Some of them are amplified reality, augmented reality, augmented virtuality, mediated reality, mixed reality, blended reality, diminished reality, virtualised reality, etc. Augmented reality is one of the prime tasks to execute virtual reality project. The first step to achieve VR is to prepare model in AR. Head mounted display (HMD) and heads-up display (HUD) are two common types of AR. HMD means AR, which includes a display system built around the technology and includes hardware. In this, as shown

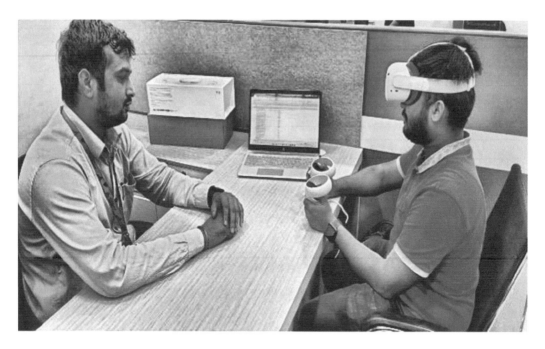

FIGURE 24.1 Head Mounted Devices (HMD) devices application.

in Figure 24.1, the user has to wear hardware like a helmet by which the user can observe projection of prepared model. The second one is HUD, which means a user has to wear hardware like a helmet, but this time a projection of the prepared model is superimposed on the user's eyesight. Thus, HUD helps in coordinating between a prepared digital model and actual physical model for better coordination.

In VR, a researcher produces a virtual environment where real objects are placed virtually and gives a feel of actual reality when user uses HMD or HUD. As 2D or 3D objects, information related to sound, light or scent is used in VR (Abdelhamid 2012). The most basic advantage in VR over a physical model is a requirement of minimal cost for preparation of virtual reality. In the construction industry, however, little efforts were made in comparison with other industries. As an example, Bae et al. (2012) and Bae et al. (2013) developed hybrid four-dimensional augmented reality (HD4AR), which uses site photography to identify the location and orientation of field personnel to allow them to query and access semantically rich 3D cyber information and see it precisely overlaid on real-world imagery. Irizarry et al. (2013) used KHARMA Hill et al. (2010) to develop an information surveyed point for observation and tracking (Info-POST), as a mobile AR tool for facilities managers for accessing information about the facilities they maintain. But these little efforts are not sufficient for a dynamic industry like construction. It is high time now that researchers should explore the applicability of VR in a wider extent for the construction industry like perfect coordination, reducing risks, reducing delays and improving quality and safety. This paper presents an approach by which coordination between client and contractor increases as an outcome of AR- and VR-based system applications. Nowadays, it becomes easy to use AR and VR with the availability of smartphones and tablets. In India, for the AEC industry, the popularity of AR and VR is slowly growing (Al-Adhami et al. 2018). Research into the use of such applications is still few and far between. It is now important to produce integrated basic HMD units and smartphones to offer a compact, affordable and agile yet multi-purpose process that should solve basic problems on construction projects like coordination and construction activity execution delay issues.

24.2 CHALLENGES AND OPPORTUNITIES IN AR AND VR

24.2.1 OPPORTUNITIES

24.2.1.1 Scheduling and Project Tracking

In construction management, during the early stage of evolution of management systems, the most popular system was as plan as built system. Researchers tried to implement this system by coordinating a planned schedule and actual construction. But tools used were not producing effective solutions. AR- and VR-led technology actually produce the best form of a plan as a built system can be truly understood by different stakeholders of construction projects without any ambiguity. Figure 24.2 shows use of VR glasses for project tracking. Some of the primary efforts made by researchers are explained below. Zaher et al. (2018) used mobile AR technologies integrated with Microsoft Project and Primavera for construction project scheduling and schedule monitoring. Zang et al. (2017) showed in their research that AR could be effectively used for safety task scheduling in a construction project.

24.2.1.2 Total Quality Management

In construction management, two important features are quality management and defect control management. Many times, the client accepted completed projects from a contractor but due to not getting required desired quality, a dispute occurred in between them at the completion of the projects. AR- and VR-led technology produces a solution that can effectively be applicable in improving quality. Automation is possible by using AR and VR technology to determine the defects in the early stage and can be removed during the initial stages of construction itself. Also, along with quality control, quality assurance also plays an important role to maintain client faith on contractor work. By producing a spatial walkthrough, the client can get an exact idea about his project and also, he can see if something is missing in a digitally produced walkthrough by using AR and VR. This actually helps contractors to understand visually the exact demand from the client and he can produce the structure accordingly. In this way, quality assurance can be maintained. In construction,

FIGURE 24.2 Virtual Reality (VR) glass used for project tracking.

FIGURE 24.3 Use of AI in every aspect of Total Quality Management (TQM.)

the defect liability period is the most important period w.r.t quality in construction. AR- and VR-led visual demonstrations can take care of the defect liability period and help to understand the quality of the structure in a better way during that period.

Current construction industry practices to check quality are inspection, checklist preparations, brainstorming meetings, etc. All these techniques include limitations because of the involvement of the human factor. It is not possible to cover each and every aspect of construction quality merely by inspection. AR- and VR-led tools can produce a visual walkthrough of a structure. The defects that are unnoticed during inspection can easily be detected by a visually produced walkthrough using BIM and AR and VR combined frameworks (Blinn et al. 2015). You can easily detect defects and workers' mistakes wherever the digitally produced walkthrough does not match with the actually constructed structure. Most importantly, a project manager can detect these defects daily and at the initial stage, enabling a project manager to mitigate it at an early stage and produce a very satisfied output for client. The parameter shown in Figure 24.3 can be monitored by using a AR-VR-led system in the best possible way.

24.2.1.3 Time and Cost Management

A project manager deals with time and cost management on a construction project. Undoubtedly, managing cost and time are major challenges in front of any construction manager. Currently, as per the planning commission report, the Indian construction sector faces around 34% cost overruns and around 40% time overruns on infrastructure projects. A developing country like India will not be able to afford this much cost overrun in the future. A project manager's main objective is to complete the construction project in a minimum time and cost with maximum quality in work. Basically, a construction delay in executing activity is the main cause of all problems. Effective coordination is the remedial solution to reduce delay. In the early stage of management evolution, the contractor is relying on the primitive techniques like daily hurdle meetings, inspections, etc. The author is presenting an AR- and VR-based framework that can reduce the delay in coordination and thus reduction in construction delay. AR- and VR-based applications on construction can be further used for the purpose of monitoring and controlling functions for projects.

Also, one of the major advantages of AR and VR frameworks is that it can be used significantly for construction data acquisition. A project manager can easily identify excess allocation

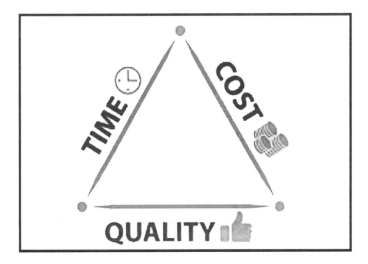

FIGURE 24.4   Time and cost managed by Building Information Modeling (BIM) software.

of resources, flaws in construction procedures, faulty work, etc. This data acquisition is accurate compared to a conventional method and also less cost. This reduces the cost of data acquisition in a significant manner. With the help of a smartphone, nowadays it becomes extremely easy to use an AR-VR-led framework in construction, leading to mitigate delay and excessive cost usage on construction projects. AR and VR also help in implementing safety on-site by providing visual insights on possible risk-prone locations in construction. Figure 24.4 reflects analogy of time, cost and quality in BIM philosophy.

24.2.2 CHALLENGES

1. Stakeholder Engagement:
   Stakeholder of the project finds the following challenges while adopting AR-VR technology:
   i.   The interface of AR-VR is dynamic in character. A technocratic user is needed to use this technology. This becomes issue when AR-VR is implemented on civil engineering projects without competencies of AR and VR technology.
   ii. The common belief of people is with connecting people for coordination and execution of project-related activities. AR-VR-led technology provides better coordination but with that it also provided a sense of isolation from people.
   iii. VR experiences is new to industry and as a new technology nobody shares it easily.
   iv. Also, stakeholders experience problems wherever augmentation does not see sufficiently real and Also, while preparing model, data feeding is inaccurate, and unrealistic luminance of virtual objects. In such cases, a stakeholders like contractor finds it difficult to convince a client based on an AR-VR setup.
2. Design support:
   Design support finds the following challenges:
   i.   It is very difficult to analyse changes made in AR-VR supporting software into BIM software as it is required mostly for the purpose of cost and time analysis with new changes.
   ii. Though AR-VR-led technology is based on a digital generated model, it is very difficult to take measurable feedbacks from customers for later review and in the process of continuous improvement. One can set a designed manual feedback system, but it does not provide the useful output to that extent by which original technology can improve.

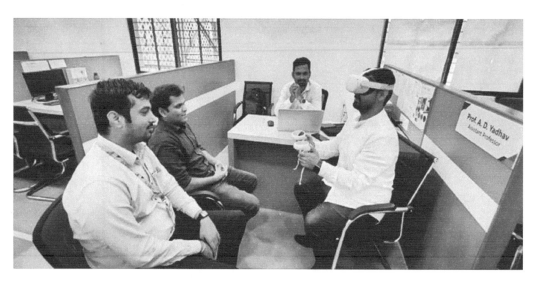

FIGURE 24.5　Stakeholder discussion.

3. Design Review:
   AR-VR-led technology, though technology savvy, there are different technical challenges in end use, as below:
   i.　The polygon meshes which helps in the development of an AR-VR-led model is not sufficient for producing better visual output. Now it's time to think about automatic 3D reconstruction of building parts.
   ii.　A marker is always guiding in a VR set for the user to see VR model. But this actually creates a problem to the user when he wants to compare products in structure. So, unfailing marker-less orientation is the need of the time.
   iii.　In construction, delay analysis plays an important role to minimize delay and thus reduce indirect cost associated with the project. For this to happen, a AR-VR-led technology should compare the planned schedule with the actual schedule, which is not yet there.
   VR Challenges:
   i.　Maintaining accuracy in data collection for preparation in a 3D model is extremely important, which is the biggest challenge in front of VR.
   ii.　There are no standard forms that guide data integration of BIM, photogrammetry and VR platforms.
   iii.　Validation of VR-produced models in terms of whether it reflects the true reality of the structure is absent on the technology front.
4. Operations and Management:
   In construction support, major operational issues are tracking the project schedule, collaborative interfaces, etc. AR-VR led technology is also facing the same issue. A major challenge is the context awareness and information visualization methods. AR-VR-led technology is a solo working technique and if the operator tries to integrate other facility management systems, then it becomes little bit difficult as such ready-made interfaces are not available. It is now the need of the hour to produce interfaces that can be integrated with other facility management systems. The quality of AR-VR-led output depends on information feed. Low-accuracy information always results in poor model production in the AR-VR method.

FIGURE 24.6 Design review using mobile applications.

FIGURE 24.7 O&M by using AI in civil engineering.

24.3 CHALLENGES AND OPPORTUNITIES IN AI ML

24.3.1 Opportunities in AI ML

Machine learning is a new technique in the field of civil engineering. Machine learning makes software learn from previous data based on a different algorithm assigned to it. Take a, example of air quality monitoring. In air quality monitoring, if the algorithm is developed to study air quality throughout a month, then based on that study, software can easily predict the future air quality data. Machine learning is used to solve such different problems in civil engineering. One can speed up the identification of design possibilities while designing the structure of the algorithm is developed for it. Several more examples are there, like environment impact assessment, water treatment plant monitoring, traffic monitoring, quality predictions, etc.

As an additional trend, now civil engineers are developing decision-making capacity for machine learning applications. After studying the material stress strain curve for different materials, a machine learning algorithm–based application can make a decision about the usability of material in different construction conditions. It can also predict maintenance cost or material corrosion behaviour over the significant time frame. Some of the machine learning applications in civil engineering are below:

1. Machine Learning Can Produce Stronger and Less Corrosive Metals:
   Researchers studied grain boundaries that actually improve the properties of metals. An algorithm was prepared by using a grain boundary study that gives decisions while metal production giving improved metal strength, corrosion resistance and improved conductivity. Better strength helps in building construction, corrosion resistance helps in bridge construction while improved conductivity helps in electricity transmission.
2. Machine Learning in Structural Health in Civil Infrastructure:
   Another important breakthrough is to describe structural health. King's College London produced an algorithm that produced suitable changes in structure to improve its structural stability (Azuma et al. 2001). They set up an impact hammer test that could easily describe changes to ten damage steel frame, which improves its structural stability if implemented. They accomplished and trained four repeated data sets and had 100% of identification records in their tests.
3. Reducing Injury in Construction Industry:
   Artificial intelligence is allowing scientists at the University of Waterloo new awareness to reduce wear-and-tear injuries and boost the productivity of skilled construction workers. Their research showed master masons don't follow the standard ergonomic rules taught to novices. Instead, they create their method of working quickly and safely. Examples

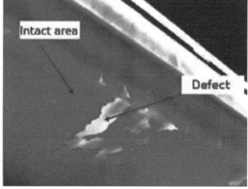

FIGURE 24.8 Defect identification using infrared thermography.

FIGURE 24.9   Quantifying deformation of bridge using ML tool.

include more swinging than the lifting of blocks and less bending of backs. Studies demonstrated that using sensors and artificial intelligence have disclosed that expert bricklayers use previously unidentified techniques to limit loads on their joints; information can now be passed on to apprentices in training sessions.

4. Quality of Designs Improvements:

Machine learning can improve designs overall to make spaces better for their ultimate human end users. In one exemplary example, WeWork wanted its meeting spaces to match the ways people would be using the spaces. The workspace start-up used machine learning to help understand and predict the frequency of use for these meeting rooms, and the company was able to design the space to best fit the needs of the people before starting construction on it.

The benefits of machine learning in design do not end there. Machine learning can also help workers figure out mistakes and omissions that might be present in the design before going forward with building. Instead, you can leave that to machine learning, which ultimately saves teams critical time that can be used for more productive tasks. With machine learning, you can even test various environmental conditions and situations in the model. The technology can help to determine if a particular element of the design is optimal, or can predict if it could create an issue down the road.

5. Create a Safer Workplace:

Increased safety is a priority for construction sites and machine learning provides a high-tech way to achieve this goal.

Let's look at an example of what machine learning is capable of accomplishing. For its annual Year in Construction Photo Contest, *Engineering News-Record* uses experts to check for safety in the submitted photos. In 2016, it looked at the difference between human safety experts and VINNIE artificial intelligence for checking safety within the images. VINNIE was able to find safety concerns, such as a person not wearing a hard hat. Also, it required less time compared to human effort. VINNIE required 10 minutes, while

FIGURE 24.10   Surface cutting with Artificial Intelligence and Machine Learning (AI-ML.)

humans took 4.5 hours. Also, VINNIE accurately identified 446 images while the human team identified 414 images.

Just like humans, machine learning also requires time to learn techniques. The potential of this test is that a tool like VINNIE could quickly sort through data and provide relevant results to humans, who can then look more closely at the results. Hence, machine learning can change the construction sector to a great context.

6. Increase the Project's Life Cycle:

Beyond design and construction, machine learning can even be instrumental in facility management to extend the total life cycle of an asset. Due to lack of important information, it's difficult to efficiently and cost-effectively manage renovations and repairs on-site.

Machine learning can help streamline the process by collecting and utilising information and data information better. It can do this by classifying documents and data like work orders and assessing pertinent conditions in real time, with surprising accuracy. This takes away these tedious and time-consuming administrative duties from people and allows them to focus on the real problem at hand. Furthermore, if machine learning is integrated into a BIM model in operations and maintenance, it can also determine the best way to carry out maintenance and repairs by visualising when and where problems will occur. Figure 24.10 shows surface cutting with Artificial Intelligence and Machine Learning.

24.4   CASE STUDY OF AR AND VR

Augmented and virtual reality improves coordination between planning and execution segments of construction. To validate this statement, a case study of the bungalow is taken. Virtual reality (VR) is one new technology that is changing the construction industry by solving old problems. Virtual reality in construction is the next level in 3D modelling. Like 3D modelling, it involves a detailed virtual model of the project. But, it places the user directly inside the virtual environment, so that the user experiences a full immersion into the virtual space. VR has also been integrated with other enabling technologies to further enhance the performance of construction education and training.

The case study is for a bungalow villa located in Pune. The bestowed case study project is officially undertaken for a Pune-based real-estate firm as a part of the validation of AR and VR use. The project is mainly focused on developing an alternative to the traditional method of buying-selling or marketing products in the real estate industry. The construction industry today is moving towards a new market and would profusely incorporate new modern-day technology to deliver new incentives. To save time and money at the same time is not everybody's cup of tea, but giving a try to an upcoming method of transitioning the whole idea of site-seeing, customising, buying or selling on the virtual reality platform by making use of augmented and virtual reality is like a dream come true. Nevertheless, this technology tends to deliver results with the new potential eventually saving the time of both parties involved in the business. As a plunge into this case study, the initial decision-making and abstract model-designing process were taken up as a challenge. The initial week of work was to get all the pointers and brainstorm ideas of how efficiently we as engineers can utilise the 3D modelling platform and virtual reality platforms to make something worthwhile and use it most productively. To advocate this project, a keen observation on how the real-estate companies operate, market and sell their finished products like villas, bungalows, flats or even township, is the most prior part of setting out the format of the research. It is fascinating that typically how a sales and marketing firm gets into the depth of field and displays the mock villas to the clients. This traditional approach, according to us, could be modernised. The challenge now was to bestow the same idea to the industry professionals to get to know their point of view.

So to begin with the model development phase, a 2D plan was required, which was readily available. The same 2D plan devised an imaginary plan that eventually took the shape of a 3D model later in the end. 3D modelling is a part of graphics that deals with creating 3D models. 3D modelling requires, like everything else, a certain amount of experience and gradual training. A 3D constraint modeller lies only in his skills and imagination. 3D modelling software allows you to easily create any view or section or later detail. It also allows you to use the "walkthrough" tool to create an animated visualization of a project that has the task of simplifying communication with the customer or just for the sake of a better idea of the virtual buildings. In this way, a structural model of the building is prepared. Figure 24.11 shows the structural model of the villa.

Now the model at its early stage looked monotonous and needed an interior upgrade, so the next task was to design and develop the interior for the same. Here, the BIM objects and REVIT-based 3D model is prepared. Figure 24.12 shows a fully developed 3D model of the villa.

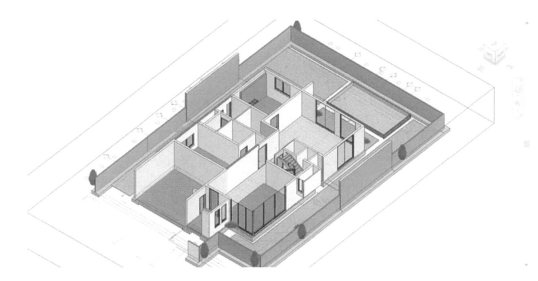

FIGURE 24.11   Structural model of villa.

FIGURE 24.12 Fully developed 3D model of the villa.

The walkthrough or rendering has conglomerate tools that can help us complete our desired task. Different software platforms demand their hardware as well as software requirements, and so does this software. When it comes to developing a walkthrough of a three-dimensional model, Lumion and Sketchup are the software that tends to deliver exceptional results. Due to hardware constraints, Revit was the only software that helped render and develop the walkthrough of the 3D model.

Once the walkthrough was ready, the same now was to be exported onto a platform that can create an environment where design alternatives are populated in a generative fashion. VR can assist the user to organize and navigate solution spaces in a 3D environment. This can be seen as an immersive

FIGURE 24.13 View of the entrance hall of the villa in virtual reality.

data visualization tool, in addition to the fact that each solution can be viewed and assessed independently. Generative design can also play a major role in mass-populating future virtual worlds in VR, such as a tool called IrisVR. As virtual environments hold fewer physical constraints than our real world, generative design techniques can assist designers and even end users to generate and customize their surrounding virtual environments in a fast, creative and emergent mode. Being a prominent developer tool, it helped transition the 3D model onto a virtual reality platform, which can be viewed using several VR gears. Oculus Pro VR headsets are found to be the most compatible pair for getting the immersive VR experience, but because of their unavailability, used the Play-Station 4 VR gear to achieve desired results. Interconnecting the 3D Revit model with augmented reality software (Kubity Go) and virtual reality software (Iris Vr) enabled us to achieve the desired result.

The complete process of getting to know what it takes to develop a 3D model, link it to a VR platform, search for the most compatible developer tool and finally obtain a result out of all this is a complete learning package at the back end. While on the front end this method of simulation and immersive experience can make the workflow of the client and firm unchallenging and quite effortless. Figure 24.13 shows view of the entrance hall of the villa in virtual reality.

The methodology followed for AR and VR is as follows.

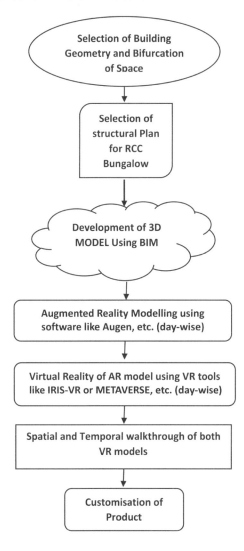

FIGURE 24.14 Framework used for spital and temporal walkthrough.

With the above methodology, step-by-step models were developed. The fully designed villa in Revit Architecture and IRIS-VR is as follows.

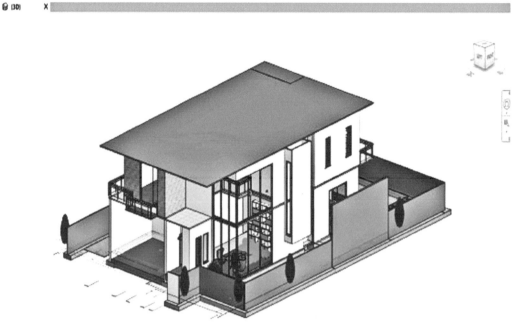

FIGURE 24.15   3D model of a villa in Revit Architecture.

FIGURE 24.16   Living room of a villa in virtual reality (IRIS-VR).

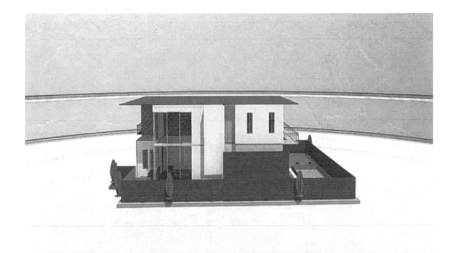

FIGURE 24.17   3D model in virtual reality (IRIS-VR).

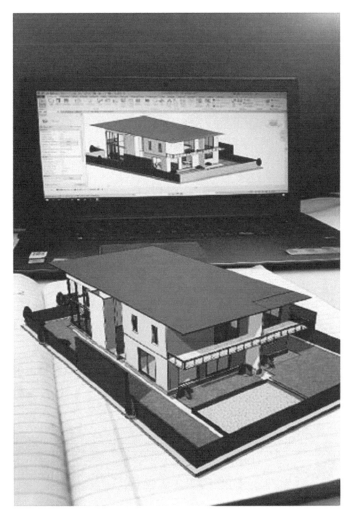

FIGURE 24.18   Augmented reality on table top.

FIGURE 24.19 Wall specification in IRIS-VR.

To validate the use of AR- and VR-led walkthroughs, a cost comparison for the sample bungalow is done as per the following table. From the table and techniques used, it is proved that customers will get maximum assurance of the quality of bungalows in AR- and VR-led framework as compared to conventional sample flat construction. Figure 24.20 shows sofa specification in IRIS-VR.

FIGURE 24.20 Sofa specification in IRIS-VR.

TABLE 24.1

Operational Cost Comparison of Traditional Method and VR

Resources (Traditional Method)	Cost (Traditional Method)	Resources (VR Technology)	Cost (VR Technology)
Vehicle cost	1000 INR	Oculus Quest VR headset	38,000 INR
Villa maintenance with security	1500 INR	–	–
Electricity	50 INR	–	–
Other utilities	500 INR	–	–
Total cost per day	3050 INR	–	–
Total cost per month	91,500 INR	Total cost (depends on quantity)	**38,000 INR**

Table 24.1 shows total cost per month requirement for operating a traditional sample bungalow and a modern VR technology–led approach. The cost required for a traditional method is Rs. 91,500 and for VR it is Rs. 38,000. The percentage difference in both is around 58.46%. So, in Indian context, approximately 58% cost is saved by using VR technology. The cost comparison omits the construction cost of the sample bungalow as well. Here, the benefits of AR- and VR-led frameworks are considered, which are maximum compared to a conventional sample bungalow construction. Thus, the use of AR- and VR-led technology validates their use over conventional methods (Dunston & Wang 2005). Although the AEC industry is far behind other industries such as healthcare and retail in adopting AR/VR technologies in the research literature, the results of this project showed that the AEC industry is changing its previous path towards utilising these technologies. This paper presents two case studies of a survey that were conducted at two different periods with about a year apart. The results were analysed to assess the current state, growth and savings opportunities for AR/VR technologies in the AEC industry. Furthermore, the results show a significant increase in AR/VR utilisation in the AEC industry over the past year and potential opportunities.

The case study considers demonstrated that:

1. Reduced on-site visits
2. Fixed problems during the pre-construction phase
3. Improved customer experience
4. Immersive walkthrough in VR
5. Real-time changes in VR
6. Cost comparison

24.5   FUTURE TRENDS

Researchers have focused on the implementation of AR and VR technologies concerning CPM aspects in recent years, where these technologies have proven to be promising for CPM advancements in many areas, including the management of costs, the management of defects, the management of safety, the management of education, the management of progress and the access to on-site information. The purpose of this study is to provide a comprehensive overview of the available literature about AR and VR technologies in CPM. This will help us understand the size, scope and nature of research activity in this area. Moreover, the objective of this study was to identify AR and VR's main application areas and characteristics in the CPM. This provides a comprehensive platform for researchers and practitioners to explore potential application areas, understand effective uses of these technologies and propose new applications.

Recent studies indicate that AR and VR systems are being developed for use in construction safety management. It has been demonstrated that these technologies are efficient for the purpose of safety planning, safety training and education, as well as operator training for construction equipment. In the advanced CPM domain, AR and VR implementations have great potential to enhance traditional methods for project scheduling and construction progress tracking. It has been proven that AR and VR can be effective tools for enhancing communication and data collection. The use of augmented reality has demonstrated the capability of improving on-site information retrieval, improving worker productivity and prompting field reports, while VR technology has been proposed as a useful tool for multidisciplinary communications during the early stages of a project. Among researchers, AR is highly valued for its efficiency in defects and quality management, as it differs from traditional methods by providing proactive prevention of many defects at a construction site and enabling inspectors and site managers to check errors remotely and efficiently.

It has been proven that AR and VR technologies can both be used to teach CPM concepts and aspects, with some of these technologies proving to be more efficient than traditional learning methods in teaching CPM concepts. It was possible to use VR and AR for visualization because of their advantages. Technology can be used in both design and constructability reviews to enhance decision-making in the review process and enhance communications among stakeholders. By combining VR and AR for visualization of building projects, throughout their various phases of life, significant waste can be avoided and value can be generated. By utilising it at each phase of the project and at the beginning, it can provide transparency and reliability in decision-making, avoiding future issues such as negative iterations, waste, delays, and rework (Dodevska & Mihić 2018). Depending on the context, each technique is used differently. VR is appropriate for desktop work such as the project definition phase and the early stages of the design phase. In fieldwork, however, AR can be useful since it only requires a target and a smartphone or tablet. As a result of the sudden increase in research into these cutting-edge technologies, the construction sector is poised to undergo a radical change in how visual management is conducted.

REFERENCES

Abdelhamid, W. A. (2012). "Virtual reality applications in project management scheduling." Computer-Aided Design and Applications, 9(1), pp. 71–78. doi: 10.3722/cadaps.2012.71-78.
Ahmed, S. (2019). "A Review on Using Opportunities of Augmented Reality and Virtual Reality in Construction Project Management." Organization, Technology and Management in Construction: An International Journal, 11(1), pp. 1839–1852. doi: 10.2478/otmcj-2018-0012.
Al-Adhami, M., L. Ma and S. Wu (2018). "Exploring virtual reality in construction, visualization and building performance analysis." ISARC 2018 - 35th International Symposium on Automation and Robotics in Construction and International AEC/FM Hackathon: The Future of Building Things, (ISARC). doi: 10.22260/isarc2018/0135.
Avhad, A. A. and G. A. Hinge (2017). "Implementation of Virtual Reality in Construction Industry." International Journal of Innovative Research in Science, 6(6), pp. 67–69. doi: 10.15680/IJIRSET.2017.0606276. http://www.ijirset.com/upload/2017/june/276_ashok%20avhad%20pgcon%20paper_IEEEE.pdf
Azuma, R., Y. Baillot, R. Behringer, S. Feiner, S. Julier and B. MacIntyre (2001). "Recent advances in augmented reality." IEEE Computer Graphics and Applications, 21(6): 34–47.
Bae, H., M. Golparvar-Fard and J. White (2012). "Enhanced HD4AR (Hybrid 4-Dimensional Augmented Reality) for Ubiquitous Context-aware AEC/FM Applications." 12th international conference on construction applications of virtual reality (CONVR 2012), National Taiwan University, Taipei, Taiwan.
Bae, H., M. Golparvar-Fard and J. White (2013). "High-precision vision-based mobile augmented reality system for context-aware architectural, engineering, construction and facility management (AEC/FM) applications." Visualization in Engineering, 1(1): 3.
Blinn, N. et al. (2015). "Using Augmented Reality to Enhance Management Educational Experiences Construction." Proc. of the 32nd CIB W78 Conference 2015, 27th-29th October 2015, Eindhoven, The Netherlands, pp. 69–78.
Brandon, P., H. Li and Q. Shen (2005). "Construction IT and the 'tipping point'." Automation in Construction, 14(3): 281–286.
Dodevska, Z. A. and M. M. Mihić (2018). "Augmented Reality and Virtual Reality Technologies in Project Management: What Can We Expect?" European Project Management Journal, 8(1), pp. 17–24. doi: 10.18485/epmj.2018.8.1.3

Dunston, P. S. and X. Wang (2005). "Mixed Reality-Based Visualization Interfaces for Architecture, Engineering, and Construction Industry." Journal of Construction Engineering and Management, 131(12): 1301–1309.

Hill, A., B. MacIntyre, M. Gandy, B. Davidson and H. Rouzati (2010). "KHARMA: An open KML/HTML architecture for mobile augmented reality applications." 2010 IEEE International Symposium on Mixed and Augmented Reality.

Irizarry, J., M. Gheisari, G. Williams and B. N. Walker (2013). "InfoSPOT: A mobile Augmented Reality method for accessing building information through a situation awareness approach." Automation in Construction, 33(Supplement C): 11–23

Zaher, M., Greenwood, D., & Marzouk, M. (2018). Mobile augmented reality applications for construction projects. *Construction Innovation*, 18(2), 152–166.

Zhang, W., Han, B., & Hui, P. (2017). "On the networking challenges of mobile augmented reality." In *Proceedings of the Workshop on Virtual Reality and Augmented Reality Network, August 2017* (pp. 24–29).

25 Smart Control of Flutter of Suspension Bridges Using Optimized Passive Tuned Mass Damper

*Hamed Alizadeh, H.H. Lavasani, Vahidreza Gharehbaghi,
Tony T.Y. Yang, and Ehsan Noroozinejad Farsangi*

CONTENTS

25.1 INTRODUCTION

In-plane stretched suspension bridges form renders them vulnerable to dynamic loads, especially as their length increases. The air flow around the bridge deck can produce fundamental challenges, leading to serviceability issues or total collapse. As a result, the aerodynamic behavior of bridges should be studied as thoroughly as possible. Vortex-shedding, buffeting, galloping, static divergence, and flutter are the most prominent aeroplane phenomena in suspension bridges.

The disastrous failure of the Tacoma Narrows bridge in 1940 due to high cable vibrations prompted numerous studies in the field of cable bridge aerodynamics. The main reason for this disaster was the one degree of freedom torsional flutter reported by Scanlan. During the stated sort of flutter, the amplitude of one of the torsional modes steadily develops until it diverges and leads to total collapse. The principal reason for amplitude divergence is negative aerodynamic damping. So, when the bridge is subjected to wind velocity, the negative aerodynamic damping grows, and at a certain threshold, it neutralizes the mechanical damping known as flutter velocity (Alizadeh et al., 2017). The flutter can be avoided from the bridge by attaching the instruments contributing to stability (Li et al., 2015).

Many studies have been conducted in the previous few decades on vibration control systems that include passive, active, semi-active, and hybrid components (Javadinasab Hormozabad and Gutierrez Soto, 2021, Jadhav et al., 2022). Passive ones, due to straightforward design and low cost of maintenance, were widely used in practice. Tuned mass damper (TMD) as a passive device includes three main parameters: mass ratio, tuning frequency, and damping ratio (Elias and Matsagar 2017). Larsena et al. (1995) suggested the aspects design of TMD for the Great-Belt suspension bridge. Pourzeynali and Datta (2002) investigated the effects of TMD's parameter on the flutter velocity during a parametric study. The results indicated that three TMDs with any arbitrary values led to an increase of flutter velocity. In another study, Pourzeynali and Esteki (2009) addressed the seismic response of suspension bridge controlled by TMDs which their parameters changed until the effect of each of them was clear. Casciati and Giuliano (2009) established a novel parameter called frequency range instead of tuning frequency to reduce the susceptibility of TMDs to mis-tuning in an MTMD system used to regulate the suspension bridge towers under gust loading. Alizadeh et al. (2018) evaluated the influence of TMD location and gyration radius on the Vincent Thomas suspension bridge's flutter velocity.

When the parameters of TMD devices are set to their optimum values, they can improve the overall response of the structure. Optimization is the process of obtaining a set of parameter values to satisfy the requirements of a problem under specified conditions (Chen et al., 2018). The gradient search method and the meta-heuristic algorithm can be employed in this case. The meta-heuristic algorithm searches a feasible and complex discontinues space for the best value of variables in the shortest period (Pourzeynali et al., 2007). Carpineto et al. (2010) attempted to optimize the TMD parameters of the footbridge suspension bridge under a live load. Miguel et al. (2016) improved the TMD parameters in the vehicle bridge. The results demonstrated that three tuned TMDs can control the vertical response of the bridge.

In this chapter, a truss-type suspension bridge's flutter analysis is first done in the timed domain. After that, the meta-heuristic optimization algorithm is utilized to optimize the TMD parameters in order to minimize the bridge's flutter. The Vincent Thomas suspension bridge and the sun and leaf optimization (SLO) algorithm are employed in the case study and optimization process. Additionally, based on the side and center spans, the mass ratio, gyration radius, tuning frequency, and damping ratio are calculated, and the optimal values resulting in the greatest flutter velocity are determined.

25.2   FLUTTER

Generally, increasing the wind velocity intensifies the structure's static and dynamic responses. In some situations, the responses can result in unstable behaviors that occur when a particular limit is exceeded. In general, cases of structural behavior near a stability limit can be classified according to the displacement reaction that emerges. There are four distinct forms of such behaviors in a bridge section. First, let's consider the probability of static divergence occurring in the torsional mode dubbed static divergence. Second, the occurrence of dynamic instability in the vertical direction perpendicular to the wind is referred to as galloping. Thirdly, the appearance of dynamic instability in pure torsional mode, and lastly, the occurrence of dynamic instability involving vertical and torsional motion, referred to as flutter. The last two items indicate that flutter instability can be classified into two distinct types: coupled flutter or classical flutter, which is prevalent in suspension bridges with airfoils and streamlined girders, and torsional flutter, which is more likely in traditional suspension bridges with truss-type girders (Larsen and Larose, 2015, Strommen, 2016).

Scanlan and Tomko noticed that numerous modes of the model control and constrain their amplitude except for one whose frequency corresponded to the frequency of the bridge's collapse during an experimental test. They demonstrated that the bridge fell as a result of torsional flutter in one degree of freedom. They subsequently added flutter derivatives and specified the aeroelastic forces of lift, drag, and torque employing them. According to Scanlan's formula, the self-excited linear lift and torque imposed on the bridge's deck are as follows (Ge and Xiang, 2008):

$$L_{se} = \frac{1}{2}\rho U^2 B \left(kH_1^* \frac{\dot{h}}{U} + kH_2^* \frac{B\dot{\theta}}{U} + k^2 H_3^* \theta + k^2 H_4^* \frac{h}{B} \right) \tag{25.1}$$

$$M_{se} = \frac{1}{2}\rho U^2 B^2 \left(kA_1^* \frac{\dot{h}}{U} + kA_2^* \frac{B\dot{\theta}}{U} + k^2 A_3^* \theta + k^2 A_4^* \frac{h}{B} \right) \tag{25.2}$$

in which, ρ, U, and B are the air density, wind velocity, and width of the deck, respectively. Also, H_i^* and A_i^* are the flutter derivatives. In addition, h and θ signify the vertical and torsional responses.

Their result indicated that the main reason of torsional instability was the abrupt increase of values of A_2^* derivatives.

Here, H_1^* represents the response of structure, while the torsional motion is perfectly constrained. H_2^* and H_3^* show the effects of torsional oscillation on the decrease of vertical response. A_1^* takes part in the flutter when a close coupling between vertical and torsional modes exists. Generally, the role of A_1^* in the flutter of the suspension bridge is negligible proved by many pieces of research and experimental tests. A_2^* is a sign of dynamic stability of bridge's deck while the vertical degree of freedom is constrained. This derivative is related to the aerodynamic stability of the structure and the possibility of the occurrence of torsional divergence. A_3^* derivative represents the difference of the dominant mode and the flutter frequency common to be 1 to 3% (Scanlan and Tomko, 1971).

While modern fluid mechanics tools can be used to evaluate flutter derivatives, the primary technique of evaluation is the wind tunnel test. For this purpose, four main combinational methods are used as follows: (i) time domain with free vibration's data, (ii) time domain with forced vibration's data, (iii) frequency domain with free vibration's data, and (iv) frequency domain with forced vibration's data. In this chapter, the used flutter derivatives are according to the experimental reports of Scanlan and Tomko (1971).

25.3 EQUATION OF MOTION

Several integral assumptions are taken into account during computation:

   i. Linear behavior is considered and any nonlinearity, whether due to material or geometric, are neglected.
  ii. The hangers are completely vertical and inexorable.
 iii. The main cables carry out the dead load, and the bridge's deck does not experience any stress.

The equation of motion using potential and kinetic energies and applying Hamilton principal can be obtained as follows:

$$[M]_{2n\times2n}\{\ddot{x}\}_{2n\times1}+[C]_{2n\times2n}\{\dot{x}\}_{2n\times1}+[K]_{2n\times2n}\{x\}_{2n\times1}=\{F\}_{2n\times1} \qquad (25.3)$$

where M, C, and K are the mass, damping, and stiffness matrices, respectively. Also, x and F are the response and force vectors. n denotes the number of degrees of freedom. Each structural property matrix is formed from two independent parts related to the vertical and torsional modes.

$$[M]=\begin{bmatrix} M_V & 0 \\ 0 & M_\theta \end{bmatrix}.[C]=\begin{bmatrix} C_V & 0 \\ 0 & C_\theta \end{bmatrix}.[K]=\begin{bmatrix} K_V & 0 \\ 0 & K_\theta \end{bmatrix} \qquad (25.4)$$

Also, F is defined as follows:

$$\{F\}=\begin{Bmatrix} L_{se} \\ M_{se} \end{Bmatrix} \qquad (25.5)$$

In the time domain, the equation can be solved by using the mode superposition approach. As a corollary, the modal displacement relationship should be expressed as follows:

$$\{x\}=[\Phi]_{m\times m}\{y\}_{m\times1} \qquad (25.6)$$

where Φ and y are the mode shape matrix and modal displacement, respectively. Also, sub-index m determine the number of modes which should be considered for computation. By replacing Equation (25.6) in Equation (25.3) and pre-multiplying in transposed mode shape matrix, the following relationships are obtained:

$$\tilde{M}\{\ddot{y}\}+\tilde{C}\{\dot{y}\}+\tilde{K}\{y\}=\tilde{F} \tag{25.7}$$

$$\tilde{M}=[\Phi]^{T}[M][\Phi]\cdot\tilde{C}=[\Phi]^{T}[C][\Phi]\cdot\tilde{K}=[\Phi]^{T}[K][\Phi]\cdot\tilde{F}=[\Phi]^{T}[F][\Phi] \tag{25.8}$$

Equation (25.7) by pre-multiplying in the inverse of modal mass is as follows:

$$\{\ddot{y}\}+\tilde{M}^{-1}\tilde{C}\{\dot{y}\}+\tilde{M}^{-1}\tilde{K}\{y\}=\tilde{M}^{-1}\tilde{F} \tag{25.9}$$

Flutter response of bridge can be considered sinusoidal harmonic. Therefore, the final response can be represented by the following relation:

$$\{y\}=\{\lambda e^{i\omega t}\} \tag{25.10}$$

Now, the final equation which can provide the flutter condition is written as follows:

$$\left(\left[\tilde{M}^{-1}\tilde{K}\right]-\omega^{2}\left[\tilde{M}^{-1}\tilde{K}_{ae}\right]\right)+i\left(\left[\tilde{M}^{-1}\tilde{C}\right]-\omega\left[\tilde{M}^{-1}\tilde{C}_{ae}\right]\right)\{a\}=0 \tag{25.11}$$

Equation (25.11) is a well-known eigenvalues problem, the solution of which should result in the zero determinant of the complex matrix's real and imaginary components. A trial-and-error approach is required to arrive at the correct solution. For this aim, the flutter derivatives should be assigned different values until a unique ω be appears. Finally, the flutter velocity can be computed by the following relation:

$$v_{f}=\frac{B\omega_{f}}{k_{f}} \tag{25.12}$$

25.4 STRUCTURAL PROPERTIES MATRICES

As previously stated, structural property matrices are required to examine the response of a bridge to any dynamic loading. They can be computed using the finite element approach in this case. To accomplish this, the bridge should be separated into the finite elements.

Each element compresses the girder, primary cables, and a minimum of two hangers. Four nodes may appear at the ends of the main cables and girder, but because the vertical displacement of the nodes of the main cables and girder will be identical due to inextensible hangers, only two nodes are considered placed at the ends of the girder's centerline, as illustrated in Figure 25.1.

V and B are the vertical displacement and bending rotation, while W and T represent the warping and torsional rotation, respectively. More details are explained by Abdel-Ghaffar (1979 and 1980). The stiffness matrix of a suspended structure is composed of two components that correspond to the primary cables and girder parameters. Cables are composed of two distinct components: gravity and elastic stiffness matrices:

$$\begin{bmatrix}k_{cg}\end{bmatrix}=\frac{-H_{w}}{30L}\begin{bmatrix} 36 & -3L & -36 & -3L \\ -3L & 4L^{2} & 3L & -L^{2} \\ -36 & 3L & 36 & 3L \\ -3L & -L^{2} & 3L & 4L^{2} \end{bmatrix} \tag{25.13}$$

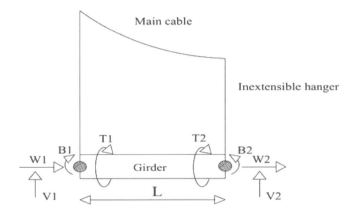

FIGURE 25.1 Arrangement of degree of freedom.

$$[k_{ce}]_e = \frac{E_c A_c}{L_E} * \left(\sum_{i=1}^{3} \frac{W_d}{H_w} \{\hat{f}\}_{N_i} \right) * \left(\sum_{i=1}^{3} \frac{W_d}{H_w} \{\hat{f}\}^T_{N_i} \right) \cdot \{\hat{f}\}^T = \left[\frac{L}{2} \cdot \frac{-L^2}{12} \cdot \frac{L}{2} \cdot \frac{L^2}{12} \right] \quad (25.14)$$

Equations (25.13) and (25.14) are common in the vertical and torsional modes. In the vertical mode, the girder contains the main part related to its elastic stiffness:

$$[k_{ge}]_e = \frac{E_g I_g}{L^3} \begin{bmatrix} 12 & -6L & -12 & -6L \\ -6L & 4L^2 & 6L & 2L^2 \\ -12 & 6L & 12 & 6L \\ -6L & 2L^2 & 6L & 4L^2 \end{bmatrix} \quad (25.15)$$

In the torsional mode, the girder has two stiffness components:

$$[k_{sd}] = \frac{2GJ}{15b^2} \begin{bmatrix} 36 & -3L & -36 & -3L \\ -3L & 4L^2 & 3L & -L^2 \\ -36 & 3L & 36 & 3L \\ -3L & -L^2 & 3L & 4L^2 \end{bmatrix} \cdot [k_{sc}]_e = \frac{4E_g\Gamma}{b^2 L^3} \begin{bmatrix} 12 & -6L & -12 & -6L \\ -6L & 4L^2 & 6L & 2L^2 \\ -12 & 6L & 12 & 6L \\ -6L & 2L^2 & 6L & 4L^2 \end{bmatrix} \quad (25.16)$$

Also, mass matrices of the vertical (m_t) and torsional (I_θ) modes are as follows:

$$[m_t]_e = \frac{m_t L}{420} \begin{bmatrix} 156 & -22L & 54 & 13L \\ -22L & 4L^2 & -13L & -3L^2 \\ 54 & -13L & 156 & 22L \\ 13L & -3L^2 & 22L & 4L^2 \end{bmatrix} \quad (25.17)$$

$$[I_\theta]_e = \frac{I_m L}{105b^2} \begin{bmatrix} 156 & -22L & 54 & 13L \\ -22L & 4L^2 & -13L & -3L^2 \\ 54 & -13L & 156 & 22L \\ 13L & -3L^2 & 22L & 4L^2 \end{bmatrix} \quad (25.18)$$

TABLE 25.1

Parameters of TMD

Degree of freedom	Mass	Damping	Stiffness
Vertical	$m_h = m_r m_s$	$c_h = 2m_h \xi_h \omega_h$	$k_h = m_h \omega_h^2$
Torsional	$I_\theta = m_h r^2$	$c_\theta = 2I_\theta \xi_\theta \omega_\theta$	$k_\theta = I_\theta \omega_\theta^2$

25.5   TMD DEVICE

TMD is a passive control system tuned to a specific structure mode to dissipate dynamic energy (Debbarma and Das 2016). In flutter with one degree of freedom, one of the torsional modes dominates the final response. As a result, inserting TMDs with a tuning frequency close to the dominant mode can help decrease vibration. The TMD can be mounted to the bridge deck and distributed along the bridge's cross section. The TMD's primary properties are calculated using the formulas in Table 25.1.

m_r, m_s, and r are the mass ratios, the mass of the structure and gyration radius. Also, ξ and ω are the damping ratio and tuning frequency. Furthermore, sub-index h and θ represent the vertical and torsional degrees of freedom. The configuration of structural properties matrices of structure with the presence of TMD will have a few changes. The component of stiffness matrix that the placement of TMD corresponds to its row (i) and column (j) alters as follows:

$$k_{ij}^{new} = k_{ij}^{old} + k_T \cdot k_{i(n+1)} = -k_T \cdot k_{(n+1)i} = -k_T \cdot k_{(n+1)(n+1)} = k_T \qquad (25.19)$$

Damping matrix has similar changes. Also, for mass matrix,

$$m_{ij}^{new} = m_{ij}^{old} + m_T \cdot m_{(n+1)(n+1)} = m_T \qquad (25.20)$$

where m and k show the composition of the mass and stiffness matrices. The equation can be readily extended for more TMDs.

25.6   SLO ALGORITHM

The design of search algorithms for locating global optima is classified into two broad categories: classical mathematical algorithms and meta-heuristic algorithms. Meta-heuristic algorithms are non-gradient-based approaches frequently used to solve mathematical models inspired by physical or social laws or natural phenomena. (Mortazavi. et al., 2018). They begin with an initial population of solutions and progress incrementally toward the optimal solution with each generation.

SLO is a meta-heuristic algorithm inspired by the shining of sunlight and the growth of leaves to find optimum solutions (Hosseini and Kaedi, 2018). The existence of green leaves in some areas of trees is not fortuitous and is due to some appropriate condition that one of them and maybe the most important one is the sunlight. To locate the greenest leaves, it makes sense to hunt for greener areas. Additionally, wind can alter the direction of leaf growth, causing them to be denser on the opposite side of the sun or scattered on the side that receives the most sunlight. These are two techniques for searching within a population without querying at local minima.

A tree represents all possible solutions to a problem, while the leaves indicate the solutions. The germination and growth of leaves serve as a model for the optimization and search processes. As a result, areas with more green leaves indicate a more desirable solution, including more acceptable alternatives. Additionally, places with low-quality leaves harbor lower-quality solutions. Thus, depending on the quality of the leaves, different portions of the tree may be spread or concentrated. Finally, to account for other natural impacts on leaf density, random wind effects will be created. More details are expressed by (Hosseini and Kaedi, 2018).

25.7   NUMERICAL ANALYSIS

The Vincent Thomas suspension bridge in Los Angeles has been chosen as a case study. The length of the main span is 457.5 m while the length of symmetric side spans is 155.5 m. the bridge has a truss-type stiffening suspended structure two hinged and its width is 18 m. the supplementary structural and geometrical properties are provided by the following data: (i) $m_t = 5354$ kg/m , $I_m = 453,010$ kg.m; (ii) $H_w = 300,37.5$ kN; (iii) $A_c = 781.28$ cm^2, $L_E = 1055.3$ m, $J = 4243.3$ cm^2.m^2; (iv) $E_c = 200,028$ MN/m^2 , $E_g = 186,232$ MN/m^2 , $E_c = 80,011$ MN/m^2 , $\Gamma_{1.3} = 45.45$ m^6, $\Gamma_2 = 44$ m^6. Note that sub-index 1.3 denote side spans while 2 signifies main span. In this study, the primary and side spans are divided into 28 and 11 finite elements, respectively. The modal data and shapes of the bridge are provided in Figure 25.2 and Table 25.2.

According to former parts, for the considered bridge, the reduced frequency and flutter frequency are 1 and 2.94 rad/s providing the 53 m/s flutter velocity. Modal participation of each mode in five different conditions containing 6, 8, 9, 10, and 12 initial modes in the flutter are summarized in Table 25.3.

As shown in the table, the initial torsional mode is crucial during the flutter situation, which many torsional and vertical modes can accompany. Figure 25.3 illustrates the bridge's response to flutter velocity, both below and above it.

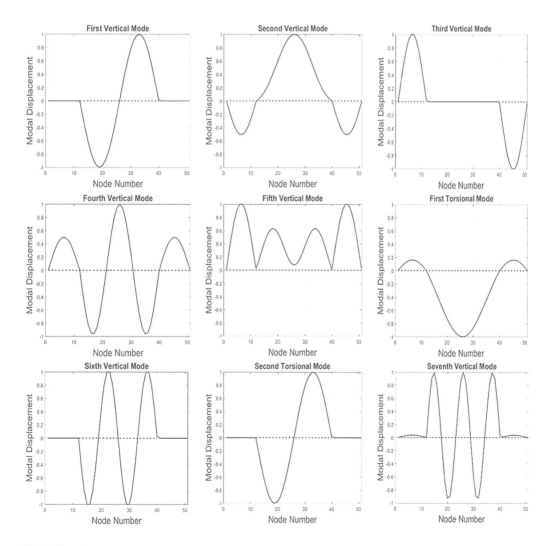

FIGURE 25.2   Mode shapes of bridge.

TABLE 25.2
Modal Properties of the Vincent Thomas Bridge

Mode No.	Period (s)	Frequency (rad/s)	Mode Type
1	5.07	1.24	V-AS
2	4.55	1.38	V-S
3	2.93	2.14	V-AS
4	2.87	2.19	V-S
5	2.17	2.90	V-S
6	2.08	3.02	T-S
7	1.82	3.46	V-AS
8	1.54	4.08	T-AS
9	1.24	5.06	V-S
10	1.01	6.21	T-AS
11	1	6.27	T-S
12	0.95	6.64	T-S

When the wind velocity is 53 m/s, the bridge due to 0.01-radian initial condition vibrate with constant amplitude is equal to the initial condition. However, when the wind velocity surpasses the specified value marginally, the amplitude of the vibration steadily diverges, resulting in the bridge collapsing.

25.8   OPTIMIZED TMD

The cost function in this study is defined as the absolute value of the difference between the eigenvalues of the real and imaginary parts:

$$f = |\omega_{re} - \omega_{im}| \qquad (25.21)$$

TABLE 25.3
Modal Participation of Bridge's Modes in the Flutter

$|\lambda_i|$ Mode No.	Case 1	2	3	4	5
1	2×10^{-8}	2×10^{-8}	3×10^{-8}	3×10^{-8}	2×10^{-8}
2	6×10^{-2}	7×10^{-2}	6×10^{-2}	6×10^{-2}	6×10^{-2}
3	3×10^{-17}	3×10^{-17}	4×10^{-17}	4×10^{-17}	3×10^{-17}
4	9×10^{-5}	8×10^{-5}	9×10^{-5}	8×10^{-5}	8×10^{-5}
5	2×10^{-2}	3×10^{-2}	2×10^{-2}	1×10^{-2}	3×10^{-2}
6	1	1	1	1	1
7	–	5×10^{-9}	5×10^{-9}	6×10^{-9}	6×10^{-9}
8	–	9×10^{-6}	9×10^{-6}	9×10^{-6}	9×10^{-6}
9	–	–	9×10^{-5}	8×10^{-5}	8×10^{-5}
10	–	–	–	3×10^{-9}	4×10^{-9}
11	–	–	–	–	5×10^{-10}
12	–	–	–	–	6×10^{-12}

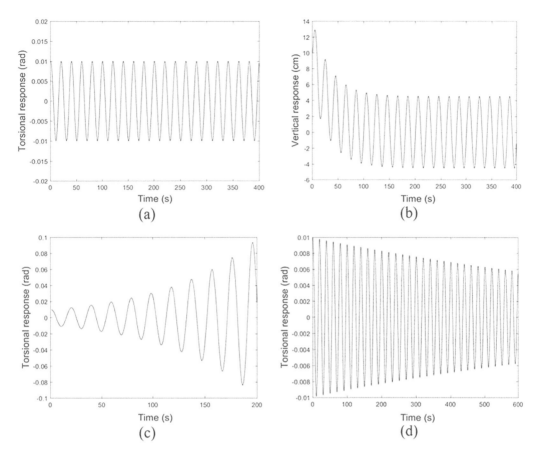

FIGURE 25.3 Response of (a) first torsional and (b) second vertical modes at the 53 m/s; response of first torsional mode at the (c) 55 m/s and (d) 48 m/s.

TMD devices should be developed following specific parameters that safeguard the bridge against flutter. For the present bridge, the TMD system has been optimized to take into consideration the qualities of the side and central spans. Table 25.4 summarizes the variables that were chosen and their optimum values.

The mass ratio, gyration radius, damping ratio, and frequency ratio are designed individually for side and center spans. In contrast to the typical technique, a novel metric termed the *frequency ratio* is utilized in place of computing the tuning frequency. Because the predominant mode has

TABLE 25.4

Optimized Values of TMD Devices

Parameter	Side Span	Center Span
Mass ratio	0.07	0.05
Gyration radius (m)	1.2	1.17
Vertical frequency ratio	1.1	0.98
Torsional frequency ratio	0.99	0.95
Damping ratio	0.12	0.21
Host node	6, 47	27

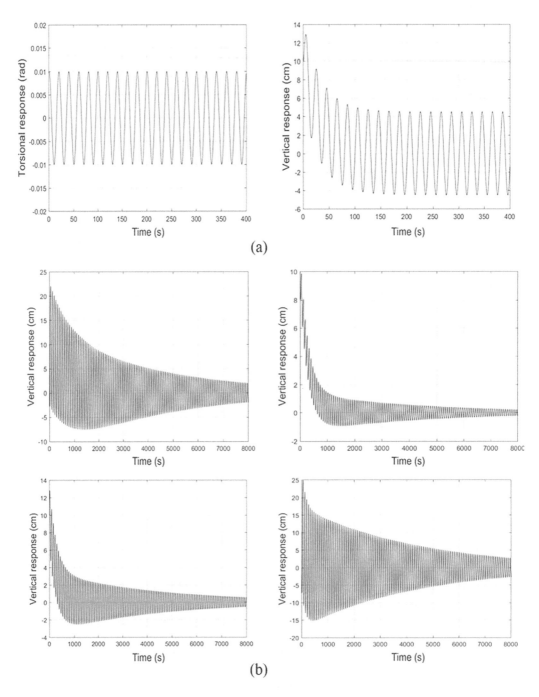

FIGURE 25.4 Response of bridge (a) in flutter and (b) under flutter velocity.

already been found, finding the optimal tuning frequency is unnecessary; however, the frequency ratio of the prominent mode can be tuned to get the greatest reduction in the ultimate response. Additionally, the geometry of the first mode indicates that the center and side spans will suffer the most significant responses at their midpoints. As a result, it appears appropriate to place one TMD at the aforementioned location.

Finally, three TMDs are implanted at the computed host nodes with the optimum values determined during analysis. The flutter frequency and reduced frequency of the bridge-TMD system are 3.6 rad/s and 0.77, respectively, providing 84 m/s flutter velocity indicating 58% increase in compare to uncontrolled condition. Figure 25.4 shows the controlled response of the bridge.

As Figure 25.4(a) illustrates, the torsional and vertical responses have the same magnitude as the vibration. In Figure 25.4(b), the ultimate response of the first four vertical degrees of freedom is decreased independently throughout the analysis by decreasing the wind velocity.

25.9 CONCLUSION

In this study, the flutter analysis and control of a conventional truss type suspension bridge were performed. Case studies on the Vincent Thomas suspension bridge and passive TMD were chosen. The TMD parameters were tuned using the SLO algorithm to maximize the increase in flutter velocity.

When a suspension bridge with a short span truss structure vibrates, the first torsional mode dominates. The TMD's characteristics should be examined in connection with its properties.

The longer center span length compared to the side spans requires lighter mass and lower stiffness values while maintaining a higher damping ratio.

In terms of the maximal response occurring in the main span, the 0.05 mass of each span is the most optimal value. The optimal value for all spans is 0.065 r/B.

REFERENCES

A.M. Abdel-Ghaffar, Free Torsional vibrations of suspension bridges, Journal of Structural Division, 106(10) (1979) 2053–2075.

A.M. Abdel-Ghaffar, Vertical vibration analysis of suspension bridges, Journal of Structural Division, 106(10) (1980) 2053–2075.

H. Alizadeh, H.H. Lavasani, S. Pourzeynali, Flutter instability control in suspension bridge by TMD, Proceedings of the 11th International Congress on Civil Engineering, Tehran, 2018.

N. Carpineto, W. Lacarbonara, F. Vestroni, Mitigation of pedestrian-inducted vibrations in suspension footbridges via multiple tuned mass damper, Journal of Vibration and Control, 16(5) (2010) 749–776.

F. Casciati, F. Giuliano, Performance of multi-TMD in the towers of suspension bridges, Journal of Vibration and Control, 15(6) (2009) 821–847.

J. Chen, H. Cai, W. Wang, A new metaheuristic algorithm: car tracking optimization algorithm, Soft Computing, 22(12) (2018) 38–57.

R. Debbarma, D. Das, Vibration control of building using multiple tuned mass dampers considering real earthquake time history, International journal of civil and environmental engineering, 10(6) (2016) 694–704.

S. Elias, V. Matsagar, Research developments in vibration control of structures using passive tuned mass dampers, Annual Reviews in Control, 44 (2017) 1–28.

Y.J. Ge, H.F. Xiang, Computational models and methods for aerodynamic flutter of long-span bridges, Journal of Wind Engineering and Industrial Aerodynamics, 96 (2008) 1912–1924.

F. Hosseini, M. Kaedi, A metaheuristic optimization algorithm inspired by the effect of sunlight on the leaf germination, International Journal of Applied Metaheuristic Computing, 9(1) (2018) 40–48.

R. Jadhav, P. Digambar, and S. Prachi. Seismic response elimination of structure by using passive devices: An overview. Recent Advances in Earthquake Engineering, 2022. DOI: 10.1007/978-981-16-4617-1_25

S. Javadinasab Hormozabad, and M. Gutierrez Soto. Optimal replicator dynamic controller via load balancing and neural dynamics for semi-active vibration control of isolated highway bridge structures. Sensors and Instrumentation, Aircraft/Aerospace, Energy Harvesting & Dynamic Environments Testing, Volume 7. Springer, Cham, 2021.

A. Larsena, E. Svensson, H. Andersen, Design aspects of tuned mass dampers for the great belt east bridge approach spans, Wind Engineering and Industrial Aerodynamics, 54–55 (1995) 413–426.

A. Larsen, G.L. Larose, Dynamic wind effects on suspension and cable-stayed bridges, Journal of Sound and Vibration, 334 (2015) 2–28.

K. Li, Y.J. Ge, Z.W. Gue, L. Zhao, Theoretical framework of feedback aerodynamic control of flutter oscillation for long-span suspension bridges by the twin-winglet system, Journal of Wind Engineering and Industrial Aerodynamics, 145 (2015) 166–177.

L.F.F. Miguel, R.H. Lopez, A.J. Torii, L.F.F. Miguel, A.T. Beck, Robust design optimization of TMDs in vehicle–bridge coupled vibration problems, Engineering Structures, 126 (2016) 703–711.

A. Mortazavi, V. Togan, A. Nuhoglu, Interactive search algorithm: A new hybrid metaheuristic optimization algorithm, Engineering Applications of Artificial Intelligence, 71 (2018) 275–292.

S. Pourzeynali, T.K. Datta, Control of flutter of suspension bridge deck using TMD, Wind and Structure, 5(5) (2002) 407–422.

S. Pourzeynali, H.H. Lavasani, A.H. Modarayi, Active control of high rise building structures using fuzzy logic and genetic algorithms, Engineering Structures, 29(3) (2007) 346–357.

S. Pourzeynali, S. Esteki, Optimization of the TMD parameters to suppress the vertical vibration of the suspension bridges subjected to earthquake excitation, IJE Transactions B: Applications, 22(1) (2009) 23–34.

R.H. Scanlan, J.J. Tomko, Airfoil and bridge deck flutter derivatives, Journal of the Engineering Mechanics, 97 (1971) 1717–1737.

E.N. Strommen, Theory of Bridge Aerodynamics, 2nd Edition, Springer, Netherland, 2016.

26 A Chronological Review of Construction Progress Monitoring Using Various Sensors and Remote Sensing Techniques

Saurabh Gupta

CONTENTS

26.1   INTRODUCTION

The building market is expanding because of demographic shifts, increased urbanisation and general economic prosperity. Global construction volume increased to 55% in 2020, up from 46% over the past decade (Roumeliotis, 2010). By 2030, experts anticipate a global return of 85%, or $15.5 trillion (Global Construction Perspectives and Oxford Economics, 2015). As an outcome of this expansion, manual management in construction projects has been supplanted by automation and robotics such as computer vision, the Internet of Things (IoT), machine learning, artificial intelligence, neural networks and more. These technologies are used to monitor the status of projects, oversee construction site activities, conduct safety inspections and manage databases. With the continued commoditisation of technologies like sensor networks and the IoT, the construction industry can only expect to deal with even greater volumes of diverse data. Unlike the industrial sector, the construction industry relies on individualised approaches to each project, indicating holistic construction monitoring.

A construction project is a unique undertaking in which sets of tasks are performed in a logical sequence to achieve predefined goals (Gupta et al., 2022). A construction project may be split up into four phases, as shown in Figure 26.1; the first phase, project initiation, involves the project's definition, tendering, contract and design phases (Gupta & Nair, 2021). Followed by a detailed design process, the second phase in the construction project is project planning which involves team formation, breaking the project into manageable chunks known as work breakdown structure (WBS), and breaking WBS into even smaller chunks of the activities. These WBS and activities are

DOI: 10.1201/9781003325246-26

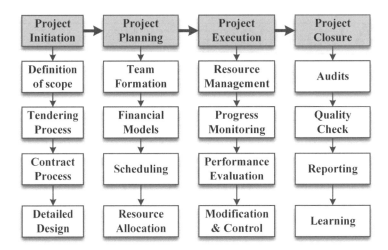

FIGURE 26.1 Life cycle of a construction project.

then arranged into a logical sequence, called scheduling and resource allocation. The third phase is the most important, starting with the project's execution. This phase includes progress and performance monitoring. The last and final phase is project closure involving audits and reporting as a major activity. This chapter emphasiseses the progress monitoring of the activities required during project execution time.

Construction progress monitoring is generally acknowledged as an important factor in the overall success of building projects. Monitoring the progress of construction allows for the prompt implementation of corrective measures and other activities, bringing the result as close as possible to initially planned milestones. Construction monitoring costs approximately 1% of the total budget, according to the Construction Industry Institute (Jaselskis & El-Misalami, 2017). Current methods of collecting this information through direct observation require a great deal of time investment from the site manager. On-site work, such as data gathering and analysis, accounts for 30–49% of a site manager's time (Son & Kim, 2010).

Furthermore, manual techniques of measuring progress are often erroneous. The present practice in construction progress monitoring is tedious and has many challenges. It is time- and resource-consuming. Progress monitoring is often not periodical, which may cause inaccurate project decisions. Project management is currently limited to manually collecting data using construction drawings and monitoring schedules by manually measuring work progress (Navon, 2007). The site engineer's data collected from the construction sites is submitted to the project manager (PM) in the form of drawings, spreadsheets, bar charts, etc. or the last decades' photographs and videos and manual data. These methods are discrete and time-consuming and may or may not cover the required level of detail (LOD). The project management team must sort, prioritise and interpret data (Song et al., 2005).

26.2 KEY ASPECTS OF CONSTRUCTION PROGRESS MONITORING

The project manager has a challenging task in distributing the detailed design of construction, schedules and the progress of construction activities. A project manager must sometimes communicate every detail with clients, engineers, site engineers, or even crews. This communication is periodical and requires whenever alternation in design and schedule needed. The report must be a comprehensive and easy deliverable that ensures proper understanding, which leads to a quick and correct decision. Improper collection or misinterpretation of field data leads to poor judgment and will result in time and cost overruns for the project.

According to Barrie and Paulson (1992), an effective construction project management system must include efficient measurements, data collection and quantification of construction progress concerning planned project milestones. After collecting data from construction sites, the project staff needs to report information precisely. This database helps in identifying the progress of the project and assess the critical issue. The next step is reporting the data to the PM within the given time. The PM interpret the data and take corrective actions if required. The corrective actions and decision-making process entirely rely on the facts and information collected from the site. Thus, the activities and progress data collection from the active construction site is one of the most crucial pursuits for successful project management.

Poor planning and progress management are both potential causes of schedule overruns. If a progress deviation is not detected until it has already caused significant damage to the Aproject's timeline, corrective measures can no longer be implemented in time to prevent the delay. Another critical point in construction progress monitoring, is the low quality of the progress data obtained manually from the construction sites. Data usually collected by the supervisor, site engineers and other staff is subjective based on status visualised on the construction site and manual measurement, which may not be accurate, and therefore impacts project progress time (Navon & Sacks, 2007). Since the data collected is not reliable, it affects the interpretation and decision for the project.

26.3 CONVENTIONAL METHODS IN CONSTRUCTION PROGRESS MONITORING

Conventional methods of measuring construction progress are disorganised. Accurate measurements are difficult in big projects, and gathering that information is also difficult (Meredith et al., 2017); for example, a subcontractor reports 50% completion of work. This can be interpreted in many ways, like 50% of volume of work finished or 50% of the planned inventory has been utilised or 50% of man hours spent. It may be given correct interpretation for a small projects but may or may not be correct for big projects. Misinterpretation of data by different role players in construction could be very misleading. One of the most common approaches to displaying a project is in percentile, which is based on the experience of the project manager. This representation is inefficient in delivering actual physical progress to different project players (Song et al., 2005). One other way of monitoring progress is based on budget; the percentage of the budget spent also gives an idea about the progress of the project. This method is also sometimes misleading and based on judgement by field managers (Shih & Wang, 2004). With the improper comparison of project progress with the actual plan, overpaying or delay in expected time may be observed.

The conventional method involves progress tracking with the direct and indirect approaches. The indirect approach is to check with the remaining stores and inventories to calculate the quantities used. The used quantities give the estimates of work completed and indicate the project progress status. For example, in a construction project, if x quantity of concrete is required and 40% of x is consumed, thus, one can have a rough estimate of the progress. This approach does not always reflect the actual progress and status of the project. On the other hand, in the direct method, manual measurements are made and maintained in the progress register, which is later used to check the progress of the project using different techniques such as bar chart, Gantt chart, network diagrams and line of balance method. The approach is presented in Figure 26.2; with effective progress monitoring, the achieved milestone can be monitored, and any deviation from the planned schedule can be assessed and corrected with the addition of extra resources and time.

Visualisation of project monitoring data is complex. Kerzner (2010) presented a different types of data representation practised in the construction industry. Representing data graphically helps show, analyse and communicate between different role players in the construction project. This method involves drawings, graphs, charts etc. The choice of method depends on the LOD required.

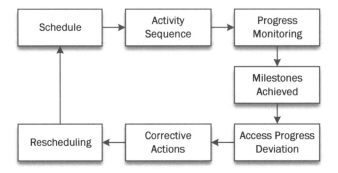

FIGURE 26.2 General approach for construction progress monitoring.

Before the last two decades, the prevailing reporting method was ineffective. The multi-dimensional information was not presented on a single platform to show the performance of the ongoing projects holistically, such as activities schedule and project cost. The previous techniques could not provide visualisation for as-planned activities vs. as-built activities in construction projects, along with the complexities involved. As a result, it impacts the ability to provide significant progress control, which it is a must for successful project management. Presently, visualisation techniques are widely accepted and practised around the globe for construction progress monitoring. These techniques take leverage of building information modelling (BIM) (Omar et al., 2018) (Reja et al., 2022) (Golparvar-Fard et al., 2010), unmanned aerial vehicles (UAVs) (Braun et al., 2018) (Srivastava et al., 2020), good quality cameras and various other remote sensing technologies (Tezel & Aziz, 2017). Different combinations of sensors and techniques used are explained in the next section.

26.4   A SYSTEMATIC REVIEW OF STATE-OF-THE-ART SOLUTIONS

To gain an understanding of the domain under consideration, this research uses a scientific mapping strategy to sift through the massive body of scholarly literature in the field. By establishing links between concepts in the realm of writing that have been missed in manual surveys, the scientific mapping method paves the way for researchers to steer orderly disclosures in this area. Using vast bibliometric data sources as raw material, the method constructs a comprehensive image of the study topic, shown with perceptive visuals. See Figure 26.3 for an overview of the research process, which consists of a scientometric analysis of the literature search and qualitative deliberation.

Google Scholar, an academic database covering most of the relevant peer-reviewed literature, met the criteria for this survey because it contains reports from a wide variety of academic journals (e.g., type of document, keywords and language of the articles). Literature consists solely of peer-reviewed, English-language journal articles due to journals' status as definitive and authoritative sources in their respective fields (Mahami et al., 2019). In addition to the journal articles, book chapters and editorials, we consulted proceedings from conferences because they include novel methods and continuing lines of inquiry.

A study of author co-authorship is performed to determine which scientists are the most productive and with whom they collaborate. Researchers (nodes) and the collaboration between them (links) are represented as nodes and edges in the co-author co-occurrence network graph, as shown in Figure 26.4. When plotted, a node's size is proportional to the author's total number of publications, while a link's thickness shows the strength of the relationship between the authors. The total number of links for a certain author shows the breadth and depth of their scholarly collaborations, while the colour of the links indicates the research group to which that author belongs.

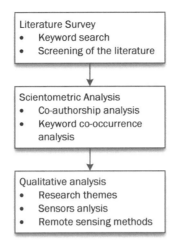

FIGURE 26.3 General approach for construction progress monitoring.

The literature review focuses on keywords because they show the most important issues and current directions in the field. In the keyword co-occurrence network graph, the gap between the nodes' size represents how strongly the individual study topics or subjects are related to one another. When a large distance separates two keywords, their connections are less strong (research topics), as shown in Figure 26.5. The extent to which a node in the resulting network represents a keyword that has been the subject of study in the appropriate field is indicated by the node's size. The number of links between terms is a proxy for the frequency with which they have been used together, and the varying hues of clusters stand for distinct sub-fields of study within the larger research domain.

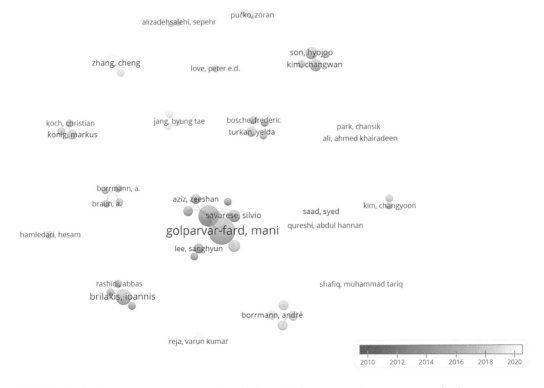

FIGURE 26.4 Co-occurrence of authors in articles related to construction progress monitoring.

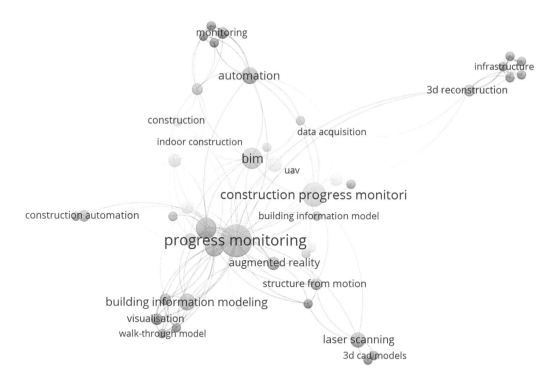

FIGURE 26.5 Co-occurrence of keywords in articles related to construction progress monitoring.

26.4.1 SENSORS USED IN CONSTRUCTION PROGRESS MONITORING

As illustrated in Table 26.1, several sensors are utilised for image data collection, such as cameras (single/stereo/arrays), smart devices, UAVs with cameras, LIDAR scanners, depth-IR cameras, satellite images and so on. With the evolution of IoT-based solutions and the development of sensors, many sensors and combinations are used for different purposes at construction sites. In recent decades, most of the research has used camera-based solutions for monitoring interior construction, building construction, cable-stayed bridge construction, scaffolding construction and erection of steel or concrete structures, as shown in Table 26.1.

The use of radio frequency identification (RFID) is very popular in steel construction and prefabricated construction sites and managing different precast members RFID is widely used (Azimi et al., 2011; Giretti et al., 2017; Soman & Whyte, 2017). During inspections, RFID technology has been utilised to acquire active site data and incorporate it into a building information model (BIM) (Ergen et al., 2007; Song et al., 2006; X. Wang et al., 2014; Changyoon Kim et al., 2017). The inspector can instantly get the data they need by simply scanning the tag with a mobile device, such as a smartphone or tablet computer. Even though this method makes it easier to collect valuable data and is compatible with commercially available BIM-based inspection tools, it still necessitates the setup and upkeep of RFID tags. As RFID requires it to be physically passed to every member, it is quite a costly process; the other approach is to manage construction progress CCTV images and video used as for managing bridge construction (Kim et al., 2017) and cannot be continuously implemented in the constantly shifting construction industry.

As-built data collection by laser scanning is another common automated progress monitoring technique. Information gleaned from a LIDAR is represented as a point cloud consisting of the geometric location of each point (X, Y and Z values). Despite their excellent level of precision and accuracy, these LIDAR scanners are not widely preferred due to their high initial cost, ongoing maintenance costs and the necessity for specially trained operators. Several drawbacks include a

TABLE 26.1

Different Sensor Combinations for Various Progress Monitoring Applications.

	Sensor Combinations	Progress Monitoring Applications	Authors
1.	RFID	Steel construction	(Azimi et al., 2011), (Giretti et al., 2017)
2.	RGB Camera	Interior construction Building construction Cable-stayed bridge construction Construction sites Scaffolding construction	(Kropp et al., 2018) (S. Lee & Pena-Mora, 2006) (Fard & Peña-Mora, 2007) (Mani Golparvar-Fard et al., 2009) (Mani Golparvar-Fard et al., 2009) (Golparvar-Fard et al., 2010) (Lukins & Trucco, 2007) (Peña-Mora et al., 2011) (Mani Golparvar-Fard et al., 2011) (Roh et al., 2011), (Golparvar-Fard et al., 2012)\ (Changwan Kim et al., 2013) (Kropp et al., 2013) (Changyoon Kim et al., 2013), (Dimitrov & Golparvar-Fard, 2014) (Kevin K. Han & Golparvar-Fard, 2014) (Tuttas et al., 2014) (Braun et al., 2015) (Kevin K. Han & Golparvar-Fard, 2015) (Kevin K. Han et al., 2015) (Braun et al., 2017), (Chi et al., 2017) (Ratajczak et al., 2019)
3.	CCTV	Construction site. Bridge Cons.	(Liu et al., 2018), (Kim et al., 2017)
6.	RGBD Camera (Infrared Camera)	Steel structure Construction site Building elements	(Choi et al., 2008), (Son & Kim, 2010), (Rebolj et al., 2017)' (Pučko et al., 2018) (Lei et al., 2019), (Pour Rahimian et al., 2020), (Pazhoohesh & Zhang, 2015)
7.	Laser Scanners	Erection of structure Steel structure Construction site	(Turkan et al., 2010), (Bosché, 2010) (Yelda et al., 2011), (Turkan et al., 2011) (Zhang & Arditi, 2013), (Maalek et al., 2019)
8.	Camera + Laser Scanners	Construction site Masonry block	(C. Wang et al., 2014), (Mani Golparvar-Fard et al., 2011)
9.	Camera + GPS + Electronic compass	Erection of concrete structures	(Kamat et al., 2010)
10.	UAV + Camera	Construction site Indoor construction	(Freimuth & König, 2015), (Jacob & Golparvar-Fard, 2016), (Hamledari et al., 2017), (Cheng & Deng, 2017), (Braun et al., 2018), (J. H. Lee et al., 2018), (Asadi et al., 2019), (J. H. Lee et al., 2019), (Braun & Borrmann, 2019)
11.	Camera + VR device, RFID + Stereo IR	Building construction	(Soman & Whyte, 2017)

lack of continuity in the spatial data, the requirement for mixed pixel restoration, the regularity with which sensor calibrations must be performed and a lengthy warmup period (Golparvar-Fard et al., 2012). Machine and human motion both contribute to data noise. In addition, laser scanners are bulky and lose quality at greater distances (Golparvar-Fard et al., 2012). To improve the efficiency and precision of data retrieval from building sites, El-Omari and Moselhi (2008) introduced a system that combines photogrammetry with laser scanning. However, the manual point-by-point selection is used to merge the photo images and scanned data requirements.

Digital stills and moving pictures are an alternative methods of recording as-built conditions. It's a typical approach that can supply data directly from the site by monitoring development, facilitating communication and recording the many stages of the building. Image-based systems are low in cost and simple in comparison to laser scanners. The advantages and disadvantages of employing cameras to monitor progress are discussed by Bohn and Teizer (Bohn & Teizer, 2010), while an overview of imaging applications in construction is provided by (Bayrak & Kaka, 2004). Several methods exist for gathering photographic evidence. Monocular (Lukins & Trucco, 2007) and stereo (both possible) cameras could be used (Son & Kim, 2010). Multiple cameras were proposed by (Abeid Neto et al., 2002) and (Mani Golparvar-Fard et al., 2009) for usage on a construction site, whereas (Rebolj et al., 2008; Zhang et al., 2009) and (Ibrahim et al., 2009) all employed a stationary camera at a well-established, fixed location. Occlusions, obstructions and weather can all hinder the view from a stationary camera. That's why it's impossible to provide a full picture of development. (Golparvar-Fard et al., 2011) attempted to address these shortcomings by collecting a series of photographs taken at and near the construction site. In some of the recent studies, (Rebolj et al., 2017), (Pučko et al., 2018), (Lei et al., 2019), (Pour Rahimian et al., 2020), (Pazhoohesh & Zhang, 2015) and (Pazhoohesh et al., 2021) research used an infrared camera in place of the ordinary optical camera, which includes a fourth element depth in the collected data; the depth factors help in conducting volumetric analysis to find out the progress in the construction activities.

Spatial properties of public works projects can be recorded in 3D point clouds using video (Brilakis et al., 2011). The possibility of employing visual data for as-built data gathering is growing as camera and performance processing unit developments raise the precision of the received data, decrease the time required for processing and speed up the data acquisition process. Different types of information are needed for indoor and outdoor settings. The principal architectural elements are columns, beams and walls on the exterior. However, interior scenes include a wide range of construction aspects and schedules connected to a large number of subcontractors. Work inside often involves modifying the wall surface (by painting, tiling, laying wood etc.) or mounting something on the wall (e.g. windows, doors etc). Although specific exterior-focused methods (Bosché, 2010; Golparvar-Fard et al., 2012; Lukins & Trucco, 2007) are adaptable for usage in interior settings, they fail to solve the difficulties just highlighted. As a result, the problems of the interior environment are still not being adequately addressed by the current state of research.

26.5 ANALYSING REMOTE SENSING TECHNIQUES

To infer the existence of unplanned model objects in as-built images, some studies have turned to image processing and computer vision methods. The majority of these systems rely on component detection based on temporal as-built pictures taken by a single static camera. One such method is that presented in Lukins and Trucco (2007). Initially, a 3D BIM is aligned with the camera, and then template masks are generated for the 2D photos based on the 3D model components that should be in view (Lukins & Trucco, 2007). Next, a derivative filter is applied to successive photos to find changes in pixel intensity within the masked regions, with results above a threshold considered successful. These methods are vulnerable to the expected occlusions and shifting light conditions of a construction site. In addition, because 3D data cannot be reconstructed from a series of 2D photos, developments at varying depths or in various parts of an item may go unnoticed.

Despite these limitations, image processing techniques show promise in determining which resources can be used to aid in object recognition. These methods often characterise texture and colour/intensity within uniform image regions by employing image filtering using morphology, edge and corner detectors and statistical analysis. Afterwards, material identification is achieved by analysing pixels properties generated by above-mentioned methods by studying photographs of these materials collected from construction sites (e.g., I. K. Brilakis et al., 2006). Improvements and additional automation of the material identification decision-making process using machine-learning classifiers have also been implemented (e.g. Dimitrov & Golparvar-Fard, 2014). These methods require an extensive library of annotated pictures of recognised substances. However, it might be difficult for such methods to account for the fact that there is often noticeable variation in the texture and colour of materials from a different sample collected from a different construction site. For building materials to be identified accurately and consistently, large photographic databases with labels would be required.

Many researchers attempted to record the 3D nature of development by collecting detailed point cloud data from the construction site of the as-built scene, aligning the digitised model after registering it with a design model and then determining which points agree to the things that were originally envisioned. Most of these studies gathered their information through LiDAR (see also: Bosché, 2010). Most recently, (Golparvar-Fard et al., 2012) and (Tuttas et al., 2014) have used photogrammetric point clouds-based visualisation techniques. Progress is recognised by contrasting the as-built point cloud with either a sampled point cloud (point-to-point) or flat surface segments (point-to-plane) from the as-planned model. For example, Bosché (2010) describes a point-to-point method that looks for correspondences within a certain distance threshold to determine whether a planned point exists in the scene (Bosché, 2010). Point-to-plane methods, such as those developed by Tuttas et al. (2014), look for as-built points that fall inside the orthogonal distance tolerances of the planar surface segments. Classifiers based on machine learning, such as decision trees (Tuttas et al., 2014) and support vector machines (Golparvar-Fard et al., 2012), have been used to learn complex boundaries for making the object recognition decision. Again, most studies in this field have concentrated on bettering the construction of structural components. Some new research has focused on developments in circular cross-sectioned mechanical, electrical and plumbing (MEP) components (Bosché et al., 2015); secondary or temporary construction objects (such as formwork or scaffolding) (Turkan et al., 2014); and operational level activities (such as reinforcing bar, insulation etc.) in building construction (K K Han & Golparvar-Fard, 2015).

Automation is monitoring construction activity progress with the help of 3D and 4D civil information modelling (CIM) (as-built to as-planned comparison). Some researchers looked at how contractors and transportation agencies use CIM, and they came to the conclusion that 4D models were the most beneficial tool for modelling construction and traffic control plans, as well as for transmitting design intent to project stakeholders and the general public (Guo et al., 2014). There is a hole in the current body of research about the practical applications of these models in easing the burden of project control throughout the construction phase.

Emerging UAVs' data processing techniques can perform various surveying outputs with high accuracy and data points at meager cost. This was achieved using photogrammetric post-processing, SLAM, and compact LiDAR sensors. Due to these advantages, UAVs present an exciting new data collection platform for tracking the progress of road and underground infrastructure projects. Recently, Nex and Remondino (2014) released a comprehensive literature analysis on applying UAV data to 3D modelling and photogrammetric mapping (Nex & Remondino, 2014). The accuracy of DTMs generated by sensors installed on UAVs (LiDAR or camera) has been assessed by several researchers (Hugenholtz et al., 2014; Lin et al., 2011; Mancini et al., 2013) by comparing the results to those from other established methodologies. According to (Greenwood et al., 2019), 5 cm or less is an acceptable level of accuracy for most engineering applications, which was generally reported. There have also been suggestions made for how to generate digital terrain models for use in massive earthwork projects (Kim et al., 2015; Uysal et al., 2015). A UAV is also used for progress monitoring of paved and unpaved pavements and flaws detection as experimentally tested by the researcher (Chunsun Zhang & Elaksher, 2012; Ellenberg et al., 2015). UAV-created photogrammetric terrain models were recommended for use in earthwork project monitoring in one study (Siebert & Teizer, 2014). While this method does show how UAV-acquired photogrammetric DTMs might aid in building projects, it is only limited to earthwork progress but has more emphasis on controlling the quality than on accurate progress monitoring.

The study that contributed the most pertinent information for this chapter (Kivimäki & Heikkilä, 2015) presented a real-time system that uses CIM surface models to do cloud-based as-built assessments on construction projects. This is the only study to date to compare the measured site surfaces with the various CIM design surfaces as per knowledge and research. It is a quality assurance system in its purest form, with no capacity for measuring actual advancement. There is no thought about how different project surfaces should be elevated at different times. Furthermore, for unlabelled data, it is necessary to manually annotate as-built data points over as-planned layers. The authors

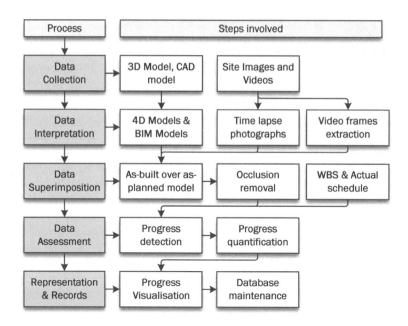

FIGURE 26.6 Holistic framework for the construction progress monitoring.

see this as a limitation to the method's overall effectiveness. Existing research hasn't looked at the possibility of automatically assigning unlabelled as-built data points to suitable CIM design layers.

26.6   A GENERAL FRAMEWORK FOR CONSTRUCTION PROGRESS MONITORING

The general framework for any progress monitoring involves the key concept of comparing the as-planned vs as-built status of undergoing construction. Figure 26.6 shows a holistic approach that needs to be followed for construction progress monitoring using the image and video data along with remote sensing technology. The very first stage is data collection; the 3D CAD models, rendered drawing and BIM model are collected for the same. For as-built data, site images and videos are recorded using the cameras, drones and other sources. These data serve as raw information for the second process of data interpretation, in the second stage, the 3D model is then converted into 4D models, and initial processing of images and videos. The processed data is then superimposed over each other for comparison in the third stage of data superimposition. The occlusion and any kind of irregularities of data are then rectified. After all data is ratified, the progress is examined against the actual schedule for the project progress detection and its quantification. With the use of visualisation techniques, the data is then shared among all stakeholders for remediation of any delays in construction activities.

26.7   CONCLUSION

This chapter presented a review of the previous research and an analysis of the most recent findings in automated progress tracking. As seen in Table 26.1, none of the recommended solutions achieves the performance of the ideal instance, and the most appropriate method varies according to the situation's specifics.

Augmented reality–based systems present an improvement to the other proposed alternatives because they satisfy many requirements. Many alternatives are proposed by researchers, which are illustrated in the literature review, and use the fundamental of as-built vs. as-planned comparison;

these sensors also require cost and time as an initial cost but later helps in effective progress management. However, they are inexpensive and easy to use in any established environment. Although model-based augmented reality algorithms for superimposing as-built models and as-planned have been developed, their performance within the restrictions necessary for successful real-time operation on a construction site has not been examined. This is even though these algorithms have been developed.

Techniques such as laser scanning, image processing and computer vision were utilised in most of the study conducted to achieve automatic progress assessment in outside settings. Because of the high level of accuracy it provides, laser scanning has the potential to be an effective technique for the collection of data for 3D models. Nevertheless, it is pricey and time-consuming, particularly in interior contexts where a high number of scans are necessary to obtain information for each room in a building. This makes it difficult to use in indoor environments. Even if the point cloud is reconstructed from pictures (Golparvar-Fard et al., 2012), point clouds do not contain information linked to the items they contain, and the process of characterising the objects inside them is still in its infancy. Because of this, it is impossible to measure the progress made in interior activities such as tiling, painting and so on.

Photogrammetry suffers from the same restrictions. For indoor contexts, a large number of overlapping photos captured from various locations within a building are required. Processing the photos to rebuild the 3D interior scene incurs additional computational and time costs. In the realm of computer vision and image processing, algorithms have been created to recognise things of interest in buildings; however, each can only detect a single object at a time (concrete or steel elements, bricks, windows, doors). As a direct consequence of this, there is no standard procedure. Only a modest amount of automation has been produced due to the several research projects focused on tracking development inside. Either the user must manually allocate the information, or the user must manually compare the as-planned model to the model produced. In addition, a convoluted internal environment is a recipe for disaster for the system.

REFERENCES

Abeid Neto, J., Arditi, D., & Evens, M. W. (2002). Using Colors to Detect Structural Components in Digital Pictures. *Computer-Aided Civil and Infrastructure Engineering*, *17*(1), 61–67. https://doi.org/10.1111/1467-8667.00253

Asadi, K., Ramshankar, H., Noghabaei, M., & Han, K. (2019). Real-Time Image Localisation and Registration with BIM Using Perspective Alignment for Indoor Monitoring of Construction. *Journal of Computing in Civil Engineering*, *33*(5), 1–15. https://doi.org/10.1061/(ASCE)CP.1943-5487.0000847

Azimi, R., Lee, S., Abourizk, S. M., & Alvanchi, A. (2011). A Framework for an Automated and Integrated Project Monitoring and Control System for Steel Fabrication Projects. *Automation in Construction*, *20*(1), 88–97. https://doi.org/10.1016/j.autcon.2010.07.001

Barrie, D. S., & Paulson, B. C. (1992). *Professional Construction Management : Including CM, Design-Construct, and General Contracting* (3rd ed.). New York: McGraw-Hill.

Bayrak, T., & Kaka, A. (2004). Evaluation of Digital Photogrammetry and 3D Cad Modelling Applications in Construction Manangement. *20th Annual ARCOM Conference, Heriot Watt University, Association of Researchers in Construction Management*, *1*(September), 613–621. http://www.arcom.ac.uk/-docs/proceedings/ar2004-0613-0619_Bayrak_and_Kaka.pdf

Bohn, J. S., & Teizer, J. (2010). Benefits and Barriers of Construction Project Monitoring Using High-Resolution Automated Cameras. *Journal of Construction Engineering and Management*, *136*(6), 632–640. https://doi.org/10.1061/(ASCE)CO.1943-7862.0000164

Bosché, F. (2010). Automated Recognition of 3D CAD Model Objects in Laser Scans and Calculation of as-Built Dimensions for Dimensional Compliance Control in Construction. *Advanced Engineering Informatics*, *24*(1), 107–118. https://doi.org/10.1016/j.aei.2009.08.006

Bosché, F., Ahmed, M., Turkan, Y., Haas, C. T., & Haas, R. (2015). The Value of Integrating Scan-to-BIM and Scan-vs-BIM Techniques for Construction Monitoring Using Laser Scanning and BIM: The Case of Cylindrical MEP Components. *Automation in Construction*, *49*, 201–213. https://doi.org/10.1016/j.autcon.2014.05.014

Braun, Alex, & Borrmann, A. (2019). Combining Inverse Photogrammetry and BIM for Automated Labeling of Construction Site Images for Machine Learning. *Automation in Construction*, *106*(June), 102879. https://doi.org/10.1016/j.autcon.2019.102879

Braun, Alexander, Tuttas, S., Borrmann, A., & Stilla, U (2015). A Concept for Automated Construction Progress Monitoring Using BIM-Based Geometric Constraints and Photogrammetric Point Clouds. *Journal of Information Technology in Construction*, *20*(November 2014), 68–79.

Braun, Alexander, Tuttas, S., Borrmann, A., & Stilla, U. (2017). Automated Progress Monitoring Based on Photogrammetric Point Clouds and Precedence Relationship Graphs. *Proceedings of the 32nd International Symposium on Automation and Robotics in Construction and Mining (ISARC 2015)*. https://doi.org/10.22260/isarc2015/0034

Braun, Alexander, Tuttas, S., Stilla, U., & Borrmann, A. (2018). Process- and Computer Vision-Based Detection of As-Built Components on Construction Sites. In *Proceedings of the 35th International Symposium on Automation and Robotics in Construction (ISARC)* (Issue Isarc). https://doi.org/10.22260/isarc2018/0091

Brilakis, I., Fathi, H., & Rashidi, A. (2011). Progressive 3D Reconstruction of Infrastructure with Videogrammetry. *Automation in Construction*, *20*(7), 884–895. https://doi.org/10.1016/j.autcon.2011.03.005

Brilakis, I. K., Soibelman, L., & Shinagawa, Y. (2006). Construction Site Image Retrieval Based on Material Cluster Recognition. *Advanced Engineering Informatics 20 (2006)*, *20*, 443–452. https://doi.org/10.1016/j.aei.2006.03.001

Cheng, J, & Deng, C. P., Y. (2017). Proactive Construction Project Controls via Predictive Visual Data Analytics. *Computing in Civil Engineering 2015*, 667–674. https://doi.org/10.1061/9780784479247.083

Chi, H. L., Chai, J., Wu, C., Zhu, J., Liu, C., & Wang, X. (2017). Scaffolding progress monitoring of LNG plant maintenance project using BIM and image processing technologies. *International Conference on Research and Innovation in Information Systems, ICRIIS*, 1–6. https://doi.org/10.1109/ICRIIS.2017.8002505

Choi, N., Son, H., Kim, C., Kim, C., & Kim, H. (2008). *Rapid 3D object recognition for automatic project progress monitoring using a stereo vision system.* 58–63. https://doi.org/10.3846/isarc.20080626.58

Dimitrov, A., & Golparvar-Fard, M. (2014). Vision-Based Material Recognition for Automated Monitoring of Construction Progress and Generating Building Information Modeling from Unordered Site Image Collections. *Advanced Engineering Informatics*, *28*(1), 37–49. https://doi.org/10.1016/j.aei.2013.11.002

Ellenberg, A., Branco, L., Krick, A., Bartoli, I., & Kontsos, A. (2015). Use of Unmanned Aerial Vehicle for Quantitative Infrastructure Evaluation. *Journal of Infrastructure Systems*, *21*(3), 1–8. https://doi.org/10.1061/(ASCE)IS.1943-555X.0000246

El-Omari, S., & Moselhi, O. (2008). Integrating 3D Laser Scanning and Photogrammetry for Progress Measurement of Construction Work. *Automation in Construction*, *18*(1), 1–9. https://doi.org/10.1016/J.AUTCON.2008.05.006

Ergen, E., Akinci, B., & Sacks, R. (2007). Tracking and Locating Components in a Precast Storage Yard Utilising Radio Frequency Identification Technology and GPS. *Automation in Construction*, *16*(3), 354–367. https://doi.org/10.1016/j.autcon.2006.07.004

Fard, M. G., & Peña-Mora, F. (2007). *Application of Visualization Techniques for Construction Progress Monitoring.* 40937(July 2007), 216–223. https://doi.org/10.1061/40937(261)27

Freimuth, H., & König, M. (2015). Generation of waypoints for UAV-assisted progress monitoring and acceptance of construction work. *15th International Conference on Construction Applications of Virtual Reality*.

Giretti, A., Carbonari, A., Vaccarini, M., Robuffo, F., & Naticchia, B. (2017). Interoperable Approach in Support of Semi-Automated Construction Management. *28th International Symposium on Automation and Robotics in Construction (ISARC 2011)*, June, 267–272. https://doi.org/10.22260/isarc2011/0048

Global Construction Perspectives and Oxford Economics. (2015). Global Construction Market to Grow $8 Trillion by 2030: Driven by China, US and India. *Global Construction 2030*, *44*, 8–10.

Golparvar-Fard, Mani, Bohn, J., Teizer, J., Savarese, S., & Peña-Mora, F. (2011). Evaluation of Image-Based Modeling and Laser Scanning Accuracy for Emerging Automated Performance Monitoring Techniques. *Automation in Construction*, *20*(8), 1143–1155. https://doi.org/10.1016/j.autcon.2011.04.016

Golparvar-Fard, Mani, Peña-Mora, F., Arboleda, C. A., & Lee, S. (2009). Visualization of Construction Progress Monitoring With 4D Simulation Model Overlaid on Time-Lapsed Photographs. *Journal of Computing in Civil Engineering*, *23*(6), 391–404. https://doi.org/10.1061/(ASCE)0887-3801(2009)23:6(391)

Golparvar-Fard, M, Peña-Mora, F., & Savarese, S (2010). Model-Based Detection of Progress Using D 4 AR Models Generated by Daily Site Photologs and Building Information Models. *Proc, 1*, 2–7.

Golparvar-Fard, Mani, Peña-Mora, F., & Savarese, S. (2011). Integrated Sequential As-Built and As-Planned Representation With D4AR Tools in Support of Decision-Making Tasks in the AEC/FM Industry. *Journal of Construction Engineering and Management, 137*(12), 1099–1116. https://doi.org/10.1061/(asce)co.1943-7862.0000371

Golparvar-Fard, Mani, Peña-Mora, F., & Savarese, S. (2012). Automated Progress Monitoring Using Unordered Daily Construction Photographs and IFC-Based Building Information Models. *Journal of Computing in Civil Engineering, 29*(1), 04014025. https://doi.org/10.1061/(asce)cp.1943-5487.0000205

Golparvar-Fard, Mani, Peña Mora, F., & Silvio, S (2009). D4AR-A 4-Dimensional Augmented Reality Model for Automating Construction Progress Monitoring Data Collection, Processing and Communication. *Electronic Journal of Information Technology in ConstrucTion, 14*(June), 129–153.

Greenwood, W. W., Lynch, J. P., & Zekkos, D. (2019). Applications of UAVs in Civil Infrastructure. *Journal of Infrastructure Systems, 25*(2), 1–21. https://doi.org/10.1061/(ASCE)IS.1943-555X.0000464

Guo, F., Turkan, Y., Jahren, C. T., & David Jeong, H. (2014). Civil information modeling adoption by Iowa and Missouri DOT. *Computing in Civil and Building Engineering – Proceedings of the 2014 International Conference on Computing in Civil and Building Engineering*, 463–471. https://doi.org/10.1061/9780784413616.058

Gupta, S., George, R. C., Philip, D., & Nair, S. (2022). Activity Time Variations and Its Influence on Realisation of Different Critical Paths in a PERT Network: An Empirical Study Using Simulations. In M. Casini (Ed.), *Proceedings of the 2nd International Civil Engineering and Architecture Conference* (pp. 674–680). Springer Nature Singapore. https://doi.org/10.1007/978-981-19-4293-8_70

Gupta, S., & Nair, S. (2021). Evaluation of Bid in Construction Industry Based on Multi-Criteria Approach Using TOPSIS. *Lecture Notes in Civil Engineering, 172*, 139–151. https://doi.org/10.1007/978-981-16-4396-5_13/COVER/

Hamledari, H., McCabe, B., & Davari, S. (2017). Automated Computer Vision-Based Detection of Components of Under-Construction Indoor Partitions. *Automation in Construction, 74*, 78–94. https://doi.org/10.1016/j.autcon.2016.11.009

Han, Kevin K., Cline, D., & Golparvar-Fard, M. (2015). Formalised Knowledge of Construction Sequencing for Visual Monitoring of Work-in-Progress via Incomplete Point Clouds and Low-LoD 4D BIMs. *Advanced Engineering Informatics, 29*(4), 889–901. https://doi.org/10.1016/j.aei.2015.10.006

Han, K. K, & Golparvar-Fard, M. (2015). Appearance-Based Material Classification for Monitoring of Operation-Level Construction Progress Using 4D BIM and Site Photologs. In *Automation in Construction*. Elsevier. https://www.sciencedirect.com/science/article/pii/S0926580515000266

Han, Kevin K., & Golparvar-Fard, M. (2014). Automated monitoring of operation-level construction progress using 4D bim and daily site photologs. *Construction Research Congress 2014: Construction in a Global Network – Proceedings of the 2014 Construction Research Congress*, 1033–1042. https://doi.org/10.1061/9780784413517.106

Han, Kevin K., & Golparvar-Fard, M. (2015). Appearance-Based Material Classification for Monitoring of Operation-Level Construction Progress Using 4D BIM and Site Photologs. *Automation in Construction, 53*, 44–57. https://doi.org/10.1016/j.autcon.2015.02.007

Hugenholtz, C. H., Walker, J., Owen, Brown, & Myshak, S. (2014). Earthwork Volumetrics with an Unmanned Aerial Vehicle and Softcopy Photogrammetry. *Journal of Surveying Engineering, 141*(1), 06014003. https://doi.org/10.1061/(ASCE)SU.1943-5428.0000138

Ibrahim, Y. M., Lukins, T. C., Zhang, X., Trucco, E., & Kaka, A. P. (2009). Towards Automated Progress Assessment of Workpackage Components in Construction Projects Using Computer Vision. *Advanced Engineering Informatics, 23*(1), 93–103. https://doi.org/10.1016/J.AEI.2008.07.002

Jacob, L., & Golparvar-Fard, M. (2016). Web-Based 4D Visual Production Models for Decentralised Work Tracking and Information Communication on Construction Sites. *PROCEEDINGS Construction Research Congress 2016*, 2039–2049. https://doi.org/10.1061/9780784479827.203

Jaselskis, E. J., & El-Misalami, T. (2017). Radio Frequency Identification Application for Constructors. *Proceedings of the 17th IAARC/CIB/IEEE/IFAC/IFR International Symposium on Automation and Robotics in Construction*. https://doi.org/10.22260/ISARC2000/0021

Kamat, V. R., Martinez, J. C., Fischer, M., Golparvar-Fard, M., Peña-Mora, F., & Savarese, S. (2010). Research in Visualization Techniques for Field Construction. *Journal of Construction Engineering and Management, 137*(10), 853–862. https://doi.org/10.1061/(asce)co.1943-7862.0000262

Kerzner, H. (2010). Project Management: A systems approach to planning, scheduling and control. In Wiley (Tenth Edit). https://doi.org/10.1115/1.859643

Kim, Changyoon, Kim, H., & Ju, Y. (2017). Bridge Construction Progress Monitoring Using Image Analysis. *Proceedings of the 2009 International Symposium on Automation and Robotics in Construction (ISARC 2009), Isarc*, 2007–2010. https://doi.org/10.22260/isarc2009/0023

Kim, Changyoon, Kim, B., & Kim, H. (2013). 4D CAD Model Updating Using Image Processing-Based Construction Progress Monitoring. *Automation in Construction*, 35, 44–52. https://doi.org/10.1016/j.autcon.2013.03.005

Kim, Changwan, Kim, C., & Son, H. (2013). Fully Automated Registration of 3D Data to a 3D CAD Model for Project Progress Monitoring. *Automation in Construction*, 35, 587–594. https://doi.org/10.1016/j.autcon.2013.01.005

Kim, H., Sim, S. H., & Cho, S. (2015). Unmanned aerial vehicle (UAV)-powered concrete crack detection based on digital image processing. *International Conference on Advances in Experimental Structural Engineering, 2015-Augus*.

Kivimäki, T., & Heikkilä, R. (2015). Infra BIM based real-time quality control of infrastructure construction projects. *32nd International Symposium on Automation and Robotics in Construction and Mining: Connected to the Future, Proceedings*. https://doi.org/10.22260/ISARC2015/0117

Kropp, C., Koch, C., & König, M. (2018). Interior Construction State Recognition With 4D BIM Registered Image Sequences. *Automation in Construction*, 86(October 2017), 11–32. https://doi.org/10.1016/j.autcon.2017.10.027

Kropp, C., König, M., & Koch, C. (2013). Object Recognition in BIM Registered Videos for Indoor Progress Monitoring. *Proceedings 20th International Workshop: Intelligent Computing in Engineering, November 2015*, 1–10.

Lee, J. H., Park, J. H., & Jang, B. T. (2018). Design of Robot based Work Progress Monitoring System for the Building Construction Site. *9th International Conference on Information and Communication Technology Convergence: ICT Convergence Powered by Smart Intelligence, ICTC 2018*, 1420–1422. https://doi.org/10.1109/ICTC.2018.8539444

Lee, J. H., Park, J. H., & Jang, B. T. (2019). Progress Monitoring system based on Volume Comparison for the Construction Site. *ICTC 2019 – 10th International Conference on ICT Convergence: ICT Convergence Leading the Autonomous Future, C*, 986–989. https://doi.org/10.1109/ICTC46691.2019.8939969

Lee, S., & Pena-Mora, F. (2006). Visualisation of construction progress monitoring. *Joint International Conference on Computing and Decision Making in Civil and Building Engineering*, 2527–2533. https://doi.org/10.1061/(ASCE)0887-3801(2009)23:6(391)

Lei, L., Zhou, Y., Luo, H., & Love, P. E. D. (2019). A CNN-Based 3D Patch Registration Approach for Integrating Sequential Models in Support of Progress Monitoring. *Advanced Engineering Informatics*, 41(May), 100923. https://doi.org/10.1016/j.aei.2019.100923

Lin, Y., Hyyppä, J., & Jaakkola, A. (2011). Mini-UAV-borne LIDAR for Fine-Scale Mapping. *IEEE Geoscience and Remote Sensing Letters*, 8(3), 426–430. https://doi.org/10.1109/LGRS.2010.2079913

Liu, C. W., Wu, T. H., Tsai, M. H., & Kang, S. C. (2018). Image-Based Semantic Construction Reconstruction. *Automation in Construction*, 90(February), 67–78. https://doi.org/10.1016/j.autcon.2018.02.016

Lukins, T. C., & Trucco, E. (2007). Towards Automated Visual Assessment of Progress in Construction Projects. *Proceedings of the British Machine Vision Conference 2007, University of Warwick*, 18.1–18.10. https://doi.org/10.5244/c.21.18

Maalek, R., Lichti, D. D., & Ruwanpura, J. Y. (2019). Automatic Recognition of Common Structural Elements from Point Clouds for Automated Progress Monitoring and Dimensional Quality Control in Reinforced Concrete Construction. *Remote Sensing*, 11(9). https://doi.org/10.3390/rs11091102

Mahami, H., Nasirzadeh, F., Ahmadabadian, A. H., & Nahavandi, S. (2019). Automated Progress Controlling and Monitoring Using Daily Site Images and Building Information Modelling. *Buildings*, 9(3). https://doi.org/10.3390/buildings9030070

Mancini, F., Dubbini, M., Gattelli, M., Stecchi, F., Fabbri, S., & Gabbianelli, G. (2013). Using Unmanned Aerial Vehicles (UAV) for High-Resolution Reconstruction of Topography: The Structure from Motion Approach on Coastal Environments. *Remote Sensing*, 5(12), 6880–6898. https://doi.org/10.3390/rs5126880

Meredith, J. R., Shafer, S. M., & Mantel, S. J. (2017). *Project management: a strategic managerial approach*.

Navon, R. (2007). Research in Automated Measurement of Project Performance Indicators. *Automation in Construction*, 16(2), 176–188. https://doi.org/10.1016/j.autcon.2006.03.003

Navon, Ronie, & Sacks, R. (2007). Assessing Research Issues in Automated Project Performance Control (APPC). *Automation in Construction*, 16(4), 474–484. https://doi.org/10.1016/j.autcon.2006.08.001

Nex, F., & Remondino, F. (2014). UAV for 3D Mapping Applications: a Review. *Applied Geomatics*, 6(1), 1–15. https://doi.org/10.1007/s12518-013-0120-x

Omar, H., Mahdjoubi, L., & Kheder, G. (2018). Towards an Automated Photogrammetry-Based Approach for Monitoring and Controlling Construction Site Activities. *Computers in Industry, 98,* 172–182. https://doi.org/10.1016/j.compind.2018.03.012

Pazhoohesh, M., & Zhang, C. (2015). Automated Construction Progress Monitoring Using Thermal Images and Wireless Sensor Networks. *Building on Our Growth Opportunities Miser Sur Nos Opportunités de Croissance May 27–30, 2015 AND INFER FROST PENETRATION Building on Our Growth Opportunities Miser Sur Nos Opportunités de Croissance May 27 – 30, 2015, September,* 1–10.

Pazhoohesh, M., Zhang, C., Hammad, A., Taromi, Z., & Razmjoo, A. (2021). Infrared Thermography for a Quick Construction Progress Monitoring Approach in Concrete Structures. *Architecture, Structures and Construction, 1*(2–3), 91–106. https://doi.org/10.1007/s44150-021-00008-7

Peña-Mora, F., Golparvar-Fard, M., Aziz, Z., & Roh, S. (2011). Design Coordination and Progress Monitoring During the Construction Phase. *Collaborative Design in Virtual Environments,* 89–99. https://doi.org/10.1007/978-94-007-0605-7_8

Pour Rahimian, F., Seyedzadeh, S., Oliver, S., Rodriguez, S., & Dawood, N. (2020). On-Demand Monitoring of Construction Projects Through a Game-Like Hybrid Application of BIM and Machine Learning. *Automation in Construction, 110* (November 2019), 103012. https://doi.org/10.1016/j.autcon.2019.103012

Pučko, Z., Šuman, N., & Rebolj, D. (2018). Automated Continuous Construction Progress Monitoring Using Multiple Workplace Real Time 3D Scans. *Advanced Engineering Informatics, 38*(April), 27–40. https://doi.org/10.1016/j.aei.2018.06.001

Ratajczak, J., Marcher, C., Schimanski, C. P., Schweikopfler, A., Riedl, M., & Matt, D. T. (2019). BIM-based augmented reality tool for the monitoring of construction performance and progress. *Proceedings of the 2019 European Conference for Computing in Construction, 1*(July), 467–476. https://doi.org/10.35490/ec3.2019.202

Rebolj, D., Babič, N. Č., Magdič, A., Podbreznik, P., & Pšunder, M. (2008). Automated Construction Activity Monitoring System. *Advanced Engineering Informatics, 22*(4), 493–503. https://doi.org/10.1016/J.AEI.2008.06.002

Rebolj, D., Pučko, Z., Babič, N. Č., Bizjak, M., & Mongus, D. (2017). Point Cloud Quality Requirements for Scan-vs-BIM Based Automated Construction Progress Monitoring. *Automation in Construction, 84*(August), 323–334. https://doi.org/10.1016/j.autcon.2017.09.021

Reja, V. K., Varghese, K., & Ha, Q. P. (2022). Computer Vision-Based Construction Progress Monitoring. *Automation in Construction, 138*(April), 104245. https://doi.org/10.1016/j.autcon.2022.104245

Roh, S., Aziz, Z., & Peña-Mora, F. (2011). An Object-Based 3D Walk-Through Model for interior Construction Progress Monitoring. *Automation in Construction, 20*(1), 66–75. https://doi.org/10.1016/j.autcon.2010.07.003

Roumeliotis, G. (2010). *Global Construction Growth to Outpace GDP This Decade - PwC - Reuters.* https://in.reuters.com/article/idINIndia-55293920110303

Shih, N. J., & Wang, P. H. (2004). Point-Cloud-Based Comparison between Construction Schedule and as-Built Progress: Long-Range Three-Dimensional Laser scanner's Approach. *Journal of Architectural Engineering, 10*(3), 98–102. https://doi.org/10.1061/(ASCE)1076-0431(2004)10:3(98)

Siebert, S., & Teizer, J. (2014). Mobile 3D Mapping for Surveying Earthwork Projects Using an Unmanned Aerial Vehicle (UAV) System. *Automation in Construction, 41,* 1–14. https://doi.org/10.1016/j.autcon.2014.01.004

Soman, R. K., & Whyte, J. K. (2017). *A Framework for Cloud-Based Virtual and Augmented Reality Using Real-Time Information for Construction Progress Monitoring. 0*(July), 833–840. https://doi.org/10.24928/jc3-2017/0273

Son, H., & Kim, C. (2010). 3D Structural Component Recognition and Modeling Method Using Color and 3D Data for Construction Progress Monitoring. *Automation in Construction, 19*(7), 844–854. https://doi.org/10.1016/j.autcon.2010.03.003

Song, J., Haas, C. T., & Caldas, C. H. (2006). Tracking the Location of Materials on Construction Job Sites. *Journal of Construction Engineering and Management, 132*(9)(August), 911–918. https://doi.org/10.1061/(ASCE)0733-9364(2006)132

Song, K., Pollalis, S. N., & Pena-Mora, F. (2005). Project Dashboard: Concurrent Visual Representation Method of Project Metrics on 3D Building Models. *Proceedings of the 2005 ASCE International Conference on Computing in Civil Engineering,* 1561–1572. https://doi.org/10.1061/40794(179)147

Srivastava, S., Gupta, S., Dikshit, O., & Nair, S. (2020). A Review of UAV Regulations and Policies in India. In *Lecture Notes in Civil Engineering* (Vol. 51, pp. 315–325). Springer. https://doi.org/10.1007/978-3-030-37393-1_27

Tezel, A., & Aziz, Z (2017). From Conventional to It Based Visual Management: A Conceptual Discussion for Lean Construction. *Journal of Information Technology in Construction*, *22*, 220–246.

Turkan, Y., Bosche, F., Haas, C. T., & Haas, R. (2010). Towards Automated Progress Tracking Of Erection Of Concrete Structures. *Proc. of 6th International AEC Innovation Conference.*

Turkan, Y., Bosche, F., Haas, C. T., & Haas, R. (2011). Automated Progress Tracking Using 4D Schedule and 3D Sensing Technologies. *Automation in Construction*, *22*, 414–421. https://doi.org/10.1016/j. autcon.2011.10.003

Turkan, Y., Bosché, F., Haas, C. T., & Haas, R. (2014). Tracking of Secondary and Temporary Objects in Structural Concrete Work. *Construction Innovation: Information, Process, Management, 14*(2), 145–167. https://doi.org/10.1108/CI-12-2012-0063

Tuttas, S., Braun, A., Borrmann, A., & Stilla, U. (2014). Comparision of Photogrammetric Point Clouds With BIM Building Elements for Construction Progress Monitoring. *International Archives of the Photogrammetry, Remote Sensing and Spatial Information Sciences - ISPRS Archives, 40*(3), 341–345. https://doi.org/10.5194/isprsarchives-XL-3-341-2014

Uysal, M., Toprak, A. S., & Polat, N. (2015). DEM Generation with UAV Photogrammetry and Accuracy Analysis in Sahitler Hill. *Measurement*, *73*, 539–543. https://doi.org/10.1016/J.MEASUREMENT. 2015.06.010

Wang, C., Cho, Y. K., & Park, J. W. (2014). Performance Tests for Automatic 3D Geometric Data Registration Technique for Progressive As-Built Construction Site Modeling. *International Conference on Computing in Civil and Building Engineering (ICCCBE), ASCE, June*, 1053–1061. https://doi.org/10. 1061/9780784413616.131

Wang, X., Truijens, M., Hou, L., Wang, Y., & Zhou, Y. (2014). Integrating Augmented Reality With Building Information Modeling: Onsite Construction Process Controlling for Liquefied Natural Gas Industry. *Automation in Construction*, *40*, 96–105. https://doi.org/10.1016/j.autcon.2013.12.003

Yelda, T., Bosche, F. N., Carl, H., & Ralph, H. (2011). Automated Progress Tracking of Erection of Concrete Structures. *Annual Conference of the Canadian Society for Civil Engineering*, 2746–2756 TS-BibTeX. Yelda, Bosche et al 2011 – Automated Progress Tracking of Erection.pdf

Zhang, Chengyi, & Arditi, D. (2013). Automated Progress Control Using Laser Scanning Technology. *Automation in Construction*, *36*, 108–116. https://doi.org/10.1016/j.autcon.2013.08.012

Zhang, Chunsun, & Elaksher, A. (2012). An Unmanned Aerial Vehicle-Based Imaging System for 3D Measurement of Unpaved Road Surface Distresses. *Computer-Aided Civil and Infrastructure Engineering, 27*(2), 118–129. https://doi.org/10.1111/j.1467-8667.2011.00727.x

Zhang, X., Bakis, N., Lukins, T. C., Ibrahim, Y. M., Wu, S., Kagioglou, M., Aouad, G., Kaka, A. P., & Trucco, E. (2009). Automating Progress Measurement of Construction Projects. *Automation in Construction*, *18*(3), 294–301. https://doi.org/10.1016/j.autcon.2008.09.004

27 Review of Novel Seismic Energy Dissipating Techniques Toward Resilient Construction

Naida Ademović and Ehsan Noroozinejad Farsangi

CONTENTS

27.1   INTRODUCTION

During their exploitation and life duration, structures and infrastructure are exposed to various actions, and ones that are of particular interest are dynamic, like traffic and wind actions, as well as natural hazards like earthquakes. Historically, structures experienced small to heavy damage and even collapse due to seismic forces caused by ground motion. In order to minimize structural damage caused by earthquake actions, seismic engineering moved from force-based design toward performance-based design (PBD). One of the ways to fulfill the PBD's requirements is to connect energy dissipation systems to the structures. The main idea is that the main structures take over the gravity loads, and during the earthquake activity, they should remain in the elastic range. The special structural elements known as energy dissipation devices (EDDs), or so-called dampers, are designed in such a way as to absorb the energy from the earthquake and release the main structure from plastic deformations, leaving it undamaged. The concentration of the damage is located in the EDDs that can be easily checked and replaced after earthquakes. As the structure remains intact, its operation and functionality are unquestionable, enabling its unremitting use and thus enhancing resilience (Aydin et al. 2019, Benavent-Climent et al. 2021).

The first systems that have been used for the mitigation of vibrations caused by seismic actions in civil engineering buildings and long bridges were passive control systems. This type of control system opposed to active devices does not involve the usage of an external power supply for its operation, and it provides additional stiffness to the structure. This means that this kind of system cannot be controlled, and it is not possible to make any modification to the control forces or device behavior once exposed to ground movement. Once exposed to earthquake movement, the reaction of devices is manifested through the conversion of mechanical energy and its transformation into heat, enabling energy dissipation, which affects the response of the structure. Advantages are seen in their low maintenance needs and the fact that these systems are not affected by power outages during seismic events. Various dampers can be classified here, from viscous dampers, viscoelastic dampers, and friction dampers, to hysteretic/yielding dampers; each of them has its area of application and limitations (Katayama et al. 2000, Kim et al. 2012, Xu et al. 2017, Armali et al. 2019). Metallic and friction dampers are representatives of the hysteresis devices, which are not a function of the loading rate, but displacement-dependent EDD, while representatives of velocity-dependent EDDs are viscous and viscoelastic dampers. Friction dampers are effective only if exposed to minor earthquake intensities. Lee et al. (2017) overcame this deficiency by producing a hybrid damper connected in parallel made of steel silt damper (SD) and rotational friction damper (RFD), which is efficient if exposed to minor and minor ground motions. This is a displacement-dependent device. In the case of minor earthquakes, only the RFD gets activated, while for extensive ground motion, simultaneous activation of both dampers is evident.

Another possibility was proposed by Ghorbani and Rofooei (2020), where instead of a single slip load, a double slip loads (DSL) friction damper was manufactured. The double slip load consists of two parts one, which is activated for minor ground movements (smaller slip load), and after a certain amount of slippage, a larger slip load is activated when exposed to moderate to intense ground motions. These dampers considerably suppressed the seismic response of structures of various heights exposed to various ground motions in terms of story drifts, residual drifts, absolute floor accelerations, and base shear forces as to single slip load friction dampers. The best performance was noted in mid-rise buildings.

Metallic dampers dissipate energy through the inelastic deformation of their constitutive material (Javanmardi et al. 2020). As well, metallic dampers are resistant to ambient temperatures and reliability, show stable hysteretic behavior, and their usage is acceptable in practice as the behavior of a material is well known to civil engineers. Usually, dampers are made of mild steel due to their stable hysteretic behavior and a large inherent plastic deformation capacity (Benavent-Climent et al. 2021). A hybrid system consisting of a brace restrained buckle (BRB) and viscous damper (VD) was proposed by Li et al. (2021) for resistance of a structure exposed simultaneously to earthquakes and strong winds. In the analysis, each component was analyzed separately as well as jointly. Regarding the effects of the natural hazards, four combinations were investigated: minor earthquake and weak wind, minor earthquake, and strong wind, strong earthquake and weak wind, a strong earthquake, and strong wind. The hybrid system was more effective in all cases except in the case of minor earthquakes and weak wind, where individual components performed more efficiently Li et al. (2021).

A tuned mass damper (TMD) is a device that is connected to the structure and its aim is to reduce mechanical vibrations, while in a tuned liquid damper (TLD) the mass is replaced by a liquid (usually water), tuned liquid column damper (TLCD) (Chenaghlou et al. 2021). On the other hand, base isolations alter the dynamic characteristics of the structure, reducing the frequency of the system and indirectly the seismic forces to which the structure will be exposed once an earthquake hits.. Base isolators that have been implemented in the structures are friction pendulum systems, laminated rubber bearings, and laminated rubber bearings with lead core (Torunbalci 2004, Pokhrel et al. 2016).

As mentioned above, for the operation of the active control system, which has progressed from the passive control systems, a power source is required for the operation of the external control actuators. The function of the actuators is to employ forces on the structures in a predefined manner that may increase or decrease the energy in the structure. Active control systems include active tuned mass dampers (ATMD) or active mass drivers (AMD), distributed actuators, active tendon systems, and active coupled building systems Fisco and Adeli (2011).

The third system is a combination of the two above-named semiactive control systems. This system operates on a battery and, in this way, resolves the biggest disadvantage of the active control system requirement for a high power supply. In the case of power loss still, some protection of the structure is feasible thanks to the passive component of the control system. Dampers that can be classified in this category are magnetorheological (MR) fluid dampers, semi-active stiffness dampers (SASD), electrorheological dampers (ER), semi-active tuned liquid column dampers, piezoelectric (PZT) dampers, and semi-active TMD (Fisco and Adeli 2011, Xu et al. 2017). A combination of passive and active or semi-active devices is defined as a hybrid control system. As well, the bases for the development of semiactive and hybrid control systems were passive control technologies.

The usage of EDD is twofold in the sense that they can be installed in existing buildings as a retrofit measure or in new structures with the aim of mitigating seismic hazards. This is one of the justifications why a significant amount of research has been devoted to passive energy dissipation systems.

27.2 NOVEL SEISMIC ENERGY DISSIPATING TECHNIQUES

The main purpose of EDD is to absorb the energy and dissipate the seismic forces, which will result in the reduction of the deformation of the structures, that are exposed to ground motions. Passive EDDs have been used for centuries and have found their application throughout the world. Ibrahim (2008) presented the advances in nonlinear passive vibration isolators. It is necessary to utilize an overall nonlinear control design for the improvement of the passive isolator performances (Wagner and Liu 2000, Heertjes and van de Wouw 2006, Lee and Kawashima 2006, 2007). Parulekar and Reddy (2009) gave an overview of all systems providing special emphasis on the pros of the passive systems.

In moment-resistant frames application of friction or yielding EDDs in the location where it is expected to have high ductility requirements during an earthquake, a motion has been seen as a beneficial mitigation technique. These EDDs were firstly applied to reinforced concrete frames (Martinez Rueda and Elnashai 1995, Martinez Rueda 1998) and then applied to steel frames (Mulas and Martinez Rueda 2003). Martinelli and Mulas (2010) in their work proposed new rotational friction devices for precast industrial frames, which are located around the beam-column connections. The new EDD had to be of small dimension and the structural damping had to be boosted to a high extent while the increase of the structural stiffness should be limited. Very detailed calculations have been conducted from limit analysis, capacity design approach, and two numerical models. In order to determine the optimum value for the "plastic" moment of the EDD, a sensitivity time-history analysis was conducted. Once the EDDs were installed on the structures, the top displacement has been reduced by 37% and the velocity by 22%, as well as a reduction of the stresses in the concrete and steel layers. It was noticed that there was an increase in the structural stiffness, which, as a result, increases the seismic input, as well as the base shear. However, as the increase in the structural damping is significant, this effect has been largely relaxed (Martinelli and Mulas (2010). They have recommended further research in the experimental campaign as well as numerical analysis.

The structural system that has been seen as very effective in resisting seismic forces for low-rise buildings is a concentrically braced frame (CBF). High stiffness and lateral strength required for dissipation of the seismic energy are performed by the steel braces, and low lateral displacement is seen as a benefit of this system. If exposed to moderate or high-intensity ground movement, there is a high chance that compression diagonal braces will buckle, and this will decrease the ductility of the CBF and reduce the ability of hysteretic behavior (Inamasu et al. 2017, Skalomenos et al. 2017). For the improvement of hysteretic behavior, various EDDs have been developed from added damping and stiffness (ADAS) developed by Bergman and Goel (1987), which was used by (Tsai et al. 1993) as the basis for the development of triangular added damping and stiffness (TADAS) (Tehranizadeh 2001, Bakre et al. 2006, Sajjadi Alehashem el al. 2008, Mohammadi et al. 2017, Tahamouli Roudsari et al. 2018, Saiedi 2021). ADAS dampers in the rhombic configuration, which uses low-strength steel with pinned joints at each end, were proposed by Shih et al. (2004) and Shih and Sung (2005). It has shown that the prevention of local cracks in the damper is connected to the low-yield strength of steel strain hardening property. The yield displacement is minimized, on one hand, and on the other, the damper's energy dissipation capacity and ductility are improved, thanks to the mechanical properties of low-yield strength steel (Shih and Sung 2005).

Buckling-restrained brace (BRB) was first proposed by Takeda et al. (1972); four years later, Kimura et al. (1976) constructed a typical type of BRBs, where an external restraint mechanism was formed of square steel pipe and the space between the pipe and the steel core was filled with mortar.

The local buckling on the steel core was noted due to the cracking of the infill mortar and this was identified as the major disadvantage. In order to overcome this all-steel, BRBs were developed (ABRB) (Jiang et al. 2020). This EDD was then further elaborated by researchers among which are Black et al. (2002), Tremblay et al. (2006), Fahnestock et al. (2007a, b), Palmer et al. (2011), Atlayan and Charney (2014), Budaházy and Dunai (2015), and Ebadi Jamkhaneh et al. (2018). The steel core is made of a ductile material and designed in such a way as to yield in both tension and compression. The benefit of the BRBs is that once exposed to cyclic loading, its behavior is almost symmetrical, possesses high ductility, exhibits high energy dissipation, and they are cost-effective and lightweight compared to traditional BRB. However, one of the major flaws of the BRBs is local bucking. In order to prevent buckling, once the core goes in compression, it is enveloped with steel casting and then filled either with concrete or mortar.

The form of the core for the BRB can have various forms from a linear core, circular core, cross, and crosswise core. A new device was proposed by Tsai et al. (2009), which, besides the core plate, had multiple neck portions connected together in order to form several segments that would be capable to dissipate energy. This multi-curve buckling restrained brace (MC-BRB) exhibited good behavior during the experimental testing campaign and as such could enable good protection of structures exposed to seismic motions. A double circular steel tube was proposed by Kuwahara et al. (1993), while Takeita et al. (2005) proposed a triple circular steel tube as a means to prevent local buckling. To solve these problems, Ma et al. (2010) used a steel tube to restrain the cruciform steel core. As steel welding led to a major drop in fatigue performance for steel, this led to the introduction of bolts in the BRBs. Zhao et al. (2011) proposed an angle steel buckling-restrained brace (ABRB), which showed stable cyclic behavior and acceptable cumulative plastic ductility capacity. A design method for the stability verification for a four-angle ABRB was conducted by Guo and Wnag (2013), and experimental tests and finite element modeling (FEM) were performed by Ghowsi and Sahoo (2019). The idea was to use the four angles to restrain the core in order to resist global flexure on one hand, and on the other to provoke higher mode bucking of the core. In their work, Ghowsi and Sahoo (2019) investigated the hysteretic response, energy dissipation response, and displacement ductility. The FEM results were in good consistency with the experimental results regarding the fracture behavior of ABRB specimens and the hysteretic response. A quite new and promising mechanism to be used in structures exposed to ground shaking is the self-centering buckling-restrained brace technique. Miller et al. (2012) developed a self-centering buckling-restrained brace (SCBRB), with nickel-titanium (NiTi) and shape memory alloy (SMA) as the steel core. The requirement for the structure to go back to its initial position has been developed parallel with the various damper types. For recentering bars of steel of NiTi and other alloy combinations, prestressed strands are used (Dizaji and Dizaji 2016). A combination of steel slit plates and SMA gave a new hybrid damper created by Naeem et al. (2017). A substantial decrease in the residual and maximum deformations was noted once the SMA was employed in comparison with the traditional steel slit dampers. Once the SMA bars were installed, the reduction in the maximum inter-story and roof displacement was 68% and 48%, respectively. The reduction of the residual deformation at the roof level was 80% in relation to the structure where traditional steel slit dampers were installed. This opens the possibilities for the application of such material not only for strengthening and rehabilitation of existing buildings, but as well for mitigation strategies of new structures.

NiTi SMA possesses superelasticity as very large strains going to 8% are recovered elastically after load removal. Besides the energy dissipation, this enables the resisting system to return to its initial position. In this system, it is the braced-frame systems with self-centering braces that return to their initial position which is seen as an advantage (Garlock and Li 2008). Previous work on the usage of NiTi SMA was done by (Dolce and Cardone 2006, Zhu and Zhang 2007, and Yang et al. 2010). Qian et al. (2016) proposed a new superelastic shape memory alloy friction damper (SSMAFD), which represents a combination of a friction damper used for energy dissipation and SMA wires for the purpose of recentering of the structure once exposed to earthquake motion. Energy dissipation is provided as well by the SMA wires, thanks to the intrinsic damping property.

Once exposed to various earthquake motions, the structures with integrated SSMAFD showed suppressed dynamic behavior and a large amount of energy was dissipated.

Application of BRBs in civil engineering practice is going very slow, regardless of the above-mentioned research; the main reason lies in the various types of configurations and complexity of their production. As well, the problem of local bucking, multi-wave buckling of the core plate, etc. still remains an open issue that needs to be understood and resolved in an acceptable manner. Jiang et al. (2020) proposed a new BRB in which T-shaped steel is welded to the angle steel to ensure lateral constraint on the core plate which is made of steel and ingot iron which showed notable ductility and showed stable energy dissipation. Besides performing an experimental test, a FEM model was created and a nonlinear finite element analysis was conducted for validation of the experimental results.

Steel ring dampers (SRD) were investigated by Abbasnia et al. (2008), Bazzaz et al. (2012), and Bazzaz et al. (2014) and their work investigated the best position of these dampers in the steel frames and well as their linear and nonlinear behavior. The advantages of the application of the steel rings and validation finite element modeling were done by Bazzaz et al. (2015). Azandariani et al. (2020) conducted a very comprehensive investigation of the steel ring dampers's (SRS's) cyclic behavior utilizing analytical and numerical methods. They have conducted a very detailed parametric study investigating the influence of thickness (t), diameter (D), and length (L) of the steel ring. Yield force, elastic stiffness, yield displacement, ductility coefficient, and total dissipated energy were considered as results of conducted parametric investigations by Azandariani et al. (2020). It was indicative that all these paraments have a major influence on the cyclic behavior of SRD.

In recent years, much attention has been devoted to the research of the eccentrically braced frame with vertical shear link (V-EBF). This kind of dampers improved the behavior of the structural system, and reduced the damage to the load-bearing structure; they are easily replaced after earthquakes. It is the shear behavior that is the main issue in V-EBFs. Double vertical links were introduced by Shayanfar et al. (2008) and showed to have a more appropriate behavior compared with single vertical links. New details for the vertical shear link in the form of partially encase I or H shape profiles was proposed by Shayanfar et al. (2012) and an experimental campaign of these composite vertical shear links was conducted. Experiments have revealed an increase in ductility and shear capacity of these links and validation of results was conducted numerically. Bouwkamp et al. (2016) in the experimental campaign noted that this damping system is very ductile, capable of dissipating a large amount of energy having a stable hysteretic behavior. Additionally, they proposed an analytical model that was in good consistency with the obtained experimental results. From the results, Vetr et al. (2017) Vetr and Ghamari (2019) realized that the concentration of the inelastic deformation was seen in the vertical link, and the performance of all specimens is governed by this link. This means that the vertical shear link represents a ductile fuse that absorbs the energy. The analytical model proposed by Bouwkamp et al. (2016) was modified in the work of Vetr et al. (2017) Vetr and Ghamari (2019). A new damper that can be easily constructed on the site, replaced after a severe earthquake, and is more cost-effective was developed by Ghamari et al. (2019), who proposed an I-shaped steel damper with a shear yield mechanism. This system besides being a ductile fuse that absorbed the energy additionally prevents the bucking of the diagonal elements. Installing such a damper results in no stiffness degradation of the CBF, and the energy dissipation increases by 16%. A new damper that can be easily constructed on the site, replaced after severe earthquakes, and is more cost-effective was developed by Ghamari et al. (2021). In order to improve the behavior of the damper the main plate was connected to the cross-flexural plate and boundary plates in a form of an octagon. The boundary plates have to remain in the elastic range as they are supporting the main plates, and this enables the damper to dissipate energy. For improvement of the strength (an increase of 84%) and stiffness (an increase of 3.9), it is important to connect the cross plate to the web plate; however, this will cause some reduction in the ultimate displacement. If cross plates were not included, the damper did not experience nonlinear zoning, making it inadequate for usage as seismic dampers.

Bakhshayesh et al. (2021) proposed a steel damper with a shear yield mechanism. This system enabled the concentration of damage and most of the plasticity in the EDDs dissipating more than 99% of the seismic energy and in this way leaving the structure intact and in the elastic range. Additional benefits of this innovation were simple construction, low cost, and its replacement. An additional benefit is seen in the fact that there is no need for complex numerical modeling, and Bakhshayesh et al. (2021) proposed a very simple procedure for the calculation of the nonlinear behavior. A vertical link system with double-stage yielding (VLDY) creating a new hybrid damper was proposed by (Kiani and Hashemi 2021). This system is able to deliver a significant energy dissipation covering a wide range of inter-story drift levels with suitable low-cycle fatigue (LCF) resistance. The system is known under the name vertical links system with double-stage yielding (VLDY). In this case, when the moment resisting frame undergoes lateral deformation, this activated the minor link that yields in shear. The inelastic shear rotation rises in the minor link all until the lockout mechanism is activated. Once this happens, the occurrence of the inelastic deformations is noted in the major shear link, and then the complete system undergoes the second yielding stage. Adjustment of the links can be done in such a way so that the minor link dissipated the input energy in relatively small inter-story drift levels during moderate earthquakes, and the major link preserves the structural integrity under severe earthquake ground motions (Kiani and Hashemi 2021). Stable dissipation capacity is obtained at inter-story drift levels between 0.3% and 4%. The benefits of this system are ease of construction and cost-effectiveness; however, during the experimental campaign, numerical modeling limitations have been noted, requiring additional experimental research and FEM modeling before such systems are implemented in practice.

In most urban centers buildings of different total heights are constructed one next to the other, or very close to each other, and unreinforced masonry buildings are mostly suspectable to pounding effects. Once exposed to earthquake excitation, these buildings may collide due to their different dynamic properties and may cause substantial or even severe damage to the buildings. To prevent the pounding of the adjacent high-rise buildings, connected control method (CCM) has been utilized (Christenson et al. 2006, Lim et al. 2011, Zhu et al. 2011). However, CCM implementation of this method has not been widely used due to its complexity and high financial requirements. The CCM can be active, semi-active, and passive. An inertial mass damper (IMD) is a novel type of response control damper that is capable of producing an inertial force on a structure having an inertial/gyro mass component. IMD with a ball-screw amplifying mechanism can yield large inertial forces and a corresponding negative stiffness effect (Sarlis et al. 2013). Combining an energy-dissipation element with elements that possess negative stiffness provides a larger capacity for energy absorption from the vibrating structure. VIMD is a passive damper exhibiting pseudo-negative stiffness. Lu et al. (2021) combined the benefits of a viscous damper (VD) and an inertial mass damper (IMD), creating a viscous inertial mass damper (VIMD) in the CCM strategy. After a very detailed study, it was concluded that the VIMD has a much higher damping efficiency and much better behavior in relation to the viscous damper. It was noted that VIMD is able to deliver substantially larger damping ratios for the dominant modes and reduce the dominant frequencies as well Lu et al. (2021). Another way of dealing with the pounding effect is the application of the fluid viscous (FV) dampers (Pratesi et al. 2014), which have a twofold effect; one to dissipate energy (damper) and the other act as a strategy for mitigation of pounding. It is believed that this system could be widely used in a case where there is an inadequate gap between tall and short buildings in urban areas; however, the case should be taken into account as this system alters the dynamic characteristic of buildings (Sravan Ashwin et al. 2021). Mitigation of vibration on buildings and bridges due to various actions (earthquakes, winds, etc.) can be realized with the application of tuned mass dampers (TMDs). The TMD is mounted on the structure and is made of a mass, spring, and damper. The highest kinetic energy absorption is obtained when the primary structure is opposite to the direction of the TMD (phase angle $\pi/2$ radians). This means that once the fundamental natural frequency of the main structure is close to the TMD frequency there will be a growth in the equivalent effective damping quantity thanks to the energy dissipation. The effectiveness of the TMD is highly dependent on the duration of the excitation input, and in order for it to be effective, the input duration should be longer. This means that its effectiveness is discouraging and should be avoided in the near-field zones (Lukkunaprasit and Wanitkorkul 2001). Soto-Brito and Ruiz (1999) investigated the influence of the TMD on a high-rise nonlinear shear building and concluded its efficiency if the building is not exposed to high-intensity earthquakes.

Salvi et al. (2018) analyzed the influence of the soil-structure interaction on the effectiveness of TMD. It was determined that the optimum TMD configuration is dependent on the soil-structure interaction, and especially when soft soils are in question. In order to upgrade the standard vibration mitigation performance of TMD, Cheng et al. (2020) proposed an inertial amplification mechanism (IAM). It is the geometric amplification effects of a triangular-shaped mechanical system that causes the mass amplification effect of the new IAT-TMD system. The benefit of the IAM is that the response of the absorber and the primary structure is much smaller compared to a standard TMD Cheng et al. (2020). The mass amplification effect is associated with the obtained thanks to the geometrical configuration and mass distribution. The apparent mass of the absorber by application of the IAT-TMD system is amplified without increasing the actual mass of the system. It has to be kept in mind that according to the numerical simulations conducted by Cheng et al. (2020), the responses of the absorber are big (relative displacement), while the primary structure displacements are blocked and reduced. Further research for multi-degree systems is envisaged. Stiffness and damping parameters (Warburton and Ayorinde 1980, Vickery et al. 1983) of TMD have been the subject of most studies. However, Ozturk et al. (2022) for the first time investigated TMD mass parameters with the goal of optimizing the mass quantity, even though the mass ratio factor was investigated by (Bekdas and Nigdeli 2013), and minimal TMD mass was mandatory for the activation of the nonlinear response was analyzed by (Zhang and Xu 2022).

Aluminum dampers in the form of a shear link in the form of an I beam have been proposed by (Rai and Wallace 1998) for enhanced seismic resistance. The aluminum has low yield strength, which is identified as a benefit. The buckling issue is prevented with the utilization of thicker webs. Once aluminum yields in shear it shows a very ductile behavior, excellent energy dissipation capacity over a wide range of strains, and excellent stiffness capacity (Rai and Wallace 1998), and exhibits very big inelastic deformations without material tearing and buckling (Jain et al. 2008). (Rai and Wallace 1998) constructed a shear-link braced frame (SLBF) system. An experimental campaign revealed that this kind of damper has high initial stiffness, reduced base shear, larger energy dissipation capacity per unit drift, and a more uniform distribution of story drifts. Like in the case of steel dampers, replacement of dampers after an earthquake event is an easy task. Energy is greatly absorbed by the aluminum shear link and to a large extent, it limits the energy demand on the primary structure (Sahoo and Rai 2009). The device proposed by (Sahoo and Rai 2009) envisaged the usage of the external steel caging for enhancement of the flexural/shear strength of columns while the goal of the aluminum shear yielding damper (Al-SYD) was to further improve lateral strength, stiffness, and overall energy dissipation capacity of non-ductile reinforced concrete frames. Very stable hysteresis behavior was observed indicating the high ability of energy dissipation of this device. Equivalent damping potential increased five times, reducing the inelastic demand and in this way protecting the RC frame from damage once exposed to earthquakes. Al-SYD was tested and investigated by (Sachan and Rai 2012) and has been applied to the truss moment frames (TMFs). During the simulations, it was noted that the TMFs with the installation of Al-SYD attracted less base shear leading to smaller design forces in columns as compared with the traditional TMFs. Testing the effectiveness of the aluminum shear-link enabled braced frame (SLBF) on ordinary concentric braced frame (OCBF) was done by (Rai et al. 2013). The reduction of the base shear, in this case, was up to 64%; at the large value of the peak ground acceleration, the acceleration demands were reduced thanks to the shift in the natural periods which has been caused by yielding and/or buckling of shear-links at lower lateral loads. Further experimental campaigns on large-scale specimens and analytical solutions are required.

Compilation of pure aluminum energy devices was shown by De Matteis et al. (2011). Slightly better performance of geometrically the same aluminum dampers compared to steel dampers has been noted.

Additionally, researchers have investigated lead as a potential material, a material whose function is dependent on temperature. Lead is characterized by softness and high ductility, behaving like an ideal plastic material. Thanks to its excellent hysteretic characteristics, lead is capable of absorbing large amounts of energy in the process of deformations. One of the benefits is its recrystallization at room temperatures, meaning that after an earthquake event there is no need for repair or replacement of such dampers. However, its high cost and weight compared to steel devices are identified as disadvantages of these systems. Due to this, and its poor connection at lower intensities, lead is combined with other materials, like steel, for the production of energy devices, leading to a simple process, low cost, and excellent hysteretic properties. Soydan et al. (2014) investigated experimentally and conducted analytical studies on the performance of the steel connection equipped with the lead extrusion damper (LED) exposed to seismic actions. Energy dissipation of these devices was significant, going up to 175% larger dissipation, and displacement decreases by approximately 50% compared to a system without dampers. The effective stiffness has increased by 1.39 times in relation to the bare connection, and the increase of the equivalent damping ratio went up to 6.5 times compared to the bare connection. Cheng et al. (2017) constructed clapboard-type lead dampers (CLDs) and conducted experimental testing and numerical modeling with the application of the finite element method. This specific device solved the problem of steel-lead connection where cracks usually occurred. The benefits of this CLD are cost-effective production, simple structure, good deformability, minor yield displacement and stable symmetric hysteretic behavior, adequate energy dissipation capacity, and stable working performance. A hybrid damper combining lead extrusion and friction composite damper (LEFCD) was created by Yan et al. (2018) for multi-level seismic protection. Small ground motions trigger the operation of the lead extrusion, and as the ground motion intensifies in the case of stronger earthquakes it mobilizes at the same time both lead extrusion and friction dampers to dissipate the seismic energy.

In general advantages and disadvantages may be presented in Table 27.1.

27.3   CONCLUSION

Various types of dampers were presented in the paper. It is seen that the application of steel dampers stands out in comparison with other types of dampers. This is connected to their ease of construction, cost-effectiveness, and acceptance by the practice engineers as a familiar type of device. Besides the material type that has a large impact on the hysteresis performance of the damper, the geometrical configuration should be considered. Aluminum devices showed slightly better performance compared to steel dampers; however, steel as a material is still massively used due to the other benefits. Lead dampers are characterized by recrystallization; however, high price and weight are seen as disadvantages. As a result, a combination of lead and steel are utilized for the production of new damping devices to be used in the seismic-prone zones. Shape memory alloy proved to be an effective energy dissipation system due to several characteristics among which are its shape memory effect, high damping, superelasticity, resistance to corrosion, low and high fatigue life, and ability to bear big strains with no signs of residual deformation once unloaded. Compared to other metallic dampers, the SMA dampers have fairly higher life cycles. However, the price of SMA is higher compared to steel energy devices. The application of steel energy devices is the most economical solution as they can be used for strengthening and rehabilitation of existing structures as well as for placement on new structures, and buildings or infrastructures like bridges. Further investigations and upgrading of the dampers are necessary and envisaged by many researchers.

TABLE 27.1

Pros and Cons of Some Types of Dampers and Intelligent Control Systems

Type of Damper	Advantages	Disadvantages	References
Viscous damper (passive)	Avoidance of excessive additional force in a structural frame	Difficulty in responding to impulsive loading	(Kim et al. 2012)
Viscous fluid damper (passive)	Activated at low displacements Minimal restoring force For linear dampers, the modeling of a damper is simplified Properties are largely frequency and temperature independent Proven record of performance in military applications	Possible fluid seal leakage (reliability concern)	(Symans et al. 2008)
General viscoelastic damper (passive)	Cost-effective Excellent energy dissipation capacity Simple construction Easily manufactured Good durability Fine energy dissipation capacity	Introduction of excessive additional force in a structural frame Temperature, frequency, amplitude-dependence	(Kim et al. 2012) (Xu et al. 2017)
Viscoelastic solid damper (passive)	Activated at low displacements Provides restoring force Linear behavior, therefore simplified modeling of the damper	Limited deformation capacity Properties are frequency and Temperature-dependent Possible debonding and tearing of VE material (reliability concern)	(Symans et al. 2008)
High-hardness rubber damper (passive)	Low temperature and frequency dependence Large initial stiffness and large deformation capacity	Introduction of excessive additional Force in a structural frame	(Kim et al. 2012)
Hysteretic damper (shear, buckling-restrained brace) (passive)	Cost-effective	Introduction of excessive additional Force in a structural frame	(Kim et al. 2012)
Metallic damper (passive)	Stable hysteretic behavior Long-term reliability Insensitivity to ambient temperature Materials and behavior familiar to practicing engineers	Device damaged after earthquake-may require replacement Nonlinear behavior-may require nonlinear analysis	(Symans et al. 2008) (Javanmardi et al. 2020)

(Continued)

TABLE 27.1 (*Continued*)
Pros and Cons of Some Types of Dampers and Intelligent Control Systems

Type of Damper	Advantages	Disadvantages	References
Friction damper (passive)	Large energy dissipation per cycle Insensitivity to ambient temperature Damping force can be adjusted by tightening or Loosening pressure adjustment elements	Sliding interface conditions may change with time (reliability concern) Strongly nonlinear behavior may excite higher modes and require nonlinear analysis Permanent displacements if no restoring force mechanism is provided Corrosion of the friction surface, which will change the damping force of the damper No self-reset capability and it must rely on structure stiffness to reset	(Symans et al. 2008) (Xu et al. 2017)
TMD (mass damper)	Response to small levels of excitation Properties can be adjusted in the field Low maintenance Cost-effective Can be designed to add damping to two orthogonal modes of vibration Effective across all typical tall building periods Control higher building accelerations than TLCDs Effectively reduce structural dynamic responses, The device is of simple construction Easy to manufacture Little effect on the function of the main structure	A large mass and large space are required for installation, but smaller than TLCDs of equivalent performance Effectiveness depends on the maximum mass that can be utilized Effectiveness depends on the tuning accuracy Mass, no other functional use Affected by the fundamental frequency and the modes of the structure	(Chenaghlou et al. 2021) (Xu et al. 2017)
TLCD (mass damper)	Response to small levels of excitation Mass can be utilized as water supply/storage/firefighter Can be designed to add damping to two orthogonal modes of vibration Small sloshing damping and the level can be restored after vibration	Damping depending on the screens provided Water can freeze at low temperature TLCD typically suffers a change in active mass upon tuning Performance in periods beyond 8 s and/or controlling very high accelerations can be challenging Possible leakage	(Chenaghlou et al. 2021) (Xu et al. 2017)

(Continued)

TABLE 27.1 (*Continued*)
Pros and Cons of Some Types of Dampers and Intelligent Control Systems

Type of Damper	Advantages	Disadvantages	References
AMD/ATMD (active damper)	AMD system can provide additional damping to different modes by a single system An excellent mitigation effect in anti-earthquake or wind resistance applications.	May destabilize the structural system if the parameters change Sensitive to the stiffness of the structure Cost is high	(Higashino et al. 1998)
Semiactive dampers	SA control with variable dampers is the ability to modify the mechanical properties of the devices Reduction of the overall structural seismic response Limiting the bending moment at the base of the columns and displacement at the top of the building		(Caterino et al. 2022) (Xu et al. 2017)
Magnetorheological damper (MR)	Continuously adjustable control force Fast response Low power consumption Excellent reliability Wide dynamic range High-frequency response Ideal devices for reducing structural responses induced by wind or earthquake		(Xu et al. 2017)
Electrorheological damper (ER)	Advantage of a continuously adjustable control force Fast response Low power consumption Excellent reliability Wide dynamic range A high-frequency response	Heat yield strength of the ER fluid is too low and the required electric field strength is too high	(Xu et al. 2017)
Piezoelectric friction damper	Self-adaptive vibration control	Instability of the friction parameters	(Xu et al. 2017)
Semi-active variable stiffness damper	Adaptive to the structures with small stiffness	Range of natural frequency of the structure is narrow.	(Xu et al. 2017)
Semi-active variable damping damper	Controls are unconditionally stable and have good robustness		(Xu et al. 2017)

(Continued)

TABLE 27.1 (*Continued*)
Pros and Cons of Some Types of Dampers and Intelligent Control Systems

Type of Damper	Advantages	Disadvantages	References
Magnetorheological elastomer device	No particle sedimentation problem Controllable Reversible Fast-response characteristics Fine stability Simple design and low cost		(Xu et al. 2017)
Shape memory alloy dampers	High damping High resilience High driving characteristics Good fatigue resistance Corrosion resistance Durability	Sensitive to temperature	(Xu et al. 2017)

REFERENCES

Abbasnia, R.; Vetr, M.G.H.; Ahmadi, R.; Kafi, M.A. Experimental and analytical investigation on the steel ring ductility. *Sharif J. Sci. Technol.* 2008, 52, 41–48.

Armali, M.; Damerji, H.; Hallal, J.; Fakih, M. Effectiveness of Friction Dampers on the Seismic Behavior of High Rise Building VS Shear Wall System, Engineering Reports, 2019. DOI: 10.1002/eng2.12075.

Atlayan, O.; Charney, F.A. Hybrid buckling-restrained braced frames. *J. Constr. Steel Res.* 2014, 96, 95–105. https://doi.org/10.1016/j.jcsr.2014.01.001.

Aydin, E.; Noroozinejad Farsangi, E.; Öztürk, B.; Bogdanovic, A.; Dutkiewicz, M. Improvement of Building Resilience by Viscous Dampers. In: Noroozinejad Farsangi, E., Takewaki, I., Yang, T., Astaneh-Asl, A., Gardoni, P. (eds) *Resilient Structures and Infrastructure.* Springer: Singapore, 2019. https://doi.org/10.1007/978-981-13-7446-3_4.

Azandariani, M.G.; Abdolmaleki, H.; Azandariani, A.G. Numerical and analytical investigation of cyclic behavior of steel ring dampers (SRDs). *Thin-Walled Struct.* 2020, 151, 106751. https://doi.org/10.1016/j.tws.2020.106751.

Bakhshayesh, Y.; Shayanfar, M.; Ghamari, A. Improving the performance of concentrically braced frame utilizing an innovative shear damper. *J. Constr. Steel Res.* 2021, 182, 106672. https://doi.org/10.1016/j.jcsr.2021.106672.

Bakre, S.V.; Jangid, R.S.; Reddy, G.R. Optimum X-plate dampers for seismic response control of piping systems. *Int. J. Press. Vessel. Pip.* 2006, 83, 672–685. https://doi.org/10.1016/j.ijpvp.2006.05.003.

Bazzaz, M.; Andalib, Z.; Kheyroddin, A.; Kafi, M.A. Numerical comparison of the seismic performance of steel rings in off-centre bracing system and diagonal bracing system. *J. Steel Compos. Struct.* 2015, 19, 917–937. https://doi.org/10.12989/scs.2015.19.4.917.

Bazzaz, M.; Kheyroddin, A.; Kafi, M.A.; Andalib, Z. Evaluation of the seismic performance of off-centre bracing system with ductile element in steel frames. *Steel Compos. Struct., Int. J.* 2012, 12(5), 445–464. https://doi.org/10.12989/scs.2012.12.5.445.

Bazzaz, M.; Kheyroddin, A.; Kafi, M.A.; Andalib, Z.; Esmaeili, H. Evaluating the seismic performance of off-centre bracing system with circular element in optimum place. *Int. J. Steel Struct.* 2014, 14(2), 293–304. https://doi.org/10.1007/s13296-014-2009-x.

Bekdas, G.; Nigdeli, S.M. Mass ratio factor for optimum tuned mass damper strategies. *Int. J. Mech. Sci.* 2013, 71, 68–84. https://doi.org/10.1016/j.ijmecsci.2013.03.014.

Benavent-Climent, A.; Escolano-Margarit, D.; Arcos-Espada, J.; Ponce-Parra, H. New metallic damper with multiphase behavior for seismic protection of structures. *Metals.* 2021, 11, 183. https://doi.org/10.3390/met11020183.

Bergman, D.; Goel, S. *Evaluation of Cyclic Testing of Steel-Plate Devices for Added Damping and Stiffness.* Department of Civil Engineering, University of Michigan: Michigan, MI, USA, 1987.

Black, C.J.; Makris, N.; Aiken, I.D. Component Testing, Stability Analysis and Characterization of Buckling Restrained 'Unbonded' Braces. Technical Report PEER 2002/08, Pacific Earthquake Engineering Research Center, University of California, Berkeley, CA. 2002.

Bouwkamp, J.; Vetr, M.G.; Ghamari, A. An analytical model for inelastic cyclic response of eccentrically braced frame with vertical shear link (V-EBF). *Case Stud. Struct. Eng.* 2016, 6, 31–44. https://doi.org/10.1016/j.csse.2016.05.002.

Budaházy, V.; Dunai, L. Numerical analysis of concrete filled buckling restrained braces. *J. Constr. Steel Res.* 2015, 115, 92–105. https://doi.org/10.1016/j.jcsr.2015.07.028.

Caterino, N.; Spizzuoco, M.; Piccolo, V.; Magliulo, G. A semi-active control technique through MR fluid dampers for seismic protection of single-story RC precast buildings. *Materials.* 2022, 15, 759. https://doi.org/10.3390/ma15030759.

Chenaghlou, M.R.; Gharabaghi, A.R.; Mohasel, M.H. Dynamic response control of offshore jacket platforms. In Proceedings of 12th International Congress on Civil Engineering, Mashhad, Iran, 12–14 July 2021, pp. 1–8.

Cheng, S.; Du, S.; Yan, X.; Guo, Q.; Xin, Y. Experimental study and numerical simulation of clapboard lead damper. *Proc. Inst. Mech. Eng. Part. C. J. Mech. Eng. Sci.* 2017, 231, 1688–1698. https://doi.org/10.1177/09544 06215 62133 9.

Cheng, Z.; Palermo, A.; Shi, Z.; Marzani, A. Enhanced tuned mass damper using an inertial amplification mechanism. *J. Sound Vib.* 2020, 475, 115267. https://doi.org/10.1016/j.jsv.2020.115267.

Christenson, R.E.; Spencer, B.F. Jr; Johnson, E.A.; Seto, K. Coupled building control considering the effects of building/connector configuration. *J. Struct. Eng.* 2006, 132, 853–863. https://doi.org/10.1061/(ASCE)0733-9445(2006)132:6(853).

De Matteis, G.; Brando, G.; Mazzolani, F.M. Hysteretic behaviour of bracing-type pure aluminium shear panels by experimental tests. *Earthq. Eng. Struct. Dyn.* 2011, 40, 1143–1162. https://doi.org/10.1002/eqe

Dizaji, F.S.; Dizaji, M.S. A Novel Smart Memory Alloy Recentering Damper for Passive Protection of Structures Subjected to Seismic Excitations Using High Performance NiTiHfPd Material. 2016, arXiv:2105.04081 [nlin. AO]. https://arxiv.org/ftp/arxiv/papers/2105/2105.04081.pdf.

Dolce, M.; Cardone, D. Theoretical and experimental studies for the application of shape memory alloys in civil engineering. *J. Eng. Mater. Technol., Trans. ASME.* 2006, 128(3), 302–311. https://doi.org/10.1115/1.2203106.

Ebadi Jamkhaneh, M.; Homaioon Ebrahimi, A.; Shokri Amiri, M. Seismic performance of steel-braced frames with an all-steel buckling restrained brace. *Pract. Period. Struct. Des. Constr.* 2018, 23(3), 04018016. https://doi.org/10.1061/(ASCE)SC.1943-5576.0000381.

Fahnestock, L.A.; Sause, R.; Ricles, J.M. Experimental evaluation of a large-scal buckling-restrained braced frame. *J. Struct. Eng.* 2007a, 33(9), 1205–1214. https://doi.org/10.1061/(ASCE)0733-9445(2007)133: 9(1205).

Fahnestock, L.A.; Sause, R.; Ricles, J.M. Seismic response and performance of buckling-restrained braced frames. *J. Struct. Eng.* 2007b, 133(9), 1195–1204. https://doi.org/10.1061/(ASCE)0733-9445(2007)133: 9(1195).

Fisco, N.R.; Adeli, H. Smart structures: Part I—Active and semi-active control. *Sci. Iran Trans.* 2011, 18(3), 275–284. https://doi.org/10.1016/j.scient.2011.05.034.

Garlock, M.E.M.; Li, J. Steel self-centering moment frames with collector beam floor diaphragms. *J. Construct. Steel. Res.* 2008, 64, 526–538. https://doi.org/10.1016/j.jcsr.2007.10.006.

Ghamari, A.; Almasi, B.; Kim, C.-H.; Jeong, S.-H.; Hong, K.-J. An innovative steel damper with a flexural and shear–flexural mechanism to enhance the CBF system behavior: An experimental and numerical study. *Appl. Sci.* 2021, 11, 11454. https://doi.org/10.3390/app112311454.

Ghamari, A.; Haeri, H.; Khaloo, A.; Zhu, Z. Improving the hysteric behavior of concentrically braced frame (CBF) by a proposed shear damper. *Steel Compos. Struct.* 2019, 30, 383–392. https://doi.org/10.12989/scs.2019.30.4.383

Ghamari, A.; Kim, Y.; Bae, J. Utilizing an I-shaped shear link as a damper to improve the behaviour of a concentrically braced frame. *J. Construct. Steel Res.* 2021, 186, 1–13. https://doi.org/10.1016/j.jcsr.2021.106915.

Ghorbani, H.R.; Rofooei, F.R. A novel double slip loads friction damper to control the seismic response of structures. *Eng. Struct.* 2020, 225, 111273. https://doi.org/10.1016/j.engstruct.2020.111273.

Ghowsi, A.F.; Sahoo, D.R. Experimental study of all-steel buckling restrained braces under cyclic loading. In Proceedings of the International Conference on Earthquake Engineering and Structural Dynamics, Geotechnical, Geological and Earthquake Engineering 47, 1st ed.; Rupakhety, R., Olafsson, S., Bessason, B., Eds.; Springer, Cham, 2019, pp 67–80. https://doi.org/10.1007/978-3-319-78187-7_6.

Guo, Y-L.; Wnag, X-A. Study on the restrain ratio of a four-angle assembled steel buckling-restrained brace[J]. *Eng. Mech.* 2013, 30(10), 35–45. https://doi.org/10.6052/j.issn.1000-4750.2012.06.0458.

Heertjes, M.; Van de Wouw, N. Nonlinear dynamics and control of a pneumatic vibration isolator. *ASME J. Vib. Acoust.* 2006, 28, 439–448. https://doi.org/10.1115/1.2128642.

Higashino, M.; Aizawa, S.; Yamamoto, M.; Toyama, K. Application of active mass damper (AMD) system, and earthquake and wind observation results. In The Proceedings of 2nd World Conference on Structural Control, Kyoto, Japan, June 28–July 1 1998.

Ibrahim, R.A. Recent advances in nonlinear passive vibration isolators. *J. Sound Vib.* 2008, 314, 371–452. https://doi.org/10.1016/j.jsv.2008.01.014.

Inamasu, H.; Skalomenos, K.A.; Hsiao, P.C.; Hayashi, K.; Kurata, M.; Nakashima, M. Gusset plate connections for naturally buckling braces, *J. Struct. Eng.* 2017, 143(8) 04017065. https://doi.org/10.1061/(ASCE)ST.1943-541X.0001794.

Jain, S.; Rai, D.C.; Sahoo, D.R. Post yield cyclic buckling criteria for aluminum shear panels. *J. Appl. Mech.* 2008, 75, 021015-1-1 to 021015-1-8. https://doi.org/10.1115/1.2793135.

Javanmardi, A.; Ibrahim, Z.; Ghaedi, K. et al. State-of-the-art review of metallic dampers: Testing, development and implementation. *Arch. Computat. Methods Eng.* 2020, 27, 455–478. https://doi.org/10.1007/s11831-019-09329-9

Jiang, T.; Dai, J.; Yang, Y.; Liu, Y.; Bai, W. Study of a new-type of steel buckling-restrained brace. *Earthq. Eng. Eng. Vib.* 2020, 19(1), 239–256. https://doi.org/10.1007/s11803-020-0559-9.

Katayama, T.; Ito, S.; Kamura, H.; Ueki, T.; Okamoto, H. Experimental study on hysteretic damper with low yield strength steel under dynamic loading. In Proceedings of the 12th World Conference on Earthquake Engineering, Auckland, New Zealand, 30 January–4 February 2000, Paper 1020, pp. 1–8.

Kiani, B.K.; Hashemi, B. H. Development of a double-stage yielding damper with vertical shear links. *Eng. Struct.* 2021, 246, 112959. https://doi.org/10.1016/j.engstruct.2021.112959.

Kim, H.-G.; Yoshitomi, S.; Tsuji, M.; Takewaki, I. Ductility inverse-mapping method for SDOF systems including passive dampers for varying input level of ground motion. *Eng. Struct.* 2012, 3, 59–81, https://doi.org/10.12989/eas.2012.3.1.059.

Kimura, K.; Yoshioka, K.; Takeda, T. Tests on braces encased by mortar in-filled steel tubes. Summaries of technical papers of annual meeting, Architectural Institute of Japan, 1976, 1041, pp. 1–42.

Kuwahara, S.; Tada, M.; Yoneyama, T.; Imai, K. A study on stiffening capacity of double-tube members. *J. Struct. Constr. Eng.* 1993, 445(3), 151–158. https://doi.org/10.3130/aijsx.445.0_151.

Lee, J.; Kang, H.; Kim, J. Seismic performance of steel plate slit-friction hybrid dampers. *J. Constr. Steel Res.* 2017, 136, 128–139. https://doi.org/10.1016/j.jcsr.2017.05.005.

Lee, T.Y.; Kawashima, K. Effectiveness of seismic displacement response control for nonlinear isolated bridge. *Struct. Eng./Earthq. Eng.* 2006, 23, 1s–15s.

Lee, T.Y.; Kawashima, K. Semiactive control of nonlinear isolated bridges with time delay. *ASCE J. Struct. Eng.* 2007, 133, 235–241. https://doi.org/10.1061/(ASCE)0733-9445(2007)133:2(235).

Li, C.; Liu, Y.; Li, H.-N. Fragility assessment and optimum design of a steel–concrete frame structure multi-hazards of earthquake and wind. *Eng. Struct.* 2021, 245, 112878. https://doi.org/1010.1016/j.engstruct.2021.112878.

Lim, J.; Bienkiewicz, B.; Richards, E. Modeling of structural coupling for assessment of modal properties of twin tall buildings with a skybridge. *J. Wind Eng. Ind. Aerodyn.* 2011, 99, 615–623. https://doi.org/10.1016/j.jweia.2011.02.010.

Lu, L.; Xu, J.; Zhou, Y.; Lu, W.; Spencer, B.F. Viscous inertial mass damper (VIMD) for seismic responses control of the coupled adjacent buildings. *Eng. Struct.* 2021, 233. https://doi.org/10.1016/j.engstruct.2021.111876.

Lukkunaprasit, P.; Wanitkorkul, A. Inelastic buildings with tuned mass dampers under moderate ground motions from distant earthquakes. *Earthq. Eng. Struct. Dyn.* 2001, 30, 537–551. 10.1002/eqe.22.

Ma, N.; Wu, B.; Zhao, J.X. "Full scale uniaxial and subassemblage tests on the seismic behavior of all-steel buckling-resistant brace." *China Civ. Eng. J.* 2010, 43(4), 1–7. (in Chinese).

Martinelli, P.; Mulas, M.G. An innovative passive control technique for industrial precast frames. *Eng. Struct.* 2010, 32, 1123–1132. https://doi.org/10.1016/j.engstruct.2009.12.038.

Martinez Rueda, J.E.; Elnashai, A.S. A novel technique for the retrofitting of reinforced concrete structures. *Eng. Struct.* 1995, 17(5), 359–371. https://doi.org/10.1016/0141-0296(95)00019-4.

Martinez Rueda, J.E. Seismic redesign of RC frames by local incorporation of energy dissipation devices. In Proceedings of the 6th US National Conference on Earthquake Engineering, Seattle, Washington, May 31–June 4 1998.

Miller, D.J.; Fahnestock, L.A.; Eatherton, M.R. Development and experimental validation of a Nickel–Titanium shape memory alloy self-centering buckling-restrained brace." *Eng. Struct.* 2012, 40, 288–298. https://doi.org/10.1016/j.engstruct.2012.02.037.

Mohammadi, R.K.; Nasri, A.; Ghaffary, A. TADAS dampers in very large deformations. *Int. J. Steel Struct.* 2017, 17, 515–524. https://doi.org/10.1007/s13296-017-6011-y.

Mulas, M.G.; Martinez Rueda, J.E. Passive control in seismic retrofitting of steel MRFs. In Proceedings of 3rd World Conference on Structural Control, Wiley, Chichester, 2003, Vol. 3, pp. 223–234.

Naeem, A.; Eldin, M.N.; Kim, J.; Kim, J. Performance evaluation of a structure retrofitted using steel slit dampers with shape memory alloy bars. *Int. J. Steel Struct.* 2017, 17(4), 1627–1638. https://doi.org/10.1007/s13296-017-1227-4.

Ozturk, B.; Cetin, H.; Dutkiewicz, M.; Aydin, E.; Noroozinejad Farsangi, E. On the efficacy of a novel optimized tuned mass damper for minimizing dynamic responses of cantilever beams. *Appl. Sci.* 2022, 12, 7878. https://doi.org/10.3390/app12157878.

Palmer, K.; Roeder, C.; Okazaki, T.; Shield, C.; Lehman, D. Three-dimensional tests of two-story, one-bay by one-bay, steel concentric-braced frames. In The Proceedings of Structures Congress, ASCE, Las Vegas, NV, 2011, pp. 3057–3067. https://doi.org/10.1061/41171(401)266.

Parulekar, Y.M.; Reddy, G.R. Passive response control systems for seismic response reduction: A state-of-the-art review. *Int. J. Struct. Stab. Dyn.* 2009, 9, 151–177. https://doi.org/10.1142/S0219455409002965.

Pokhrel, A.; Li, J.C.; Li, Y.C.; Maksis, N.; Yu, Y. Comparative studies of base isolation systems featured with lead rubber bearings and friction pendulum bearings. *Appl. Mech. Mater.* 2016, 846, 114–119. https://doi.org/10.4028/www.scientific.net/AMM.846.114.

Pratesi, F.; Sorace, S.; Terenzi, G. Analysis and mitigation of seismic pounding of a slender R/C bell tower. *Eng. Struct.* 2014, 71, 23–34. https://doi.org/10.1016/j.engstruct.2014.04.006.

Qian, H.; Li, H.; Song, G. Experimental investigations of building structure with a superelastic shape memory alloy friction damper subject to seismic loads. *Smart Mater. Struct.* 2016, 25, 125026. https://doi.org/10.1088/s13296-017-1227-4.

Rai, D.C.; Annam, P.K.; Pradhan, T. Seismic testing of steel braced frames with aluminum shear yielding dampers. *Eng Struct.* 2013, 46, 737–747. https://doi.org/10.1016/j.engstruct.2012.08.027.

Rai, D.C.; Wallace, B.J. Aluminum shear link for enhanced seismic resistance. *J. Earthq. Eng. Struct. Dyn.* 1998, 27, 315–342. https://doi.org/10.1002/(SICI)1096-9845(199804)27:4<315::AID-EQE703>3.0.CO;2-N.

Sachan, A.; Rai, D.C. Aluminium shear yielding damper (Al-SYD) as an energy dissipation device in truss moment frames (TMFs). In Proceedings of the 15th World Conference on Earthquake Engineering, Lisbon, Portugal, 24–28 September 2012, pp. 1–10.

Sahoo, D.R.; Rai, D.C. A novel technique seismic strengthening of RC frame using steel caging and aluminum shear yielding device. *Earthq Spectra (EERI).* 2009, 25(2), 415–437. https://doi.org/10.1193/1.3111173.

Saiedi, A. A Review of Triangle Yielding Metal Dampers (TADAS) in Brace, PACE 2021- Ataturk University, Engineering Faculty, Department of Civil Engineering, Erzurum, 25030, Turkey, 20–23 June 2021

Sajjadi Alehashem, S.M.; Keyhani, A.; Pourmohammad, H. Behavior and performance of structures equipped with ADAS & TADAS dampers a comparison with conventional structures. In Proceedings of the 14th World Conference on Earthquake Engineering, Beijing, China, 12–17 October 2008, pp. 1–8.

Salvi, J.; Pioldi, F.; Rizzi, E. Optimum tuned mass dampers under seismic soil-structure interaction, *Soil Dyn. Earthq. Eng.* 2018, 114, 576–597. https://doi.org/10.1016/j.soildyn.2018.07.014.

Sarlis, A.A.; Pasala, D.T.R.; Constantinou, M.C.; Reinhorn, A.M.; Nagarajaiah, S.; Taylor, D.P. Negative stiffness device for seismic protection of structures. *Ascelibrary Org.* 2013, 139, 1124–33. https://doi.org/10.1061/(ASCE)ST.1943-541X.0000616.

Shayanfar, M.A.; Barkhordari, M.A.; Rezaeian, A.R. Experimental study of cyclic behavior of composite vertical shear link in eccentrically braced frames. *Steel Compos. Struct.* 2012, 12, 13–29. https://doi.org/10.12989/scs.2011.12.1.013.

Shayanfar, M.A.; Rezaeian, A.; Taherkhani, S. Assessment of the seismic behavior of eccentrically braced frame with double vertical link (DV-EBF). In Proceedings of the 14th World Conference on Earthquake Engineering, Beijing, China, 12-18 October 2008, pp. 1–8.

Shih, M.-H.; Sung, W.-P. A model for hysteretic behavior of rhombic low yield strength steel added damping and stiffness. *Comput. Struct.* 2005, 83, 895–908. https://doi.org/10.1016/j.compstruc.2004.11.012.

Shih, M.-H.; Wen-pei, S.; Cheer-Germ, G.O. Investigation of newly developed added damping and stiffness device with low yield strength steel. *J. Zhejiang Univ. A.* 2004, 5, 326–334. https://doi.org/10.1007/BF02841018.

Skalomenos, K.A.; Inamasu, H.; Shimada, H.; Nakashima, M. Development of a steel brace with intentional eccentricity and experimental validation. *J. Struct. Eng.* 2017, 143(8) 04017072. https://doi.org/10.1061/(ASCE)ST.1943-541X.0001809.

Soto-Brito, R.; Ruiz, S.E. Influence of ground motion intensity on the effectiveness of tuned mass dampers. *Earthq. Eng. Struct. Dyn.* 1999, 28, 1255–1271. https://doi.org/10.1002/(SICI)1096-9845(199911)28:11<1255::AID-EQE865>3.0.CO;2-C.

Soydan, C.; Yuksel, E.; Irtem, E. The behavior of a steel connection equipped with the lead extrusion damper. *Adv. Struct. Eng.* 2014, 17, 25–39. https://doi.org/10.1260/1369-4332.17.1.25.

Sravan Ashwin, A.; Arunachalam, P.; Sreenivas, M.K.; Rahima Shabeen, S. Seismic energy dissipation systems – A review. *Aust. J. Struct. Eng.* 2021, 23(1), 1–25. https://doi.org/10.1080/13287982.2021.1989167.

Symans, M.D.; Charney, F.A.; Whittaker, A.S.; Constantinou, M.C.; Kircher, C.A.; Johnson, M.W.; McNamara, R.J. Energy dissipation systems for seismic applications: Current practice and recent developments. *J. Struct. Eng. ASCE.* 2008, 134(1), 3–21. https://doi.org/10.1061/(ASCE)0733-9445(2008)134:1(3).

Tahamouli Roudsari, M.; Eslamimanesh, M.B.; Entezari, A.R.; Noori, O.; Torkaman, M. Experimental assessment of retrofitting RC moment resisting frames with ADAS and TADAS yielding dampers. *Structures.* 2018, 14, 75–87. https://doi.org/10.1016/j.istruc.2018.02.005.

Takeda, T.; Takemoto, Y.; Furuya, Y. An experimental study on moment frame with steel braces part 3. *Annu. Meet. AIJ.* 1972, 47, 1389–1390. (in Japanese)

Takeita, K.; Nagao, T.; Taguti, T. Studies on Buckling-Restrained Brace Using Triple Steel Tubes Part 2: Consideration on Experimental Results and Finite Element Method Analysis. In: *Summaries of Technical Papers of Annual Meeting.* Architectural Institute of Japan: Tokyo, 2005, pp. 1013–1014.

Tehranizadeh, M. Passive energy dissipation device for typical steel frame building in Iran. *Eng. Struct.* 2001, 23, 643–655. https://doi.org/10.1016/S0141-0296(00)00082-1.

Torunbalci, N. Seismic isolation and energy dissipating stems in earthquake resistant design. In Proceedings of the 13th World Conference on Earthquake Engineering, Vancouver, B.C., Canada 1–6 August 2004 Paper No. 3273, pp. 1–12.

Tremblay, R.; Bolduc, P.; Neville, R.; DeVall, R. Seismic testing and performance of buckling-restrained bracing systems. *Can J. Civil. Eng.* 2006, 33, 183–198. https://doi.org/10.1139/l05-103.

Tsai, K.; Chen, H.; Hong, C.; Su, Y. Design of steel triangular plate energy absorbers for seismic-resistant construction. *Earthq. Spectra.* 1993, 9, 505–528. https://doi.org/10.1193/1.1585727.

Tsai, C.S.; Lin, Y.; Chen, W.; Su, H.C. Mathematical modeling and full-scale shaking table tests for multi-curve buckling restrained braces. *Earthq. Eng. Eng. Vib.* 2009, 8(3), 359–371. https://doi.org/10.1007/s11803-009-9004-9.

Vetr, M.G.; Ghamari, A. Experimentally and analytically study on eccentrically braced frame with vertical shear links. *Struct. Des. Tall Spec. Build.* 2019, 28, e1587. https://doi.org/10.1002/tal.1587.

Vetr, M.G.; Ghamari, A.; Bouwkamp, J. Investigating the nonlinear behavior of eccentrically braced frame with vertical shear links (V-EBF). *J. Build. Eng.* 2017, 10, 47–59. https://doi.org/10.1016/j.jobe.2017.02.002.

Vickery, B.J.; Isyumov, N.; Davenport, A.G. The role of damping. *MTMD Accel. J. Wind. Eng. Ind. Aerodyn.* 1983, 11, 285–294.

Wagner, J.; Liu, X. Nonlinear modeling and control of automotive vibration isolation systems. In Proceedings of the American Control Conference (ACC), (IEEE Cat. No. 00CH36334), Chicago, IL, 2000, Vol 1, pp. 564–568.

Warburton, G.B.; Ayorinde, E.O. Optimum absorber parameters for simple systems. *Earthq. Eng. Struct. Dyn.* 1980, 8, 197–217. https://doi.org/10.1002/eqe.4290080302.

Xu, Z.-D.; Guo, Y.-Q.; Zhu, J.-T.; Xu, F.-H. Introduction. In: *Intelligent Vibration Control in Civil Engineering Structures*, 2017, pp. 1–20. https://doi.org/10.1016/B978-0-12-405874-3.00001-1.

Yan, X.; Chen, Z.; Qia, A.; Wanga, X.; Shi, S. Experimental and theoretical study of a lead extrusion and friction composite damper. *Eng. Struct.* 2018, 177, 306–317. https://doi.org/10.1016/j.engstruct.2018.09.080.

Yang, C.S.W.; DesRoches, R.; Leon, R.T. Design and analysis of braced frames with shape memory alloy and energy-absorbing hybrid devices. *Eng. Struct.* 2010, 32, 408–507. https://doi.org/10.1016/j.engstruct.2009.10.011.

Zhang, M.; Xu, F. Tuned mass damper for self-excited vibration control: Optimization involving non-linear aeroelastic effect. *J. Wind. Eng. Ind. Aerodyn.* 2022, 220, 104836. https://doi.org/10.1016/j.jweia.2021.104836.

Zhao, J.; Wu, B.; Ou, J. A novel type of angle steel buckling-restrained brace: Cyclic behavior and failure mechanism. *Earthq. Eng. Struct. Dyn.* 2011, 40(10), 1083–1102. https://doi.org/10.1002/eqe.1071.

Zhu, H.; Ge, D.D.; Huang, X. Optimum connecting dampers to reduce the seismic responses of parallel structures. *J. Sound Vib.* 2011, 330, 1931–1949. https://doi.org/10.1016/j.jsv.2010.11.016.

Zhu, S.; Zhang, Y. Seismic behaviour of self-centering braced frame buildings with reusable hysteretic damping brace. *Earthq. Eng. Struct. Dynam.* 2007, 36, 1329–1346. https://doi.org/10.1002/eqe.683.

28 Application of Machine Learning in Design of Steel Plate Shear Walls

Arsalan Mousavi, Arman Mamazizi,
and Vahidreza Gharehbaghi

CONTENTS

28.1 INTRODUCTION

Buildings are key assets for each nation, since they represent the development and culture of the country. Every year, lateral and horizontal forces from natural disasters such as earthquakes, floods, and violent storms risk the safety and integrity of structures. Thus, building codes have devised numerous lateral force resisting methods, such as different shear walls, moment frames, and braced systems. Because the building codes are commonly updated annually, designers must keep their knowledge up to date in order to provide reliable recommendations for structural design.

Steel plate shear walls (SPSWs) have been utilized to build new structures or retrofitting existing ones, notably in seismic areas, such as the United States and Japan. Excellent energy dissipation capacity, steady hysteresis characteristics, strong lateral stiffness, and substantial ductility make SPSWs one of the most dependable lateral resisting structures [1]. Practically for mid-rise and high-rise structures, these systems have showed significant performance based on previous studies [2–4]. Several advantages have made SPSWs attractive for designers and constructors. Some of the main merits are stated below [5]:

- **Thickness:** SPSWs provide a thinner wall structure than concrete shear walls of equivalent strength.
- **Weight:** A building using SPSW is approximately 18% lighter than one with concrete shear walls.
- **Rapid Installation:** Constructing takes less time when an SPSW system is used. There is no required curing time; thus, it may be assembled quickly.

DOI: 10.1201/9781003325246-28

- **Ductility:** In resisting post-buckling, even a thin steel plate holds up very well. Studies into the SPSW system show that the system can tolerate up to 4% drift without encountering severe damage.
- **Resiliency:** Several strong earthquakes have shaken at least two structures to their foundations, proving the effectiveness of SPSW as a key lateral force resisting device.

A typical SPSWs comprise three major components, including beams (horizontal boundary elements), columns (vertical boundary elements), and thin web panels that infill frames. Design of these systems requires considering various parameters from the geometry of components, thickness of plates, and materials properties. Considering these factors requires complicated efforts. Also, nonlinear response history analysis is required to and impacts of all variables such as web plate's tension are not precisely considered in buildings codes. For example, American Institute of Steel Construction (AISC) [6] and the Canadian Standards Association (CSA) [7] do not explicitly account for the strength of the wall provided by the moment resisting action of the boundary frame, which can lead to a more conservative, and therefore more expensive, design [8]. Hence, considering more empirical-based design methods could be more reliable and practical in some cases.

Structure analysis and design processes cause a time-consuming calibration process and could be difficult to accomplish in practice when applied to large structural systems subjected to severe reactions with nonstationary behavior. Previously, many different types of variables utilized in structural design, response analysis, and performance evaluation have been predicted using statistical models based on experimental or field data. When physics-based models, which are often simplified for practical reasons, do not even accurately reflect known physical mechanisms, these alternative models have been extremely beneficial. For example, the ACI-318 Building Code Requirements for structural concrete features a variety of analytical connections that are either wholly empirical (based on regression utilizing experimental data) or hybrid (i.e., engineering equations incorporating empirically determined parameters). As a result, it is worthwhile to investigate if alternative solutions have the potential to improve the prediction performance of these empirical correlations and, if so, what trade-offs are involved [9].

The present level of influence of artificial intelligence (AI) on every aspect of human existence is unparalleled. Since the 1980s, when logic-based AI systems were initially used, structural engineers, for example, have used AI technology to undertake design explorations. AI-based engineering solutions have been available for a while, but their popularity has lately increased because of the broad availability of low-cost data collection and processing capabilities. Machine learning (ML) is a subset of AI that can detect patterns in data even when they are masked by noise or other sources of uncertainty. This uncertainty is caused by the tiny sample sizes employed, but it is also due to errors committed during data collection (including measurements) and a lack of clear epistemic certainty [10].

In terms of complex engineering problems, ML offers potentially practical alternates that will help to save time and energy [11]. Seven sub-classes of ML methods that are popular for structural predictions and design application are: NN methods (i.e., ANN, ANFIS, and CNN), boosting algorithms, regression analysis, support vector machines, random forest, decision trees, and other techniques like Naïve-Bayes and K-nearest neighbor. According to a comprehensive survey in 2022 [11], over 50% of ML approaches used in the structural engineering domain apply NN, and more than 80% of studies have focused on ANN, highlighting the importance of ANN in this field.

One of the earliest applications of ML in structural engineering was carried out in 1989 by Adeli and Yeh [12], who used ANNs to design steel beams. Subsequent research in the 1990s utilized ANNs [13] to the challenges of performance assessment and structural design, expanding on the foundation laid by the aforementioned pioneering work. Since then, ANN methods have been widely used in innovative structural engineering research and predictive problems for structural analysis and design; as a case in point, for predication of structural members in shear resistance, axial resistance, buckling strength, flexural resistance, and serviceability.

Although there are numerous studies of the application of ANN in the design of RC and masonry shear walls, limited studies have been focused on SPSWs. To name some, in a study by Vahidi and Roshani [14], load-carrying capacity of these systems is investigated under static load with the aid of nonlinear geometrical and material analysis in FEM software. The thickness of the plate, the location of the opening, the width of the diagonal stiffeners, and the diameter of the circular stiffeners are the variables that have been examined. Then, an ANN is proposed to model the effects of these parameters. Using the verified simulated data, the obtained findings show that the suggested ANN model has achieved has achieved good agreement, with a correlation coefficient of greater than 0.99. Moradi et al. [15] employed FEM to investigate how the presence of rectangular openings affects the lateral load-bearing performance of SPSWs. An ANN makes predictions on SPSW behavior based on the FEM's outputs. Effects of the rectangular opening at different plate thicknesses are modeled using a radial basis function (RBF) network.

With the above review, in this chapter, the authors attempt to design an ANN for prediction of the variables in a SPSW with one story. To feed the network, a FEM verified by an empirical specimen is used to simulate the required input data. Different indices are then employed to demonstrate the accuracy of the model. First, the SPSW sample and FEM configuration are described. The next step is the introduction of an empirical specimen to validate the generated numerical data. The next section illustrates the employed ANN's architecture, input parameters, and performance indexes. The next part includes results and discussion, followed by the conclusion.

28.2   MATERIALS AND DATA

In this section, the numerical model and the creation of the data set are provided. A typical SPSW with and components is shown in Figure 28.1. These systems are classified into two types, stiffened and unstiffened, each of which has its own mechanisms [16].

In order to do finite element analysis and collect numerical data for this investigation, a sample model is created. The reference model is a non-stiffened SPSW with one level and one-third the scale of the laboratory specimen described in the reference papers [16, 17]. In these papers, three samples of stiffened steel shear wall (DS-PSW-0%), unstiffened steel shear wall (DS-PSW), and surrounding frame (frame) were tested and the corresponding behavior was explored. The primary purpose of this study was to evaluate the effect of stiffeners on the response of the specimen, and

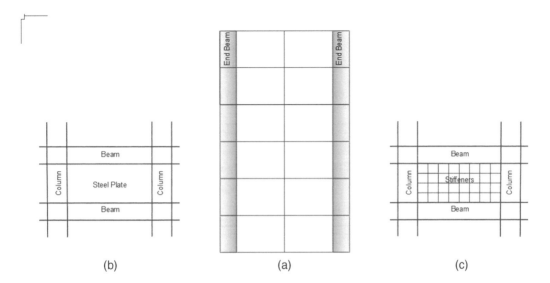

FIGURE 28.1   Schematic of SPSWs. (a) Frame elevation; (b) SPSW without stiffeners; (c) SPSW with stiffeners.

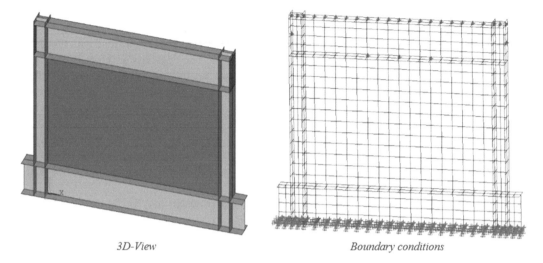

3D-View   Boundary conditions

FIGURE 28.2 Configuration of numerical model.

the authors concluded stiffeners could assess the seismic behavior of the SPSW. Herein, we utilize the same specimen to validate the numerical model used to generate numerical data.

The final data set consists of 60 unstiffened samples based on DS-PSW models that are subjected to uniform lateral loading. Several parameters influence the maximum lateral load of the walls. In this study, a set of parameters is assumed to be constant across all samples in order to evaluate the effect of other variable parameters. All the samples are subjected to the same loads and have similar connections, boundary constraints, and material properties. In this investigation, the only variables are the geometric dimensions of the components. To analyze the samples, we employ ANSYS software and the SHELL181 elements with four nodes and six degrees of freedom at each node [18]. Figure 28.2 presents the configuration of the numerical model used in this study.

The goal is to attain the maximal lateral load capacity provided by uniform loading of the samples. Hence, the samples are loaded equally and at the level of the upper beam's center line. Rigid plates are employed in the modeling procedure to prevent stress concentrations in the areas of applied load. Also, all the beam-to-column connections are fully rigid. The displacement of the bottom flange of the bottom beam is constrained in all directions, including the middle of the upper beam, three locations in the middle of the upper beam, and two points of the loading position in the direction perpendicular to the plane (Z). Details of the mechanical properties of the materials are represented bilinearly in Table 28.1.

The section that follows validates the numerical model by comparing it to the empirical specimen.

TABLE 28.1
Mechanical Properties of Finite Element Samples

Parameter	Numerical Sample	Empirical Sample
Yield stress of beam and column (Mpa)	414.9	414.9
Yield strain of beam and column	0.0020745	-
Ultimate stress of beam and column (MPa)	551.8	551.8
Ultimate strain of beam and column	0.18	-
Yield stress of plate	192.4	192.4
Ultimate stress of plate (Mpa)	0.0009431	-
Ultimate strain of plate	277.2	277.2

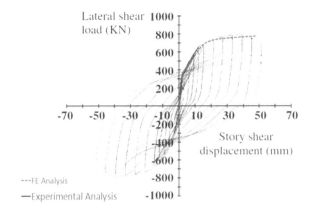

FIGURE 28.3 Force-displacement graph for DS-PSW and reference specimen.

28.3 VERIFICATION OF FEM

As previously stated, the DS-PSW reference sample is employed to ensure optimum accuracy in modeling for validation. The mechanical properties of the DS-PSW laboratory specimen with limited components are shown in Table 28.1. In the FEM, loading is applied uniformly in 100 stages with 0.5 mm increments. As depicted in Figure 28.3, the comparison of models demonstrates adequate convergence. Moreover, the shape of the plate and its out-of-plane buckling in the numerical sample and the laboratory specimen confirm the accuracy of the modeling done (see Figure 28.4).

28.4 RESEARCH METHODOLOGY

Three steps are required to create a ML model, which includes preparing the data set, learning, and performance assessment. Figure 28.5 depicts the flow chart of ML for prediction applications.

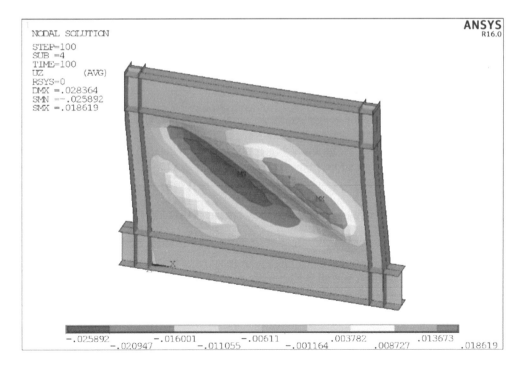

FIGURE 28.4 Deformation of DS-PSW sample.

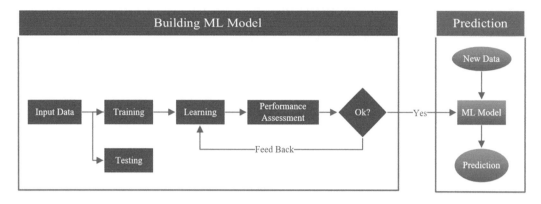

FIGURE 28.5   ML typical flow chart for prediction application (adopted from [11]).

After reaching adequate accuracy in training and resting procedure, the trained model can use with new input data to predict a new problem. In the following sections, a ML model is implemented to predict the SPSW parameters.

28.4.1   Step 1: Data Preparation

Many parameters are effective in determining the maximum lateral load of SPSW; as previously stated, the geometric dimensions of the sample components are considered variables, whereas the remaining parameters, such as boundary conditions, type of connections, loading, and mechanical properties, are constant in all samples in the modeling.

Based on the reference specimen, 60 one-story and one-span samples with varied geometric dimensions are subjected to uniform lateral loading. The final data set comprises the geometric dimensions as well as the maximum lateral force. For training the ML model, 80% of the samples are chosen at random as training data and 20% as test data. Table 28.2 displays 19 geometric variables, which comprise the geometric dimensions of beams, columns, and steel plates. These variables are fed into the network's input layer, which consists of 19 neurons. Figure 28.6 illustrates the histogram regarding the geometric dimensions of the frame.

28.4.2   Building Model

The ANN employed in this study comprises three layers: an input layer with 19 neurons, a hidden layer with a variable number of neurons ranging from 1 to 200 neurons, and the output layer with one neuron (see Figure 28.7). Three accuracy metrics are used to determine the number of hidden layer neurons. The number of neurons in the hidden layer is chosen based on the highest coefficient of determination (R^2), lowest mean square error (MSE), and root-mean-square error (RMSE) for the training and test data.

Accordingly, the objective of employing accuracy criteria is to achieve the following:

- Number of neurons
- The largest amount of neurons in the hidden layer
- Monitoring the network's overall performance.

28.4.3   Performance Metrics

As previously stated, the criteria for determining the accuracy of the ML model are determined in accordance with the architecture of the ANN, which has a fixed input and output layer and a hidden

TABLE 28.2
Parameters Used in This Study

Number	Parameter	Description
1	Btop	Flange width (upper beam)
2	bttop	Flange thickness (upper beam)
3	Htop	Distance of the ends of flanges (upper beam)
4	Dtop	Distance of centerline of flanges (upper beam)
5	httop	Thickness of web (upper beam)
6	Bbot	Width of flange (upper beam)
7	btbot	Thickness of flange (upper beam)
8	Hbot	Distance of ends of flanges (upper beam)
9	Dbot	Distance of centerlines of flanges (lower beam)
10	htbot	Thickness of web (lower beam)
11	Hcol	Distance of ends of flanges of columns
12	Bcol	Width of column's flange
13	tfcol	Thickness of column's flange
14	twcol	Thickness of column's web
15	Lcol	Length of column
16	Lbeam	Length of beam
17	Dplate	Width of plate
18	Lplate	Length of plate
19	tplate	Thickness of plate

layer with a variable number of neurons, to obtain the number of neurons with the highest accuracy and the lowest error. Herein, we utilize three measures to assess the accuracy of the proposed ANN: mean square error (MSE), root-mean-square error (RMSE), and coefficient of determination (R^2).

MSE: Mean squared error is a method for estimating error, which is the difference between actual and estimated values. The mean square error is one of the most common criteria for both training and comparing models. This criterion is multi-dimensional and always positive. Another essential aspect of this criterion is that it gives more weight to larger errors. The following are the relationships used to calculate the mean square error.

$$MSE(x, y) = \frac{1}{n}\sum_{i=1}^{n}(x_i - y_i)^2 \tag{28.1}$$

RMSE: Using the following relationship, the square root of the error is obtained by calculating the root of the average square of the errors. This index is used because its dimension and scale are the same as the target variable.

$$RMSE(x, y) = \sqrt{MSE(x, y)} \tag{28.2}$$

R^2: Unlike the previous criteria, the coefficient of determination shows the highest sensitivity in the model as it increases. This index is always less than one and is determined as follows:

$$R^2 = 1 - \frac{\sum_{i=1}^{n}(y_i - x_i)^2}{\sum_{i=1}^{n}(x_i - \overline{y_i})^2} \tag{28.3}$$

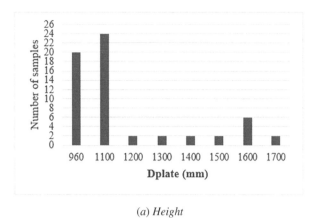

(a) Height

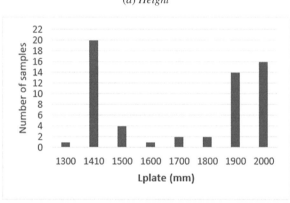

(b) Length of plate

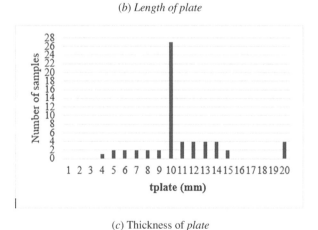

(c) Thickness of plate

FIGURE 28.6   Histogram of number of input variables.

where x represents the predicted output, y represents the data set output, $\overline{y_i}$ indicates the average of data set output, and n refers to the number of data set samples.

28.5   RESULTS AND DISCUSSIONS

In this section, the proposed model is used to predict the SPSW's parameters and the results are compared with the reference specimen. To build the ML model, coding is carried out by Python 3 [19].

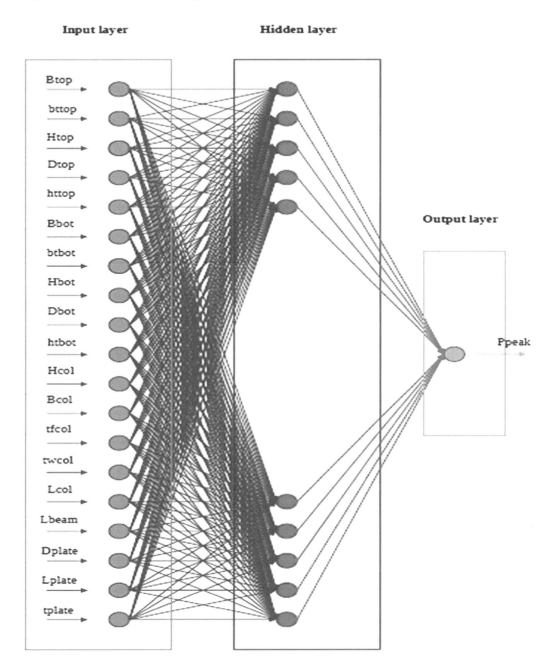

FIGURE 28.7 Architecture of ANN.

Python has surpassed other ML languages as the preferred language for coding ML models because of its availability of pre-built libraries that include hundreds of methods. Below are the libraries used in this study [11].

Pandas: This is the most widely used ML library in Python. In 2008, Pandas made its debut. It provides choices for manipulating several types of data at a high level, including matrix data, tabular data, and time series. It also offers several effective features for effectively managing big data. This library is used in this study's coding to read the input data file.

TABLE 28.3

System Specifications

CPU	Intel Core i7-5500U
Ram	8GB DDR3
GPU	2GB- NVIDIA GeForce 940M
HDD	750 GB

NumPy: This is a library for handling arrays in a wide variety of situations. NumPy stands for "Numerical Python," and is built upon the basis of the much-older "Numeric" library. Because of its extensive mathematical functions for working with massive multi-dimensional arrays and matrices, it has rapidly become one of the most important and widely used libraries for ML. It works well for random simulations, the Fourier transform, and elementary algebra. We utilize NumPy and the min-max method to normalize the input data.

Matplot: This Python visualization package contains a wide variety of charting tools. NumPy serves as the foundation for this program. Line plots, scatter plots, bar charts, a histogram, and many other charting tools are only some of the many that can be found in Matplotlib. It is straightforward and easily understandable to employ.

Scikit-learn: This library was developed in 2007 using the Python numerical and scientific libraries NumPy and SciPy as a framework. The scikit-learn package includes several useful algorithms for dimensionality reduction, clustering, and classification. With its focus on data mining and analysis, it has quickly become one of the most widely used ML libraries. This library is employed in this chapter to design and implement the ML model.

All the codes and scripts are executed through a single system with the specifications listed in Table 28.3.

The hyperparameters of a learning algorithm are the parameters whose values regulate the learning procedure and define the values of the model parameters that the algorithm learns. The hyper_ prefix indicates that these parameters govern both the learning procedure and the resulting model parameters. It is stated that hyperparameters are "external to the model" if their values cannot be modified by the model during the training and development phases.

Various methods exist for selecting a model's hyperparameters, such as random search, grid search, manual search, Bayesian optimizations. As part of our research, we used GridSearchCV model from scikit-learn library, that uses the grid search approach to get the best values for model hyperparameters. After calculating the performance for each combination of the hyperparameters and their values, grid search chooses the optimal value for the hyperparameters. In this study, the learning rate is set to 0.001, and the number of hidden layer neurons is 100. Also, we assumed a maximum repeat of 1000 and one hidden layer in total. To optimize the network, all hyperparameters are considered being fixed except the number of neurons in the hidden layer, and we aim to identify the most suitable number of neurons in the hidden layer by altering the number of neurons In the hidden layer and checking the accuracy of measurement criteria.

By comparing the results of ANN accuracy evaluation criteria based on the number of hidden layer neurons, we concluded a general similar tendency between training and test data. To make it easy to inspect the ANN results, we divide the maximum lateral load of the steel shear wall by 10^6, so that the results are in mega-newton (MN). According to Equation (28.1), which describes MSE, the unit of MSE is $10^{12} N^2$ due to the square power of the outputs, and the unit of RMSE is 10^6 N according to Equation (28.2).

Regardless of the limited number of results from the number of hidden layer neurons, the range of changes of the MSE criterion for the training data is between 0 and 2, which is equally true for

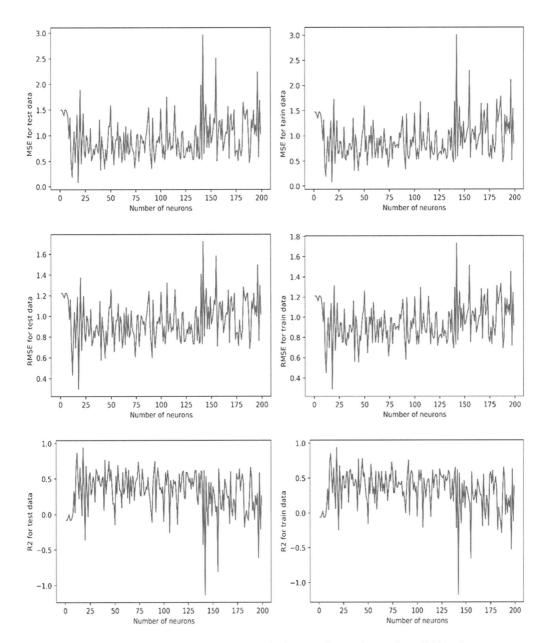

FIGURE 28.8 Results of accuracy measurement criteria according to the number of hidden layer neurons.

the test data. In addition, for training and test data, the standard range of RMSE is 0 and 1.4, and the standard range of R^2 is –0.3 to 1.

Figure 28.8 depicts the accuracy of measurement criterion for training and test data, demonstrating that the optimal measurement criteria values occur when the number of hidden neurons is less than 25. MSE values vary between 0.5 and 1.5 without taking into consideration the best and worst results of the measurement criteria for both training and test data sets. RMSE values range between 0.6 and 1.2, and R^2 values range between 0 and 0.7. Furthermore, we concluded that having 18 neurons leads is the best performance in both training and test data sets across all three accuracy assessment criteria.

TABLE 28.4

Results of the Accuracy of Measurement Criteria for the Number of 18 Neurons in the Hidden Layer

Performance Metrics	Train Data	Test Data
MSE	0.083	0.088
RMSE	0.289	0.296
R^2	0.939	0.936

Table 28.4 and Figure 28.9 present the results of measurement criteria for the number of 18 neurons in the hidden layer for the training and test data, with MSE, RMSE, and R^2 values of approximately 0.083, 0.289, and 0.939 for the training data and 0.088, 0.296, and 0.936 for the test data, respectively.

Figure 28.10 depicts the distribution of FEM outputs and the ML model with 18 neurons in the hidden layer for training and test data, which can estimate the accuracy of the ANN. It substantiates the ML model's proper performance, which differed little from the actual results obtained from FEM.

28.6 CONCLUSION

Steel shear wall is one sort of lateral restraining systems, and its design should account for the maximum lateral load. SPSW parameters can be determined using a variety of computational and laboratory approaches. Our goal of this study is to predict the maximum lateral load of a steel shear wall based on the geometric dimensions through an ANN. Like other machine learning algorithms, ANN needs a series of data for training. The input data of the neural network consists of finite element analysis of 60 unstiffened SPSW samples with different geometric dimensions, 80% of these 60 samples are employed for training the ML model and 20% for testing.

The designed ANN comprises three layers, an input layer with 19 neurons consisting of the geometric dimensions of steel shear wall components, a hidden layer with 18 neurons, and an output layer with one neuron, which is defined as the maximum lateral load.

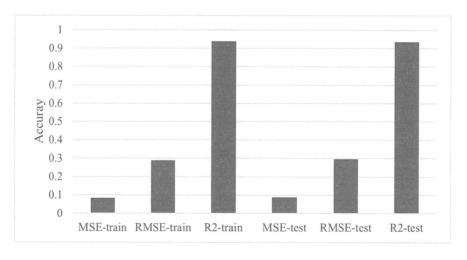

FIGURE 28.9 Accuracy measurement criteria for the network with optimum numbers of neurons in the hidden layer.

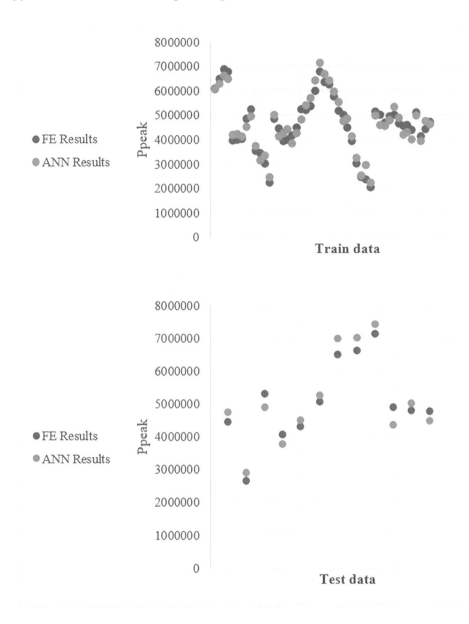

FIGURE 28.10 Distribution of outputs of finite element analysis and artificial neural network: (a) training data and (b) test data.

The following are the most important results gained after training the ML model and comparing the results with the reference specimen:

- By studying the measurement results of the number of neurons in the hidden layer, it was concluded that the ANN with the hidden layer comprising 18 neurons had the best performance, with R^2 values of 0.939 and 0.936, respectively, for the training and test data.
- According to the results for 18 neurons in the hidden layer, the difference between the training and test data is 5.2% in the MSE criterion, and 2.5% and 0.3% in the RMSE and R^2 criterion, respectively, showing strong convergence between the two data.

- Considering the configuration of ANN, there is no clear relationship between the number of neurons in the hidden layer and the accuracy of measurement criterion, and the most vital point that can be stated about the measurement criteria is that the network performs best when the number of hidden layer neurons is less than 25.

REFERENCES

1. Mamazizi, A.; Khani, S.; Gharehbaghi, V.; Farsangi, E. N., Modified plate frame interaction method for evaluation of steel plate shear walls with beam-connected web plates. *Journal of Building Engineering* 2022, 45, 103682.
2. Nie, J.; Fan, J.; Liu, X.; Huang, Y., Comparative study on steel plate shear walls used in a high-rise building. *Journal of Structural Engineering* 2013, 139 (1), 85–97.
3. Usefi, N.; Ronagh, H.; Sharafi, P., Lateral performance of a new hybrid CFS shear wall panel for mid-rise construction. *Journal of constructional Steel Research* 2020, 168, 106000.
4. Zhang, J.; Liu, J.; Li, X.; Cao, W., Seismic behavior of steel fiber-reinforced high-strength concrete mid-rise shear walls with high-strength steel rebar. *Journal of Building Engineering* 2021, 42, 102462.
5. Seilie, I. F.; Hooper, J. D., Steel plate shear walls: practical design and construction. *Modern steel construction* 2005, 45 (4), 37–43.
6. American Institute of Steel Construction, *Steel construction manual*. American Institute of Steel Construction: 2005.
7. Home, M.; No, C. N., Canadian standards association. 1980.
8. Qu, B.; Bruneau, M., Design of steel plate shear walls considering boundary frame moment resisting action. *Journal of Structural Engineering* 2009, 135 (12), 1511–1521.
9. Burton, H.; Mieler, M., Emerging technology machine learning applications hope, hype, or hindrance for structural engineering. *Struct Magazine June* 2021, 16–20.
10. Málaga-Chuquitaype, C., Machine learning in structural design: an opinionated review. *Frontiers in built environment* 2022, 8, 6.
11. Thai, H.-T. *Machine learning for structural engineering: A state-of-the-art review*, Structures, 2022. Elsevier: 2022; pp. 448–491.
12. Adeli, H.; Yeh, C., Perceptron learning in engineering design. *Computer-Aided Civil and Infrastructure Engineering* 1989, 4 (4), 247–256.
13. Hopfield, J. J., Neural networks and physical systems with emergent collective computational abilities. *Proceedings of the National Academy of Sciences* 1982, 79 (8), 2554–2558.
14. Khalilzadeh Vahidi, E.; Roshani, M., Prediction of load-carrying capacity in steel shear wall with opening using artificial neural network. *Journal of Engineering* 2016, 2016. http://dx.doi.org/10.1155/2016/4039407
15. Moradi, M. J.; Roshani, M. M.; Shabani, A.; Kioumarsi, M., Prediction of the load-bearing behavior of SPSW with rectangular opening by RBF network. *Applied Sciences* 2020, 10 (3), 1185.
16. Sabouri-Ghomi, S.; Kharrazi, M. H.; Mam-Azizi, S. E. D.; Sajadi, R. A., Buckling behavior improvement of steel plate shear wall systems. *The structural Design of Tall and Special Buildings* 2008, 17 (4), 823–837.
17. Sabouri-Ghomi, S.; Sajjadi, S. R. A., Experimental and theoretical studies of steel shear walls with and without stiffeners. *Journal of Constructional Steel Research* 2012, 75, 152–159.
18. Stolarski, T.; Nakasone, Y.; Yoshimoto, S., *Engineering analysis with ANSYS software*. Butterworth-Heinemann: 2018.
19. Sheridan, C., *The Python language reference manual*. Lulu Press, Inc: 2016.

29 Blast Mitigation of Irregular Buildings Equipped with Resilient Passive Control Systems

Muhammed Zain Kangda, Rohan Raikar, and Ehsan Noroozinejad Farsangi

CONTENTS

29.1   INTRODUCTION

The civil engineering structures are most commonly plagued by an issue to classify them as regular and irregular when subjected to earthquake induced vibrations. Under the seismic action, the regular buildings perform better as compared to irregular structures. Various international seismic codes of practice [1–4] have defined regular structures as one which is free from any horizontal and vertical discontinuities or irregularities in the structure. Some of the essential characteristics of regular buildings are rectangular structural plan, uniform mass distribution over the height and continuous placement of shear walls in each storey [5]. An irregular structure is one that is afflicted by any one or the combination of the discontinuity as tabulated in Table 29.1 in its configuration or in its lateral force resisting system. The damages incurred to irregular civil engineering structures in past earthquakes are primarily due to asymmetry in plan and elevation, sudden changes in strength and stiffness resulting in weak and soft storeys, respectively, and sudden changes in floor plan area leading to setbacks. The earthquake loadings generate torsion and local deformations in irregular buildings due to asymmetry and presence of re-entrant corners and excessive openings. These effects generate additional stress concentration in the structural members of buildings resulting in large deformation and damages. The study presented by [6] highlights the advantages of irregular buildings as compared to a regular-shaped building in the form of architectural, environmental, aesthetics and urbanization aspects with a case study. Thus, the construction of civil engineering irregular structures is inevitable. Researchers have developed and extensively studied methods to safeguard and retrofit the severely damaged irregular structures during the past earthquakes.

Many researchers have employed various passive control techniques namely tuned mass dampers [7–9], viscoelastic dampers [10–12], friction dampers [13–15], base isolation [16–18] and fluid viscous damper [19–22] to enhance the seismic performance of irregular buildings. Various active control techniques such as magnetorheological damper [23–25], active tuned dampers [26–28] and semi-active dampers [29–31] have also been incorporated to protect these asymmetrical structures

DOI: 10.1201/9781003325246-29

TABLE 29.1
Types of Irregularities as per Various International Codes of Practice

Sr No.	Name of Country	Code of Practice	Types of Irregularities	
			Plan Irregularity	Vertical Irregularity
1.	India	IS:1893 – 2016	• Torsional irregularity • Re-entrant corners • Floor slab having excessive openings • Out of plane offset in vertical element • Non-parallel lateral force system	• Stiffness irregularity (soft storey) • Mass irregularity • Vertical geometric irregularity • In-plane discontinuity in vertical elements resisting lateral force • Strength irregularity (weak storey) • Floating or stub column • Irregular modes of oscillation in two principal plane directions
2.	America	ASCE 7-2021	• Torsional irregularity • Extreme torsional irregularity • Re-entrant corners • Diaphragm discontinuity • Out of plane offset • Non-parallel systems	• Stiffness irregularity (soft storey) • Extreme stiffness irregularity • Weight (mass) irregularity • Vertical geometric irregularity • In-plane discontinuity in vertical elements resisting lateral force • Discontinuity in lateral strength (weak storey)
3.	China	GB 50001 – 2010	• Torsional irregularity • Uneven irregularity • Partial discontinuity of floor slab	• Irregularity of lateral rigidity • Discontinuity of vertical lateral force resisting component • Discontinuity of storey-bearing capacity
4.	UK	Eurocode 8	• Torsional irregularity • Re-entrant corners • Diaphragm discontinuity	• Stiffness irregularity (soft storey) • Mass irregularity • Vertical geometry irregularity • In-plane discontinuity in vertical elements resisting lateral force

in the past in the field of earthquake engineering. Thus, the present structural engineers have developed methods to mitigate and protect structures against the disastrous damages caused to the irregular buildings during the earthquake phenomenon. In the recent events of bombings and attacks on civil engineering structures, namely The World Trade Centre, New York (2001); Imam Ali Mosque, Saudi Arabia (2003); The Taj Mahal Hotel, Mumbai (2008); Badr Mosque, Yemen (2015) and the apartment building in Chasiv Yar, Ukraine (2022) has urged the structural engineering community to design blast-resistant structures. Extensive research on the input blast load calculations [32–34] is available and methods to safeguard structures by providing reinforced steel stud walls [35–37], securing building envelopes [38–40] and keeping the glass intact [41–43] and improving the building façade [44–46] have been studied extensively in the past. Researchers have also studied the performance of irregular structures subjected to air and mine blast loading. Pachla et al. [47] compared the performance of irregular buildings subjected to different earthquakes and mine blast durations and observed that long-duration high ground acceleration earthquake phenomena are detrimental to structural safety as compared to medium duration high peak ground accelerations generated due to mining activities. Kiakojouri et al. [48] presented a detailed review on various strengthening and retrofitting techniques namely reinforced polyurethane bricks, sandwich panels, sacrificial cladding, steel jacketing and bracing, fibre-reinforced plastic and energy absorber devices against fire and blast and impact loadings. Ebrahimi et al. [49] concluded that the energy absorption capacity of regular structures is higher as compared to irregular structures due to column removal and the force demand to capacity ratio of columns in irregular structures is in the range of 1.5 to 2 times that of regular structures. Nica et al. [50] validated the collapse mechanism of irregular RC structure using applied element method and experimentally using a one-fourth scale model with an objective to identify the non-progressive collapse scenario on column removal. The performance of a geometrically irregular steel building designed for seismic loads is compared under different blast loads with concentrically and eccentrically braced frames by Coffield and Adeli [51]. The blast fragility curves were developed for regular and irregular braced steel frame structures and the effect of blast standoff distance on the presented results was also discussed by Kumar and Matsagar [52].

Thus, it is observed that a limited study investigating the performance of irregular RC structures subjected to blast loading is available and it is the prime objective of the present study. The study implements the earthquake mitigation techniques to the irregular structures against blast-induced vibrations primarily due to blasting near mine sites. The present study investigates the issue of plan irregularity in buildings subjected to blast-induced vibrations and the effectiveness of passive control techniques namely fluid viscous dampers and X-plate dampers in controlling the damages incurred to the irregular RC buildings under the short-duration blast loads. The blast charge weight is kept constant in the present whereas the distance of the blast from the building position is varied to understand the effect of the standoff distance (R) on the building performance. The study also reviews the damper placement techniques to optimize the number of the installed dampers in mitigating the blast-induced responses.

29.2 METHODOLOGY

Blasting is routinely conducted in mines to fragment huge rocks. This causes motion on the ground and have an impact on the nearby structures. The ground motions generated are similar to seismic excitations and cannot be ignored since it causes significant structural damage. The blast-induced acceleration ($\ddot{x}_g$) is a function of peak particle velocity (v) in m/s and time of arrival (t_a) calculated as ratio of radial distance (R) in meters and wave propagation velocity of soil (c) in m/s of soil. The study incorporated the blast load parameters studied by Kangda and Bakre [53] to develop the blast time history, as shown in Figure 29.1 and given by Equation (29.1).

$$\ddot{x}_g(t) = -\frac{1}{t_a} v e^{\frac{-t}{t_a}} \tag{29.1}$$

The peak particle velocity of the rock particles is calculated using Equation (29.2) proposed by Kumar et al. [54] and formulation is a function of average mass density (γ_d) of rock particles, uniaxial compressive strength (fc) of rock particles and scaled distance (D) in m/kg$^{1/2}$ and determined as the ratio of distance from charge point, R (m) to the square root of charge mass Q (kg). The granite rock particles parameters examined in the present study are taken from the study by [55].

$$v = \frac{f_c^{0.642} \, SD^{-1.463}}{\gamma_d} \tag{29.2}$$

In the present study, the blast load of intensity (Q) of 50 tonnes is applied to the structure by varying the standoff distance (R) in the range of 100, 200, 300 and 400 m. The input blast history is produced at a step time of 0.0005 sec and 20,001 input steps.

It is thus crucial to implement structural control techniques namely viscous and viscoelastic damper, metallic damper, friction damper and tuned mass damper to minimize the structural responses and protect them under such erratic loading conditions. Fluid viscous dampers are now widely employed worldwide. This damper operates on the energy dissipation principle. The system consists of a stainless-steel piston, a steel cylinder with two chambers and silicon fluid to provide

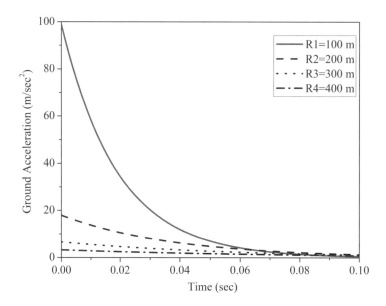

FIGURE 29.1   Time history of blast-induced ground acceleration.

smooth fluid circulation. The fluid viscous damper is modelled as a nonlinear damper called a
Maxwell element. The behaviour of a viscous damper is modelled as pure stiffness-free damping. The study takes into account zero stiffness to obtain a pure damping in a nonlinear analysis.
The fluid viscous damper's energy dissipator per cycle depends on several parameters. The ideal
damping force of viscous damper is given by Equation (29.3). The parameter F represents damping force, C is the damping coefficient, V is the velocity of piston relative to cylinder and α is the
damping exponent. In the present study, the damping coefficient (C) is evaluated from Table 29.2
as presented by [56] and given input to link damper exponential element in SAP 2000 finite element software.

$$F = C \ V^\alpha \tag{29.3}$$

The present study also evaluates the effectiveness of X-plate dampers in mitigating the structural
responses of regular and irregular RC buildings subjected to blast ground acceleration. One of the
main properties of X-plate dampers (XPD) is that, during cyclic deformations, the metal plates are
subjected to hysteretic mechanism and the plasticization of these plates consumes a substantial

TABLE 29.2
Properties of Viscous Damper [56]

Damper Notation	Group A		Group B	
	Damping Coefficient (kNs/m)	Damping Exponent (α)	Damping Coefficient (kNs/m)	Damping Exponent (α)
A1/B1	420	0.8	330	0.36
A2/B2	330	0.56	280	0.4
A3/B3	330	0.51	280	0.39
A4/B4	310	0.52	260	0 45
A5/B5	290	0.55	260	0.5

TABLE 29.3
Properties of X-Plate Damper [57]

1	Modulus of Elasticity (E)	1.94E+05 N/mm^2
2	Yield Stress (σ_y)	235 N/mm^2
3	Strain Hardening Rate (H)	5,000 N/mm^2
4	Thickness of Plate (t)	2, 4 and 8 mm
5	Breadth of Triangular Portion (b)	20, 40, 60, 80, 100, 120, 140, 160
6	Height of Triangular Portion (a)	20, 40, 60, 80, 100, 120, 140, 160
7	Yielding Exponent (n)	10

portion of the structural vibration energy. Moreover, the additional stiffness introduced by the metallic elements increase the lateral strength of the building, with the consequent reduction in deformations and damage in the main structural members. The cyclic response of yielding metallic devices is strongly nonlinear accompanied by changes in element stiffness due to loading, unloading and reloading of yielding elements.

The introduction of these devices in a structure will render it to behave nonlinearly, even if the other structural elements that support these devices remain linear when they are subjected to the real ground motions. A hysteretic forcing model such as Bouc Wen model is used to replicate the parameters involved in the design of a typical metallic element as studied by Bakre et al. [57]. The properties of XPD detailed in Table 29.3 are input to SAP 2000 software using Plastic Wen link element to compare the effectiveness with viscous dampers in mitigating the blast structural responses.

The present study examines four different G+10 story reinforced concrete structures to compare the performance of regular and irregular plan geometries in buildings subjected to blast load. The regular plan buildings have square and rectangular geometries, whereas plan shapes, namely C-shaped and L-shaped are grouped as irregular buildings. The sizes of the various building shapes chosen for the study along with the building data assumed in the present study are tabulated in Tables 29.4 and 29.5, respectively. All the selected G+10 buildings have a plinth height of 1.5 m above the foundation level with a floor-to-floor height of 3 m. All the buildings are residential-style structures with a Poisson's ratio of 0.2 and elastic modulus of 250 MPa. The concrete unit weight is 25 kN/m^3, masonry unit weight is 22 kN/m^3, footing depth is 1.5 m and slab thickness is 125 mm with thickness of external and internal walls kept at 230 and 150 mm, respectively. In the present study, the following three models are analysed in SAP 2000 software.

a. G + 10 regular and irregular RC buildings subjected to blast loading without damper.
b. G + 10 regular and irregular RC buildings subjected to blast loading installed with viscous damper at all locations, Alternate floors and zig-zag pattern.
c. G + 10 Regular and irregular RC buildings subjected to blast loading installed with XPD damper at all locations, alternate floors and zig-zag pattern.

Initially, a linear static analysis is carried out for all building shapes, namely square, rectangle, C-shaped and L-shaped using FE software SAP 2000 under dead load and live load combinations and it is observed that all the member sections pass under the conventional loading conditions. Next, the volatile blast loading is input to all the selected buildings and it is observed that member sections of both regular and irregular building plans fail and a need to retrofit the structures with passive control techniques is evident to protect the structures from failure. A schematic representation of the methodology incorporated in the present study to enhance the structural performance

TABLE 29.4
Selected Plan Dimensions with Member Properties and Structural Output Parameters

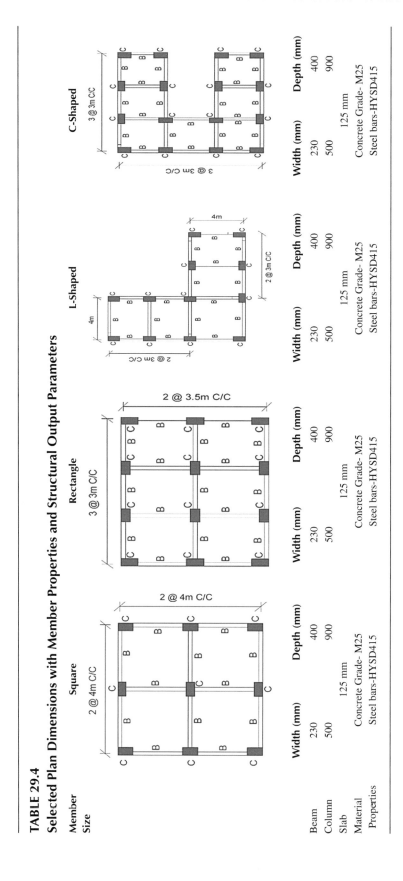

TABLE 29.5

Summary of Building Data, Load Calculations and X-Plate Damper Properties

1	Number of Storeys	G + 10
2	Plinth Height	1.5m
3	Floor-to-Floor Height	3m
4	Type of Structure	Residential Structure
5	Poisson's Ratio	0.2
6	Elastic Modulus of Concrete	250 MPa
7	Unit Weight of Concrete (IS 875 Part – I)	25 kN/m^3
8	Unit Weight of Masonry (IS 875 Part – I)	22 kN/m^3
9	Footing Depth	1.5 m
10	Slab Thickness	125 mm
11	External Wall Thickness	230 mm
12	Internal Wall Thickness	150 mm
13	Live Load (IS 875 Part – II)	2 kN/m^2
14	Live Load on Terrace (IS 875 Part – II)	0.75 kN/m^2
15	Floor Finish (IS 875 Part – II)	1.5 kN/m^2
16	Wall Load Calculation	
	• Main Wall Load	$= (3 - 0.4) \times 0.23 \times 22$
		$= 13.156$ kN/m
	• Partition Wall Load	$= (3 - 0.4) \times 0.15 \times 22$
		$= 8.58$ kN/m
	• Parapet Wall Load	$= (1 \times 0.15 \times 22)$
		$= 3.3$ kN/m
17	X-Plate Damper	
	• Effective Stiffness of XPD	$= K_d = \dfrac{Ebt^3}{12a^3} \, n$
		$= \dfrac{1.94E + 05 \times 20 \times 2^3}{12 \times 20^3} \times 10$
		$= 3233.33$ kN/m
	• Yield Force of XPD	$= F_y = \dfrac{\sigma_y bt^2}{6a} \, n$
		$= \dfrac{235 \times 20 \times 2^2}{6 \times 20} \times 10$
		$= 1566.67$ N $= 1.56667$ kN
	• Yield Displacement of XPD	$= q = \dfrac{2 \times 235 \times 20^2}{1.94E + 05 \times 2}$
		$= 0.4845$ mm
	• Post Yield Strength Ratio	$= 0.025$ (Ratio of plastic stiffness to elastic stiffness of X-plate ADAS element)
	• Yielding Exponent	$= 10$

(Continued)

TABLE 29.5 (*Continued*)

Summary of Building Data, Load Calculations and X-Plate Damper Properties

18

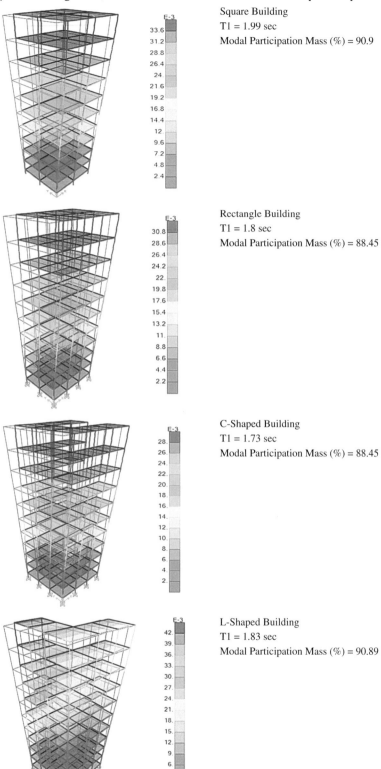

Square Building
T1 = 1.99 sec
Modal Participation Mass (%) = 90.9

Rectangle Building
T1 = 1.8 sec
Modal Participation Mass (%) = 88.45

C-Shaped Building
T1 = 1.73 sec
Modal Participation Mass (%) = 88.45

L-Shaped Building
T1 = 1.83 sec
Modal Participation Mass (%) = 90.89

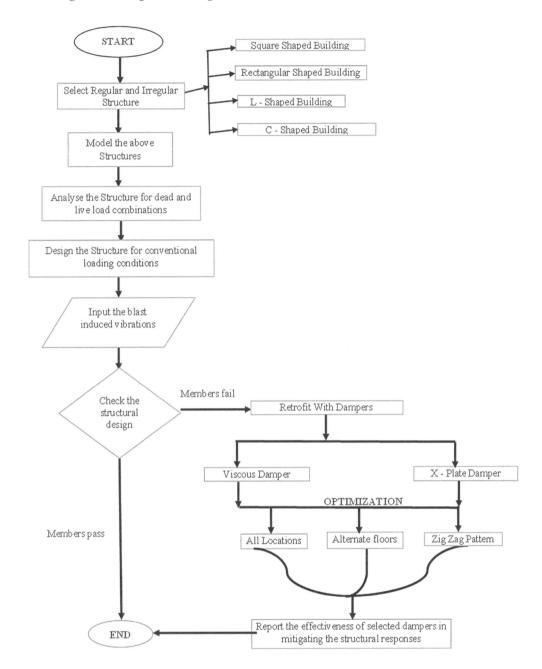

FIGURE 29.2 Schematic representation of the retrofitting methodology adopted for blast protection.

of all selected buildings is depicted by Figure 29.2. The study also compares the structural output parameters, namely natural period and percentage mass excited for all the selected regular and irregular buildings.

29.3 RESULTS AND DISCUSSIONS

The nonlinear dynamic analysis of regular and irregular buildings equipped without and with fluid viscous and XPD subjected to four different blast histories is performed in the present study. The plan area of all the selected plan geometries is kept constant in the range of 64 m² to compare the blast effects of regular and irregular structures. The selected plan geometry buildings are square, rectangle, C-shaped and L-shaped. The modal time period of the selected buildings is found to be 1.99, 1.8, 1.83 and 1.73 sec, respectively. The buildings are designed using IS 456:2000 [58] for a load combination 1.5 (dead load + live load) located in a low seismic zone of Chakki Khapa village near Nagpur City, Maharashtra, India. The design results show that the buildings are inadequately designed for ground vibrations produced by the mining activities nearby. The design results obtained from blast analysis show that many columns fail and the need to retrofit the selected buildings is essential to prevent them from collapse. In the present study, two resilient passive control techniques, namely fluid viscous dampers and X-plate dampers, are installed in the regular and irregular buildings susceptible to the mine blast loads. The viscous dampers are further classified into two categories, namely class A damper and class B damper. The blast load is assumed to be 50 tonnes, located at standoff distances of 100, 200, 300 and 400 m, respectively. The analysis results show that the peak top storey displacement of square, rectangular, C-shaped and L-shaped buildings are found be 89.57, 71.22, 77.32 and 69.58 mm when subjected to the blast at a standoff distance of 400 m. Thus, unlike seismic loads, the blast load causes maximum destruction to regular buildings in comparison to irregular buildings. The results show that the viscous dampers show a nominal reduction in the peak storey displacement responses of square, rectangle, L-shaped and C-shaped buildings in the range of 7–26%, 6–24%, 7–27% and 7–32% for the selected viscous damper properties for blast load generated at 400 m. The efficiency of the blast load further reduces for closely denoted blast loads at 100 m and the reductions are in the range of 5–7%, 4–6%, 5–7% and 6–9% for square, rectangle, L-shaped and C-shaped buildings, respectively. The other structural responses, namely peak top storey acceleration, bottom-corner column shear force and bending moments generated due to blast charge of 50 tonnes located at a standoff distance of 100 (R1), 200 (R2), 300 (R3) and 400 m (R4) are enhanced with the viscous damper passive control technique and tabulated in Table 29.6. The maximum reductions observed in absolute acceleration for square, rectangle, L-shaped and C-shaped buildings are 41%, 40%,41% and 46%, respectively, under far considered blast (R = 40 0m). A similar range in reductions is observed for bending moments in bottom-corner columns. The most effective and least effective damper in class A has a damping coefficient (kNs/m) of 330 and 420, respectively, with damping exponent values of 0.51 and 0.8, respectively. For class B dampers, the maximum and minimum reductions are achieved for damping coefficient (kNs/m) of 330 and 260, respectively, with damping exponent values 0.36 and 0.5, respectively. It must be noted that plan irregular buildings equipped with viscous dampers show better response mitigation ability as compared to the regular buildings. The square-shaped buildings are the worst affected under the blast loads, whereas C-shaped buildings are the least affected under various selected blast loads. It is concluded that damper B class is the most effective damper in mitigating the structural responses under all selected blast loads. Next, the efficiency of XPD is investigated in mitigating the structural responses of regular and irregular buildings subjected to blast-induced vibrations. The XPD parameters, namely thickness of plate, width and height of the triangular portion of the plate, are varied to determine the optimum damper properties in mitigating the blast effects in RC regular and irregular building. The maximum reductions in structural responses observed in XPD at t = 8 mm, a = 20 mm and b = 160 mm. It is important to note that XPD is inefficient in mitigating the top storey acceleration responses for all the selected buildings subjected to near and far occurring blasts. However, the reduction in shear forces and bending moments are in the range of 22–54% and 37–44% for square-shaped buildings, 31–55% and 35–43% for rectangular buildings, 23–42% and 40–45% for L-shaped buildings and 42–50% and 35–46% for C-shaped buildings for four different blast loads. Like viscous dampers, XPD improved the performance of irregular buildings better in comparison to regular buildings. In XPD the thickness of plate is critical parameter to obtain maximum reductions as the results obtained with thickness 2 mm are ineffective and are in the range of 1–10% for all the buildings subjected to selected blast loads. Thus, it is concluded that XPD have low efficiency to reduce top storey acceleration responses of the selected buildings and a high thickness value along with a low value of a and high value b yields reductions at par with that

TABLE 29.6

Performance of Regular and Irregular RCC Structure Equipped with Viscous Damper

Geometry	Fluid Viscous Damper Properties	Top Acceleration				Shear Force				Bending Moment			
		R1	R2	R3	R4	R1	R2	R3	R4	R1	R2	R3	R4
Square-Shaped Building	Without Damper	29.18	8.121	3.95	2.365	1315	453.5	236.8	147.5	6330	2215	1175	742
	$C = 330, \alpha = 0.51$	20.65	6.094	2.784	1.554	1137	347.7	168.5	100.2	4992	1595	822.7	527.1
	$C = 420, \alpha = 0.8$	19.46	6.122	3.012	1.773	1139	347.6	177.2	111.6	5191	1782	944.3	600.6
	$C = 330, \alpha = 0.36$	21.15	5.982	2.598	1.394	1131	348.1	170.8	100.9	4711	1476	794.1	512.3
	$C = 260, \alpha = 0.5$	21.24	6.354	2.924	1.635	1144	348.9	168.6	100.6	5189	1666	850.4	534.9
Rectangular-Shaped Building	Without Damper	32.22	10.03	4.771	2.665	1292	438	226.9	141.2	5632	1940	1020	642.5
	$C = 330, \alpha = 0.51$	25.27	7.272	3.32	1.853	1027	329.5	164.7	99.37	4519	1442	748.6	478.3
	$C = 420, \alpha = 0.8$	25.69	7.861	3.709	2.12	1039	348	181.5	114.2	4726	1611	851.1	541.1
	$C = 330, \alpha = 0.36$	24.38	6.698	2.962	1.605	1017	327.2	163.2	97.57	4251	1378	710.4	449.8
	$C = 260, \alpha = 0.5$	26.4	7.66	3.502	1.96	1035	331.4	165.1	99.62	4694	1501	768.9	485.9
L-Shaped Building	Without Damper	29.3	7.978	4.086	2.449	1258	438.4	230.6	144.5	5663	1990	1058	670
	$C = 330, \alpha = 0.51$	22.91	6.552	2.968	1.65	1102	338.6	165.5	100.7	4459	1428	753.8	488.7
	$C = 420, \alpha = 0.8$	21.7	6.666	3.212	1.872	1104	337.1	173.2	109.7	4668	1599	848.8	541.7
	$C = 330, \alpha = 0.36$	23.06	6.304	2.714	1.448	1097	340.1	167	99.55	4173	1369	741.6	473.9
	$C = 260, \alpha = 0.5$	23.71	6.891	3.133	1.747	1110	339.6	165.5	101.2	4647	1486	768.1	491.9
C-Shaped Building	Without Damper	33.1	10.22	4.772	2.688	1222	413.1	213.6	132.7	5429	1866	980.3	617
	$C = 330, \alpha = 0.51$	24.3	6.839	3.099	1.713	960	307.9	154	92.61	4114	1321	703	451.1
	$C = 420, \alpha = 0.8$	24.7	7.504	3.537	2.026	966	310.6	162.4	102.7	4353	1483	786.3	501.4
	$C = 330, \alpha = 0.36$	23.34	6.2	2.698	1.44	951.4	305.5	150.9	90.39	3940	1311	670.5	418.4
	$C = 260, \alpha = 0.5$	25.62	7.286	3.309	1.834	968.3	309.8	154.7	93.14	4303	1367	713.3	456.8

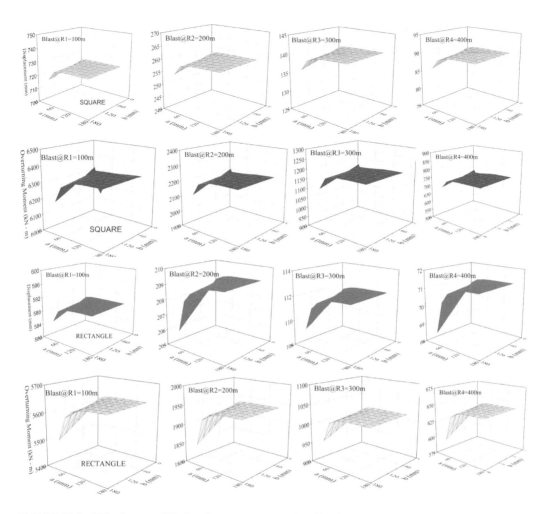

FIGURE 29.3 Effectiveness of X-plate damper parameters in mitigating the performance of regular buildings.

obtained using fluid viscous dampers. The study also investigates the structural responses, namely top storey displacement and bending moments for different plan buildings (regular and irregular) equipped with XPD having a thickness of plate equal to 2 mm and varying other parameters of XPD and plotted in Figures 29.3 and 29.4. It is observed that for all the plan geometries, the height of the triangular portion of XPD is more critical compared to the width of the XPD. The maximum reductions are obtained at lower value of "a" and higher values of "b" and higher thickness values. The maximum reductions become constant at a higher value of "b", keeping the value of "a" constant.

In the next section, the force deformation behaviour of fluid viscous dampers and X-plate dampers in controlling the blast vibration of regular and irregular buildings is discussed. It is evident from Figures 29.5 and 29.6 that damper B class with a damping coefficient equal to 330 and damping exponent 0.36 yields maximum axial force at lower deformations to produce maximum reduction in responses. The comparison of class A dampers shows that in case of square buildings, damper A1 shows the best response reduction ability and is evident from Table 29.6 by generating maximum axial force and undergoing comparable deformations. In the case of rectangular-, L- and C-shaped buildings the axial force deformation behaviour of A3 damper is best suited to yield maximum reduction in responses for all considered blasts. The comparison of class B dampers suggests that damper B1 is the best suited damper for all selected buildings subjected to different blast loads. It must be noted that regular buildings resulted in higher axial forces and deformations

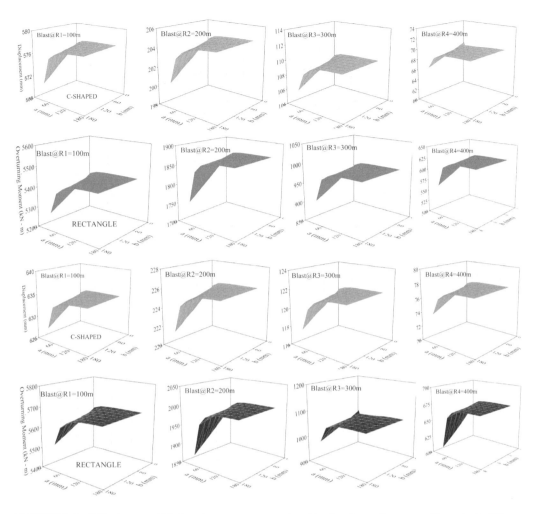

FIGURE 29.4 Effectiveness of X-plate damper parameters in mitigating the performance of irregular buildings.

to mitigate the blast effects in comparison to irregular buildings equipped with viscous dampers. The axial force deformation behaviour of XPD is depicted in Figure 29.7, comparing the effect of width of plate thickness in mitigating the blast loads. The regular buildings yield larger axial forces and deformations in comparison to irregular-shaped buildings. The axial force deformation study is extended further to generate the energy dissipated by the selected dampers in mitigating the structural responses. The energy dissipated by class B damper is 13%, 15%, 14% and 13% more than that dissipated by class A damper for square, rectangular, L-shaped and C-shaped buildings. The fluid viscous damper installed in C-shaped buildings dissipated maximum energy to mitigate the blast responses and is validated by Figures 29.8 and 29.9. The study also compares the energy dissipation ability of XPD by varying the thickness of plate and keeping the other parameters the same as shown in Figure 29.10. The XPD installed in C-shaped buildings dissipated the maximum hysteretic energy in mitigating the blast responses. Increasing the thickness of plate from 2 to 8 mm results in energy dissipation increases by 84%, 94%, 71% and 48% for square, rectangle, L-shaped and C-shaped buildings when subjected to far occurring blasts. For closely occurring blasts (R = 100 m), the energy dissipation capability of XPD with 8 mm thickness is six times for regular buildings (square and rectangular plan) and five times for irregular buildings installed with a 2 mm thick plate.

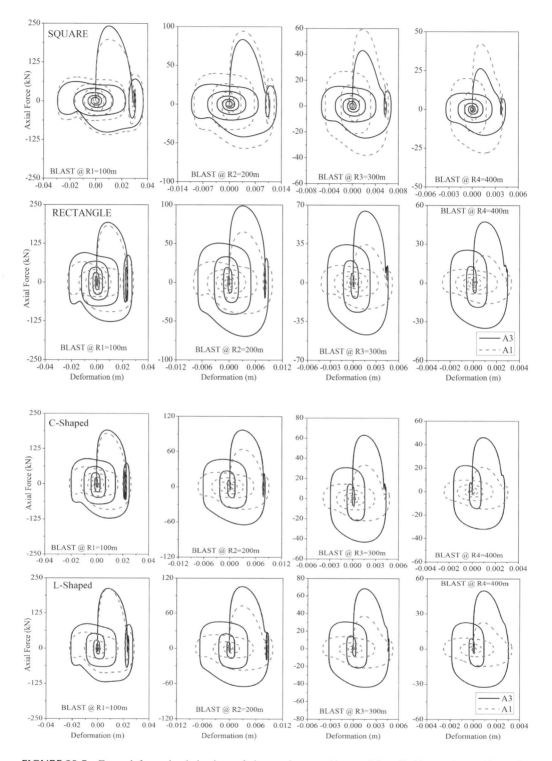

FIGURE 29.5   Force deformation behaviour of viscous dampers (A group) installed in regular and irregular buildings.

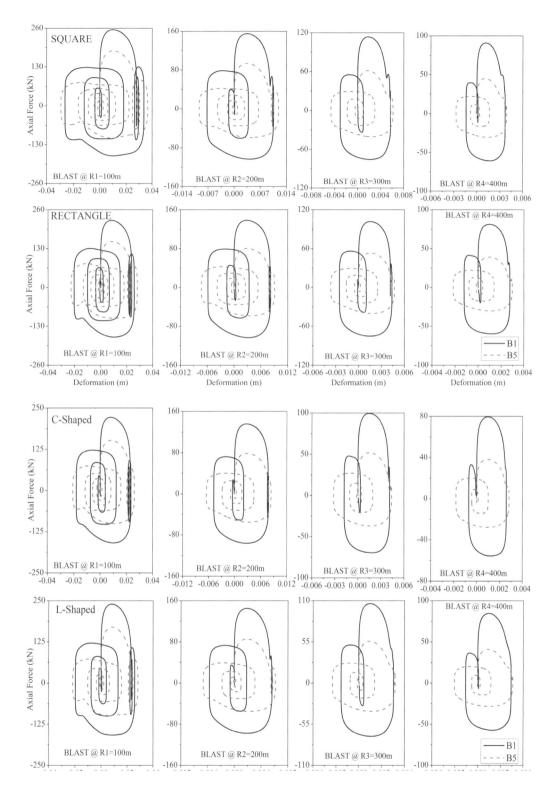

FIGURE 29.6 Force deformation behaviour of viscous dampers (B group) installed in regular and irregular buildings.

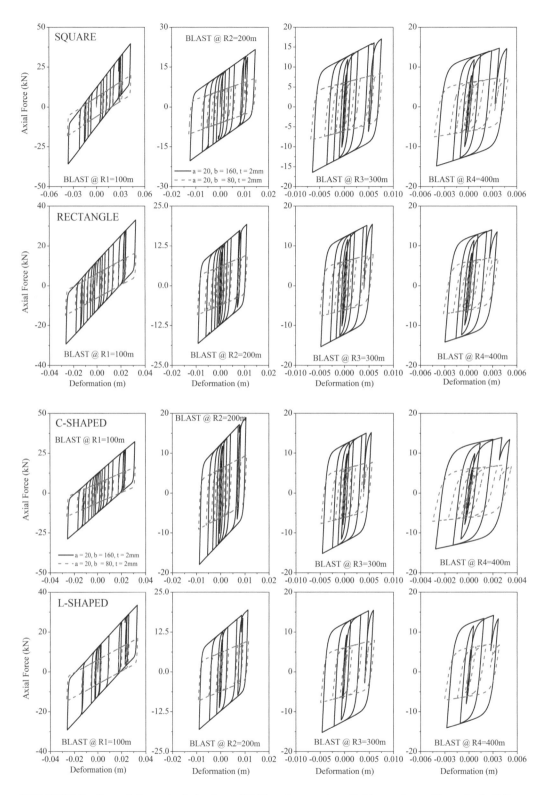

FIGURE 29.7 Force deformation behaviour of X-Plate dampers installed in regular and irregular buildings.

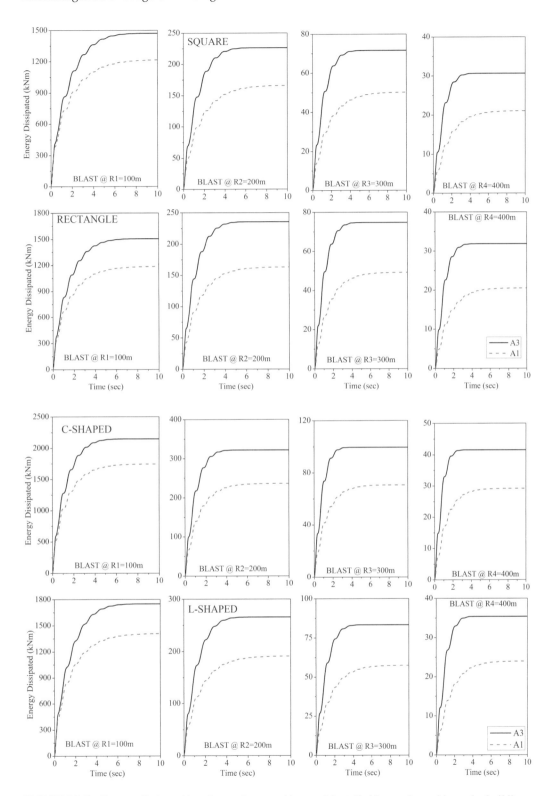

FIGURE 29.8 Energy dissipated by viscous dampers (A group) installed in regular and irregular buildings.

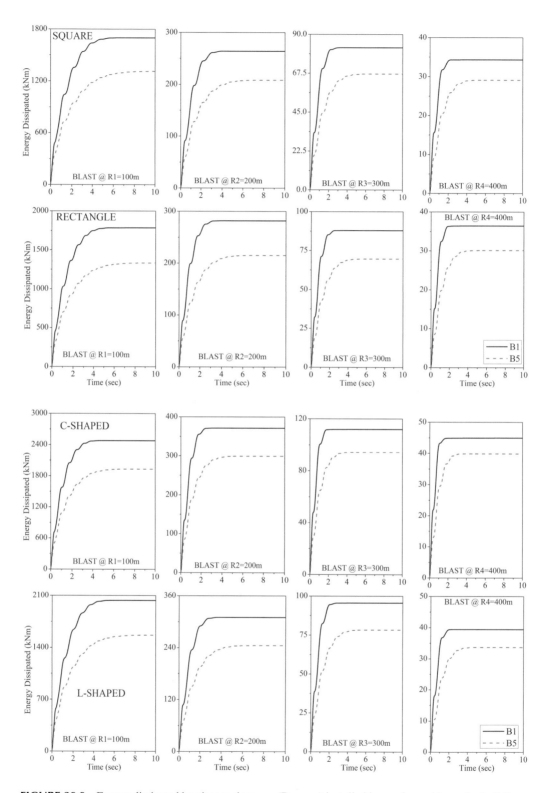

FIGURE 29.9 Energy dissipated by viscous dampers (B group) installed in regular and irregular buildings.

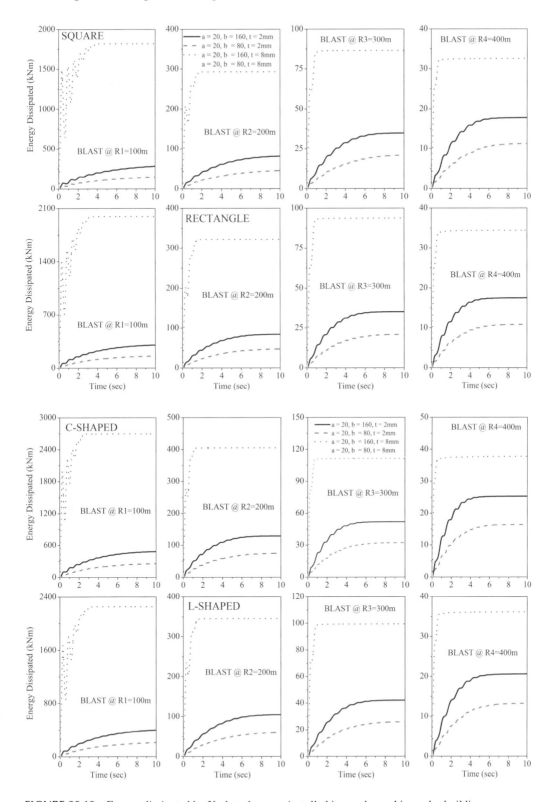

FIGURE 29.10 Energy dissipated by X-plate dampers installed in regular and irregular buildings.

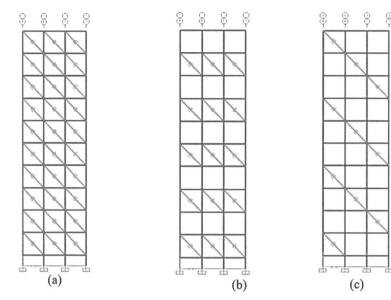

FIGURE 29.11   Optimum damper positions to control the blast responses of regular and irregular buildings: (a) damper all locations, (b) damper alternate floors and (c) zig-zag pattern.

Three cases of placement techniques are compared in the present study, namely dampers placed at all locations, alternate floors and zig-zag pattern to optimize the number of dampers in mitigating the blast effects, as shown in Figure 29.11. The results obtained from the placement of viscous dampers are tabulated in Table 29.7 and show that in case of square buildings the zig-zag pattern

TABLE 29.7

Blast Responses of Different Building Configuration at Optimum Placement of Viscous Damper

Geometry	Damper Placement	Top Acceleration (m/sec²)				Bending Moment (kNm)			
Damper Property (C = 330, α = 036)		R1	R2	R3	R4	R1	R2	R3	R4
Square-Shaped	Without Damper	29.18	8.12	3.95	2.36	6330	2215	1175	742
Building	All Location	21.15	5.98	2.59	1.39	4711	1476	794.1	512.3
	Alternate Floor	22.5	6.90	3.13	1.70	5402	1685	831.2	517.7
	Zig-Zag Pattern	22.34	6.73	3.04	1.66	5312	1650	820.8	510.2
Rectangle-Shaped	Without Damper	32.22	10.03	4.71	2.66	5632	1940	1020	642.5
Building	All Location	24.38	6.69	2.96	1.60	4251	1378	710.4	449.8
	Alternate Floor	27.66	7.89	3.60	2.01	4844	1503	747.4	467.9
	Zig-Zag Pattern	29.01	8.48	3.88	2.16	5078	1620	806.6	492.7
L-Shaped	Without Damper	29.3	7.97	4.08	2.44	5663	1990	1058	670
Building	All Location	23.06	6.30	2.71	1.44	4173	1369	741.6	473.9
	Alternate Floor	25.27	7.46	3.35	1.83	4842	1490	752	478.4
	Zig-Zag Pattern	25.65	7.57	3.44	1.90	4973	1563	780.2	486.7
C-Shaped	Without Damper	33.1	10.22	4.77	2.68	5429	1866	980.3	617
Building	All Location	23.34	6.2	2.69	1.44	3940	1311	670.5	418.4
	Alternate Floor	27.05	7.60	3.43	1.88	4462	1358	693.3	440.9
	Zig-Zag Pattern	28.1	8.02	3.61	2.00	4626	1430	718.8	454.5

TABLE 29.8

Blast Responses of Different Building Configuration at Optimum Placement of X-Plate Damper

Geometry	Damper Placement	Top Acceleration (m/sec²)				Bending Moment (kNm)			
Damper Property									
(a = 20 mm, b = 160 mm, t = 8 mm)		R1	R2	R3	R4	R1	R2	R3	R4
Square-Shaped	Without Damper	29.18	8.121	3.95	2.365	6330	2215	1175	742
Building	All Location	34.11	9.797	4.249	2.507	4020	1296	698.9	417.8
	Alternate Floor	30.97	9.081	4.133	2.242	4880	1355	704.9	435.4
	Zig-Zag Pattern	31.27	8.993	4.272	2.149	4599	1256	635.1	384.1
Rectangle-Shaped	Without Damper	32.22	10.03	4.711	2.665	5632	1940	1020	642.5
Building	All Location	32.44	9.098	4.301	2.413	3656	1173	594.7	367.7
	Alternate Floor	32.05	9.107	4.252	2.162	4312	1262	638.9	387.3
	Zig-Zag Pattern	32.07	9.242	4.247	2.283	4719	1364	687.8	423.4
L-Shaped	Without Damper	29.3	7.978	4.086	2.449	5663	1990	1058	670
Building	All Location	34.84	9.429	4.25	2.414	3674	1182	620.1	371.4
	Alternate Floor	32.94	9.472	3.899	2.075	4202	1250	645.2	382.4
	Zig-Zag Pattern	31.81	8.765	3.988	2.093	4434	1271	650.1	400.6
C-Shaped	Without Damper	33.1	10.22	4.772	2.688	5429	1866	980.3	617
Building	All Location	33.92	9.359	4.521	2.331	3392	1065	550.5	336.1
	Alternate Floor	32.55	9.259	4.18	2.338	3839	1176	574.3	345.8
	Zig-Zag Pattern	32.73	9.457	4.196	2.122	4145	1238	616.8	377.4

is most efficient in improving the building performance under far blasts. For rectangular, L-shaped and C-shaped buildings, placing the dampers at alternate floors yielded responses at par with that obtained by placing dampers at all floors subjected to near and far occurring blasts. The optimal results obtained by the XPD support the results obtained from a fluid viscous damper with a zig-zag pattern suitable for square-shaped buildings, whereas the alternate damper placement technique is conducive for rectangular-, L- and C-shaped buildings are tabulated in Table 29.8.

29.4 CONCLUSIONS

In the present study, an attempt is made to evaluate the effectiveness of resilient passive control dampers namely fluid viscous and X-plate in mitigating the blast effects of regular (square and rectangular) and irregular (C-shaped and L-shaped) plan buildings. The buildings are located at the mine blasting zone of the Indian subcontinent designed for the conventional loading conditions. The study incorporates four blast histories data generated from the available empirical formulae for granite rock conditions. Both regular and irregular buildings are analysed and designed for with and without damper conditions subjected to a blast charge of intensity 50 tonnes located at standoff distances of 100, 200, 300 and 400 m. The study identifies the most efficient vibration control in mitigating the blast effects in regular and irregular buildings. The study also presents an optimum damper placement technique for regular and irregular buildings in mitigating the mine blast effects. The main outcomes from the present study are summarized as follows:

1. The fluid viscous dampers installed within regular and irregular buildings having a damping coefficient of 330 and damping exponent of 0.36 yielded maximum reduction in blast responses as compared to the ten other selected damper properties. The irregular buildings

equipped with and without viscous dampers performed better as compared to regular buildings when subjected to various blast-induced vibrations.

2. The X-plate dampers controlled the performance of regular buildings better in comparison to irregular buildings and maximum reduction in responses are obtained at thickness of 8 mm, width of plate equal to 160 mm and height of triangular portion equal to 20 mm. The maximum reductions are obtained at high thickness and width of plate and minimum height of damper.

3. The results evaluated from the axial force deformation behaviour of the selected dampers show that regular buildings generate more axial forces and deformations in comparison to irregular buildings in mitigating the structural responses. The damper class B1 performed better in comparison to all selected dampers in mitigating the structural responses of selected buildings subjected to four selected near and far occurring blasts.

4. The energy dissipation capability of B1 damper is maximum compared with class A dampers and XPD installed in regular and irregular buildings. The blast mitigation by energy dissipation of selected dampers shows that dampers installed in C-shaped buildings dissipated maximum hysteretic energy.

5. The study also investigates various damper placement techniques, namely dampers placed at all locations, alternated floors and dampers in a zig-zag pattern. The buildings equipped with dampers at all floors produced maximum reduction in responses. To minimize the number of dampers, the study incorporates alternate and zig-zag pattern placement producing comparable responses. It is interesting to note that the zig-zag pattern is best suited for square-shaped buildings in mitigating the blast responses, whereas in all other shaped buildings, alternate placement of both fluid viscous and XPD generated maximum reduction in responses.

REFERENCES

1. Eurocode 8: 2020 (2020) Design of Structures for Earthquake Resistance - Part 1-1: General Rules and Seismic Action and Part 1-2: Rules for New Buildings, prEN 1998-1-1 & 1-2. CEN European Committee for Standardization, Draft, Brussels.
2. ASCE (2021) ASCE/SEI 7 - 22 Standard: Minimum Design Loads and Associated Criteria for Buildings and Other Structures. American Society of Civil Engineers (ASCE), Reston, VA.
3. IS 1893:2016 Part 1 (2016) Criteria for Earthquake Resist Design of Structures. Bureau of Indian Standards, New Delhi, India.
4. GB50011 (2010) Code for Seismic Design of Buildings. China Building Industry Press, Beijing.
5. Breyer, D. E., Cobeen, K. E., Fridley, K. J., & Pollock, D. G. (2015). Design of Wood Structures—ASD/LRFD. McGraw-Hill Education, New York, NY.
6. Raven, E. L., & López, O. A. (1996). Regular and irregular plan shape buildings in seismic regions; approaching to an integral evaluation. In Eleventh World Conference on Earthquake Engineering, Paper (No. 619).
7. Lin, C. C., Ueng, J. M., & Huang, T. C. (2000). Seismic response reduction of irregular buildings using passive tuned mass dampers. Engineering Structures, 22(5), 513–524.
8. Gutierrez Soto, M. (2012). Investigation of passive control of irregular building structures using bidirectional tuned mass damper (Doctoral dissertation, The Ohio State University).
9. Bigdeli, Y., & Kim, D. (2016). Damping effects of the passive control devices on structural vibration control: TMD, TLC and TLCD for varying total masses. KSCE Journal of Civil Engineering, 20(1), 301–308.
10. Sophocleous, A. A. (2001). Seismic control of regular and irregular buildings using viscoelastic passive dampers. Journal of Structural Control, 8(2), 309–325.
11. Gong, S., Zhou, Y., & Ge, P. (2017). Seismic analysis for tall and irregular temple buildings: A case study of strong nonlinear viscoelastic dampers. The Structural Design of Tall and Special Buildings, 26(7), e1352.
12. Alam, Z., Zhang, C., & Samali, B. (2020). The role of viscoelastic damping on retrofitting seismic performance of asymmetric reinforced concrete structures. Earthquake Engineering and Engineering Vibration, 19(1), 223–237.

13. Daniel, Y., Lavan, O., & Levy, R. (2013). A simple methodology for the seismic passive control of irregular 3D frames using friction dampers. In Seismic Behaviour and Design of Irregular and Complex Civil Structures (pp. 285–295). Springer, Dordrecht.

14. Pekau, O. A., & Guimond, R. (1991). Controlling seismic response of eccentric structures by friction dampers. Earthquake Engineering & Structural Dynamics, 20(6), 505–521.

15. Armali, M., Damerji, H., Hallal, J., & Fakih, M. (2019). Effectiveness of friction dampers on the seismic behavior of high rise building VS shear wall system. Engineering Reports, 1(5), e12075.

16. Cancellara, D., & De Angelis, F. (2017). Assessment and dynamic nonlinear analysis of different base isolation systems for a multi-storey RC building irregular in plan. Computers & Structures, 180, 74–88.

17. Fujii, K., & Masuda, T. (2021). Application of mode-adaptive bidirectional pushover analysis to an irregular reinforced concrete building retrofitted via base isolation. Applied Sciences, 11(21), 9829.

18. De Stefano, M., & Pintucchi, B. (2008). A review of research on seismic behaviour of irregular building structures since 2002. Bulletin of Earthquake Engineering, 6(2), 285–308.

19. Puthanpurayil, A. M., Lavan, O., & Dhakal, R. P. (2020). Multi-objective loss-based optimization of viscous dampers for seismic retrofitting of irregular structures. Soil Dynamics and Earthquake Engineering, 129, 105765.

20. Lavan, O. (2015). Optimal design of viscous dampers and their supporting members for the seismic retrofitting of 3D irregular frame structures. Journal of Structural Engineering, 141(11), 04015026.

21. Hwang, J. S., Lin, W. C., & Wu, N. J. (2013). Comparison of distribution methods for viscous damping coefficients to buildings. Structure and Infrastructure Engineering, 9(1), 28–41.

22. Huang, X., & Bae, J. (2022). Evaluation of genetic algorithms for optimizing the height-wise viscous damper distribution in regular and irregular buildings. Arabian Journal for Science and Engineering, 47(10), 12945–12962.

23. Zafarani, M. M., & Halabian, A. M. (2020). A new supervisory adaptive strategy for the control of hysteretic multi-story irregular buildings equipped with MR-dampers. Engineering Structures, 217, 110786.

24. Yoshida, O., Dyke, S. J., Giacosa, L. M., & Truman, K. Z. (2003). Experimental verification of torsional response control of asymmetric buildings using MR dampers. Earthquake Engineering & Structural Dynamics, 32(13), 2085–2105.

25. Bagherkhani, A., & Baghlani, A. (2021). Reliability assessment and seismic control of irregular structures by magnetorheological fluid dampers. Journal of Intelligent Material Systems and Structures, 32(16), 1813–1830.

26. Nazarimofrad, E., & Zahrai, S. M. (2018). Fuzzy control of asymmetric plan buildings with active tuned mass damper considering soil-structure interaction. Soil Dynamics and Earthquake Engineering, 115, 838–852.

27. Azimi, M., Pan, H., Abdeddaim, M., & Lin, Z. (2017). Optimal design of active tuned mass dampers for mitigating translational–torsional motion of irregular buildings. In International Conference on Experimental Vibration Analysis for Civil Engineering Structures (pp. 586–596). Springer, Cham.

28. Kim, H., & Adeli, H. (2005). Hybrid control of irregular steel highrise building structures under seismic excitations. International Journal for Numerical Methods in Engineering, 63(12), 1757–1774.

29. Jung, H. J., Spencer, B. F. Jr, Ni, Y. Q., & Lee, I. W. (2004). State-of-the-art of semiactive control systems using MR fluid dampers in civil engineering applications. Structural Engineering and Mechanics, 17(3–4), 493–526.

30. Pnevmatikos, N. G., & Hatzigeorgiou, G. D. (2014). Seismic response of active or semi active control for irregular buildings based on eigenvalues modification. Earthquakes and Structures, 6(6), 647–664.

31. Rather, F., & Alam, M. (2022). Optimal placement of active tendons to control seismic structural response with dual irregularities-mass and stiffness. In International Conference on Advances in Structural Mechanics and Applications (pp. 72–84). Springer, Cham.

32. Filice, A., Mynarz, M., & Zinno, R. (2022). Experimental and empirical study for prediction of blast loads. Applied Sciences, 12(5), 2691.

33. Smith, P. D., & Rose, T. A. (2002). Blast loading and building robustness. Progress in Structural Engineering and Materials, 4(2), 213–223.

34. Sun, W. B., Jiang, Y., & He, W. Z. (2011). An overview on the blast loading and blast effects on the RC structures. Applied Mechanics and Materials, 94, 77–80.

35. Dinan, R. J. (2005). Blast Resistant Steel Stud Wall Design. Columbia, Missouri:University of Missouri-Columbia.

36. Viau, C., & Doudak, G. (2016). Investigating the behavior of light-frame wood stud walls subjected to severe blast loading. Journal of Structural Engineering, 142(12), 04016138.

37. King, K. W., Wawclawczyk, J. H., & Ozbey, C. (2009). Retrofit strategies to protect structures from blast loading. Canadian Journal of Civil Engineering, 36(8), 1345–1355.

38. Figuli, L., & Štaffenova, D. (2017). Practical aspect of methods used for blast protection. In Key Engineering Materials (Vol. 755, pp. 139–146). Trans Tech Publications Ltd, Zurich, Switzerland.

39. Malhotra, A., Carson, D., & Stevens, T. (2014). Blast resistant design of an embassy building. In Applied Mechanics and Materials (Vol. 566, pp. 415–419). Trans Tech Publications Ltd, Zurich, Switzerland.

40. Dusenberry, D. O. (Ed.). (2010). Handbook for Blast Resistant Design of Buildings. John Wiley & Sons, Hoboken, NJ.

41. Hidallana-Gamage, H. D., Thambiratnam, D. P., & Perera, N. J. (2014). Failure analysis of laminated glass panels subjected to blast loads. Engineering Failure Analysis, 36, 14–29.

42. Larcher, M., Solomos, G., Casadei, F., & Gebbeken, N. (2012). Experimental and numerical investigations of laminated glass subjected to blast loading. International Journal of Impact Engineering, 39(1), 42–50.

43. Zhang, X., Hao, H., & Ma, G. (2013). Parametric study of laminated glass window response to blast loads. Engineering Structures, 56, 1707–1717.

44. Lori, G., Morison, C., Larcher, M., & Belis, J. (2019). Sustainable facade design for glazed buildings in a blast resilient urban environment. Glass Structures & Engineering, 4(2), 145–173.

45. Zobec, M., Lori, G., Lumantarna, R., Ngo, T., & Nguyen, C. (2014). Innovative design tool for the optimization of blast-enhanced facade systems. Journal of Facade Design and Engineering, 2(3–4), 183–200.

46. Santos, F. A., Cismasiu, C., & Bedon, C. (2016). Smart glazed cable facade subjected to a blast loading. Proceedings of the Institution of Civil Engineers-Structures and Buildings, 169(3), 223–232.

47. Pachla, F., Kowalska-Koczwara, A., Tatara, T., & Stypuła, K. (2019). The influence of vibration duration on the structure of irregular RC buildings. Bulletin of Earthquake Engineering, 17(6), 3119–3138.

48. Kiakojouri, F., De Biagi, V., Chiaia, B., & Sheidaii, M. R. (2022). Strengthening and retrofitting techniques to mitigate progressive collapse: A critical review and future research agenda. Engineering Structures, 262, 114274.

49. Ebrahimi, H. A. (2018). The investigation of the effect of plan irregularities on the progressive collapse response of low to medium rise steel structures (Doctoral dissertation, University of Birmingham).

50. Nica, G. B., Lupoae, M., Pavel, F., & Baciu, C. (2018). Numerical analysis of RC column failure due to blast and collapse scenarios for an irregular RC-framed structure. International Journal of Civil Engineering, 16(9), 1125–1136.

51. Coffield, A., & Adeli, H. (2016). Irregular steel building structures subjected to blast loading. Journal of Civil Engineering and Management, 22(1), 17–25.

52. Kumar, A., & Matsagar, V. (2018). Blast fragility and sensitivity analyses of steel moment frames with plan irregularities. International Journal of Steel Structures, 18(5), 1684–1698.

53. Kangda, M. Z., & Bakre, S. (2021). Performance of linear and nonlinear damper connected buildings under blast and seismic excitations. Innovative Infrastructure Solutions, 6(2), 1–19.

54. Kumar, R., Choudhury, D., & Bhargava, K. (2016). Determination of blast-induced ground vibration equations for rocks using mechanical and geological properties. Journal of Rock Mechanics and Geotechnical Engineering, 8(3), 341–349.

55. Kangda, M. Z., & Bakre, S. (2019). Response control of adjacent structures subjected to blast-induced vibrations. Proceedings of the Institution of Civil Engineers-Structures and Buildings, 172(12), 902–921.

56. Narkhede, D. I., & Sinha, R. (2014). Behavior of nonlinear fluid viscous dampers for control of shock vibrations. Journal of Sound and Vibration, 333(1), 80–98.

57. Bakre, S. V., Jangid, R. S., & Reddy, G. R. (2006). Optimum X-plate dampers for seismic response control of piping systems. International Journal of Pressure Vessels and Piping, 83(9), 672–685.

58. IS: 456-2000 (2000) Plain and Reinforced Concrete–Code of Practice. Bureau of Indian Standards, New Delhi.

Index

Note: Locators in *italics* represent figures and **bold** indicate tables in the text

For Product Safety Concerns and Information please contact our
EU representative GPSR@taylorandfrancis.com Taylor & Francis
Verlag GmbH, Kaufingerstraße 24, 80331 München, Germany